Green Energy and Technology

Climate change, environmental impact and the limited natural resources urge scientific research and novel technical solutions. The monograph series Green Energy and Technology serves as a publishing platform for scientific and technological approaches to "green"—i.e. environmentally friendly and sustainable—technologies. While a focus lies on energy and power supply, it also covers "green" solutions in industrial engineering and engineering design. Green Energy and Technology addresses researchers, advanced students, technical consultants as well as decision makers in industries and politics. Hence, the level of presentation spans from instructional to highly technical.

Indexed in Scopus.

Indexed in Ei Compendex.

Luis Baringo · Miguel Carrión · Ruth Domínguez

Electric Vehicles and Renewable Generation

Power System Operation and Planning Under Uncertainty

Springer

Luis Baringo
Electrical Engineering
University of Castilla-La Mancha
Ciudad Real, Spain

Miguel Carrión
Electrical Engineering
University of Castilla-La Mancha
Toledo, Spain

Ruth Domínguez
Department of Economics and Management
University of Brescia
Brescia, Italy

ISSN 1865-3529 ISSN 1865-3537 (electronic)
Green Energy and Technology
ISBN 978-3-031-09078-3 ISBN 978-3-031-09079-0 (eBook)
https://doi.org/10.1007/978-3-031-09079-0

This Springer imprint is published by the registered company Springer Nature Switzerland AG
The registered company address is: Gewerbestrasse 11, 6330 Cham, Switzerland

To Carmen, María, and Luis
To Sofía and Mayte
To Felipa, Jesús, Doroteo, and Maruja

Preface

Power systems are experiencing relevant changes that are modifying their structures, participants, and functioning procedures. The ever-increasing installation of renewable generation technologies, which most of them are characterized by uncertain and variable power outputs, is changing the traditional generation mixes of power systems, typically composed of thermal and hydropower units. Additionally, the continuous electrification of energetic processes, such as heating and transportation sectors, may change significantly the demand profiles in the future. In particular, the foreseen massive replacement of combustion-engine vehicles by electric vehicles is expected to cause different operational problems due to possible high peak consumption at specific time periods. Consequently, new tools are required to assist in the operation of current and future power systems. Moreover, planning models considering these aspects are also needed to gather a technical and economical proper transition from current power systems to future ones.

This book provides a set of operation and planning models for power systems that explicitly take into account the presence of a large capacity of renewable generation technologies and a high number of electric vehicles. Then, innovative short- and long-term decision-making models are comprehensively described considering the points of view of power system operators, planners, and aggregators of electric vehicles. The presented models take into account different sources of uncertainty, which are handled using different mathematical tools.

This book comprises twelve chapters. Chapters 1–3 describe the situation of current and future power systems, as well as the models and tools that are used throughout the book. Chapters 4 and 5 provide short-term decision-making tools taking the perspective of the power system operator. Chapters 6 and 7 face problems dealt by electric vehicle aggregators and energy suppliers, respectively. Chapters 8–12 describe planning models. Chapters 8–10 are devoted to the determination of the generation and storage capacity investments, whereas Chaps. 11 and 12 take the points of view of transmission and distribution network planners, respectively. The contents of this book are summarized below.

Chapter 1 describes briefly the operating and planning problems that power system operators and participants will face in future power systems. The main challenges of future power systems are also described in this chapter, focusing on the role that renewable energies and electric vehicles are expected to play. Finally, some motivating examples are described.

Chapter 2 presents relevant modeling information about the operation of power systems. It also describes the energy production process of wind and solar photovoltaic power plants and the modeling of the energy consumption of electric vehicles. Additionally, this chapter analyzes the temporal characterization of long-term planning horizons.

Chapter 3 provides the fundamentals of some mathematical tools that are used throughout this book. The basics of stochastic programming and robust optimization are presented. The practical implementation of the linear decision rules approach and the Benders' decomposition technique are described in detail.

Chapter 4 models the day-ahead energy and reserve scheduling of a renewable-dominated power system considering the uncertainty in the balancing market. The presented model is based on stochastic programming and adopts the perspective of the system operator. The uncertainty of the hourly demand, and wind and solar power is modeled through a set of scenarios.

Chapter 5 studies the day-ahead scheduling problem of a power system with a high presence of electric vehicles. The presented model considers that electric vehicles can provide reserve services when they are connected to the grid. The system demand, the intermittent renewable power, and the battery status of electric vehicles are considered uncertain parameters. This model is formulated as a stochastic programming problem.

Chapter 6 analyzes the bidding strategy problem of an aggregator of electric vehicles. Different models are presented in this chapter that take into account the uncertainty in market prices and driving needs. The impact of the bidding decisions of the electric vehicle aggregator on market prices is also addressed in the decision-making tool.

Chapter 7 describes the selling price determination problem faced by an electricity supplier that provides the energy demand of a set of electric vehicle users. The presented procedure considers the uncertainty of the electricity prices, the consumption patterns of electric vehicles, and the selling prices offered by rival suppliers.

Chapter 8 presents the static generation expansion planning problem of a power system considering renewable energies. This problem is modeled taking the perspective of a central planner that aims at determining the generation expansion plans that minimize both the investment and operation costs. Short- and long-term uncertainties are modeled using both scenarios and confidence bounds, which allows formulating this problem using stochastic programming and robust optimization approaches, respectively.

Chapter 9 extends the model presented in Chap. 8 considering that investment decisions can be made at different stages along the planning horizon. The demand growth and the investment costs of the units are considered uncertain parameters.

The presented model is formulated using both stochastic programming and linear decision rules approaches.

Chapter 10 describes the generation expansion problem considering explicitly the presence of electric vehicles. The presented model assumes that electric vehicle users may be willing to leave the charge of their vehicles in hands of the power system operator if high-enough financial incentives are offered to them. The model analyzed in this chapter considers uncertainties such as the annual demand growth, the capital costs of generating and storage units, and the number of electric vehicles. The proposed model is formulated using a two-stage stochastic programming problem.

Chapter 11 analyzes the transmission expansion problem considering the presence of renewable energies and electric vehicles. The presented model determines the transmission lines to reinforce and the storage facilities to install. A two-stage stochastic programming formulation is proposed considering the uncertainties of the generation capacity installed and the number of electric vehicles.

Chapter 12 studies the distribution expansion problem considering a large number of chargers for electric vehicles. The presence of electric vehicle chargers may change substantially the energy load at some locations and may require the reinforcement of the distribution system to preserve its adequate operation. A non-linear programming formulation is proposed to determine the reinforcement of the distribution grid.

This book is intended to be useful for energy engineering and economics communities. In particular, graduate students and practitioners may profit from the different models proposed in this book. The large number of illustrative examples included throughout each chapter, as well as the analyzed illustrative and realistic case studies, intend to facilitate the comprehension of the theoretical concepts included in this book.

Many colleagues have contributed with their views, comments, and research to inspire the contents of this book. We are grateful to Prof. Conejo from the Ohio State University; Prof. Oggioni from the University of Brescia; Profs. Morales and Pineda from the University of Málaga; Prof. Pandžić from the University of Zagreb; Prof. Dvorkin from the Johns Hopkins University; Prof. Vitali from the University of Bergamo; Prof. Ruiz from the University Carlos III de Madrid; Profs. Alguacil-Conde, Arroyo, Cañas-Carretón, and García-Bertrand from the University of Castilla-La Mancha; and in particular, we would like to acknowledge the work of Prof. Zárate-Miñano from the University of Castilla-La Mancha that helped the authors to elaborate the contents of this book.

We are grateful to Anthony Doyle for encouraging us to write this book, and to Chandra Sekaran and Padma Subbaiyan for their support in administrative paperwork.

We are thankful to the National Research Agency of Spain for their financial support during the last years.

Finally, we would like to state that the writing of this book has been really hard because of the difficult events that we all have experienced during years 2020 and

2021. However, we believe that the time and efforts we have dedicated in writing this book can be considered well spent if it contributes, only a little, to increase the interest of readers in the operation and planning of power systems.

Ciudad Real, Spain Luis Baringo
Toledo, Spain Miguel Carrión
Brescia, Italy Ruth Domínguez
April 2022

Contents

Part I
Introduction and Mathematical Characterization

Chapter 1
Introduction

This chapter provides an overview of the operation and planning problems that power system operators, planners, and participants will face in future power systems. The decision framework of each problem is also discussed. The main challenges of power systems are pointed out, focusing on the role of renewable energies and electric vehicles.

This chapter is organized as follows. Section 1.1 describes the main challenges that future power systems must face. Sections 1.2 and 1.3 describe the basic aspects of renewable energies and electric vehicles, respectively. Section 1.4 analyzes the presence of uncertainty in the different problems associated with the operation and planning of power systems. Section 1.5 outlines the different decision-making problems considered in this book and provides some motivating examples. Finally, Sect. 1.6 concludes this chapter providing a summary and some relevant remarks.

1.1 Challenges of Future Electricity Systems

Power systems have been continuously evolving since the first systems dedicated to the supply of electric light were developed at the end of the nineteenth century. However, we can observe that, nowadays, power systems are facing new challenges that call into question the procedures used to operate these systems so far.

The first challenge that power systems must handle is the fast incorporation of large capacities of new generation technologies, especially those based on wind and solar photovoltaic (PV) power. These technologies are based on renewable and weather-dependent sources that provide an intermittent power output. For this reason, the power plants belonging to these technologies are usually referred to as intermittent units. The installation of renewable technologies worldwide during the

L. Baringo et al., *Electric Vehicles and Renewable Generation*,
Green Energy and Technology, https://doi.org/10.1007/978-3-031-09079-0_1

last decades has been massive and obeys to two main reasons. The first one is due to the increasing environmental concern of the societies. The desire to revert or reduce the consequences of the global warming by cutting off the emission of greenhouse gases (GHGs) has placed the focus of governments on the power sector, which is responsible of about one third of the global emissions of CO_2. In this sense, renewable technologies, which are characterized by very low GHG emissions during their life cycle, have been strongly promoted by governments to replace thermal units fed by fossil fuels, as coal and gas power plants.

The second reason that explains the rapid growth of the installation of renewable power plants is the technical development of these technologies and the strong reduction of their capital costs. The development of the power electronics at the end of the twentieth century increased the reliability and the capacity of power electronic devices and has boosted the installation of wind and solar PV technologies by private investors. Although the investments in renewable generating units required financial support by governments to ensure their profitability in the past, these technologies are currently economically competitive with respect to mature generation technologies and financial aids are not longer needed, specially for wind and solar PV technologies.

The performance of intermittent power plants is substantially different to that of traditional technologies, basically hydro and thermal power plants. As stated above, the production of intermittent plants is variable in time, but it should not be forgotten that this production is also uncertain. In other words, it is not possible to forecast with absolute precision the power output of an intermittent power plant in a future time period. These two characteristics, intermittency and uncertainty of the production, cause that 1 MW of a thermal power plant that is decommissioned from the system cannot be straightforwardly replaced by 1 MW of an intermittent generation technology. The variability and the uncertainty of the production of intermittent units, as well as the spatial correlation of the production of power plants located at different locations, must be carefully considered in the operation of an existing power system. Nonetheless, these issues should also be considered in the design process of new power systems, or in the transformation of existing power systems based on hydro and thermal generation technologies into new ones dominated by intermittent power plants.

Another important difference between the performance of conventional and intermittent power plants concerns the provision of inertia to the system. The inertia provided by the rotating elements of hydro, steam, or gas turbines is very useful to limit frequency variations produced by the imbalances between generation and consumption in a power system. Therefore, if an energy imbalance is produced, the excess or defect of energy can be partly or totally compensated by increasing or reducing the kinetic energy stored in the rotating elements of the turbines. However, the lack of turbines in solar PV plants or the existence of an electronic layer between the wind turbine and the network connection in wind power units prevents the provision of this inertia support. As a result, high frequency variations have been observed in systems with high penetration of intermittent units. To solve this problem, lots of efforts have been placed in the design and implementation of the provision of the so-called synthetic inertia by intermittent units. In this manner, the power output of intermittent units is adjusted electronically to counteract frequency variations. The results obtained from the provision of synthetic inertia in small power systems are promising. In the same vein, the presence of intermittent generation has an impact in the scheduling of ancillary services of the system. The ancillary services are a set of mechanisms that intend to balance the production and consumption at every moment maintaining grid stability and security. In this manner, higher amounts of reserve are scheduled when a large intermittent and uncertain generation is scheduled. However, nowadays, intermittent plants, specially wind power units, are also able to participate in the ancillary services to provide different types of reserves.

The economic impacts of the incorporation of intermittent units in the power systems are also non negligible. Unlike thermal units, and similarly to hydro units, intermittent units do not use fuel to produce energy and they are characterized by high investment costs but very low operation costs. However, contrary to those hydro units with water reservoirs that are able to decide in which periods they produce, intermittent units are able to produce energy only in those periods in which the renewable source is available. As a result, the operation of hydro and intermittent power units is essentially different. In this sense, given that water resources are limited, hydro units can decide to produce only in those periods in which electricity prices are high. On the other hand, considering that production costs of intermittent units are almost negligible and that the renewable resources as wind and solar irradiation cannot be stored in general, these technologies are interested in producing as long as the renewable resource is available.

The intermittent operation of most of renewable technologies has a high impact in the electricity prices of electricity markets, especially in the day-ahead market. The day-ahead market is typically the market in which most part of the electricity is traded by producers and consumers in a power system. This market is specially designed for the operation of hydro and thermal units and it schedules the production of each power plant one-day ahead in one-hour or half-hour basis. In this manner, thermal units can be configured to supply the scheduled power in each hour of the next day. Then, one-day ahead, producers submit generation offers stating for each hour of the next day the energy that they are willing to supply and the selling price for each offer. On the other hand, consumers (or electricity retailers) submit consumption bids with information about the energy that they are interested to buy and the purchasing price for each bid in every hour of the next day. Usually, the day-ahead market is settled using a marginal pricing mechanism that incentives producers to participate by offering selling prices equal to their operation costs. This market configuration is not beneficial for intermittent plants because it forces them to know the production of the plant one-day ahead. Since the accurate forecast of the renewable source of the intermittent plant is not possible, intermittent units are forced to be penalized for production deviations or to participate in adjustment markets to modify their production forecasts. The adjustment markets have an energy volume much lower than the day-ahead market and the participation in them is not always economically advantageous for intermittent power producers. Another consequence of the participation of intermittent technologies in the day-ahead market is the reduction of the electricity prices, which is good news for consumers but it can constitute a profitability problem for producers, including also intermittent power plants. Considering the marginal pricing used in most of day-ahead electricity markets, intermittent power plants are incentivized to offer prices close to zero that place out of the market those plants with higher operation costs. Then, if a large enough capacity of intermittent units participate in the day-ahead market, it is probable that electricity prices be close to zero in those periods in which the intermittent production be able to supply the system demand. Therefore, a cannibalism risk may exist if the amount of installed capacity of intermittent units is high enough to cause a decrease of prices avoiding the profitability of intermittent power plants. In order to avoid this situation, new trading platforms or remuneration mechanisms may be in order. For instance, the renewable auctions for installing new renewable plants that are used in many countries ensure economic incomes out of the market for new intermittent plants may be interesting to remunerate the production of these units. For these reasons, the adequate operation and planning of power systems dominated by intermittent power plants is currently an open research topic.

The second challenge that future power systems must face is the electrification of energetic processes that do not use electricity so far. It is expected that, in the incoming years, heating and low-temperature industry processes be electrified, as well as the transportation sector. The reason of the electrification of these processes is to decrease GHG emissions. The efficiency of electric engines is higher than that of thermal ones, and the emission of pollutants associated with electric devices may be low if the generation mix used to supply the demand of these vehicles is based on renewable energies. Consequently, the electrification of energetic processes is advised if GHG emissions are desired to be reduced.

The lead actor in this electrification process is the electric vehicle (EV). We can define an EV as a vehicle that is powered by electricity and is equipped with an electrochemical battery that can be charged from the grid. EVs are intended to replace combustion-engine vehicles progressively during the next decades. Governments around the globe have boosted decisively the usage of EVs for environmental and strategic reasons during the last years. From an environmental point of view, EVs are considered to be more efficient than combustion-engine vehicles since they do not emit pollutant gases and do not generate noise when they are driven. As stated above, the primary energy used to charge EVs may emit a low amount of GHGs if it is based on renewable energies. From an energy policy point of view, the increase of EVs may be useful to reduce the energetic dependence of some countries with respect to fossil fuels. This may be of special interest in those countries without fossil fuel reserves and with high potentials of renewable energies.

The massive integration of EVs is a challenge from several points of view. Although the increase of the annual system demand needed to charge the vehicles is technically affordable for most power systems, some problems may happen at some specific hours. For instance, cloudy days with low production of wind power may be problematic for renewable-dominated systems if the capacity of other generation technologies or storage units is not available. The ability of the system to ramp up and down the production in some specific hours has also to be adequate to face the simultaneous charge of a large number of EVs.

Notwithstanding the above, the main difficulties of the incorporation of EVs are expected at the distribution level. The charge of an EV may double the energy consumption of a home and triple the peak power demanded. Therefore, if a large number of EVs are simultaneously charged in a distribution network, voltage and distribution line congestion problems may happen. To avoid this situation, the reinforcement of the distribution system and smart charging procedures must be implemented to avoid the simultaneous charge of a high number of vehicles.

In this sense, demand-side management actions including the direct control of the charge of vehicles by an external entity may be advisable to reduce or delay the investments in new distribution assets. Observe that since vehicles are parked most of the time, EVs are ideal for participating in energy management actions. Additionally, they are equipped with a large battery that can be even discharged into the grid using vehicle-to-grid capabilities. Therefore, EVs can be used as energy storage units that can be useful to perform energy arbitrage and provide ancillary services to the grid. However, it should be also mentioned that this type of demand-side management procedures must handle concerns about privacy, security, and communication failures. For this reason, the energy management of EVs is also an open research topic. In another vein, flexible voltage and power flow control devices can be placed at strategic locations in distribution grids to optimize the operation of the distribution facilities.

The massive integration of renewable energies in power systems and the electrification of the transport sector and other energetic processes are the main challenges that power system operators and planners must face in the incoming years. It should not be forgotten that, despite of the technical complexities associated with the generation, transmission, and distribution of electricity, power systems can be considered so far as an example of a reliable and secure system with high quality standards. The average interruption time of power systems is almost null in most developed countries and the cost of the power supply for consumers is comparatively lower than other energy types. Therefore, the good performance of power systems should not be deteriorated as a consequence of the challenges that have been described above. Consequently, power system operators and planners must anticipate the changes that power systems are about to experience with the objective of maintaining the quality of the service. Fortunately, there are several tools that are available and can be used to ease the expected transformation of power systems. First, new electricity storage technologies are expected to play a relevant role in future power systems. Specially, electrochemical batteries are experiencing a technological revolution that make them suitable for providing different services as energy arbitrage, and local and grid support. Note that, so far, only hydro pumping units have been used to storage high amounts of energy. However, this type of facilities requires a specific location and resources that do not exist in all power systems. On the other hand, electrochemical batteries have a modular structure that make them suitable to be installed in any location of the system in a wide range of sizes. The usage of new net-zero emission fuels, as hydrogen, is also promising. Hydrogen is a fuel that may be burnt in gas turbines, as natural gas, to produce power without the emission of CO_2. The problem of obtaining hydrogen is that it requires the usage of electricity.

Electrolyzers are the devices used to obtain hydrogen from water and they require significant amounts of power. Then, hydrogen power plants may be of high interest in renewable dominated power systems. Renewable power surpluses can be used to produce hydrogen using electrolyzers, and hydrogen power plants can be used to produce electricity when renewable power is not available. The key factors for the future development of hydrogen-based power plants are to reduce the capital costs of electrolyzers, as well as to increase their efficiency and lifetime, and develop economical hydrogen storage units. The increasing participation of consumers in the operation of power systems is also another tool that can be used to facilitate the operation of power systems. So far, only large consumers have traditionally participated in the provision of specific services, as voluntary demand shed, or directly in the electricity markets. However, small-size consumers are expected to participate in demand-side management programs to modify their consumption patterns according to the needs of the operation of the system. In this manner, indirect or direct control procedures may be implemented to modify consumption profiles in exchange for a economical compensation. The services that power electronics can offer are also very interesting for the technical and economical operation of power systems. It is well known that Flexible AC Transmission Systems (FACTS) are able to improve significantly the performance of the transmission system. In this sense, the optimal operation of transmission lines and voltage control will be key in future power systems, where it is expected that some specific transmission lines be subject to a higher operation level. Finally, power system operators and planners have also available sophisticated mathematical-based decision-making tools and higher computational resources that will allow them to make informed decisions in this uncertain framework. Big data and decision-making under uncertainty models are examples of mathematical tools that can be used to operate and design future power systems.

As a conclusion, it can be stated that there is still a lot of work to be done to overcome the great challenges that power systems must face during next years. However, the power system community is working hard to develop new models, procedures, and devices that will allow us to increase the reliability, security, and quality of future power systems.

1.2 Renewable Energies

Renewable energies have significantly increased their presence in electric energy systems around the world because of their numerous advantages with respect to fossil-fuel and conventional generating technologies. Wind, solar, hydraulic, biomass, or geothermal energy, among others, represent natural endless resources available around the planet. This implies, firstly, a solution to generate electricity with a much lower carbon footprint than fossil fuel technologies [21]; secondly, the possibility of generating electricity independently of the willingness of third parties to supply fossil fuels or other scarce resources; thirdly, a distributed and local generation system

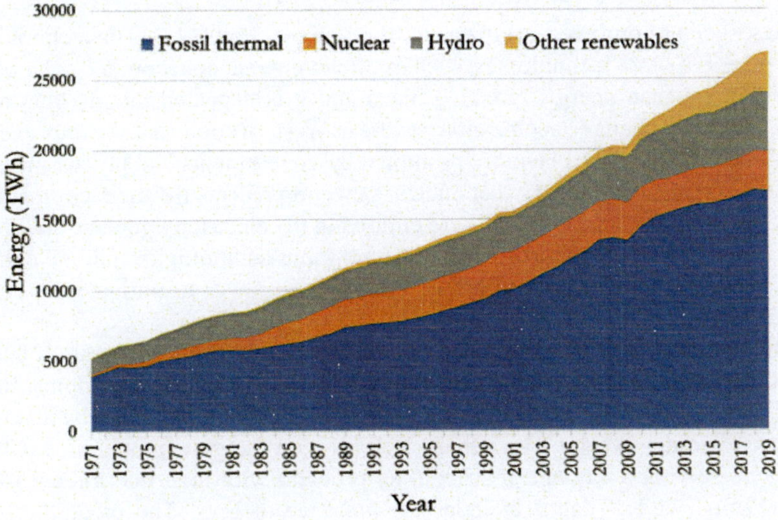

Fig. 1.1 World electricity generation by sources from 1971 to 2019

against a centralized and concentrated system; and fourthly, the growth of the local economy and the possibility of becoming autonomous in terms of energy supply.

The problem of climate change caused by anthropogenic GHG emissions has impulsed the use of renewable sources for electricity generation. According to the International Energy Agency (IEA) [12], the new capacity built in 2019 based on solar PV and wind power was 108 and 60 GW, respectively, and the forecast is a growth of 4% in the net new renewable capacity built in 2020. It is expected a global share of renewable sources in electricity generation of 27% in 2020. As it can be observed in Fig. 1.1, the electricity generation from renewable sources has grown 2.5 times from 2010 to 2019 [11]. Still, it is evident that a long way remains to effectively reduce the fossil-fuel dependency of this sector.

The main drawback of renewable sources such as wind or solar radiation is their lack of dispatchability. It is not possible to store these resources and it is difficult to predict their availability with an acceptable precision until a few hours in advance. This complicates the operation of a traditional power system where most of the energy and reserve capacity is scheduled at least one day prior to the energy delivery. Therefore, modifications in the market structures have been implemented in the last years to ease the integration of intermittent renewable electricity generation. In the European electricity market, for instance, the integration of the continuous and auction-based intraday markets is taken place among the countries inside the coordinated market [8]. Intraday markets happening along the delivery day allow a better adjustment of the intermittent generation to the actual renewable source availability, which entails a reduction in the ancillary service costs.

The technology of onshore wind and solar PV has reached the maturity level and their investment costs can nowadays compete with those of conventional technologies, such as combined-cycle gas turbines or coal units [13]. The technological development of other technologies, such as offshore wind, concentrated solar power, bioenergy, or ocean energy, is taking place and it is expected that as long as the effects of climate change become more severe in Western countries, the improvement of those technologies will be more promoted by governments and private investors. Finally, it is worth mentioning that nuclear power represents the main competitor to renewable energies since no GHGs are emitted in the electricity generation process. The technological evolution of emergent renewable technologies may be compromised if governments decide to bet for the nuclear power to mitigate the climate change effects.

In this book, we tackle different problems related with the integration and the impact of intermittent renewable generation in conventional power systems. In particular, Sect. 2.3 of Chap. 2 explains how to model the power production by renewable units using wind and solar PV sources. Related with the system operation, Chap. 4 focuses on solving a day-ahead scheduling problem with high intermittent renewable capacity, while Chap. 5 includes the presence of EVs. The modeling of the short-term uncertainties is specially crucial in these type of problems to obtain technically and economically feasible solutions. The other important problem related with renewable energies is the long-term planning. Therefore, Chaps. 8–12 are devoted to solve different planning situations. Specifically, Chaps. 8–9 study the single-stage and multi-stage capacity expansion problems in renewable capacity considering different uncertainty sources and modeling approaches.

1.3 Electric Vehicles

EVs are expected to play an important role in electric energy systems since their penetration is foreseen to significantly increase in the near future. This increase is mainly motivated by the fact that EVs can facilitate the reduction of CO_2 and other GHG emissions, especially in renewable-dominated power systems [15]. Moreover, the maturity of the technologies used for EVs makes them available for long-distance trips, which makes EVs more attractive for consumers.

Despite of the above possible benefits, a high penetration of EVs may have a great impact on the planning and operation of electric energy systems. Most EV users have similar driving patterns, arriving and departing from home and work at similar times. Therefore, if all users charge their EVs at the same time, this would translate into an increase in the peak demand in the system, which may make it necessary to build new generation and/or transmission facilities [5].

Nevertheless, EVs are parked and, thus, available for charging, most of the time. It is not necessary to charge them as soon as they are parked; instead the charging can be shifted to a more suitable time, e.g., to those hours with low prices, with low demand, or with high renewable power availability. This would result in a reduction of the charging costs or a reduction in the peak demand.

Another relevant aspect of EVs is that their batteries may be also used to inject energy into the network, i.e., the charging of EVs can be bidirectional, also known as vehicle-to-grid (V2G). In such a case, EVs can be seen as mobile energy storage units, which would increase the flexibility of power systems.

However, managing individual EVs has two main problems. On the one hand, an individual EV has limited flexibility since the owner may not be willing to change the timing of the charging unless it receives an incentive to do that. On the other hand, the high computation effort that would be needed to manage the charging and/or discharging of a high number of EVs. To deal with these two issues, the figure of an aggregator, which can be seen as an energy management system in charge of an EV fleet, is generally considered in the technical literature [14, 26]. The aggregator manages the charging of these EVs with the aim of minimizing the charging costs and guaranteeing that the driving requirements of EV users are satisfied.

Within this context, this book addresses some of the most relevant problems in electric energy systems with a high penetration of EVs. Firstly, Sect. 2.4 of Chap. 2 describes how to model the working of EVs in decision-making problems in power systems. Secondly, Chaps. 5–7 analyze operation problems with EVs, including the day-ahead market scheduling in Chap. 5, the bidding strategy of an EV aggregator in Chap. 6, and the pricing strategy of an electricity supplier for EVs in Chap. 7. Finally, Chaps. 8–12 deal with planning problems in power systems with EVs, including the generation expansion planning problem in Chaps. 8–10 and the transmission and distribution expansion planning problems in Chaps. 11 and 12, respectively.

1.4 Uncertainty

Decision-making problems in power systems are generally solved within an uncertain environment. For example, a power producer deciding its self-scheduling decisions for the next day faces the uncertainty in the market power prices, which are unknown at that time. Similarly, the owner of a renewable generating unit does not know in advance the available production levels since these depend on a meteorological phenomena such as the solar radiation or wind speed. Lastly, if an expansion planning problem is considered, the entity in charge of making these decisions, such as the building of new transmission lines or new generating units, must anticipate the power system conditions in the future, which is a difficult task since it involves forecasting the future demand or the future number of EVs in the system.

In order to obtain informed decisions, it becomes relevant to model these uncertainties in the decision-making problems. For this purpose, two relevant methods are generally considered in the technical literature and also addressed in this book, namely, scenario-based stochastic programming [3] and robust optimization [2].

Scenario-based stochastic programming models uncertain parameters through a set of scenarios. These scenarios characterize the possible realizations of uncertain parameters and each one has an associated probability of occurrence. In general, scenario-based stochastic programming problems can be classified into two-stage and multi-stage stochastic programming problems.

Two-stage stochastic programming models are used to represent decisions in two stages, namely, before and after knowing the uncertainty realization. The decision sequence in two-stage stochastic programming problems is as follows. Firstly, the decision-maker takes a decision before knowing the actual realization of the uncertain parameters. These decisions, which are also known as first-stage or here-and-now decisions, are independent on the future scenario realization. Secondly, the decision-maker is informed or gets information about the actual realization of the uncertain parameters. Finally, the decision-maker takes some corrective actions after knowing the actual uncertainty realization. These decisions, also known as second-stage or wait-and-see decisions, do depend on the scenario realization.

Multi-stage stochastic programming models are similar to two-stage stochastic programming problems but the decision sequence is repeated several times.

A key issue of scenario-based stochastic programming is an appropriate generation of scenarios, which must accurately represent the uncertain parameters. For this, it is key to have information about the probability distribution function of these parameters, which may be not available to the decision maker in some situations. Moreover, it is generally needed to consider a large enough number of scenarios in order to have a good representation of the uncertainty, which may translate into an intractable problem, mainly in complex decision-making problems. Even if the problem can be solved, it may involve a high computation time, which may not be acceptable for practical applications that need decisions to be made in a short time. Therefore, it is generally needed to achieve a trade-off between modeling accuracy and problem tractability when stochastic programming is considered to model decision-making problems under uncertainty.

On the other hand, robust optimization can be also used to model decision-making problems under uncertainty in power systems [25]. The main difference with scenario-based stochastic programming is that robust optimization guarantees the feasibility and optimality of the decisions provided that the uncertain parameters take values within a pre-specified uncertainty set, while scenario-based stochastic programming only guarantees the feasibility of the solution for the considered scenarios. This means that if the actual uncertainty realization is different to one of the scenarios considered, the solution may be infeasible.

Uncertainty sets for robust optimization are generally based on confidence bounds, which are usually easier to define even if the probability distribution function of the uncertain parameters is unknown. As a disadvantage of robust optimization, it can be mentioned that solutions may be too conservative.

Robust optimization problems can be also classified into static and dynamic problems, depending on the sequence of decisions. In a static robust optimization framework, decisions are made by anticipating that, once made, the worst-case uncertainty realization will occur. Within a dynamic robust optimization framework, after the uncertainty realization occurs, it is possible to apply some corrective actions, which provides the decision maker with additional flexibility.

Further information about stochastic programming and robust optimization models can be found in Sects. 3.2 and 3.3 of Chap. 3 of this book, respectively. Moreover, stochastic programming and robust optimization are used in this book to model decision making-problems under uncertainty, including operation problems in Chaps. 4–7 and planning problems in Chaps. 8–12.

1.5 Decision Making and Decision Framework. Motivating Examples

The main objective of this book is to provide scheduling and planning tools for power system operators and planners considering the presence of renewable energies and EVs. Different decision frameworks are considered depending on the characteristics of the particular problem to solve. Then, we can find throughout this book from day-ahead scheduling procedures to multi-year planning tools, spanning from short-term to long-term planning horizons.

The main decision-making problems analyzed in this book are summarized below.

1.5.1 Day-Ahead Market Scheduling Considering Renewable Energies

We consider a power system operator that aims at developing a day-ahead scheduling model to co-optimize energy and reserve capacity. The appropriate scheduling of energy and reserve becomes particularly important in power systems with high presence of renewable and intermittent power sources. This model can be formulated as a two-stage mixed-integer stochastic programming problem, where the first stage represents the day-ahead market and the second stage the balancing market. This problem is analyzed and formulated in Chap. 4. The uncertainty related to the demand, as well as to the wind and solar power is modeled. The decision framework of this problem is represented in Fig. 1.2.

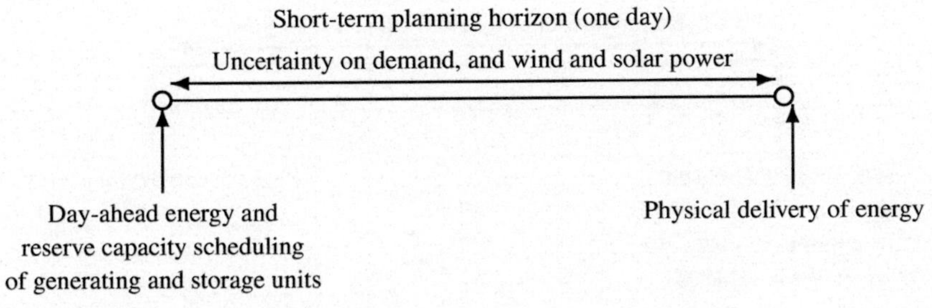

Fig. 1.2 Decision framework of the day-ahead market scheduling considering renewable energies

Illustrative Example 1.1 Decision making: day-ahead market scheduling considering renewable energies

Let us consider a power system comprised by the following generation technologies: 20 MW of hydro power, 15 MW of gas units, 50 MW of wind power, and 25 MW of solar PV power. Additionally, the power system has available a storage capacity of 5 MW/30 MWh. The objective of the power system operator is to schedule the day-ahead market for each hour of next day considering the availability of intermittent power sources (wind and solar PV units) to minimize the total cost. Additionally, it has to schedule the reserve capacity so that the power system be ready to counteract unexpected variations of consumptions and power output of intermittent power units. The power system operator has to solve an optimization problem similar to that described in Chap. 4 to schedule the power production and the reserve provided by each generation technology and each storage unit.

□

1.5.2 Day-Ahead Market Scheduling Considering Electric Vehicles

We consider a day-ahead energy market considering a large presence of EVs. A two-stage mixed-integer problem is proposed to represent the day-ahead market and the real-time operation. In this problem it is considered that, together with dispatchable generating units, EVs can participate actively in the day-ahead energy market, and they can provide reserve services when connected to the network.

The decision framework of this problem is represented in Fig. 1.3. This problem is analyzed in Chap. 5 and extends the model described in Chap. 4 modeling the participation of EVs in the electricity markets.

Illustrative Example 1.2 Decision making: day-ahead market scheduling considering electric vehicles

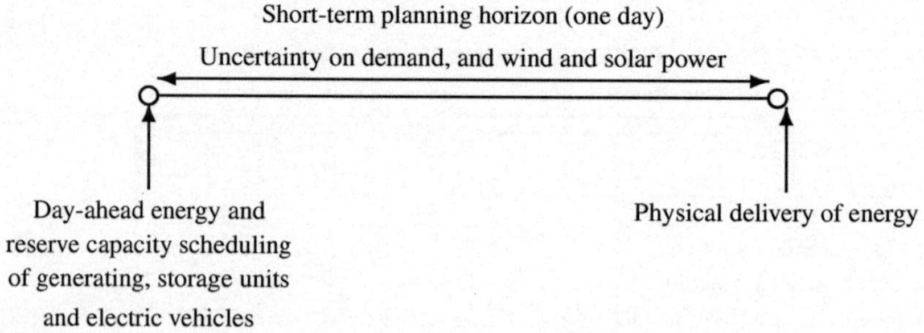

Fig. 1.3 Decision framework of the day-ahead market scheduling considering electric vehicles

Let us consider a power system similar to that described in Illustrative Example 1.1 comprised by 20 MW of hydro power, 15 MW of gas units, 50 MW of wind power, 25 MW of solar PV power, and storage capacity of 5 MW/30 MWh. It is also considered that 15 million of EVs use the power system to charge their batteries. Considering that 70% of these vehicles can be treated as inflexible and uncertain demand, the market system operator can control the charge of the rest of 30% of EVs to facilitate the operation of the system. The objective of the power system operator is to schedule the day-ahead market and the reserve capacity market for each day considering the availability of intermittent power sources (wind and solar PV units) and the flexibility of controlled EVs to minimize the total cost. Then, the power system operator may solve an optimization problem similar to that described in Chap. 5 to schedule the power production and the reserve provided by generators, storage units, and EVs. □

1.5.3 Bidding Strategy of an Electric Vehicle Aggregator

We consider the bidding strategy problem of an aggregator of EVs. The objective of EV aggregators is to determine the optimal bidding to submit to the day-ahead market minimizing the charging costs of EVs meeting also the driving requirements of EV users. This problem is analyzed in Chap. 6. The models developed in this chapter consider simultaneously the uncertainty in both market prices and driving needs. The decision framework of this problem is represented in Fig. 1.4.

Illustrative Example 1.3 Decision making: bidding strategy determination of an electric vehicle aggregator

We consider an EV aggregator aiming at deciding its self-scheduling strategy in the day-ahead energy market. The number of EVs belonging to the aggregation is 2,000.

The aggregator has to decide the quantities of energy to purchase from and sell to the day-ahead market, as well as the purchasing and selling prices to submit. The aggregator may consider whether its participation in the market can influence the

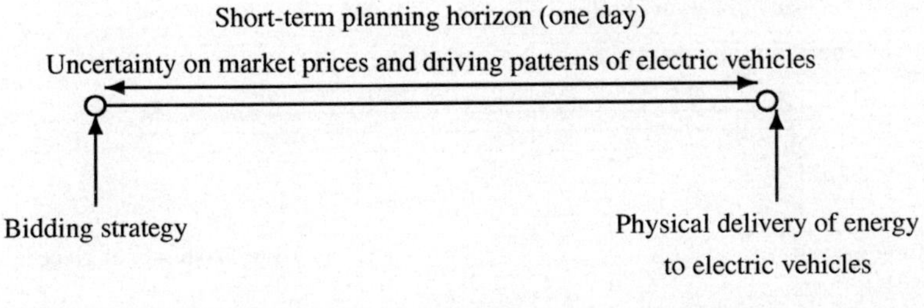

Fig. 1.4 Decision framework of the bidding strategy determination of an electric vehicle aggregator

resulting market prices, or not. Additionally, it should also take into account that its generation offers and consumption bids may not be scheduled if they are not competitive enough. The EV aggregator may solve one of the optimization problems described in Chap. 6 to determine optimally the self-scheduling strategy in the day-ahead energy market.

□

1.5.4 Pricing Strategy of Electricity Suppliers for Electric Vehicles

Electricity suppliers or retailers participate in the electricity markets to purchase energy that is afterwards delivered to their clients. We consider the problem of a supplier that provides the power demand of a set of EV users and desires to determine the offered selling price. This electricity supplier must anticipate the response of their potential clients to the offered selling price. This is a medium-term decision-making problem that is analyzed in Chap. 7.

The electricity supplier has to handle a number of uncertainties: (i) the electricity price in the pool market, (ii) the consumption patterns of EVs, and (iii) the selling prices offered by rival suppliers. Additionally, the electricity supplier can offer a more competitive price to those clients that are willing to leave the decision about the charge of their vehicles to the electricity supplier. The decision framework of this problem is represented in Fig. 1.5.

Illustrative Example 1.4 Decision making: pricing strategy determination by electricity suppliers of electric vehicles

Let us consider an electricity supplier that may sell electricity to a set of potential clients comprising EV users. It is considered that the supplier has currently 20,000 clients. However, the number of potential clients is 85,000. Then, this supplier must decide what is the selling price to offer to their potential clients, including their current clients, to maximize its profit. For that, the electricity supplier must forecast

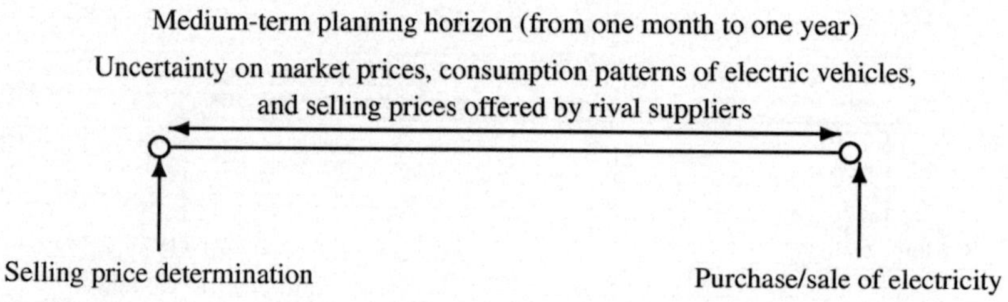

Fig. 1.5 Decision framework of pricing strategy determination by electricity suppliers for electric vehicles

the selling price offered by rival suppliers and the answer of the potential clients to the price offered. This supplier may solve a problem similar to that described in Chap. 7 to determine the best selling price that it can offer to maximize profits.

□

1.5.5 Generation Expansion Planning Considering Renewable Energies

We assume the point of view of a central planner that aims at determining the generation expansion plans that minimize both the investment and operation costs. This central planner may be interested in promoting the investments of the most suitable generating units among private profit-oriented investors. Short- and long-term uncertainties can be modeled using both scenarios and confidence interval approaches, which allow formulating the generation expansion planning problem using stochastic programming and robust optimization approaches, respectively. This problem is analyzed in Chap. 8. A static approach is adopted, where investment decisions are made at the beginning of the planning horizon. The decision framework of this problem is represented in Fig. 1.6.

Illustrative Example 1.5 Decision making: generation expansion planning considering renewable energies

Let us consider a power system planner that intends to decide the optimal configuration of the power system for 2060. This power system planner has an estimation of the capacity potentials of renewable energies that are equal to 60 MW of wind power and 40 MW of solar PV power.

The power system planner has to decide the capacity to install of each available technology facing the uncertainties related to: (i) the annual demand growth and (ii) the evolution of the natural gas prices. The power system planner may solve an optimization problem similar to that described in Chap. 8 to determine the capacity to install from each available technology.

□

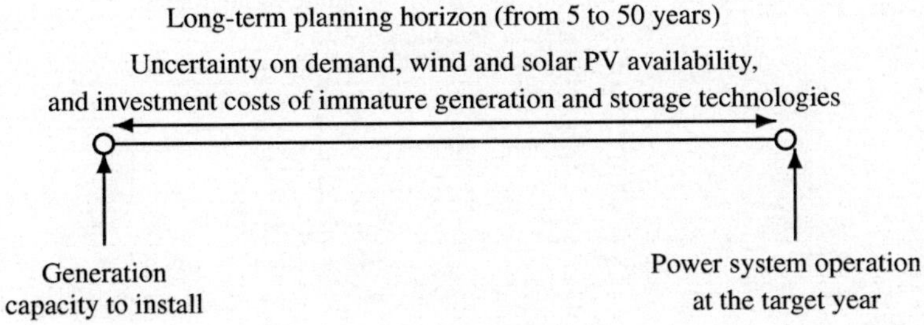

Fig. 1.6 Decision framework of the generation expansion planning considering renewable energies

1.5.6 Multi-Stage Modeling of the Generation Expansion Planning

Central planners can be interested in either determining the optimal configuration of the power system at a particular target year or in deciding the transformation process of a current power system into a desired one considering multiple decision stages. We consider a generation expansion problem in which investment decisions are made at different stages along the planning horizon. This planning problem usually consists in a long-term decision process that is subject to multiple uncertain sources throughout time. This problem is analyzed in Chap. 9 and expands the model described in Chap. 8. In this chapter, a multi-stage investment model in generation capacity considering long-term uncertainties. This problem is formulated using different approaches which differ in how the uncertain parameters are represented. The decision framework of this problem is represented in Fig. 1.7.

Illustrative Example 1.6 Decision making: multi-stage modeling of the generation expansion planning

Let us consider a power system planner that aims at deciding the optimal investments in generation and storage technologies to make in years 2030, 2040, and 2050 to obtain the best possible configuration of the power system for year 2060. This power system planner has an estimation of the capacity potentials of renewable energies that are equal to 60 MW of wind power and 40 MW of solar PV power. The estimation of maximum storage capacity that can be installed is 20 MW/120 MWh.

The power system planner has to decide the capacity to install in each decision stage of each available technology considering the uncertainties related to the annual demand growth, the evolution of the investment costs of solar, wind, and storage technologies, and the evolution of the natural gas prices. The power system planner may solve one of the optimization problems described in Chap. 9 to determine the capacity to install in each stage from each available technology.

□

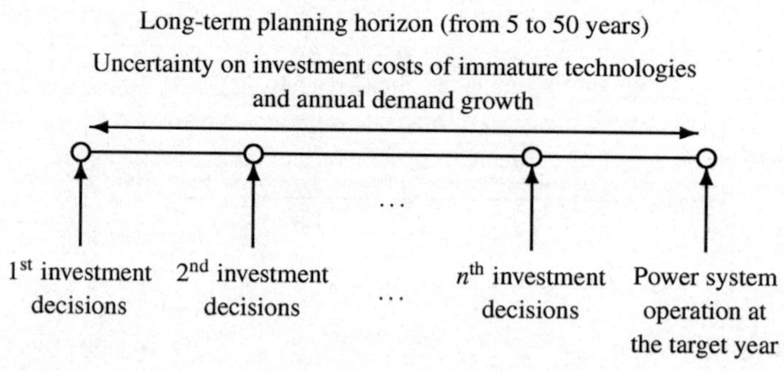

Fig. 1.7 Decision framework of the multi-stage modeling of the generation expansion plannings

1.5.6.1 Generation Expansion Planning Considering Electric Vehicles

The flexibility of the charge of EVs may help to reduce the capacity to install of generating units in future power systems. Considering this, we analyze the generation expansion planning considering the possibility of using financial incentives to modify the consumption patterns of EV users. Then, we consider that EV users may leave the charge of their vehicles in hands of the power system operator if a high-enough financial incentive is offered to them. The willingness of EV users to give the control of the charge/discharge of their vehicles to the system operator must be formulated mathematically. This problem is analyzed in Chap. 10 and expands the model described in Chap. 8.

The model analyzed in this chapter considers uncertainties such as the annual demand growth, the capital costs of generating and storage units, and the number of EVs. The Benders' decomposition technique is applied to solve the resulting mathematical problem. The decision framework of this problem is represented in Fig. 1.8.

Illustrative Example 1.7 Decision making: generation expansion planning considering electric vehicles

Let us consider a power system planner that intends to decide the optimal configuration of the power system for 2060. This power system planner has an estimation of the capacity potentials of renewable energies that are equal to 60 MW of wind power and 40 MW of solar PV power. The estimation of the storage capacity that can be installed is 20 MW/120 MWh. It has also available an estimation of 15 million of EVs that will use the power system to charge their batteries in 2060.

The power system planner has to decide the capacity to install of each available technology facing different sources of uncertainties and the possibility of offering a financial incentive of EV users to let the power system operator to decide the charging patterns of their vehicles to facilitate the operation of the system. Then, the power system planner has also to decide the incentives to offer to EV users so they allow

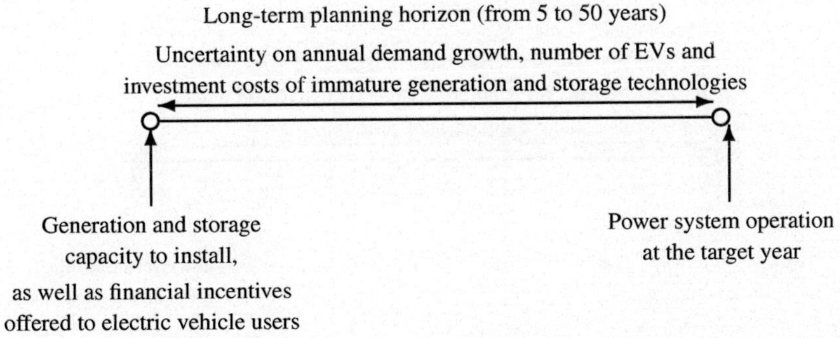

Fig. 1.8 Decision framework of the generation expansion planning considering electric vehicles

the direct control of the charge of their vehicles. An optimization problem similar to that described in Chap. 10 can be solved by the power system planner to determine the capacity to install from each available technology and the incentives to offer to EV users.

□

1.5.7 *Transmission Expansion Planning Considering Renewable Energies and Electric Vehicles*

The transmission expansion planner must decide the optimal investments in transmission assets to upgrade existing networks to facilitate the incorporation of renewable energies and EVs. In this problem, the transmission expansion planer must face uncertainties as the future renewable energy capacity installed and the number of EVs. This problem is formulated using a two-stage stochastic programming approach that is formulated using a mixed-integer linear programming formulation. This problem is analyzed in Chap. 11. The proposed model considers investments in new transmission lines and storage units. The decision framework of this problem is represented in Fig. 1.9.

Illustrative Example 1.8 Decision making: transmission expansion planning considering renewable energies and electric vehicles

Let us consider a transmission expansion planner that desires to determine the reinforcement of the transmission network to be ready to accommodate a large capacity of renewable energies and EVs in 2060. The transmission network comprises 55 lines of 400 kV and 120 lines of 220 kV connecting 210 buses. Then, the investment decisions that can be made comprise the building of new transmission lines, the reinforcement of some existing lines, and the installation of storage units.

The transmission expansion planner must face the uncertainty of the capacity installed from each generation technology in each bus of the system, as well as

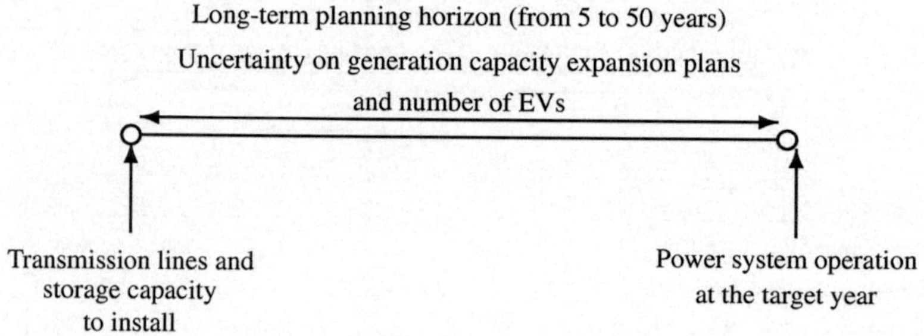

Fig. 1.9 Decision framework of the transmission expansion planning considering renewable energies and electric vehicles

the number of EVs. The transmission network planner may solve the optimization problem described in Chap. 11 to determine the investment decisions on transmission lines and storage units to be at operation for year 2060.

□

1.5.8 Distribution Expansion Planning Considering Electric Vehicles

We consider the problem faced by a distribution system operator that desires to determine the optimal investment decisions to incorporate the presence of a high number of EV chargers. This distribution system operator has to decide the installation of new distribution lines in existing or new corridors. Observe that the presence of a large number of EV chargers may cause low-voltage and line congestion problems in distribution networks. Power flows must be characterized using the non-linear expressions of the AC model of the network. As a result, a mixed-integer non-linear programming problem must be formulated. This problem is analyzed in Chap. 12. The decision framework of this problem is represented in Fig. 1.10.

Illustrative Example 1.9 Decision making: distribution expansion planning considering electric vehicles

A distribution network operator aims at upgrading an urban distribution network to accommodate the installation of a large number of EV chargers. The considered distribution network is an urban grid that comprises five feeders connected to a transformer station of 200 kVA. The distribution network delivers energy to 55 households. The owner of the distribution network desires to determine the distribution lines to reinforce to allow the incorporation of EV chargers avoiding undesirable voltage drops and line congestion.

The distribution network operator must consider the non-linear formulation of the power flows to model properly the voltage magnitude value in each bus of the network. The considered distribution network operator may solve the optimization problem described in Chap. 12 to select the distribution lines to reinforce.

□

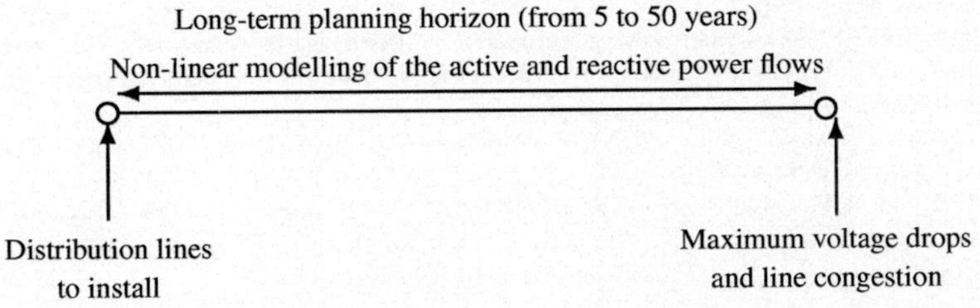

Fig. 1.10 Decision framework of the distribution expansion planning considering electric vehicle

1.6 Summary and Further Reading

This chapter provides an overview of future power systems, focusing on the presence of renewable energies and EVs. A large penetration of both renewable energies and EVs may lead to the increase of operation problems linked to the variability and uncertainty of renewable energies and to the increase of demand caused by the charge of EVs from the power grid. However, it has to be also considered that the dispatchability of renewable energies is significantly increasing and that new coordinated control schemes may favor the integration of EVs. Considering this, the main problems that power system operators and planners will face in future power systems are introduced and briefly analyzed. Finally, some motivating examples are described.

Significant references on the description of future power systems are [1, 9, 10, 16, 22, 23]. Relevant works describing the operation and planning of power systems are [4–7, 15, 17–20, 24–26].

References

1. Arghandeh, R., Zhou, Y.: Big Data Application in Power Systems. Elsevier, Amsterdam, The Netherlands (2018)
2. Bertsimas, D., Sim, M.: The price of robustness. Oper. Res. **52**, 35–53 (2004)
3. Birge, J.R., Louveaux, F.: Introduction to Stochastic Programming, 2nd edn. Springer, New York, NY (2011)
4. Conejo, A.J., Carrión, M., Morales, J.M.: Decision Making Under Uncertainty in Electricity Markets. Springer, New York, US (2010)
5. Conejo, A.J., Baringo, L., Kazempour, S.J., Siddiqui, A.S.: Investment in Electricity Generation and Transmission: Decision Making Under Uncertainty. Springer, Cham (2016)
6. Domínguez, R., Oggioni, G., Smeers, Y.: Reserve procurement and flexibility services in power systems with high renewable capacity: effects of integration on different market designs. Int. J. Electric. Power Energy Syst. **113** (2019)
7. Domínguez, R., Carrión, M., Conejo, A.J.: Influence of the number of decision stages on multi-stage renewable generation expansion models. Int. J. Electric. Power Energy Syst. **126**(A), 106588 (2021)
8. ENTSOE. Single Intraday Coupling (SIDC) (2022). https://www.entsoe.eu/
9. Götz, M., Lefebvre, J., Mörs, et al.: Renewable power-to-gas: a technological and economic review. Renewab. Energy **85**, 1371–1390 (2016)
10. Heinen, S., Mancarella, P., O'Dwyer, C., O'Malley, M.: Heat electrification: the latest research in Europe. IEEE Power Energy Mag. **16**(4), 69–78 (2018)
11. International Energy Agency (IEA): World electricity generation by fuel 1971-2019. IEA, Paris (2022). https://www.iea.org/data-and-statistics
12. International Energy Agency (IEA): Renewables 2020. IEA, Paris (2020). https://www.iea.org/reports/renewables-2020
13. International Renewable Energy Agency (IRENA) (2022). https://www.irena.org/
14. Kristoffersen, T.K., Capion, K., Meibom, P.: Optimal charging of electric drive vehicles in a market environment. Appl. Energy. **88**(5), 1940–1948 (2011)
15. Li, Z., Guo, Q., Sun, H., Wang, Y., Xin, S.: Emission-concerned wind-EV coordination on the transmission grid side with network constraints: concept and case study. IEEE Trans. Smart Grid. **4**(3), 1692–1704 (2013)

16. Milano, F., Dörfler, F., Hug. G., Hill, D. J., Verbič, G.: Foundations and challenges of low-inertia systems. In: Power Systems Computation Conference (PSCC) (2018)
17. Morales, J.M., Conejo, A.J., Liu, K., Zhong, J.: Pricing electricity in pools with wind producers. IEEE Trans. Power Syst. **27**(3), 1366–1376 (2012)
18. Morales, J.M., Pineda, S.: On the inefficiency of the merit order in forward electricity markets with uncertain supply. Eur. J. Oper. Res. **261**, 789–799 (2017)
19. Ortega-Vazquez, M.A., Kirschen, D.S.: Estimating the spinning reserve requirements in systems with significant wind power generation penetration. IEEE Trans. Power Syst. **24**(1), 114–124 (2009)
20. Papavasiliou, A., Oren, S.S., O'Neill, R.P.: Reserve requirements for wind power integration: a scenario-based stochastic programming framework. IEEE Trans. Power Syst. **26**(4), 2197–2206 (2011)
21. Parliamentary Office of Science and Technology UK: Carbon footprint of electricity generation. Parliamentary Copyright (2006). https://www.parliament.uk/globalassets/documents/post/postpn268.pdf
22. Peças Lopes, J.A., Guimarães Madureira, A., et al.: The future of power systems: challenges, trends, and upcoming paradigms. In: WIREs Energy and Environment, pp. 1–16 (2019)
23. Pereirinha, P.G., González, M., Carrilero, I., Anseán, D., Alonso, J., Viera, J.C.: Main trends and challenges in road transportation electrification. Transp. Res. Proc. **33**, 235–242 (2018)
24. Pritchard, G., Zakeri, G., Philpott, A.: A single-settlement, energy-only electric power market for unpredictable and intermittent participants. Oper. Res. **58**(4), 1210–9 (2010)
25. Sun, X.A., Conejo, A.J.: Robust Optimization in Electric Energy Systems. Springer, Cham (2021)
26. Vandael, S., Claessens, B., Hommelberg, M., Holvoet, T., Deconinck, G.: A scalable three-step approach for demand side management of plug-in hybrid vehicles. IEEE Trans. Smart Grid. **4**(2), 720–728 (2013)

Chapter 2
Modeling

This chapter describes some modeling aspects that are recommended to be known by the readers of this book to fully understand the decision-making models that are presented in Chaps. 4–12. In particular, this chapter describes the modeling of the operation of power systems, as well as the energy production processes of wind and solar photovoltaic power plants and the energy consumption of electric vehicles. The temporal characterization of long-term planning horizons is also analyzed in this chapter. In order to better understand the contents presented in this chapter, a basic knowledge about mathematical optimization theory is helpful. In summary, this chapter aims at providing some basic concepts related to the modeling of operation and planning problems of power systems with high presence of renewable energies and electric vehicles. Then, those readers familiar with these topics may skip this chapter.

2.1 Introduction

The technical and economical operation of power systems is a complex task that requires techniques and tools belonging to different knowledge areas as electrical engineering, energy, economics, mathematics, operations research, etc. For instance, the full management of power systems comprises topics ranging from power system dynamics to electricity markets, or from power system operations to power system planning. Therefore, it is not easy for practitioners to have an in-depth understanding of the wide range of topics related to the management of power systems. This chapter tries to clarify some basic concepts in the context of the modeling of the operation and planning of power systems with high presence of renewable energies and electric vehicles (EVs).

L. Baringo et al., *Electric Vehicles and Renewable Generation*,
Green Energy and Technology, https://doi.org/10.1007/978-3-031-09079-0_2

The operation of power systems is firstly analyzed, including the description of the day-ahead energy market and the market clearing procedure. Additionally, the ancillary services market is also defined. The energy scheduling based on a day-ahead energy market is present in most of liberalized power systems around the world and it is also used to determine the energy sold and purchased by generators and consumers in each time period, typically one hour. Electricity prices are also settled down using the market clearing procedure described in this chapter.

In a framework in which the capacity installed from renewable energies is continuously increasing, this chapter also analyzes the energy production process of the most usual renewable energy technologies used for electricity production: solar photovoltaic (PV) and wind power. The power output of these two technologies are dependent on weather variables as the solar radiation and the wind speed, respectively. Some illustrative examples are analyzed to compute quantitatively the power produced by these technologies from actual values of solar radiation and wind speed.

Considering the topic of this book, it is mandatory to establish a modeling framework of EVs and their aggregation. Because of computational reasons, it is not possible to model each EV individually, and it is necessary to aggregate them. Then, this chapter provides a basic formulation of the charging and discharging processes of EV groups in a power system framework.

Finally, the modeling of the temporal characterization for planning analyses is addressed. Observe that planning problems are long-term decision making problems, spanning up to several years, in which the operation of the system must be simulated in order to make informed planning decisions. How to model the operation of the system in such a large optimization problem is an open research question that is described in this chapter.

The remaining parts of this chapter are organized as follows. The modeling of the operation of power systems is described in Sect. 2.2. Section 2.3 provides the physical modeling of the power output of wind and solar PV power units. The mathematical characterization of charging and discharging processes of EVs is provided in Sect. 2.4. The modeling of the temporal characterization of long-term planning problems is analyzed in Sect. 2.5. Section 2.6 summarizes this chapter. The notation used in this chapter is listed in Sect. 2.7. Finally, Sect. 2.8 provides the GAMS [11] codes to solve some of the illustrative examples presented in this chapter.

2.2 Operation of Power Systems

The operation of power systems comprises the decisions made within one day of the actual power delivery. These decisions can be classified into those done one day ahead, some hours before, and just some minutes before power delivery.

Firstly, the commitment of conventional generating units, also known as dispatchable or controllable generating units, is decided one day in advance for the next-day operation. Moreover, the required reserves are also scheduled one day in advance, i.e., the scheduling of production capacity that is not used but is ready to be deployed if needed. This reserve capacity may be needed due to the uncertainty in demand, production from renewable generating units, or contingencies in generation and transmission facilities. Day-ahead operation can be handled in a centralized or a market framework. Within a centralized framework, the system operator solves a unit commitment problem to identify the status of generating units, as well as the production levels and reserves allocated to each unit that minimize the total production costs. In a market framework, the day-ahead operation is mainly based on a market clearing algorithm that determines the energy production and consumption levels, as well as the reserves with the aim of maximizing the social welfare.

Secondly, some hours before the power delivery and to ensure an economic and secure power supply, the operator deploys the required reserves to adjust the energy balance between production and consumption. In addition and if needed, the operator procures additional reserves. The main objective of these adjustments is the security of the supply, although they are done at minimum cost. For this purpose, an economic dispatch problem is generally solved.

Finally, some minutes before the power delivery, the operator solves an optimal power flow or a security-constrained optimal power flow problem. These problems focus on guarantying the security in the actual power system operation. These problems represent precisely the operation of the power system and incorporate appropriate active and reactive power adjustments to guarantee maximum security, even if one or several contingencies occur. This type of problems are out of the scope of this book.

This book focuses on the day-ahead operation of power systems considering a market framework. Additional details about the operation of power systems can be found in the monograph by Conejo and Baringo [6].

2.2.1 *Day-Ahead Energy Market*

This section reviews the day-ahead energy market. This market is the one with the largest volume of energy traded and is cleared once a day, one day in advance, and generally on an hourly basis.

2.2.1.1 Production Offers and Consumption Bids

Market participants submit their offers and bids the day before the actual energy delivery. On the one hand, producers submit their production offers comprising energy blocks at increasing prices. These prices represent the minimum prices that producers are willing to accept for their production. On the other hand, consumers submit their consumption bids comprising energy blocks at decreasing prices. These prices represent the maximum prices that consumers are willing to pay for consuming energy. Then, the market operator uses these offers and bids to clear the market and

Table 2.1 Illustrative Example 2.1: production offers

Producer	Energy (MWh)	Price (€/MWh)
1	80	50
2	100	80
3	20	70
4	50	60

Table 2.2 Illustrative Example 2.1: consumption bids

Consumer	Energy (MWh)	Price (€/MWh)
1	40	85
2	100	65
3	100	75

determines the power schedule of each participant in the day-ahead energy market, as well as the market prices.

Illustrative Example 2.1 Production offers and consumption bids

The production offers of four producers are provided in Table 2.1, while the consumption bids of three consumers are provided in Table 2.2. For the sake of simplicity it is considered that producers and consumers submit a single price-quantity block.

These production offers and consumption bids are used to build the so-called stepwise increasing production offer and stepwise decreasing consumption bid curves depicted in Figs. 2.1 and 2.2, respectively. □

2.2.1.2 Market Clearing

After receiving the production offers and consumption bids, the market operator clears the market with the purpose of maximizing a measure of the business welfare. In this sense, most market clearing auctions aim at maximizing the social welfare [4]. The social welfare is the sum of the producer surplus and the consumer surplus. On the one hand, the producer surplus is equal to the difference between the actual revenue of producers (energy times market clearing price) and the revenue that they are willing to accept (energy times offer price). On the other hand, the consumer surplus is equal to the difference between the amount that the consumers are willing to pay (energy times bid price) and their payment (energy times market clearing price). The market clearing price is determined by the intersection of the offer and bid curves. All production offers with a price lower than the market clearing price are accepted, as well as all consumption bids with a price greater than the market clearing prices. The rest of production offers and consumption bids are rejected.

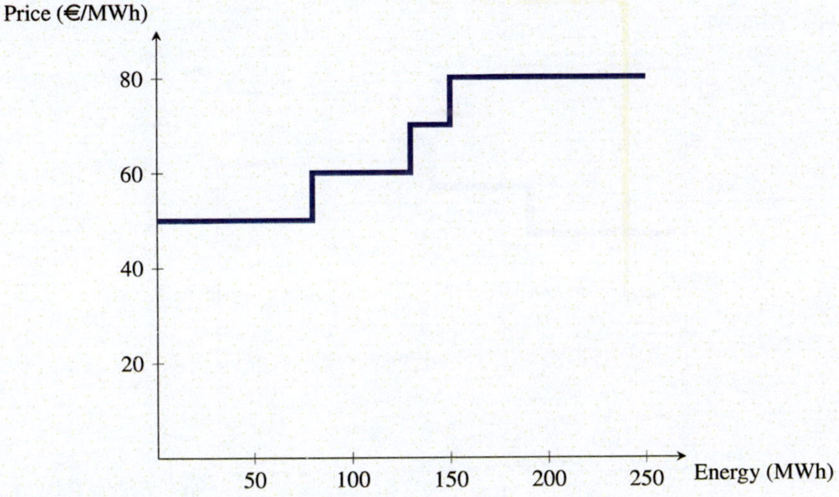

Fig. 2.1 Illustrative Example 2.1: production offer curve

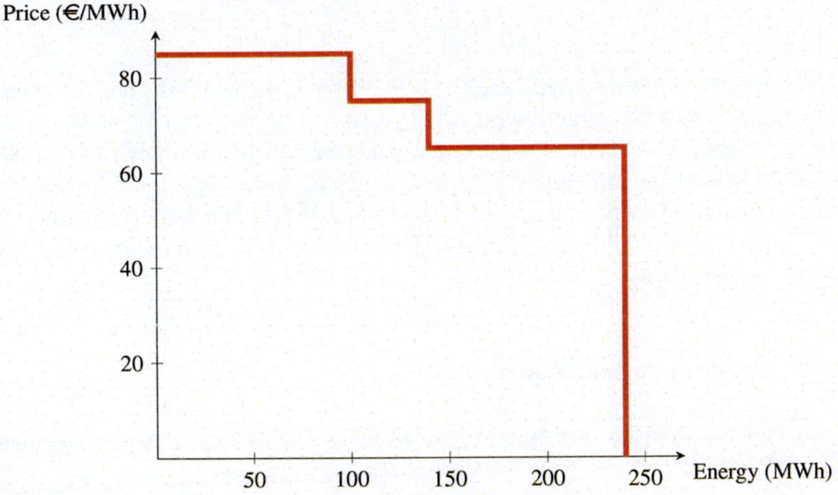

Fig. 2.2 Illustrative Example 2.1: consumption bid curve

Illustrative Example 2.2 Day-ahead market clearing price

The data of Illustrative Example 2.1 are considered again. Figure 2.3 depicts both the production offer (in blue) and consumption bid (in red) curves. Thick lines represent accepted offers and bids, whereas thin lines are related to non accepted ones. The crossing of these two curves determines the accepted production and consumption levels (total of 140 MWh) in the market, as well as the market clearing price (70 €/MWh).

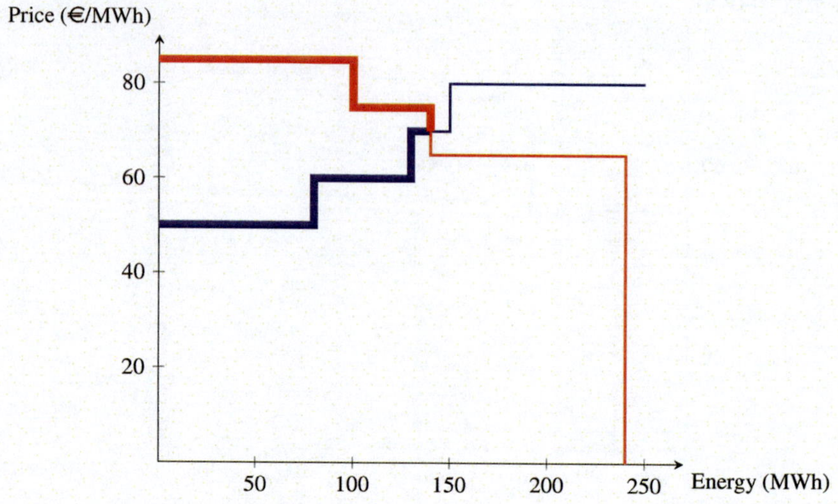

Fig. 2.3 Illustrative Example 2.2: accepted production/consumption levels and market clearing price

Note that the offers of producers 1 (80 MWh) and 4 (50 MWh) are completely accepted since they have the lowest offer prices, while the offer of producer 3 is partially accepted (10 MWh). On the other hand, the bids of consumers 1 (40 MWh) and 3 (100 MWh) are completely accepted, which have the highest bid prices. Nevertheless, the bid of consumer 2 (100 MW at 65 €/MWh) is not accepted since this would require the use of the generating unit owned by producer 3, whose offer price is higher (70 €/MWh). □

2.2.1.3 Problem Formulation

Considering the previous descriptions, the DA market clearing problem is formulated as follows:

$$\text{Maximize}_{\Phi^{\mathrm{DA}}} \sum_{t \in T} \left(\sum_{d \in D} \chi^{\mathrm{D}}_{dt} p^{\mathrm{D}}_{dt} - \sum_{g \in G} \chi^{\mathrm{G}}_{gt} p^{\mathrm{G}}_{gt} \right) \tag{2.1a}$$

subject to:

$$\sum_{g \in G} p^{\mathrm{G}}_{gt} = \sum_{d \in D} p^{\mathrm{D}}_{dt}, \quad \forall t \in T, \tag{2.1b}$$

$$0 \le p^{\mathrm{G}}_{gt} \le \overline{P}^{\mathrm{G}}_{gt}, \quad \forall g \in G, \forall t \in T, \tag{2.1c}$$

$$0 \le p^{\mathrm{D}}_{dt} \le \overline{P}^{\mathrm{D}}_{dt}, \quad \forall d \in D, \forall t \in T, \tag{2.1d}$$

where variables in set $\Phi^{\mathrm{DA}} = \left\{ p^{\mathrm{G}}_{gt}, \forall g \in G, \forall t \in T; \, p^{\mathrm{D}}_{dt}, \forall d \in D, \forall t \in T \right\}$ are the optimization variables of problem (2.1). Note that this formulation considers that each producer and consumer submits a single price-quantity block.

Model (2.1) is a linear programming (LP) problem [23]. The objective function (2.1a) represents the maximization of the social welfare. Constraints (2.1b) impose the power balance between generation and demand, while constraints (2.1c) and (2.1d) bound the power production and consumption levels to the offer and bid quantities, respectively.

Illustrative Example 2.3 Day-ahead market clearing

The data of Illustrative Example 2.2 are considered again. Using these data, the market clearing problem is formulated below.

First, the objective function to be maximized, representing the social welfare, is as follows:

$$\begin{aligned} &85p^{\mathrm{D}}_1 + 65p^{\mathrm{D}}_2 + 75p^{\mathrm{D}}_3 \\ &\qquad - 50p^{\mathrm{G}}_1 - 80p^{\mathrm{G}}_2 - 70p^{\mathrm{G}}_3 - 60p^{\mathrm{G}}_4. \end{aligned}$$

Next, the constraints of the problem are formulated:

- Power balance:

$$p^{\mathrm{G}}_1 + p^{\mathrm{G}}_2 + p^{\mathrm{G}}_3 + p^{\mathrm{G}}_4 = p^{\mathrm{D}}_1 + p^{\mathrm{D}}_2 + p^{\mathrm{D}}_3.$$

- Bounds on production levels:

$$\begin{aligned} &0 \le p^{\mathrm{G}}_1 \le 80, \\ &0 \le p^{\mathrm{G}}_2 \le 100, \\ &0 \le p^{\mathrm{G}}_3 \le 20, \\ &0 \le p^{\mathrm{G}}_4 \le 50. \end{aligned}$$

- Bounds on consumption levels:

$$\begin{aligned} &0 \le p^{\mathrm{D}}_1 \le 40, \\ &0 \le p^{\mathrm{D}}_2 \le 100, \\ &0 \le p^{\mathrm{D}}_3 \le 100. \end{aligned}$$

Finally, the optimization variables of the problem are those included in set $\Phi^{\text{IE2.3}} = \{p^{\text{G}}_1, p^{\text{G}}_2, p^{\text{G}}_3, p^{\text{G}}_4, p^{\text{D}}_1, p^{\text{D}}_2, p^{\text{D}}_3\}$. Note that the market clearing is solved for a single time period. Therefore, the previous optimization variables include a single subscript that denotes the power producer/consumer, while the subscript related to the time period is not included for the sake of clarity.

The above model is an LP problem [23] that is solved using CPLEX [27] under a GAMS environment [11].

The optimal solution of this problem is as follows:

$$\begin{aligned} p^{\text{G}*}_1 &= 80, \\ p^{\text{G}*}_2 &= 0, \\ p^{\text{G}*}_3 &= 10, \\ p^{\text{G}*}_4 &= 50, \\ p^{\text{D}*}_1 &= 40, \\ p^{\text{D}*}_2 &= 0, \\ p^{\text{D}*}_3 &= 100. \end{aligned}$$

Note that this solution is identical to that provided in Illustrative Example 2.2.

Finally, the optimal value of the objective function, which represents the social welfare, is 3200 €. □

A GAMS [11] code for solving Illustrative Example 2.3 is provided in Sect. 2.8.

Illustrative Example 2.4 Day-ahead market prices

The data and results of Illustrative Example 2.3 are considered again. The optimal value of the dual variable associated with the power balance equation is equal to 70 €/MWh. This dual variable represents the market clearing price and is used in many electricity markets as the price paid to power producers for their production and the price paid by power consumers for their energy consumption. Note that it is a marginal price that represents the incremental cost of marginally supplying an additional unit of electrical energy [19].

Note that this market price is identical to that achieved in Illustrative Example 2.2 as the crossing of offer and bid curves. □

2.2.2 *Ancillary Services Market*

The system operator must ensure a reliable energy supply. Therefore, it schedules some capacity that is not used but is ready to be used in case any contingency occurs, e.g., the failure of a generating unit or an unexpected increase in the demand. This reserve capacity is generally scheduled one day in advance in the so-called ancillary services market, also known as reserve market. As in the day-ahead energy market described in Sect. 2.2.1, participants in the ancillary services market submit their offers, which in this case indicate the price they are willing to accept to reserve some of their capacity and be ready to inject some additional energy if needed. Then, once all the offers are collected, the reserve capacity is scheduled to those units that provide it at minimum cost. Therefore, the clearing of the ancillary services market can be formulated as the following LP problem:

$$\text{Minimize}_{\Phi^{\text{AS}}} \quad \sum_{t\in T}\sum_{o\in O} \chi^{\text{R}}_{ot} p^{\text{R}}_{ot} \tag{2.2a}$$

subject to:

$$\sum_{o\in O} p^{\text{R}}_{ot} = R_t, \quad \forall t \in T, \tag{2.2b}$$

$$0 \leq p^{\text{R}}_{ot} \leq \overline{P}^{\text{R}}_{ot}, \quad \forall o \in O, \forall t \in T, \tag{2.2c}$$

where variables in set $\Phi^{\text{AS}} = \left\{p^{\text{R}}_{ot}, \forall o \in O, \forall t \in T\right\}$ are the optimization variables of problem (2.2).

The objective function (2.2a) minimizes the reserve scheduling costs. Constraints (2.2b) impose the balance between scheduled capacity and reserve needed in each time period. Note that these reserve requirements represented by parameters R_t generally depend on the demand forecast and the expected production of intermittent power units in each time period. Finally, constraints (2.2c) bound the capacity scheduled to each producer.

Illustrative Example 2.5 Ancillary services market

The capacity offers of four producers participating in the ancillary services market are provided in Table 2.3, while the required reserve capacity is 60 MW.

Considering the above data, the formulation of the ancillary services market clearing problem is as follows:

$$\text{Minimize}_{\Phi^{\text{IE2.5}}} \quad 30p^{\text{R}}_1 + 20p^{\text{R}}_2 + 40p^{\text{R}}_3 + 10p^{\text{R}}_4$$

Table 2.3 Illustrative Example 2.5: capacity offers

Producer	Capacity (MW)	Price (€/MW)
1	50	30
2	40	20
3	20	40
4	10	10

subject to:

$$p_1^R + p_2^R + p_3^R + p_4^R = 60,$$
$$0 \leq p_1^R \leq 50,$$
$$0 \leq p_2^R \leq 40,$$
$$0 \leq p_3^R \leq 20,$$
$$0 \leq p_4^R \leq 10,$$

where $\Phi^{\text{IE2.5}} = \left\{p_1^R, p_2^R, p_3^R, p_4^R\right\}$.

The solution of this problem is trivial, namely, $p_1^R = 10$ MW, $p_2^R = 40$ MW, $p_3^R = 0$, and $p_4^R = 10$ MW. □

2.3 Modeling of the Power Production by Renewable Units

The modeling of the production obtained from renewable generating units is necessary for planning and operating power systems. The intermittency, variability, and uncertainty of the power output of most of renewable technologies cause that the modeling of this production be more complex than that of conventional thermal units. Additionally, the dependency of the power production with respect to weather conditions make it necessary to characterize mathematically the physical phenomena underlying the power production of renewable technologies.

In this section we focus on the power production of wind power and solar PV technologies. These technologies have been widely installed during the last years, and it is expected that a large number of power systems all around the globe be based on these two technologies in the future.

A monograph describing the fundamentals of renewable energies is [24]. Relevant monographs focused on wind and solar PV power technologies are [1, 3, 17, 30], respectively.

2.3.1 *Solar PV Power*

A utility-scale solar PV facility comprises basically PV panels, inverters, and a transformation center. The PV panels capture the solar radiation and generate direct current power. The PV panels consist of a set of interconnected solar cells that are able to transform solar radiation into electrical power. The absorption of light by the solar cells generates, as a consequence of the photoelectric effect, an electron-hole pair. The electron and hole are then separated by the structure of the device generating electrical power. Solar cells are mainly made of a semiconducting material in the

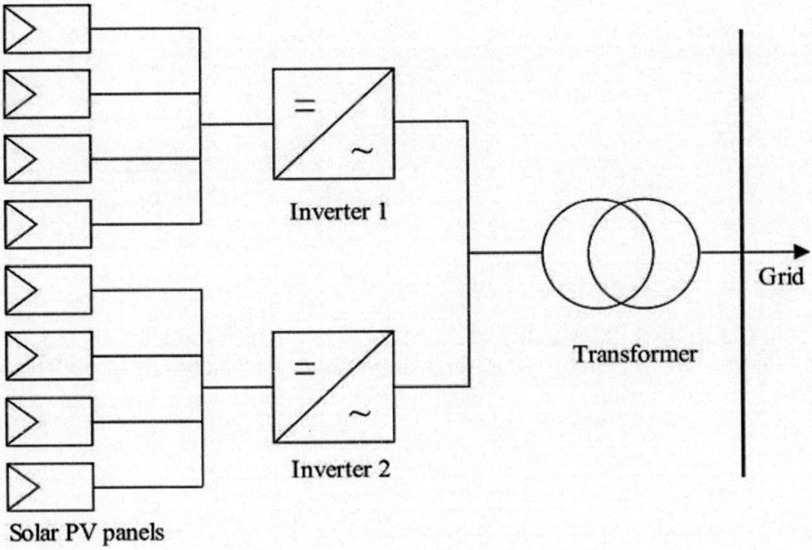

Fig. 2.4 Example of single-line chart of a utility-scale solar PV unit

form of a p-n junction. Silicon is the most used semiconductor in solar cells. An inverter is an electronic device whose purpose is to convert direct current power into alternating current. The voltage of the power output of the inverter is increased in the transformation center to inject the power production into the grid. As an example, Fig. 2.4 represents the single-line schema of a utility-scale solar PV facility.

The power generated by PV panels depends on the solar radiation that the panel receives, G (W/m^2), and the temperature of the panel, T (K). Therefore, the production of the panel will be maximum if the position of the sun is perpendicular to the surface of the panel, the atmosphere is clear of clouds, the temperature is low, and the panel is completely clean.

The equivalent circuit of an ideal photovoltaic panel is represented in Fig. 2.5 and consists of an ideal current source connected in parallel with a real diode. The current produced by a solar panel, I, is equal to the difference between the current generated by the source, I_{ph}, and the diode current, I_{d}. Current I_{ph} is generated by the solar radiation of the panel and its value is proportional to the radiation according to the following expression:

$$I_{\text{ph}} = \frac{G}{G_{STC}} \left(I_{\text{ph,ref}} + \mu_{sc} \Delta T \right), \tag{2.4}$$

where G_{STC} is the solar radiation at Standard Test Conditions (STC), typically 1000 W/m^2. The STC are used by manufacturers to rate their products in lab conditions with prefixed values of solar radiation, G_{STC}, and temperature, T_{STC}. ΔT is the

Fig. 2.5 Equivalent circuit of an ideal solar panel

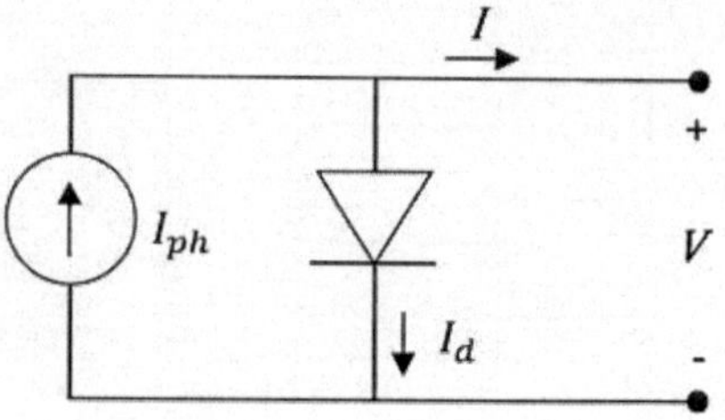

temperature variation of the panel with respect to the temperature at STC (K), while μ_{sc} is a thermal coefficient that represents the variation of the current I_{ph} with respect to the temperature variation of the panel, (A/K). The diode current is equal to:

$$I_{\text{d}} = I_0 \left(e^{\frac{V \cdot q}{k_B \cdot A \cdot N_S \cdot T}} - 1 \right), \tag{2.5}$$

where I_0 es the reverse current of the diode (A), k_B is the Boltzmann constant ($1.381 \cdot 10^{-23}$ J/K), V is the voltage of the panel (V), A is an ideal factor that depends on the material of the cell (between 1.2 and 1.3 for Si, and 1.3 for AsGa), N_S is the number of cells of the panel, and q is the electron charge ($1.602176634 \cdot 10^{-19}$ C). The reverse current of the diode, I_0, can be computed as a function of the reverse current at STC, $I_{0,\text{STC}}$, as follows:

$$I_0 = I_{0,\text{STC}} \left(\frac{T}{T_{\text{STC}}} \right)^3 e^{\left(\frac{q \cdot \Xi_{\text{G}}}{A \cdot k_{\text{B}}} \cdot \left(\frac{1}{T} - \frac{1}{T_{\text{STC}}} \right) \right)}, \tag{2.6}$$

where Ξ_{G} is the band-gap energy of the semiconductor used in the cell (eV). Current $I_{0,\text{STC}}$ can be computed as a function of the short-circuit current of the panel, $I_{\text{sc,STC}}$, and the open-circuit voltage, $V_{\text{oc,STC}}$, as STC:

$$I_{0,\text{STC}} = \frac{I_{\text{sc,STC}}}{e^{\frac{q \cdot V_{oc,STC}}{A \cdot N_S \cdot k_B \cdot T_{STC}}} - 1}. \tag{2.7}$$

Note that $I_{\text{sc,STC}}$ and $V_{\text{oc,STC}}$ are known data available in the data sheet of the considered panel.

Illustrative Example 2.6 Determination of the power production of a solar panel

Let us consider a solar monocrystalline-silicon photovoltaic panel with the following characteristics:

- Short-circuit current, $I_{sc,STC} = 10.36$ A.
- Open-circuit voltage, $V_{oc,STC} = 49.80$ V.
- Number of solar cells, $N_S = 72$.
- Ideal factor, $A = 1.3$.

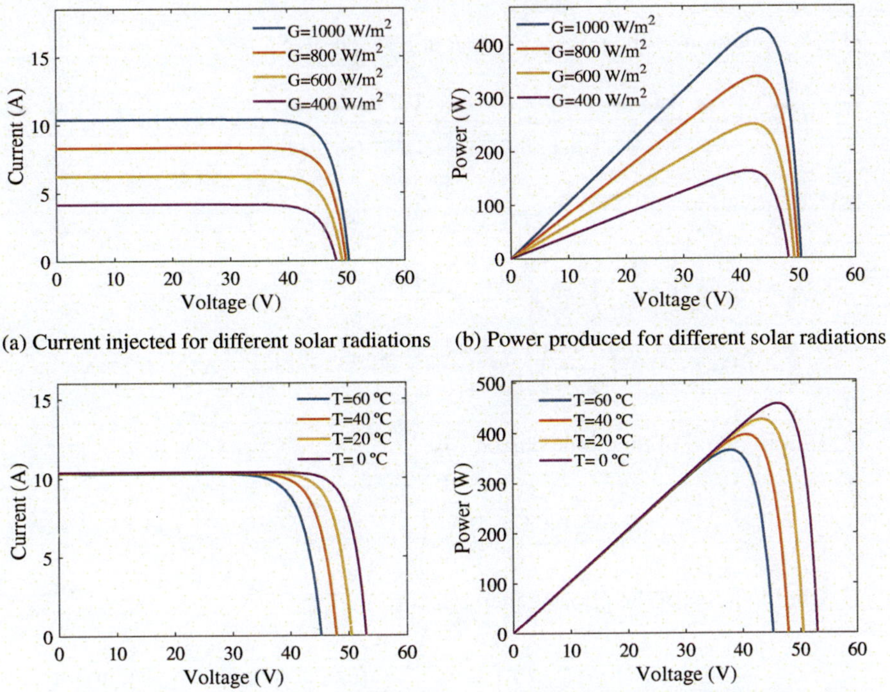

(a) Current injected for different solar radiations
(b) Power produced for different solar radiations
(c) Current injected for different temperatures
(d) Power produced for different temperatures

Fig. 2.6 Illustrative Example 2.6: current injected and power produced by a solar photovoltaic panel for different solar radiations and temperatures

- Thermal coefficient, $\mu_{sc} = 0.0013$ A/K.
- Band-gap energy, $\Xi_G = 1.12$ eV.

Considering that, at STC, $G_{STC} = 1000$ W/m^2, $T_{STC} = 295$ K, and $I_{ph,STC} = I_{sc,STC}$, Fig. 2.6 represents the current and the power generated by the considered panel as a function of the voltage of the panel for different solar radiations and temperatures obtained using expressions (2.4)–(2.7).

Figure 2.6a shows that the solar radiation has a great influence on the value of the current injected by the panel. The increment of the current causes that the power produced by the panel grows. In this manner, in Fig. 2.6b it can be observed that the maximum power obtained increases from 161.9 W for $G = 400$ W/m^2 to 426.1 W for $G = 1000$ W/m^2.

The decrease of the temperature of the panel allows to increase the voltage of the panel, as shown in Fig. 2.6c. Figure 2.6d reveals that the maximum power generated by the panel increases from 369.2 W for $T = 60$ °C to 454.9 W for $T = 0$ °C.

As an example, the power generated by the PV panel for $G = 800$ W/m^2, $T = 20$ °C, and $V = 42$ V is computed as follows:

(1) Computation of the reverse current at STC, $I_{0,\mathrm{STC}}$:

$$I_{0,\mathrm{STC}} = \frac{I_{\mathrm{sc,STC}}}{e^{\frac{q \cdot V_{oc,STC}}{A \cdot N_S \cdot k_B \cdot T_{STC}}} - 1} = \frac{10.36}{e^{\frac{1.602 \cdot 10^{-19} \cdot 49.8}{1.3 \cdot 72 \cdot 1.380649 \cdot 10^{-23} \cdot 295}} - 1} = 1.043 \cdot 10^{-8}\,\mathrm{A}.$$

(2) Computation of the reverse current, I_0:

$$I_0 = I_{0,\mathrm{STC}} \left(\frac{T}{T_{\mathrm{STC}}}\right)^3 e^{\left(\frac{q \cdot \Xi_G}{A \cdot k_B} \cdot \left(\frac{1}{T} - \frac{1}{T_{\mathrm{STC}}}\right)\right)}$$

$$= 1.043 \cdot 10^{-8} \left(\frac{293}{295}\right)^3 e^{\left(\frac{1.602 \cdot 10^{-19} \cdot 1.12}{1.3 \cdot 1.380649 \cdot 10^{-23}} \cdot \left(\frac{1}{293} - \frac{1}{295}\right)\right)} = 6.032 \cdot 10^{-8}\,\mathrm{A}.$$

(3) Computation of the diode current, I_{d}:

$$I_{\mathrm{d}} = I_0 \left(e^{\frac{V \cdot q}{k_B \cdot A \cdot N_S \cdot T}} - 1\right) = 6.032 \cdot 10^{-8} \left(e^{\frac{42 \cdot 1.602 \cdot 10^{-19}}{1.380649 \cdot 10^{-23} \cdot 1.3 \cdot 72 \cdot 293}} - 1\right) = 1.011\,\mathrm{A}.$$

(4) Computation of the photo-generated current, I_{ph}:

$$I_{\mathrm{ph}} = \frac{G}{G_{STC}} \left(I_{\mathrm{ph,STC}} + \mu_{sc} \cdot \Delta T\right) = \frac{800}{1000}(10.36 + 0.0013 \cdot 5) = 8.304\,\mathrm{A}.$$

(5) Computation of the current generated by the photovoltaic panel, I:

$$I = I_{\mathrm{ph}} - I_{\mathrm{d}} = 8.304 - 1.011 = 7.293\,\mathrm{A}.$$

(6) Computation of the power generated by the photovoltaic panel, P:

$$P = V \cdot I = 42 \cdot 7.293 = 306.3\,\mathrm{W}.$$

□

Figure 2.7 represents the power production of an actual solar PV unit located in Toledo (Spain) during three different days in October. These days are referred to as sunny, partly cloudy, and cloudy depending on the availability of solar radiation. Panels are disposed with a declination angle equal to 30° and the latitude of the location is 39.86° N. As observed, the energy production differs significantly in each day and depends on the sky conditions. For instance, the energy production during partly cloudy day is 48.2% less than in the sunny day. In the same manner, the energy production in the cloudy day represents only 14.1% of the energy production during the sunny day.

Despite of the cloud conditions, the production of a solar PV facility varies also significantly during the year. For instance, the number of daylight hours during summer is larger and the sun is higher in the sky than in the rest of seasons. Opposite to this, the number of daylight hours in winter is small and the sun is lower in the sky.

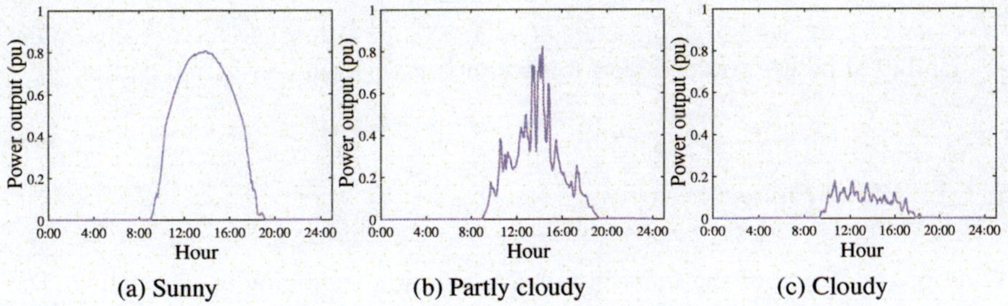

Fig. 2.7 Actual power production of a solar photovoltaic facility

Fig. 2.8 Comparison of the power production of a solar photovoltaic panel in two different sunny days of the year

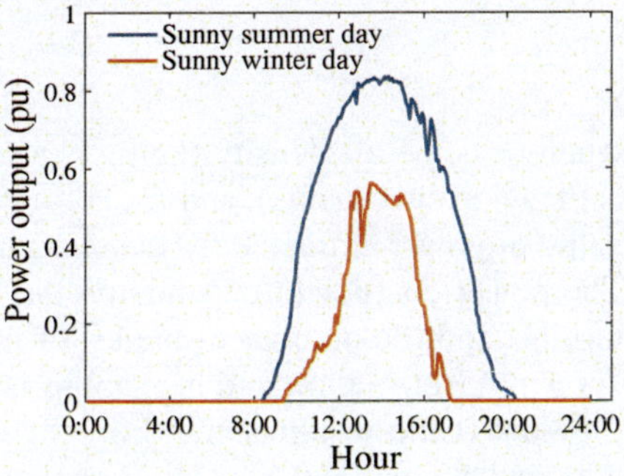

The number of daylight hours and the position of the sun depends on the period of the year but also on the latitude in which the solar PV facility is located. Locations close to the equator benefit from a number of daylight hours stable during the year, which are close to 12. On the contrary, locations close to the North and South Poles present great differences between summer and winter with a large number of daylight hours in summer, but quite low exploitable hours in winter. Therefore, the location is also a key aspect that determines the power production of a solar PV facility. Figure 2.8 represents the actual power production of a solar PV facility located in two different sunny days in two different times of the year, 4th September (summer) and 31th December (winter).

Due to the variability of the position of the sun with respect to the panels, some facilities use PV panels equipped with a tracking system that allows them to orient perpendicularly to the sun position in every time period. In this manner, the power production of the panel is maximized in each instant. However, the high decrease of the costs of PV panels during the second decade of 21st century has caused that, in some locations, tracking systems be less economically competitive that increasing the number of installed fixed panels.

As a summary, the power production of solar PV units is variable in time and in space. This reality causes that the mathematical characterization of the available power output of solar PV units be more complex than that of dispatchable generating units, as nuclear, coal, or gas units. Therefore, the planning and operation models of power systems with high presence of solar PV units must carefully represent the availability of power production of this technology in each considered period.

2.3.2 *Wind Power*

Wind power is one of the most installed generation technologies worldwide. The kinetic energy of wind is transformed into electrical power in wind turbines. The power output of a wind turbine largely depends on the wind speed as expressed by Eq. (2.8):

$$P = \frac{1}{2}\rho A C_p u^3, \tag{2.8}$$

where ρ is the air density (kg/m^3), A is the rotor area of the wind turbine (m^2), u is the wind speed (m/s), and C_p is the power coefficient that represents the per-unit value of power extracted by the wind turbine rotor to the available power of wind. The power coefficient depends on the wind speed, the rotation speed of the wind turbine, and the number of blades of the rotor. The theoretical maximum of C_p is known as Betz's limit and is equal to $16/27 = 0.593$, [3].

Wind is the result of the different thermal conditions of air masses. However, the wind at a particular location and time also depends on the terrain configuration. Therefore, the mathematical modeling of the wind is a complex task and measures on the ground are usually recommended to characterize the wind at specific locations. Figure 2.9 depicts a wind rose that is used to display graphically the orientation and frequency of different wind speeds in the cardinal directions (North, East, South, and West). In this figure it is observed that, for the considered location, wind blows mainly in North-West direction and that the most frequent wind speeds are within 5–15 m/s interval. As stated above, the most relevant magnitude that characterizes the wind is the wind speed. The wind speed determines the kinetic energy associated with a mass of air. For instance, Fig. 2.10 represents the hourly wind speed at four meters height during a week in a location in the center of Spain. This figure shows that the wind speed is variable, and also volatile.

It is also interesting to note that the wind speed varies with altitude. The Hellman's equation establishes that the wind speed profile follows a power law. According to this, assuming that the wind speed at height h_0 is u_{h0}, the wind speed at height h_1, u_{h1}, can be computed as:

$$u_{h1} = u_{h0}\left(\frac{h_1}{h_0}\right)^{\alpha}, \tag{2.9}$$

where α, representing the friction coefficient or Hellman's exponent, is used to specify the type of terrain. This coefficient may vary from 0.1, for lakes or smooth grounds, to 0.4 for city areas with high rise buildings. Figure 2.11 plots an example of the value of wind speed for different heights according to the Hellman's law.

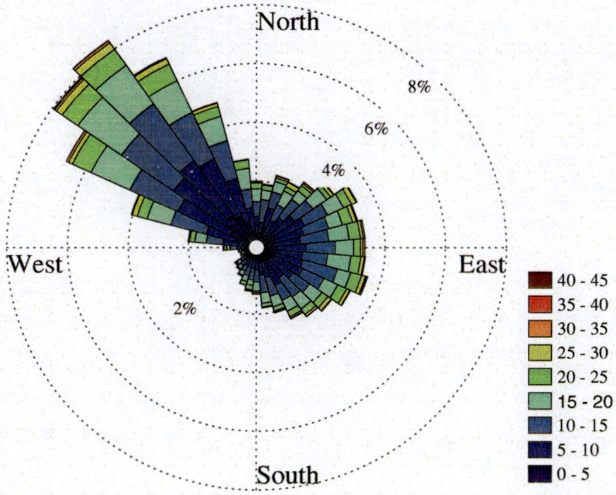

Fig. 2.9 Example of wind rose

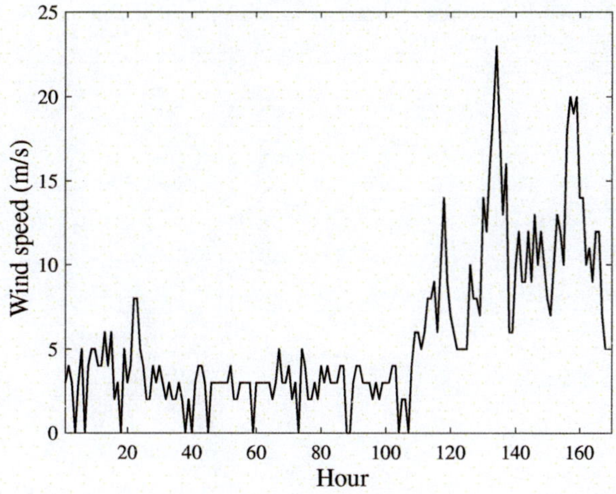

Fig. 2.10 Example of wind speed during one week

Then, using expression (2.9) it is possible to obtain wind speed values at different heights based on measures made at a specific altitude. This is of interest if it is desired to obtain the wind speed values at a wind turbines altitude (typically higher than 90 m) based on the wind speed measures obtained from a measuring center located at a much lower height (typically below 10 m).

For instance, Fig. 2.12 represents the wind speeds represented in Fig. 2.10 at 5 and 110-m height. Wind speeds at 110-m high are obtained from using Hellman's law with exponent $\alpha = 0.1$.

Illustrative Example 2.7 Computation of wind speed at a different height using Hellman's law

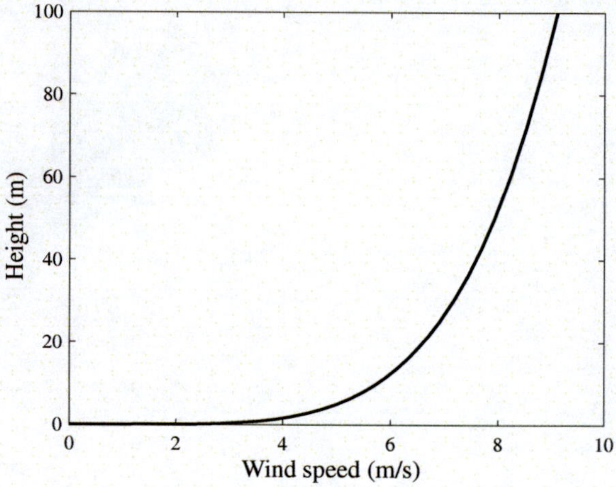

Fig. 2.11 Example of wind speed profile for different heights

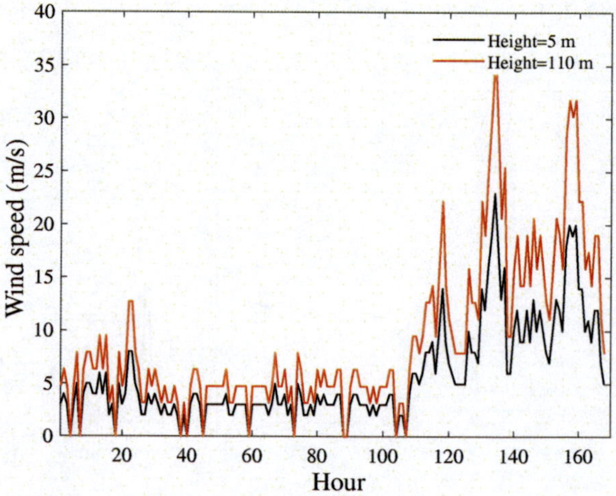

Fig. 2.12 Example of wind speed during one week at different heights

Table 2.4 Illustrative Example 2.7: wind speed at 5 m

Hour	Wind speed (m/s)
1	1
2	5
3	3

Let us consider the set of wind speed measures included in Table 2.4. These measures refer to hourly average values of wind speed at 5 m high. Considering these values it is desired to obtain the corresponding wind speeds at 100-m high using the Hellman's law formulated in (2.9) considering that the ground is either (i) smooth or (ii) rough. These two types of terrains are represented by Hellman's exponents equal to $\alpha = 0.1$ and 0.4, respectively.

The wind speeds at 100-m high for each time period and terrain type are computed below:

- Smooth ground ($\alpha = 0.1$):

$$u^{\text{Smooth}}_{100m,t=1} = u_{5m,t=1}\left(\frac{100}{5}\right)^{0.1} = 1\cdot\left(\frac{100}{5}\right)^{0.1} = 1.35\ \text{m/s},$$

$$u^{\text{Smooth}}_{100m,t=2} = u_{5m,t=2}\left(\frac{100}{5}\right)^{0.1} = 5\cdot\left(\frac{100}{5}\right)^{0.1} = 6.75\ \text{m/s},$$

$$u^{\text{Smooth}}_{100m,t=3} = u_{5m,t=3}\left(\frac{100}{5}\right)^{0.1} = 3\cdot\left(\frac{100}{5}\right)^{0.1} = 4.05\ \text{m/s}.$$

- Rough ground ($\alpha = 0.4$):

$$u^{\text{Rough}}_{100m,t=1} = u_{5m,t=1}\left(\frac{100}{5}\right)^{0.4} = 1\cdot\left(\frac{100}{5}\right)^{0.4} = 3.31\ \text{m/s},$$

$$u^{\text{Rough}}_{100m,t=2} = u_{5m,t=2}\left(\frac{100}{5}\right)^{0.4} = 5\cdot\left(\frac{100}{5}\right)^{0.4} = 16.57\ \text{m/s},$$

$$u^{\text{Rough}}_{100m,t=3} = u_{5m,t=3}\left(\frac{100}{5}\right)^{0.4} = 3\cdot\left(\frac{100}{5}\right)^{0.4} = 9.94\ \text{m/s}.$$

Figure 2.13 represents graphically the obtained wind speed values. Observe that, for identical wind speeds at 5 m high, wind speeds are higher at 100 m for the rough terrain. The reason of this result is that the increment of wind speed with altitude is higher in rough terrains than in smooth ones. This fact is observed more clearly in Fig. 2.14, where wind speed-height pairs are represented for both types of terrains.

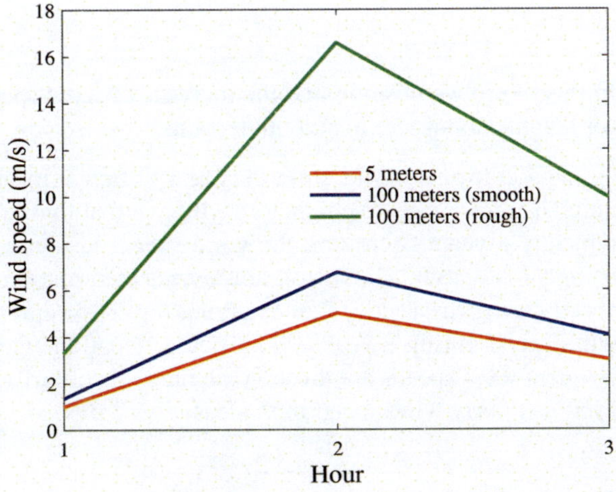

Fig. 2.13 Illustrative Example 2.7: wind speeds for different heights and ground conditions

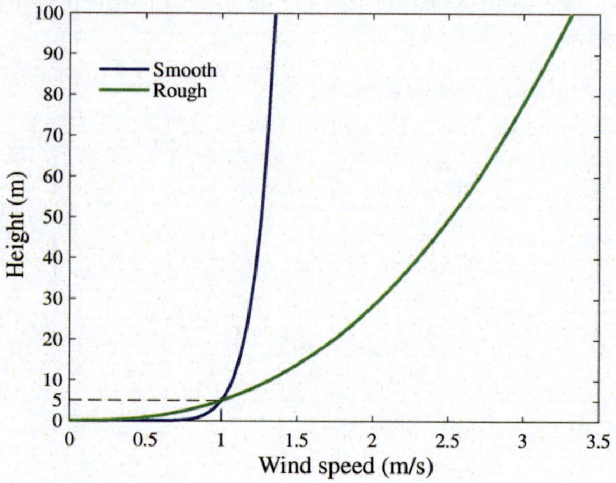

Fig. 2.14 Illustrative Example 2.7: wind speed profiles for different heights

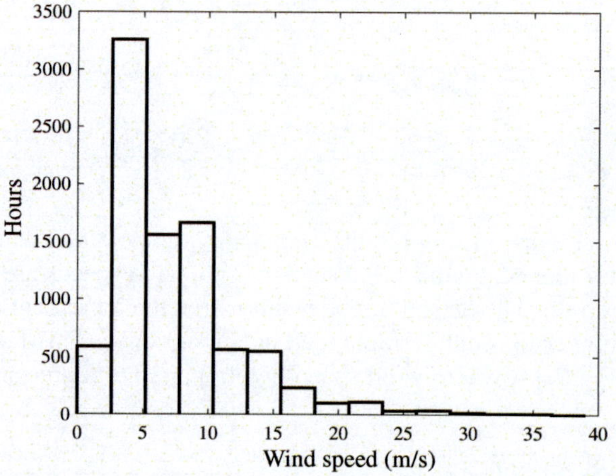

Fig. 2.15 Example of wind speed histogram

In this figure we observe that the increase of wind speed with altitude is very small for heights about 5 m in smooth terrains. □

In order to analyze the suitability of a particular location for installing wind power units, it is necessary to characterize the wind at that location for a long time period, typically a year. Therefore, the wind speed during a year is usually characterized using a histogram. The histogram reveals information about the frequency of wind speed during the year. As an example, Fig. 2.15 represents the histogram of hourly wind speed during a year at a particular location. This figure shows that the most frequent wind speeds are those in the interval comprised between 2.5 and 5 m/s. The average hourly wind speed in this case is 6.2 m/s.

Illustrative Example 2.8 Wind speed histogram

Let us consider a location with the wind speed data at 5 m height listed in Table 2.5. The total number of observations is 8760.

The histogram representing the values listed in Table 2.5 is depicted in Fig. 2.16.

The expected wind speed is equal to:

$$\bar{u} = \frac{\sum_{i=1}^{n_B} n_i u_i}{n_T} = \frac{657 \cdot 0 + 1205 \cdot 3 + 1971 \cdot 6 + 1752 \cdot 9 + \cdots + 109 \cdot 30}{8760} = 9.56 \text{ m/s},$$

where u_i is the wind speed associated to block i, n_i is the number of observations associated to block i, and n_T is the total number of observations.

Table 2.5 Illustrative Example 2.8: wind speed data at 5 m high

Wind speed (m/s)	Frequency (h)
0	657
3	1205
6	1971
9	1752
12	1095
15	876
18	438
21	328
24	219
27	110
30	109

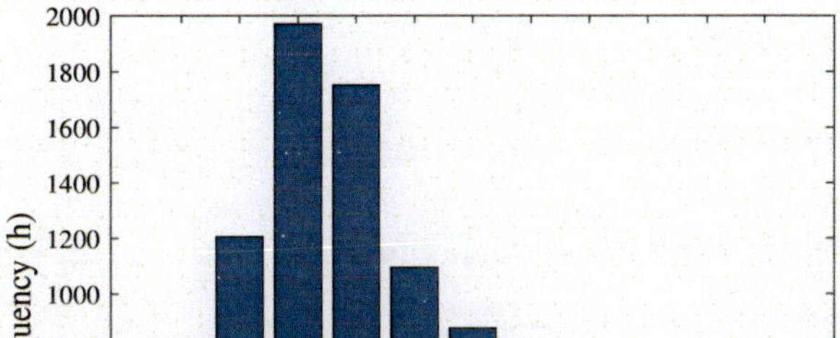

Fig. 2.16 Illustrative Example 2.8: histogram of wind speeds

The standard deviation of wind speed is computed as follows:

$$\sigma = \sqrt{\frac{\sum_{i=1}^{n_B} n_i (u_i - \bar{u})^2}{n_T}} = \sqrt{\frac{657 \cdot (0 - 9.56)^2 + \cdots + 109 \cdot (30 - 9.56)^2}{8760}} = 6.49 \text{ m/s}.$$

The probability of wind speed to be greater than 21 m/s is the following:

$$P(u > 21) = \frac{\sum_{\{i=1:n_B/(u_i>21)\}} n_i}{n_T} = \frac{219 + 110 + 109}{8760} = 0.05.$$

The probability of wind speed to be less than or equal to 12 m/s is computed as follows:

$$P(u \leq 12) = \frac{\sum_{\{i=1:n_B/(u_i\leq 12)\}} n_i}{n_T} = \frac{657 + 1205 + 1971 + 1752 + 1095}{8760} = 0.76.$$

□

For some applications, it is desirable to use a continuous distribution function to characterize wind speed. In these cases, the Weibull distributions function is typically used to characterize wind speeds. The density and cumulative probability functions of the Weibull distributions, $f(x)$ and $F(x)$, are mathematically expressed as follows:

$$f(x) = \frac{k}{c}\left(\frac{x}{c}\right)^{k-1} e^{-\left(\frac{x}{c}\right)^k}, \quad \forall x \geq 0, \tag{2.10}$$

$$F(x) = 1 - e^{-\left(\frac{x}{c}\right)^k}, \quad \forall x \geq 0, \tag{2.11}$$

where c and k are the parameters of the function and named as scale and shape factors, respectively. These parameters define the particular shape of the distribution.

The mean value of a Weibull distribution with parameters c and k may be computed as follows:

$$E(x) = \frac{\Gamma\left(1 + \frac{1}{k}\right)}{c}, \quad \forall x \geq 0, \tag{2.12}$$

where $\Gamma(x)$ is the gamma function. If x is a real number, the gamma function is equal to:

$$\Gamma(x) = \int_0^\infty t^{x-1} e^{-t} dt, \quad \forall x \in R. \tag{2.13}$$

As an example, Fig. 2.17 represents a Weibull distribution with parameters $c = 7.86$ and $k = 1.34$ to characterize the wind speed values shown in Fig. 2.15.

Figure 2.18 represents different Weibull distribution with different scale and shape factor values. From Fig. 2.18a we observe that, if the shape factor is fixed, different scale factor values lead to distributions with different mean and standard deviation

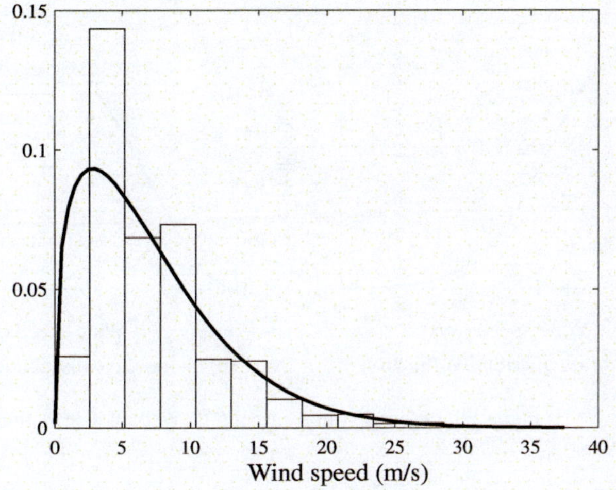

Fig. 2.17 Example of wind speed adjusted by a Weibull distribution

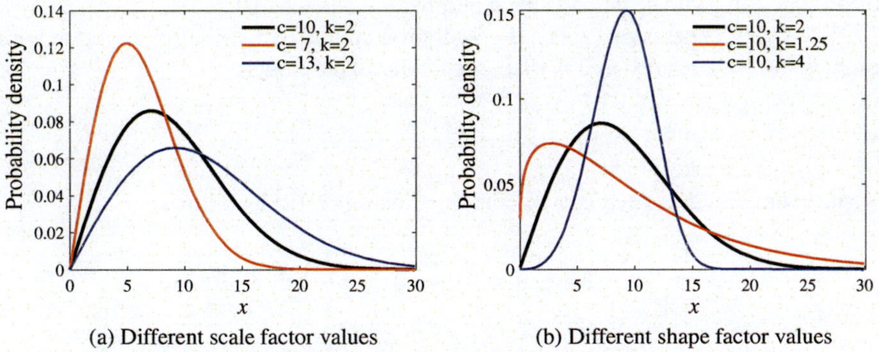

Fig. 2.18 Weibull distributions with different scale and shape factor values

values, but similar shapes. On the contrary, Fig. 2.18b shows that decreasing the shape factor k causes asymmetry and a widening of the distribution.

Illustrative Example 2.9 Weibull distribution

Let us consider a location where the wind speed follows a Weibull distribution with scale and shape parameters equal to 10 and 2, respectively. Considering this, it is desired to compute the following values:

- Average wind speed.
- The probability of wind speed to be less than 20 m/s.
- The probability of wind speed to be greater than 3 m/s.
- The probability of wind speed to be comprised in the interval between 10 and 20 m/s.

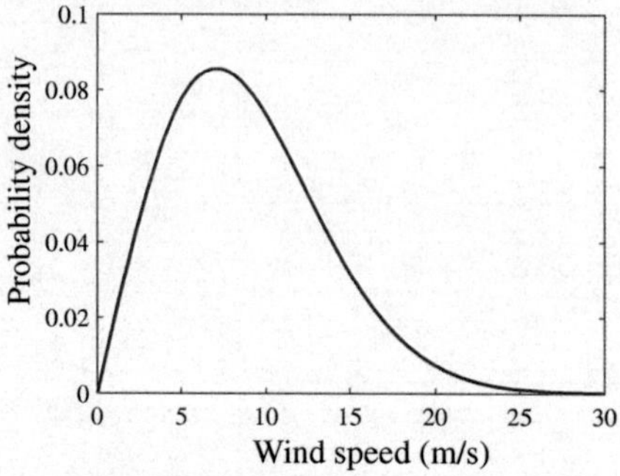

(a) Probability distribution function

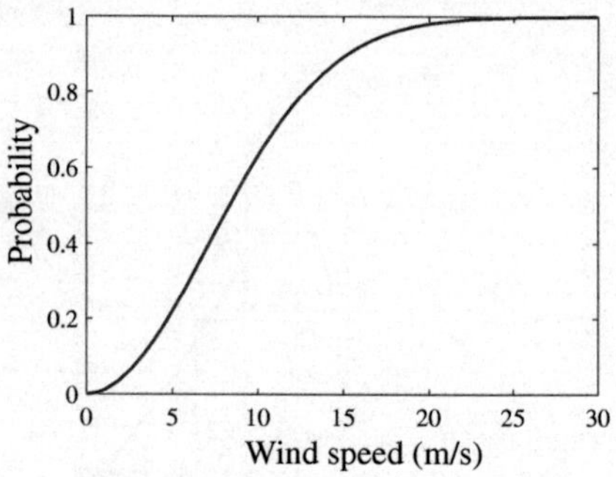

(b) Cumulative distribution function

Fig. 2.19 Illustrative Example 2.9: probability and cumulative Weibull distribution functions with parameters $c = 10$ and $k = 2$

Figure 2.19 represents respectively the probability and cumulative Weibull distribution functions with scale and shape parameters equal to 10 and 2.

Considering expression (2.11), the Weibull cumulative distribution function with parameters $c = 10$ and $k = 2$ is mathematically expressed as:

$$F(x) = 1 - e^{-\left(\frac{x}{c}\right)^k} = 1 - e^{-\left(\frac{x}{10}\right)^2}.$$

The average wind speed can be computed using (2.12) as follows:

$$\bar{u} = c \cdot \Gamma\left(1 + \frac{1}{k}\right) = 10 \cdot \Gamma\left(1 + \frac{1}{2}\right) = 8.86\,\text{m/s}.$$

The probability of wind speed to be lower than 20 m/s is computed as:

$$P(x \leq 20) = F(20) = 1 - e^{-\left(\frac{20}{10}\right)^2} = 0.98.$$

The probability of wind speed to be greater than 3 m/s is computed as:

$$P(x \geq 3) = 1 - P(x \leq 3) = 1 - F(3) = 1 - \left(1 - e^{-\left(\frac{3}{10}\right)^2}\right) = 0.92.$$

Finally, the probability of wind speed to be comprised between 10 and 20 m/s is computed as follows:

$$P(10 \leq x \leq 20) = P(x \leq 20) - P(x \leq 10) = F(20) - F(10)$$
$$= 1 - e^{-\left(\frac{20}{10}\right)^2} - \left(1 - e^{-\left(\frac{10}{10}\right)^2}\right) = 0.35.$$

□

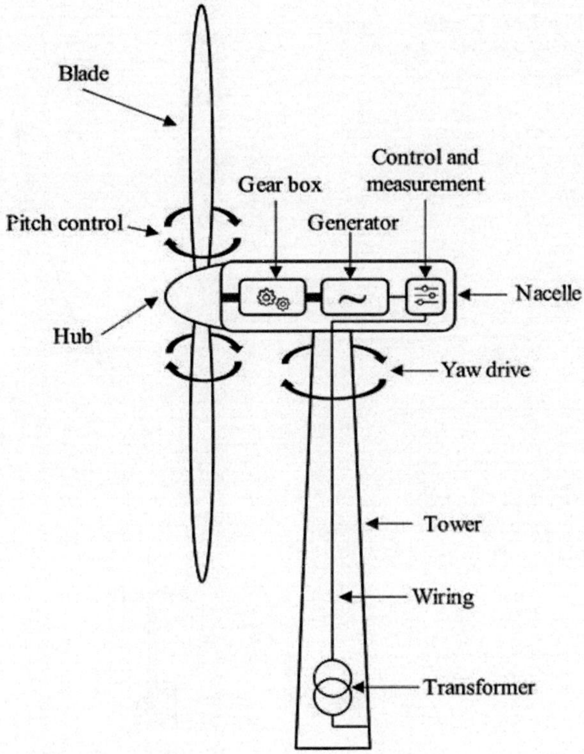

Fig. 2.20 Schema of a wind turbine

Wind turbines are used to transform the kinetic power of wind into electrical power. Figure 2.20 represents the schema of a wind turbine. Utility-scale wind turbines usually have three blades to optimize the value of the power factor C_p included in Eq. (2.8). The blades are connected to the rotor of the turbine through the hub. The pitch control allows the rotation of the blades over their longitudinal axis to modify the aerodynamic behavior of the blades and, therefore, the power produced. Traditionally, wind turbines have used a gear box to increase the rotation speed of the axis of the generator and reduce its number of poles and size. The generator is responsible of converting mechanical power into electrical power. There are different types of generators that are typically used in wind power applications, which are described below. However, new generators with variable number of poles have been used successfully to eliminate the need of using a gear box with the objective of reducing costs and increasing efficiency. The nacelle accommodates the gear box, the generator, and control and measurement elements. The tower sustains and elevates the rotor of the wind turbine and may contain the transformer, wiring, and several measuring and protection elements. The nacelle of the wind turbine may rotate over the tower to orientate according to the wind direction thanks to the yaw drive located in the tower.

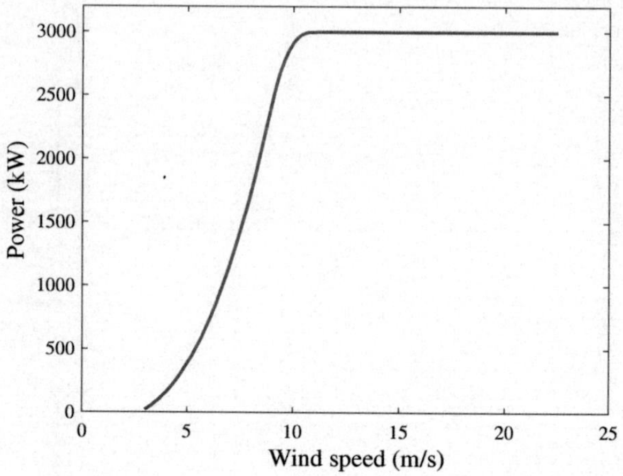

Fig. 2.21 Example of wind power curve (3 MW)

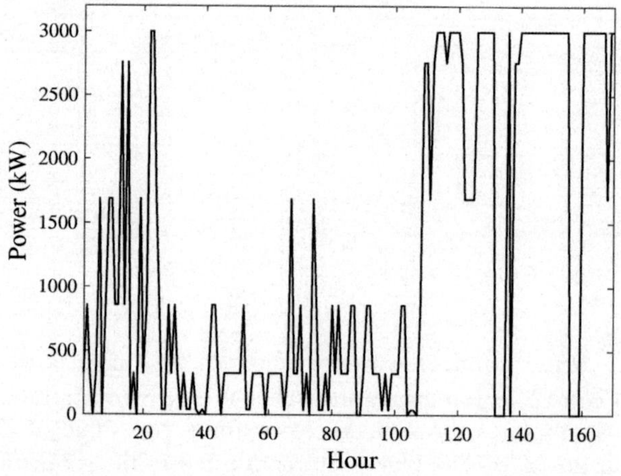

Fig. 2.22 Example of power production of a wind turbine during one week (3 MW)

The power produced by a specific wind turbine for different wind speeds is represented by the so-called power curve. Figure 2.21 represents a typical power curve corresponding to a 3-MW wind turbine.

Figure 2.22 represents the average power obtained in each hour of one week considering the wind speed at 110-m high depicted in Fig. 2.12 and the power curve represented in Fig. 2.21.

Illustrative Example 2.10 Example of wind power production

Let us consider a 3-MW wind turbine placed in a plain terrain with a hub height of 120 m and a power curve described by expression (2.14):

$$p(u) = \begin{cases} 0 & 0 \leq u \leq 3 \\ 0.88888u^3 & 3 \leq u \leq 15 \\ 3 \cdot 10^3 & 15 \leq u \leq 28 \\ 0 & u > 28 \end{cases} \quad \text{(kW)}. \tag{2.14}$$

Suppose that it is desired to compute the energy produced by this wind turbine during a year considering the wind profile included in Illustrative Example 2.8. The process to compute this energy is the following:

- For each block of the histogram depicted in Fig. 2.16:
  - Transform the wind speed measured at 10-m height to the hub height, 120 m, using the Hellman's law (2.9).
  - Compute the power associated with the wind speed at hub height using the power curve expression (2.14).
  - Compute the energy associated with each block by multiplying for each block the power times the frequency listed in Table 2.5.
- To conclude, sum the energy associated with each block to obtain the total energy generated by the wind turbine.

As an example, the process for computing the energy associated with the second block of Table 2.5 is the following:

- The wind speed of the second block at 10-m height is 3 m/s.
- Using the Hellman's law, and considering $\alpha = 0.1$ for plain terrains, the wind speed at 120 m/s is computed as:

$$u_{120} = u_{10} \left(\frac{120}{10}\right)^{0.1} = 3 \cdot \left(\frac{120}{10}\right)^{0.1} = 3.846 \text{ m/s}.$$

- According to expression (2.14), the power associated to 3.846 m/s is equal to $0.88888 \cdot 3.846^3 = 50.578$ kW.
- Finally, considering that the second block has a frequency equal to 1205 h (Table 2.5), the energy associated to this block is equal to $50.578 \cdot 1205 = 60946.4$ kWh $= 60.9464$ MWh.

If this process is repeated for each block defined in Table 2.5, the energy generated for each block is listed in Table 2.6 and depicted in Fig. 2.23. The highest production is obtained for a wind speed equal to 12 m/s at 10-m height, and 15.385 m/s at hub height. At this wind speed, the wind power unit operates at maximum capacity and it has a relatively high frequency of 1095 h per year.

The number of hours equivalent of nominal output of the wind power unit can be computed as the energy produced divided by the power capacity, which in this case is equal to $11462.0/3 = 3820.7$ h.

Finally, considering an average electricity price equal to 70 €/MWh, the revenue of this wind power unit during the year is $11462 \cdot 70 = 802340$ €. □

Table 2.6 Illustrative Example 2.10: energy produced per block

Wind speed at 10 m (m/s)	Wind speed at 120 m (m/s)	Power (kW)	Frequency (h)	Energy (MWh)
0	0.000	0.0	657	0.0
3	3.846	50.6	1205	60.95
6	7.693	404.6	1971	797.51
9	11.539	1365.6	1752	2392.54
12	15.385	3000.0	1095	3285.0
15	19.231	3000.0	876	2628.0
18	23.078	3000.0	438	1314.0
21	26.924	3000.0	328	984.0
24	30.770	0.0	219	0.0
27	34.616	0.0	110	0.0
30	38.463	0.0	109	0.0
Total:			8760	11462.0

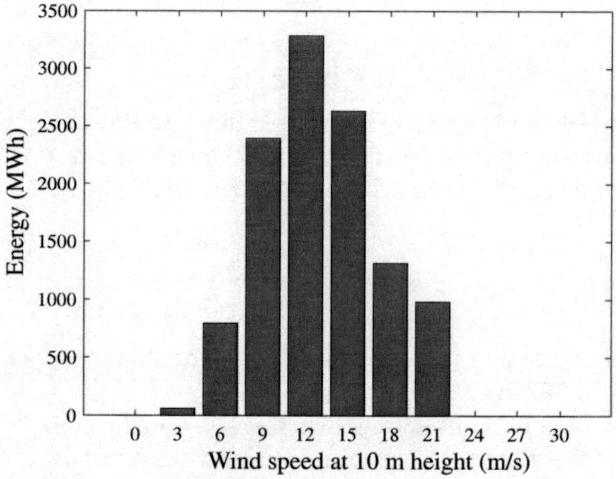

Fig. 2.23 Illustrative Example 2.10: energy produced for each wind speed value

The first wind turbines were the so-called fixed-speed wind turbines equipped with induction (asynchronous) generators, which were directly connected to the grid through a transformer. This type of wind turbine is depicted in Fig. 2.24. The reasons of using these generators for wind power applications are their robust and reliable operation as well as their low cost. Additionally, the frequency of the electrical power generated by induction generators is enforced by the grid at which they are connected, which simplifies its power control. Induction generators are designed to work at a constant speed close to the maximum torque speed that maximizes the energy production. Rotation speeds different to that value cause a high decrease of the power output of induction generators. Considering that the wind speed is variable and

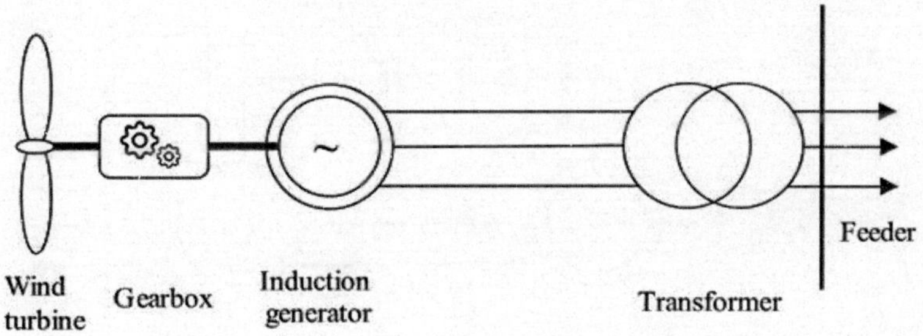

Fig. 2.24 Schema of a wind turbine with an induction generator

the generation rotational speed is nearly constant, fixed-speed wind turbines suffer from mechanical stress that derives to fluctuation on the electrical power injected into the grid. Furthermore, induction generators need the installation of soft starters to reduce the high startup currents of induction machines and capacitor banks to reduce the quantity of reactive power demanded by these generators.

In addition to the mechanical stress caused by the fixed-speed rotation of the induction generator, the inability of induction generators to adapt its rotational speed to different wind speeds causes a high reduction of power production specially when the wind speed is greater than the nominal one for which the induction generator is designed. One solution to increase the variation range of the rotational speed of inductions generators is to add a variable resistance in the rotor of the generator controlled electronically. If the value of this resistance is modified, the rotational speed of the generator at which the maximum torque is reached is also modified. Then it is possible to vary the rotational speed of the generator up to 8% maintaining the maximum torque speed for different wind speed situations.

The solution used to change substantially the rotational speed of a generator to take advantage of the different wind speeds is to use power electronic converters. The development of power electronics during the first decades of the 21st century has allowed the usage of reliable and not very expensive electronic devices that are able to handle high power quantities. A power converter may enable an efficient conversion of the variable frequency outputs of induction or synchronous generators, driven by a variable speed wind turbine, to a fixed frequency adequate for the grid.

A slightly different application of power converters to wind turbines is the so-called Double Fed Induction Generator (DFIG) represented in Fig. 2.25. Basically, the objective of this power converter is to feed an adjustable frequency AC power to the rotor windings to get a desired variation in the generator speed. The advantage of this configuration is that only 30% of the power generated passes through the power converter. However, it is possible to modify the rotational speed around 30% with respect to the nominal one and to control the reactive power production. This reduces the mechanical stress of the wind turbine, increases the production, and eliminates the need of soft starters and capacity banks.

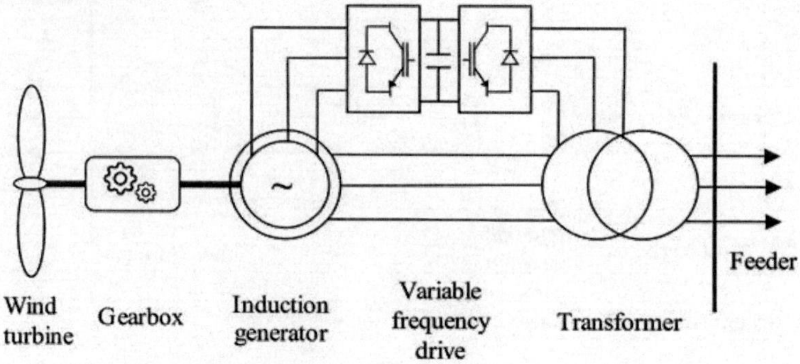

Fig. 2.25 Schema of a wind turbine with a Double-Feed Induction Generator (DFIG)

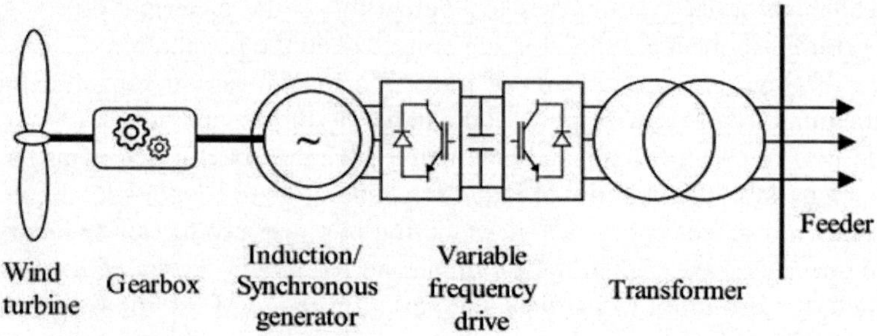

Fig. 2.26 Schema of a wind turbine with a induction/synchronous generator and variable frequency drive

The last model of variable generator is represented in Fig. 2.26. In this case a power converter that comprises a variable frequency drive that is connected directly to the output of the generator. This configuration allows the usage of either induction or synchronous generators. Unlike the DFIG, the whole production of the wind power turbine passes through the power converter. This requires the installation of high capacity power converters that increase the total cost of the wind turbine. But in return, this wind turbine configuration allows a total active and reactive power control and, in some cases, allows to exclude the usage of a gearbox if a multiple pole generator is used.

Typically, wind turbines at the same location are grouped together to constitute a so-called wind farm. Wind farms can be installed either onshore or offshore and they can comprise from only one wind turbine to hundreds. Offshore wind farms have higher capital and operation and maintenance costs, but they usually profit from higher and more stable wind speeds. Figure 2.27 represents the single-line chart of a wind farm comprising four wind turbines.

Similarly to the case of solar PV units, the power production of wind turbines is variable in time and in space. As a consequence of this, the planning and oper-

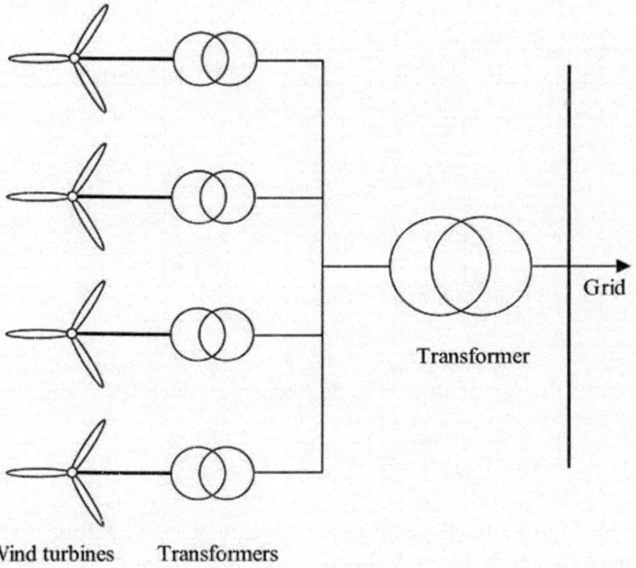

Fig. 2.27 Example of single-line chart of a wind farm

ation models of future power systems must represent carefully the availability and uncertainty of power production of wind power.

2.4 Electric Vehicles

This section describes the mathematical modeling of the charging and discharging processes of EVs.

2.4.1 Modeling of Individual Electric Vehicles

The increasing penetration of EVs in electric energy systems requires a precise characterization of their charging patterns. The most simple model is to consider that the EV charging is uncontrolled, i.e., EVs start charging as soon as they are parked and until the battery is completely filled or they depart for their next trip. This can be modeled by using the following equations:

$$p_t^{\mathrm{I,C}} = \begin{cases} 0 & \text{if } e_{t-1}^{\mathrm{I}} = \overline{E}^{\mathrm{I}} \\ U_t^{\mathrm{I}} \overline{P}^{\mathrm{I,C}} & \text{if } e_{t-1}^{\mathrm{I}} < \overline{E}^{\mathrm{I}} \end{cases}, \quad \forall t \in T. \tag{2.15}$$

Table 2.7 Illustrative Example 2.11: status of EV

Time period	Parked	Consumption [kWh]
1	Yes	–
2	Yes	–
3	No	3
4	No	4
5	Yes	–
6	No	2.5

In addition, the energy evolution in the battery of each EV is defined as:

$$e_t^{\mathrm{I}} = \min\left\{e_{t-1}^{\mathrm{I}} + p_t^{\mathrm{I,C}}\eta^{\mathrm{I,C}}\Delta_t - E_t^{\mathrm{I,C}}, \overline{E}^{\mathrm{I}}\right\}, \quad \forall t \in T. \tag{2.16}$$

Equations (2.16) impose that the energy level at time period t is equal to the minimum among the energy level in the previous time period, plus the charging energy in the current time period, minus the energy consumed by the EV in that time period, and the maximum energy level of the battery.

Illustrative Example 2.11 Uncontrolled individual EV charging

The uncontrolled charging of an individual EV during a 6-h planning horizon is analyzed below. The minimum and maximum energy levels of the battery are 2 and 10 kWh, respectively, being 5 kW the charging power level and 90% the charging efficiency rate. The initial energy level is 5 kWh. Table 2.7 provides the status of the EV in each time period, indicating whether it is parked and available for charging, and its energy consumption.

Considering Eqs. (2.15) and (2.16), the power and energy profiles of the EV are obtained as follows:

- Time period 1:

$$p_1^{\mathrm{I,C}} = 5,$$
$$e_1^{\mathrm{I}} = \min\{5 + 5 \cdot 0.9 - 0, 10\} = 9.5.$$

- Time period 2:

$$p_2^{\mathrm{I,C}} = 5,$$
$$e_2^{\mathrm{I}} = \min\{9.5 + 5 \cdot 0.9 - 0, 10\} = 10.$$

- Time period 3:

$$p_3^{\mathrm{I,C}} = 0,$$
$$e_3^{\mathrm{I}} = \min\{10 + 0 \cdot 0.9 - 3,\, 10\} = 7.$$

- Time period 4:

$$p_4^{\mathrm{I,C}} = 0,$$
$$e_4^{\mathrm{I}} = \min\{7 + 0 \cdot 0.9 - 4,\, 10\} = 3.$$

- Time period 5:

$$p_5^{\mathrm{I,C}} = 5,$$
$$e_5^{\mathrm{I}} = \min\{3 + 5 \cdot 0.9 - 0,\, 10\} = 7.5.$$

- Time period 6:

$$p_6^{\mathrm{I,C}} = 0,$$
$$e_6^{\mathrm{I}} = \min\{7.5 + 0 \cdot 0.9 - 2.5,\, 10\} = 5.$$

The price paid for the energy consumption in each hour of the planning horizon is provided in Fig. 2.28. Therefore, the total charging cost is:

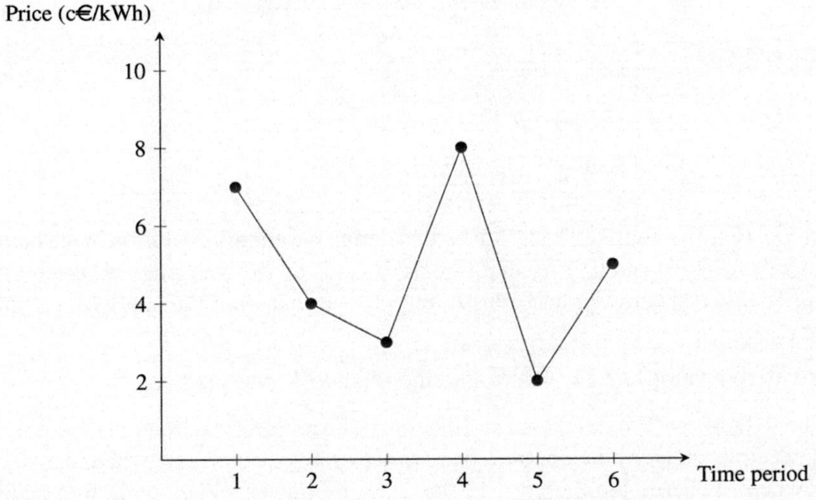

Fig. 2.28 Illustrative Example 2.11: prices

$$\text{Cost} = 7 \cdot \frac{4.5}{0.9} + 4 \cdot \frac{0.5}{0.9} + 3 \cdot 0 + 8 \cdot 0 + 2 \cdot \frac{4.5}{0.9} + 5 \cdot 0 = 47.22\,\text{c€}.$$

Note that in the above equation the energy increase in each time period, obtained as the difference of energy levels in two consecutive time periods, is divided by the corresponding charging efficiency rate to account for the charging losses. □

2.4.2 Controlled Charging and Vehicle-to-Grid Capabilities

The uncontrolled charging model characterized by Eqs. (2.15) and (2.16) has three important drawbacks, namely:

1. EVs start charging as soon as they are parked without taking into account the charging price. Therefore, charging costs may be too high depending on the starting point of the charging process.
2. EV users have similar driving patterns. For example, many EVs arrive at home at similar times in the evening after work. Thus, most EVs may start charging at the same time, which would result in a high increase in the peak demand.
3. EVs are expected to be equipped with V2G technologies in the near future, i.e., EVs may be also used to inject energy into the network if needed. This aspect is not considered in Eqs. (2.15) and (2.16).

The following controlled charging model overcomes the above issues [14]:

$$e_t^{\mathrm{I}} = e_{t-1}^{\mathrm{I}} + p_t^{\mathrm{I,C}} \eta^{\mathrm{I,C}} \Delta_t - \frac{p_t^{\mathrm{I,D}} \Delta_t}{\eta^{\mathrm{I,D}}} - E_t^{\mathrm{I,C}}, \quad \forall t \in T, \tag{2.17a}$$

$$0 \le p_t^{\mathrm{I,C}} \le U_t^{\mathrm{I}} \overline{P}^{\mathrm{I,C}}, \quad \forall t \in T, \tag{2.17b}$$

$$0 \le p_t^{\mathrm{I,D}} \le U_t^{\mathrm{I}} \overline{P}^{\mathrm{I,D}}, \quad \forall t \in T, \tag{2.17c}$$

$$\underline{E}^{\mathrm{I}} \le e_t^{\mathrm{I}} \le \overline{E}^{\mathrm{I}}, \quad \forall t \in T. \tag{2.17d}$$

Model (2.17) comprises Eqs. (2.17a) that define the energy evolution in the battery, constraints (2.17b) and (2.17c) that impose bounds on the charging and discharging power levels, respectively, and constraints (2.17d) that limit the energy level in the battery.

Illustrative Example 2.12 Controlled individual EV charging

The data of the EV described in Illustrative Example 2.11 are considered again. This example analyzes the controlled charging strategy of this EV, which can decide both when to charge the battery and the charging power level, taking into account the driving requirements as well as the charging costs. Moreover, the EV has the capability of injecting energy into the network if appropriate, being the maximum discharging power level 5 kW and being paid for this at the prices provided in Fig. 2.28. The discharging efficiency rate is 90%.

Under these conditions, the EV determines its charging strategy by solving the following optimization problem.

First, the objective function to be minimized is as follows:

$$7\left(p_1^{\mathrm{I,C}} - p_1^{\mathrm{I,D}}\right) + 4\left(p_2^{\mathrm{I,C}} - p_2^{\mathrm{I,D}}\right) + 3\left(p_3^{\mathrm{I,C}} - p_3^{\mathrm{I,D}}\right)$$
$$+ 8\left(p_4^{\mathrm{I,C}} - p_4^{\mathrm{I,D}}\right) + 2\left(p_5^{\mathrm{I,C}} - p_5^{\mathrm{I,D}}\right) + 5\left(p_6^{\mathrm{I,C}} - p_6^{\mathrm{I,D}}\right).$$

This objective function represents the charging costs minus the revenues achieved by discharging the EV.

Next, the constraints of the problem are formulated:

- Energy evolution in the battery:

$$e_1^{\mathrm{I}} = 5 + 0.9p_1^{\mathrm{I,C}} - \frac{p_1^{\mathrm{I,D}}}{0.9} - 0,$$
$$e_2^{\mathrm{I}} = e_1^{\mathrm{I}} + 0.9p_2^{\mathrm{I,C}} - \frac{p_2^{\mathrm{I,D}}}{0.9} - 0,$$
$$e_3^{\mathrm{I}} = e_2^{\mathrm{I}} + 0.9p_3^{\mathrm{I,C}} - \frac{p_3^{\mathrm{I,D}}}{0.9} - 3,$$
$$e_4^{\mathrm{I}} = e_3^{\mathrm{I}} + 0.9p_4^{\mathrm{I,C}} - \frac{p_4^{\mathrm{I,D}}}{0.9} - 4,$$
$$e_5^{\mathrm{I}} = e_4^{\mathrm{I}} + 0.9p_5^{\mathrm{I,C}} - \frac{p_5^{\mathrm{I,D}}}{0.9} - 0,$$
$$e_6^{\mathrm{I}} = e_5^{\mathrm{I}} + 0.9p_6^{\mathrm{I,C}} - \frac{p_6^{\mathrm{I,D}}}{0.9} - 2.5.$$

- Maximum charging power:

$$0 \le p_1^{\mathrm{I,C}} \le 5,$$
$$0 \le p_2^{\mathrm{I,C}} \le 5,$$
$$0 \le p_3^{\mathrm{I,C}} \le 0,$$
$$0 \le p_4^{\mathrm{I,C}} \le 0,$$
$$0 \le p_5^{\mathrm{I,C}} \le 5,$$
$$0 \le p_6^{\mathrm{I,C}} \le 0.$$

- Maximum discharging power:

$$0 \le p_1^{\mathrm{I,D}} \le 5,$$
$$0 \le p_2^{\mathrm{I,D}} \le 5,$$
$$0 \le p_3^{\mathrm{I,D}} \le 0,$$
$$0 \le p_4^{\mathrm{I,D}} \le 0,$$
$$0 \le p_5^{\mathrm{I,D}} \le 5,$$
$$0 \le p_6^{\mathrm{I,D}} \le 0.$$

- Energy bounds:

$$2 \le e_1^{\mathrm{I}} \le 10,$$
$$2 \le e_2^{\mathrm{I}} \le 10,$$
$$2 \le e_3^{\mathrm{I}} \le 10,$$
$$2 \le e_4^{\mathrm{I}} \le 10,$$
$$2 \le e_5^{\mathrm{I}} \le 10,$$
$$2 \le e_6^{\mathrm{I}} \le 10.$$

Finally, the optimization variables of the problem are those included in set $\Phi^{\mathrm{IE2.12}} = \{p_1^{\mathrm{I,C}}, p_2^{\mathrm{I,C}}, p_3^{\mathrm{I,C}}, p_4^{\mathrm{I,C}}, p_5^{\mathrm{I,C}}, p_6^{\mathrm{I,C}}, p_1^{\mathrm{I,D}}, p_2^{\mathrm{I,D}}, p_3^{\mathrm{I,D}}, p_4^{\mathrm{I,D}}, p_5^{\mathrm{I,D}}, p_6^{\mathrm{I,D}}, e_1^{\mathrm{I}}, e_2^{\mathrm{I}}, e_3^{\mathrm{I}}, e_4^{\mathrm{I}}, e_5^{\mathrm{I}}, e_6^{\mathrm{I}}\}$.

The above model is a linear programming (LP) problem [23] that is solved using CPLEX [27] under a GAMS environment [11].

The optimal charging and discharging profile is provided in Fig. 2.29, while the energy evolution in the battery is depicted in Fig. 2.30. Positive values in Fig. 2.29 denote that the EV is charging the battery, while negative values indicate that the battery is discharged. Note that the EV decides both its charging/discharging timing and power level in each time period. Therefore, it prefers to charge the EV battery in time periods 2 and 5, which has the lowest prices among the periods in which the EV is parked and available for charging, and to discharge in time period 1, with a comparatively higher price.

The total charging cost minus discharging revenue is 22.41 c€, which is 52.54% lower than in the case of the uncontrolled charging analyzed in Illustrative Example 2.11. This result highlights the importance of a controlled and smart charging strategy for EVs. □

A GAMS [11] code for solving Illustrative Example 2.12 is provided in Sect. 2.8.

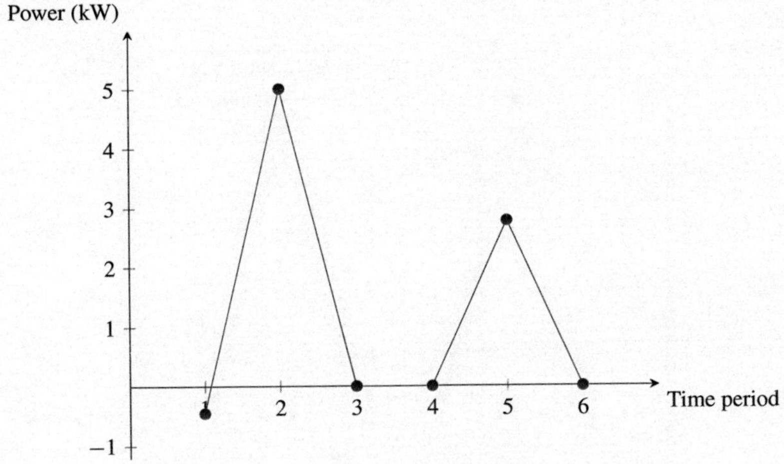

Fig. 2.29 Illustrative Example 2.12: charging and discharging power levels

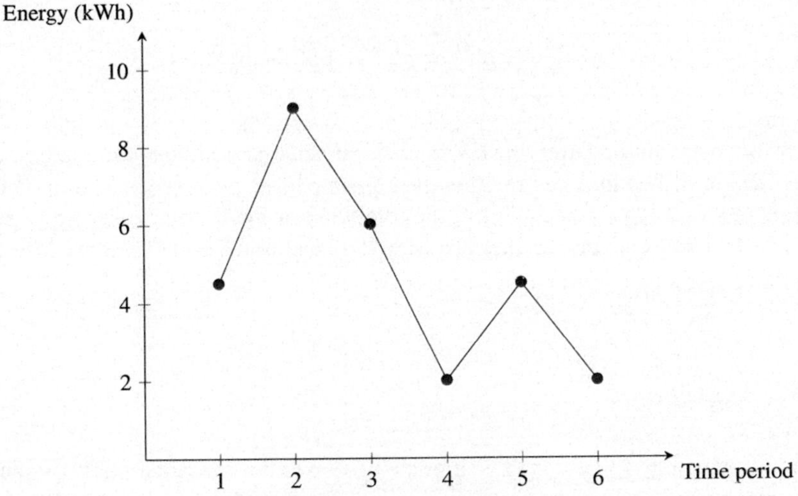

Fig. 2.30 Illustrative Example 2.12: energy level in the battery

Illustrative Example 2.13 Impact of the V2G technology

The data of Illustrative Example 2.12 are considered again. Here, the impact of the V2G technology is analyzed. With this purpose, the problem formulated in Illustrative Example 2.12 is solved again but considering in this case that the EV does not have the capability of injecting energy into the network, i.e., it is imposed that:

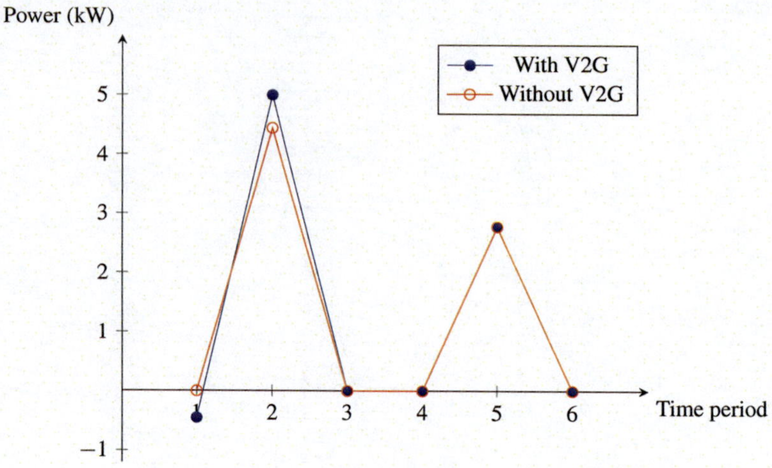

Fig. 2.31 Illustrative Example 2.13: charging and discharging power levels

$$p_1^{I,D} = p_2^{I,D} = p_3^{I,D} = p_4^{I,D} = p_5^{I,D} = p_6^{I,D} = 0.$$

Figure 2.31 shows the charging profiles of both options, namely, with and without V2G technology. In the latter, the EV does not discharge nor charge in the first time period, which also influences the charging power level in the second time period, which is reduced. As a consequence, the total cost of the EV increases up to 23.33 c€, i.e., a 4.11% higher than in the case with V2G technology analyzed in Illustrative Example 2.12. □

2.4.3 *Aggregator*

Managing EVs individually may be unrealistic due to the computational complexity of the problem if the number of EVs is large enough. Therefore, the figure of an EV aggregator is defined next. This aggregator manages a fleet of EVs, which are modeled as a *virtual battery*, whose charging and discharging model is defined using the following constraints [14]:

$$e_t^{EV} = e_{t-1}^{EV} + p_t^{EV,C}\eta^{EV,C}\Delta_t - \frac{p_t^{EV,D}\Delta_t}{\eta^{EV,D}} + E_t^{EV,IO}, \quad \forall t \in T, \tag{2.18a}$$

$$0 \le p_t^{EV,C} \le \overline{P}_t^{EV,C}, \quad \forall t \in T, \tag{2.18b}$$

$$0 \le p_t^{EV,D} \le \overline{P}_t^{EV,D}, \quad \forall t \in T, \tag{2.18c}$$

$$\underline{E}_t^{EV} \le e_t^{EV} \le \overline{E}_t^{EV}, \quad \forall t \in T. \tag{2.18d}$$

Equations (2.18a) model the energy evolution in the virtual battery representing the EV aggregation. The energy level in a given time period is equal to the energy level in the previous time period, plus the charging energy, minus the discharging energy, plus the energy contribution/drop of EVs arriving to/departing from the EV aggregation in that time period. Constraints (2.18b) and (2.18c) impose bounds on the charging and discharging power levels, respectively, while constraints (2.18d) limit the energy level of the virtual battery. The maximum charging ($\overline{P}_t^{\mathrm{EV,C}}$) and discharging ($\overline{P}_t^{\mathrm{EV,D}}$) power levels are related to the number of EVs parked and available to charge/discharge in each time period, while the minimum ($\underline{E}_t^{\mathrm{EV}}$) and maximum ($\overline{E}_t^{\mathrm{EV}}$) energy bounds are related to the driving requirements of EV users. For example, if a large number of EVs departs from the EV aggregation in a given time period, the lower energy bound in the previous time period will be comparatively higher so that EVs depart with a high energy level in their batteries. On the other hand, during night hours in which a very low number of EVs departs from the EV aggregation, this lower energy bound will be comparatively lower since there is no need to have EV batteries fully charged.

Illustrative Example 2.14 Charging strategy of an EV aggregator

The charging strategy throughout a 6-h planning horizon of an EV aggregator is analyzed below. The EV aggregation comprises 10,000 individual EVs whose driving requirements are represented using the maximum charging and discharging power levels depicted in Fig. 2.32, the minimum and maximum energy levels provided in Fig. 2.33, as well as the energy contributions/drops of EVs arriving to/departing from the virtual battery representing the EV aggregation depicted in Fig. 2.34. The maximum charging and discharging power levels are related to the number of EVs parked

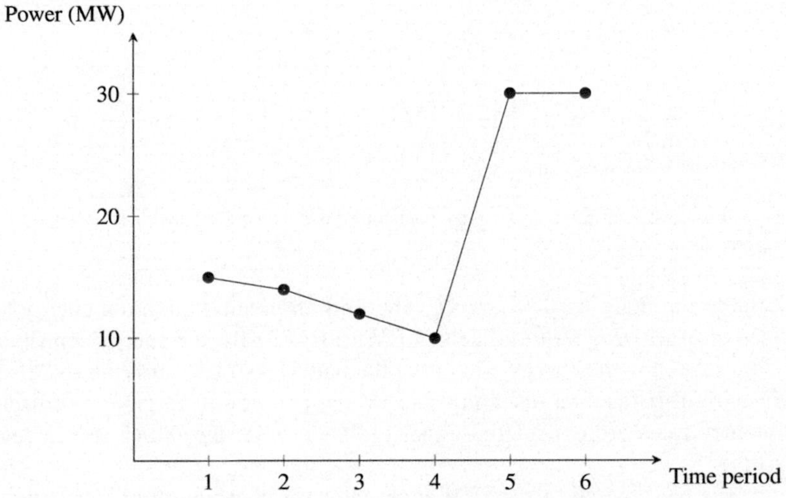

Fig. 2.32 Illustrative Example 2.14: maximum charging and discharging power levels

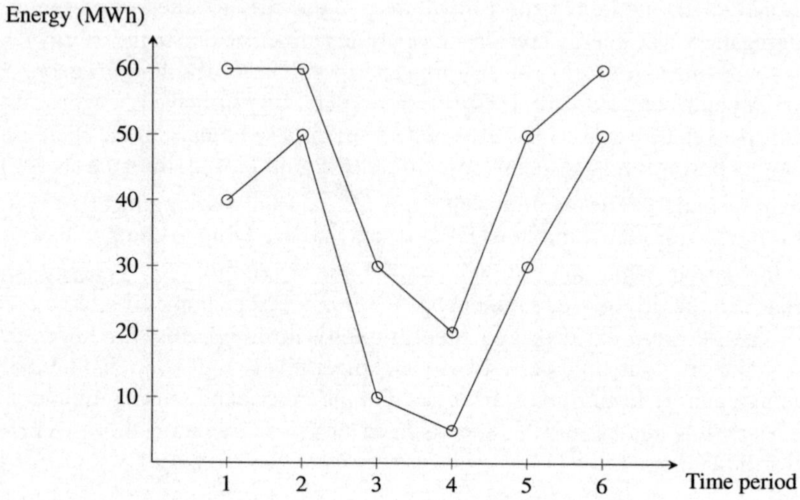

Fig. 2.33 Illustrative Example 2.14: minimum and maximum energy levels

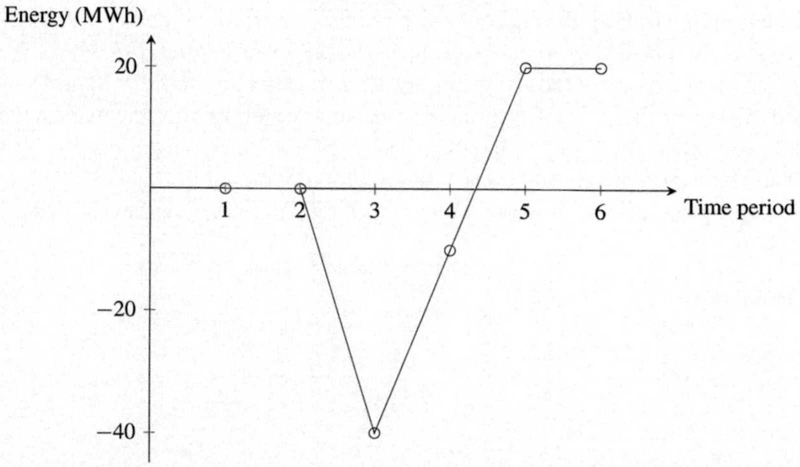

Fig. 2.34 Illustrative Example 2.14: energy contributions/drop of EVs arriving to/departing from the EV aggregation

and available for charging/discharging. The minimum and maximum energy levels are related to the driving requirements of EV users. Finally, the energy contributions/drops are related to the energy content of the batteries of EVs arriving to/departing from the virtual battery representing the EV aggregation. Charging and discharging efficiency rates are considered equal to 90%, while the initial energy level is 30 MWh.

The prices paid by/paid to the EV aggregator for charging/discharging the EVs are provided in Fig. 2.35.

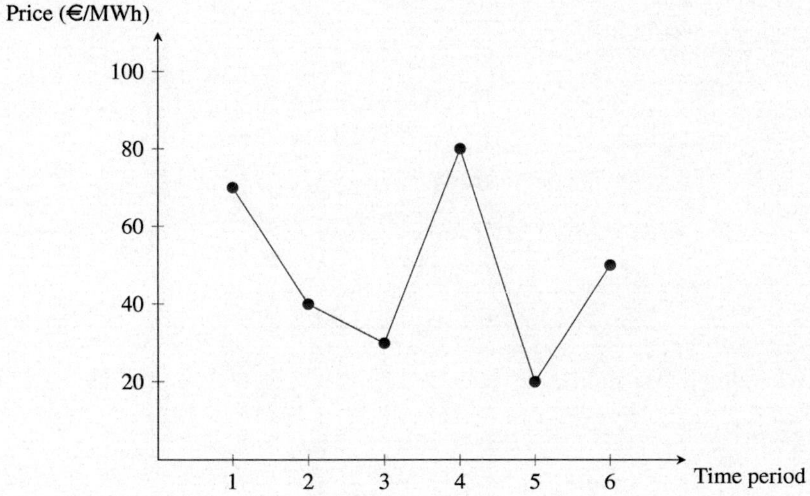

Fig. 2.35 Illustrative Example 2.14: prices

Considering these data, the problem is formulated below.

First, the objective function of the problem is the minimization of purchasing costs minus selling revenues:

$$70\left(p_1^{\mathrm{EV,C}} - p_1^{\mathrm{EV,D}}\right) + 40\left(p_2^{\mathrm{EV,C}} - p_2^{\mathrm{EV,D}}\right) + 30\left(p_3^{\mathrm{EV,C}} - p_3^{\mathrm{EV,D}}\right)$$
$$+ 80\left(p_4^{\mathrm{EV,C}} - p_4^{\mathrm{EV,D}}\right) + 20\left(p_5^{\mathrm{EV,C}} - p_5^{\mathrm{EV,D}}\right) + 50\left(p_6^{\mathrm{EV,C}} - p_6^{\mathrm{EV,D}}\right).$$

Next, the constraints of the problem are formulated:

- Energy evolution of the virtual battery representing the EV aggregation:

$$e_1^{\mathrm{EV}} = 30 + 0.9 p_1^{\mathrm{EV,C}} - \frac{p_1^{\mathrm{EV,D}}}{0.9} + 0,$$
$$e_2^{\mathrm{EV}} = e_1^{\mathrm{EV}} + 0.9 p_2^{\mathrm{EV,C}} - \frac{p_2^{\mathrm{EV,D}}}{0.9} + 0,$$
$$e_3^{\mathrm{EV}} = e_2^{\mathrm{EV}} + 0.9 p_3^{\mathrm{EV,C}} - \frac{p_3^{\mathrm{EV,D}}}{0.9} - 40,$$
$$e_4^{\mathrm{EV}} = e_3^{\mathrm{EV}} + 0.9 p_4^{\mathrm{EV,C}} - \frac{p_4^{\mathrm{EV,D}}}{0.9} - 10,$$
$$e_5^{\mathrm{EV}} = e_4^{\mathrm{EV}} + 0.9 p_5^{\mathrm{EV,C}} - \frac{p_5^{\mathrm{EV,D}}}{0.9} + 20,$$
$$e_6^{\mathrm{EV}} = e_5^{\mathrm{EV}} + 0.9 p_6^{\mathrm{EV,C}} - \frac{p_6^{\mathrm{EV,D}}}{0.9} + 20.$$

- Charging power limits:

$$0 \leq p_1^{\mathrm{EV,C}} \leq 15,$$
$$0 \leq p_2^{\mathrm{EV,C}} \leq 14,$$
$$0 \leq p_3^{\mathrm{EV,C}} \leq 12,$$
$$0 \leq p_4^{\mathrm{EV,C}} \leq 10,$$
$$0 \leq p_5^{\mathrm{EV,C}} \leq 30,$$
$$0 \leq p_6^{\mathrm{EV,C}} \leq 30.$$

- Discharging power limits:

$$0 \leq p_1^{\mathrm{EV,D}} \leq 15,$$
$$0 \leq p_2^{\mathrm{EV,D}} \leq 14,$$
$$0 \leq p_3^{\mathrm{EV,D}} \leq 12,$$
$$0 \leq p_4^{\mathrm{EV,D}} \leq 10,$$
$$0 \leq p_5^{\mathrm{EV,D}} \leq 30,$$
$$0 \leq p_6^{\mathrm{EV,D}} \leq 30.$$

- Energy bounds:

$$40 \leq e_1^{\mathrm{EV}} \leq 60,$$
$$50 \leq e_2^{\mathrm{EV}} \leq 60,$$
$$10 \leq e_3^{\mathrm{EV}} \leq 30,$$
$$5 \leq e_4^{\mathrm{EV}} \leq 20,$$
$$30 \leq e_5^{\mathrm{EV}} \leq 50,$$
$$50 \leq e_6^{\mathrm{EV}} \leq 60.$$

Finally, the optimization variables of the problem are those included in set $\Phi^{\mathrm{IE2.14}} = \{p_1^{\mathrm{EV,C}}, p_2^{\mathrm{EV,C}}, p_3^{\mathrm{EV,C}}, p_4^{\mathrm{EV,C}}, p_5^{\mathrm{EV,C}}, p_6^{\mathrm{EV,C}}, p_1^{\mathrm{EV,D}}, p_2^{\mathrm{EV,D}}, p_3^{\mathrm{EV,D}}, p_4^{\mathrm{EV,D}}, p_5^{\mathrm{EV,D}}, p_6^{\mathrm{EV,D}}, e_1^{\mathrm{EV}}, e_2^{\mathrm{EV}}, e_3^{\mathrm{EV}}, e_4^{\mathrm{EV}}, e_5^{\mathrm{EV}}, e_6^{\mathrm{EV}}\}$.

The above model is an LP problem [23] that is solved using CPLEX [27] under a GAMS environment [11].

The optimal charging and discharging profile of the virtual battery representing the EV aggregation is provided in Fig. 2.36, while the energy evolution in this virtual battery is depicted in Fig. 2.37. Note that both the charging and energy levels are within the considered lower and upper bounds. Moreover, the cheapest price corresponds to time period 5, where the EV aggregator charges the virtual battery until the upper energy level is reached. On the other hand, the time period with the highest

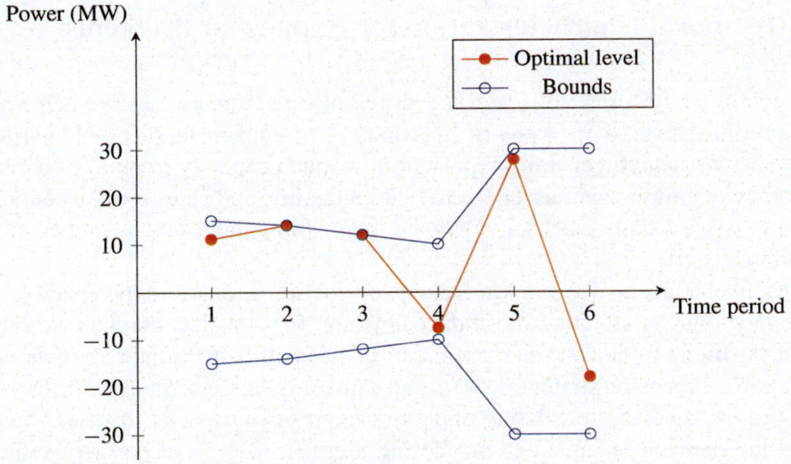

Fig. 2.36 Illustrative Example 2.14: charging and discharging power levels

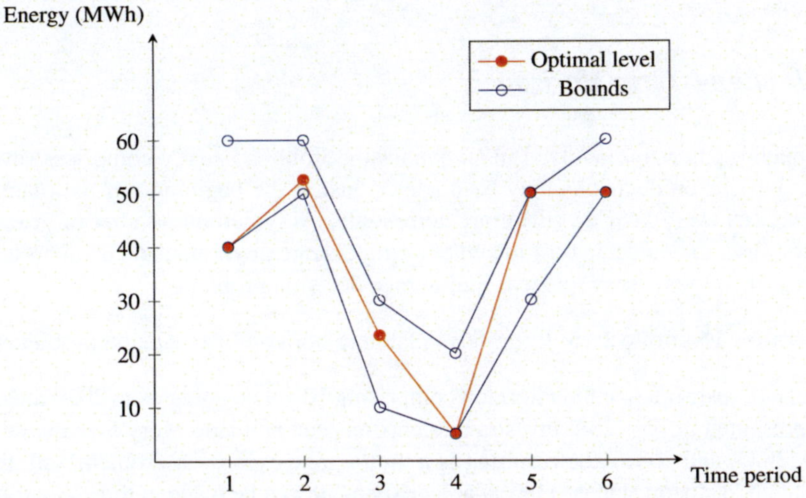

Fig. 2.37 Illustrative Example 2.14: energy level in the virtual battery representing the EV aggregation

price is time period 4, in which the EV aggregator discharges the virtual battery and is paid for this.

The total charging cost minus discharging revenue is 748.53€. If the problem is solved again but considering that EVs are not equipped with the V2G technology, i.e., they cannot discharge the batteries, then this cost increases up to 1500€. □

A GAMS [11] code for solving Illustrative Example 2.14 is provided in Sect. 2.8.

2.5 Temporal Characterization for Expansion Problems

This section studies how to model the different time frameworks we can find in expansion problems. Two types of decisions are found in these problems: long-term decisions, which generally represent how much capacity to build of a certain technology or simply whether or not to build a facility, and short-term decisions, or how to satisfy the daily electricity demand maintaining the technical and economic feasibility.

Capacity expansion decisions are taken considering a long-term perspective. The uncertainty sources affecting those decisions are, for instance, the annual demand growth, the future capital and operating costs of the facilities to be built, the national or international expansion strategies and the sustainable goals, etc. Short-term decisions imply, for instance, the scheduling of energy and reserve capacity in the DA market so that the demand is satisfied considering a certain degree of security, either to counteract the prediction errors of the demand and intermittent renewable power, or against the failure of some facilities.

2.5.1 Long-Term Horizon

The planning horizon involved in an expansion problem usually comprises several years. Expansion decisions may be made (i) just at the beginning of the planning horizon, (ii) every year or, (iii) more commonly, every certain number of years. In the first case, the model is formulated as a single-stage investment decision problem, whereas the other cases represent multi-stage decision problems.

Illustrative Example 2.15 Long-term planning horizon in expansion problems

Let us consider a planning horizon comprising 10 years, going from 2020 to 2030, as represented in Fig. 2.38. Investment decisions can be made every 5 years, i.e., in 2020, 2025, and 2030. The resulting is a multi-stage expansion problem with three investment decision stages. The capacity expansion can be made in three generating technologies, namely thermal fossil fuel, wind, and/or solar PV units.

Assuming that there is a limit on the capacity that can be built of each technology, this problem can be formulated in a very simple way as follows:

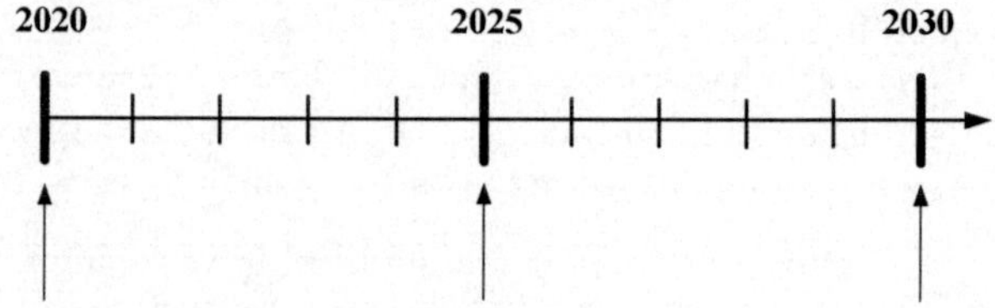

Fig. 2.38 Illustrative Example 2.15: schema of a long-term planning horizon

$$\text{Minimize} \quad \Big(C^{\text{In,Th}}(p_1^{\text{Th}} + p_2^{\text{Th}} + p_3^{\text{Th}}) + C^{\text{In,W}}(p_1^{\text{W}} + p_2^{\text{W}} + p_3^{\text{W}})$$

$$+ C^{\text{In,PV}}(p_1^{\text{PV}} + p_2^{\text{PV}} + p_3^{\text{PV}}) \Big) \tag{2.19a}$$

subject to:

$$0 \le p_1^{\text{Th}} + p_2^{\text{Th}} + p_3^{\text{Th}} \le P^{\text{Th,max}}, \tag{2.19b}$$

$$0 \le p_1^{\text{W}} + p_2^{\text{W}} + p_3^{\text{W}} \le P^{\text{W,max}}, \tag{2.19c}$$

$$0 \le p_1^{\text{PV}} + p_2^{\text{PV}} + p_3^{\text{PV}} \le P^{\text{PV,max}}, \tag{2.19d}$$

where p_1^{Th}, p_2^{Th}, and p_3^{Th} represent the capacity to be built of thermal power in each stage, p_1^{W}, p_2^{W}, and p_3^{W} represent the capacity to be built of wind power in each stage, and equivalently, p_1^{PV}, p_2^{PV}, and p_3^{PV} represent the capacity to be built of solar PV in each stage; $C^{\text{In,Th}}$, $C^{\text{In,W}}$, and $C^{\text{In,PV}}$ correspond to the investment cost of each technology, which are assumed to be equal along the planning horizon; and $P^{\text{Th,max}}$, $P^{\text{W,max}}$ and $P^{\text{PV,max}}$ are the maximum capacity that can be built of each technology throughout the 10-years planning horizon.

The objective of the planner (2.19a) is to minimize the total investment cost. Constraints (2.19b)–(2.19d) limit the total capacity that can be built of each technology. As it can be observed, no short-term technical constraints have been considered in this example. □

In this type of problems, it is important to correctly introduce the investment cost of the facilities to be built, specially when there are investment and operating decisions involved. To do this, it is common to find that investment or capital costs are annualized so that the total cost is distributed along the lifetime of the facility. To annualize the capital cost of a facility, the cost is multiplied by the capital recovery factor (a), which is defined as follows [25]:

$$a = \frac{i(1+i)^Y}{(1+i)^Y - 1}, \tag{2.20}$$

where i represents the interest rate and Y the lifetime of the facility in years. It is important to note that the annualized costs must be considered as many times as number of years are simulated. For instance, if three years are simulated to model the effect of a given decision, the investment cost has to be annualized and multiplied by three. In that case, the capital recovery factor will be $3 \cdot a$, where a is obtained from Eq. (2.20).

Illustrative Example 2.16 Annualization of the investment cost of wind and solar PV power units

Let us consider that it is desired to annualize the investment costs of two generating units: a wind and a solar PV power unit. The investment cost of the wind power unit is 1200 €/kW, whereas the solar PV power unit has an investment cost of 700 €/kW. The expected life time of these units are 25 and 20 years, respectively. If the interest rate is 5%, the annualized investment costs of these units are:

- Annualized cost of the wind power unit: $1200 \cdot \frac{(0.05(1+0.05)^{25})}{((1+0.05)^{25}-1)} = 1200 \cdot 0.071 =$ 85.20 €/kW.
- Annualized cost of the solar PV power unit: $700 \cdot \frac{(0.05(1+0.05)^{20})}{((1+0.05)^{20}-1)} = 700 \cdot 0.0802 =$ 56.17 €/kW. □

2.5.2 *Modeling of the Short-Term Variability in a Long-Term Planning Problem*

In order to make informed long-term decisions for a given power system, it is needed to model properly the influence of these decisions on the short-term operation of that system. The representation of the short-term variability in an expansion problem has been and is widely studied among the scholars. Simply along the day, we find multiple operating conditions in terms of the demand level, the intermittent power available, or the state of the interconnection in the local or international network, for instance. However, due to the mathematical complexity of expansion problems, often it is not possible to consider all the possible states of the power system because the problem results computationally intractable. Therefore, a trade-off between computing tractability and short-term variability representation must be found.

There are multiple papers in the technical literature proposing techniques to represent the daily variability with a restricted set of operating conditions [2, 5, 9, 12, 13, 16, 18, 20–22, 26, 28, 29]. Initially, it was common to use the so called load-duration or wind-duration curves [2]. Figure 2.39 represents an example of a load-duration curve, where the demand power of the 8760 h of the year is sorted in descendent order.

Then, techniques based on the *k-means* or hierarchical clustering where proposed to generate or select representative operating conditions. Figure 2.40 depicts an example of how the *k-means* algorithm clusters the input data into four clusters (filled dots). It is common in capacity expansion problems to select representative hours, days, or weeks of a typical meteorological year [8].

A different approach for modeling the variability of the demand and the power output of wind and solar PV units consists in selecting a set of representative days from historical data. The advantage of this procedure is that the correlation among the different series is maintained for the 24 h of each day. For instance, let us consider a year represented by 365 days with hourly data of system demand, wind power, and solar PV production. To consider the whole set of 365 days comprising 8760 hourly values in a planning model may lead to computational infeasibility. Then, the objective can be to select the fewest number of days that are able to represent adequately the original set of 365 days. One manner of selecting this number of days is through the minimization of the probability distance between the original

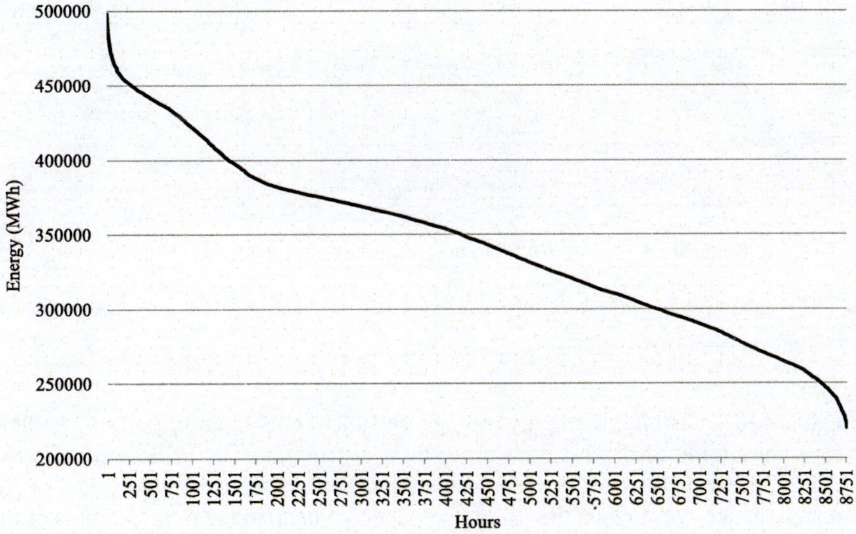

Fig. 2.39 Example of a load-duration curve

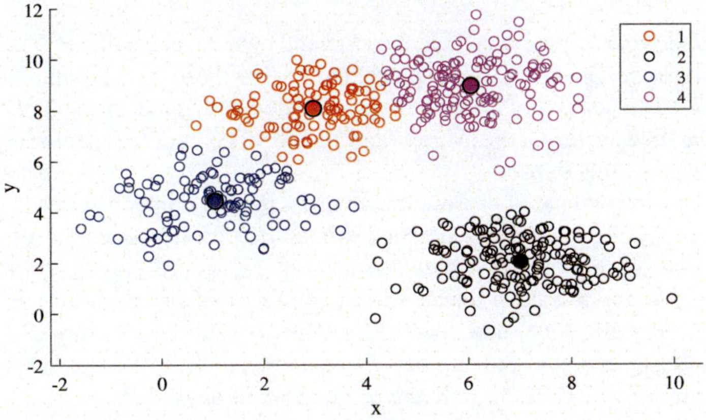

Fig. 2.40 Example of the *k-means* clustering algorithm

and reduced sets of days. For instance, the Kantorovich distance is typically used to characterize the distance between two discrete probability distributions [10]. Based on this distance it is possible to implement an heuristic iterative algorithm that is able to select in each iteration the day that minimizes the distance between the original and reduced sets of days. This procedure assigns a weight to each selected day as a function of the distance between that day and non-selected ones. Further details on this procedure can be found in [7, 15].

Currently, the problem of how to maintain the temporal correlation among the representative operating conditions is under study because this aspect conditions

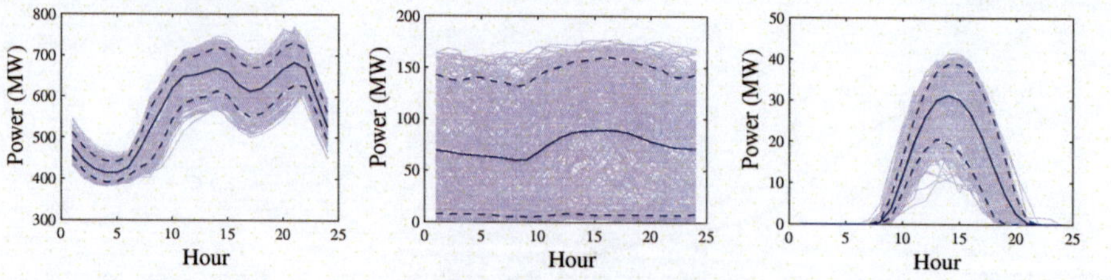

(a) Demand (b) Wind power production (c) Solar PV power production

Fig. 2.41 Illustrative Example 2.17: hourly demand and wind and solar PV power values during one year

the expansion decisions on energy storage capacity. This is an important problem in renewable-dominated power systems since storage capacity will be required to efficiently integrate intermittent renewable generation in a sustainable way. For the interested reader, we suggest the following papers to deep in the main features of the proposed algorithms [8, 12, 20].

Illustrative Example 2.17 Selection of representative days

As an example, let us consider the demand, wind, and solar PV productions depicted in Fig. 2.41. The series represented in this figure correspond to one year comprising 365 instances of 24-hourly values of demand, wind, and solar PV production. For the sake of clarity, the average and 10 and 90 percentiles are also represented for each series in this figure.

In order to select a set of representative days, the net demand series is used. The net demand is computed as the demand minus the power production of wind and solar PV. Considering the net demand series, Fig. 2.42 represents in dark blue color the set of representative days selected using the fast forward algorithm described in [15] considering different numbers of days. The original set of values is depicted in light blue. Observe that as the number of considered days increases, the reduced set of days represents more accurately the original set of days.

Finally, Fig. 2.43 represents the selected representative days in the original series of demand, wind, and solar PV production. Since in this case the values of demand are much higher than those of wind and solar PV production, the demand series is quite similar to the net demand series. As a consequence of this, the demand series is the one that is better represented in this example. □

Illustrative Example 2.18 Expansion problem with long- and short-term decisions

Let us consider the planning horizon and generating system of Illustrative Example 2.15, but in this case investment decisions can be made in 2020 and 2025. The strategic objective is to achieve a minimum renewable share of 50% in 2030 of electricity generation. The annual demand growth is considered as a deterministic parameter, assuming an increment of 0.5% every year.

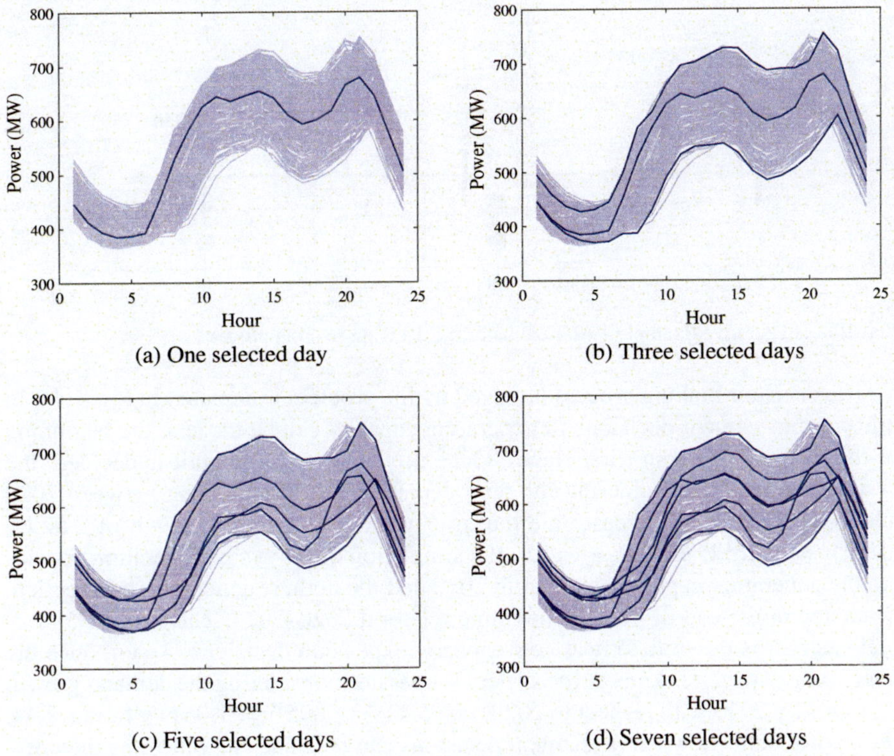

(a) One selected day (b) Three selected days

(c) Five selected days (d) Seven selected days

Fig. 2.42 Illustrative Example 2.17: net demand for different number of selected days

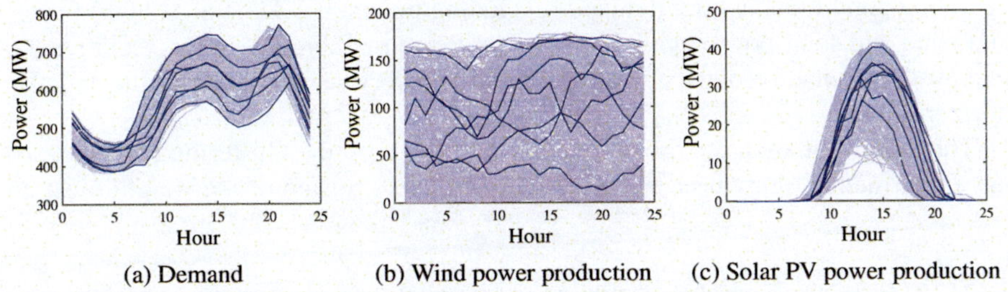

(a) Demand (b) Wind power production (c) Solar PV power production

Fig. 2.43 Illustrative Example 2.17: selection of 7 representative days of demand and wind and solar PV power production

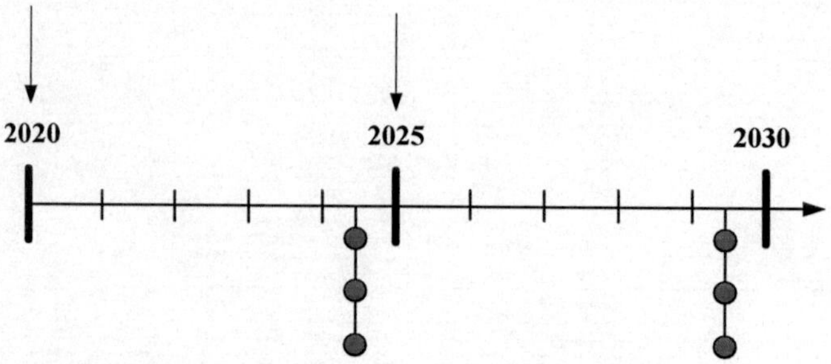

Fig. 2.44 Illustrative Example 2.18: schema of the decision-making process

The decision-making process involved in this problem is depicted in Fig. 2.44. In this way, investment decisions in generating capacity can be made at the beginning of the planning horizon (first stage). Once the capacity to be built is decided, the system is operated considering the total demand growth happening between 2020 and 2025, i.e., 0.5% increase rate during 5 years gives a total growth of 2.525% ($1.005^5 = 1.02525$). In other words, the simulation of the system operation considers the generating capacity decided in 2020 and the highest demand to be supplied, occurring in the end of the first investment period 2020–2025. Then, in 2025 (second stage) it is possible to take new capacity expansion decisions. Again, once the generating capacity is known, the system is operated considering the demand growth expected by 2030 with respect to 2020, i.e., 5.11% ($1.005^{10} = 1.0511$). In Fig. 2.44, the arrows indicate when investment decisions can be made and blue dots the operating decisions.

In this example, we assume that the variability along the year is represented by 3 operating conditions, whose demand (excluding the annual demand growth) power and wind and solar power availability are provided in Table 2.8. In this example we assume the replication of the operating conditions every year comprising the planning horizon.

The weight of each operating condition is given by the clustering technique or any other methodology used to select those points. The sum of the weight must be

Table 2.8 Illustrative Example 2.18: demand and renewable power availability in each operating condition

Operating condition	Demand (MWh)	Wind (pu)	Solar PV (pu)
t1	100	0.5	0.6
t2	120	0.3	0.8
t3	70	0.7	0.0

equal to the total number of hours considered in the short-term framework. In this simple example, we consider that the 8760 h of the year are equivalently distributed among the 3 points, resulting in a weight of 2920 h per operating condition.

This problem can be formulated as follows:

$$\begin{aligned}
\text{Minimize} \quad & C^{\text{In,Th}}(a_1 \cdot p_1^{\text{Th}} + a_2 \cdot p_2^{\text{Th}}) \\
& + C^{\text{In,W}}(a_1 \cdot p_1^{\text{W}} + a_2 \cdot p_2^{\text{W}}) \\
& + C^{\text{In,PV}}(a_1 \cdot p_1^{\text{PV}} + a_2 \cdot p_2^{\text{PV}}) \\
& + 2920 \cdot C^{\text{O,Th}}(p_{11}^{\text{G,Th}} + p_{12}^{\text{G,Th}} + p_{13}^{\text{G,Th}} + p_{21}^{\text{G,Th}} + p_{22}^{\text{G,Th}} + p_{23}^{\text{G,Th}}) \\
& + 2920 \cdot C^{\text{O,W}}(p_{11}^{\text{G,W}} + p_{12}^{\text{G,W}} + p_{13}^{\text{G,W}} + p_{21}^{\text{G,W}} + p_{22}^{\text{G,W}} + p_{23}^{\text{G,W}}) \\
& + 2920 \cdot C^{\text{O,PV}}(p_{11}^{\text{G,PV}} + p_{12}^{\text{G,PV}} + p_{13}^{\text{G,PV}} + p_{21}^{\text{G,PV}} + p_{22}^{\text{G,PV}} + p_{23}^{\text{G,PV}})
\end{aligned} \tag{2.21}$$

subject to:

$$0 \le p_1^{\text{Th}} + p_2^{\text{Th}} \le P^{\text{Th,max}}, \tag{2.22a}$$

$$0 \le p_1^{\text{W}} + p_2^{\text{W}} \le P^{\text{W,max}}, \tag{2.22b}$$

$$0 \le p_1^{\text{PV}} + p_2^{\text{PV}} \le P^{\text{PV,max}}, \tag{2.22c}$$

$$p_{11}^{\text{G,Th}} + p_{11}^{\text{G,W}} + p_{11}^{\text{G,PV}} = 100 \cdot 1.02525, \tag{2.23a}$$

$$p_{12}^{\text{G,Th}} + p_{12}^{\text{G,W}} + p_{12}^{\text{G,PV}} = 120 \cdot 1.02525, \tag{2.23b}$$

$$p_{13}^{\text{G,Th}} + p_{13}^{\text{G,W}} + p_{13}^{\text{G,PV}} = 70 \cdot 1.02525, \tag{2.23c}$$

$$p_{21}^{\text{G,Th}} + p_{21}^{\text{G,W}} + p_{21}^{\text{G,PV}} = 100 \cdot 1.511, \tag{2.23d}$$

$$p_{22}^{\text{G,Th}} + p_{22}^{\text{G,W}} + p_{22}^{\text{G,PV}} = 120 \cdot 1.511, \tag{2.23e}$$

$$p_{23}^{\text{G,Th}} + p_{23}^{\text{G,W}} + p_{23}^{\text{G,PV}} = 70 \cdot 1.511, \tag{2.23f}$$

$$0 \le p_{11}^{\text{G,Th}} \le p_1^{\text{Th}}, \tag{2.24a}$$

$$0 \le p_{21}^{\text{G,Th}} \le p_1^{\text{Th}}, \tag{2.24b}$$

$$0 \le p_{31}^{\text{G,Th}} \le p_1^{\text{Th}}, \tag{2.24c}$$

$$0 \le p_{11}^{\text{G,W}} \le 0.5 \cdot p_1^{\text{W}}, \tag{2.24d}$$

$$0 \le p_{12}^{\text{G,W}} \le 0.3 \cdot p_1^{\text{W}}, \tag{2.24e}$$

$$0 \le p_{13}^{\text{G,W}} \le 0.7 \cdot p_1^{\text{W}}, \tag{2.24f}$$

$$0 \le p_{11}^{\text{G,PV}} \le 0.6 \cdot p_1^{\text{PV}}, \tag{2.24g}$$

$$0 \le p_{12}^{\text{G,PV}} \le 0.8 \cdot p_1^{\text{PV}}, \tag{2.24h}$$

$$0 \le p_{13}^{\text{G,PV}} \le 0.0 \cdot p_1^{\text{PV}}, \tag{2.24i}$$

$$0 \le p_{21}^{\text{G,Th}} \le p_1^{\text{Th}} + p_2^{\text{Th}}, \tag{2.24j}$$

$$0 \leq p_{22}^{\mathrm{G,Th}} \leq p_1^{\mathrm{Th}} + p_2^{\mathrm{Th}}, \tag{2.24k}$$
$$0 \leq p_{23}^{\mathrm{G,Th}} \leq p_1^{\mathrm{Th}} + p_2^{\mathrm{Th}}, \tag{2.24l}$$
$$0 \leq p_{21}^{\mathrm{G,W}} \leq 0.5(p_1^{\mathrm{W}} + p_2^{\mathrm{W}}), \tag{2.24m}$$
$$0 \leq p_{22}^{\mathrm{G,W}} \leq 0.3(p_1^{\mathrm{W}} + p_2^{\mathrm{W}}), \tag{2.24n}$$
$$0 \leq p_{23}^{\mathrm{G,W}} \leq 0.7(p_1^{\mathrm{W}} + p_2^{\mathrm{W}}), \tag{2.24o}$$
$$0 \leq p_{21}^{\mathrm{G,PV}} \leq 0.6(p_1^{\mathrm{PV}} + p_2^{\mathrm{PV}}), \tag{2.24p}$$
$$0 \leq p_{22}^{\mathrm{G,PV}} \leq 0.8(p_1^{\mathrm{PV}} + p_2^{\mathrm{PV}}), \tag{2.24q}$$
$$0 \leq p_{23}^{\mathrm{G,PV}} \leq 0.0(p_1^{\mathrm{PV}} + p_2^{\mathrm{PV}}), \tag{2.24r}$$

$$(p_{21}^{\mathrm{G,W}} + p_{21}^{\mathrm{G,PV}} + p_{22}^{\mathrm{G,W}} + p_{22}^{\mathrm{G,PV}} + p_{23}^{\mathrm{G,W}} + p_{23}^{\mathrm{G,PV}}) \geq 0.50 \cdot 1.5(100 + 120 + 70), \tag{2.25}$$

where p_1^{Th} and p_2^{Th} represent the capacity to be built of thermal power in each stage, p_1^{W} and p_2^{W} the capacity to be built of wind power in each stage, and equivalently, p_1^{PV} and p_2^{PV} the capacity to be built of solar PV in each stage. Parameters a_1 and a_2 are used to annualize the investment costs of thermal, wind, and solar PV units in stages 1 and 2. $p_{11}^{\mathrm{G,Th}}, ..., p_{23}^{\mathrm{G,Th}}$ represent the power generated by the thermal unit in each operating condition at each stage; $p_{11}^{\mathrm{G,W}}, ..., p_{23}^{\mathrm{G,W}}$ represent the power generated by the wind unit in each operating condition at each stage; $p_{11}^{\mathrm{G,PV}}, ..., p_{23}^{\mathrm{G,PV}}$ represent the power generated by the solar PV unit in each operating condition at each stage; $C^{\mathrm{In,Th}}$, $C^{\mathrm{In,W}}$, and $C^{\mathrm{In,PV}}$ correspond to the investment cost of each technology, whereas $C^{\mathrm{O,Th}}$, $C^{\mathrm{O,W}}$, and $C^{\mathrm{O,PV}}$ correspond to the operating cost; $P^{\mathrm{Th,max}}$, $P^{\mathrm{W,max}}$ and $P^{\mathrm{PV,max}}$ are the maximum capacity that can be built of each technology.

The objective of the planner, formulated through (2.21), is to minimize the total cost, comprising the investment and operating costs. As defined by Eq. (2.20) and assuming an interest rate of 3% and a lifetime of 25 years, the capital recovery factor amounts to 5.74%. Considering that the capacity installed in stage 1 will be also used between years 2025 and 2030, the annualized cost of the investment decisions made at stage 1 must be considered twice, one for each year in which the operation of the year is considered.Then, $a_1 = 2 \cdot a_2$, and $a_1 = 2 \cdot 5.74\ \%$ and $a_2 = 5.74\ \%$. Moreover, note that in (2.21) the weight of the operating conditions is considered, being equal to 2920 h. Therefore, the total cost comprises (i) the annualized investment cost attained in 2020 and 2025 for building new generating capacity, and (ii) the operating cost of supplying the demand in 2024, with the highest demand growth between 2020 and 2025, and the demand in 2029, with the highest demand growth of the whole planning period. As stated previously, this is based on the assumption of replicating the operating conditions every year. Alternatively, we could multiply the investment cost of each stage by the number of years comprising such stage. The same should be done with the operating cost. The investment and operating decisions would be the same as with the formulation proposed in this example.

Constraints (2.22) limit the capacity that can be built of each technology. Constraints (2.23) impose the power balance in each operating condition, meaning that the total power generated must be equal to the total demand. Constraints (2.24) limit the power generated of each generating unit in each operating condition. The availability factor of the wind and solar PV units is taken into account. Finally, constraint (2.25) enforces that the generating share from renewable sources must be at least 50% of the total demand needs at the end of the planning horizon. □

2.6 Summary

This chapter describes modeling aspects of interest for power systems with high presence of renewable energies and EVs. First, the day-ahead operation of power systems is described, including the market clearing procedure and the ancillary services market. Next, the modeling of the power output of wind and solar PV units is described. From wind speed and solar radiation values, the power output of wind and solar PV units is mathematically determined. The modeling of the charge of EVs is also described in this chapter. Concepts as uncontrolled and controlled charge of EVs are introduced, and the notion of aggregator is also explained. Finally, the temporal characterization of long-term planning horizons has been described.

2.7 Notation

The main notation used in he mathematical formulations included in this chapter is provided below for quick reference. Other symbols are defined as needed throughout the chapter.

Sets and Indices

d Index of power consumers participating in the day-ahead energy market.
D Set of power consumers participating in the day-ahead energy market.
g Index of power producers participating in the day-ahead energy market.
G Set of power producers participating in the day-ahead energy market.
o Index of power producers participating in the ancillary services market.
O Set of power producers participating in the ancillary services market.
t Index of time periods.
T Set of time periods.

Parameters

$\underline{E}_t^{\text{EV}}$ Minimum energy that must contain the virtual battery representing the EV aggregation in time period t (MWh).

$\overline{E}_t^{\text{EV}}$ Maximum energy that can contain the virtual battery representing the EV aggregation in time period t (MWh).

$E_t^{\text{EV,IO}}$ Energy drop/contribution of the EVs departing from/arriving to the virtual battery representing the EV aggregation in time period t (MWh).

$E_t^{\text{I,C}}$ Energy consumption of the individual EV in time period t (kWh).

$\overline{E}^{\text{I}}$ Maximum energy level of the battery of the individual EV (kWh).

$\underline{E}^{\text{I}}$ Minimum energy level of the battery of the individual EV (kWh).

$\overline{P}_{dt}^{\text{D}}$ Power bid quantity of power consumer d in time period t (MW).

$\overline{P}_{gt}^{\text{G}}$ Power offer quantity of power producer g in time period t (MW).

$\overline{P}^{\text{I,C}}$ Maximum charging power rate of the individual EV (kW).

$\overline{P}^{\text{I,D}}$ Maximum discharging power rate of the individual EV (kW).

$\overline{P}_{ot}^{\text{R}}$ Capacity offer in the ancillary services market of power producer o in time period t (MW).

R_t Reserve capacity required in the ancillary services market in time period t (MW).

U_t^{I} Binary parameter that is equal to 1 if the individual EV is parked and available for charging in time period t, being 0 otherwise.

χ_{dt}^{D} Bid price of power consumer d in time period t (€/MWh).

χ_{gt}^{G} Offer price of power producer g in time period t (€/MWh).

χ_{ot}^{R} Capacity offer price of power producer o in time period t (€/MW).

Δ_t Duration of time period t (h).

$\eta^{\text{EV,C}}$ Charging efficiency of the virtual battery representing the EV aggregation (pu).

$\eta^{\text{EV,D}}$ Discharging efficiency of the virtual battery representing the EV aggregation (pu).

$\eta^{\text{I,C}}$ Charging efficiency of the individual EV (pu).

$\eta^{\text{I,D}}$ Discharging efficiency of the individual EV (pu).

Variables

e_t^{EV} Energy stored in the virtual battery representing the EV aggregation in time period t (MWh).

e_t^{I} Energy stored in the battery of the individual EV in time period t (kWh).

p_{dt}^{D} Consumption level of power consumer d in time period t (MW).

$p_t^{\text{EV,C}}$ Charging power level of the virtual battery representing the EV aggregation in time period t (MW).

$p_t^{\mathrm{EV,D}}$ Discharging power level of the virtual battery representing the EV aggregation in time period t (MW).
p_{gt}^{G} Generation level of power producer g in time period t (MW).
$p_t^{\mathrm{I,C}}$ Charging power level of the individual EV in time period t (MW).
$p_t^{\mathrm{I,D}}$ Discharging power level of the individual EV in time period t (MW).
p_{ot}^{R} Capacity scheduled in the ancillary services market of power producer o in time period t (MW).

2.8 GAMS Codes

This section provides the GAMS [11] codes used to solve some of the illustrative examples of this chapter.

2.8.1 Day-Ahead Market Clearing Problem

A GAMS code for solving Illustrative Example 2.3 is provided below:

```
$title DAY-AHEAD MARKET CLEARING PROBLEM

******************************************************************
* DATA
******************************************************************

SETS
d           'Consumers'                 /d1*d3/
g           'Producers'                 /g1*g4/
;

SCALARS
Deltat      'Duration of time steps' /1/
;

TABLE GDATA(g,*)
            PGmax       XiG
g1          80          50
g2          100         80
g3          20          70
g4          50          60;

TABLE DDATA(d,*)
            PDmax       XiD
d1          40          85
d2          100         65
d3          100         75;

******************************************************************
* DECLARATION OF VARIABLES AND MATHEMATICAL CHARACTERIZATION
******************************************************************

VARIABLES
```

```
**          OBJECTIVE FUNCTION
z               'Objective function'
;

POSITIVE VARIABLES

**          OPERATIONAL VARIABLES
pG(g)           'Production level'
pD(d)           'Consumption level'
;

*******************************************************************
*   EQUATIONS
*******************************************************************

EQUATIONS

**          OBJECTIVE FUNCTION
ObjectiveFunction    'Objective function'

**          CONSTRAINTS
Balance                'Power balance'
MaxPG                  'Maximum production levels'
MaxPD                  'Maximum consumption levels'
;

ObjectiveFunction..                      z=e=sum(d,DDATA(d,'XiD')*pD(d))
     -sum(g,GDATA(g,'XiG')*pG(g));

Balance..                                sum(g,pG(g))=e=sum(d,pD(d));
MaxPG(g)..                               pG(g)=l=GDATA(g,'PGmax');
MaxPD(d)..                               pD(d)=l=DDATA(d,'PDmax');

*******************************************************************
*   MODEL AND SOLVING
*******************************************************************

MODEL MarketClearing /all/;
MarketClearing.optcr=0;
Option lp=cplex;

SOLVE MarketClearing USING lp MAXIMIZING z;
```

2.8.2 Controlled Individual EV Charging

A GAMS code for solving Illustrative Example 2.12 is provided below:

```
$title CONTROLLED INDIVIDUAL EV CHARGING

*******************************************************************
* DATA
*******************************************************************

SETS
t          'Time periods'              /t1*t6/
;
```

```
SCALARS
Deltat    'Duration of time steps' /1/
Emin      'Minimum energy level'   /2/
Emax      'Maximum energy level'   /10/
E0        'Initial energy level'   /5/
PCmax     'Charging capacity'      /5/
PDmax     'Discharging capacity'   /5/
etaC      'Charging efficiency'    /0.9/
etaD      'Discharging efficiency' /0.9/
;

PARAMETER lambda(t) 'market price'
/t1       7
t2        4
t3        3
t4        8
t5        2
t6        5/
;

PARAMETER Econs(t) 'energy consumption'
/t1       0
t2        0
t3        3
t4        4
t5        0
t6        2.5/
;

PARAMETER UC(t) 'parking status'
/t1       1
t2        1
t3        0
t4        0
t5        1
t6        0/
;

******************************************************************
* DECLARATION OF VARIABLES AND MATHEMATICAL CHARACTERIZATION
******************************************************************

VARIABLES

**        OBJECTIVE FUNCTION
z             'Objective function'
;

POSITIVE VARIABLES

**        OPERATIONAL VARIABLES
e(t)          'Energy level'
pch(t)        'Charging power'
pd(t)         'Discharging power'
;

******************************************************************
*  EQUATIONS
******************************************************************

EQUATIONS
```

```
**        OBJECTIVE FUNCTION
ObjectiveFunction   'Objective function'

**        CONSTRAINTS
EnergyEvolution       'Energy evolution in the battery'
EnergyEvolution0      'Energy evolution in the battery in the first
    period'
ChargingLimits        'Charging power limits'
DischargingLimits     'Discharging power limits'
EnergyLimitsL         'Lower energy limits'
EnergyLimitsU         'Upper energy limits'
;

ObjectiveFunction..                         z=e=sum(t,lambda(t)*(pch(t)-pd(
    t)));

EnergyEvolution(t)$(ord(t) gt 1)..  e(t)=e=e(t-1)+pch(t)*etaC*
    Deltat-pd(t)*Deltat/etaD-Econs(t);
EnergyEvolution0..                  e('t1')=e=E0+pch('t1')*etaC*
    Deltat-pd('t1')*Deltat/etaD-Econs('t1');
ChargingLimits (t)..                pch(t)=l=PCmax*UC(t);
DischargingLimits(t)..              pd(t)=l=PDmax*UC(t);
EnergyLimitsL(t)..                  Emin=l=e(t);
EnergyLimitsU(t)..                  e(t)=l=Emax;

*****************************************************************
*  MODEL AND SOLVING
*****************************************************************

MODEL IndividualEV /all/;
IndividualEV.optcr=0;
Option lp=cplex;

SOLVE IndividualEV USING lp MINIMIZING z;
```

2.8.3 EV Aggregator Charging

A GAMS code for solving Illustrative Example 2.14 is provided below:

```
$title EV AGGREGATOR CHARGING

*****************************************************************
* DATA
*****************************************************************
SETS
t          'Time periods'            /t1*t6/
;

SCALARS
Deltat     'Duration of time steps' /1/
E0         'Initial energy level'   /30/
etaC       'Charging efficiency'    /0.9/
etaD       'Discharging efficiency' /0.9/
;
```

```
PARAMETER lambda(t) 'market price'
/t1       70
t2        40
t3        30
t4        80
t5        20
t6        50/
;

PARAMETER PCmax(t) 'maximum charging power level'
/t1       15
t2        14
t3        12
t4        10
t5        30
t6        30/
;

PARAMETER PDmax(t) 'maximum discharging power level'
/t1       15
t2        14
t3        12
t4        10
t5        30
t6        30/
;

PARAMETER Emin(t) 'minimum energy level'
/t1       40
t2        50
t3        10
t4        5
t5        30
t6        50/
;

PARAMETER Emax(t) 'maximum energy level'
/t1       60
t2        60
t3        30
t4        20
t5        50
t6        60/
;

PARAMETER EIO(t) 'energy input/output'
/t1       0
t2        0
t3        -40
t4        -10
t5        20
t6        20/
;

*******************************************************************
* DECLARATION OF VARIABLES AND MATHEMATICAL CHARACTERIZATION
*******************************************************************

VARIABLES
```

```
**        OBJECTIVE FUNCTION
z             'Objective function'
;

POSITIVE VARIABLES

**        OPERATIONAL VARIABLES
e(t)          'Energy level'
pch(t)        'Charging power'
pd(t)         'Discharging power'
;


*****************************************************************
*  EQUATIONS
*****************************************************************


EQUATIONS

**        OBJECTIVE FUNCTION
ObjectiveFunction    'Objective function'

**        CONSTRAINTS
EnergyEvolution      'Energy evolution in the battery'
EnergyEvolution0     'Energy evolution in the battery in the first period'
ChargingLimits       'Charging power limits'
DischargingLimits    'Discharging power limits'
EnergyLimitsL        'Lower energy limits'
EnergyLimitsU        'Upper energy limits'
;

ObjectiveFunction..                         z=e=sum(t,lambda(t)*(pch(t)-pd(t)));

EnergyEvolution(t)$(ord(t) gt 1)..  e(t)=e=e(t-1)+pch(t)*etaC*Deltat-pd(t)*Deltat/etaD+EIO(t);
EnergyEvolution0..                          e('t1')=e=E0+pch('t1')*etaC*Deltat-pd('t1')*Deltat/etaD+EIO('t1');
ChargingLimits(t)..                        pch(t)=l=PCmax(t);
DischargingLimits(t)..                      pd(t)=l=PDmax(t);
EnergyLimitsL(t)..                          Emin(t)=l=e(t);
EnergyLimitsU(t)..                          e(t)=l=Emax(t);

*****************************************************************
*  MODEL AND SOLVING
*****************************************************************

MODEL AggregatorEV /all/;
AggregatorEV.optcr=0;
Option lp=cplex;

SOLVE AggregatorEV USING lp MINIMIZING z;
```

References

1. Ackerman, T.: Wind Power in Power Systems, 2nd edn. Wiley (2012)
2. Baringo, L., Conejo, A.J.: Correlated wind-power production and electric load scenarios for investment decisions. Appl. Energy **101**, 475–482 (2013)
3. Burton T., Jenkins, N., Sharpe, D., Bossanyi, E.: Wind Energy Handbook, 2nd edn. Wiley (2011)
4. Chang, H.-J.: Economics. In: The User's Guide. Bloomsbury Press, New York (2014)
5. Chicco, G.: Overview and performance assessment of the clustering methods for electrical load pattern grouping. Energy **42**(1), 68–80 (2012)
6. Conejo, A.J., Baringo, L.: Power System Operations. Springer, Cham (2018)
7. Conejo, A.J., Carrión, M., Morales, J.M.: Decision Making Under Uncertainty in Electricity Markets. Springer, New York, NY (2010)
8. Domínguez, R., Carrión, M., Oggioni, G.: Planning and operating a renewable-dominated European power system under uncertainty. Appl. Energy **258**, 113989 (2020)
9. Domínguez, R., Vitali, S.: Multi-chronological hierarchical clustering to solve capacity expansion problems with renewable sources. Energy **227**, 120491 (2021)
10. Dupačová, J., Gröwe-Kuska, N., Römisch, W.: Scenario reduction in stochastic programming. Math. Program. **95**, 493–511 (2003)
11. GAMS (2021). www.gams.com
12. García-Cerezo, Á., García-Bertrand, R., Baringo, L.: Enhanced representative time periods for transmission expansion planning problems. IEEE Trans. Power Syst. **36**(4), 3802–3805 (2021)
13. García-Cerezo, Á., García-Bertrand, R., Baringo, L.: Priority chronological time-period clustering for generation and transmission expansion planning problems with long-term dynamics. IEEE Trans. Power Syst. (2022). (In press)
14. González-Vayá, M., Baringo, L., Krause, T., Andersson, G., Almeida, P., Geth, F., Rapoport, S.: EV aggregation models for different charging scenarios. In: Proceedings of the 23rd International Conference on Electricity Distribution (CIRED) (2015)
15. Gröwe-Kuska, N., Heitsch, H., Römisch, W.: Scenario reduction and scenario tree construction for power management problems. In: Proceedings of IEEE Bologna Power Technology Conferences Bologna, Italy (2003)
16. Liu, Y., Sioshansi, R., Conejo, A.J.: Hierarchical clustering to find representative operating periods for capacity-expansion modeling. IEEE Trans. Power Syst. **33**(3), 3029–3039 (2017)
17. McEvoy, A., Markvart, T., Castañer, L.: Practical Handbook of Photovoltaics Fundamentals and Applications, 2nd edn. Academic Press (2011)
18. Nahmmacher, P., Schmid, E., Hirth, L., Knopf, B.: Carpe diem: a novel approach to select representative days for long-term power system modeling. Energy **112**, 430–442 (2016)
19. Pérez-Arriaga, J.I., Meseguer, C.: Wholesale marginal prices in competitive generation markets. IEEE Trans. Power Syst. **12**(2), 710–717 (1997)
20. Pineda, S., Morales, J.M.: Chronological time-period clustering for optimal capacity expansion planning with storage. IEEE Trans. Power Syst. **33**(6), 7162–7170 (2018)
21. Poncelet, K., Hoschle, H., Delarue, E., Virag, A., Dhaeseleer, W.: Selecting representative days for capturing the implications of integrating intermittent renewables in generation expansion planning problems. IEEE Trans. Power Syst. **32**(3), 1936–1948 (2016)
22. Scott, I.J., Carvalho, P.M., Botterud, A., Silva, C.A.: Clustering representative days for power systems generation expansion planning: capturing the effects of variable renewables and energy storage. Appl. Energy **253**, 113603 (2019)
23. Sioshansi, R., Conejo, A.J.: Optimization in Engineering. Models and Algorithms. Springer, New York (2017)
24. Sørensen, B.: Renewable Energy: Its Physics, Engineering, Use, Environmental Impacts, Economy, and Planning Aspects, 3rd edn. Editor Elsevier Academic Press (2004)
25. Steiner, H.M.: Engineering Economic Principles. McGraw-Hill Companies (1992)

26. Teichgraeber, H., Brandt, A.R.: Clustering methods to find representative periods for the optimization of energy systems: an initial framework and comparison. Appl. Energy. **239**, 1283–1293 (2019)
27. The ILOG CPLEX (2021). www.ilog.com/products/cplex/
28. Tso, W.W., Demirhan, C.D., Heuberger, C.F., Powell, J.B., Pistikopoulos, E.N.: A hierarchical clustering decomposition algorithm for optimizing renewable power systems with storage. Appl. Energy **270**, 115190 (2020)
29. Ward, J.J.H.: Hierarchical grouping to optimize an objective function. J. Am. Stat. Assoc. **58**(301), 236e44 (1963)
30. Winter, C.J., Sizmann, R.L., Vant-HullSolar, L.L.: Power Plants: Fundamentals, Technology, Systems, Economics. Springer (2011)

Chapter 3
Mathematical Tools

This chapter describes some specific mathematical tools that are used throughout this book. First, two decision-making under uncertainty tools are analyzed: stochastic programming and robust optimization. Next, the linear decision rules approach is presented. This technique allows to establish linear relationships between the optimization variables and the uncertain parameters, which allows to reduce the computational size of optimization problems. The linear decision rules approach is specially convenient in multi-stage programming problems under uncertainty. Finally, the practical implementation of the Benders' decomposition is described in detail. This decomposition technique is of special interest for those large-scale optimization problems that can be decomposed easily if the values of some variables are fixed. Note that this chapter intends to provide basic concepts of some mathematical tools of interest. Those readers familiar with these tools may skip this chapter.

3.1 Introduction

Decision-making problems are present in different knowledge areas such as economics, engineering, etc. They are usually formulated as optimization problems, where the optimization variables constitute the decisions that are desired to be determined. The mathematical formulation and the practical resolution of these problems may be complex due to many factors. First, the modeling of the physical aspects of a particular decision-making problem may be hard. This is the case, for instance, of the mathematical formulation of a decision-making problem including the dynamic behavior of a power system that requires the usage of nonlinear mathematical expressions. Additionally, unknown data are frequent in real-world problems. This requires the use of decision-making tools under uncertainty, as for instance stochastic pro-

L. Baringo et al., *Electric Vehicles and Renewable Generation*,
Green Energy and Technology, https://doi.org/10.1007/978-3-031-09079-0_3

gramming or robust optimization [5]. An example of unknown data in a decision-making problem may be the value of the system demand at a future year when a power system planner desires to determine the optimal configuration of the power system for that year. Other complexities fall on the resolution process of the resulting optimization problem. For instance, it is well known that nonlinear programming problems are much difficult to solve than linear problems. Furthermore, an excessive computational size of the resulting optimization problems may difficult the process of obtaining the problem solutions. In this case, decomposition techniques can be used, which are iterative procedures based on dividing the original problem in comparatively smaller and solvable ones.

In this chapter, we describe three procedures for modeling decision-making problems under uncertainty: stochastic programming, robust optimization, and linear decision rules. Stochastic programming is a widely-used tool for modeling optimization problems with uncertain data [3]. Stochastic programming is based on the knowledge of the probability distribution functions of the uncertain parameters. If these distribution functions are known, it is possible to generate plausible realizations of the uncertain parameters, also called scenarios. On the contrary, robust optimization and linear decision rules techniques do not require the knowledge of the probability distribution of the uncertain parameters. Instead of using scenarios to represent the uncertain parameters, confidence intervals are considered. In particular, robust optimization is able to find the optimal decision for the worst occurrence of the uncertain parameters [2]. This worst case is a byproduct of the robust optimization technique. Therefore, the usage of robust optimization is recommended if either the probability distributions of the uncertain parameters are unknown, or if it is desired to obtain a robust solution against the worst realization of the uncertain parameters. The linear decision rules approach is based on the assumption that optimization variables may be linearly expressed as a function of the uncertain parameters. If the uncertain parameters are defined by a confidence interval, specifying upper and lower limits, the assumption of linearity of the optimization variables reduces significantly the computational size of the optimization problem. In fact, this approach is specially devised to be used in multi-stage decision making problems that are characterized by a high computational size. An example of multi-stage decision-making problem is the multi-year generation expansion problem of a power system. This problem consists in determining for each considered year the optimal generation investment decisions to make to minimize the total cost of the system while the demand is satisfied.

Finally, this chapter also describes the basics of the Benders' decomposition. As stated above, decomposition techniques are useful to reduce the size of an optimization problem that is solved using an iterative procedure. Roughly speaking, a large-size problem can be solved by solving many small-size problems. The Benders' decomposition technique is of special interest for those optimization problems that are easy to decompose if the values of some optimization variables are fixed. This is the case of many stochastic programming problems that can be decomposed by scenario if those optimization variables depending of all scenarios are fixed.

The contents of this chapter are organized as follows. The fundamentals of stochastic programming are described in Sect. 3.2. Section 3.3 provides the basic concepts of the robust optimization. The mathematical implementation of the linear decision rule approach is carefully explained in Sect. 3.4. The Benders' decomposition technique is described in Sect. 3.5. Section 3.6 provides some relevant conclusions. Finally, Sect. 3.7 provides the GAMS codes to solve some of the illustrative examples presented in this chapter.

3.2 Stochastic Programming

Uncertainty is present in most decision-making problems faced by power systems planners, operators, and participants. For example, the future demand and generating mix are unknown when the transmission network planner has to decide the new investments in transmission assets. Similarly, at the time of operating a power system, the power system operator does not know accurately the next-day demand and power output of renewable generating units.

However, decisions need to be made even with lack of perfect information. This is what motivates the use of stochastic programming for dealing with decision-making problems under uncertainty. Stochastic programming is based on the knowledge or estimation of the probability distribution functions of the unknown data. In practice, the probability distributions of uncertain input data are approximated by sets of plausible realizations of the input data together with their probabilities of occurrence. In this manner, the decision-making problem is solved accounting for a set of possible realizations of the input data considered.

Some relevant monographs about stochastic programming are [3, 12]. The practical application of stochastic programming to different decision-making problems under uncertainty in electricity markets can be consulted in [5].

3.2.1 Random Variables and Stochastic Processes

Within a stochastic programming framework, each uncertain parameter is modeled as a random variable. For practical purposes, random variables are usually represented by a finite set of realizations or scenarios. For instance, random variable a can be represented by $\{a_\omega, \forall \omega \in \Omega\}$, where ω is the scenario index and Ω is the set of scenarios. Therefore, a_ω is a possible realization of random variable a. Each scenario a_ω is associated with a probability π_ω, where $\sum_{\omega \in \Omega} \pi_\omega = 1$.

Illustrative Example 3.1 Random variable: production of a wind farm at a given hour

Let us consider a wind farm with nominal capacity equal to 20 MW. The power output of this wind farm at a future hour is unknown and it can be characterized as a random variable.

As an example, let us assume that the distribution of the power output of the wind farm at 10:00 am can be characterized by the following discrete distribution:

$$\begin{aligned}&8 \text{ MW with probability } 0.25,\\&10 \text{ MW with probability } 0.50,\\&12 \text{ MW with probability } 0.25.\end{aligned}$$

Then, the expected production of this wind farm at 10:00 am is $0.25 \cdot 8 + 0.5 \cdot 10 + 0.25 \cdot 12 = 10$ MW. □

A stochastic process is a random variable whose value evolves over time. For instance, the hourly power output of a wind farm over one day is an example of a stochastic process. Therefore, stochastic processes can be identified as a collection of random variables for different time periods. For the interested reader, a formal description of stochastic processes is given in [15].

Similarly to random variables, stochastic processes spanning a given time horizon can be represented by scenarios. Let us assume that B contains the set of dependent random variables constituting the stochastic process, $\{b_1, b_2, \ldots, b_{N_\mathrm{T}}\}$. For instance, stochastic process B can be represented by the set of scenarios $\{B_\omega, \forall \omega \in \Omega\}$, where ω is the scenario index and Ω is the set of scenario indexes. Again, each realization B is associated with a probability π_ω, where $\sum_{\omega \in \Omega} \pi_\omega = 1$.

For example, if B represents the 24 hourly productions of a wind farm for the next day, B_ω is a 24×1 vector representing one possible realization of these productions.

Illustrative Example 3.2 Stochastic process: wind power production at two consecutive hours

We consider that the random variable representing the wind farm production at 10:00 am is:

$$\begin{aligned}&8 \text{ MW with probability } 0.25,\\&10 \text{ MW with probability } 0.50,\\&12 \text{ MW with probability } 0.25.\end{aligned}$$

The random variable representing the wind power production at the next hour, 11:00 am, can be described as follows:

1. If the production of the wind farm at 10:00 am is 8 MW, then the production at 11:00 am can be 10, 12, or 14 MW with probabilities 0.25, 0.50, and 0.25, respectively.

Table 3.1 Illustrative Example 3.2: scenarios

Scenario	Production at 10:00 am (MW)	Production at 11:00 am (MW)	Probability
1	8	10	0.0625
2	8	12	0.1250
3	8	14	0.0625
4	10	12	0.1250
5	10	14	0.2500
6	10	16	0.1250
7	12	14	0.0625
8	12	16	0.1250
9	12	18	0.0625

2. If, the production of the wind farm at 10:00 am is 10 MW, then the production at 11:00 am can be 12, 14, or 16 MW with probabilities 0.25, 0.50, and 0.25, respectively.
3. Finally, if the production of the wind farm at 10:00 am is 12 MW, then the production at 11:00 am can be 14, 16, or 18 MW with probabilities 0.25, 0.50, and 0.25, respectively.

According to the description included above, the production of the wind farm at 10:00 am and 11:00 am constitutes a stochastic process comprising two random variables that can be characterized by 9 scenarios. Table 3.1 lists the considered scenarios and the probability of each scenario. The probability of each scenario is computed as the product of the probabilities of the elements in each scenario. For instance, the probability of scenario 1 is equal to the product of the probability of the production of the wind farm at 10:00 am to be equal to 8 MW (0.25) and the probability of the production of the wind farm at 11:00 am to be equal to 10 MW (0.25), which is equal to $0.25 \cdot 0.25 = 0.0625$. Note that the sum of the probabilities of all scenarios is equal to 1.

□

3.2.2 *Scenarios and Scenario Trees*

As stated above, random variables are represented in practice using scenarios, where a scenario is a single realization of the uncertain parameter. To adequately characterize a random variable, it is required that the set of scenarios represents faithfully the probability distribution of the considered variable. Different techniques have been proposed in the technical literature to build scenario trees, [6, 11]. Two typically used methods are *path-based* and *moment-matching* approaches. Path-based methods generate scenarios using econometric and time series models. Typically, a large number of scenarios is randomly generated using this method. On the other hand, moment-matching methods are used to build discrete distributions that satisfy a pre-fixed set of statistical properties of the original distributions of the random variables, e.g., moments, correlation matrix, etc. In other situations, scenarios can be generated using either historical data or based on expert judgement and existing technical reports.

The set of scenarios used to model a random variable is usually arranged in a so-called *scenario tree*. The main characteristic of scenario trees is that they provide information about both the decision-making points and the realizations of the uncertain parameters. Graphically, a scenario tree is made up of nodes and arcs. The nodes constitute states of the world at a particular instant, and they also represent the points where decisions are made. Each node has a single predecessor and can have several successors. The first node is called the root node, and it corresponds to the beginning of the planning horizon. In the root node, the first-stage decisions are made. The nodes connected to the root node are the second-stage nodes, and so on. The nodes in the last stage are called leaves. Therefore, a scenario is the path between the root node and a leave.

The nodes are linked together by a set of arcs, representing different realizations of the uncertain parameters. Each arc has associated a probability of occurrence. In this manner, the probability of a scenario is the product of all arcs probabilities associated with that scenario.

Illustrative Example 3.3 Scenario tree

Figure 3.1 shows an example of a scenario tree representing a stochastic process, e.g. the power production of a wind farm at two consecutive hours, 10 am and 11 am. This scenario tree comprises four scenarios and three stages. Each arc has associated a realization of the stochastic process and a probability of occurrence. For instance, observe that two arcs are leaving the root node. These two arcs have associated the values 12 and 10 MW, with probabilities 0.4 and 0.6, respectively.

As an example, Scenario 1 comprises the values 12 and 16 MW for 10 and 11 am, and its probability of occurrence is equal to $0.4 \cdot 0.3 = 0.12$. Observe that the sum of the probabilities over all scenarios is equal to 1: $0.4 \cdot 0.3 + 0.4 \cdot 0.7 + 0.6 \cdot 0.2 + 0.6 \cdot 0.8 = 1$. □

Usually, the number of scenarios generated to represent adequately the probability distribution of an uncertain parameter can be very large if scenarios are generated randomly. For instance, this is the case of path-based procedures. In this situation, considering that the size of a stochastic programming problem increases significantly with the number of scenarios, it is advisable to reduce the cardinality of the set of scenarios to maintain the computational tractability of the related optimization problem. To that end, the usage of scenario reduction techniques is recommended. The objective of a scenario-reduction technique is to reduce the number of scenarios maintaining the statistical description of the uncertain parameters as close as possible to that provided by the original set of scenarios. The description of scenario-reduction techniques is outside of the scope of this book, but relevant information about these techniques can be found in [5, 7, 10].

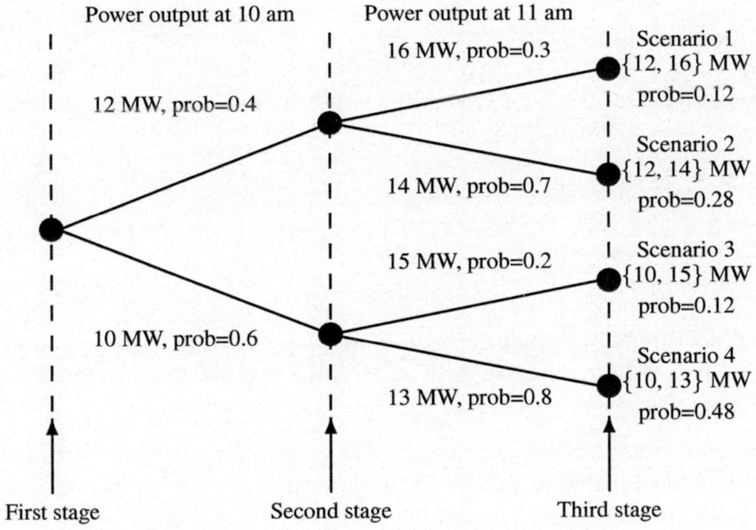

Fig. 3.1 Illustrative Example 3.3: scenario tree

3.2.3 *Two- and Multi-stage Stochastic Programming Problems*

Considering a stochastic programming framework, decision makers desire to make optimal decisions throughout a specified decision horizon considering incomplete information. The points in time where decisions are made or where uncertainty is totally or partially disclosed are named stages. The types of decisions and the amount of information available to the decision maker is usually different from stage to stage. According to the number of stages considered, we can distinguish between two-stage and multi-stage stochastic programming problems.

Let us start defining a classical two-stage stochastic programming problem. We consider a decision-making problem where decisions are made at two stages and there exists a stochastic process $\boldsymbol{\lambda}$ represented by a set of scenarios $\boldsymbol{\lambda}_{\omega}, \forall \omega \in \Omega$. We assume that two different decision variable vectors, namely $\mathbf{x}$ and $\mathbf{y}$, are involved in this problem. Decision $\mathbf{x}$ is made before knowing the actual value of the stochastic process $\boldsymbol{\lambda}$, while $\mathbf{y}$ is determined after the realization of $\boldsymbol{\lambda}$. Consequently, decision $\mathbf{y}$ depends on the decision $\mathbf{x}$ previously made and on the realization $\boldsymbol{\lambda}_{\omega}$ of the stochastic process $\boldsymbol{\lambda}$. Thus, we can express $\mathbf{y}$ as $\mathbf{y}_{\omega}$. The decision-making process is represented in Fig. 3.2.

According to Fig. 3.2, it is observed that:

1. First, decision $\mathbf{x}$ is made.
2. Then, stochastic process $\boldsymbol{\lambda}$ is realized as $\boldsymbol{\lambda}_{\omega}$.
3. Finally, decision $\mathbf{y}_{\omega}$ is made.

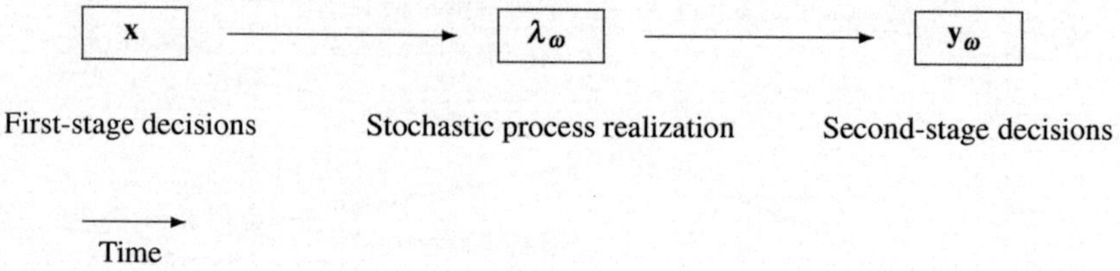

Fig. 3.2 Two-stage stochastic programming problem

Note that decision $\mathbf{y}_\omega$ depends on the decision $\mathbf{x}$ previously made and on the realization of the stochastic process $\boldsymbol{\lambda}$ in $\boldsymbol{\lambda}_\omega$. Considering this, two different types of decisions are observed:

1. *First-stage* or *here-and-now* decisions. These decisions are made before the realization of the stochastic process. Hence, variables representing here-and-now decisions do not depend on each realization of the stochastic process. In other words, first-stage decisions are made considering all possible realizations of the stochastic process.
2. *Second-stage* or *wait-and-see* decisions. These decisions are made after knowing the actual realization of the stochastic process. Consequently, these decisions depend on each realization vector of the stochastic process. If the stochastic process is represented by a set of scenarios, a second-stage decision variable is defined for each single scenario considered.

The decision framework above can be conveniently arranged through a scenario tree, as the one in Fig. 3.3. This scenario tree comprises n scenarios and two stages.

The compact expression of a two-stage stochastic linear programming problem is the following:

$$\text{Minimize}_{\mathbf{x},\mathbf{y}_\omega} \quad z = \sum_{\omega \in \Omega} \pi_\omega \left(\mathbf{c}_\omega^\top \mathbf{x} + \mathbf{q}_\omega^\top \mathbf{y}_\omega \right) \tag{3.1a}$$

$$\text{subject to:} \quad \mathbf{A}_\omega \mathbf{x} = \mathbf{b}_\omega, \quad \forall \omega \in \Omega, \tag{3.1b}$$

$$\mathbf{T}_\omega \mathbf{x} + \mathbf{W}_\omega \mathbf{y}_\omega = \mathbf{h}_\omega, \quad \forall \omega \in \Omega, \tag{3.1c}$$

$$\mathbf{x} \in \mathbf{X}, \ \mathbf{y}_\omega \in \mathbf{Y}, \quad \forall \omega \in \Omega, \tag{3.1d}$$

where $\mathbf{A}_\omega$, $\mathbf{B}_\omega$, $\mathbf{c}_\omega$, $\mathbf{q}_\omega$, $\mathbf{b}_\omega$, $\mathbf{h}_\omega$, $\mathbf{A}$, $\mathbf{T}_\omega$, and $\mathbf{W}_\omega$ are known vectors and matrices of appropriate size. Note that any of the vectors and matrices in problem (3.1) representing input data may depend (or not) on the set of stochastic process realizations $\boldsymbol{\lambda}_\omega$.

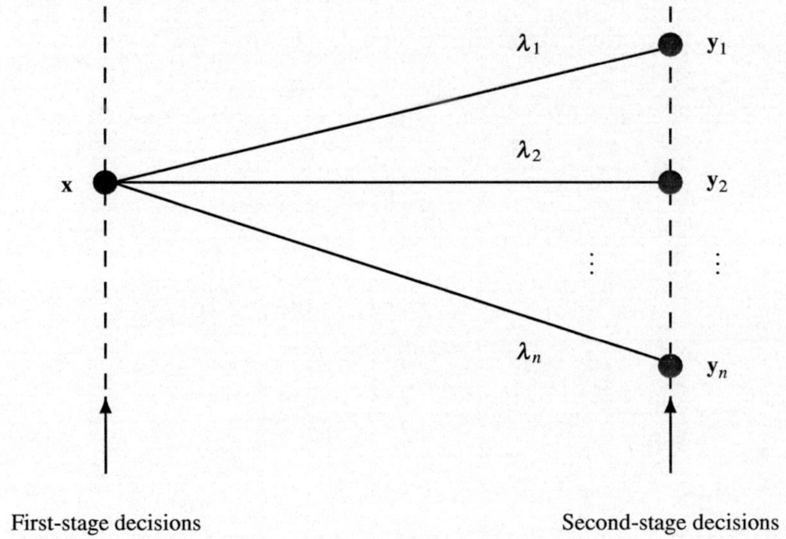

Fig. 3.3 Scenario tree of a two-stage stochastic programming problem

Table 3.2 Illustrative Example 3.4: demand scenarios

Demand (MWh)	Low		Medium		High	
	Value	Prob.	Value	Prob.	Value	Prob.
Period 1	800	1/3	900	1/3	1000	1/3
Period 2	1100	1/3	1300	1/3	1400	1/3

Illustrative Example 3.4 Formulation of a two-stage stochastic programming problem

Let us consider a power system planner that desires to determine the capacity to install from two candidate technologies: wind power (x_1) and combined-cycle gas (x_2). The capacity installed must be able to supply the demand during one year represented by two hourly periods at minimum cost. The annualized capital costs of these technologies are 200 €/kW and 100 €/kW, respectively. The operation cost of wind power is negligible in this example, while it is equal to 75 €/MWh for combined-cycle gas units. The availability of the wind power in each period is 20% and 40% of the installed capacity, respectively. The demand in each time period is an uncertain parameter that is modeled using three equiprobable scenarios (low, medium, and high). Table 3.2 provides the demand in each period and scenario. Then, the summation of the energy produced by each technology in each period $t \in \{1, 2\}$ and scenario $\omega \in \{1, 2, 3\}$, $y_{1,t,\omega}$ and $y_{2,t,\omega}$, must satisfy the demand in that period and scenario.

The problem described above can be formulated as follows:

$$\text{Minimize}_{x_1,x_2,y_{1,1,1},y_{1,1,2},y_{1,1,3},y_{1,2,1},y_{1,2,2},y_{1,2,3},y_{2,1,1},y_{2,1,2},y_{2,1,3},y_{2,2,1},y_{2,2,2},y_{2,2,3}}$$

$$(200 \cdot x_1 + 100 \cdot x_2) + \frac{1}{3} \cdot \frac{8760}{2} \cdot 75 \cdot \left(y_{2,1,1} + y_{2,1,2} + y_{2,2,1} + y_{2,2,2} + y_{2,1,3} + y_{2,2,3}\right)$$

subject to:

$$\begin{aligned}
&0 \le y_{1,1,1} \le 0.2 \cdot x_1,\\
&0 \le y_{1,2,1} \le 0.4 \cdot x_1,\\
&0 \le y_{1,1,2} \le 0.2 \cdot x_1,\\
&0 \le y_{1,2,2} \le 0.4 \cdot x_1,\\
&0 \le y_{1,1,3} \le 0.2 \cdot x_1,\\
&0 \le y_{1,2,3} \le 0.4 \cdot x_1,\\[1ex]
&0 \le y_{2,1,1} \le x_2,\\
&0 \le y_{2,1,2} \le x_2,\\
&0 \le y_{2,1,3} \le x_2,\\
&0 \le y_{2,2,1} \le x_2,\\
&0 \le y_{2,2,2} \le x_2,\\
&0 \le y_{2,2,3} \le x_2,\\[1ex]
&y_{1,1,1} + y_{2,1,1} = 800,\\
&y_{1,2,1} + y_{2,2,1} = 1100,\\
&y_{1,1,2} + y_{2,1,2} = 900,\\
&y_{1,2,2} + y_{2,2,2} = 1300,\\
&y_{1,1,3} + y_{2,1,3} = 1000,\\
&y_{1,2,3} + y_{2,2,3} = 1400,\\[1ex]
&x_1, x_2 \ge 0.
\end{aligned} \tag{3.2}$$

The objective function is equal to the summation of the investment cost of wind and combined-cycle gas units plus the expected operation cost of operating the system during one year. Note that the operation cost is computed assigning a probability equal to $1/3$ for each scenario and a weight of $8760/2$ for each period, where 8760 is the duration of one year in terms of hours.

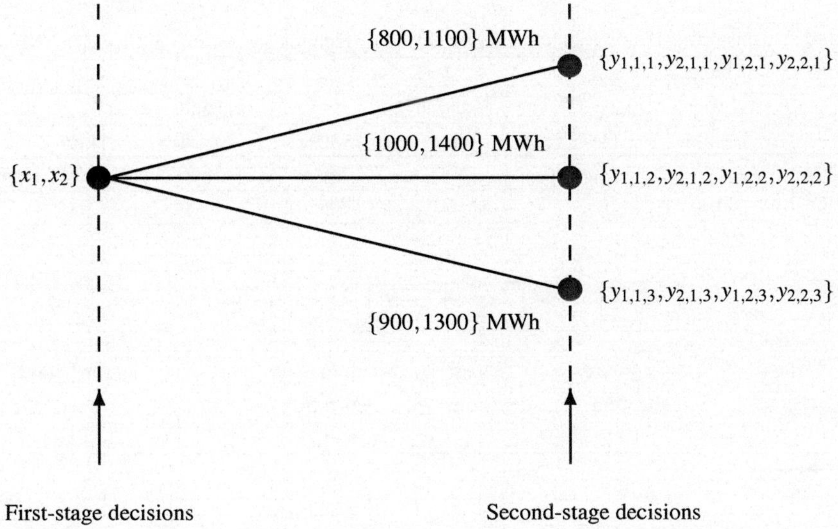

Fig. 3.4 Illustrative Example 3.4. Scenario tree

The first set of constraints establishes that the energy produced by wind units is constrained by the installed capacity and the availability of the wind resources in each period and scenario. The power generated by combined-cycle gas units is bounded by the second set of constraints. The third set of constraints enforces the energy balance for each period and scenario. Finally, the last set of constraints establishes the non-negative nature of the installed capacity. Observe that variables x_1 and x_2 are first-stage variables that do not depend on the specific realized scenario. On the other hand, second-stage variables $y_{1,t,\omega}$ and $y_{2,t,\omega}$ depend on the scenario index ω and they can be different for each realization of the uncertain parameters. Figure 3.4 depicts the corresponding scenario tree.

The resulting expected cost obtained from solving the above problem is 764.72 million €. The annual investment cost is 595 million €, whereas the expected operation cost is 169.72 million €. The optimal capacities installed from wind and gas units, x_1 and x_2, are 2750 and 450 MW, respectively. Table 3.3 lists the energy produced in each period by wind and gas units for each scenario. For the sake of clarity, the energy values provided in this table refer to the energy produced in hourly periods 1 and 2, which are denoted in the formulation of the problem by variable y. In order to obtain the energy produced during the whole year it is necessary to multiply the values provided in this table in each hour by $8760/2$ and add them. Observe that the energy produced in each period and scenario by wind units is always the maximum available, which is computed as the product of the wind power availability times the capacity installed. In this manner, the energy produced in period 1 is $0.2 \cdot 2750 = 550$ MWh, while it is equal to $0.4 \cdot 2750 = 1100$ MWh in period 2. The energy produced by gas units in each period and scenario is the difference between

Table 3.3 Illustrative Example 3.4: energy produced by technology, period, and scenario (MW)

	Scenario					
	Low		Medium		High	
	Wind	Gas	Wind	Gas	Wind	Gas
Period 1	550	250	550	350	550	450
Period 2	1100	0	1100	200	1100	300

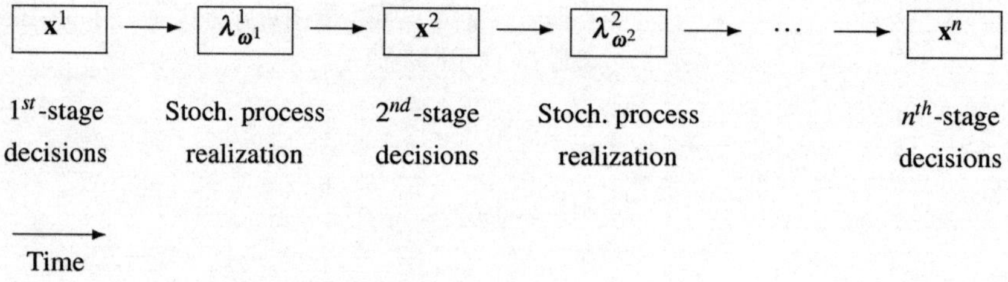

Fig. 3.5 Multi-stage stochastic programming problem

the demand and the energy produced by wind units. Note that the capacity installed by gas units coincides with the maximum produced energy by this technology, which is equal to 450 MW in period 1 and scenario High.

A GAMS [8] code for solving Illustrative Example 3.4 is provided in Sect. 3.7. □

Note that not all decision-making problems can be formulated considering only two stages. In those cases, a multi-stage stochastic programming problem can be formulated to model properly the decisions sequence. As an example, Fig. 3.5 represents graphically a decision-making problem with n stages.

The decision-making process depicted in Fig. 3.5 is the following:

1. Decisions $\mathbf{x}^1$ are made.
2. Stochastic process $\boldsymbol{\lambda}^1$ is realized as $\boldsymbol{\lambda}^1_{\omega^1}$.
3. Decisions $\mathbf{x}^2$ are made. Observe that these decisions depend on decisions $\mathbf{x}^1$ previously made and on the realization $\boldsymbol{\lambda}^1_{\omega^1}$.
4. Stochastic process $\boldsymbol{\lambda}^2$ is realized as $\boldsymbol{\lambda}^2_{\omega^2}$.
5. Decisions $\mathbf{x}^3$ are made. These decisions depend on previous decisions $\mathbf{x}^1$ and $\mathbf{x}^2$, and on the realizations $\boldsymbol{\lambda}^1_{\omega^1}$ and $\boldsymbol{\lambda}^2_{\omega^2}$.

...

2n-2. Stochastic process $\boldsymbol{\lambda}^{n-1}$ is realized as $\boldsymbol{\lambda}^{n-1}_{\omega^{n-1}}$.

2n-1. Decisions $\mathbf{x}^n$ are made, which depend on decisions $\mathbf{x}^1, \cdots, \mathbf{x}^{n-1}$ and on realizations $\boldsymbol{\lambda}^1_{\omega^1}, \cdots, \boldsymbol{\lambda}^{n-1}_{\omega^{n-1}}$.

As an example, Fig. 3.6 shows a three-stage scenario tree. Observe that in this example each node branches in two successors and that a total number of four scenarios are considered. In the first node (root node), the first-stage decisions, x^1,

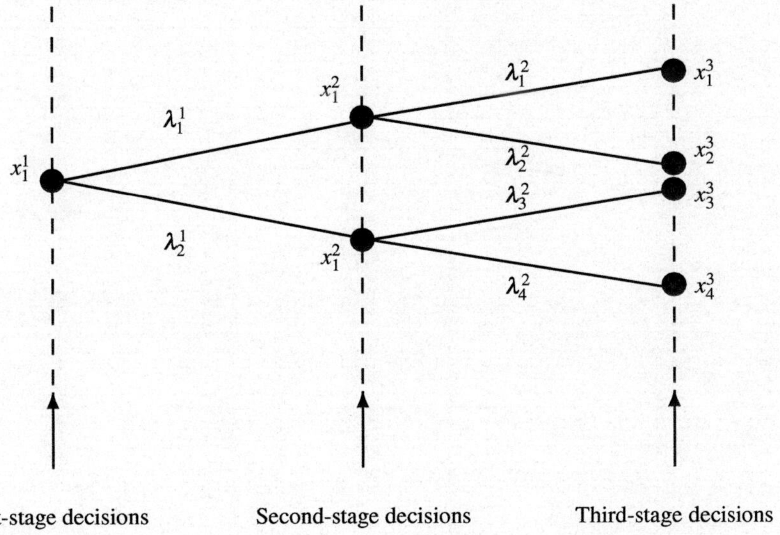

Fig. 3.6 Example of three-stage stochastic tree

are made. Note that these decisions are the same for all possible values of the random variables. The nodes connected to the root node represent the points where the second-stage decisions are taken. Considering that random variables can have two different values in this example, two different decisions can be made at the second-stage. Then, if random variable λ^1 materializes in λ_1^1, decision x_1^2 is made. However, if λ^1 takes the value λ_2^1, then decision x_2^2 is made. Observe that decisions x_1^2 and x_2^2 can be different. Then, it can be said that decisions x_1^2 and x_2^2 are wait-and-see with respect to random variable λ^1. However, they are here-and-now decisions over the future realizations of λ^2. In other words, x_1^2 has to be the same for realizations λ_1^2 and λ_2^2, and x_2^2 for λ_3^2 and λ_4^2. This decision-making process continues up to reach the number of stages, that is three in this case.

The compact formulation of a multistage-stage stochastic linear programming problem is presented below.

$$\text{Minimize}_{\mathbf{x}_\omega^n} \quad z = \sum_{n\in N}\sum_{\omega\in\Omega} \pi_\omega^n \left(\mathbf{c}_\omega^{n,\top}\mathbf{x}_\omega^n\right) \tag{3.3a}$$

$$\text{subject to:} \quad \mathbf{A}_\omega^n\mathbf{x}_\omega^n = \mathbf{b}_\omega^n, \quad \forall n \in N, \forall \omega \in \Omega, \tag{3.3b}$$

$$\mathbf{x}_\omega^n = \mathbf{x}_{\omega'}^n, \quad \forall n \in N, \forall\{\omega, \omega'\} \in \Omega^n, \tag{3.3c}$$

$$\mathbf{x}_\omega^n \in \mathbf{X}, \quad \forall n \in N, \forall \omega \in \Omega, \tag{3.3d}$$

where $\mathbf{A}_\omega^n$, $\mathbf{B}_\omega^n$, and $\mathbf{c}_\omega^n$ are known vectors and matrices of appropriate size. Any of the vectors and matrices in problem (3.3) representing input data can depend (or not) on the set of stochastic process realizations $\boldsymbol{\lambda}_\omega^n$. Observe that (3.3) is a scenario-based formulation in which all variables are dependent on the set of scenarios Ω.

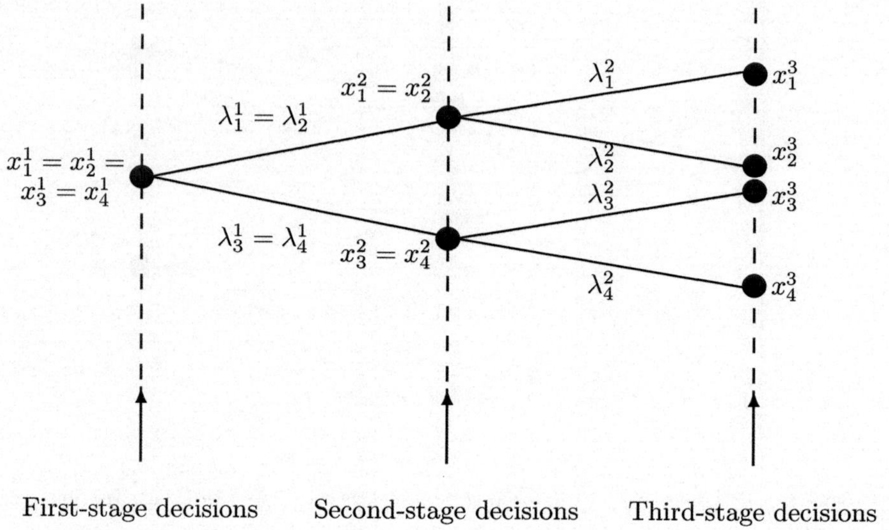

Fig. 3.7 Example of three-stage stochastic tree using scenario-dependent notation

This formulation requires the explicit formulation of non-anticipativity constraints, which establish that if the realizations of the random variables are equal in two scenarios up to stage n, then the decisions in stage n must be the same. In this manner, constraints (3.3c) enforce that if scenarios ω and ω' are equal up to stage n, $\{\omega, \omega'\} \in \Omega^n$, then the decisions in those scenarios in stage n have to be the same, $x_\omega^n = x_{\omega'}^n$. The main advantages of this formulation are the clarity of the resulting formulation and the possibility of applying decomposition techniques for solving the resulting mathematical problem. Alternative formulations, as the node-dependent formulation, can be also used [17]. Figure 3.7 represents the scenario tree depicted in Fig. 3.6 using a scenario-dependent notation for referring the decision variables, x_ω^n, and random variables, λ_ω^n. Observe that non-anticipativity constraints are applied to first- and second-stage variables.

Illustrative Example 3.5 provides the formulation of a three-stage stochastic programming approach.

Illustrative Example 3.5 Formulation of a three-stage stochastic programming problem

Let us consider an aggregator of electric vehicles that desires to determine the energy procurement of the vehicles during a planning horizon comprising two periods. The expected demand of the aggregator in each period is 500 and 400 MWh. This aggregator can sign a forward contract at fixed price for a quantity to determine at the beginning of each period, x^1, x^2. Additionally, it can participate in the pool to purchase energy in each period, y^1, y^2. The purchasing prices for the forward contracts applicable to periods 1 and 2 are 54 €/MWh and 45 €/MWh, respectively. The pool price and the demand of EVs are uncertain parameters that are represented using four scenarios. Table 3.4 shows the values of these scenarios.

Table 3.4 Illustrative Example 3.5: pool price and demand scenarios

Scenario	Probability	Pool price (€/MWh)		Demand (MWh)	
		Period 1	Period 2	Period 1	Period 2
1	0.25	60	52	600	450
2	0.25	60	48	600	400
3	0.25	50	42	500	350
4	0.25	50	38	500	300

At the beginning of period 1, the aggregator has to decide the energy to contract from forward contract 1, x^1. This is a here-and-now decision with respect to all possible scenarios of pool prices. The demand that is not satisfied by the forward contract during the first period is purchased from the pool, y^1, at the pool price. This is a wait-and-see decision that depends on the resulting pool price. At this point, the aggregator has also to determine the energy to contract from forward contract 2. This decision is a here-and-now decision with respect to pool prices in period 2, but it can be considered wait-and-see regarding the pool prices in period 1. Finally, the aggregator has to decide its participation in the pool in period 2. This decision-making problem can be arranged as a three-stage stochastic programming problem as follows:

- First stage: Forward contracting decisions for period 1.
- Second stage: Pool participation in period 1 and forward contracting decisions for period 2.
- Third stage: Pool participation in period 2.

The mathematical formulation of this problem is the following:

$$\text{Minimize}_{x_1^1,x_2^1,x_3^1,x_4^1,x_1^2,x_2^2,x_3^2,x_4^2,y_1^1,y_2^1,y_3^1,y_4^1,y_1^2,y_2^2,y_3^2,y_4^2}$$

$$\begin{aligned}0.25 \cdot \frac{8760}{2}\Big(&54 \cdot \left(x_1^1 + x_2^1 + x_3^1 + x_4^1\right) + 45 \cdot \left(x_1^2 + x_2^2 + x_3^2 + x_4^2\right)\\ &+ 60 \cdot \left(y_1^1 + y_2^1\right) + 50 \cdot \left(y_3^1 + y_4^1\right)\\ &+ 52 \cdot y_1^2 + 48 \cdot y_2^2 + 42 \cdot y_3^2 + 38 \cdot y_4^2\Big)\end{aligned}$$

subject to:

$$\begin{aligned}x_1^1 + y_1^1 &= 600,\\ x_2^1 + y_2^1 &= 600,\\ x_3^1 + y_3^1 &= 500,\\ x_4^1 + y_4^1 &= 500,\end{aligned}$$

$$x_1^2 + y_1^2 = 450,$$
$$x_2^2 + y_2^2 = 400,$$
$$x_3^2 + y_3^2 = 350,$$
$$x_4^2 + y_4^2 = 300,$$

$$x_1^1 = x_2^1 = x_3^1 = x_4^1,$$
$$x_1^2 = x_2^2,$$
$$x_3^2 = x_4^2,$$
$$y_1^1 = y_2^1,$$
$$y_3^1 = y_4^1,$$

$$x_1^1, x_2^1, x_3^1, x_4^1, x_1^2, x_2^2, x_3^2, x_4^2 \geq 0,$$
$$y_1^1, y_2^1, y_3^1, y_4^1, y_1^2, y_2^2, y_3^2, y_4^2 \geq 0.$$

The objective function is equal to the summation of the expected cost of the purchase of energy through forward contracts plus the expected pool cost. Total expected costs are computed assigning a probability equal to 0.25 to each scenario. It has also assigned a weight of 8760/2 for each period, where 8760 is the duration of one year in terms of hours.

The first set of constraints establishes the energy balance for each period and scenario. In this manner, the demand of the electric vehicles is satisfied in each period and scenario by the energy coming from the forward contracts and the pool. The non-anticipativity of the decisions is formulated through the second set of constraints. The last set of constraints defines the positive nature of the energy purchased through forward contracts. Observe that the sign of the transactions of energy in the pool is not constrained. Then, selling and purchasing in the pool is allowed.

The decision-making problem described above can be arranged in the scenario tree depicted in Fig. 3.8.

The resulting expected cost is 2.378 million €. Table 3.5 provides the energy purchased from forward contracts and the pool in each period and scenario. Observe that 500 MWh are signed from the forward contract in period 1. Note that, compared with pool prices, the price of this contract is very competitive. Since this is a first stage decision, this decision is equal for all scenarios. In period 2, the forward contract is not as competitive as in period 1, and it is only used in scenarios 1 and 2, where pool prices are higher than in scenarios 3 and 4. Observe that the demand not provided by the forward contract is supplied by the pool in each period and scenario.

The GAMS codes used to solve this problem are included in Sect. 3.7.

□

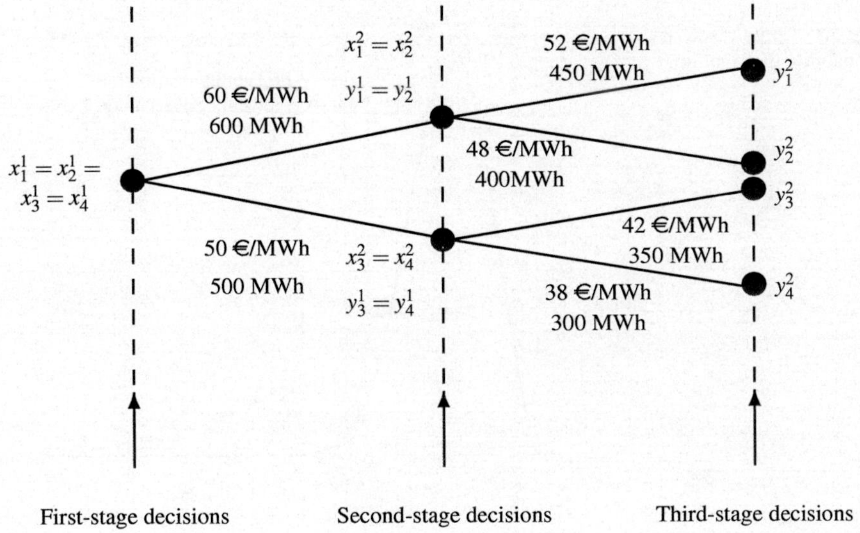

Fig. 3.8 Illustrative Example 3.5: scenario tree

Table 3.5 Illustrative Example 3.5: energy purchased from forward contracts and the pool (MWh)

Scenario	Forward contracts (x_ω^t)		Pool (y_ω^t)	
	Period 1	Period 2	Period 1	Period 2
1	500	400	100	50
2	500	400	100	0
3	500	0	0	350
4	500	0	0	300

3.2.4 *Risk and Conditional Value-at-Risk*

The perceptive reader may have observed that the formulations of problems (3.1) and (3.3) have the particularity that they optimize the expected value of a probability distribution. Those problems optimizing the expected value of a probability distribution are usually referred to as *risk-neutral* problems. The term risk neutral indicates that decision makers only focus on the expected value of the distribution, neglecting other information of the probability distribution. For instance, if the desire of a decision maker is to minimize costs, it should be also worried of not suffering from high costs in some negative scenarios, even though the expected cost be acceptable. In the same manner, if the objective of a decision maker is to maximize the profit, it should try to obtain a high expected profit but also to avoid experiencing low profits in some scenarios. As an example, Fig. 3.9 shows two continuous probability distributions representing costs. Observe that both distributions have a similar expected value.

Fig. 3.9 Comparison of two probability distributions

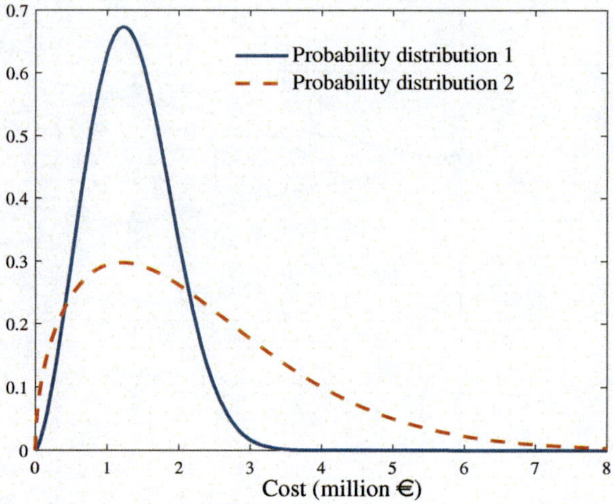

However, the probability distribution 2 exhibits a much higher risk of experiencing higher costs than probability distribution 1. For this reason, a rational decision maker may prefer probability distribution 1 instead of probability distribution 2.

A *risk-averse* formulation is that in which the occurrence of experiencing undesirable outcomes is quantified using a metric. There are many risk measures that can be used to quantify risk. For instance, the variance has been traditionally used for this purpose. A high variance indicates that the probability of experiencing an outcome different to the expected one is high. One of the drawbacks of this measure is that it penalizes also the positive deviations from the expected value (high profits or low costs). Nowadays, the most used risk measure in decision making is the so-called *Conditional Value-at-Risk* (CVaR) [16]. Considering a cost distribution, the CVaR can be roughly defined as the expected value of the costs in worst scenarios. Given its wide usage among decision makers, next section describes in detail the formulation of the CVaR.

3.2.4.1 Conditional Value-at-Risk

For a given $\alpha \in (0, 1)$ and considering a cost distribution, the CVaR is the expected value of the cost greater than the $(1 - \alpha)$-quantile of the cost distribution, [16]. Considering a profit distribution, the CVaR is the expected value of the profit smaller than the $(1 - \alpha)$-quantile of the profit distribution.

As an example, Fig. 3.10 represents a cost probability distribution. The expected value of this distribution of is 1.96 million €. In dark blue color is represented the area below the distribution whose cost is greater than the 0.9-quantile of the distribution, 4.35 million €. Observe that a rational decision maker is interested in reducing as

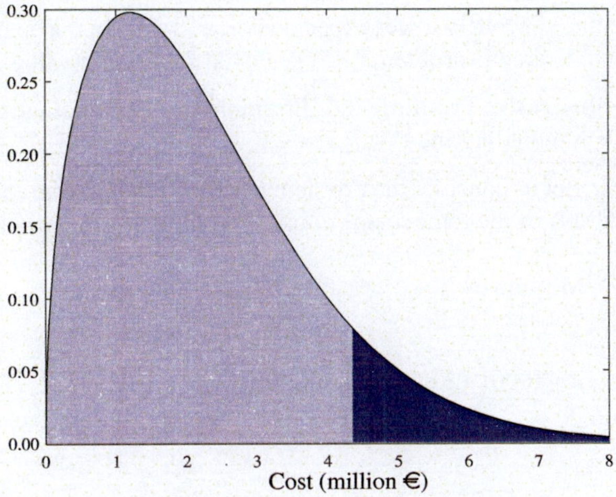

Fig. 3.10 CVaR of a cost distribution function

much as possible the expected value of that part of the distribution, which is equal to the CVaR.

Mathematically, the CVaR for $\alpha \in (0, 1)$, CVaR(α), of a cost distribution, C_ω, is defined as:

$$\text{CVaR}(\alpha) = \min\left\{\eta + \frac{1}{1-\alpha} E_\omega\left\{\max\{C_\omega - \eta, 0\}\right\}\right\},\ \forall \alpha \in (0, 1), \tag{3.4}$$

where $E_\omega\{\cdot\}$ is the expectation operator and η is a real number.

The CVaR can be easily incorporated in the formulation of a stochastic programming problem. Next, the formulation of a two-stage stochastic programming problem including the CVaR is provided:

$$\text{Minimize}_{\mathbf{x},\mathbf{y}_\omega,\xi,\eta_\omega} \quad z = \beta \sum_{\omega\in\Omega} \pi_\omega \left(\mathbf{c}_\omega^\top \mathbf{x} + \mathbf{q}_\omega^\top \mathbf{y}_\omega\right) + (1-\beta)\left(\xi + \frac{1}{1-\alpha}\pi_\omega \eta_\omega\right) \tag{3.5a}$$

$$\text{subject to:} \quad \mathbf{A}_\omega \mathbf{x} = \mathbf{b}_\omega, \tag{3.5b}$$

$$\mathbf{T}_\omega \mathbf{x} + \mathbf{W}_\omega \mathbf{y}_\omega = \mathbf{h}_\omega, \quad \forall \omega \in \Omega, \tag{3.5c}$$

$$\mathbf{c}_\omega^\top \mathbf{x} + \mathbf{q}_\omega^\top \mathbf{y}_\omega - \xi \leq \eta_\omega, \quad \forall \omega \in \Omega, \tag{3.5d}$$

$$\mathbf{x} \in \mathbf{X},\ \mathbf{y}_\omega \in \mathbf{Y}, \quad \eta_\omega \geq 0,\ \forall \omega \in \Omega. \tag{3.5e}$$

where β is a parameter used to make a tradeoff between expected cost and risk. A risk-neutral decision maker will prefer to minimize the expected cost, independently of the value of the CVaR and he/she will select a value of β close to 1. However, a risk-averse decision maker will try to reduce the cost in worst scenarios and he/she will prefer values of β close to 0. The optimal solution of problem (3.5) for a given vlaue of β is denominated *efficient point*, which is a pair of expected cost/risk in such a a way that it is impossible to find a set of optimization variables obtaining simultaneously lower expected cost and risk. In this manner, an *efficient frontier* may be derived if problem (3.5) is solved for several values of β.

Illustrative Example 3.6 Formulation of a two-stage stochastic programming problem including the CVaR

Let us consider the problem described in Illustrative Example 3.4. In this case, the CVaR of the total cost is considered in the formulation of the problem as follows:

$$\text{Minimize}_{x_1,x_2,y_{1,1,1},y_{1,1,1},y_{1,1,2},y_{1,1,3},y_{1,2,1},y_{1,2,2},y_{1,2,3},y_{2,1,1},y_{2,1,2},y_{2,1,3},y_{2,2,1},y_{2,2,2},y_{2,2,3},\xi,\eta_1,\eta_2,\eta_3}$$

$$\begin{aligned}
&\beta \cdot 1000\Big((200 \cdot x_1 + 100 \cdot x_2) \\
&\qquad + \frac{1}{3} \cdot \frac{8760}{2} \cdot 75 \left(y_{2,1,1} + y_{2,1,2} + y_{2,2,1} + y_{2,2,2} + y_{2,1,3} + y_{2,2,3}\right)\Big) \\
&\qquad + (1-\beta)\Big(\xi + \frac{1}{1-0.9} \cdot \frac{1}{3} (\eta_1 + \eta_2 + \eta_3)\Big)
\end{aligned}$$

subject to:

$$\begin{aligned}
&0 \le y_{1,1,1} \le 0.2 \cdot x_1, \\
&0 \le y_{1,2,1} \le 0.4 \cdot x_1, \\
&0 \le y_{1,1,2} \le 0.2 \cdot x_1, \\
&0 \le y_{1,2,2} \le 0.4 \cdot x_1, \\
&0 \le y_{1,1,3} \le 0.2 \cdot x_1, \\
&0 \le y_{1,2,3} \le 0.4 \cdot x_1,
\end{aligned}$$

$$\begin{aligned}
&0 \le y_{2,1,1} \le x_2, \\
&0 \le y_{2,1,2} \le x_2, \\
&0 \le y_{2,1,3} \le x_2, \\
&0 \le y_{2,2,1} \le x_2, \\
&0 \le y_{2,2,2} \le x_2, \\
&0 \le y_{2,2,3} \le x_2,
\end{aligned}$$

$$\begin{aligned}
&y_{1,1,1} + y_{2,1,1} = 800, \\
&y_{1,2,1} + y_{2,2,1} = 1100, \\
&y_{1,1,2} + y_{2,1,2} = 1000, \\
&y_{1,2,2} + y_{2,2,2} = 1400, \\
&y_{1,1,3} + y_{2,1,3} = 900, \\
&y_{1,2,3} + y_{2,2,3} = 1300,
\end{aligned}$$

$$
\begin{aligned}
& x_1, x_2 \geq 0, \\
& (200 \cdot x_1 + 100 \cdot x_2) + \frac{8760}{2} \cdot 50 \left(y_{2,1,1} + y_{2,2,1}\right) - \xi \leq \eta_1, \\
& (200 \cdot x_1 + 100 \cdot x_2) + \frac{8760}{2} \cdot 50 \left(y_{2,1,2} + y_{2,2,2}\right) - \xi \leq \eta_2, \\
& (200 \cdot x_1 + 100 \cdot x_2) + \frac{8760}{2} \cdot 50 \left(y_{2,1,3} + y_{2,2,3}\right) - \xi \leq \eta_3, \\
& \eta_1, \eta_2, \eta_3 \geq 0.
\end{aligned} \tag{3.6}
$$

Comparing formulation (3.6) with respect to risk-neutral formulation (3.2), it can be observed that the weighting parameter β and the term computing the CVaR are incorporated in the objective function of formulation (3.6). Note that a value of $\alpha = 0.9$ has been used in this example. Additionally, the last four set of constraints in (3.6) are needed to calculate the value of auxiliary variables ξ and η_1, η_2, and η_3, that are used to compute the CVaR.

In this example, three different values of β, $(1, 0.2$, and $0)$ are considered. Note that if $\beta = 1$, the risk-neutral case is considered and the results provided in Illustrative Example 3.4 are obtained. The reason of this result is that if $\beta = 1$ the term modeling the CVaR in the objective function of problem (3.6) is not accounted for. On the other hand, if $\beta = 0$ (risk-averse case) the expected cost is neglected in the objective function and only the CVaR is minimized.

Table 3.6 shows the total expected costs and the CVaR for the three different values of β. Additionally, for the sake of clarity, the resulting investment and expected operation costs are also listed. The results provided in this table show that the expected total cost increases as the decision maker becomes more risk averse and the value of parameter β decreases. On the contrary, the value of the CVaR (the expected cost in most unfavourable scenarios) decreases if β also diminishes. This behavior can be also graphically observed in the efficient frontier represented in Fig. 3.11. As β decreases, it is noted that the investment costs increase, whereas the expected operation costs decrease. The explanation of this result can be obtained from Table 3.7. This table provides the capacity installed from wind and gas units for each value of β. It is observed that the wind capacity installed increases as β decreases. The opposite effect happens for gas units. Note that, while the operation cost of wind units is null, gas units have associated an operation cost of 75 €/MWh. Then, considering that the demand is the only uncertain parameter, the randomness of the total costs is solely caused by the operation cost of gas units. Therefore, if it is desired to reduce the CVaR (costs higher than the expected one), then it is necessary to reduce the production of gas units in those scenarios with a high demand. Then, if β is low, the system planner will be interested in reducing the CVaR and it will prefer to produce a greater amount of energy from wind units than from gas units. Considering that wind units have higher investment costs and lower operation costs than gas units, a low value of β implies high investment costs and low operation costs, as observed in Table 3.6.

Table 3.6 Illustrative Example 3.6: expected costs and CVaR (million €)

β	Investment cost	Expected operation cost	Expected total cost	CVaR
1	595.0	169.7	764.7	841.4
0.2	685.0	93.1	778.1	832.0
0	730.0	76.6	806.6	828.5

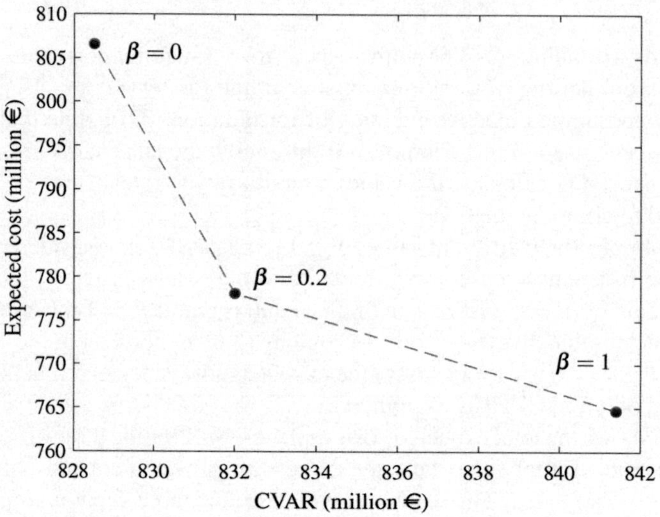

Fig. 3.11 Illustrative Example 3.6: efficient frontier

Table 3.7 Illustrative Example 3.6: capacity installed (MW)

β	Wind	Gas
1	2750	450
0.2	3250	350
0	3500	300

Table 3.8 provides the energy generated by wind and gas units for different values of β for each period and scenario. Considering that the capacity installed of wind units increases as the system planner becomes more risk averse (low values of β), it is observed that the production of wind units increases as the value of β decreases. Observe that the capacity installed from gas units coincides with the maximum production of this type of technology in each case. For example, if $\beta = 1$ the capacity installed of gas units is 450 MW (Table 3.7) and the maximum production of gas units

Table 3.8 Illustrative Example 3.6: energy produced by technology, period, and scenario for each value of parameter β (MWh)

		Scenario					
β	Period	Low		Medium		High	
		Wind	Gas	Wind	Gas	Wind	Gas
1	1	550	250	550	350	550	450
	2	1100	0	1100	200	1100	300
0.2	1	650	150	650	250	650	350
	2	1100	0	1300	0	1300	100
0	1	700	100	700	200	700	300
	2	1100	0	1300	0	1400	0

is also 450 MWh (period 1, scenario high). It is interesting to note that the production of gas units is null in the second time period if $\beta = 0$.

A GAMS [8] code for solving Illustrative Example 3.6 is provided in Sect. 3.7. □

3.3 Robust Optimization

The scenario-based stochastic programming model described in Sect. 3.2 aims at minimizing the expected cost over the set of scenarios used to represent the uncertain variables. In order to obtain informed decisions, it is essential:

1. To generate accurate scenarios for which it is required to have information about the probability distribution functions of uncertain variables.
2. To consider a large enough number of scenarios since the feasibility of the problem is only guaranteed for the considered scenarios.

These two issues are not trivial in many applications. On the one hand, information about the probability distribution is not always available, e.g., due to the lack of historical data. On the other hand, considering a very large number of scenarios may result in computationally intractable problems.

In addition, implementing the solution that minimizes the expected cost is not always the *best* strategy. For instance, if the system operator determines the new generation capacity to be built in a power system with the aim of minimizing the expected costs, this would mean that these decisions are optimal for this objective, but this may imply that part of the demand is not supplied if some specific scenarios realize, or if the actual scenario realization is different to one of the scenarios considered. In this type of problems, instead of focusing on the expected value over scenarios, it may be more convenient to make the decisions for the *worst-case* realizations of uncertain variables.

Robust optimization deals with the above issues. Firstly, it represents uncertain variables using uncertainty sets, which may comprise, for example, confidence bounds for a given confidence level. In this sense, it is not needed to have a complete information about the probability distribution of uncertain variables, and thus, it is generally easier to generate these confidence bounds than accurate scenarios. Secondly, the feasibility of the problem is guaranteed provided that the uncertain variables take values within the considered confidence bounds so that the computational tractability of the resulting problem is not affected by the number of input scenarios. Finally, decisions are made with the aim of minimizing the impact of the *worst-case* uncertainty realizations. In other words, the decisions are those that minimize the maximum cost.

This section describes the basic theory of robust optimization. For the sake of simplicity, a two-stage decision making problem is considered in which the decision maker implements first-stage decisions before knowing the actual uncertain realization. In this context, robust optimization can be classified into *static* robust optimization, also known as robust optimization without recourse, and *adaptive* robust optimization, also known as robust optimization with recourse. The former makes first-stage decisions by anticipating that, once they are made, the *worst-case* uncertainty realization will occur. The latter follows the same principle; however, it identifies the *worst-case* uncertainty realizations by anticipating that, after they occur, it is possible to make some corrective actions, which is represented by the recourse problem. This section focuses on adaptive robust optimization since it is the approach implemented in the book, while the interested reader is referred to [2] for the basis of static robust optimization.

3.3.1 Problem Formulation

Two-stage decision making problems under uncertainty generally follow the decision sequence below:

1. The decision maker determines the first-stage decisions before knowing the actual uncertainty realization.
2. The actual uncertainty realization is known.
3. The decision maker makes the second-stage decisions after knowing the actual uncertainty realization.

Under a robust optimization approach, the first-stage decisions are made by anticipating that once they are made, the worst uncertainty realization will occur. In addition, the worst uncertainty realization is identified by assuming that, after it occurs, it is possible to make corrective actions, i.e., the second-stage decisions. Considering this framework, it is possible to formulate the following three-level optimization problem:

$$\text{Minimize}_{\mathbf{x},c^{\text{WC}}} \ \mathbf{e}^{\text{T}}\mathbf{x} + c^{\text{WC}} \tag{3.7a}$$

subject to:

$$\mathbf{Kx} \leq \mathbf{f}, \tag{3.7b}$$

$$c^{\text{WC}} = \Bigg\{ \max_{\mathbf{u},c} \ c \tag{3.7c}$$

subject to:

$$\mathbf{u} \in \mathscr{U}, \tag{3.7d}$$

$$c = \Bigg[\min_{\mathbf{y}} \ \mathbf{b}^{\text{T}}\mathbf{y} \tag{3.7e}$$

subject to:

$$\mathbf{Ax} + \mathbf{By} + \mathbf{Du} = \mathbf{a}, \tag{3.7f}$$

$$\mathbf{Fx} + \mathbf{Gy} + \mathbf{Lu} \geq \mathbf{d} \Bigg] \Bigg\}, \tag{3.7g}$$

where vector $\mathbf{x}$ comprises first-stage decision variables, variables c and c^{WC} refer to second-stage and worst-case costs, respectively, vector of variables $\mathbf{u}$ represents worst-case uncertainty realizations, and vector $\mathbf{y}$ includes second-stage variables. Furthermore, $\mathbf{A}$, $\mathbf{B}$, $\mathbf{D}$, $\mathbf{F}$, $\mathbf{G}$, and $\mathbf{L}$ are coefficient matrices, while $\mathbf{a}$, $\mathbf{b}$, $\mathbf{d}$, $\mathbf{e}$, and $\mathbf{f}$ are coefficient vectors. Finally, all possible materializations of the uncertain parameters are represented by the uncertainty set $\mathscr{U}$, which is further described in Sect. 3.3.2.

Problem (3.7) is a three-level model comprising three optimization levels:

1. The upper level (3.7a)–(3.7b) identifies the first-stage decision variables that minimize first-stage and worst-case second-stage costs.
2. The middle level (3.7c)–(3.7d) determines the worst-case uncertainty realization that maximizes the second-stage costs for the first-stage decisions identified by the upper level.
3. The lower level (3.7e)–(3.7g) determines the second-stage decisions that minimize the second-stage costs for given upper- and middle-level decision variables.

Illustrative Example 3.7 Robust optimization: motivating example

A power producer owns a generating unit with a capacity of 100 MW and a marginal production cost of 50 €/MWh. The power producer has to supply two uncertain demands so that it has to decide if it is necessary or not to invest and build a new generating unit, with an investment cost of 2000 €, a capacity of 100 MW, and a marginal production cost of 40 €/MWh. The power producer has to make its investment decisions before knowing the actual demand. The load-shedding cost is 1000 €/MWh.

Considering the above data, the problem can be formulated as the following three-level optimization problem:

$$\text{Minimize}_{x,c^{\text{WC}}} \; 2000x + c^{\text{WC}}$$

subject to:

$$x \in \{0, 1\},$$

$$c^{\text{WC}} = \Bigg\{ \max_{d_1,d_2,c} \; c$$

$$\text{subject to:}$$

$$d_1, d_2 \in \mathscr{U},$$

$$c = \Bigg[\min_{p_1,p_2,d^{\text{LS}}} \; 50p_1 + 40p_2 + 1000d^{\text{LS}}$$

$$\text{subject to:}$$

$$p_1 + p_2 = d_1 + d_2 - d^{\text{LS}},$$

$$0 \leq p_1 \leq 100,$$

$$0 \leq p_2 \leq 100x,$$

$$0 \leq d^{\text{LS}} \leq d_1 + d_2 \Bigg] \Bigg\},$$

The above problem comprises three optimization levels:

1. The upper level determines whether it is optimal to build the candidate generating unit ($x = 1$) or not ($x = 0$) to minimize investment costs and worst-case operating costs.
2. The middle level determines the worst case uncertainty realization of demands (d_1, d_2) that maximizes the operating costs for given investment decisions. Note that random variables d_1 and d_2 take values within a given uncertainty set represented by $\mathscr{U}$.
3. The lower level determines the production levels of the existing and candidate generating units (p_1, p_2) that minimize the operation costs for given investment and uncertain decision variables. □

3.3.2 Uncertainty Set

Uncertain variables represented by vector **u** take values within known confidence bounds for a given confidence level:

$$\mathbf{u} \in \left[\tilde{\mathbf{u}} - \underline{\mathbf{u}}, \tilde{\mathbf{u}} + \overline{\mathbf{u}}\right], \tag{3.9}$$

where $\tilde{\mathbf{u}}$ is a vector representing the forecast values of variables in vector $\mathbf{u}$, while parameters in vectors $\underline{\mathbf{u}}$ and $\overline{\mathbf{u}}$ represent the maximum negative and positive deviations of variables in vector $\mathbf{u}$ from the forecast values, respectively.

As shown in [14], the worst case uncertainty realization of uncertain variables in an adaptive robust optimization problem occurs in a vertex of the polyhedron. Therefore, uncertainty set $\mathscr{U}$ used in constraints (3.7d) can be defined as follows:

$$\mathbf{u} = \tilde{\mathbf{u}} + \mathbf{v}^{+}\overline{\mathbf{u}} - \mathbf{v}^{-}\underline{\mathbf{u}}, \tag{3.10a}$$

$$\mathbf{v}^{+}, \mathbf{v}^{-} \in \{\mathbf{0}, \mathbf{1}\}, \tag{3.10b}$$

$$\mathbf{v}^{+} + \mathbf{v}^{-} \leq \mathbf{1}, \tag{3.10c}$$

$$\mathbf{1}^{\mathrm{T}}\left(\mathbf{v}^{+} + \mathbf{v}^{-}\right) \leq \Gamma, \tag{3.10d}$$

where $\mathbf{v}^{+}$ and $\mathbf{v}^{-}$ are vectors of binary variables that indicate if the corresponding uncertain variables take values at their upper or lower bounds, respectively, while Γ is the so-called uncertainty budget and is an integer that represents the maximum number of uncertain variables in vector $\mathbf{u}$ that can deviate from their forecast values.

Constraints (3.10a) define variables in vector $\mathbf{u}$ in terms of their forecast and maximum deviation values. Constraints (3.10b) define binary variables $\mathbf{v}^{+}$ and $\mathbf{v}^{-}$. Constraints (3.10c) impose that variables in vector $\mathbf{u}$ cannot be at its lower and upper bound simultaneously. Finally, constraints (3.10d) limit the number of variables that can deviate from their forecast values. If Γ is equal to 0, then all variables are equal to their forecast values, i.e., $\mathbf{u} = \tilde{\mathbf{u}}$. However, if Γ is equal to the number of uncertain variables, then all the uncertain variables can take values different from their forecasts, which leads to a conservative solution. Intermediate values allow the decision maker to control the robustness in the solution.

Illustrative Example 3.8 Robust optimization: uncertainty sets

The data of Illustrative Example 3.7 are considered again. Uncertain demands have a forecast level of 40 MW and 50 MW, respectively, but these values can be deviated up to 10 and 5 MW, respectively, i.e.:

$$d_1 \in [30, 50],$$
$$d_2 \in [45, 55].$$

The worst-case uncertainty realization of uncertain variables in an adaptive robust optimization problem occurs in a vertex of the polyhedron. Therefore, the uncertainty set $\mathscr{U}$ can be formulated as follows:

$$
\begin{aligned}
&d_1 = 40 + 10d_1^+ - 10d_1^-,\\
&d_2 = 50 + 5d_2^+ - 5d_2^-,\\
&d_1^+, d_1^-, d_2^+, d_2^- \in \{0, 1\},\\
&d_1^+ + d_1^- \le 1,\\
&d_2^+ + d_2^- \le 1,\\
&d_1^+ + d_1^- + d_2^+ + d_2^- \le \Gamma.
\end{aligned}
$$

Note that Γ is a parameter that controls the level of robustness in the problem and determines the maximum number of demand levels that can deviate from their forecast value. In the case analyzed in this example, Γ can be equal to 0, 1, and 2, i.e., the number of demands. If $\Gamma = 0$, then $d_1 = 40$ and $d_2 = 50$. If $\Gamma = 1$, then only one of the demands can take values different to their forecast level. Finally, if $\Gamma = 2$, then the two demand levels can deviate from the forecast values. Selecting the value of Γ depends on the preferences of the decision maker. A value of $\Gamma = 2$ would represent a conservative strategy since it considers all possible situations, while a value of $\Gamma = 0$ may be seen as a risky strategy since it assumes that demands will take their forecast values. □

3.3.3 *Solution Methodology*

Problem (3.7) is a three-level optimization problem that cannot be solved directly. Typically, this type of problems are handled by means of the column-and-constraint generation algorithm described in [18]. This method comprises two problems, namely, a master problem and a subproblem that are iteratively solved until convergence to the optimal solution. The formulation of these two problems and the aforementioned algorithm are described below.

3.3.3.1 Master Problem

The master problem is a relaxed version of the three-level problem (3.7) in which the second-stage problem is approximated in each iteration through a set of feasible constraints and provides the values of first-stage decision variables in vector **x**. The master problem at iteration υ is formulated as:

$$\text{Minimize}_{\mathbf{x},\mathbf{y}_{\upsilon'},\theta} \mathbf{e}^{\mathrm{T}}\mathbf{x} + \theta \tag{3.12a}$$

subject to:

$$\text{Constraints (3.7b)}, \tag{3.12b}$$

$$\theta \ge \mathbf{b}^{\mathrm{T}}\mathbf{y}_{\upsilon'}, \quad \upsilon' = 1, \dots, \upsilon, \tag{3.12c}$$

$$\mathbf{A}\mathbf{x} + \mathbf{B}\mathbf{y}_{\upsilon'} + \mathbf{D}\mathbf{u}^{(\upsilon')} = \mathbf{a}, \quad \upsilon' = 1, \ldots, \upsilon, \tag{3.12d}$$

$$\mathbf{F}\mathbf{x} + \mathbf{G}\mathbf{y}_{\upsilon'} + \mathbf{L}\mathbf{u}^{(\upsilon')} \geq \mathbf{d}, \quad \upsilon' = 1, \ldots, \upsilon, \tag{3.12e}$$

where the additional vectors of second-stage operating decision variables $\mathbf{y}_\upsilon$ associated with $\mathbf{y}$ are related to the uncertain realizations identified by the subproblem at iteration υ through $\mathbf{u}^{(\upsilon)}$.

The objective function (3.12a) is identical to (3.7a) except for the term θ, which is an approximation of the worst-case second-stage cost. Constraints (3.12b) include the constraints of the upper-level problem in (3.7). Constraints (3.12c) represent a lower bound for θ defined by the worst-case second-stage costs associated with the uncertainty realizations identified at previous iterations υ'. Constraints (3.12d)–(3.12e) are equivalent to lower-level constraints in (3.7).

Illustrative Example 3.9 Robust optimization: master problem

The master problem associated to the three-level optimization problem formulated in Illustrative Example 3.7 is as follows:

$$\text{Minimize}_{x,\theta,p_{1\upsilon'},p_{2\upsilon'},d^{\text{LS}}_{\upsilon'}} \; 2000x + \theta$$

subject to:

$$x \in \{0, 1\},$$
$$\theta \geq 50p_{1\upsilon'} + 40p_{2\upsilon'} + 1000d^{\text{LS}}_{1\upsilon'}, \quad \upsilon' = 1, \ldots, \upsilon,$$
$$p_{1\upsilon'} + p_{2\upsilon'} = d_1^{(\upsilon')} + d_2^{(\upsilon')} - d^{\text{LS}}_{\upsilon'}, \quad \upsilon' = 1, \ldots, \upsilon,$$
$$0 \leq p_{1\upsilon'} \leq 100, \quad \upsilon' = 1, \ldots, \upsilon,$$
$$0 \leq p_{2\upsilon'} \leq 100x, \quad \upsilon' = 1, \ldots, \upsilon,$$
$$0 \leq d^{\text{LS}}_{\upsilon'} \leq d_1^{(\upsilon')} + d_2^{(\upsilon')}, \quad \upsilon' = 1, \ldots, \upsilon.$$

□

3.3.3.2 Subproblem

At each iteration υ, the two lowermost levels of (3.7) are solved under the following max-min subproblem:

$$\text{Maximize}_{\mathbf{u}} \; \min_{\mathbf{y}} \; \mathbf{b}^{\text{T}}\mathbf{y} \tag{3.14a}$$

subject to:

$$\mathbf{u} = \tilde{\mathbf{u}} + \mathbf{v}^{+}\overline{\mathbf{u}} - \mathbf{v}^{-}\underline{\mathbf{u}}, \tag{3.14b}$$

$$\mathbf{v}^{+}, \mathbf{v}^{-} \in \{\mathbf{0}, \mathbf{1}\}, \tag{3.14c}$$

$$\mathbf{v}^+ + \mathbf{v}^- \leq \mathbf{1}, \tag{3.14d}$$

$$\mathbf{1}^\mathrm{T}\left(\mathbf{v}^+ + \mathbf{v}^-\right) \leq \Gamma, \tag{3.14e}$$

subject to:

$$\mathbf{Ax} + \mathbf{By} + \mathbf{Du} = \mathbf{a} \,:\, \lambda, \tag{3.14f}$$

$$\mathbf{Fx} + \mathbf{Gy} + \mathbf{Lu} \geq \mathbf{d} \,:\, \mu, \tag{3.14g}$$

where the dual variables associated to lower-level constraints are provided following a colon.

Subproblem (3.14) is a bi-level problem. Constraints (3.14b)–(3.14e) represent the uncertainty set, while constraints (3.14f)–(3.14g) are the lower-level constraints of problem (3.7).

The solution of the subproblem provides the worst-case uncertainty realizations linked to the maximum value of the minimum second-stage cost for given first-stage decisions attained by the previous master problem (3.12). In other words, subproblem (3.14) is parameterized in terms of first-stage decision variables in vector $\mathbf{x}$.

Bi-level subproblem (3.14) can be recast as a single-level problem by replacing its lower-level problem by its dual formulation. Note that this transformation is possible because the lower-level problem in (3.14) is continuous and linear on its decision variables. The resulting problem is as follows:

$$\text{Maximize}_{\mathbf{u},\lambda,\mu}\ \lambda^\mathrm{T}\left(\mathbf{a} - \mathbf{Ax}^{(\nu)} - \mathbf{Du}\right) + \mu^\mathrm{T}\left(\mathbf{d} - \mathbf{Fx}^{(\nu)} - \mathbf{Lu}\right) \tag{3.15a}$$

subject to:

$$\mathbf{u} = \tilde{\mathbf{u}} + \mathbf{v}^+\overline{\mathbf{u}} - \mathbf{v}^-\underline{\mathbf{u}}, \tag{3.15b}$$

$$\mathbf{v}^+, \mathbf{v}^- \in \{\mathbf{0}, \mathbf{1}\}, \tag{3.15c}$$

$$\mathbf{v}^+ + \mathbf{v}^- \leq \mathbf{1}, \tag{3.15d}$$

$$\mathbf{1}^\mathrm{T}\left(\mathbf{v}^+ + \mathbf{v}^-\right) \leq \Gamma, \tag{3.15e}$$

$$\mathbf{B}^\mathrm{T}\lambda + \mathbf{G}^\mathrm{T}\mu = \mathbf{b}, \tag{3.15f}$$

$$\mu \geq \mathbf{0}. \tag{3.15g}$$

Illustrative Example 3.10 Robust optimization: subproblem

The subproblem associated to the three-level optimization problem formulated in Illustrative Example 3.7 and the uncertainty set described in Illustrative Example 3.8 is as follows:

$$\text{Maximize}_{d_1,d_2,d_1^+,d_1^-,d_2^+,d_2^-}\ \text{Minimize}_{p_1,p_2,d^{\mathrm{LS}}}\ 50p_1 + 40p_2 + 1000d^{\mathrm{LS}}$$

subject to:

$$
\begin{aligned}
& d_1 = 40 + 10d_1^+ - 10d_1^-, \\
& d_2 = 50 + 5d_2^+ - 5d_2^-, \\
& d_1^+, d_1^-, d_2^+, d_2^- \in \{0, 1\}, \\
& d_1^+ + d_1^- \leq 1, \\
& d_2^+ + d_2^- \leq 1, \\
& d_1^+ + d_1^- + d_2^+ + d_2^- \leq \Gamma, \\
& \qquad \text{subject to:} \\
& \qquad\qquad p_1 + p_2 = d_1 + d_2 - d^{\text{LS}} \ : \ \lambda, \\
& \qquad\qquad 0 \leq p_1 \leq 100 \ : \ \underline{\mu}_1, \overline{\mu}_1, \\
& \qquad\qquad 0 \leq p_2 \leq 100x^{(\nu)} \ : \ \underline{\mu}_2, \overline{\mu}_2, \\
& \qquad\qquad 0 \leq d^{\text{LS}} \leq d_1 + d_2 \ : \ \underline{\phi}, \overline{\phi},
\end{aligned}
$$

where the dual variables associated to the lower-level problem are provided following a colon.

This bi-level model can be recast as a single-level optimization problem by replacing the lower-level problem by its dual formulation:

$$
\text{Maximize}_{d_1, d_2, d_1^+, d_1^-, d_2^+, d_2^-} \lambda \left(d_1 + d_2\right) - 100\overline{\mu}_1 - 100x^{(\nu)}\overline{\mu}_2 - \overline{\phi}\left(d_1 + d_2\right)
$$

subject to:

$$
\begin{aligned}
& d_1 = 40 + 10d_1^+ - 10d_1^-, \\
& d_2 = 50 + 5d_2^+ - 5d_2^-, \\
& d_1^+, d_1^-, d_2^+, d_2^- \in \{0, 1\}, \\
& d_1^+ + d_1^- \leq 1, \\
& d_2^+ + d_2^- \leq 1, \\
& d_1^+ + d_1^- + d_2^+ + d_2^- \leq \Gamma. \\
& \lambda + \underline{\mu}_1 - \overline{\mu}_1 = 50, \\
& \lambda + \underline{\mu}_2 - \overline{\mu}_2 = 40, \\
& \lambda + \underline{\phi} - \overline{\phi} = 1000, \\
& \underline{\mu}_1, \overline{\mu}_1, \underline{\mu}_2, \overline{\mu}_2, \underline{\phi}, \overline{\phi} \geq 0.
\end{aligned}
$$

Note that the above problem is a mixed integer nonlinear programming problem since it includes the products of continuous variables $\lambda \left(d_1 + d_2\right)$ and $\overline{\phi} \left(d_1 + d_2\right)$ in the objective function. However, these terms can be replaced by equivalent mixed-integer linear expressions as explained below.

Using the first two constraints, term $\lambda \left(d_1 + d_2\right)$ can be rewritten as:

$$\lambda (d_1 + d_2) = \lambda \left(40 + 10d_1^+ - 10d_1^- + 50 + 5d_2^+ - 5d_2^-\right).$$

This equation is still nonlinear. However, it includes the products of continuous and binary variables that can be linearized using exact mixed-integer linear expressions. In particular, this nonlinear equation can be replaced by the following constraints:

$$\begin{aligned}
&\lambda (d_1 + d_2) = 90\lambda + 10z_1^+ - 10z_1^- + 5z_2^+ - 5z_2^-,\\
&-Md_1^+ \le z_1^+ \le Md_1^+,\\
&-M\left(1 - d_1^+\right) \le z_1^+ - \lambda \le M\left(1 - d_1^+\right),\\
&-Md_1^- \le z_1^- \le Md_1^-,\\
&-M\left(1 - d_1^-\right) \le z_1^- - \lambda \le M\left(1 - d_1^-\right),\\
&-Md_2^+ \le z_2^+ \le Md_2^+,\\
&-M\left(1 - d_2^+\right) \le z_2^+ - \lambda \le M\left(1 - d_2^+\right),\\
&-Md_1^- \le z_2^- \le Md_2^-,\\
&-M\left(1 - d_2^-\right) \le z_2^- - \lambda \le M\left(1 - d_2^-\right),
\end{aligned}$$

where M is a large enough positive constant. Note that the above expressions include auxiliary variables z_1^+, z_1^-, z_2^+, and z_2^- that are respectively equal to λd_1^+, λd_1^-, λd_2^+, and λd_2^-. The working of these equations is as follows. Let us consider that $d_1^+ = 0$, then $z_1^+ = 0$ as imposed by the second constraint. Instead, if $d_1^+ = 1$, then the third constraint imposes that $z_1^+ = \lambda$, while the second constraint implies that $-M \le z_1^+ \le M$, which is a non-active constraint provided that M is fixed to a large enough value. The working of the remaining constraints is very similar. Similarly, term $\overline{\phi}\,(d_1 + d_2)$ can be also replaced by exact mixed-integer linear equations.

As a result of the above linearization, the subproblem is finally formulated as a mixed-integer linear programming problem. □

3.3.3.3 Algorithm

The master problem (3.12) and subproblem (3.15) are iteratively solved following the column-and-constraint generation algorithm described below:

1. Initialize the iteration counter, $\upsilon \leftarrow 0$, select the convergence tolerance, ε, and set the lower bound (LB) and upper bound (UB) to $-\infty$ and $+\infty$, respectively.
2. Solve the master problem (3.12).
3. Update the lower bound as:
$$\mathrm{LB} = z^{\mathrm{M}*}, \tag{3.18}$$
where $z^{\mathrm{M}*}$ is the optimal value of the objective function (3.12a).
4. Set $\mathbf{x}^{(\upsilon)} = \mathbf{x}^*$, where $\mathbf{x}^*$ denotes the optimal values of variables in vector $\mathbf{x}$ obtained in the master problem (3.12) in step 2.

5. Solve the subproblem (3.15).
6. Update the upper bound:

$$\mathrm{UB} = \min\Big\{\mathrm{UB}, \mathbf{e}^{\mathrm{T}}\mathbf{x} + z^{\mathrm{S}*}\Big\}, \tag{3.19}$$

where $z^{\mathrm{S}*}$ is the optimal value of the objective function (3.15a).
7. Check if $\mathrm{UB} - \mathrm{LB} < \varepsilon$. If so, the algorithm stops and the optimal first-stage decisions are those in vector $\mathbf{x}^*$. Otherwise, go to step 8.
8. Update the iteration counter $\upsilon \leftarrow \upsilon + 1$.
9. Set $\mathbf{v}^{(\upsilon)} = \mathbf{v}^*$, where $\mathbf{v}^*$ denotes the optimal values of variables in vector $\mathbf{y}$ obtained in the subproblem (3.15) in step 5.
10. Go to step 2.

Illustrative Example 3.11 Robust optimization: solution

The data of Illustrative Examples 3.7 and 3.8 are considered again. Control parameter Γ is fixed equal to 1, which means that only one of the demands can deviate from its forecast value. The adaptive robust optimization problem formulated in Illustrative Example 3.7 is solved using the column-and-constraint generation algorithm that comprises the steps below:

- *Step 1*. Initialize the iteration counter, $\upsilon \leftarrow 0$, select the convergence tolerance, $\varepsilon = 0$, and set the lower bound (LB) and upper bound (UB) to $-\infty$ and $+\infty$, respectively.
- *Step 2*. Solve the master problem. This problem is formulated in Illustrative Example 3.9 and its solution for $\upsilon = 0$ is trivial, namely, $x = 0$ and $\theta = -\infty$.
- *Step 3*. Update the lower bound:

$$\mathrm{LB} = -\infty.$$

- *Step 4*. Set $x^{(0)} = 0$.
- *Step 5*. Solve the subproblem. This problem is formulated in Illustrative Example 3.9 and it identifies the worst-case uncertainty realization for given first-stage decision variables. In this case, the optimal solution of this problem is $d_1 = 50$ and $d_2 = 50$.
- *Step 6*. Update the upper bound:

$$\mathrm{UB} = \min\Big\{\infty, 5000\Big\} = 5000.$$

- *Step 7*. As $\mathrm{UB} - \mathrm{LB} = 5000 - (-\infty) > \varepsilon$, go to step 8.
- *Step 8*. Update the iteration counter $\upsilon \leftarrow 1$.
- *Step 9*. Set $d_1^{(1)} = 50$ and $d_2^{(1)} = 50$.
- *Step 2*. Solve the following master problem:

$$\mathrm{Minimize}_{x,\theta,p_{11},p_{21},d_1^{\mathrm{LS}}}\ 2000x + \theta$$

subject to:

$$
\begin{aligned}
& x \in \{0, 1\}, \\
& \theta \geq 50 p_{11} + 40 p_{21} + 1000 d_{11}^{\mathrm{LS}}, \\
& p_{11} + p_{21} = 50 + 50 - d_{1}^{\mathrm{LS}}, \\
& 0 \leq p_{11} \leq 100, \\
& 0 \leq p_{21'} \leq 100x, \\
& 0 \leq d_{1'}^{\mathrm{LS}} \leq 50 + 50.
\end{aligned}
$$

The optimal solution of this problem is $x = 0$ and $\theta = 5000$.

- *Step 3*. Update the lower bound:

$$\mathrm{LB} = 5000.$$

- *Step 4*. Set $x^{(1)} = 0$.
- *Step 5*. Solve the subproblem. This problem is formulated in Illustrative Example 3.9 and it identifies the worst-case uncertainty realization for given first-stage decision variables. In this case, the optimal solution of this problem is $d_1 = 50$ and $d_2 = 50$.
- *Step 6*. Update the upper bound:

$$\mathrm{UB} = \min\left\{\infty, 5000\right\} = 5000.$$

- *Step 7*. As $\mathrm{UB} - \mathrm{LB} = 5000 - 5000 = 0$, the algorithm stops and the optimal solution is $x = 0$, i.e., the candidate generating unit is not built.

The problem is also solved for the other two possible values of control parameter Γ, namely, 0 and 2. The optimal solutions are provided in Table 3.9. Note that for both $\Gamma = 0$ and $\Gamma = 1$, it is not optimal to build the candidate generating unit since the existing unit can supply the maximum demand in both cases. The worst-case cost for $\Gamma = 0$ is lower since demands are assumed to take their forecast value. However, if $\Gamma = 2$, both demands can deviate from their forecasts and the worst case is $d_1 = 50$ and $d_2 = 55$. In such a case, it becomes optimal to build the candidate generating unit. Otherwise, it would result in load shedding with its high associated cost.

□

Table 3.9 Illustrative Example 3.11: results

Γ	x	Worst-case cost (€)
0	0	4500
1	0	5000
2	1	6250

3.4 Linear Decision Rules

As explained in Sect. 3.3, stochastic programming presents some drawbacks that may lead to computational intractability when a more precise representation of the decision-making problem is required. For instance, capacity expansion exercises are usually developed in a long-term planning horizon and decision makers update the investment decisions as long as additional information becomes available. This can be mathematically modelled using multi-stage stochastic programming problems since they better represent the decision-making process involved. However, the larger the number of decision stages, the larger the set of scenarios, and the higher the risk of computing intractability. To face this issue, decomposition techniques, such as Benders' decomposition (see Sect. 3.5 of this chapter), can be applied to efficiently solve the problem. In this section, the methodology proposed is based on establishing linear relationships between the random parameters and the decision variables. This implies that the optimal value of the variables is limited to that holding a linear dependency with respect to the random parameters. Here-in-after this approach is called Linear Decision Rule (LDR) approach. In this approach, the random parameters are characterized by confidence intervals and their expected value in such intervals. These assumptions lead to much less computational expensive problems, while the numerical experiments show that the solution obtained with the LDR approach is very similar to that obtained from the equivalent stochastic programming problem.

This section explains the mathematical fundamentals of LDRs and how to apply this methodology to solve a multi-stage stochastic programming problem. This section takes into consideration a multi-stage decision framework, but it can be easily extended to a two-stage decision problem. In particular, we focus on a multi-stage capacity expansion problem because it is one of the problems analyzed in this book, but a more general approach is explained in [9, 13].

3.4.1 Multi-Stage Decision Problem Formulation

In a multi-stage model, decisions are taken at different time points or stages along the planning horizon. The stages divide the planning horizon into periods. Thus, the decisions are made according to the following sequence:

1. The decision maker takes the first decision at the beginning of the planning horizon with lack of information about the evolution of the uncertain parameters;
2. The values of the uncertain parameters realize along the first period;
3. The decision maker can take new decisions, this time knowing the realization of the uncertain parameters in the previous period.

This procedure of uncertainty realization and decision making repeats as many times as stages considered in the planning horizon.

Considering a problem of minimization, a multi-stage model can be generally formulated as follows:

$$\text{Minimize}_{\mathbf{x}_n} \quad \mathbb{E}_{\Xi}\left\{\sum_{n\in N} \mathbf{c}_n^{\mathrm{N}}\mathbf{x}_n\right\} \tag{3.21a}$$

subject to:

$$\mathbf{A}\mathbf{x}_n \geq \mathbf{b}_n, \quad \forall n \in N, \tag{3.21b}$$

$$\mathbf{x}_n \geq 0, \quad \forall n \in N, \tag{3.21c}$$

where vector $\mathbf{x}_n$ comprises the decision variables for each decision stage n; vector $\mathbf{c}_n$ represents the uncertain parameters considered in the objective function in stage n, while vector $\mathbf{b}_n$ comprises the uncertain parameters involved in the constraints of the problem for each stage n; and $\mathbf{A}$ is a coefficient matrix with all its elements constant. E_{Ξ} is the expectation operator with respect to all uncertain parameters. As previously stated, the value of decisions $\mathbf{x}_n$ depend on the decisions made in stage $n-1$ and on the uncertain parameters realized until stage n.

It is assumed that the uncertain parameters $\mathbf{b}_n$ and $\mathbf{c}_n$ take values within confidence intervals defined as follows:

$$\bar{\mathbf{b}}_n(1-\theta_n^{\mathrm{B}}) \leq \mathbf{b}_n \leq \bar{\mathbf{b}}_n(1+\theta_n^{\mathrm{B}}), \quad \forall n \in N, \tag{3.22a}$$

$$\bar{\mathbf{c}}_n(1-\theta_n^{\mathrm{C}}) \leq \mathbf{c}_n \leq \bar{\mathbf{c}}_n(1+\theta_n^{\mathrm{C}}), \quad \forall n \in N, \tag{3.22b}$$

being $\bar{\mathbf{b}}_n$ and $\bar{\mathbf{c}}_n$ the average value of the uncertain parameters $\mathbf{b}_n$ and $\mathbf{c}_n$, respectively, in each stage, and θ_n^{B} and θ_n^{C} the uncertainty bounds.

Illustrative Example 3.12 LDR approach: multi-stage decision problem

Let us consider a capacity expansion problem in which investment decisions can be made at 3 different stages ($N = 3$) along the planning horizon. Figure 3.12 depicts the planning schema used as example. For instance, if the planning horizon lasts 15 years, investment decisions can be made at the beginning of the planning horizon (stage 1), after 5 years (stage 2), and finally after 10 years (stage 3). The stages divide the planning horizon into 3 periods. Note that at each stage decisions of capacity expansion of different technologies can be made.

In a capacity expansion model, operating decisions related to how to supply the electricity consumption in each period must be included. Once the generating, transmission, and/or storage capacity of the system is known, the decision maker must determine which resources to use to supply the demand.

If limits on the capacity that can be built are imposed, the formulation of the problem results in the following mathematical program:

$$\text{Minimize}_{\mathbf{x}_1,\mathbf{x}_2,\mathbf{x}_3,\mathbf{y}_1,\mathbf{y}_2,\mathbf{y}_3} \quad \mathbb{E}_{\Xi}\left\{\mathbf{c}_1^{\mathrm{T}}\mathbf{x}_1 + \mathbf{c}_2^{\mathrm{T}}\mathbf{x}_2 + \mathbf{c}_3^{\mathrm{T}}\mathbf{x}_3 + \mathbf{K}_1\mathbf{y}_1 + \mathbf{K}_2\mathbf{y}_2 + \mathbf{K}_3\mathbf{y}_3\right\} \tag{3.23a}$$

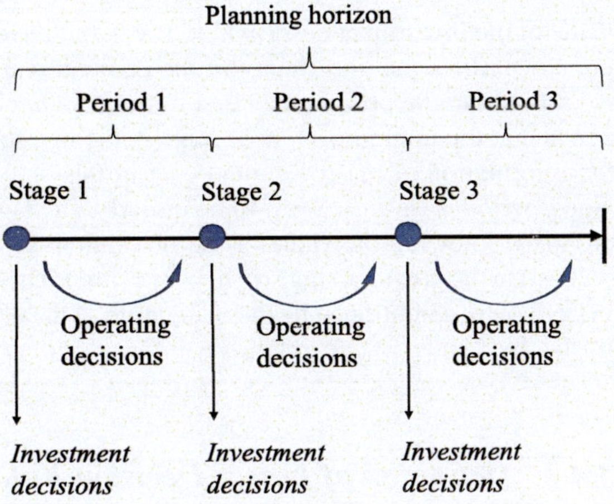

Fig. 3.12 Illustrative Example 3.12: planning schema of multi-stage capacity expansion model

subject to:

$$\mathbf{X} - \mathbf{x}_1 - \mathbf{x}_2 - \mathbf{x}_3 \geq 0, \tag{3.23b}$$

$$\mathbf{x}_1 - \mathbf{y}_1 \geq 0, \tag{3.23c}$$

$$\mathbf{x}_1 + \mathbf{x}_2 - \mathbf{y}_2 \geq 0, \tag{3.23d}$$

$$\mathbf{x}_1 + \mathbf{x}_2 + \mathbf{x}_3 - \mathbf{y}_3 \geq 0, \tag{3.23e}$$

$$\mathbf{y}_1 - \mathbf{b}_1 = 0, \tag{3.23f}$$

$$\mathbf{y}_2 - \mathbf{b}_2 = 0, \tag{3.23g}$$

$$\mathbf{y}_3 - \mathbf{b}_3 = 0, \tag{3.23h}$$

$$\mathbf{x}_1, \mathbf{x}_2, \mathbf{x}_3, \mathbf{y}_1, \mathbf{y}_2, \mathbf{y}_3 \geq 0, \tag{3.23i}$$

where $\mathbf{x}_1, \mathbf{x}_2$, and $\mathbf{x}_3$ represent the capacity to be built of each technology in stages 1, 2, and 3, respectively; $\mathbf{y}_1, \mathbf{y}_2$, and $\mathbf{y}_3$ represent the power generated by each technology in each period; $\mathbf{c}_1$, $\mathbf{c}_2$, and $\mathbf{c}_3$ correspond to the investment cost of building new capacity of the considered technologies; $\mathbf{K}_1$, $\mathbf{K}_2$, and $\mathbf{K}_3$ represent the operating cost of the technologies in each stage; $\mathbf{X}$ is a vector comprising the maximum capacity that can be built of each technology; and $\mathbf{b}_1$, $\mathbf{b}_2$, and $\mathbf{b}_3$ represent the demand in each period. Note that $\mathbf{c}_1$, $\mathbf{c}_2$, $\mathbf{c}_3$, and $\mathbf{b}_1$, $\mathbf{b}_2$, and $\mathbf{b}_3$ are uncertain parameters, whereas $\mathbf{X}$ and $\mathbf{K}_1$, $\mathbf{K}_2$, $\mathbf{K}_3$ are considered as known parameters.

The objective function (3.23a) represents the expected cost of building new capacity and supplying the demand. Constraints (3.23b) limit the capacity that can be built along the whole planning horizon for each technology. Constraints (3.23c)–(3.23e) impose limits on the power that can be generated according to the capacity built up to stages 1, 2, and 3, respectively. Constraints (3.23f)–(3.23h) set the power balance in each period, meaning that the power generated must be equal to the demand. Finally, inequalities (3.23i) establish the non negativity of the variables.

Note that in this problem the investment decisions at stage 1, $\mathbf{x}_1$, are taken considering the information known at the beginning of the planning horizon, i.e., the current value of the investment cost (in $n = 1, \mathbf{c}_1$), but the demand growth along period 1 is unknown; then, the uncertainty of the demand realizes, the demand along period 2 is known, and the power generated by each technology is determined to supply the demand at the minimum cost; at stage 2, the investment decisions $\mathbf{x}_2$ are made with the information of the investment cost in this stage and of the demand realization along period 1 (in $n = 2, \mathbf{c}_2, \mathbf{b}_1$); similarly, in stage 3 the investment decisions $\mathbf{x}_3$ are made knowing the value of the investment costs at that stage and the demand realized in the previous stage (in $n = 3, \mathbf{c}_3, \mathbf{b}_2$). The operating decisions in each stage $\mathbf{y}_1, \mathbf{y}_2, \mathbf{y}_3$ are conditioned by the realization of the demand in the corresponding stage $\mathbf{b}_1, \mathbf{b}_2, \mathbf{b}_3$. □

3.4.2 Definition of Linear Decision Rules

To facilitate the formulation of the decision rules, it is convenient to define a vector that comprises all the uncertain parameters of the problem, being $\tilde{\xi}$. Without loss of generality and for mathematical purposes, it is assumed that the first element of this vector is equal to 1. Thus,

$$\tilde{\xi} = [1, \mathbf{c}_1, \mathbf{b}_1, \mathbf{c}_2, \mathbf{b}_2, ..., \mathbf{c}_n, \mathbf{b}_n, ..., \mathbf{c}_N, \mathbf{b}_N]. \tag{3.24}$$

In particular, $\tilde{\xi}_m$ would refer to the element located in position m in vector $\tilde{\xi}$, i.e., $\tilde{\xi}_1 = [1], \tilde{\xi}_2 = [\mathbf{c}_1], \tilde{\xi}_3 = [\mathbf{b}_1]$, and so on. We define as M_n the set indices of elements considered in $\tilde{\xi}$ up to stage n, i.e., $M_1 = [1, 2, 3]$, $M_2 = [1, 2, 3, 4, 5]$, and so on.

A linear decision rule expresses the linear dependency between the uncertain parameters and the decision variables. For variables $\mathbf{x}_n$, the LDRs are written as follows:

$$\mathbf{x}_n = \sum_{m \in M_n} \hat{\mathbf{x}}_{nm} \tilde{\xi}_m, \quad \forall n \in N, \tag{3.25}$$

where $\hat{\mathbf{x}}_{nm}$ are the new decision variables. Note that index m counts for the indices in set M_n for each stage, i.e., the indices of uncertain parameters up to stage n in vector $\tilde{\xi}$.

In the same manner, LDRs can be established for deterministic parameters. For instance, parameter $\mathbf{c}_n$ of the objective function can be rewritten as follows:

$$\mathbf{c}_n = \sum_{m \in M_n} \hat{\mathbf{c}}_{nm} \tilde{\xi}_m, \quad \forall n \in N, \tag{3.26}$$

being $\hat{\mathbf{c}}_{nm}$ the new parameter to be considered in the reformulation of the initial multi-stage problem.

Illustrative Example 3.13 LDR approach: definition of linear decision rules

Considering Illustrative Example 3.12, vector $\tilde{\xi}$ is defined as follows:

$$\tilde{\xi} = [1, \mathbf{c}_1, \mathbf{b}_1, \mathbf{c}_2, \mathbf{b}_2, \mathbf{c}_3, \mathbf{b}_3].$$

In this way, the LDRs establishing a linear dependency between the decision variables $\mathbf{x}_{nm}$ and $\mathbf{y}_{nm}$, and the uncertain parameters can be written as follows:

$$\mathbf{x}_n = \sum_{m \in M_n^{\mathrm{I}}} \hat{\mathbf{x}}_{nm} \tilde{\xi}_m, \quad \forall n \in N,$$

$$\mathbf{y}_n = \sum_{m \in M_n^{\mathrm{O}}} \hat{\mathbf{y}}_{nm} \tilde{\xi}_m, \quad \forall n \in N,$$

where $\hat{\mathbf{x}}_{nm}$ and $\hat{\mathbf{y}}_{nm}$ are the new investment and operating variables, respectively; M_n^{I} is the set of indices of uncertain elements known when the investment decisions of stage n are made; and M_n^{O} is the set of indices of uncertain elements known when the operating decisions of stage n are made. In particular, index sets M_n^{I} and M_n^{O} for each stage comprise the following elements of $\tilde{\xi}$:

$$\tilde{\xi}_{M_1^{\mathrm{I}}} = \{1, \mathbf{c}_1\},$$
$$\tilde{\xi}_{M_2^{\mathrm{I}}} = \{1, \mathbf{c}_1, \mathbf{b}_1, \mathbf{c}_2\},$$
$$\tilde{\xi}_{M_3^{\mathrm{I}}} = \{1, \mathbf{c}_1, \mathbf{b}_1, \mathbf{c}_2, \mathbf{b}_2, \mathbf{c}_3\},$$
$$\tilde{\xi}_{M_1^{\mathrm{O}}} = \{1, \mathbf{c}_1, \mathbf{b}_1\},$$
$$\tilde{\xi}_{M_2^{\mathrm{O}}} = \{1, \mathbf{c}_1, \mathbf{b}_1, \mathbf{c}_2, \mathbf{b}_2\},$$
$$\tilde{\xi}_{M_3^{\mathrm{O}}} = \{1, \mathbf{c}_1, \mathbf{b}_1, \mathbf{c}_2, \mathbf{b}_2, \mathbf{c}_3, \mathbf{b}_3\}.$$

Note that the investment cost of new capacity installed at the beginning of the planning horizon $\mathbf{c}_1$ is known, or in other words, the first-stage investment decisions $\mathbf{x}_1$ do not depend on any uncertain parameter. □

3.4.3 Approximate Multi-Stage Problem

Therefore, if the LDRs presented in Sect. 3.4.2 are applied to multi-stage decision problem (3.21), this problem can be rewritten as follows:

$$\text{Minimize}_{\mathbf{x}_n} \quad \mathbb{E}_{\Xi} \left\{ \sum_{n \in N} \sum_{m \in M_n} \hat{\mathbf{c}}_{nm} \tilde{\xi}_m \left(\sum_{r \in M_n} \hat{\mathbf{x}}_{nr} \tilde{\xi}_r \right) \right\} \tag{3.27a}$$

subject to:

$$\mathbf{A}\sum_{m\in M_n}\hat{\mathbf{x}}_{nm}\tilde{\xi}_m-\sum_{m\in M_n}\hat{\mathbf{b}}_{nm}\tilde{\xi}_m\geq 0,\quad \forall n\in N, \tag{3.27b}$$

$$\sum_{m\in M_n}\hat{\mathbf{x}}_{nm}\tilde{\xi}_m\geq 0,\quad \forall n\in N. \tag{3.27c}$$

Observe that all the variables and parameters of problem (3.21) are defined regarding a linear relationship with the uncertain parameters represented by vector $\tilde{\xi}$.

We call problem (3.27) *approximate* because the space of optimal variables is limited to that with variables holding a linear dependency with the uncertain parameters.

Illustrative Example 3.14 LDR approach: approximate problem formulation

Illustrative Example 3.12 is considered again. Applying the LDRs described in Illustrative Example 3.13, problem (3.23) can be rewritten as follows:

$$\begin{aligned}\text{Minimize}_{\hat{\mathbf{x}}_{nm},\hat{\mathbf{y}}_{nm},\forall n\in N,\forall m\in M_n}\quad &\mathbb{E}_{\Xi}\left\{\sum_{m=1}^{2}\hat{\mathbf{c}}_{1m}^{\mathrm{T}}\tilde{\xi}_m\left(\sum_{r=1}^{2}\hat{\mathbf{x}}_{1r}\tilde{\xi}_r\right)\right.\\
&+\sum_{m=1}^{4}\hat{\mathbf{c}}_{2m}^{\mathrm{T}}\tilde{\xi}_m\left(\sum_{r=1}^{4}\hat{\mathbf{x}}_{2r}\tilde{\xi}_r\right)+\sum_{m=1}^{6}\hat{\mathbf{c}}_{3m}^{\mathrm{T}}\tilde{\xi}_m\left(\sum_{r=1}^{6}\hat{\mathbf{x}}_{3r}\tilde{\xi}_r\right)\\
&+\sum_{m=1}^{3}\hat{\mathbf{K}}_{1m}\tilde{\xi}_m\left(\sum_{r=1}^{3}\hat{\mathbf{y}}_{1r}\tilde{\xi}_r\right)+\sum_{m=1}^{5}\hat{\mathbf{K}}_{2m}\tilde{\xi}_m\left(\sum_{r=1}^{5}\hat{\mathbf{y}}_{2r}\tilde{\xi}_r\right)\\
&\left.+\sum_{m=1}^{7}\hat{\mathbf{K}}_{3m}\tilde{\xi}_m\left(\sum_{r=1}^{7}\hat{\mathbf{y}}_{3r}\tilde{\xi}_r\right)\right\}\end{aligned} \tag{3.28}$$

subject to:

$$\sum_{m=1}^{6}\hat{\mathbf{X}}_{3m}\tilde{\xi}_m-\sum_{m=1}^{2}\hat{\mathbf{x}}_{1m}\tilde{\xi}_m-\sum_{m=1}^{4}\hat{\mathbf{x}}_{2m}\tilde{\xi}_m-\sum_{m=1}^{6}\hat{\mathbf{x}}_{3m}\tilde{\xi}_m\geq 0,$$

$$\sum_{m=1}^{2}\hat{\mathbf{x}}_{1m}\tilde{\xi}_m-\sum_{m=1}^{3}\hat{\mathbf{y}}_{1m}\tilde{\xi}_m\geq 0,$$

$$\sum_{m=1}^{2}\hat{\mathbf{x}}_{1m}\tilde{\xi}_m+\sum_{m=1}^{4}\hat{\mathbf{x}}_{2m}\tilde{\xi}_m-\sum_{m=1}^{5}\hat{\mathbf{y}}_{2m}\tilde{\xi}_m\geq 0,$$

$$\sum_{m=1}^{2}\hat{\mathbf{x}}_{1m}\tilde{\xi}_m+\sum_{m=1}^{4}\hat{\mathbf{x}}_{2m}\tilde{\xi}_m+\sum_{m=1}^{6}\hat{\mathbf{x}}_{3m}\tilde{\xi}_m-\sum_{m=1}^{7}\hat{\mathbf{y}}_{3m}\tilde{\xi}_m\geq 0,$$

$$\sum_{m=1}^{3}\hat{\mathbf{y}}_{1m}\tilde{\xi}_m-\sum_{m=1}^{3}\hat{\mathbf{b}}_{1m}\tilde{\xi}_m=0,$$

$$\sum_{m=1}^{5} \hat{\mathbf{y}}_{2m}\tilde{\xi}_m - \sum_{m=1}^{5} \hat{\mathbf{b}}_{2m}\tilde{\xi}_m = 0,$$

$$\sum_{m=1}^{7} \hat{\mathbf{y}}_{3m}\tilde{\xi}_m - \sum_{m=1}^{7} \hat{\mathbf{b}}_{3m}\tilde{\xi}_m = 0,$$

$$\sum_{m=1}^{2} \hat{\mathbf{x}}_{1m}\tilde{\xi}_m \geq 0,$$

$$\sum_{m=1}^{4} \hat{\mathbf{x}}_{2m}\tilde{\xi}_m \geq 0,$$

$$\sum_{m=1}^{6} \hat{\mathbf{x}}_{3m}\tilde{\xi}_m \geq 0,$$

$$\sum_{m=1}^{3} \hat{\mathbf{y}}_{1m}\tilde{\xi}_m \geq 0,$$

$$\sum_{m=1}^{5} \hat{\mathbf{y}}_{2m}\tilde{\xi}_m \geq 0,$$

$$\sum_{m=1}^{7} \hat{\mathbf{y}}_{3m}\tilde{\xi}_m \geq 0.$$

Let us clarify that to apply the mathematical procedure explained above, it is required to express every element of the optimization problem as a function of vector $\tilde{\xi}$. Consequently, parameter **X**, representing the maximum capacity that can be built per technology, is now defined as stated in the first constraints of the previous problem and its elements take the following values:

$$\hat{\mathbf{X}}_{nm} = \begin{bmatrix} \mathbf{X}\,0\,0\,0\,0\,0\,0 \\ \mathbf{X}\,0\,0\,0\,0\,0\,0 \\ \mathbf{X}\,0\,0\,0\,0\,0\,0 \end{bmatrix}.$$

□

3.4.4 Uncertainty Bounds

The uncertainty bounds (3.22), where the random parameters are defined, can be rewritten using parameters **J** and **h** as follows:

$$\sum_{m \in M_n} J_{km} \tilde{\xi}_m \geq h_k, \quad \forall k \in K, \tag{3.29}$$

where the number of elements in K is twice the size of $\tilde{\xi}$ because the upper and lower bounds of the interval must be defined.

Illustrative Example 3.15 LDR approach: uncertainty bounds

Considering Illustrative Example 3.12 and according to (3.22), the intervals where the uncertain parameters take values can be written as follows:

$$\begin{aligned}
\bar{\mathbf{b}}_n(1-\theta_n^{\mathrm{B}}) \leq \mathbf{b}_n \leq \bar{\mathbf{b}}_n(1+\theta_n^{\mathrm{B}}), \quad n = 1, 2, 3, \\
\bar{\mathbf{c}}_n(1-\theta_n^{\mathrm{C}}) \leq \mathbf{c}_n \leq \bar{\mathbf{c}}_n(1+\theta_n^{\mathrm{C}}), \quad n = 1, 2, 3.
\end{aligned}$$

The previous uncertainty bounds can be also established regarding vector $\tilde{\xi} = [1, \mathbf{c}_1, \mathbf{b}_1, \mathbf{c}_2, \mathbf{b}_2, \mathbf{c}_3, \mathbf{b}_3]$ as done in (3.29). Consequently, $\mathbf{J}$ and $\mathbf{h}$ are expressed as:

$$\mathbf{J} = \begin{bmatrix}
1 & 0 & 0 & 0 & 0 & 0 & 0 \\
-1 & 0 & 0 & 0 & 0 & 0 & 0 \\
-\bar{\mathbf{c}}_1(1-\theta_1^{\mathrm{C}}) & 1 & 0 & 0 & 0 & 0 & 0 \\
\bar{\mathbf{c}}_1(1+\theta_1^{\mathrm{C}}) & -1 & 0 & 0 & 0 & 0 & 0 \\
-\bar{\mathbf{b}}_1(1-\theta_1^{\mathrm{B}}) & 0 & 1 & 0 & 0 & 0 & 0 \\
\bar{\mathbf{b}}_1(1+\theta_1^{\mathrm{B}}) & 0 & -1 & 0 & 0 & 0 & 0 \\
-\bar{\mathbf{c}}_2(1-\theta_2^{\mathrm{C}}) & 0 & 0 & 1 & 0 & 0 & 0 \\
\bar{\mathbf{c}}_2(1+\theta_2^{\mathrm{C}}) & 0 & 0 & -1 & 0 & 0 & 0 \\
-\bar{\mathbf{b}}_2(1-\theta_2^{\mathrm{B}}) & 0 & 0 & 0 & 1 & 0 & 0 \\
\bar{\mathbf{b}}_2(1+\theta_2^{\mathrm{B}}) & 0 & 0 & 0 & -1 & 0 & 0 \\
-\bar{\mathbf{c}}_3(1-\theta_3^{\mathrm{C}}) & 0 & 0 & 0 & 0 & 1 & 0 \\
\bar{\mathbf{c}}_3(1+\theta_3^{\mathrm{C}}) & 0 & 0 & 0 & 0 & -1 & 0 \\
-\bar{\mathbf{b}}_3(1-\theta_3^{\mathrm{B}}) & 0 & 0 & 0 & 0 & 0 & 1 \\
\bar{\mathbf{b}}_3(1+\theta_3^{\mathrm{B}}) & 0 & 0 & 0 & 0 & 0 & -1
\end{bmatrix},$$

$$\mathbf{h} = \begin{bmatrix} 1 & -1 & 0 & 0 & 0 & 0 & 0 & 0 & 0 & 0 & 0 & 0 & 0 & 0 \end{bmatrix}.$$

As it can be observed, the number of elements in K is equal to 14 that is twice the size of $\tilde{\xi}$. Recall that $\bar{\mathbf{c}}_1$, $\bar{\mathbf{c}}_2$ and $\bar{\mathbf{c}}_3$ represent the expected value of parameter $\mathbf{c}_n$ in each stage. The same applies to parameter $\mathbf{b}_n$. Parameters $\theta_n^{\mathrm{B/C}}$ set the size of the uncertainty bound in each stage. □

3.4.5 Reformulation of the Objective Function

The next step to obtain a formulation of a multi-stage decision problem based on LDRs that can be solved using linear programming consists in reformulating the problem such that vector $\tilde{\xi}$ disappears from the equations.

Firstly, let us focus on the objective function (3.27a). In this equation, the only element affected by the expectation operator is vector $\tilde{\xi}$. Applying basic statistics, the objective function results in:

$$\mathbb{E}_{\Xi}\left\{\sum_{n\in N}\sum_{m\in M_n}\hat{\mathbf{c}}_{nm}\tilde{\xi}_m\left(\sum_{r\in M_n}\hat{\mathbf{x}}_{nr}\tilde{\xi}_r\right)\right\}=\mathbb{E}_{\Xi}\left\{\sum_{n\in N}\sum_{m\in M_n}\sum_{r\in M_n}\hat{\mathbf{c}}_{nm}\ \hat{\mathbf{x}}_{nr}\ \tilde{\xi}_m\ \tilde{\xi}_r\right\}$$
$$=\sum_{n\in N}\sum_{m\in M_n}\sum_{r\in M_n}\hat{\mathbf{c}}_{nm}\hat{\mathbf{x}}_{nr}\ \mathbb{E}_{\Xi}\left\{\tilde{\xi}_m\tilde{\xi}_r\right\}=\sum_{n\in N}\sum_{m\in M_n}\sum_{r\in M_n}\hat{\mathbf{c}}_{nm}\hat{\mathbf{x}}_{nr}\ \mathrm{M}^{\mathrm{M}}_{mr},\tag{3.30}$$

being $\mathbf{M}^{\mathrm{M}}$ the second-order moment matrix of vector $\tilde{\xi}$. The elements of matrix $\mathbf{M}^{\mathrm{M}}$ are computed as:

$$\mathrm{M}^{\mathrm{M}}_{mr}=\mathbb{E}_{\Xi}\left\{\tilde{\xi}_m\tilde{\xi}_r\right\}=\bar{\xi}_m\bar{\xi}_r+var\left\{\tilde{\xi}_m,\tilde{\xi}_r\right\},\quad\forall m,r\in M_n,\forall n.$$

This means that the computation of matrix $\mathbf{M}^{\mathrm{M}}$, and consequently the objective function (3.30), depends on the expected value of the uncertain parameters in each stage and their variance.

Illustrative Example 3.16 LDR approach: reformulation of the objective function

Illustrative Example 3.14 is considered again. The objective function (3.28) can be rewritten as follows:

$$\mathbb{E}_{\Xi}\left\{\sum_{m=1}^{2}\hat{\mathbf{c}}^{\mathrm{T}}_{1m}\tilde{\xi}_m\left(\sum_{r=1}^{2}\hat{\mathbf{x}}_{1r}\tilde{\xi}_r\right)+\sum_{m=1}^{4}\hat{\mathbf{c}}^{\mathrm{T}}_{2m}\tilde{\xi}_m\left(\sum_{r=1}^{4}\hat{\mathbf{x}}_{2r}\tilde{\xi}_r\right)\right.$$
$$+\sum_{m=1}^{6}\hat{\mathbf{c}}^{\mathrm{T}}_{3m}\tilde{\xi}_m\left(\sum_{r=1}^{6}\hat{\mathbf{x}}_{3r}\tilde{\xi}_r\right)+\sum_{m=1}^{3}\hat{\mathbf{K}}_{1m}\tilde{\xi}_m\left(\sum_{r=1}^{3}\hat{\mathbf{y}}_{1r}\tilde{\xi}_r\right)$$
$$\left.+\sum_{m=1}^{5}\hat{\mathbf{K}}_{2m}\tilde{\xi}_m\left(\sum_{r=1}^{5}\hat{\mathbf{y}}_{2r}\tilde{\xi}_r\right)+\sum_{m=1}^{7}\hat{\mathbf{K}}_{3m}\tilde{\xi}_m\left(\sum_{r=1}^{7}\hat{\mathbf{y}}_{3r}\tilde{\xi}_r\right)\right\}$$
$$=\mathbb{E}_{\Xi}\left\{\sum_{m=1}^{2}\sum_{r=1}^{2}\hat{\mathbf{c}}^{\mathrm{T}}_{1m}\hat{\mathbf{x}}_{1r}\tilde{\xi}_m\tilde{\xi}_r+\sum_{m=1}^{4}\sum_{r=1}^{4}\hat{\mathbf{c}}^{\mathrm{T}}_{2m}\hat{\mathbf{x}}_{2r}\tilde{\xi}_m\tilde{\xi}_r\right.$$
$$+\sum_{m=1}^{6}\sum_{r=1}^{6}\hat{\mathbf{c}}^{\mathrm{T}}_{3m}\hat{\mathbf{x}}_{3r}\tilde{\xi}_m\tilde{\xi}_r+\sum_{m=1}^{3}\sum_{r=1}^{3}\hat{\mathbf{K}}_{1m}\hat{\mathbf{y}}_{1r}\tilde{\xi}_m\tilde{\xi}_r$$

$$+\sum_{m=1}^{5}\sum_{r=1}^{5}\hat{\mathbf{K}}_{2m}\hat{\mathbf{y}}_{2r}\tilde{\xi}_m\tilde{\xi}_r+\sum_{m=1}^{7}\sum_{r=1}^{7}\hat{\mathbf{K}}_{3m}\hat{\mathbf{y}}_{3r}\tilde{\xi}_m\tilde{\xi}_r\Bigg\}$$

$$=\sum_{m=1}^{2}\sum_{r=1}^{2}\hat{\mathbf{c}}_{1m}^{\mathrm{T}}\hat{\mathbf{x}}_{1r}\mathbb{E}_{\Xi}\{\tilde{\xi}_m\tilde{\xi}_r\}+\sum_{m=1}^{4}\sum_{r=1}^{4}\hat{\mathbf{c}}_{2m}^{\mathrm{T}}\hat{\mathbf{x}}_{2r}\mathbb{E}_{\Xi}\{\tilde{\xi}_m\tilde{\xi}_r\}$$

$$+\sum_{m=1}^{6}\sum_{r=1}^{6}\hat{\mathbf{c}}_{3m}^{\mathrm{T}}\hat{\mathbf{x}}_{3r}\mathbb{E}_{\Xi}\{\tilde{\xi}_m\tilde{\xi}_r\}+\sum_{m=1}^{3}\sum_{r=1}^{3}\hat{\mathbf{K}}_{1m}\hat{\mathbf{y}}_{1r}\mathbb{E}_{\Xi}\{\tilde{\xi}_m\tilde{\xi}_r\}$$

$$+\sum_{m=1}^{5}\sum_{r=1}^{5}\hat{\mathbf{K}}_{2m}\hat{\mathbf{y}}_{2r}\mathbb{E}_{\Xi}\{\tilde{\xi}_m\tilde{\xi}_r\}+\sum_{m=1}^{7}\sum_{r=1}^{7}\hat{\mathbf{K}}_{3m}\hat{\mathbf{y}}_{3r}\mathbb{E}_{\Xi}\{\tilde{\xi}_m\tilde{\xi}_r\}$$

$$=\sum_{m=1}^{2}\sum_{r=1}^{2}\hat{\mathbf{c}}_{1m}^{\mathrm{T}}\hat{\mathbf{x}}_{1r}\mathbf{M}_{mr}^{\mathrm{M}}+\sum_{m=1}^{4}\sum_{r=1}^{4}\hat{\mathbf{c}}_{2m}^{\mathrm{T}}\hat{\mathbf{x}}_{2r}\mathbf{M}_{mr}^{\mathrm{M}}$$

$$+\sum_{m=1}^{6}\sum_{r=1}^{6}\hat{\mathbf{c}}_{3m}^{\mathrm{T}}\hat{\mathbf{x}}_{3r}\mathbf{M}_{mr}^{\mathrm{M}}+\sum_{m=1}^{3}\sum_{r=1}^{3}\hat{\mathbf{K}}_{1m}\hat{\mathbf{y}}_{1r}\mathbf{M}_{mr}^{\mathrm{M}}$$

$$+\sum_{m=1}^{5}\sum_{r=1}^{5}\hat{\mathbf{K}}_{2m}\hat{\mathbf{y}}_{2r}\mathbf{M}_{mr}^{\mathrm{M}}+\sum_{m=1}^{7}\sum_{r=1}^{7}\hat{\mathbf{K}}_{3m}\hat{\mathbf{y}}_{3r}\mathbf{M}_{mr}^{\mathrm{M}}.$$

As an example, the elements of matrix $\mathbf{M}^{\mathrm{M}}$ are computed as follows:

$$\mathrm{M}_{11}^{\mathrm{M}}=\bar{\xi}_1\bar{\xi}_1+var\left\{\tilde{\xi}_1,\tilde{\xi}_1\right\}=1,$$

$$\mathrm{M}_{21}^{\mathrm{M}}=\bar{\xi}_2\bar{\xi}_1+var\left\{\tilde{\xi}_2,\tilde{\xi}_2\right\}=\bar{\mathbf{c}}_1,$$

$$\mathrm{M}_{12}^{\mathrm{M}}=\mathrm{M}_{21}^{\mathrm{M}}=\bar{\mathbf{c}}_1,$$

$$\mathrm{M}_{23}^{\mathrm{M}}=\bar{\xi}_{23}\bar{\xi}_{32}+var\left\{\tilde{\xi}_{23},\tilde{\xi}_{32}\right\}=\bar{\mathbf{c}}_1\bar{\mathbf{b}}_1.$$

Note that parameters **c** and **b** are assumed to be uncorrelated. Hence, the variance between both parameters is null. □

3.4.6 Reformulation of Constraints

Secondly, to reformulate the constraints of problem (3.27) it is needed to consider the definition of the uncertainty bounds (3.29) and Proposition 1 of [13]. This proposition states that being Z_{nm} a variable holding

$$\sum_{m\in M_n} Z_{nm}\tilde{\xi}_m \geq 0, \quad \forall n \in N, \tag{3.31}$$

and $\tilde{\xi}$ taking values in the uncertainty bounds (3.29), there exists a variable Λ_{nk} that allows to rewrite (3.31) as:

$$Z_{nm} = \sum_{k \in K} J_{km} \lambda_{nk}, \quad \forall n \in N, \forall m \in M_n, \tag{3.32a}$$

$$\sum_{k \in K} h_k \Lambda_{nk} \geq 0, \quad \forall n \in N, \tag{3.32b}$$

$$\Lambda_{nk} \geq 0, \quad \forall n \in N, \forall k \in K. \tag{3.32c}$$

As a result, the constraints become deterministic, i.e., there is no dependency of random parameters.

In this way, applying this proposition to problem (3.27), constraints (3.27b) are reformulated as follows:

$$\mathbf{A}\hat{\mathbf{x}}_{nm} - \mathbf{b}_{nm} = \sum_{k \in K} J_{km} \Lambda_{nk}, \quad \forall n \in N, \forall m \in M_n, \tag{3.33a}$$

$$\sum_{k \in K} h_k \Lambda_{nk} \geq 0, \quad \forall n \in N, \tag{3.33b}$$

$$\Lambda_{nk} \geq 0, \quad \forall n \in N, \forall k \in K, \tag{3.33c}$$

being Λ_{nk} an auxiliary variable.

3.4.7 Final LDR-Based Formulation

Finally, the formulation of a standard multi-stage decision problem as (3.21) based on LDRs is as follows:

$$\text{Minimize}_{\hat{\mathbf{x}}_n, \forall n \in N, \Lambda_{nk}, \forall n \in N, \forall k \in K} \quad \sum_{n \in N} \sum_{m \in M_n} \sum_{r \in M_n} \hat{\mathbf{c}}_{nm} \hat{\mathbf{x}}_{nr} \, \mathrm{M}^{\mathrm{M}}_{mr} \tag{3.34a}$$

subject to:

$$\mathbf{A}\hat{\mathbf{x}}_{nm} - \mathbf{b}_{nm} = \sum_{k \in K} J_{km} \Lambda^1_{nk}, \quad \forall n \in N, \forall m \in M_n \tag{3.34b}$$

$$\sum_{k \in K} h_k \Lambda^1_{nk} \geq 0, \quad \forall n \in N, \tag{3.34c}$$

$$\Lambda^1_{nk} \geq 0, \quad \forall n \in N, \forall k \in K, \tag{3.34d}$$

$$\hat{\mathbf{x}}_{nm} = \sum_{k \in K} J_{km} \Lambda^2_{nk}, \quad \forall n \in N, \forall m \in M_n, \tag{3.34e}$$

$$\sum_{k \in K} h_k \Lambda_{nk}^2 \geq 0, \quad \forall n \in N, \tag{3.34f}$$

$$\Lambda_{nk}^2 \geq 0, \quad \forall n \in N, \forall k \in K. \tag{3.34g}$$

Illustrative Example 3.17 LDR approach: final formulation of a multi-stage decision problem

Considering the formulation presented in Illustrative Example 3.14, the final formulation based on LDRs reads as follows:

$$\text{Minimize}_{\hat{\mathbf{x}}_{nm}, \hat{\mathbf{y}}_{nm}, \forall n \in N, \forall m \in M_n} \quad \sum_{m=1}^{2}\sum_{r=1}^{2} \hat{\mathbf{c}}_{1m}^{\mathrm{T}} \hat{\mathbf{x}}_{1r} \mathrm{M}_{mr}^{\mathrm{M}} + \sum_{m=1}^{4}\sum_{r=1}^{4} \hat{\mathbf{c}}_{2m}^{\mathrm{T}} \hat{\mathbf{x}}_{2r} \mathrm{M}_{mr}^{\mathrm{M}}$$
$$+ \sum_{m=1}^{6}\sum_{r=1}^{6} \hat{\mathbf{c}}_{3m}^{\mathrm{T}} \hat{\mathbf{x}}_{3r} \mathrm{M}_{mr}^{\mathrm{M}} + \sum_{m=1}^{3}\sum_{r=1}^{3} \hat{\mathbf{K}}_{1m} \hat{\mathbf{y}}_{1r} \mathrm{M}_{mr}^{\mathrm{M}}$$
$$+ \sum_{m=1}^{5}\sum_{r=1}^{5} \hat{\mathbf{K}}_{2m} \hat{\mathbf{y}}_{2r} \mathrm{M}_{mr}^{\mathrm{M}} + \sum_{m=1}^{7}\sum_{r=1}^{7} \hat{\mathbf{K}}_{3m} \hat{\mathbf{y}}_{3r} \mathrm{M}_{mr}^{\mathrm{M}}$$

subject to:

$$\hat{\mathbf{X}}_{3m} - \hat{\mathbf{x}}_{1m} - \hat{\mathbf{x}}_{2m} - \hat{\mathbf{x}}_{3m} = \sum_{k=1}^{14} J_{km} \Lambda_{3k}^1, \quad m = 1, ..., 6,$$

$$\sum_{k=1}^{14} h_k \Lambda_{3k}^1 \geq 0,$$

$$\Lambda_{3k}^1 \geq 0, \quad k = 1, ..., 14,$$

$$\hat{\mathbf{x}}_{1m} - \hat{\mathbf{y}}_{1m} = \sum_{k=1}^{14} J_{km} \Lambda_{1k}^2, \quad m = 1, ..., 3,$$

$$\sum_{k=1}^{14} h_k \Lambda_{1k}^2 \geq 0,$$

$$\Lambda_{1k}^2 \geq 0, \quad k = 1, ..., 14,$$

$$\hat{\mathbf{x}}_{1m} + \hat{\mathbf{x}}_{2m} - \hat{\mathbf{y}}_{2m} = \sum_{k=1}^{14} J_{km} \Lambda_{2k}^3, \quad m = 1, ..., 5,$$

$$\sum_{k=1}^{14} h_k \Lambda_{2k}^3 \geq 0,$$

$$\Lambda_{2k}^3 \geq 0, \quad k = 1, ..., 14,$$

$$\hat{\mathbf{x}}_{1m} + \hat{\mathbf{x}}_{2m} + \hat{\mathbf{x}}_{3m} - \hat{\mathbf{y}}_{3m} = \sum_{k=1}^{14} J_{km} \Lambda_{3k}^4, \quad m = 1, ..., 7,$$

$$\sum_{k=1}^{14} h_k \Lambda_{3k}^4 \geq 0,$$

$$\Lambda_{3k}^4 \geq 0, \quad k = 1, ..., 14,$$

$$\hat{\mathbf{y}}_{1m} - \hat{\mathbf{b}}_{1m} = \sum_{k=1}^{14} J_{km} \Lambda_{1k}^5, \quad m = 1, ..., 3,$$

$$\sum_{k=1}^{14} h_k \Lambda_{1k}^5 \geq 0,$$

$$\Lambda_{1k}^5 \geq 0, \quad k = 1, ..., 14,$$

$$\hat{\mathbf{y}}_{2m} - \hat{\mathbf{b}}_{2m} = \sum_{k=1}^{14} J_{km} \Lambda_{2k}^6, \quad m = 1, ..., 5,$$

$$\sum_{k=1}^{14} h_k \Lambda_{2k}^6 \geq 0,$$

$$\Lambda_{2k}^6 \geq 0, \quad k = 1, ..., 14,$$

$$\hat{\mathbf{y}}_{3m} - \hat{\mathbf{b}}_{3m} = \sum_{k=1}^{14} J_{km} \Lambda_{3k}^7, \quad m = 1, ..., 7,$$

$$\sum_{k=1}^{14} h_k \Lambda_{3k}^7 \geq 0,$$

$$\Lambda_{3k}^7 \geq 0, \quad k = 1, ..., 14,$$

$$\hat{\mathbf{x}}_{1m} = \sum_{k=1}^{14} J_{km} \Lambda_{1k}^8, \quad m = 1, 2,$$

$$\sum_{k=1}^{14} h_k \Lambda_{1k}^8 \geq 0,$$

$$\Lambda_{1k}^8 \geq 0, \quad k = 1, ..., 14,$$

$$\hat{\mathbf{x}}_{2m} = \sum_{k=1}^{14} J_{km} \Lambda_{2k}^9, \quad m = 1, ..., 4,$$

$$\sum_{k=1}^{14} h_k \Lambda_{2k}^9 \geq 0,$$

$$\Lambda_{2k}^9 \geq 0, \quad k = 1, ..., 14,$$

$$\hat{\mathbf{x}}_{3m} = \sum_{k=1}^{14} J_{km} \Lambda_{3k}^{10}, \quad m = 1, ..., 6,$$

$$\sum_{k=1}^{14} h_k \Lambda_{3k}^{10} \geq 0,$$

$$\Lambda_{3k}^{10} \geq 0, \quad k = 1, ..., 14,$$

$$\hat{\mathbf{y}}_{1m} = \sum_{k=1}^{14} J_{km} \Lambda_{1k}^{11}, \quad m = 1, 2, 3,$$

$$\sum_{k=1}^{14} h_k \Lambda_{3k}^{11} \geq 0,$$

$$\Lambda_{1k}^{11} \geq 0, \quad k = 1, ..., 14,$$

$$\hat{\mathbf{y}}_{2m} = \sum_{k=1}^{14} J_{km} \Lambda_{2k}^{12}, \quad m = 1, ..., 5,$$

$$\sum_{k=1}^{14} h_k \Lambda_{2k}^{12} \geq 0,$$

$$\Lambda_{2k}^{12} \geq 0, \quad k = 1, ..., 14,$$

$$\hat{\mathbf{y}}_{3m} = \sum_{k=1}^{14} J_{km} \Lambda_{3k}^{13}, \quad m = 1, ..., 7,$$

$$\sum_{k=1}^{14} h_k \Lambda_{3k}^{13} \geq 0,$$

$$\Lambda_{3k}^{13} \geq 0, \quad k = 1, ..., 14.$$

The resulting is a deterministic problem that can be solved using standard mathematical programs. □

For the interested reader, Chap. 9 applies this approach to solve a multi-stage capacity expansion problem in renewable energies considering long-term uncertainty sources. In Chap. 9, numerical examples are included to ease the understanding of the implementation of the LDR approach.

3.5 Benders' Decomposition

Decision making in power systems may involve formulating and solving mathematical programs with thousands of variables and constraints. These problems may be hard to solve in a reduced time, something that may be critical, especially when decisions must be made in a short span of time. Problems may be even intractable if they are solved with a computer with limited computation characteristics. However, some decision-making problems in power systems have the property that they can be decomposed into different subproblems if the value of some variables is fixed, allowing a reduction in the computation time. In such a case, it is possible to apply the so-called Benders' decomposition, which is a useful technique to solve problems with complicating variables [1, 4].

Illustrative Example 3.18 Benders' decomposition: motivating example

A power producer owns a generating unit with a capacity of 100 MW, a minimum production level of 20 MW, a marginal production cost of 50 €/MWh, and a fixed cost of 20 €/h. The power producer can sell its production at prices of 48 €/MWh, 65 €/MWh, 45 €/MWh, and 70 €/MWh in the following four hours. This power producer has to decide the on/off status of its unit in these four hours to maximize its profit knowing that, for technical reasons, the status must be the same for the whole 4-h period.

Considering the above data, the problem can be formulated as the following mixed-integer linear programming problem:

$$\text{Minimize}_{u,p_1,p_2,p_3,p_4}\ 50\left(p_1+p_2+p_3+p_4\right)+4\cdot 20u-48p_1-65p_2-45p_3-70p_4$$

subject to:

$$\begin{aligned}
&u\in\{0,1\},\\
&20u\le p_1\le 100u,\\
&20u\le p_2\le 100u,\\
&20u\le p_3\le 100u,\\
&20u\le p_4\le 100u.
\end{aligned}$$

The above problem minimizes the cost of the power producer minus the revenue achieved by selling its production, which is equivalent to maximize its profit. The first constraint defines the binary nature of variable u, used to represent the on/off status of the generating unit. The following constraints impose bounds on the production levels in the four time periods, according to the minimum production level and capacity of the generating unit.

The solution of this problem can be easily obtained and indicates that it is optimal to turn on the generating unit, i.e., $u=1$, and fix the production level to the minimum or maximum levels depending on whether the price is lower or higher than the marginal production cost, respectively, i.e., $p_1=20$, $p_2=100$, $p_3=20$, and $p_4=100$.

Note that the above optimization problem can be simplified if the value of variable u is fixed to a given value since the problem can be decomposed into four subproblems, namely, one for each hour. In this case, the complicating variable is u. This gives an idea of the type of problems that Benders' decomposition can be applied on. □

3.5.1 Problem Structure

Mathematical programs with complicating variables can be formulated as follows:

$$\text{Minimize}_{\mathbf{x}^{\mathrm{C}},\mathbf{x}_i^{\mathrm{D}},i=1,\ldots,n} \; f^{\mathrm{C}}\left(\mathbf{x}^{\mathrm{C}}\right) + \sum_{i=1}^{n} f^{\mathrm{D}}\left(\mathbf{x}^{\mathrm{C}}, \mathbf{x}_i^{\mathrm{D}}\right) \tag{3.37a}$$

subject to:

$$\mathbf{h}^{\mathrm{C}}\left(\mathbf{x}^{\mathrm{C}}\right) = \mathbf{0}, \tag{3.37b}$$

$$\mathbf{h}_i^{\mathrm{D}}\left(\mathbf{x}^{\mathrm{C}}, \mathbf{x}_i^{\mathrm{D}}\right) = \mathbf{0}, \quad i = 1, \ldots, n, \tag{3.37c}$$

$$\mathbf{g}^{\mathrm{C}}\left(\mathbf{x}^{\mathrm{C}}\right) \le \mathbf{0}, \tag{3.37d}$$

$$\mathbf{g}_i^{\mathrm{D}}\left(\mathbf{x}^{\mathrm{C}}, \mathbf{x}_i^{\mathrm{D}}\right) \le \mathbf{0}, \quad i = 1, \ldots, n, \tag{3.37e}$$

where vector $\mathbf{x}^{\mathrm{C}}$ comprises the complicating variables, while vectors $\mathbf{x}_i^{\mathrm{D}}$, $i = 1, \ldots, n$, include the remaining variables. Note that these vectors include subscript i that denotes the problems in which problem (3.37) can be decomposed if variables in vector $\mathbf{x}^{\mathrm{C}}$ are fixed to given values.

The objective function (3.37a) comprises two terms, namely, a term that depends exclusively on the complicating variables in vector $\mathbf{x}^{\mathrm{C}}$ and a term that depends on variables in vectors $\mathbf{x}^{\mathrm{C}}$ and $\mathbf{x}_i^{\mathrm{D}}$. Constraints (3.37b) and (3.37c) represent equality constraints that depend on the complicating variables and on both the complicating variables and the remaining ones, respectively. Similarly, constraints (3.37d) and (3.37e) represent inequality constraints that depend on the complicating variables and on both the complicating variables and the remaining ones, respectively. Note that constraints (3.37c) and (3.37e) are considered for each value of index i, i.e., these constraints can be decomposed by index i.

The optimal solution of problem (3.37) can be obtained using Benders' decomposition as explained below.

First, auxiliary variable $\alpha\left(\mathbf{x}^{\mathrm{C}}\right)$, used to express the objective function as a function exclusively of the complicating variables, is defined as follows:

$$\alpha\left(\mathbf{x}^{\mathrm{C}}\right) = \min_{\mathbf{x}_i^{\mathrm{D}},i=1,\ldots,n} \; \sum_{i=1}^{n} f^{\mathrm{D}}\left(\mathbf{x}^{\mathrm{C}}, \mathbf{x}_i^{\mathrm{D}}\right) \tag{3.38a}$$

subject to:

$$\mathbf{h}_i^{\mathrm{D}}\left(\mathbf{x}^{\mathrm{C}}, \mathbf{x}_i^{\mathrm{D}}\right) = \mathbf{0}, \quad i = 1, \ldots, n, \tag{3.38b}$$

$$\mathbf{g}_i^{\mathrm{D}}\left(\mathbf{x}^{\mathrm{C}}, \mathbf{x}_i^{\mathrm{D}}\right) \le \mathbf{0}, \quad i = 1, \ldots, n. \tag{3.38c}$$

Using function $\alpha\left(\mathbf{x}^{\mathrm{C}}\right)$, problem (3.37) can be reformulated as:

$$\text{Minimize}_{\mathbf{x}^{\text{C}}} \ \alpha\left(\mathbf{x}^{\text{C}}\right) + f^{\text{C}}\left(\mathbf{x}^{\text{C}}\right) \tag{3.39a}$$

subject to:

$$\mathbf{h}^{\text{C}}\left(\mathbf{x}^{\text{C}}\right) = \mathbf{0}, \tag{3.39b}$$

$$\mathbf{g}^{\text{C}}\left(\mathbf{x}^{\text{C}}\right) \leq \mathbf{0}. \tag{3.39c}$$

Benders' decomposition is based on an iterative procedure that allows to progressively obtain a better and better approximation of function $\alpha\left(\mathbf{x}^{\text{C}}\right)$. This procedure is based on the iterative solution of a master problem and a set of subproblems as explained in the following subsections.

3.5.2 Subproblems

Problem (3.37) is solved for fixed values of the complicating variables, i.e.:

$$\text{Minimize}_{\mathbf{x}^{\text{C}}, \mathbf{x}_i^{\text{D}}, i=1,\ldots,n} \ f^{\text{C}}\left(\mathbf{x}^{\text{C}}\right) + \sum_{i=1}^{n} f^{\text{D}}\left(\mathbf{x}^{\text{C}}, \mathbf{x}_i^{\text{D}}\right) \tag{3.40a}$$

subject to:

$$\mathbf{h}^{\text{C}}\left(\mathbf{x}^{\text{C}}\right) = \mathbf{0}, \tag{3.40b}$$

$$\mathbf{h}_i^{\text{D}}\left(\mathbf{x}^{\text{C}}, \mathbf{x}_i^{\text{D}}\right) = \mathbf{0}, \quad i = 1, \ldots, n, \tag{3.40c}$$

$$\mathbf{g}^{\text{C}}\left(\mathbf{x}^{\text{C}}\right) \leq \mathbf{0}, \tag{3.40d}$$

$$\mathbf{g}_i^{\text{D}}\left(\mathbf{x}^{\text{C}}, \mathbf{x}_i^{\text{D}}\right) \leq \mathbf{0}, \quad i = 1, \ldots, n, \tag{3.40e}$$

$$\mathbf{x}^{\text{C}} = \mathbf{x}^{\text{C},(\upsilon)} \ : \ \eta^{(\upsilon)}, \tag{3.40f}$$

where υ denotes the iteration counter.

Note that for fixed values of variables in set $\mathbf{x}^{\text{C}}$ as imposed by constraints (3.40f), $f^{\text{C}}\left(\mathbf{x}^{\text{C}}\right)$ is constant and problem (3.40) decomposes by index i, provided that the values of variables in set $\mathbf{x}^{\text{C}}$ meet constraints (3.40b) and (3.40d):

$$\text{Minimize}_{\mathbf{x}_i^{\text{D}}} \ f_i^{\text{D}}\left(\mathbf{x}^{\text{C}}, \mathbf{x}_i^{\text{D}}\right) \tag{3.41a}$$

subject to:

$$\mathbf{h}_i^{\text{D}}\left(\mathbf{x}^{\text{C}}, \mathbf{x}_i^{\text{D}}\right) = \mathbf{0}, \tag{3.41b}$$

$$\mathbf{g}_i^{\text{D}}\left(\mathbf{x}^{\text{C}}, \mathbf{x}_i^{\text{D}}\right) \leq \mathbf{0}, \tag{3.41c}$$

$$\mathbf{x}^{\text{C}} = \mathbf{x}^{\text{C},(\upsilon)} \ : \ \eta_{\text{i}}^{(\upsilon)}. \tag{3.41d}$$

Each of these problems, one for each index i, is denominated subproblem. The solutions of subproblems (3.41) provide values for variables in sets $\mathbf{x}_i^{\mathrm{D},(\nu)}$, $i = 1, \ldots, n$, as well as the vector of dual variables $\eta_i^{(\nu)}$, associated with constraints (3.41d) that are used to fix the value of complicating variables $\mathbf{x}^{\mathrm{C}}$. These dual variables are sensitivities that allow obtaining an increasingly accurate representation of function $\alpha\left(\mathbf{x}^{\mathrm{C}}\right)$ as explained in the following subsection.

Illustrative Example 3.19 Benders' decomposition: subproblems

The data of Illustrative Example 3.18 are considered again. In such a problem, the complicating variable is u. If this variable is fixed to a given value, e.g., 0, then the problem can be decomposed into the following four subproblems:

$$\text{Minimize}_{u,p_1} \quad 50p_1 - 48p_1$$

subject to:

$$20u \leq p_1 \leq 100u,$$
$$u = 0 \; : \; \eta_1^{(\upsilon)}.$$

$$\text{Minimize}_{u,p_2} \quad 50p_2 - 65p_2$$

subject to:

$$20u \leq p_2 \leq 100u,$$
$$u = 0 \; : \; \eta_2^{(\upsilon)}.$$

$$\text{Minimize}_{u,p_3} \quad 50p_3 - 45p_3$$

subject to:

$$20u \leq p_3 \leq 100u,$$
$$u = 0 \; : \; \eta_3^{(\upsilon)}.$$

$$\text{Minimize}_{u,p_4} \quad 50p_4 - 48p_4$$

subject to:

$$20u \leq p_4 \leq 100u,$$
$$u = 0 \; : \; \eta_4^{(\upsilon)}.$$

The solution of these problems is trivial, namely, $u = 0$, and $p_1 = p_2 = p_3 = p_4 = 0$. Moreover, the sensitivities are $\eta_1^{(\upsilon)} = 40$, $\eta_2^{(\upsilon)} = -1500$, $\eta_3^{(\upsilon)} = 100$, and $\eta_4^{(\upsilon)} = -2000$. □

3.5.3 Master Problem

Considering the definition of function $\alpha\left(\mathbf{x}^{\mathrm{C}}\right)$ in (3.38) and the solutions of subproblems (3.41), $i = 1, \ldots, n$, it is possible to formulate the following master problem:

$$\text{Minimize}_{\alpha,\mathbf{x}^{\mathrm{C}}} \ \alpha + f^{\mathrm{C}}\left(\mathbf{x}^{\mathrm{C}}\right) \tag{3.42a}$$

subject to:

$$\mathbf{h}^{\mathrm{C}}\left(\mathbf{x}^{\mathrm{C}}\right) = \mathbf{0}, \tag{3.42b}$$

$$\mathbf{g}^{\mathrm{C}}\left(\mathbf{x}^{\mathrm{C}}\right) \leq \mathbf{0}, \tag{3.42c}$$

$$\alpha \geq \sum_{i=1}^{n} f_i^{\mathrm{D}}\left(\mathbf{x}^{\mathrm{C},(\mathrm{m})}, \mathbf{x}_i^{\mathrm{D},(\mathrm{m})}\right) + \sum_{i=1}^{n} \eta_i^{(m)^{\mathrm{T}}}\left(\mathbf{x}^{\mathrm{C}} - \mathbf{x}^{\mathrm{C},(\mathrm{m})}\right), \quad m = 1, \ldots, \upsilon. \tag{3.42d}$$

Constraints (3.42d) are the so-called Benders' cuts and allow approximating from below the objective function (3.38) as a function of the complicating variables in vector $\mathbf{x}^{\mathrm{C}}$ and function $\alpha\left(\mathbf{x}^{\mathrm{C}}\right)$. Note that in each iteration, a new Benders' cut, i.e., a new constraint (3.42d) is considered and the approximation of the objective function (3.38) becomes increasingly accurate.

The solution of the master problem provides updated solutions of the complicating variables in vector $\mathbf{x}^{\mathrm{C}}$, which are used to solve subproblems (3.41), $i = 1, \ldots, n$.

3.5.4 Algorithm

Given the definitions of the subproblems and the master problem in the previous subsections, the Benders' decomposition algorithm is based on the following steps:

1. Initialize the iteration counter, $\upsilon \leftarrow 0$, select the convergence tolerance, ε, and set the lower bound (LB) and upper bound (UB) to $-\infty$ and $+\infty$, respectively.
2. Solve the master problem (3.42). Note that if $\upsilon = 0$, then the initial master problem does not include Benders' cuts. The solution of this problem comprises variables in vectors $\mathbf{x}^{\mathrm{C},(\upsilon)}$ and $\alpha^{(\upsilon)}$.
3. Update the lower bound as:

$$\mathrm{LB} = \alpha^{(\upsilon)} + f^{\mathrm{C}}\left(\mathbf{x}^{\mathrm{C},(\upsilon)}\right). \tag{3.43}$$

4. Set $\mathbf{x}^{\mathrm{C},(\upsilon)} = \mathbf{x}^{\mathrm{C}*}$, where $\mathbf{x}^{\mathrm{C}*}$ denotes the optimal values of variables in vector $\mathbf{x}^{\mathrm{C}}$ obtained in the master problem (3.42) in step 2.
5. Solve the subproblems (3.41), $i = 1, \ldots, n$.
6. Update the upper bound:

$$\text{UB} = f^{\text{C}}\left(\mathbf{x}^{\text{C},(\upsilon)}\right) + \sum_{i=1}^{n} f_i^{\text{C}}\left(\mathbf{x}^{\text{C},(\upsilon)}, \mathbf{x}_i^{\text{D},(\upsilon)}\right). \tag{3.44}$$

7. Check if UB − LB $< \varepsilon$. If so, the algorithm stops and the optimal solution comprises the values of variables in vectors $\mathbf{x}^{\text{C},(\upsilon)}$ and $\mathbf{x}_i^{\text{D},(\upsilon)}$. Otherwise, go to step 8.
8. Update the iteration counter $\upsilon \leftarrow \upsilon + 1$.
9. Go to step 2.

Illustrative Example 3.20 Benders'decomposition: algorithm

The data of Illustrative Example 3.18 are considered again. In this case, problem is solved using Benders' decomposition that comprises the steps below:

- *Step 1*. Initialize the iteration counter, $\upsilon \leftarrow 0$, select the convergence tolerance, $\varepsilon = 0$, and set the lower bound (LB) and upper bound (UB) to $-\infty$ and $+\infty$, respectively.
- *Step 2*. Solve the following master problem:

$$\text{Minimize}_{\alpha,u} \quad \alpha + 4 \cdot 20u$$

subject to:

$$u \in \{0, 1\}.$$

The solution of this problem is trivial: $\alpha = -\infty$ and $u = 0$.
- *Step 3*. Update the lower bound:

$$\text{LB} = -\infty + 4 \cdot 20 \cdot 0 = -\infty.$$

- *Step 4*. Set $u^{(0)} = 0$.
- *Step 5*. Solve the subproblems. Note that the subproblems considering $u = 0$ were formulated and solved in Illustrative Example 3.19.
- *Step 6*. Update the upper bound:

$$\text{UB} = 4 \cdot 20 \cdot 0 + 50\,(0 + 0 + 0 + 0) - 48 \cdot 0 - 65 \cdot 0 - 45 \cdot 0 - 70 \cdot 0 = 0.$$

- *Step 7*. As UB − LB $= 0 - (-\infty) = \infty$, go to step 8.
- *Step 8*. Update the iteration counter $\upsilon \leftarrow 1$.
- *Step 2*. Solve the following master problem:

$$\text{Minimize}_{\alpha,u} \quad \alpha + 4 \cdot 20u$$

subject to:

$$u \in \{0, 1\},$$

$$\alpha \geq 0 + 40\,(u - 0) - 1500\,(u - 0) + 100\,(u - 0) - 2000\,(u - 0)\,.$$

The solution of this problem is $\alpha = -3360$ and $u = 1$.

- *Step 3*. Update the lower bound:

$$\text{LB} = -3360 + 4 \cdot 20 \cdot 1 = -3280.$$

- *Step 4*. Set $u^{(1)} = 1$.
- *Step 5*. Solve the following subproblems:

$$\text{Minimize}_{u,p_1} \quad 50p_1 - 48p_1$$

subject to:

$$20u \leq p_1 \leq 100u,$$
$$u = 1 \;:\; \eta_1^{(1)}.$$

$$\text{Minimize}_{u,p_2} \quad 50p_2 - 65p_2$$

subject to:

$$20u \leq p_2 \leq 100u,$$
$$u = 1 \;:\; \eta_2^{(1)}.$$

$$\text{Minimize}_{u,p_3} \quad 50p_3 - 45p_3$$

subject to:

$$20u \leq p_3 \leq 100u,$$
$$u = 1 \;:\; \eta_3^{(1)}.$$

$$\text{Minimize}_{u,p_4} \quad 50p_4 - 48p_4$$

subject to:

$$20u \leq p_4 \leq 100u,$$
$$u = 1 \;:\; \eta_4^{(1)}.$$

The solution of these problems is trivial, namely, $u = 1$, $p_1 = 20$, $p_2 = 100$, $p_3 = 20$, and $p_4 = 100$. Moreover, the sensitivities are $\eta_1^{(1)} = 40$, $\eta_2^{(1)} = -1500$, $\eta_3^{(1)} = 100$, and $\eta_4^{(1)} = -2000$.

- *Step 6*. Update the upper bound:

$$\text{UB} = 4 \cdot 20 \cdot 1 + 50\,(20 + 100 + 20 + 100)$$
$$- 48 \cdot 20 - 65 \cdot 100 - 45 \cdot 20 - 70 \cdot 100 = -3280.$$

- *Step 7*. As UB − LB = −3280 − (−3280) = 0, the algorithm stops and the optimal solution is $u = 1$, $p_1 = 20$, $p_2 = 100$, $p_3 = 20$, and $p_4 = 100$.

Note that the optimal solution achieved by applying Benders' decomposition is identical to that obtained in Illustrative Example 3.18. □

3.6 Summary

This chapter provides an overview of some mathematical tools that are used throughout this book. First, the basics of stochastic programming are established. Stochastic programming requires the knowledge, or the estimation, of the probability distribution functions of uncertain parameters. The concepts of random variables, scenarios, and scenario tree are introduced. Two-stage and multi-stage stochastic programming problems are defined. Finally, the concept of risk and its quantification through the conditional value-at-risk are described.

Robust optimization has been proven as a useful tool to formulate decision-making problems under uncertainty in case of unknowing the probability distribution function of uncertain parameters. This chapter includes the fundamentals of robust optimization, including the description of the solution methodology.

The modeling of optimization problems under uncertainty using the linear decision rules approach is also described in this chapter. This technique allows to establish linear relationships between the optimization variables and the uncertain parameters aiming at reducing the computational size of optimization problems. The linear decision rules approach is specially convenient in multi-stage programming problems.

Finally, the foundations and the implementation of the Benders' decomposition technique is included in this chapter. This tool is of special interest for those large-scale optimization problems that can be decomposed easily if the values of some variables are fixed.

3.7 GAMS Codes

This section provides the GAMS [8] codes used to solve some of the illustrative examples of this chapter.

3.7.1 Two-Stage Stochastic Programming Problem

A GAMS code for solving Illustrative Example 3.4 is provided below:

```
$title Two-stage stochastic programming problem

********************************************************************
* DATA
********************************************************************

SETS
i 'generation technologies' /wind,ccgt/,
t 'time periods' /t1,t2/,
w 'scenarios' /w1*w3/
;

PARAMETERS
C_C(i)    'annualized capital cost (euro\kW)'
/wind     200
ccgt      100/,

O_C(i)    'operation cost (euro\kWh)'
/wind     0
ccgt      75/,

A(t)      'availability of wind power (pu)'
/t1       0.2
t2        0.4/;

TABLE D(t,w)    'demand (MWh)'
          w1      w2      w3
t1        800     900     1000
t2        1100    1300    1400;

********************************************************************
* DECLARATION OF VARIABLES AND MATHEMATICAL CHARACTERIZATION
********************************************************************

POSITIVE VARIABLES
x(i)      'installed power (MW)'
y(i,t,w)  'energy generated (MWh)'
;

VARIABLE
cost      'total cost (euro)'
;


********************************************************************
*  EQUATIONS
********************************************************************

EQUATIONS

**        OBJECTIVE FUNCTION
ObjectiveFunction

**        CONSTRAINTS
```

```
PowerProducedLimit_wind
PowerPorducedLimit_ccgt
EnergyBalance
;

ObjectiveFunction..                     cost =e= (1e-6)*(1000*(sum(i,C_C
    (i)*x(i)))+1/3*8760/2*sum((i,t,w),O_C(i)*y(i,t,w)));
PowerProducedLimit_wind(t,w)..          y('wind',t,w) =l= A(t)*x('wind')
    ;
PowerPorducedLimit_ccgt(t,w)..          y('ccgt',t,w) =l= x('ccgt');
EnergyBalance(t,w)..                    sum(i,y(i,t,w)) =e= D(t,w);

*******************************************************************
*  MODEL AND SOLVING
*******************************************************************

model twostage /all/;
solve twostage minimizing cost using lp;
```

3.7.2 Three-Stage Stochastic Programming Problem

A GAMS code for solving Illustrative Example 3.5 is provided below:

```
$title Three-stage stochastic programming problem

SETS

t 'time periods' /t1,t2/,
w 'scenarios' /w1*w4/
;

PARAMETERS

WT(t)    'weight of each period (hours)'
/t1      55
t2       45/,

Forw_Price(t)    'forward contract prices (euro\MWh)'
/t1      54
t2       45/,

pi(w)     'probability of scenarios'
;

pi(w)=1/card(w);

TABLE D(t,w)    demand (MWh)
           w1       w2       w3       w4
t1         600      600      500      500
t2         450      400      350      300;


TABLE Pool_Price(t,w)    'pool price (euro\MWh)'
           w1       w2       w3       w4
```

```
t1        60        60        50        50
t2        52        48        42        38;

POSITIVE VARIABLES
x(t,w)    'forward contract (MWh)'
y(t,w)    'energy purchased in the pool (MWh)';

VARIABLE
cost      'total cost (keuro)';


EQUATIONS

ObjectiveFunction
EnergyBalance
NonAnticipativity_t1
NonAnticipativity_t2a
NonAnticipativity_t2b;

ObjectiveFunction..                  cost =e= 0.001*sum(w,pi(w)*sum(t,WT(t)*(Forw_Price(t)*x(t,w)+Pool_Price(t,w)*y(t,w))));
EnergyBalance(t,w)..                 x(t,w)+y(t,w) =e= D(t,w);
NonAnticipativity_t1(w)$(ord(w) lt 4)..         x('t1',w)=e= x('t1',w+1);
NonAnticipativity_t2a..              x('t2','w1')=e= x('t2','w2');
NonAnticipativity_t2b..              x('t2','w3')=e= x('t2','w4');

model threestage /all/;

solve threestage minimizing cost using lp;

display cost.l, x.l, y.l;
```

3.7.3 Two-Stage Stochastic Programming Problem with CVaR

A GAMS code for solving Illustrative Example 3.6 is provided below:

```
$title Two-stage stochastic programming problem with CVaR

****************************************************************
* DATA
****************************************************************

SETS
i 'generation technologies' /wind,ccgt/,
t 'time periods' /t1,t2/,
w 'scenarios' /w1*w3/
;

SCALARS
beta      'weighting parameter'      /0.00001/,
*beta     'weighting parameter'      /0.2/,
```

```
*beta       'weighting parameter'      /0.99900/,

alpha       'parameter CVaR'           /0.9/;

PARAMETERS
C_C(i)      'annualized capital cost (euro\kW)'
/wind       200
ccgt        100/,

O_C(i)      'operation cost (euro\MWh)'
/wind       0
ccgt        75/,

A(t)        'availability of wind power (pu)'
/t1         0.2
t2          0.4/;

TABLE D(t,w)  'demand (MWh)'
            w1      w2      w3
t1          800     900     1000
t2          1100    1300    1400
;

*******************************************************************
* DECLARATION OF VARIABLES AND MATHEMATICAL CHARACTERIZATION
*******************************************************************

POSITIVE VARIABLES
x(i)        'installed power (MW)'
y(i,t,w)    'energy generated (MWh)'
eta(w)      'auxiliary variable used to compute the cvar (euro)'
;

VARIABLE
cost        total cost (euro)
xi          var (euro);


*******************************************************************
*  EQUATIONS
*******************************************************************

EQUATIONS

**          OBJECTIVE FUNCTION
ObjectiveFunction

**          CONSTRAINTS
PowerProducedLimit_wind
PowerPorducedLimit_ccgt
EnergyBalance
CVaR;

ObjectiveFunction..                cost =e= (1e-6)*beta*(1000*(sum(i,C_C(i)*x(i)))+1/3*8760/2*sum((i,t,w),O_C(i)*y(i,t,w)))+(1e-6)*(1-beta)*(xi+1/(1-alpha)*1/3*sum(w,eta(w)));
PowerProducedLimit_wind(t,w)..     y('wind',t,w) =l= A(t)*x('wind');
PowerPorducedLimit_ccgt(t,w)..     y('ccgt',t,w) =l= x('ccgt');
```

```
EnergyBalance(t,w)..                    sum(i,y(i,t,w)) =e= D(t,w);
CVaR(w)..                               1000*(sum(i,C_C(i)*x(i)))
     +8760/2*sum((i,t),O_C(i)*y(i,t,w))-xi =l= eta(w);

*****************************************************************
*  MODEL AND SOLVING
*****************************************************************

MODEL twostagecvar /all/;

solve twostagecvar minimizing cost using lp;
```

References

1. Benders, J.F.: Partitioning procedures for solving mixed-variables programming problems. Numer. Math. **4**, 238–252 (1962)
2. Bertsimas, D., Sim, M.: The price of robustness. Oper. Res. **52**, 35–53 (2004)
3. Birge, J.R., Louveaux, F.: Introduction to Stochastic Programming, 2nd edn. Springer, New York, US (2011)
4. Conejo, A.J., Castillo, E., Mínguez, R., García-Bertrand, R.: Decomposition Techniques in Mathematical Programming. Springer, Heidelberg (2006)
5. Conejo, A.J., Carrión, M., Morales, J.M.: Decision Making Under Uncertainty in Electricity Markets
6. Dupačová, J., Consigli, G., Wallace, S.W.: Scenarios for multistage stochastic programs. Ann. Oper. Res. **100**, 25–53 (2000)
7. Dupačová, J., Gröwe-Kuska, N., Römisch, W.: Scenario reduction in stochastic programming. Math. Progr. **95**, 493–511 (2003)
8. GAMS (2021). www.gams.com
9. Georghiou, A., Wiesemann, W., Kuhn, D.: The decision rule approach to optimisation under uncertainty: methodology and applications in operations management. Department of Computing, Imperial College London (2011)
10. Gröwe-Kuska, N., Heitsch, H., Römisch, W.: Scenario reduction and scenario tree construction for power management problems. In: Proceedings of IEEE Bologna Power Technology Conference, Bologna, Italy (2003)
11. Høyland, K., Wallace, S.W.: Generating scenario trees for mulstistage decision problems. Manag. Sci. **47**(2), 295–307 (2001)
12. Kall, P., Wallace, S.W.: Stochastic Programming. Wiley, Chichester, UK (1994)
13. Kuhn, D., Wiesemann, W., Georghiou, A.: Primal and dual linear decision rules in stochastic and robust optimization. Math. Progr. **130**(1), 199–209 (2011)
14. Mínguez, R., García-Bertrand, R.: Robust transmission network expansion planning in energy systems: improving computational performance. Eur. J. Oper. Res. **248**, 21–32 (2016)
15. Prékopa, A.: Stochastic Programming. Springer, New York, US (1995)
16. Rockafellar, R.T., Uryasev, C.: Conditional value-at-risk for general loss distributions. J. Bank. Financ. **26**(7), 1443–1471 (2016)
17. Vitali, S., Domínguez, R., Moriggia, V.: Comparing stage-scenario with nodal formulation for multistage stochastic problems. 4OR-Q J. Oper. Res. (2020)
18. Zeng, B., Zhao, L.: Solving two-stage robust optimization problems using a column-and-constraint generation method. Oper. Res. Lett. **41**(5), 457–461 (2013)

Part II
Operation Models

Chapter 4
Day-Ahead Market Scheduling Considering Renewable Energies

This chapter describes a two-stage stochastic problem that models the day-ahead energy and reserve scheduling of a renewable-dominated power system considering the uncertainty in the balancing market. The system operator perspective is adopted such that the total operating cost is minimized taking into account the technical constraints of dispatchable, intermittent renewable, and storage units, as well as of the transmission lines. The uncertainty of the hourly demand and wind and solar power is modeled through a set of scenarios. The resulting model is formulated as a mixed-integer stochastic problem that incorporates the unit commitment variables of dispatchable units. Illustrative examples allow to better understand the main ideas of the proposed model. Then, a realistic case study based on the European power system is studied and multiple numerical results show how a real power system with high penetration of renewable energies is scheduled in the short-term horizon.

4.1 Introduction

The use of intermittent renewable sources for the electricity generation is constantly growing. This is due to several reasons; first, the use of renewable sources allows to reduce the greenhouse gas (GHG) emissions in the electricity generation; second, their investment costs have strongly decreased as well as their reliability as a matter of the technology evolution; third, they allow to reduce the dependency of fossil fuels and contribute to the local economy, among others. However, the future value of the wind and solar energy availability introduces high uncertainty in the system operation. Therefore, a precise representation of this uncertainty must be introduced in the decision-making model so that the supply quality is ensured and the social welfare is maximized.

L. Baringo et al., *Electric Vehicles and Renewable Generation*,
Green Energy and Technology, https://doi.org/10.1007/978-3-031-09079-0_4

Traditional power systems with a generating system based on thermal power plants hardly faced long periods of non stability. Instead, the system operation was usually smooth and predictable. However, power systems along the world are immersed in a deep transformation, giving the leadership to renewable sources [10]. The variability of renewable sources makes the system operation difficult to predict in the medium-/long-term and, hence, instruments allowing high flexibility must be incorporated [8]. Fast thermal power plants such as gas units entail a reliable alternative, but enforcing frequent start-ups and shut-downs makes the units deteriorate faster and increase their total costs [13]. Moreover, high generation from renewable sources may cause a significant decrease in electricity prices, which may introduce high uncertainty on how to recover investments on generation capacity.

Storage units represent an alternative to thermal-based technologies. The technological improvements of new storage technologies, as utility-scale electrochemical batteries, towards increasing their autonomy and reducing their costs are making these units be considered as feasible options to electricity companies and system operators (SOs). Most storage technologies have a very fast response and their environmental impact is comparatively lower than that of fossil fuels.

Other flexible instruments being gradually introduced in the power generation are electric vehicles (EVs) and demand response. Regulators and policy markers are promoting initiatives to involve consumers in the system operation.[1] As a result, multiple demand side management tools are under development and even implemented in power systems. For instance, the battery incorporated in EVs allows consumers to provide flexible services in exchange of benefits in the electricity bill. This aspect is deeply analyzed in the next chapters of this book.

In the technical literature, multiple papers dealing with the day-ahead scheduling of power systems considering short-term uncertainties can be found [15–17, 19, 20]. In this chapter, a day-ahead scheduling model that co-optimizes energy and reserve capacity is described. The model is formulated as a two-stage mixed-integer stochastic problem where the first stage represents the day-ahead market and the second stage the balancing market. The uncertainty related to the demand and wind and solar power is included. A set of scenarios is generated to represent the possible realizations of those parameters. An illustrative example shows in a simple way the main characteristics of the proposed approach. Then, a realistic case study based on the European power system is solved and the numerical results are discussed.

The rest of this chapter is organized as follows. Section 4.2 describes the market structure adopted in this chapter. Section 4.3 explains the decision-making process involved in this problem and how the uncertain parameters are represented in the model. An example of a scenario tree is depicted to clearly illustrate the proposed approach. Section 4.4 presents the proposed formulation and provides a description of each equation. In Sect. 4.5, an illustrative case study considering a thermal, a wind and a solar PV unit is solved to easily explain how the proposed model works. Section 4.6 analyses a realistic case study based on the European power system and presents multiple numerical results showing the daily variability in the power

[1] https://ec.europa.eu/energy/topics/energy-strategy/clean-energy-all-europeans_en.

system operation. In Sect. 4.7 the summary and main conclusions of this chapter are provided. Finally, Sect. 4.8 includes the list of notation used throughout this chapter and Sect. 4.9 provides the GAMS code used to solve the illustrative case study.

4.2 Market Structure

The market structure considered in this chapter comprises two decision stages. The first stage corresponds to the day-ahead scheduling of the energy and up and down reserve capacities, and the second stage represents the balancing market. In this way, energy and up and down reserve capacities are scheduled in the day-ahead market for the 24 h of the following day. Afterwards, in the balancing market, which is settled close to the physical delivery of energy, the stochastic parameters (demand level and wind and solar power) realize and the reserves are deployed to satisfy the power balance in spite of the forecast errors.

The up and down reserves considered in this model are used to counteract the forecast error of the demand and intermittent renewable production made when the day-ahead market is cleared. These types of reserves are commonly named as secondary or spinning reserves. Since the forecast error of wind and solar photovoltaic (PV) power plants is still high one day ahead of the energy dispatching, it is important to schedule the right amount of up and down reserve capacity without overestimating or underestimating the reserve needs. For this reason, the possible realizations of the stochastic variables at the balancing market are considered in the day-ahead scheduling model, which allows the scheduling of reserve capacities such that they can cover the possible imbalances of that specific day.

The generating system comprises dispatchable and intermittent renewable units. The dispatchable technologies considered include fossil-fuel, nuclear, biomass, hydro, and concentrating solar power (CSP) units. The latter unit incorporates a thermal energy storage (TES) that enables to control the power output of the unit according to the solar radiation availability. The modeling of dispatchable units considers inter-temporal constraints such as ramping limits and minimum up/down times, where their start-up and shut-down status may be modeled using binary variables. The start-up and production costs are considered to be linear.

The power output of intermittent renewable units, i.e. onshore and offshore wind and solar PV units, depends on the resource availability. This is modeled through a set of scenarios that can be generated using historical data or stochastic models that represent the spatial and temporal correlations characterizing the resource behavior. No penalization cost is applied to the power spillage.

Storage units are also considered in the power system, which could be hydro-pumped power plants or batteries, for instance. Storage units provide high flexibility in the system operation since they can consume electricity when the demand is low or the renewable production exceeds the system needs and can generate electricity in the ramping or peak-demand hours. The inter-temporal constraints modeling the state-of-charge of storage units are imposed in the problem.

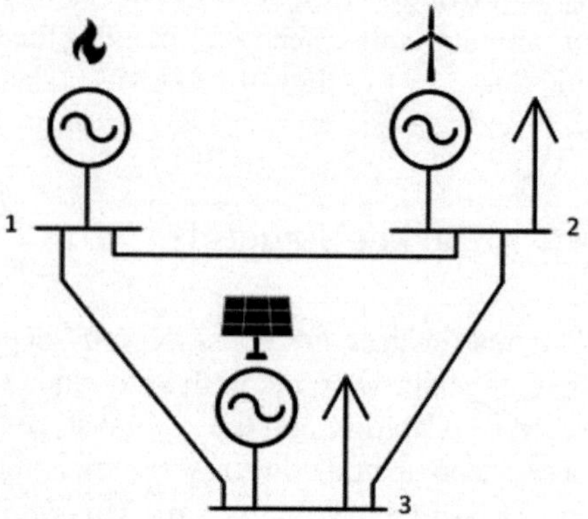

Fig. 4.1 Schema of a renewable-dominated power system

In the formulation considered in this chapter, it is considered that all generating and storage technologies, $g \in G$ and $s \in S$, may be at operation in each node of the network, $b \in B$.

It is assumed that all generation and storage units, except for solar PV units, can provide reserve capacity according to their capacity and ramping limits. In the case of wind power plants, power systems with high wind capacity like those of Spain or Denmark allow these units to participate in the secondary reserve market, achieving good results up to now. Even though it is likely that power electronics will evolve in such a way that solar PV plants will be able to provide reserve services, still in this model we assume that these plants do not participate in the reserve market. For the interested reader, the provision of reserve services by intermittent renewable technologies is described in [2].

As it is customary in this kind of problems, the transmission system is represented through a DC model. Finally, the demand is assumed to be inelastic.

Illustrative Example 4.1 Renewable-dominated power systems

Figure 4.1 illustrates an example of a renewable-dominated power system whose generating system comprises thermal (bus 1), wind (bus 2), and solar PV (bus 3) power plants. Two demands are located in buses 2 and 3, respectively, while 3 transmission lines connect the 3 buses.

□

4.3 Decision Framework and Uncertainty Characterization

The use of renewable sources such as wind and solar radiation for electricity generation introduces high uncertainty in the daily operation of power systems. For this reason, in the day-ahead scheduling it is crucial to consider the variability of intermittent sources so that up and down reserve capacities can be adequately scheduled.

Thus, the scheduling model presented in this chapter is based on the co-optimization of energy and reserve capacity in the day-ahead market considering the uncertainty related to the real-time operation.

The proposed model is formulated as a mixed-integer two-stage stochastic problem. The first stage represents the day-ahead market and the second stage represents the balancing market. In the day-ahead market, the energy and reserve capacity provided by generating and storage units are scheduled considering the technical constraints characterizing each technology and the possible realizations of the random variables, i.e. the demand and the intermittent renewable production, in the real-time operation. Unit commitment variables define the on/off status of dispatchable units. In the balancing market, the up and down reserves are deployed to ensure the demand supply for each realization of the random variables, fulfilling the technical constraints imposed by the generating and storage units, and the transmission system.

Stochastic optimization is used to represent the decision-making process involved in the scheduling problem [1]. In this way, the operating decisions are made considering the uncertainty of the available intermittent renewable sources and the demand level. The values of these random variables are represented through a set of scenarios, being a scenario a possible realization of the random variable in the real-time operation in the balancing market. Scenarios for each hour of the day may generated using time series models or historical data of the forecast error in each location. To attain computational tractability, the initial set of scenarios may be reduced using the procedure proposed in [7].

Illustrative Example 4.2 Day-ahead scheduling of renewable-dominated power systems

Figure 4.2 illustrates an example of the scenario tree representing the decision-making process for one hour. Node t_1 represents the hour of the following day in which the energy and reserve capacities must be scheduled. In other words, the decision planner is located at the day ahead of delivery $(D-1)$ to take the decision on how much energy and reserve capacity to schedule in hour t_1, taking into account different operating situations at the hour of delivery (D). In this example, we consider three random parameters, namely the demand level and the wind and solar power, which are represented by red, green and orange circles, respectively. Each random parameter is represented through two possible realizations. Assuming all the combinations among the random parameter values, it results in $2 \cdot 2 \cdot 2 = 8$ scenarios, which are denoted by $\omega_1, \omega_2, ..., \omega_8$. Observe that each scenario contains a possible value of demand level and wind and solar power.

□

4.4 Model Formulation

In this subsection, the formulation of the scheduling model described above is provided.

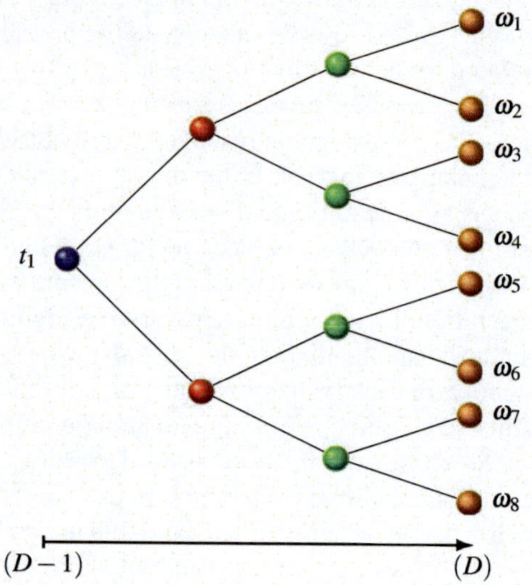

Fig. 4.2 Example of a scenario tree representing the decision-making process

4.4.1 Objective Function

The objective function to be minimized is defined by (4.1), which comprises the total expected cost. The first line in (4.1) is related to the generating units, including the start-up costs, the offering cost of energy and up and down reserve capacities in the day-ahead market. The second line comprises the cost of the up and down reserve capacity offered by the storage units and the unserved demand cost in the day-ahead market. The third line corresponds to the expected cost of the up and down reserves deployed by generating units in the balancing market. The last line includes the expected cost of deploying up and down reserves by the storage units and the unserved demand cost in the balancing market.

$$\begin{aligned}\underset{\Theta}{\text{Minimize}} \quad & \sum_{b\in B}\sum_{t\in T}\Bigg[\sum_{g\in G}\left(c_{gbt}^{\text{G,SU}} + C_{gt}^{\text{D,C}}p_{gbt}^{\text{G,D}} + C_{gt}^{\text{G,RCU}}r_{gbt}^{\text{G,CU}} + C_{gt}^{\text{G,RCD}}r_{gbt}^{\text{G,CD}}\right) \\ & + \sum_{s\in S}\left(C_{st}^{\text{S,RCU}}r_{sbt}^{\text{S,CU}} + C_{st}^{\text{S,RCD}}r_{sbt}^{\text{S,CD}}\right) + C^{\text{UD}}p_{bt}^{\text{UD,D}}\Bigg] \\ & + \sum_{b\in B}\sum_{t\in T}\sum_{\omega\in\Omega}\pi_{\omega}\Bigg[\sum_{g\in G}\left(C_{gt}^{\text{G,RDU}}r_{gbt\omega}^{\text{G,DU}} - C_{gt}^{\text{G,RDD}}r_{gbt\omega}^{\text{G,DD}}\right) \\ & + \sum_{s\in S}\left(C_{st}^{\text{S,RDU}}r_{sbt\omega}^{\text{S,DU}} - C_{st}^{\text{S,RDD}}r_{sbt\omega}^{\text{S,DD}}\right) + C^{\text{UD}}p_{bt\omega}^{\text{UD,B}}\Bigg],\end{aligned} \tag{4.1}$$

where $\Theta = \{ c^{\mathrm{G,SU}}_{gbt}, p^{\mathrm{G,D}}_{gbt}, r^{\mathrm{G,CD}}_{gbt}, r^{\mathrm{G,CU}}_{gbt}, u^{\mathrm{D}}_{gbt}, y^{\mathrm{D}}_{gbt}, z^{\mathrm{D}}_{gbt}, \forall g \in G, \forall b \in B, \forall t \in T;\ p^{\mathrm{G,B}}_{gbt\omega}, r^{\mathrm{G,DD}}_{gbt\omega}, r^{\mathrm{G,DU}}_{gbt\omega}, \forall g \in G, \forall b \in B, \forall t \in T, \forall \omega \in \Omega;\ e^{\mathrm{SD}}_{sbt}, p^{\mathrm{S,C}}_{sbt}, p^{\mathrm{S,D}}_{sbt}, r^{\mathrm{S,CD}}_{sbt}, r^{\mathrm{S,CDC}}_{sbt}, r^{\mathrm{S,CDD}}_{sbt}, r^{\mathrm{S,CU}}_{sbt}, r^{\mathrm{S,CUC}}_{sbt}, r^{\mathrm{S,CUD}}_{sbt}, \forall s \in S, \forall b \in B, \forall t \in T;\ e^{\mathrm{SB}}_{sbt\omega}, p^{\mathrm{S,BC}}_{sbt\omega}, p^{\mathrm{S,BD}}_{sbt\omega}, r^{\mathrm{S,DD}}_{sbt\omega}, r^{\mathrm{S,DDC}}_{sbt\omega}, r^{\mathrm{S,DDD}}_{sbt\omega}, r^{\mathrm{S,DU}}_{sbt\omega}, r^{\mathrm{S,DUC}}_{sbt\omega}, r^{\mathrm{S,DUD}}_{sbt\omega}, \forall s \in S, \forall b \in B, \forall t \in T, \forall \omega \in \Omega;\ p^{\mathrm{L,D}}_{\ell t}, \forall \ell \in L, \forall t \in T;\ p^{\mathrm{L,B}}_{\ell t\omega}, \forall \ell \in L, \forall t \in T, \forall \omega \in \Omega;\ p^{\mathrm{UD,D}}_{bt}, \forall b \in B, \forall t \in T;\ p^{\mathrm{UD,B}}_{bt\omega}, \forall b \in B, \forall t \in T, \forall \omega \in \Omega; \lambda^{\mathrm{D}}_{bt}, \theta^{\mathrm{D}}_{bt}, \forall b \in B, \forall t \in T; \lambda^{\mathrm{B}}_{bt\omega}, \theta^{\mathrm{B}}_{bt\omega}, \forall b \in B, \forall t \in T, \forall \omega \in \Omega \}$ is the set of optimization variables.

The problem is subject to the following constraints.

4.4.2 Power Balance in the Day-Ahead Market

Constraints (4.2) define the power balance in each bus and hour in the day-ahead market, considering the power scheduled by the generating and storage units, the power flow through the transmission lines, and the scheduled demand. Note that the dual variable of this equation, $\lambda^{\mathrm{D}}_{bt}$, represents the electricity price in each bus and hour.

$$\sum_{g \in G} p^{\mathrm{G,D}}_{gbt} + \sum_{s \in S} (p^{\mathrm{S,D}}_{sbt} - p^{\mathrm{S,C}}_{sbt}) - \sum_{\ell | O(\ell)=b} p^{\mathrm{L,D}}_{\ell t} + \sum_{\ell | F(\ell)=b} p^{\mathrm{L,D}}_{\ell t}$$
$$= (P^{\mathrm{D,D}}_{bt} - p^{\mathrm{UD,D}}_{bt}), \forall b, \forall t \ : \ \lambda^{\mathrm{D}}_{bt}. \quad (4.2)$$

4.4.3 Reserve Requirements in the Day-Ahead Market

Constraints (4.3a) and (4.3b) impose minimum down and up reserve capacities in each bus and in each hour in the day-ahead market based on the scheduled demand and the predicted intermittent power. K^{D} and K^{It} indicate the percentage of the aforementioned parameters contributing to the reserve capacity requirement. The reserve requirements may be established for the whole system or for operation zones. In this formulation, it is considered that up and down reserve requirements are defined for each bus of the system.

$$\sum_{g \in G_b} r^{\mathrm{G,CD}}_{gbt} + \sum_{s \in S_b} r^{\mathrm{S,CD}}_{sbt} \geq \left(K^{\mathrm{D}} P^{\mathrm{D,D}}_{bt} + K^{\mathrm{It}} \sum_{g \in G^{\mathrm{I}}_b} A^{\mathrm{D}}_{gbt} P^{\mathrm{C,G}}_{\max,gb} \right), \quad \forall b, \forall t, \quad (4.3a)$$

$$\sum_{g \in G_b} r^{\mathrm{G,CU}}_{gbt} + \sum_{s \in S_b} r^{\mathrm{S,CU}}_{sbt} \geq \left(K^{\mathrm{D}} P^{\mathrm{D,D}}_{bt} + K^{\mathrm{It}} \sum_{g \in G^{\mathrm{I}}_b} A^{\mathrm{D}}_{gbt} P^{\mathrm{C,G}}_{\max,gb} \right), \quad \forall b, \forall t. \quad (4.3b)$$

4.4.4 Dispatchable Units in the Day-Ahead Market

Constraints (4.4a)–(4.5i) represent the technical operation of the dispatchable generating units in the day-ahead market. Specifically, (4.4a) impose the limit on the power and up reserve scheduled by the unit to the power available in period t. Observe that most of the dispatchable units have the availability factor A^{D}_{gbt} equal to 1, but for units such as CSP or hydroelectric this factor may be lower than 1 in some hours of the day.

Constraints (4.4b) establish the minimum power output of each unit considering the down reserve scheduled. Constraints (4.4c) and (4.4d) set the up and down ramping capacity, respectively. Constraints (4.4e) define the start-up cost of the unit.

$$p^{\mathrm{G,D}}_{gbt} + r^{\mathrm{G,CU}}_{gbt} \leq u^{\mathrm{D}}_{gbt} A^{\mathrm{D}}_{gbt} P^{\mathrm{C,G}}_{\mathrm{max},gb}, \quad \forall g \in G^{\mathrm{D}}, \forall b, \forall t, \tag{4.4a}$$

$$p^{\mathrm{G,D}}_{gbt} - r^{\mathrm{G,CD}}_{gbt} \geq u^{\mathrm{D}}_{gbt} P^{\mathrm{C,G}}_{\mathrm{min},gb}, \quad \forall g \in G^{\mathrm{D}}, \forall b, \forall t, \tag{4.4b}$$

$$p^{\mathrm{G,D}}_{gbt} - p^{\mathrm{G,D}}_{gbt-1} \leq P^{\mathrm{G}}_{\mathrm{UP},g} P^{\mathrm{C,G}}_{\mathrm{max},gb}, \quad \forall g \in G^{\mathrm{D}}, \forall b, \forall t, \tag{4.4c}$$

$$p^{\mathrm{G,D}}_{gbt-1} - p^{\mathrm{G,D}}_{gbt} \leq P^{\mathrm{G}}_{\mathrm{DW},g} P^{\mathrm{C,G}}_{\mathrm{max},gb}, \quad \forall g \in G^{\mathrm{D}}, \forall b, \forall t, \tag{4.4d}$$

$$c^{\mathrm{G,SU}}_{gbt} = C^{\mathrm{G,SU}}_{gb} y^{\mathrm{D}}_{gbt}, \quad \forall g \in G^{\mathrm{D}}, \forall b, \forall t. \tag{4.4e}$$

Finally, constraints (4.5) enforce minimum up and down times in the dispatchable units, being this formulation based on that proposed in [18]. Parameters L_g and M_g are defined as $L_g = \min\left\{T, N^{\mathrm{ON}}_g\right\}$ and $M_g = \min\left\{T, N^{\mathrm{OFF}}_g\right\}$. N^{ON}_g and N^{OFF}_g are the number of hours that the generating unit has been online/offline, respectively, previous to the initial hour of the planning horizon.

$$\sum_{t=1}^{L_g} u^{\mathrm{D}}_{gbt} = L_g, \quad \forall g \in G^{\mathrm{D}}, \forall b, \tag{4.5a}$$

$$\sum_{k=t}^{t+T^{\mathrm{Umin}}_g-1} u^{\mathrm{D}}_{gbk} \geq T^{\mathrm{Umin}}_g y^{\mathrm{D}}_{gbt}, \quad \forall g \in G^{\mathrm{D}}, \forall b, t = L_g + 1, ..., T - T^{\mathrm{Umin}}_g + 1, \tag{4.5b}$$

$$\sum_{k=t}^{T} \left[u^{\mathrm{D}}_{gbk} - y^{\mathrm{D}}_{gbt}\right] \geq 0, \quad \forall g \in G^{\mathrm{D}}, \forall b, t = T - T^{\mathrm{Umin}}_g + 2, ..., T, \tag{4.5c}$$

$$\sum_{t=1}^{M_g} u^{\mathrm{D}}_{gbk} = 0, \forall g \in G^{\mathrm{D}}, \forall b, \tag{4.5d}$$

$$\sum_{k=t}^{t+T^{\mathrm{Dmin}}_g-1} (1 - u^{\mathrm{D}}_{gbk}) \geq T^{\mathrm{Dmin}}_g z^{\mathrm{D}}_{gbt}, \quad \forall g \in G^{\mathrm{D}}, \forall b, t = M_g + 1, ..., T - T^{\mathrm{Dmin}}_g + 1, \tag{4.5e}$$

$$\sum_{k=t}^{T} \left[1 - u^{\mathrm{D}}_{gbk} - z^{\mathrm{D}}_{gbt}\right] \geq 0, \quad \forall g \in G^{\mathrm{D}}, \forall b, t = T - T^{\mathrm{Dmin}}_{g} + 2, ..., T, \tag{4.5f}$$

$$u^{\mathrm{D}}_{gbt-1} - u^{\mathrm{D}}_{gbt} + y^{\mathrm{D}}_{gbt} - z^{\mathrm{D}}_{gbt} = 0, \quad \forall g \in G^{\mathrm{D}}, \forall b, \forall t > 1, \tag{4.5g}$$

$$y^{\mathrm{D}}_{gbt} + z^{\mathrm{D}}_{gbt} \leq 1, \quad \forall g \in G^{\mathrm{D}}, \forall b, \forall t, \tag{4.5h}$$

$$u^{\mathrm{D}}_{gbt}, y^{\mathrm{D}}_{gbt}, z^{\mathrm{D}}_{gbt} \in \{0, 1\}, \quad \forall g \in G^{\mathrm{D}}, \forall b, \forall t. \tag{4.5i}$$

4.4.5 Intermittent Renewable Units in the Day-Ahead Market

Constraints (4.6a) limit the power output of intermittent renewable units based on the resource availability. Constraints (4.6b) and (4.6c) establish the limit on the up and down reserve capacity that these units can provide.

$$0 \leq p^{\mathrm{G,D}}_{gbt} \leq A^{\mathrm{D}}_{gbt} P^{\mathrm{C,G}}_{\mathrm{max},gb}, \quad \forall g \in G^{\mathrm{I}}, \forall b, \forall t, \tag{4.6a}$$

$$0 \leq r^{\mathrm{G,CU}}_{gbt} \leq R^{\mathrm{G,CUmax}}_{g} A^{\mathrm{D}}_{gbt} P^{\mathrm{C,G}}_{\mathrm{max},gb}, \quad \forall g \in G^{\mathrm{I}}, \forall b, \forall t, \tag{4.6b}$$

$$0 \leq r^{\mathrm{G,CD}}_{gbt} \leq R^{\mathrm{G,CDmax}}_{g} A^{\mathrm{D}}_{gbt} P^{\mathrm{C,G}}_{\mathrm{max},gb}, \quad \forall g \in G^{\mathrm{I}}, \forall b, \forall t. \tag{4.6c}$$

4.4.6 Storage Units in the Day-Ahead Market

Constraints (4.7) characterize the operation of storage units in the day-ahead market. In particular, (4.7a) and (4.7b) limit the charged and discharged power scheduled in the day-ahead market. Constraints (4.7c) define the state-of-charge of each storage unit, considering the energy level in the previous hour and the charged and discharged power in the current hour. Constraints (4.7d) are equivalent to the previous constraints but in the initial period of the planning horizon. Constraints (4.7e) limit the energy content in the storage unit according to its hours of autonomy. Constraints (4.7f) impose that the energy level in the end of the planning horizon must be greater or equal to the initial energy level.

$$0 \leq p^{\mathrm{S,D}}_{sbt} \leq P^{\mathrm{C,SP}}_{\mathrm{max},sb}, \quad \forall s, \forall b, \forall t, \tag{4.7a}$$

$$0 \leq p^{\mathrm{S,C}}_{sbt} \leq P^{\mathrm{C,SP}}_{\mathrm{max},sb}, \quad \forall s, \forall b, \forall t, \tag{4.7b}$$

$$e^{\mathrm{SD}}_{sbt} = e^{\mathrm{SD}}_{sbt-1} - \frac{1}{\eta^{\mathrm{S}}_{s}} p^{\mathrm{S,D}}_{sbt} + \eta^{\mathrm{S}}_{s}\, p^{\mathrm{S,C}}_{sbt}, \quad \forall s, \forall b, \forall t > 1, \tag{4.7c}$$

$$e^{\mathrm{SD}}_{sbt} = E^{\mathrm{C,SE}}_{\mathrm{ini},sb} - \frac{1}{\eta^{\mathrm{S}}_{s}} p^{\mathrm{S,D}}_{sbt} + \eta^{\mathrm{S}}_{s}\, p^{\mathrm{S,C}}_{sbt}, \quad \forall s, \forall b, t = 1, \tag{4.7d}$$

$$0 \leq e^{\mathrm{SD}}_{sbt} \leq H^{\mathrm{S}}_{s} P^{\mathrm{C,SP}}_{\mathrm{max},sb}, \quad \forall s, \forall b, \forall t, \tag{4.7e}$$

$$e_{sbt}^{\text{SD}} \geq E_{\text{ini},sb}^{\text{C,SE}}, \ \forall s, \forall b, t = N_{\text{T}}. \tag{4.7f}$$

Constraints (4.8) limit the up and down reserve capacities that can be scheduled from charging and discharging. Constraints (4.8a) and (4.8d) define the up and down reserve capacities scheduled, respectively. Observe that up reserve capacity is needed for those cases in which the generation is lower than the consumption. In this case, storage units can provide this type of reserve in two ways. On one hand, storage units can reduce the charging as it implies the consumption of electricity from the system. As established by (4.8b), the maximum up reserve capacity would be limited by the power scheduled for charging. On the other hand, if the power scheduled for discharging is lower than the maximum discharging capacity, storage units can increase the discharged power and, hence, increase the electricity generation. Thus, the up reserve capacity would be limited by the difference between the maximum discharging capacity and the power scheduled for discharging, as stated by (4.8c). On the contrary, down reserve capacity is required for those situations in which the electricity generation is higher than the consumption. As previously, storage units can provide down reserve capacity in two ways. They can increase the scheduled power for charging and, hence, the electricity consumption up to the limit imposed by (4.8e). Otherwise, storage units can reduce the electricity generation by reducing the power scheduled for discharging. The maximum down reserve capacity in this case would be the power scheduled for discharging, as established by (4.8f).

$$r_{sbt}^{\text{S,CU}} = r_{sbt}^{\text{S,CUC}} + r_{sbt}^{\text{S,CUD}}, \quad \forall s, \forall b, \forall t, \tag{4.8a}$$

$$0 \leq r_{sbt}^{\text{S,CUC}} \leq p_{sbt}^{\text{S,C}}, \quad \forall s, \forall b, \forall t, \tag{4.8b}$$

$$0 \leq r_{sbt}^{\text{S,CUD}} \leq P_{\max,sb}^{\text{C,SP}} - p_{sbt}^{\text{S,D}}, \quad \forall s, \forall b, \forall t, \tag{4.8c}$$

$$r_{sbt}^{\text{S,CD}} = r_{sbt}^{\text{S,CDC}} + r_{sbt}^{\text{S,CDD}}, \quad \forall s, \forall b, \forall t, \tag{4.8d}$$

$$0 \leq r_{sbt}^{\text{S,CDC}} \leq P_{\max,sb}^{\text{C,SP}} - p_{sbt}^{\text{S,C}}, \quad \forall s, \forall b, \forall t, \tag{4.8e}$$

$$0 \leq r_{sbt}^{\text{S,CDD}} \leq p_{sbt}^{\text{S,D}}, \quad \forall s, \forall b, \forall t. \tag{4.8f}$$

Illustrative Example 4.3 Operation of storage units

Let us consider a 1 MW/3 MWh storage unit with efficiency η_s^{S} equal to 1 and a planning horizon comprising five hourly-time periods. The initial energy status of the storage is 1 MWh. A possible participation of this storage unit in these markets that satisfies constraints (4.7) and (4.8) is provided in Table 4.1. According to the data provided in this table, the storage charges during periods 1 and 2, and discharges in periods 4 and 5. The charge of the battery results from purchasing energy in the day-ahead market, $p_{sbt}^{\text{S,C}}$, whereas the energy discharged is sold in this market, $p_{sbt}^{\text{S,D}}$. In all periods, the power charged and discharged from the participation in the day-ahead markets is smaller than the power capacity of the unit, 1 MW.

The storage participates in the reserve capacity market providing up reserve $r_{sbt}^{\text{S,CU}}$ in periods 1 and 4, and down reserve $r_{sbt}^{\text{S,CD}}$ in periods 2 and 5. Because the unit is charging power in period 1, $p^{\mathrm{S,C}}_{sbt} = 0.8$ MW, the provision of up reserve is made by decreasing the charge of power, $r^{\mathrm{S,CUC}}_{sbt} = 0.1$ MW. Observe that $0.8 - 0.1 \geq 0$. On the contrary, the unit is discharging in period 4, $p^{\mathrm{S,D}}_{sbt} = 0.5$ MW. Therefore, the provision of up reserve can only be performed by increasing the discharge of power, $r^{\mathrm{S,CUD}}_{sbt} = 0.2$ MW. In this period it is also satisfied that the maximum power discharged does not exceed the power capacity of the unit, $0.5 + 0.2 \leq 1$ MW. In period 2, the storage unit is charging power, $p^{\mathrm{S,C}}_{sbt} = 0.5$ MW. In this period, the provision of down reserve can be obtained by increasing the power charged, $r^{\mathrm{S,CDC}}_{sbt} = 0.3$ MW. The total amount of power that can be charged in this period is less than the power capacity of the storage, $0.5 + 0.3 \leq 1$ MW. Finally, the storage is discharging power in period 5 , $p^{\mathrm{S,D}}_{sbt} = 0.5$ MW, and the provision of the scheduled down reserve in this period can only be achieved by decreasing the power discharged, $r^{\mathrm{S,CDD}}_{sbt} = 0.3$ MW. Observe that the total amount of power discharged is a positive value, $0.5 - 0.3 \geq 0$.

□

4.4.7 *Transmission Lines in the Day-Ahead Market*

The power flow through the transmission lines is formulated using the DC approximation of the transmission network. This approximation is useful to approximate linearly the nonlinear relationships of active power flows with respect to voltage angles. This approximation is based on the following assumptions: (i) voltage magnitudes are equal to 1 pu at all buses and (ii) the phase angle differences between adjacent buses are small. These assumptions are reasonable for transmission net-

Table 4.1 Illustrative Example 4.3: participation of a storage unit in the day-ahead and reserve capacity markets (MW)

	Periods				
	1	2	3	4	5
$p^{\mathrm{S,C}}_{sbt}$	0.8	0.5	0.0	0.0	0.0
$p^{\mathrm{S,D}}_{sbt}$	0.0	0.0	0.0	0.5	0.5
$r^{\mathrm{S,CU}}_{sbt}$	0.1	0.0	0.0	0.2	0.0
$r^{\mathrm{S,CD}}_{sbt}$	0.0	0.3	0.0	0.0	0.3
$r^{\mathrm{S,CUC}}_{sbt}$	0.1	0.0	0.0	0.0	0.0
$r^{\mathrm{S,CUD}}_{sbt}$	0.0	0.0	0.0	0.2	0.0
$r^{\mathrm{S,CDC}}_{sbt}$	0.0	0.3	0.0	0.0	0.0
$r^{\mathrm{S,CDD}}_{sbt}$	0.0	0.0	0.0	0.0	0.3

works, where the resistance/reactance ratio is low. Using this approximation, active power flows can be expressed in terms of the series reactance of the line and the difference of voltage angles of the buses at both ends of the line. The DC approximation is very useful to reduce the computational complexity of planning models. However, the disadvantage of this approximation is that reactive power flows and power losses are not modeled. For these reasons, the DC approximation is not used in simulation analyses where voltage magnitudes and reactive power flows are needed to be taken into account.

Considering the above, constraints (4.9a) define the power flow through the transmission lines in the day-ahead market using the DC model. Constraints (4.9b) and (4.9c) limit the power flow through the transmission lines and the voltage angles, respectively.

$$p^{\mathrm{L,D}}_{\ell t} = \frac{1}{X_\ell}(\theta^{\mathrm{D}}_{O(\ell),t} - \theta^{\mathrm{D}}_{F(\ell),t}), \quad \forall \ell, \forall t, \tag{4.9a}$$

$$- P^{\mathrm{L}}_{\max,\ell} \leq p^{\mathrm{L,D}}_{\ell t} \leq P^{\mathrm{L}}_{\max,\ell}, \quad \forall \ell, \forall t, \tag{4.9b}$$

$$- \pi \leq \theta^{\mathrm{D}}_{bt} \leq \pi, \quad \forall b, \forall t. \tag{4.9c}$$

4.4.8 Power Balance in the Balancing Market

Constraints (4.10) define the power balance in each bus, period and scenario in the balancing market, considering the deployment of the up and down reserves to counteract the forecast errors of the demand and the intermittent renewable power. Note that the dual variable of this equation, $\lambda^{\mathrm{B}}_{bt\omega}$, represents the weighted electricity price in the balancing market. Then, the electricity price in the balancing market at bus b, period t, and scenario ω may be computed as $\lambda^{\mathrm{B}}_{bt\omega}/\pi_\omega$.

$$\sum_{g\in G}(r^{\mathrm{G,DU}}_{gbt\omega} - r^{\mathrm{G,DD}}_{gbt\omega}) + \sum_{s\in S}(r^{\mathrm{S,DU}}_{sbt\omega} - r^{\mathrm{S,DD}}_{sbt\omega}) - (P^{\mathrm{D,B}}_{bt\omega} - P^{\mathrm{D,D}}_{bt} - p^{\mathrm{UD,B}}_{bt\omega})$$
$$- \sum_{l|O(\ell)=b}(p^{\mathrm{L,B}}_{\ell t\omega} - p^{\mathrm{L,D}}_{\ell t}) + \sum_{l|F(\ell)=b}(p^{\mathrm{L,B}}_{\ell t\omega} - p^{\mathrm{L,D}}_{\ell t}) = 0 \; : \; \lambda^{\mathrm{B}}_{bt\omega}, \quad \forall b, \forall t, \forall \omega. \tag{4.10}$$

4.4.9 Dispatchable Units in the Balancing Market

Constraints (4.11a) define the power output of the generating units in the balancing market considering the day-ahead schedule and the deployed reserves in each scenario. Constraints (4.11b) limit the minimum and maximum power output of the committed dispatchable units, while (4.11c) and (4.11d) enforce the up and down ramping limits.

$$p^{\mathrm{G,B}}_{gbt\omega} = p^{\mathrm{G,D}}_{gbt} + r^{\mathrm{G,DU}}_{gbt\omega} - r^{\mathrm{G,DD}}_{gbt\omega}, \quad \forall g \in G^{\mathrm{D}}, \forall b, \forall t, \forall \omega, \tag{4.11a}$$

$$u^{\mathrm{D}}_{gbt} P^{\mathrm{C,G}}_{\mathrm{min},gb} \le p^{\mathrm{G,B}}_{gbt\omega} \le u^{\mathrm{D}}_{gbt} A^{\mathrm{B}}_{gbt\omega} P^{\mathrm{C,G}}_{\mathrm{max},gb}, \quad \forall g \in G^{\mathrm{D}}, \forall b, \forall t, \forall \omega, \tag{4.11b}$$

$$p^{\mathrm{G,B}}_{gbt\omega} - p^{\mathrm{G,B}}_{gbt-1,\omega} \le P^{\mathrm{G}}_{\mathrm{UP},g} P^{\mathrm{C,G}}_{\mathrm{max},gb}, \quad \forall g \in G^{\mathrm{D}}, \forall b, \forall t, \forall \omega, \tag{4.11c}$$

$$p^{\mathrm{G,B}}_{gbt-1,\omega} - p^{\mathrm{G,B}}_{gbt\omega} \le P^{\mathrm{G}}_{\mathrm{DW},g} P^{\mathrm{C,G}}_{\mathrm{max},gb}, \quad \forall g \in G^{\mathrm{D}}, \forall b, \forall t, \forall \omega. \tag{4.11d}$$

4.4.10 Intermittent Renewable Units in the Balancing Market

Constraints (4.12a) define the power output of the intermittent generating units based on the deployed up and down reserves, whereas (4.12b) limit the power output according to the available power in each scenario.

$$p^{\mathrm{G,B}}_{gbt\omega} = p^{\mathrm{G,D}}_{gbt} + r^{\mathrm{G,DU}}_{gbt\omega} - r^{\mathrm{G,DD}}_{gbt\omega}, \quad \forall g \in G^{\mathrm{I}}, \forall b, \forall t, \forall \omega, \tag{4.12a}$$

$$0 \le p^{\mathrm{G,B}}_{gbt\omega} \le A^{\mathrm{B}}_{gbt\omega} P^{\mathrm{C,G}}_{\mathrm{max},gb}, \quad \forall g \in G^{\mathrm{I}}, \forall b, \forall t, \forall \omega. \tag{4.12b}$$

4.4.11 Reserves Deployment in the Balancing Market

Constraints (4.13) enforce the limits on the up and down reserves deployed by generating and storage units, respectively, considering the reserve capacities scheduled in the day-ahead market.

$$0 \le r^{\mathrm{G,DU}}_{gbt\omega} \le r^{\mathrm{G,CU}}_{gbt}, \quad \forall g, \forall b, \forall t, \forall \omega, \tag{4.13a}$$

$$0 \le r^{\mathrm{G,DD}}_{gbt\omega} \le r^{\mathrm{G,CD}}_{gbt}, \quad \forall g, \forall b, \forall t, \forall \omega, \tag{4.13b}$$

$$0 \le r^{\mathrm{S,DUC}}_{sbt\omega} \le r^{\mathrm{S,CUC}}_{sbt}, \quad \forall s, \forall b, \forall t, \forall \omega, \tag{4.13c}$$

$$0 \le r^{\mathrm{S,DUD}}_{sbt\omega} \le r^{\mathrm{S,CUD}}_{sbt}, \quad \forall s, \forall b, \forall t, \forall \omega, \tag{4.13d}$$

$$0 \le r^{\mathrm{S,DDC}}_{sbt\omega} \le r^{\mathrm{S,CDC}}_{sbt}, \quad \forall s, \forall b, \forall t, \forall \omega, \tag{4.13e}$$

$$0 \le r^{\mathrm{S,DDD}}_{sbt\omega} \le r^{\mathrm{S,CDD}}_{sbt}, \quad \forall s, \forall b, \forall t, \forall \omega. \tag{4.13f}$$

4.4.12 Storage Units in the Balancing Market

Constraints (4.14a) and (4.14b) define the up and down reserves deployed from the charging and discharging, respectively. Constraints (4.14c) and (4.14d) establish the state-of-charge of the storage units in the balancing market considering the charge and discharge power scheduled in the day-ahead market and the deployment of the reserves. Constraints (4.14e) limit the energy content in each storage unit, while (4.14f) impose the minimum energy level at the end of the planing horizon in the balancing market.

$$r^{\text{S,DU}}_{sbt\omega} = r^{\text{S,DUD}}_{sbt\omega} + r^{\text{S,DUC}}_{sbt\omega}, \quad \forall s, \forall b, \forall t, \forall \omega, \tag{4.14a}$$

$$r^{\text{S,DD}}_{sbt\omega} = r^{\text{S,DDD}}_{sbt\omega} + r^{\text{S,DDC}}_{sbt\omega}, \quad \forall s, \forall b, \forall t, \forall \omega, \tag{4.14b}$$

$$\begin{aligned} e^{\text{SB}}_{sbt\omega} &= e^{\text{SB}}_{sbt-1,\omega} - \frac{1}{\eta^{\text{S}}_s}(p^{\text{S,D}}_{sbt} + r^{\text{S,DUD}}_{sbt\omega} - r^{\text{S,DDD}}_{sbt\omega}) \\ &\quad + \eta^{\text{S}}_s(p^{\text{S,C}}_{sbt} - r^{\text{S,DUC}}_{sbt\omega} + r^{\text{S,DDC}}_{sbt\omega}), \quad \forall s, \forall b, \forall t > 1, \forall \omega, \end{aligned} \tag{4.14c}$$

$$\begin{aligned} e^{\text{SB}}_{sbt\omega} &= E^{\text{C,SE}}_{\text{ini},sb} - \frac{1}{\eta^{\text{S}}_s}(p^{\text{S,D}}_{sbt} + r^{\text{S,DUD}}_{sbt\omega} - r^{\text{S,DDD}}_{sbt\omega}) \\ &\quad + \eta^{\text{S}}_s(p^{\text{S,C}}_{sbt} - r^{\text{S,DUC}}_{sbt\omega} + r^{\text{S,DDC}}_{sbt\omega}), \quad \forall s, \forall b, t = 1, \forall \omega, \end{aligned} \tag{4.14d}$$

$$0 \le e^{\text{SB}}_{sbt\omega} \le H^{\text{S}}_s P^{\text{C,SP}}_{\text{max},sb}, \quad \forall s, \forall b, \forall t, \forall \omega, \tag{4.14e}$$

$$e^{\text{SB}}_{sbt\omega} \ge E^{\text{C,SE}}_{\text{ini},sb}, \quad \forall s, \forall b, t = 24, \forall \omega. \tag{4.14f}$$

4.4.13 Transmission Lines in the Balancing Market

Constraints (4.15a) define the power flow through the transmission lines in the balancing market using the DC model, while constraints (4.15b) limit the power flow to their capacity. Constraints (4.15c) impose limits on the voltage angles.

$$p^{\text{L,B}}_{\ell t\omega} = \frac{1}{X_\ell}(\theta^{\text{B}}_{O(\ell)t\omega} - \theta^{\text{B}}_{F(\ell)t\omega}), \quad \forall \ell, \forall t, \forall \omega, \tag{4.15a}$$

$$-P^{\text{L}}_{\text{max},\ell} \le p^{\text{L,B}}_{\ell t\omega} \le P^{\text{L}}_{\text{max},\ell}, \quad \forall \ell, \forall t, \forall \omega, \tag{4.15b}$$

$$-\pi \le \theta^{\text{B}}_{bt\omega} \le \pi, \quad \forall b, \forall t, \forall \omega. \tag{4.15c}$$

4.4.14 Limits on the Unserved Demand

Finally, constraints (4.16a) and (4.16b) establish the limits on the unserved demand in the day-ahead and balancing markets, respectively.

$$0 \le p^{\text{UD,D}}_{bt} \le P^{\text{D,D}}_{bt}, \quad \forall b, \forall t, \tag{4.16a}$$

$$0 \le p^{\text{UD,B}}_{bt\omega} \le P^{\text{D,B}}_{bt\omega}, \quad \forall b, \forall t, \forall \omega. \tag{4.16b}$$

4.5 Illustrative Case Study

In this example, a simplified version of the scheduling model described previously is solved. The power system is based on that shown in Fig. 4.1. To make it simple, we assume that there is just one bus in the system, i.e., all the generating units and demands are connected to the same bus. The capacities of thermal, wind, and solar PV units are 150 MW, 100 MW, and 50 MW, respectively. The ramping capacity of the thermal unit is 0.7 and the minimum power output is 50 MW. Neither minimum up/down times nor start-up costs are considered for the thermal unit. The operating costs of the thermal unit is 55 €/MWh, whereas the costs for the wind and solar PV units are 3 €/MWh and 1 €/MWh, respectively. The peak demand is 220 MW. Figure 4.3 represents the described system.

The factors that determine the up and down reserve requirements in the day-ahead market take the following values, $K^D = 0.03$ and $K^{In} = 0.05$. The maximum up and down reserve capacity that the wind unit can provide corresponds to 30% and 50% of the available power in each hour of the day ahead of delivery, respectively. The solar PV unit cannot provide reserve capacity. For the rest of units, the cost of providing up and down reserve capacity in the day-ahead market is equal to 5% of the operating cost. In the balancing market, the cost of deploying up and down reserves is 15% higher and 10% lower, respectively, than the operating cost.

A planning horizon of two hours is considered. The demand, and wind and solar power are random variables represented via scenarios. In this example, we consider two scenarios per random variable, resulting in eight scenarios per hour. Table 4.2 gives the factors describing the demand level and the wind and solar power availability in the day-ahead (D-1) and balancing (D) markets in each hour and scenario. Note that those factors multiply the peak demand and the power capacity per unit, respectively. The scenarios are considered to be equiprobable. The availability factor of the thermal unit is equal to 1 in all hours.

This problem is solved using the academic license of GAMS [6]. The GAMS code of this example is provided in Sect. 4.9. The results obtained from this example are described below.

Figure 4.4 shows the day-ahead scheduling of energy and up and down reserve capacity for hours t_1 and t_2. The quantities above the figures representing the generating units and the demand indicate the scheduled energy. Below the figure, the up and down reserve capacity provided by the thermal and wind units are provided.

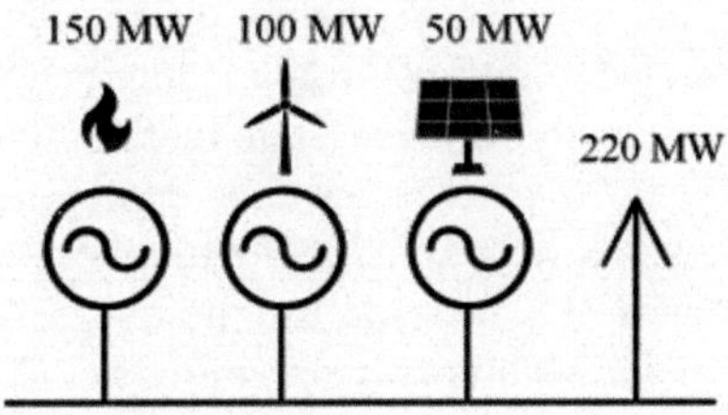

Fig. 4.3 Illustrative case study: power system

Table 4.2 Illustrative case study: demand and renewable power factors per scenario

Hour	Demand		Wind		Solar PV		Scenario
	D–1	D	D–1	D	D–1	D	
t_1	0.75	0.70	0.70	0.60	0.20	0.20	ω_1
		0.70		0.60		0.30	ω_2
		0.70		0.80		0.20	ω_3
		0.70		0.80		0.30	ω_4
		0.80		0.60		0.20	ω_5
		0.80		0.60		0.30	ω_6
		0.80		0.80		0.20	ω_7
		0.80		0.80		0.30	ω_8
t_2	0.90	0.80	0.55	0.40	0.65	0.60	ω_1
		0.80		0.40		0.70	ω_2
		0.80		0.60		0.60	ω_3
		0.80		0.60		0.70	ω_4
		1.00		0.40		0.60	ω_5
		1.00		0.40		0.70	ω_6
		1.00		0.60		0.60	ω_7
		1.00		0.60		0.70	ω_8

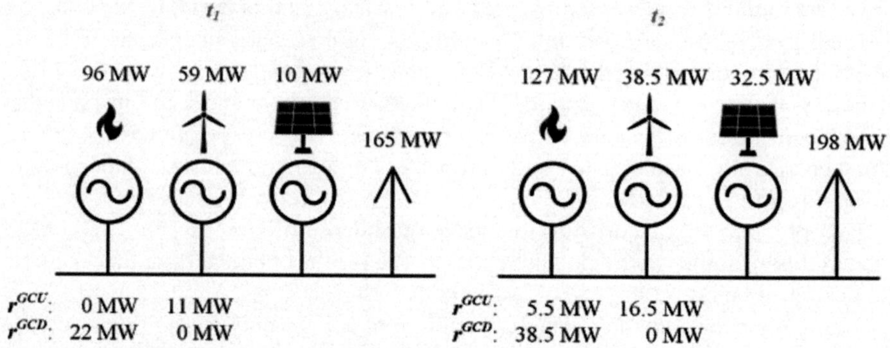

Fig. 4.4 Illustrative case study: power scheduled in the day ahead

In these results we observe that the up reserve capacity is mainly provided by the wind unit, whereas only the thermal unit schedules down reserve capacity. This is related to the costs assigned to the up and down reserve capacities, meaning that it is less costly for the system that the wind unit, with the lowest operating cost, provides the up reserve because the deployment of up reserve implies an increase in the total cost. However, the deployment of down reserve reduces the total cost of the system, hence, it is better to use the most expensive unit.

The results corresponding to the balancing market are given in Table 4.3. Let us use scenarios 4 and 5 of hour 1 to explain the results. In scenario 4, the wind power available is 80 MW, while in scenario 5 the available power is 60 MW. The wind unit schedules 69 MW in the day-ahead market and deploys 11 MW of up reserve in the scenarios with higher available power (scenario 4). However, the wind unit generates the maximum available power in the scenarios with lower available power, e.g., scenario 5, which results in deploying 9 MW of down reserve. In this situation, the thermal unit schedules 86 MW in the day ahead and deploys up and down reserves to counteract the variability of the wind unit. In hour 2, the wind power available reduces with respect to the previous hour, but the same reasoning applies. Observe that the thermal unit is at full capacity in scenarios 5 and 6.

In addition, note that solar power is spilled due to the lack of flexibility of the solar PV unit in scenarios 2, 4, 6, and 8. Obviously, real electricity markets are more complicated than the model studied in this chapter, and there exist multiple mechanisms to provide flexibility to the system so that renewable sources are not wasted.

Finally, Table 4.4 provides the day-ahead (column 2) and balancing (columns 3-10) electricity prices. From these results we can notice that the balancing price in hour 2 for scenarios 5 and 6 is equal to the unserved demand cost. This is related to what was highlighted previously, i.e., in those scenarios all the generation units are at full capacity and, even though there is no unserved demand, a possible increase in the demand could not be satisfied. Note that the day-ahead electricity price coincides with the expected value of the balancing prices.

Table 4.3 Illustrative case study: results of the balancing market (MW)

Hour	Scenario	Demand	Thermal			Wind			Solar
		$P^{D,B}$	$p^{G,D}$	$r^{G,DU}$	$r^{G,DD}$	$p^{G,D}$	$r^{G,DU}$	$r^{G,DD}$	$p^{G,B}$
t_1	ω_1	154.0	86.0	0.0	2.0	69.0	0.0	9.0	10.0
	ω_2	154.0	86.0	0.0	2.0	69.0	0.0	9.0	10.0
	ω_3	154.0	86.0	0.0	22.0	69.0	11.0	0.0	10.0
	ω_4	154.0	86.0	0.0	22.0	69.0	11.0	0.0	10.0
	ω_5	176.0	86.0	20.0	0.0	69.0	0.0	9.0	10.0
	ω_6	176.0	86.0	20.0	0.0	69.0	0.0	9.0	10.0
	ω_7	176.0	86.0	0.0	0.0	69.0	11.0	0.0	10.0
	ω_8	176.0	86.0	0.0	0.0	69.0	11.0	0.0	10.0
t_2	ω_1	176.0	124.5	0.0	18.5	43.5	0.0	3.5	30.0
	ω_2	176.0	124.5	0.0	18.5	43.5	0.0	3.5	30.0
	ω_3	176.0	124.5	0.0	38.5	43.5	16.5	0.0	30.0
	ω_4	176.0	124.5	0.0	38.5	43.5	16.5	0.0	30.0
	ω_5	220.0	124.5	25.5	0.0	43.5	0.0	3.5	30.0
	ω_6	220.0	124.5	25.5	0.0	43.5	0.0	3.5	30.0
	ω_7	220.0	124.5	5.5	0.0	43.5	16.5	0.0	30.0
	ω_8	220.0	124.5	5.5	0.0	43.5	16.5	0.0	30.0

Table 4.4 Illustrative case study: prices (€/MWh)

Hour	λ_t^{D}	$\lambda_{t\omega}^{\mathrm{B}}/\pi_\omega$							
		ω_1	ω_2	ω_3	ω_4	ω_5	ω_6	ω_7	ω_8
t_1	55.0	49.5	49.5	27.5	49.5	63.2	85.3	63.2	52.9
t_2	286.4	49.5	49.5	27.5	49.5	1000.0	1000.0	63.2	63.2

4.6 Realistic Case Study

In this section, a case study based on the European power system is described. The scheduling model presented in the previous sections is solved and multiple numerical results are explained. The optimization problems are solved using GAMS [6] in an Intel CORE i7 with 16 GB of RAM.

4.6.1 Input Data

The data used to define this case study are available at [5]. We consider a power system connecting 24 European countries, namely Austria (AT), Belgium (BE), Czech Republic (CZ), Denmark (DK), Estonia (EE), Finland (FI), France (FR), Germany (DE), Hungary (HU), Italy (IT), Latvia (LV), Lithuania (LT), Luxembourg (LU), the Netherlands (NL), Norway (NR), Poland (PL), Portugal (PT), Romania (RO), Slovakia (SK), Slovenia (SI), Spain (ES), Sweden (SE), Switzerland (CH), and the United Kingdom (UK). Each country is represented by a single bus of the network. The 24 buses are connected by 42 transmission lines, whose characteristics are provided in Table 4.5. Figure 4.5 depicts the schema of the considered power system.

The generating system comprises nuclear (Nucl), coal, gas-fired (GasF), fuel, hydroelectric (Hydr), onshore (Wons) and offshore (Woff) wind, solar PV (SoPV), CSP (SoTH), and biomass (Biom) technologies. Table 4.6 provides the total capacity and the capacity per technology installed in each country in 2018. The total capacity of the whole system is 936.6 GW.

Table 4.7 gives the technical parameters describing the generating technologies, namely the operating cost, the ramp factor, the minimum up time, the number of hours online and offline prior to the first period of the planning horizon, and the number of hours to compute the start-up cost. We assume that the start-up cost of dispatchable units is constant and corresponds to the cost of working at maximum capacity during the hours indicated in the last column of Table 4.7. These data are defined according to the reports published by the International Energy Agency [9], the International Renewable Energy Agency [11, 12].

The cost of scheduling and deploying reserve depends on the operating cost. Thus, we consider that the cost of scheduling reserve capacity in the day-ahead market is 1% of the operating cost, the cost of deploying up reserve in the balancing market is

Table 4.5 Realistic case study: characteristics of the transmission lines connecting the considered European countries

Line (#)	Origin	Destiny	Capacity (MW)	Reactance (Ω)	Line (#)	Origin	Destiny	Capacity (MW)	Reactance (Ω)
1	AT	CZ	2526	1.5	22	FR	ES	5567	1.6
2	AT	DE	4708	1.2	23	FR	CH	3769	1.9
3	AT	HU	1456	1.3	24	FR	UK	2024	1.5
4	AT	IT	300	1.4	25	DE	LU	819	1.4
5	AT	SI	1335	1.5	26	DE	NL	5134	1.3
6	AT	CH	1791	1.6	27	DE	PL	2139	1.7
7	BE	FR	4167	1.3	28	DE	SE	612	1.2
8	BE	LU	205	1.8	29	DE	CH	5352	1.3
9	BE	NL	3165	1.4	30	HU	RO	566	1.1
10	CZ	DE	3032	1.6	31	HU	SK	1807	1.7
11	CZ	PL	1749	1.2	32	IT	SI	1730	1.5
12	CZ	SK	2225	1.1	33	IT	CH	4905	1.6
13	DK	DE	2213	1.7	34	LV	LT	1315	1.8
14	DK	NR	1618	1.5	35	LT	PL	496	1.9
15	DK	SE	2269	1.4	36	LT	SE	734	1.4
16	EE	FI	1022	1.3	37	NL	NR	732	1.2
17	EE	LV	825	1.2	38	NL	UK	1103	1.3
18	FI	NR	120	1.3	39	NR	SE	4211	1.5
19	FI	SE	2770	1.6	40	PL	SK	1054	1.6
20	FR	DE	4569	1.7	41	PL	SE	601	1.7
21	FR	IT	3343	1.1	42	PT	ES	3978	1.8

10% higher than the operating cost, and the cost of deploying down reserve is 10% lower than the operating cost. The reserve capacity provided by dispatchable units is limited to their capacity and ramping limits. We assume that wind units can provide up to 20% and 30% of up and down reserve, respectively, with respect to the power available in each hour. Solar PV units cannot provide reserve capacity. In addition, we consider that parameters K^D and K^{In} of constraints (4.3a) and (4.3b) are equal to 0.03. A review of how the reserve procurement criteria is applied along different power systems is available in [14].

Two types of storage technologies are considered, i.e. interday and intraday units. Table 4.8 indicates the operating cost, the hours of autonomy and the charging rate of each technology. The capacity of each technology in each country is reported in Table 4.9. The interday capacity corresponds to the hydro pumped-storage capacity in each country in 2018. For the capacity of intraday storage units, which could be batteries or small storage devices, we take as reference the results obtained from the capacity expansion problem solved in [3] since no official data pertaining to these technologies are available. Specifically, for this case study we assume that the intraday storage capacity is 20% of the capacity resulting in 2050 as reported in [3].

Table 4.6 Realistic case study: generating capacity per technology and country (GW)

Country	Nucl	Coal	GasF	Fuel	Hydr	Wons	Woff	SoPV	SoTH	Biom	Total
AT	0.000	0.598	4.468	0.178	8.590	2.887	0.000	1.193	0.000	0.707	18.621
BE	5.931	0.470	6.545	0.330	0.179	1.979	1.178	2.953	0.000	1.061	20.626
CH	3.373	0.000	0.000	0.000	5.934	0.000	0.000	2.171	0.000	0.000	11.478
CZ	4.040	10.122	1.226	0.000	1.087	0.308	0.000	2.040	0.000	0.850	19.673
DE	9.516	46.310	31.361	4.271	5.300	51.633	5.051	42.804	0.000	11.034	207.280
DK	0.000	3.656	1.826	1.007	0.006	4.423	1.700	1.002	0.000	2.226	15.846
EE	0.000	0.085	0.119	1.976	0.008	0.487	0.000	0.000	0.000	0.126	2.801
ES	7.117	9.535	30.683	0.715	20.281	22.834	0.000	6.722	1.500	1.443	100.830
FI	2.782	2.278	1.865	1.386	3.149	1.908	0.000	0.000	0.000	3.724	17.092
FR	63.130	3.972	11.812	6.263	19.800	12.518	0.000	7.170	0.000	2.015	126.680
HU	1.899	1.049	4.107	0.412	0.056	0.325	0.000	0.336	0.000	0.355	8.539
IT	0.000	8.055	39.103	2.475	15.559	9.261	0.000	20.100	0.000	8.155	102.708
LT	0.000	0.000	1.873	0.000	0.128	0.509	0.000	0.082	0.000	0.161	2.753
LU	0.000	0.000	0.000	0.000	1.344	0.180	0.000	0.150	0.000	0.100	1.774
LV	0.000	0.000	1.124	0.000	1.539	0.053	0.000	0.000	0.000	0.105	2.821
NL	0.486	4.631	18.433	0.000	0.038	3.675	0.957	2.584	0.000	1.172	31.976
NR	0.000	0.000	0.635	0.000	28.147	1.197	0.000	0.000	0.000	0.000	29.979
PL	0.000	28.793	2.140	0.415	0.612	5.791	0.000	0.231	0.000	0.603	38.585
PT	0.000	1.756	4.621	0.000	4.349	5.073	0.000	0.559	0.000	0.634	16.992
RO	1.300	5.805	5.142	0.000	6.490	2.987	0.000	1.211	0.000	0.116	23.051
SE	8.603	0.000	0.000	0.000	15.949	6.247	0.000	0.000	0.000	7.870	38.669
SI	0.696	0.924	0.491	0.058	1.053	0.003	0.000	0.275	0.000	0.058	3.558
SK	1.940	0.566	1.111	0.257	1.626	0.003	0.000	0.531	0.000	0.778	6.812
UK	0.000	11.427	28.755	0.316	4.619	13.139	8.217	12.563	0.000	8.470	87.506
Total	110.813	140.032	197.440	20.059	145.843	147.420	17.103	104.677	1.500	51.763	**936.650**

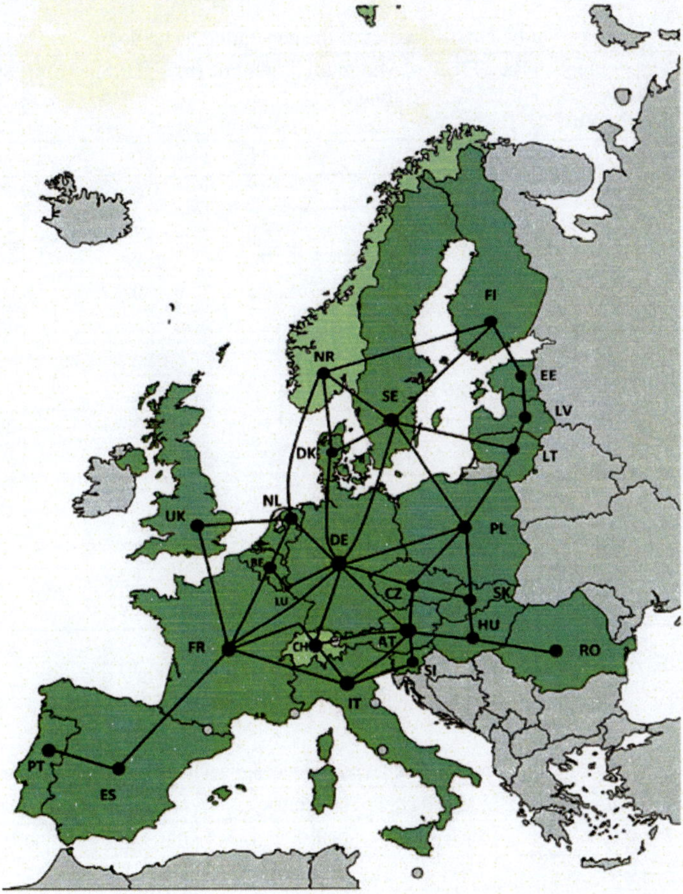

Fig. 4.5 Realistic case study: schema of the European network (created with mapchart.net)

The reason of using such small percentage is that the installed storage capacity of this type of technologies in 2018 was not relevant.

Moreover, storage units can provide reserve capacity, but it is limited to 10% of their power capacity. The cost of scheduling up and down reserve is 1% of their operating cost, while the cost of deploying up and down reserve is 1 and 5% higher than their operating cost, respectively. The cost of the unserved demand is 1000 €/MWh.

Table 4.7 Realistic case study: characteristics of the generating technologies

Technology	Operating cost (€/MWh)	Ramp factor (p.u.)	Minimum time (h)	Online (h)	Offline (h)	Start up (h)
Nucl	10.54	0.2	24	24	0	24
Coal	48.39	0.6	6	6	0	12
GasF	45.28	0.7	3	3	0	3
Fuel	35.00	0.5	6	0	2	12
Hydr	46.60	1.00	0	0	0	0
Wons	3.50	–	–	–	–	–
Woff	10.50	–	–	–	–	–
SoPV	2.70	–	–	–	–	–
SoTH	4.60	0.8	3	1	0	2
Biom	24.00	0.7	3	3	0	6

Table 4.8 Realistic case study: characteristics of storage technologies

Technology	Operating cost (€/MWh)	Autonomy (h)	Charging rate
Interday	10	6	0.7
Intraday	2	48	0.8

Table 4.9 Realistic case study: installed capacity per storage technology in each country (MW)

Country	Intraday	Interday	Country	Intraday	Interday
AT	0.0	3401	LT	0.0	900
BE	0.0	1308	LU	0.0	0
CH	47.1	6594	LV	0.0	0
CZ	0.0	1172	NL	61.7	0
DE	0.0	8918	NR	0.0	0
DK	5.7	0	PL	0.0	1780
EE	0.0	0	PT	227.8	2820
ES	684.6	5645	RO	39.4	0
FI	35.3	0	SE	0.0	0
FR	698.3	5020	SI	61.8	180
HU	0.0	0	SK	0.0	916
IT	1483.0	6453	UK	1347.0	0

4.6.2 Uncertainty Characterization

As previously stated, the random variables considered in this scheduling model are the hourly demand, onshore/offshore wind and solar PV power. A set of scenarios is generated to define the possible values that these random variables can take. The scenarios are randomly generated using the historical data of 2016 available at the transparency platform of ENTSO-E [5]. Specifically, the day-ahead forecast considered by the SO and the actual consumption/generation of the demand, wind and solar power for each country are used. Thus, the procedure applied to generate the scenarios consists on the following: (i) we consider the day-ahead forecast of the demand, wind and solar power for the 24 hours of a specific day; (ii) the forecast error between the day-ahead prediction and the actual realization of the random variable is computed and the probability distribution of the forecast error is defined; (iii) a set of 100 scenarios is generated using the forecast error distributions and the day-ahead values for each of the 24 h of the selected day; and (iv) to attain computational tractability, the initial set of scenarios is reduced by applying the scenario reduction technique proposed in [7] such that we select 10 scenarios per hour. This procedure can be applied iteratively for each day of the year. As an example, Fig. 4.6 illustrates the onshore wind and solar PV scenarios used in this case study for Denmark, Finland, Spain and Germany. The figures in the upper plots represent the day-ahead wind factor forecast in dark blue and the real-time scenarios in light blue. The lower plots represent the day-ahead solar factor forecast in magenta and the real-time scenarios in yellow.

In addition, we assume that CSP units incorporate a TES system, which implies that their available power is significantly less variable than solar PV units. Therefore, we assume that the availability factor of CSP units is equal to 1 in all hours except for 1.00 am to 8.00 am. In those hours, the availability factor is equal to 0.4 in the winter season, 0.9 in the summer season, and 0.7 in the rest of the year. Finally, the availability factor of the rest of dispatchable units is equal to 1 for all hours.

4.6.3 Results

In this section, the results obtained from this case study are described. The resulting problem comprises 297,361 variables and 526,417 equations and is solved in about 450 s.

First, we analyze the day-ahead scheduling of three different days of the year, which are January 20, August 8, and September 25. The peak demand of each selected day in the day-ahead market reaches 477.8, 292.9, and 381.1 GW, respectively.

Figure 4.7 shows the energy (left) and reserve capacity (right) scheduling resulting from the day-ahead market in the whole power system for each day. In these figures, each generating technology is represented with a different color according to the legend, while the storage participation includes both storage technologies. In the

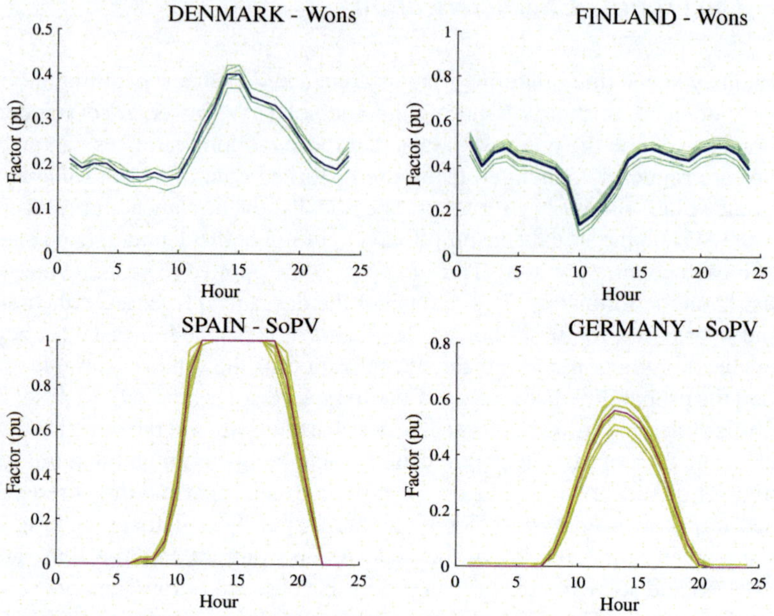

Fig. 4.6 Realistic case study: onshore wind and solar PV scenarios in different countries for a representative day of the year

left-hand plots, the demand is represented with a red line. Note that in the right-hand plots, the positive values represent the up reserve capacity, while the negative values correspond to the down reserve capacity indicating that this represents a reduction in the power output or the charge/discharge. The red lines in these plots represent the minimum up/down reserve requirements, respectively.

As general observations from these results, we can state (i) the nuclear units are at full capacity in the three days, (ii) hydro and, in a lesser extent, gas units vary their power output according to the available intermittent renewable power, (iii) biomass units maintain relatively constant production during the three days, while the production of solar PV units in warm days is meaningful, (iv) as expected due to their flexibility and low operating costs, storage units mainly supply reserve capacity, and (v) the participation of wind units in the reserve provision, especially in the down reserve, is very relevant. Note that this results are strictly related to the operating costs defined in this example.

It is also interesting to analyze what happens in specific countries. As an example, Figs. 4.8, 4.9, and 4.10 depict the day-ahead energy scheduling in the three selected days in Spain, Germany, Denmark, and France (the legend is the same for the three figures). These four countries present relevant differences among each other according to their generating system mix (for instance, in France the generating system mainly comprises nuclear power while in Denmark the wind power accumulates a high share of the generating mix); the electricity consumption, being that of Germany

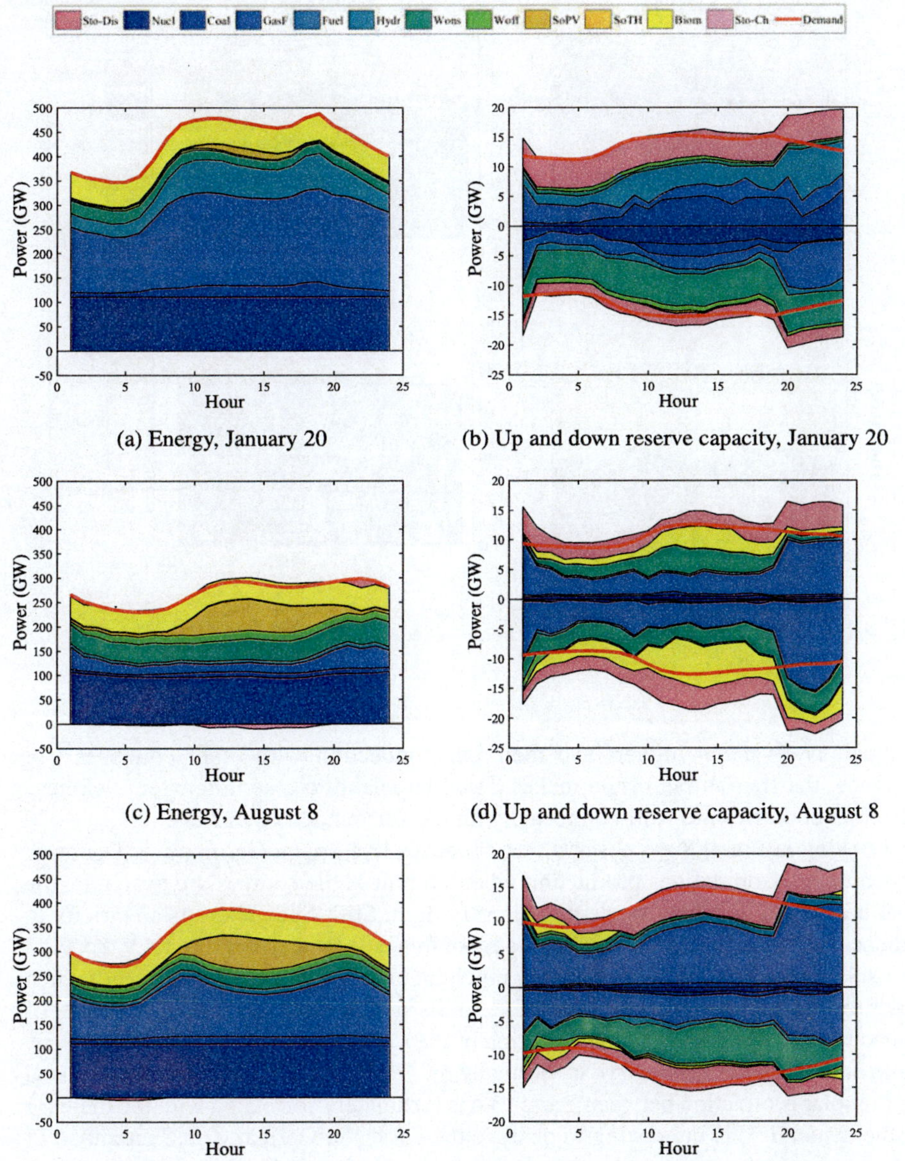

(a) Energy, January 20 (b) Up and down reserve capacity, January 20

(c) Energy, August 8 (d) Up and down reserve capacity, August 8

(e) Energy, September 25 (f) Up and down reserve capacity, September 25

Fig. 4.7 Realistic case study: energy and reserve capacity day-ahead scheduling in the whole system in three representative days

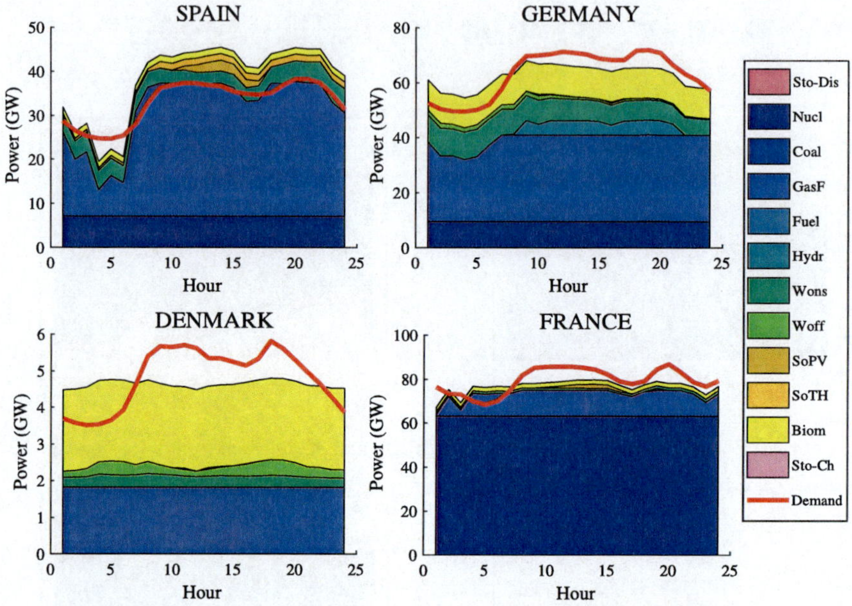

Fig. 4.8 Realistic case study: day-ahead energy scheduling in Spain, Germany, Denmark and France in January 20

comparatively much higher; and their interconnection with other countries (for instance, the Iberian peninsula makes Spain an island to a certain extent, whereas the other countries are well-connected with the surrounded countries).

Looking at Fig. 4.8 we observe that countries like Spain, Germany, or Denmark depend highly on the gas production when no renewable sources are available, the coal has a high cost and the demand is very high. Still, Spain exports electricity to France and/or Portugal during some hours of this day. However, in Fig. 4.9 the wind and solar production is comparatively much higher and the demand is lower, which leads to a insignificant participation of gas units, a high export of France to the rest of the countries, and a relevant participation of the storage units in Spain and Germany. Note how steep the solar curve in Germany on August 8 is. The strong ramp caused by the solar production between 7 and 13 h is fortunately correlated with the increase in the demand. It is interesting to observe the curves in Germany on September 25 because the gas units counteract the up ramp of the demand before the sunrise and the solar production during the central hours of the day.

As economic results, we can observe the prices obtained in the day-ahead market in each country and in each day through Fig. 4.11. In most of the hours and countries, the prices move around the operating cost of the gas and coal units. The prices on January 20 are comparatively higher because of the high demand and low renewable production, whereas on August 8 the prices are comparatively lower.

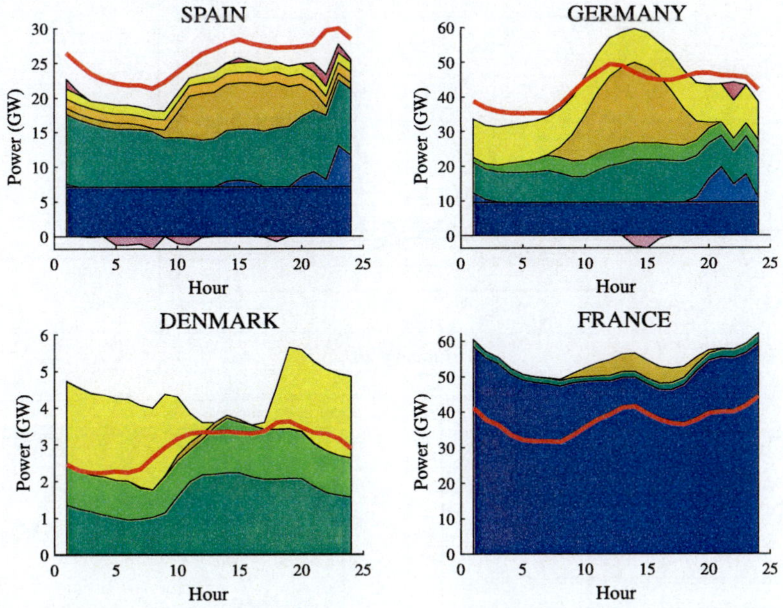

Fig. 4.9 Realistic case study: day-ahead energy scheduling in Spain, Germany, Denmark and France in August 8

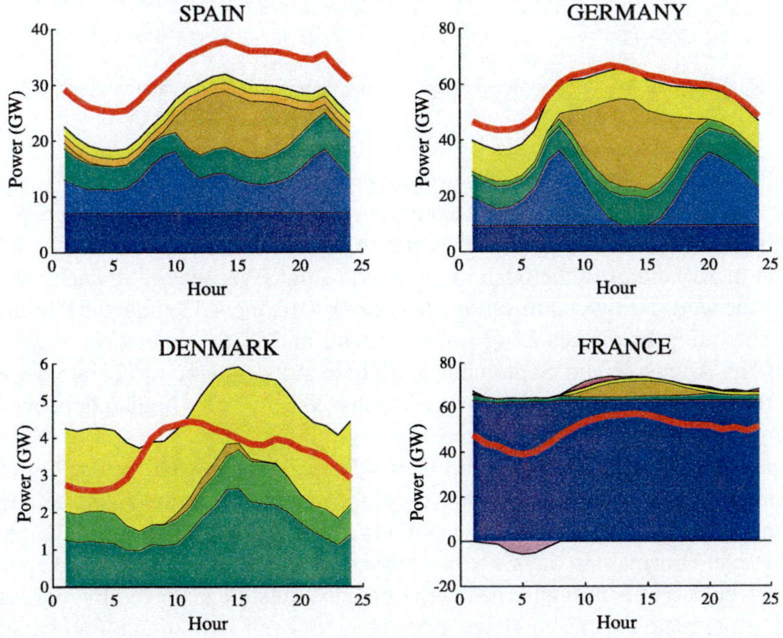

Fig. 4.10 Realistic case study: day-ahead energy scheduling in Spain, Germany, Denmark and France in September 25

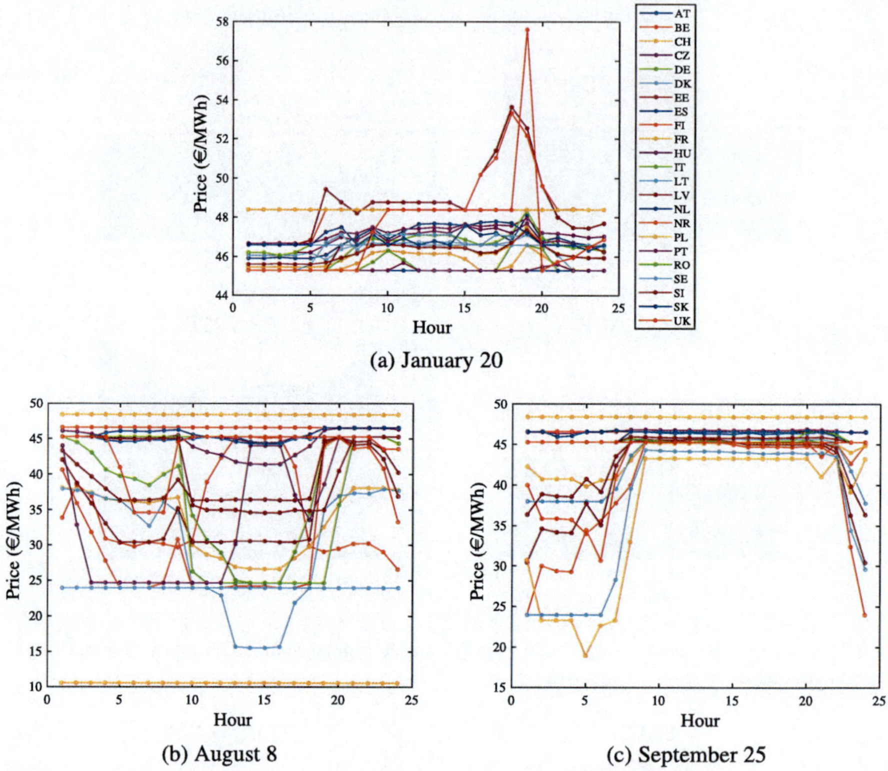

Fig. 4.11 Realistic case study: day-ahead electricity prices in three representative days

Since the global tendency goes towards increasing the wind and solar PV capacity in power systems, it becomes relevant to analyze the impact of increasing the capacity of those generation technologies. Thus, we solve again the day-ahead scheduling model but increasing 50% the onshore wind and solar PV capacity in each country regarding the values provided in Table 4.6. Figures 4.12 and 4.13 show the day-ahead energy scheduling in the resulting power system and in the four aforementioned countries for August 8 and September 25. The results of January 20 are avoided because the wind and solar production in January 20 is not very high and, hence, the differences with respect to the base case are negligible.

Comparing Figs. 4.12 and 4.13 with Figs. 4.7c, e, 4.9, and 4.10, we can highlight how important the solar PV production becomes in general. In days with relatively high demand like on September 25, gas units generally participate to balance the demand curve. However, in days with comparatively lower demand as on August 8, the gas power output is null and the energy balance must be achieved by increasing the use of storage units, spilling renewable power, or even switching off and making ramp some nuclear power plants as it happens in Spain. Moreover, the increasing of

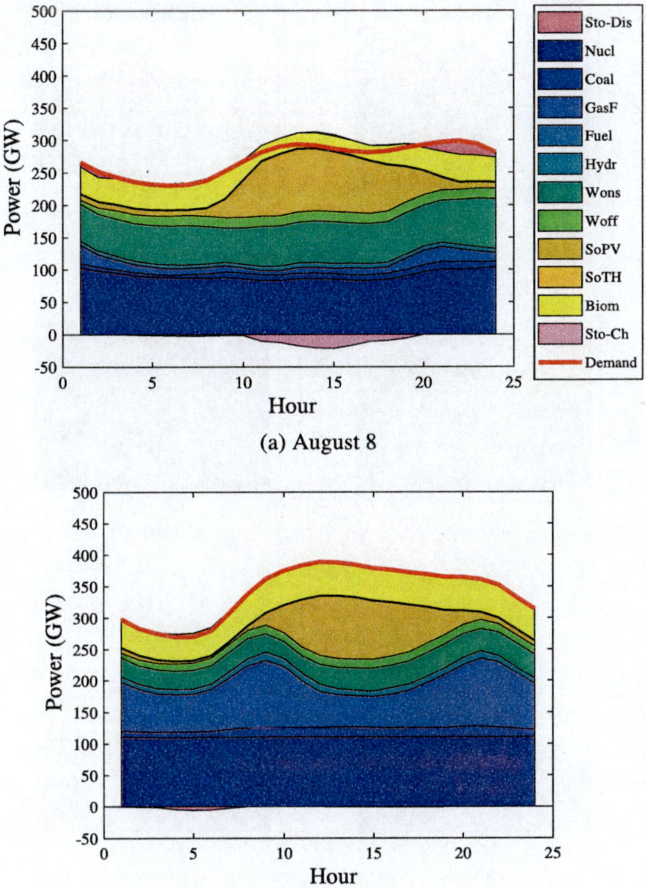

(a) August 8

(b) September 25

Fig. 4.12 Realistic case study: energy day-ahead scheduling with 50% more of intermittent renewable capacity in the whole system

the wind and solar production has an impact on the total expected cost, which can be reduced up to 20% regarding the base case.

Let us analyze now the main results of the balancing market. Going back to the initial case (no increase in the renewable capacity), Tables 4.10 and 4.11 provide the expected up and down reserve deployed in the balancing market per country and technology for September 25. The last column of these tables indicate the reserve deployed by the storage units.

The technologies deploying higher up reserve are gas and onshore wind units with a total expected energy of 39.1 and 8.3 GWh, respectively. The same occurs with the down reserve, providing 30.6 GWh the gas units and 13.1 GWh the wind units. Additionally, if we compare the up reserve deployed by wind units in the base

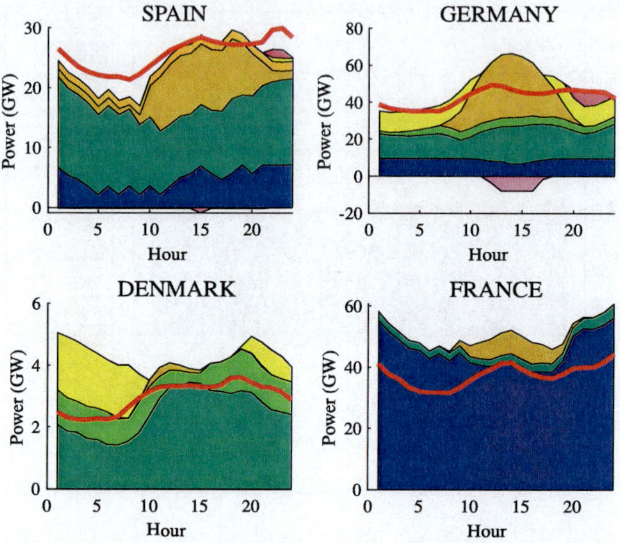

(a) August 8

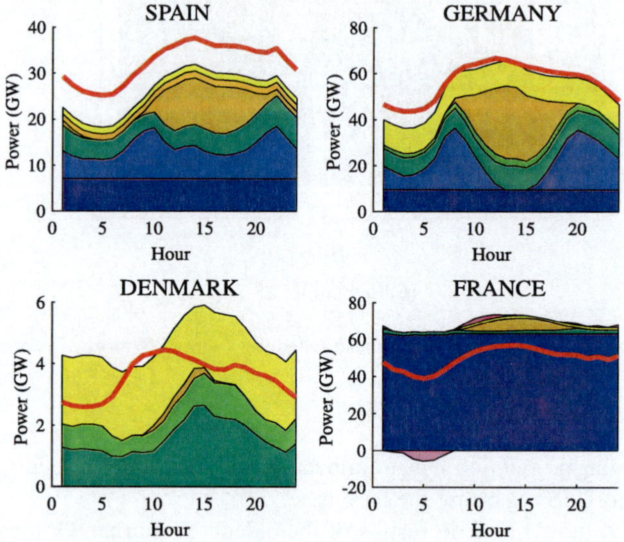

(b) September 25

Fig. 4.13 Realistic case study: energy and reserve capacity day-ahead scheduling with 50% more of intermittent renewable capacity in ES, DE, DK and FR

Table 4.10 Realistic case study: deployed up reserve on September 25 (GWh)

Country	Nucl	Coal	GasF	Fuel	Hydr	Wons	Woff	SoPV	SoTH	Biom	Sto
AT	0.00	0.00	0.00	0.00	0.00	0.24	0.00	0.00	0.00	0.00	0.00
BE	0.00	0.00	0.03	0.00	0.00	0.39	0.31	0.00	0.00	0.00	0.13
CH	0.00	0.00	0.00	0.00	0.00	0.00	0.00	0.00	0.00	0.00	0.00
CZ	0.00	0.00	0.00	0.00	0.03	0.06	0.00	0.00	0.00	0.00	0.00
DE	0.00	0.00	9.52	0.00	0.03	1.69	0.28	0.00	0.00	0.00	0.00
DK	0.00	0.00	0.60	0.00	0.00	0.28	0.01	0.00	0.00	0.00	0.00
EE	0.00	0.00	0.08	0.00	0.00	0.05	0.00	0.00	0.00	0.00	0.00
ES	0.00	0.00	0.68	0.00	0.00	0.34	0.00	0.00	0.00	0.00	0.04
FI	0.00	0.00	1.17	0.00	0.00	0.23	0.00	0.00	0.00	0.30	0.05
FR	0.87	0.00	0.00	0.00	0.00	2.36	0.00	0.00	0.00	1.87	3.36
HU	0.00	0.00	0.27	0.00	0.03	0.08	0.00	0.00	0.00	0.00	0.00
IT	0.00	0.00	4.08	0.00	0.00	0.00	0.00	0.00	0.00	0.00	0.00
LT	0.00	0.00	0.19	0.00	0.00	0.00	0.00	0.00	0.00	0.00	0.02
LU	0.00	0.00	0.00	0.00	0.00	0.04	0.00	0.00	0.00	0.00	0.00
LV	0.00	0.00	0.10	0.00	0.00	0.01	0.00	0.00	0.00	0.00	0.00
NL	0.00	0.00	1.54	0.00	0.00	0.45	0.04	0.00	0.00	0.00	0.00
NR	0.00	0.00	0.00	0.00	1.16	0.01	0.00	0.00	0.00	0.00	0.00
PL	0.00	0.74	0.00	0.00	0.00	0.52	0.00	0.00	0.00	0.00	0.00
PT	0.00	0.00	0.00	0.00	0.00	0.11	0.00	0.00	0.00	0.00	0.01
RO	0.00	0.00	0.18	0.00	0.11	0.80	0.00	0.00	0.00	0.00	0.00
SE	0.00	0.00	0.00	0.00	0.60	0.28	0.00	0.00	0.00	1.69	0.00
SI	0.00	0.00	0.01	0.00	0.01	0.00	0.00	0.00	0.00	0.00	0.00
SK	0.00	0.00	0.00	0.00	0.01	0.00	0.00	0.00	0.00	0.00	0.00
UK	0.00	0.00	20.60	0.00	0.00	0.37	0.46	0.00	0.00	0.00	0.00

case with the case with higher renewable capacity, the result is that the up reserve deployed by wind units grows 80% in the analyzed day.

Finally, to give an idea of how variable the balancing prices are, Table 4.12 indicates the minimum, average and maximum prices found in the balancing market in each country on September 25. As observed in the day-ahead market, the average prices are quite similar among the countries.

Different additional analyses could be performed. For instance, it could be evaluated the impact of (i) increasing the cost of fossil fuels, (ii) imposing a constraint on the CO_2 emissions emitted per day, (iii) limiting the hydro power available in each period of the year or considering as uncertain parameter the input to the reservoirs, (iv) increasing the operating cost of storage units or their availability to provide reserve services, etc. These are stimulating analyses that remain proposed for the interested readers.

Table 4.11 Realistic case study: deployed down reserve on September 25 (GWh)

Country	Nucl	Coal	GasF	Fuel	Hydr	Wons	Woff	SoPV	SoTH	Biom	Sto
AT	0.00	0.00	0.00	0.00	0.91	0.38	0.00	0.00	0.00	0.00	0.00
BE	0.00	0.00	0.41	0.00	0.00	0.19	0.33	0.00	0.00	0.00	0.19
CH	0.00	0.00	0.00	0.00	0.00	0.00	0.00	0.00	0.00	0.00	0.00
CZ	0.00	0.00	0.00	0.00	0.02	0.03	0.00	0.00	0.00	0.00	0.00
DE	0.00	0.00	5.76	0.00	0.00	3.88	1.25	0.00	0.00	0.00	0.00
DK	0.00	0.00	0.77	0.00	0.00	0.41	0.31	0.00	0.00	0.00	0.00
EE	0.00	0.00	0.05	0.00	0.00	0.05	0.00	0.00	0.00	0.00	0.00
ES	0.00	0.00	0.28	0.00	0.00	0.30	0.00	0.00	0.00	0.00	0.05
FI	0.00	0.00	1.18	0.00	0.00	0.43	0.00	0.00	0.00	0.29	0.06
FR	0.03	0.00	0.00	0.00	0.00	0.59	0.00	0.00	0.00	0.49	4.58
HU	0.00	0.00	0.01	0.00	0.02	0.02	0.00	0.00	0.00	0.00	0.00
IT	0.00	0.00	4.40	0.00	0.00	0.16	0.00	0.00	0.00	0.00	0.00
LT	0.00	0.00	0.37	0.00	0.00	0.17	0.00	0.00	0.00	0.00	0.02
LU	0.00	0.00	0.00	0.00	0.00	0.02	0.00	0.00	0.00	0.01	0.00
LV	0.00	0.00	0.41	0.00	0.00	0.01	0.00	0.00	0.00	0.00	0.00
NL	0.00	0.00	3.37	0.00	0.00	0.47	0.19	0.00	0.00	0.00	0.00
NR	0.00	0.00	0.00	0.00	1.83	0.13	0.00	0.00	0.00	0.00	0.00
PL	0.00	1.77	0.00	0.00	0.00	0.44	0.00	0.00	0.00	0.00	0.01
PT	0.00	0.00	0.00	0.00	0.00	0.93	0.00	0.00	0.00	0.00	0.01
RO	0.00	0.00	0.00	0.00	0.10	0.31	0.00	0.00	0.00	0.00	0.00
SE	0.00	0.00	0.00	0.00	0.00	1.05	0.00	0.00	0.00	1.35	0.00
SI	0.00	0.00	0.00	0.00	0.00	0.00	0.00	0.00	0.00	0.00	0.00
SK	0.00	0.00	0.00	0.00	0.60	0.00	0.00	0.00	0.00	0.00	0.00
UK	0.00	0.00	13.61	0.00	0.00	3.09	1.43	0.00	0.00	0.00	0.00

4.7 Summary and Conclusions

The increasing penetration of wind and solar units in power systems is deeply transforming the system operation. Renewable sources represent a sustainable alternative to polluting fossil fuels, but their intermittency implies to find units flexible enough to ensure the electricity supply when there is lack of those sources. In this chapter, we tackle the problem of how a power system with high penetration of intermittent renewable power is scheduled in the day ahead considering different sources of uncertainty. A scheduling model formulated as a two-stage mixed-integer stochastic programming model is presented. The day-ahead decisions of how much energy and reserve capacity to provide for each hour of the following day are determined considering the variability of the demand and the intermittent renewable power in the real-time operation. The perspective of the SO is adopted so that the total expected cost must be minimized. The possible realizations of the random variables, i.e. demand and wind and solar power, are represented through a set of scenarios.

Table 4.12 Realistic case study: minimum, average and maximum balancing prices on September 25 (€/MWh)

Country	Minimum	Average	Maximum
AT	37.7	45.7	53.7
BE	28.8	42.1	50.5
CH	34.6	43.5	51.9
CZ	38.3	46.4	52.0
DE	36.2	45.1	54.3
DK	24.0	42.4	53.4
EE	25.7	41.1	53.9
ES	36.6	45.0	53.3
FI	19.2	39.3	54.3
FR	8.4	37.0	49.6
HU	38.0	45.2	52.3
IT	36.4	45.0	54.3
LT	36.2	45.2	54.3
LV	30.1	42.7	53.5
NL	36.2	45.0	54.3
NR	37.3	46.2	55.9
PL	39.1	48.1	53.2
PT	36.6	45.0	53.3
RO	38.0	45.2	52.3
SE	19.2	37.0	53.1
SI	37.4	45.4	53.8
SK	37.3	46.3	52.2
UK	36.2	45.0	54.3

A case study based on the European power system is carried out. The main conclusions obtained from the numerical study are as follows:

- The high variability of the renewable sources along the year makes it necessary to have of a back-up system that ensures the energy balance under any circumstance. The results obtained show that gas and hydroelectric units provide that flexibility, specially the former one.
- In the model presented, it was assumed that wind units can provide reserve services and this was noticeable in the numerical results. Wind units in addition to gas units provide most of the up and down reserves in the balancing market in the considered power system. The fact that wind units can provide reserve services makes them increase their reliability.
- If available, nuclear units keep on generating at full capacity. However, in seasons with very high renewable production, as it can happen in renewable-dominated systems in the warm periods, they may be shut down.

- The solar production may dominate the central hours of the day in many power systems. This entails steep up and down ramps in the sunrise and sunset that requires other technologies or instruments to handle the consequences of fast changes.
- Storage units increase their participation in days with high intermittent renewable production.

4.8 Notation

This section provides the list of symbols used along this chapter.

Acronyms

CSP	Concentrating solar power.
DC	Direct current.
EV	Electric vehicle.
GHG	Greenhouse gas.
PV	Photovoltaic.
SO	System operator.
TES	Thermal energy storage.
TSO	Transmission system operator.

Sets and Indices

b	Index of buses.
B	Set of buses.
$F(\ell)$	Destination bus of line ℓ.
g	Index of generating units.
G	Set of generating units.
G^{D}	Set of dispatchable generating units.
G_b^{D}	Set of dispatchable generating units located at bus b.
G^{I}	Set of intermittent generating units.
ℓ	Index of transmission lines.
L	Set of transmission lines.
L_b^{F}	Set of transmission lines whose destination bus is b.
L_b^{O}	Set of transmission lines whose origin bus is b.
s	Index of storage units.
S	Set of storage units.
S_b	Set of storage units located at bus b.

$O(\ell)$ Origin bus of line ℓ.
t Index of time periods.
T Set of time periods.
ω Index of scenarios.
Ω Set of scenarios.

Parameters

$A^{\mathrm{B}}_{gbt\omega}$ Availability of generating unit g at bus b, period t, and scenario ω in the balancing market (pu).
A^{D}_{gbt} Availability of generating unit g at bus b and period t in the day-ahead market (pu).
$C^{\mathrm{D,C}}_{gt}$ Offering cost in the day-ahead energy market of unit g and period t (€/MWh).
$C^{\mathrm{G,RCD}}_{gt}$ Offering cost of down reserve capacity of generating unit g and period t (€/MWh).
$C^{\mathrm{G,RCU}}_{gt}$ Offering cost of up reserve capacity of generating unit g and period t (€/MWh).
$C^{\mathrm{G,RDD}}_{gt}$ Offering cost of deploying down reserve of generating unit g and period t (€/MWh).
$C^{\mathrm{G,RDU}}_{gt}$ Offering cost of deploying up reserve of generating unit g and period t (€/MWh).
$C^{\mathrm{G,SU}}_{gb}$ Startup cost parameter of generating unit g at bus b (€).
$C^{\mathrm{I,G}}_{g}$ Annualized investment cost of generating unit g (€/MW).
$C^{\mathrm{I,SE}}_{s}$ Annualized investment cost of the energy component of storage unit s (€/MW).
$C^{\mathrm{I,SP}}_{s}$ Annualized investment cost of the power component of storage unit s (€/MW).
$C^{\mathrm{S,RCD}}_{st}$ Offering cost of down reserve capacity of storage unit s and period t (€/MWh).
$C^{\mathrm{S,RCU}}_{st}$ Offering cost of up reserve capacity of storage unit s and period t (€/MWh).
$C^{\mathrm{S,RDD}}_{st}$ Offering cost of deploying down reserve of storage unit s and period t (€/MWh).
$C^{\mathrm{S,RDU}}_{st}$ Offering cost of deploying up reserve of storage unit s and period t (€/MWh).
C^{UD} Cost of unserved demand (€/MWh).
$E^{\mathrm{C,SE}}_{\mathrm{ini},sb}$ Energy content at the beginning of the planning horizon in storage unit s at bus b (MWh).
H^{S}_{s} Hours of autonomy of storage unit s (h).
N_{T} Number of time periods.
$P^{\mathrm{C,G}}_{\mathrm{max},gb}$ Capacity of generation unit g at bus b (MW).

$P^{\mathrm{C,G}}_{\min,gb}$ Minimum power output of generation unit g at bus b (MW).
$P^{\mathrm{C,SP}}_{\max,sb}$ Peak power of storage unit s at bus b (MW).
$P^{\mathrm{D,B}}_{bt\omega}$ Power demand in the balancing market at bus b, period t, and scenario ω (MW).
$P^{\mathrm{D,D}}_{bt}$ Power demand scheduled in the day-ahead market at bus b and period t (MW).
$P^{\mathrm{G}}_{\mathrm{DW},g}$ Down ramp factor of generating unit g (pu).
$P^{\mathrm{G}}_{\mathrm{UP},g}$ Up ramp factor of generating unit g (pu).
$P^{\mathrm{L}}_{\max,\ell}$ Active power capacity of transmission line ℓ (MW).
$R^{\mathrm{G,CDmax}}_{g}$ Factor limiting the down reserve capacity of intermittent generating unit g (pu).
$R^{\mathrm{G,CUmax}}_{g}$ Factor limiting the up reserve capacity of intermittent generating unit g (pu).
T^{Dmin}_{g} Minimum down time of generating unit g (h).
T^{Umin}_{g} Minimum up time of generating unit g (h).
X_{ℓ} Reactance of line ℓ (Ω).
$\alpha^{\mathrm{S,SE}}_{s}$ Factor used to model the ending status of the storage unit s (pu).
$\alpha^{\mathrm{S,SI}}_{s}$ Factor used to model the initial status of the storage unit s (pu).
$\alpha^{\mathrm{S,SMIN}}_{s}$ Factor used to model the minimum energy that must contain storage unit s (pu).
η^{S}_{s} Efficiency of charging/discharging storage unit s (pu).
$\gamma^{\mathrm{S,EP}}_{s}$ Relationship between energy and power capacities in storage unit s (h).
π_{ω} Probability of scenario ω (pu).

Variables

$c^{\mathrm{G,SU}}_{gbt}$ Startup cost of generating unit g at bus b and period t (€).
$e^{\mathrm{SB}}_{sbt\omega}$ Energy stored in the balancing market in storage unit s at bus b, period t, and scenario ω (MWh).
e^{SD}_{sbt} Energy stored in the day-ahead market in storage unit s at bus b and period t (MWh).
$p^{\mathrm{G,B}}_{gbt\omega}$ Power generated in the balancing market by generating unit g at bus b, period t, and scenario ω (MW).
$p^{\mathrm{G,D}}_{gbt}$ Power scheduled by generating unit g at bus b in period t in the day-ahead market (MW).
$p^{\mathrm{L,B}}_{\ell t\omega}$ Power flow through line ℓ in the balancing market, period t, and scenario ω (MW).
$p^{\mathrm{L,D}}_{\ell t}$ Power flow through line ℓ in the day-ahead market, and period t (MW).
$p^{\mathrm{S,BC}}_{sbt\omega}$ Power charged by storage s at bus b, period t, and scenario ω (MW).
$p^{\mathrm{S,BD}}_{sbt\omega}$ Power discharged by storage s at bus b, period t, and scenario ω (MW).

$p_{sbt}^{S,C}$ Charge power scheduled by storage unit s at bus b in period t in the day-ahead market (MW).
$p_{sbt}^{S,D}$ Discharge power scheduled by storage unit s at bus b in period t in the day-ahead market (MW).
$p_{bt\omega}^{UD,B}$ Unserved demand at bus b in the balancing market, period t, and scenario ω (MW).
$p_{bt}^{UD,D}$ Unserved demand at bus b in the day-ahead market and period t (MW).
$r_{gbt}^{G,CD}$ Down reserve capacity scheduled by generating unit g at bus b and period t (MW).
$r_{gbt}^{G,CU}$ Up reserve capacity scheduled by generating unit g at bus b and period t (MW).
$r_{gbt\omega}^{G,DD}$ Deployed down reserve by generating unit g at bus b, period t, and scenario ω (MW).
$r_{gbt\omega}^{G,DU}$ Deployed up reserve by generating unit g at bus b, period t, and scenario ω (MW).
$r_{sbt}^{S,CD}$ Down reserve capacity scheduled by storage unit s at bus b and period t (MW).
$r_{sbt}^{S,CDC}$ Down reserve capacity scheduled from the charging of storage unit s at bus b, and period t (MW).
$r_{sbt}^{S,CDD}$ Down reserve capacity scheduled from the discharging of storage unit s at bus b and period t (MW).
$r_{sbt}^{S,CU}$ Up reserve capacity scheduled by storage unit s in the day-ahead market, at bus b, and period t (MW).
$r_{sbt}^{S,CUC}$ Up reserve capacity scheduled from the charging of storage unit s at bus b, and period t (MW).
$r_{sbt}^{S,CUD}$ Up reserve capacity scheduled from the discharging of storage unit s at bus b, and period t (MW).
$r_{sbt\omega}^{S,DD}$ Deployed down reserve by storage unit s at bus b, period t, and scenario ω (MW).
$r_{sbt\omega}^{S,DDC}$ Down reserve deployed from the charging of storage unit s at bus b, period t, and scenario ω (MW).
$r_{sbt\omega}^{S,DDD}$ Down reserve deployed from the discharging of storage unit s at bus b, period t, and scenario ω (MW).
$r_{sbt\omega}^{S,DU}$ Deployed up reserve by storage unit s at bus b, period t, and scenario ω (MW).
$r_{sbt\omega}^{S,DUC}$ Up reserve deployed from the charging of storage unit s at bus b, period t, and scenario ω (MW).
$r_{sbt\omega}^{S,DUD}$ Up reserve deployed from the discharging of storage unit s at bus b, period t, and scenario ω (MW).
u_{gbt}^{D} Binary variable that is equal to 1 if generating unit g at bus b is online in period t, and 0 otherwise.
y_{gbt}^{D} Binary variable that is equal to 1 if generating unit g at bus b is started up in period t, and 0 otherwise.

z^{D}_{gbt} Binary variable that is equal to 1 if generating unit g at bus b is shut down in period t, and 0 otherwise.

$\lambda^{B}_{bt\omega}$ Dual variable used to compute the real-time energy price at bus b, period t, and scenario ω (€/MWh).

λ^{D}_{bt} Dual variable used to compute the day-ahead energy price at bus b and period t (€/MWh).

$\theta^{B}_{bt\omega}$ Voltage angle of bus b, period t, and scenario ω in the balancing market (rad).

θ^{D}_{bt} Voltage angle of bus b and period t in the day-ahead market (rad).

4.9 GAMS Codes

This section includes the GAMS code of the Illustrative Case Study solved in Sect. 4.5.

```
$title Scheduling model of renewable-dominated power systems

*****************************************************************
* DATA
*****************************************************************

SETS
g          'Generating units'         /Thrm,Wind,SoPV/
b          'Buses'                    /b1/
t          'Time periods'             /t1*t2/
w          'Scenarios'                /w1*w8/

GDis(g)   'Dispatchable units'         /Thrm/
GInt(g)   'Intermittent units'         /Wind,SoPV/
GRes(g)   'Units providing reserve'    /Thrm,Wind/
;

ALIAS(t,k);

SCALARS
Ramp       'Ramping capability (p.u.)'           /0.7/
PGmin      'Minimum power output of the thermal unit (MW)' /50/
KD         'Demand reserve factor (p.u.)'        /0.03/
KI         'Renewable reserve factor (p.u.)'     /0.05/

DemP       'Peak demand (MW)'                    /220/
CDU        'Unserved demand cost (euro/MWh)'     /1000/
;

PARAMETER
prob(w) 'Scenario probability';
prob(w)=1/card(w);

PARAMETERS
CG(g) Operating cost (euro-MWh) /
Thrm     55
Wind     3
SoPV     1
```

```
/
CRU(g) Up reserve capacity cost (euro-MWh)
CRD(g) Down reserve capacity cost (euro-MWh)
CDRU(g) Up reserve deployed cost (euro-MWh)
CDRD(g) Down reserve deployed cost (euro-MWh)
;
CRU(g)=0.05*CG(g);
CRD(g)=0.05*CG(g);
CDRU(g)=1.15*CG(g);
CDRD(g)=0.9*CG(g);

TABLE PGmax(g,b) Capacity of generating units (MW)
        b1
Thrm    150
Wind    100
SoPV    50
;

TABLE DemFD(t,b) Demand factor in day ahead (p.u.)
      b1
t1    0.75
t2    0.90
;
TABLE AFInD(g,t) Wind and solar factor in day ahead (p.u.)
        t1     t2
Wind    0.70   0.55
SoPV    0.20   0.65
;

TABLE DemFB(t,w,b) Demand factor in balancing (p.u.)
          b1
t1.w1     0.7
t1.w2     0.7
t1.w3     0.7
t1.w4     0.7
t1.w5     0.8
t1.w6     0.8
t1.w7     0.8
t1.w8     0.8
t2.w1     0.8
t2.w2     0.8
t2.w3     0.8
t2.w4     0.8
t2.w5     1.0
t2.w6     1.0
t2.w7     1.0
t2.w8     1.0
;

TABLE WindF(w,t) Wind availability (p.u.)
      t1     t2
w1    0.6    0.4
w2    0.6    0.4
w3    0.8    0.6
w4    0.8    0.6
w5    0.6    0.4
w6    0.6    0.4
w7    0.8    0.6
w8    0.8    0.6
;

TABLE SoPVF(w,t) Solar availability (p.u.)
      t1     t2
```

```
w1      0.2     0.6
w2      0.3     0.7
w3      0.2     0.6
w4      0.3     0.7
w5      0.2     0.6
w6      0.3     0.7
w7      0.2     0.6
w8      0.3     0.7
;
PARAMETER AFInB(g,t,w);
AFInB('Wind',t,w)=WindF(w,t);
AFInB('SoPV',t,w)=SoPVF(w,t);

*****************************************************************
* DECLARATION OF VARIABLES AND MATHEMATICAL CHARACTERIZATION
*****************************************************************

VARIABLES

**         OBJECTIVE FUNCTION
z          'Objective function value'
;

POSITIVE VARIABLES

**         OPERATIONAL VARIABLES
pGD(g,b,t)        'Power output in the day-ahead market'
rGCU(g,b,t)       'Up reserve capacity by generating unit'
rGCD(g,b,t)       'Down reserve capacity by generating unit'
pGB(g,b,t,w)      'Power generated in the balancing market'
rGDU(g,b,t,w)     'Up reserve deployed by generating unit'
rGDD(g,b,t,w)     'Down reserve deployed by generating unit'
dUSD(b,t)         'Unserved demand in the day-ahead market'
dUSB(b,t,w)       'Unserved demand in the balancing market'
;

rGCU.fx('SoPV',b,t)=0;
rGCD.fx('SoPV',b,t)=0;

BINARY VARIABLES
uD(g,b,t)        'On/off status'
;

*****************************************************************
*  EQUATIONS
*****************************************************************

EQUATIONS

**         OBJECTIVE FUNCTION
ObjectiveFunction    'Objective function'

**         CONSTRAINTS DAY AHEAD
BalanceDA            'Power balance in the DA'
UpResMinDA           'Up reserve capacity requirement'
DoResMinDA           'Down reserve capacity requirement'
MaxPowGenDA          'Max generated power dispatchable'
MinPowGenDA          'Min generated power dispatchable'
RampUpDA             'Up ramping limit'
RampDoDA             'Down ramping limit'
RampUpDAt1           'Up ramping limit in the first period'
RampDoDAt1           'Down ramping limit in the first period'
MaxPowGenDAIn        'Max generated power intermittent unit'
```

```
UpRerMaxDAIn          'Max up reserve capacity of inter. unit'
DoRerMaxDAIn          'Max down reserve capacity of inter. unit'
DemUnserDA            'Limit on the unserved demand'

**         CONSTRAINTS BALANCING
BalanceBA             'Power balance in the BA'
PowerGenB             'Power generated in the BA'
MaxPowGenB            'Max power generated by dispatchable unit'
MinPowGenB            'Min power generated by dispatchable unit'
RampUpB               'Up ramping limit'
RampDoB               'Down ramping limit'
RampUpBt1             'Up ramping limit in the first period'
RampDoBt1             'Down ramping limit in the first period'
MaxPowGenInB          'Max power generated by intermittent unit'
UpResDeploy           'Max up reserve deployed'
DoResDeploy           'Max down reserve deployed'
DemUnserB             'Limit on the unserved demand'
;

ObjectiveFunction..  z =e= sum((g,b,t),CG(g)*pGD(g,b,t)+CRU(g)*rGCU(g,b
     ,t)+CRD(g)*rGCD(g,b,t))+
sum((b,t),CDU*dUSD(b,t))+
sum((b,t,w),prob(w)*( sum(g,CDRU(g)*rGDU(g,b,t,w)-CDRD(g)*rGDD(g,b,t,w)
     )+ CDU*dUSB(b,t,w) ))
;

**         CONSTRAINTS DAY AHEAD
BalanceDA(b,t)..  sum(g,pGD(g,b,t)) =e= DemFD(t,b)*DemP;

UpResMinDA(b,t)..  sum(g$GRes(g),rGCU(g,b,t)) =g= KD*DemFD(t,b)*DemP+KI
     *sum(GInt(g),AFInD(g,t)*PGmax(g,b) );
DoResMinDA(b,t)..  sum(g$GRes(g),rGCD(g,b,t)) =g= KD*DemFD(t,b)*DemP+KI
     *sum(GInt(g),AFInD(g,t)*PGmax(g,b) );

MaxPowGenDA(g,b,t)$GDis(g)..  pGD(g,b,t)+rGCU(g,b,t) =l=
     uD(g,b,t)*1*PGmax(g,b);
MinPowGenDA(g,b,t)$GDis(g)..  pGD(g,b,t)-rGCD(g,b,t) =g=
     uD(g,b,t)*PGmin;
RampUpDA(g,b,t)$(ord(t)>1 and GDis(g))..
     pGD(g,b,t)- pGD(g,b,t-1) =l= uD(g,b,t)*Ramp*PGmax(g,b);
RampDoDA(g,b,t)$(ord(t)>1 and GDis(g))..
     pGD(g,b,t-1)-pGD(g,b,t) =l= uD(g,b,t)*Ramp*PGmax(g,b);
RampUpDAt1(g,b,t)$(ord(t) eq 1 and GDis(g))..
     pGD(g,b,t)-PGmin =l= uD(g,b,t)*Ramp*PGmax(g,b);
RampDoDAt1(g,b,t)$(ord(t) eq 1 and GDis(g))..
     PGmin-pGD(g,b,t) =l= uD(g,b,t)*Ramp*PGmax(g,b);

MaxPowGenDAIn(g,b,t)$GInt(g)..
     pGD(g,b,t) =l= AFInD(g,t)*PGmax(g,b);
UpRerMaxDAIn(b,t)..
     rGCU('Wind',b,t) =l= 0.30*AFInD('Wind',t)*PGmax('Wind',b);
DoRerMaxDAIn(b,t)..
     rGCD('Wind',b,t) =l= 0.50*AFInD('Wind',t)*PGmax('Wind',b);

DemUnserDA(b,t)..    dUSD(b,t) =l= DemFD(t,b)*DemP;

**         CONSTRAINTS BALANCING
BalanceBA(b,t,w).. sum(g,pGB(g,b,t,w)-pGD(g,b,t))
     +dUSB(b,t,w)-dUSD(b,t) =e= (DemFB(t,w,b)-DemFD(t,b))*DemP;

PowerGenB(g,b,t,w)..  pGB(g,b,t,w) =e= pGD(g,b,t)+rGDU(g,b,t,w)-rGDD(g,
     b,t,w);
MaxPowGenB(g,b,t,w)$GDis(g)..
```

```
     pGB(g,b,t,w) =l= uD(g,b,t)*1*PGmax(g,b);
MinPowGenB(g,b,t,w)$GDis(g)..  pGB(g,b,t,w) =g= uD(g,b,t)*PGmin;
RampUpB(g,b,t,w)$(ord(t) gt 1 and GDis(g))..
     pGB(g,b,t,w)-pGB(g,b,t-1,w) =l= uD(g,b,t)*Ramp*PGmax(g,b);
RampDoB(g,b,t,w)$(ord(t) gt 1 and GDis(g))..
     pGB(g,b,t-1,w)-pGB(g,b,t,w) =l= uD(g,b,t)*Ramp*PGmax(g,b);
RampUpBt1(g,b,t,w)$(ord(t) eq 1 and GDis(g))..
     pGB(g,b,t,w)-PGmin =l= uD(g,b,t)*Ramp*PGmax(g,b);
RampDoBt1(g,b,t,w)$(ord(t) eq 1 and GDis(g))..
     PGmin-pGB(g,b,t,w) =l= uD(g,b,t)*Ramp*PGmax(g,b);
MaxPowGenInB(g,b,t,w)$GInt(g)..
     pGB(g,b,t,w) =l= AFInB(g,t,w)*PGmax(g,b);
UpResDeploy(g,b,t,w)..  rGDU(g,b,t,w) =l= rGCU(g,b,t);
DoResDeploy(g,b,t,w)..  rGDD(g,b,t,w) =l= rGCD(g,b,t);

DemUnserB(b,t,w)..  dUSB(b,t,w) =l= DemFB(t,w,b)*DemP;


*******************************************************************
*  MODEL AND SOLVING
*******************************************************************

MODEL ExampleDispatch /all/;
ExampleDispatch.optcr=0;

SOLVE ExampleDispatch USING mip MINIMIZING z;



*******************************************************************
*  RESULTS
*******************************************************************

PARAMETER priceD(b,t) day-ahead price,
priceB(b,t,w) balancing price;
priceD(b,t)=BalanceDA.m(b,t);
priceB(b,t,w)=BalanceBA.m(b,t,w)/prob(w);
```

References

1. Birge, J.R., Louveaux, F.: Introduction to Stochastic Programming. Springer, New York (1997)
2. Denholm, P., Mai, T., Wallace Kenyon, R., Kroposki, B., O'Malley, M.: Inertia and the Power Grid: A Guide Without the Spin. National Renewable Energy Laboratory, Golden (2020). NREL/TP-6120-73856
3. Domínguez, R., Carrión, M., Oggioni, G.: Planning and operating a renewable-dominated European power system under uncertainty. Appl. Energy **258**, 113989 (2020)
4. Domínguez, R., Oggioni, G., Smeers, Y.: Reserve procurement and flexibility services in power systems with high renewable capacity: effects of integration on different market designs. Inter. J. Elect. Power Energy Syst. **113** (2019)
5. ENTSO-E transparency platform (2021). https://transparency.entsoe.eu/
6. GAMS (2021). www.gams.com
7. Gröwe-Kuska, N., Heitsch, H., Römisch, W.: Scenario reduction and scenario tree construction for power management problems. In: Proceedings IEEE Bologna Power Technology Conference, Bologna, Italy (2003)

8. Hillberg, E., et al.: Flexibility needs in the future power system. Discussion paper. Power Transmission and Distribution Systems. ISGAN Annex 6 Power T&D Systems. ISGAN (2019)
9. International Energy Agency (IEA). World Energy Outlook 2016 (2016). https://www.iea.org/weo/
10. International Energy Agency (IEA). World Energy Outlook 2019, IEA, Paris (2019). https://www.iea.org/reports/world-energy-outlook-2019
11. International Renewable Energy Agency (IRENA). Renewable Power Generation Costs in 2019 (2019). https://www.irena.org/publications/2020/Jun/Renewable-Power-Costs-in-2019
12. Kost, C., et al.: Levelized Cost of Electricity Renewable Energy Technologies, vol. 144. Fraunhofer Institute for Solar Energy Systems ISE (2013)
13. Kumar, N., Besuner, P., Lefton, S., Agan D., Hilleman D.: Power plant cycling costs. National Renewable Energy Laboratory (NREL) (2012). http://www.osti.gov/bridge
14. Milligan, et al.: Operating reserves and wind power integration: an international comparison. NREL (2010). https://www.nrel.gov/docs/fy11osti/49019.pdf
15. Morales, J.M., Conejo, A.J., Liu, K., Zhong, J.: Pricing electricity in pools with wind producers. IEEE Trans. Power Syst. **27**(3), 1366–1376 (2012)
16. Morales, J.M., Pineda, S.: On the inefficiency of the merit order in forward electricity markets with uncertain supply. Eur. J. Oper. Res. **261**, 789–799 (2017)
17. Ortega-Vazquez, M.A., Kirschen, D.S.: Estimating the spinning reserve requirements in systems with significant wind power generation penetration. IEEE Trans. Power Syst. **24**(1), 114–124 (2009)
18. Ostrowski, J., Anjos, M.F., Vannelli, A.: Tight mixed integer linear programming formulations for the unit commitment problem. IEEE Trans. Power Syst. **27**(1), 39–46 (2012)
19. Papavasiliou, A., Oren, S.S., O'Neill, R.P.: Reserve requirements for wind power integration: a scenario-based stochastic programming framework. IEEE Trans. Power Syst. **26**(4), 2197–2206 (2011)
20. Pritchard, G., Zakeri, G., Philpott, A.: A single-settlement, energy-only electric power market for unpredictable and intermittent participants. Oper. Res. **58**(4), 1210–1219 (2010)

Chapter 5
Day-Ahead Market Scheduling Considering Electric Vehicles

This chapter studies the problem of the market clearing with high presence of electric vehicles. In particular, it is presented an optimization model describing the day-ahead market, where energy and reserve capacity is scheduled, and the real-time operation, where the reserve capacity is deployed to counteract the deviations in the demand and renewable power with respect to the predictions made the day ahead of delivery. It is considered that electric vehicles can provide reserve services when connected considering the capacity of the battery and the final state requirement. The uncertainty of the demand, the intermittent renewable power, and the use of the electric vehicle batteries is modeled through a set of scenarios. Therefore, the model is formulated as a mixed-integer two-stage stochastic programming problem. Illustrative examples explain in a simple way the proposed formulations, while a realistic case study based on the power system of Lanzarote and Fuerteventura in Spain provides a deep numerical analysis.

5.1 Introduction

The urgency to drastically reduce the greenhouse gas (GHG) emissions to mitigate the climate change problem is leading to rapidly increasing the use of renewable sources to satisfy the energy needs and the electrification of sectors like the transportation or the heat production. Most of the renewable capacity being installed uses intermittent renewable sources, such as wind or solar power units [9]. This implies a deep change in the operation of electric energy systems since the lack of dispatchability of such resources requires a large enough back-up capacity being able to quickly counteract the variability of those resources [11]. Therefore, it is necessary to increase the flexibility of power systems to efficiently integrate the intermittent renewable power while avoiding fossil fuels.

L. Baringo et al., *Electric Vehicles and Renewable Generation*,
Green Energy and Technology, https://doi.org/10.1007/978-3-031-09079-0_5

Energy storage systems represent a good alternative to provide flexibility to power systems operation [19]. In particular, the improvement of the technology related to Li-ion and other types of electrochemical batteries is allowing the introduction of Electric Vehicles (EVs) in the transportation sector [12]. The availability of a large storage capacity and the electrification of the transportation sector in a renewable-dominated power system represents an extraordinary solution to the global warming.

However, a large expansion of EVs leads to important challenges to the system operation. The control system and the capacity of the distribution network must be robust enough to manage the power flows derived from charging and discharging the batteries. At the same time, smart grids should be significantly developed so that the communication between the EV owners and the distribution companies be fluid, in such a way that both parts can take advantage of the possible benefits of sharing the control of the batteries. Moreover, the investment decisions in renewable and storage capacity may be influenced by the amount and characteristics of EVs [4].

In this chapter, we focus on the operation of the daily electricity market considering a large presence of EVs. A two-stage mixed-integer problem is proposed to represent the day-ahead market and the real-time operation. In the day-ahead market, energy and reserve capacity are scheduled for the 24 h of the following day. The particularity of this problem is that, together with dispatchable generating units, EVs can provide reserve services when connected to the network. Since the publication of the pioneering work [10] that analyzes the economical advantages of using the Vehicle-to-Grid (V2G) capability, the study of the impact of EVs in the power system operation has been an active research topic. In [5], EVs are considered to participate in the day-ahead energy and the reserve markets, whereas [3] analyzes that EVs can also provide primary frequency response after contingencies. Other relevant works analyzing the participation of EVs in ancillary services are [7, 13–15, 18, 20]. Stochastic programming is used to represent the decision-making process and the uncertain parameters of this problem [2], which are the variability of intermittent renewable generation, the power demand, and the level of the EV batteries at the beginning of the charging period. A set of scenarios is generated to represent the possible realizations of those parameters.

The rest of the chapter is organized as follows. Section 5.2 presents the market structure considered in the proposed scheduling model. Section 5.3 describes the different sources of uncertainty considered in this problem and how they are modelled, while the decision framework is explained in Sect. 5.4. Section 5.5 provides a detailed description of the equations modeling the technical operation of EVs. In Sect. 5.6, the complete formulation of the proposed model is presented. Section 5.7 explains the proposed problem through an illustrative case study, while Sect. 5.8 provides numerical analyses based on a realistic case study. Section 5.9 ends the chapter with a summary and some conclusions. In Sect. 5.10 the list of symbols used in the problem formulation is explained. Finally, Sect. 5.11 provides the GAMS code used to solve the illustrative case study of this chapter.

5.2 Market Structure

The operation of a power system is a complex problem that can be divided in several stages. Firstly, a market-clearing procedure is performed. The market-clearing is usually a day-ahead market that is realized the day prior to the actual energy delivery on an hourly basis. The objective of the market-clearing is to determine for each hour of the next day: (i) the energy produced by each power unit, (ii) the energy consumed by each demand and, (iii) the market-clearing prices. Next, short-time markets (spanning from minutes to hours) as adjustment and ancillary service markets are cleared in order to adjust the power production to the actual consumption needs in each instant, while the technical constraints of the production units, loads, and transmission lines are satisfied. These markets are of high interest in power systems with either numerous uncertain production units or with uncertain loads. However, it is important to note that most part of the traded energy is negotiated in the day-ahead market.

5.3 Uncertainty Characterization

The uncertainty sources that should be taken into account in power system scheduling tasks are mainly the production of wind and solar power plants, and the power demand, whether it is conventional demand or the demand related to the charge of EVs. These uncertain parameters are represented as stochastic processes that are discretized into a set of scenarios [2]. Each scenario is a sequence of values of the parameters that represents a plausible realization of these stochastic processes.

Stochastic processes are constructed on the basis of the statistical properties of the considered parameter. Such statistical properties are inferred from historical data.

With respect to EVs, and assuming that the users of the vehicles require to have their batteries fully charged at the end of the charging periods, the uncertainty affects the time periods in which the vehicles are connected to the grid, and the initial state-of-charge (SOC) of their batteries, which determines the total energy they need and limit the energy they can provide.

The uncertainty regarding the initial SOC of the batteries is represented as follows:

- The daily distance driven by EVs is characterized by a probability distribution, which is identified by analyzing historical data. The probability distribution is discretized into a finite set of scenarios. Each scenario has information of daily distance driven and it has an occurrence probability [2].
- Accounting for both the daily driven distance and the average electricity consumption per driven kilometer, the daily energy consumption of each EV type is computed for each scenario. Using this information, the initial SOC of the batteries at the beginning of the charging period is derived.

Moreover, EVs are categorized in different types, $k \in K$, in such a way that EVs with similar usage patterns are assigned to the same type. An EV type is characterized by the daily driven distance and by the set of hours in which EVs are available to be charged. In this way, all EVs belonging to the same type are assumed to have similar initial battery SOC and connecting periods.

5.4 Decision Framework

The day-ahead market-clearing model determines:

- For each generating unit g and each period t, the day-ahead energy output, $p_{gt}^{\mathrm{G,D}}$, and the up and down reserve capacities, $r_{gt}^{\mathrm{G,CU}}$ and $r_{gt}^{\mathrm{G,CD}}$, respectively.
- For each EV group k in each bus b and period t, the energy sold and purchased in the day-ahead market, $p_{kbt}^{\mathrm{EV,DD}}$ and $p_{kbt}^{\mathrm{EV,CD}}$, and the up and down reserve capacities, $r_{kbt}^{\mathrm{EV,CU}}$ and $r_{kbt}^{\mathrm{EV,CD}}$.

The scheduled up and down reserve capacities represent the maximum quantity of power that generating units and EVs are forced to deploy upwards and downwards if the demand or the renewable power output differs from their expected values in the real-time operation. Therefore, this capacity is dedicated to deploy the operating reserve used to follow hourly deviations of intermittent production and demand.

EV users can implicitly incorporate estimations about their battery degradation costs when participating in the markets. In this manner, EVs will only participate in the different trading floors if that participation is economically profitable for them. As an example, references [16, 17] propose different procedures for determining the optimal bidding strategy of an aggregation of EVs participating in different day-ahead electricity markets. In particular, reference [16] explicitly considers the battery degradation costs for deciding the bidding strategy of EVs.

The market-clearing procedure is formulated as a two-stage stochastic programming problem, where the first-stage represents the day-ahead market in which the scheduling of energy and reserve capacities are assigned to the different market agents. The real-time operation is characterized in the second-stage in which the deployment of the reserves is determined for a number of realizations of the uncertain parameters. Co-optimizing simultaneously energy and reserve capacity in the day ahead market allows the system to have enough flexibility to face the possible deviations in the forecasted demand and intermittent renewable power.

5.5 Modeling of the Electric Vehicles Operation

For the sake of clarity, this section details the mathematical formulation describing the operation of EVs. The description of the symbols used below is included in Sect. 5.10.

5.5.1 *Initial Status of the Batteries of EVs*

The initial status of the batteries of the EVs belonging to group k is formulated through (5.1). As explained in Sect. 5.3, the initial SOC of the batteries depends on the number of kilometers driven along the day. Thus, parameter $\alpha_{k\omega}^{\text{EV,SI}}$ is a random parameter that takes a different value in each scenario. Note that this equation defines the energy level in the batteries in the time period prior to the charging period.

$$e_{kbt\omega}^{\text{EV}} = N_{kb}^{\text{EV}} \alpha_{k\omega}^{\text{EV,SI}} E_{\max,k}^{\text{EV}}, \quad \forall k \in K, \forall b \in B, \forall \omega \in \Omega, t = t_k^{\text{I}} - 1. \quad (5.1)$$

Illustrative Example 5.1 Initial status of the batteries of EVs

Let us consider two groups of EVs, $K = \{1, 2\}$, that are connected to bus $b = 1$ with 100 and 200 EVs, respectively. The battery size of all vehicles is $E_{\max,k}^{\text{EV}} = 50$ kWh, $\forall k \in \{1, 2\}$. We assume a planning horizon of four periods. EVs belonging to group $k = 1$ are connected to the grid in periods 2 and 3, whereas EVs in group $k = 2$ are connected in periods 3 and 4. Therefore, parameters $t_1^{\text{I}} = 2$ and $t_2^{\text{I}} = 3$. The initial status of the batteries of EVs is modeled by two scenarios, $\Omega = \{1, 2\}$. Then, the initial status of the batteries of EVs of group $k = 1$ may be 0.20 or 0.28, and 0.50 or 0.55 for EVs in group $k = 2$. Considering this, the initial status of the batteries of both groups of EVs is formulated as follows:

(Group $k = 1$)

$$\text{Scenario } \omega = 1 : \; e_{1111}^{\text{EV}} = 100 \cdot 0.20 \cdot 50 = 1000\,\text{kWh},$$
$$\text{Scenario } \omega = 2 : \; e_{1112}^{\text{EV}} = 100 \cdot 0.28 \cdot 50 = 1400\,\text{kWh}.$$

(Group $k = 2$)

$$\text{Scenario } \omega = 1 : e_{2121}^{\text{EV}} = 200 \cdot 0.50 \cdot 50 = 5000\,\text{kWh},$$
$$\text{Scenario } \omega = 2 : e_{2122}^{\text{EV}} = 200 \cdot 0.55 \cdot 50 = 5500\,\text{kWh}.$$

□

5.5.2 *Operation of the Batteries of EVs*

Constraints (5.2) model the SOC of the EV batteries at the real-time operation. It is considered that the batteries of the EVs can be also used to inject energy into the network using the V2G capability. In this manner, EVs can participate either as power consumers or producers depending on if they are charging o discharging power. This may happen during the periods when EVs are connected to the network, represented by set T_k. Considering this, the level of energy contained in period t is equal to the level of energy contained in the previous period, plus the energy charged in period t considering the losses during the charging, minus the energy discharged from the battery in period t taking into account the losses during the process. Note that if one hour periods are considered, power (W) and energy (Wh) units are equivalent.

$$e^{\mathrm{EV}}_{kbt\omega} = e^{\mathrm{EV}}_{kbt-1\omega} + \eta^{\mathrm{EV}} p^{\mathrm{EV,BC}}_{kbt\omega} - \frac{1}{\eta^{\mathrm{EV}}} p^{\mathrm{EV,BD}}_{kbt\omega}, \quad \forall k \in K, \forall b \in B, \forall \omega \in \Omega, t \in T_k. \tag{5.2}$$

A minimum energy level is imposed at the end of the charging period as stated by (5.3a). In this case, parameter $\alpha^{\mathrm{EV,SE}}_{k\omega}$ represents the ending status of the EV battery. Constraints (5.3b) impose limits on the minimum and maximum energy level of the battery during the charging period.

$$e^{\mathrm{EV}}_{kbt\omega} \geq N^{\mathrm{EV}}_{kb} \alpha^{\mathrm{EV,SE}}_{k\omega} E^{\mathrm{EV}}_{\max,k}, \quad \forall k \in K, \forall b \in B, \forall \omega \in \Omega, t = t^{\mathrm{F}}_k, \tag{5.3a}$$

$$N^{\mathrm{EV}}_{kb} \alpha^{\mathrm{EV,SMIN}}_{k} E^{\mathrm{EV}}_{\max,k} \leq e^{\mathrm{EV}}_{kbt\omega} \leq N^{\mathrm{EV}}_{kb} E^{\mathrm{EV}}_{\max,k}, \quad \forall k \in K, \forall b \in B, \forall \omega \in \Omega, t \in T_k. \tag{5.3b}$$

Illustrative Example 5.2 Operation of the batteries of EVs

Let us consider the input data described in Illustrative Example 5.1. Let us also consider that the efficiency of the charge and discharge of vehicles, η^{EV}, is equal to 0.9, that the final status of the batteries at the end of the charging period must be, at least, 80% of the capacity of the batteries, and that the energy stored in the batteries cannot be less than 15% of the capacity. If group $k = 1$ charges 4000 kWh in period 2 and discharges 500 kWh in period 3, in scenarios 1 and 2, the energy stored at each period and scenario is the following:
(*Group* $k = 1$)

Scenario $\omega = 1$, Period $t = 2$:

$$e^{\mathrm{EV}}_{1121} = 1000 + 0.9 \cdot 4000 - \frac{1}{0.9} \cdot 0 = 4600\,\mathrm{kWh},$$

Scenario $\omega = 1$, Period $t = 3$:

$$e^{\mathrm{EV}}_{1131} = 4600 + 0.9 \cdot 0 - \frac{1}{0.9} \cdot 500 = 4044.44\,\mathrm{kWh},$$

Scenario $\omega = 2$, Period $t = 2$:

$$e^{\mathrm{EV}}_{1122} = 1400 + 0.9 \cdot 4000 - \frac{1}{0.9} \cdot 0 = 5000\,\mathrm{kWh},$$

$$\text{Scenario } \omega = 2, \text{ Period } t = 3:$$
$$e_{1132}^{\mathrm{EV}} = 5000 + 0.9 \cdot 0 - \frac{1}{0.9} \cdot 500 = 4444.44\,\mathrm{kWh}.$$

Considering that the maximum energy that can be stored in the batteries of group $k = 1$ is equal to $N_{11}^{\mathrm{EV}} E_{\max,1}^{\mathrm{EV}} = 100 \cdot 50 = 5000$ kWh, it is observed that the energy stored at the end of the charging period, $t = 3$, in both scenarios is greater than the minimum required value, that is $N_{11}^{\mathrm{EV}} \alpha_{1\omega}^{\mathrm{EV,SE}} E_{\max,1}^{\mathrm{EV}} = 100 \cdot 0.8 \cdot 50 = 4000$ kWh, (5.3a). It is also observed that the energy stored in period $t = 2$ and scenario $\omega = 2$ is equal to the maximum value, 5000 kWh, (5.3b). Finally, it is also noted that the energy stored in each period and scenario is higher than the minimum limit $N_{11}^{\mathrm{EV}} \alpha_{1}^{\mathrm{EV,SMIN}} E_{\max,1}^{\mathrm{EV}} = 100 \cdot 0.15 \cdot 50 = 750$ kWh, (5.3b).

□

5.5.3 *Reserve Deployment by EVs*

It is assumed that EVs can provide reserve services in the real-time operation by charging and discharging the batteries regarding whether the system operator asks to reduce or increase the energy supply to the network, respectively. The up and down reserves deployed by EVs, $r_{kbt\omega}^{\mathrm{EV,DU}}$, $r_{kbt\omega}^{\mathrm{EV,DD}}$ are bounded by the reserve capacities scheduled in the day-ahead market, $r_{kbt}^{\mathrm{EV,CU}}$, $r_{kbt}^{\mathrm{EV,CD}}$. This is formulated by (5.4) as follows:

$$0 \leq r_{kbt\omega}^{\mathrm{EV,DU}} \leq r_{kbt}^{\mathrm{EV,CU}}, \quad \forall k \in K, \forall b \in B, \forall t \in T_k, \tag{5.4a}$$
$$0 \leq r_{kbt\omega}^{\mathrm{EV,DD}} \leq r_{kbt}^{\mathrm{EV,CD}}, \quad \forall k \in K, \forall b \in B, \forall t \in T_k. \tag{5.4b}$$

The power charged and discharged by the EVs of group k is defined according to (5.5). The power charged in the real-time scenario ω is equal to the power to be charged scheduled in the day-ahead market, minus the up reserve deployed by reducing the charging of the battery, plus the down reserve deployed by increasing the charging of the battery. On the other side, the power discharged in the real-time scenario ω is equal to the power to be discharged scheduled in the day-ahead market, plus the up reserve deployed by increasing the discharging of the battery, minus the down reserve deployed by reducing the discharging of the battery.

$$p_{kbt\omega}^{\mathrm{EV,BC}} = p_{kbt}^{\mathrm{EV,CD}} - r_{kbt\omega}^{\mathrm{EV,DUC}} + r_{kbt\omega}^{\mathrm{EV,DDC}}, \quad \forall k \in K, \forall b \in B, \forall \omega \in \Omega, \forall t \in T_k, \tag{5.5a}$$
$$p_{kbt\omega}^{\mathrm{EV,BD}} = p_{kbt}^{\mathrm{EV,DD}} + r_{kbt\omega}^{\mathrm{EV,DUD}} - r_{kbt\omega}^{\mathrm{EV,DDC}}, \quad \forall k \in K, \forall b \in B, \forall \omega \in \Omega, \forall t \in T_k. \tag{5.5b}$$

The power charged and discharged by EVs in the real-time operation are bounded by constraints (5.6). Parameter $P^{\text{EV}}_{\max,k}$ indicates the rate power capacity of EV chargers belonging to group k.

$$0 \leq p^{\text{EV,BC}}_{kbt\omega} \leq N^{\text{EV}}_{kb} P^{\text{EV}}_{\max,k}, \quad \forall k \in K, \forall b \in B, \forall \omega \in \Omega, \forall t \in T_k, \tag{5.6a}$$

$$0 \leq p^{\text{EV,BD}}_{kbt\omega} \leq N^{\text{EV}}_{kb} P^{\text{EV}}_{\max,k}, \quad \forall k \in K, \forall b \in B, \forall \omega \in \Omega, \forall t \in T_k. \tag{5.6b}$$

Equation (5.7a) defines the total up reserve deployed by the EVs of group k, which comprises the up reserve deployed by charging and discharging modes. Constraints (5.7b) establish that the up reserved deployed from the charging cannot be higher than the power to be charged scheduled in the day-ahead market. Constraints (5.7c) imposes that the up reserved deployed from discharging the batteries must be lower than the total battery capacity of group k minus the power to be discharged scheduled in the day-ahead market.

$$r^{\text{EV,DU}}_{kbt\omega} = r^{\text{EV,DUC}}_{kbt\omega} + r^{\text{EV,DUD}}_{kbt\omega}, \quad \forall k \in K, \forall b \in B, \forall \omega \in \Omega, \forall t \in T_k, \tag{5.7a}$$

$$0 \leq r^{\text{EV,DUC}}_{kbt\omega} \leq p^{\text{EV,CD}}_{kbt}, \quad \forall k \in K, \forall b \in B, \forall \omega \in \Omega, \forall t \in T_k, \tag{5.7b}$$

$$0 \leq r^{\text{EV,DUD}}_{kbt\omega} \leq N^{\text{EV}}_{kb} P^{\text{EV}}_{\max} - p^{\text{EV,DD}}_{kbt}, \quad \forall k \in K, \forall b \in B, \forall \omega \in \Omega, \forall t \in T_k. \tag{5.7c}$$

Finally, constraints (5.8) describe the down reserve deployed by EVs of group k. Equation (5.8a) defines the down reserve deployed as the sum of the reserve deployed from the charging and the discharging. Constraints (5.8b) limit the down reserve deployed from the charging to the maximum capacity of the batteries of group k minus the power to be charged scheduled in the day-ahead market. Constraints (5.8c) state that the down reserve deployed from the discharging cannot be higher than the power to be discharged scheduled in the day-ahead market.

$$r^{\text{EV,DD}}_{kbt\omega} = r^{\text{EV,DDC}}_{kbt\omega} + r^{\text{EV,DDD}}_{kbt\omega}, \quad \forall k \in K, \forall b \in B, \forall \omega \in \Omega, \forall t \in T_k, \tag{5.8a}$$

$$0 \leq r^{\text{EV,DDC}}_{kbt\omega} \leq N^{\text{EV}}_{kb} P^{\text{EV}}_{\max} - p^{\text{EV,CD}}_{kbt}, \quad \forall k \in K, \forall b \in B, \forall \omega \in \Omega, \forall t \in T_k, \tag{5.8b}$$

$$0 \leq r^{\text{EV,DDD}}_{kbt\omega} \leq p^{\text{EV,DD}}_{kbt}, \quad \forall k \in K, \forall b \in B, \forall \omega \in \Omega, \forall t \in T_k. \tag{5.8c}$$

Illustrative Example 5.3 Reserve deployment by EVs

Let us consider an EV group, denoted by $k = 1$, connected to bus $b = 1$ comprising $N^{\text{EV}}_{11} = 300$ EVs, with a rated power capacity per vehicle equal to $P^{\text{EV}}_{\max,1} = 7.4$ kW. This group of EVs participates in the day-ahead market charging 1000 kWh in period 1, $p^{\text{EV,CD}}_{111}$, and discharging 500 kWh in period $t = 2$, $p^{\text{EV,DD}}_{112}$. Additionally, this group of EVs provides 200 kW of up-reserve capacity, $r^{\text{EV,CU}}_{11t}$, and 500 kW of down-reserve capacity, $r^{\text{EV,CD}}_{11t}$, in both periods, $t = \{1, 2\}$.

In the case that the deployment of 200 kW of up-reserve is required during period $t = 1$, the EV group can provide it by reducing its charge, $r^{\text{EV,DUC}}_{111\omega}$, from 1000 to 800 kWh in that period, (5.7b). If this EV group has to provide 500 kW of down-reserve during the whole period $t = 1$, it can increase its charge, $r_{111\omega}^{\text{EV,DDC}}$, from 1000 to 1500 kWh, (5.8b). Observe that a charge rate equal to 1500 kW is less than the rated capacity which is equal to $N_{11}^{\text{EV}} P_{\text{max},1}^{\text{EV}} = 300 \cdot 7.4 = 2200$ kW, (5.6a).

Considering period $t = 2$, if the deployment of 200 kW of up-reserve is needed during one hour, EVs can deploy it by increasing the discharge, $r_{111\omega}^{\text{EV,DUD}}$, from 500 to 700 kWh, (5.7c). If this EV group has to provide 500 kW of down-reserve, $r_{111\omega}^{\text{EV,DDD}}$, it can stop the discharge of 500 kWh during that period, (5.8c). □

5.6 Model Formulation

This section presents the full mathematical formulation of the problem described above. The following assumptions have been considered:

- The perspective of the System Operator (SO) is considered and it is assumed that its objective consists in minimizing the total expected production cost.
- The demand is assumed to be inelastic. Therefore, it is also considered that the owners of EVs desire to charge their vehicles up to a desired level, regardless of electricity prices.
- Unit commitment variables for thermal units are included, allowing to describe the minimum power output and the start-up and shout-down costs of such units.
- The generation-side offer curves are assumed as linear.
- Reserve capacity can be provided only by dispatchable thermal units, not by intermittent renewable units. EVs can also provide reserve services when connected due to the flexibility given by their batteries.
- Reserve capacity needs are computed endogenously as a function of the considered scenarios. Therefore, unlike the model presented in Chap. 4, constraints formulating minimum reserve requirements are not considered.
- The transmission network is represented using a DC formulation, as usually done in this type of problems [6].
- For simplicity, storage units are not considered in this formulation. The description of the modeling of storage units can be consulted in Chap. 4.

5.6.1 Objective Function

The objective function to be minimized is expressed by (5.9). As a consequence of considering an inelastic system demand, maximizing the expected social welfare is equivalent to minimizing the expected operation cost. Based on this, objective function (5.9) comprises (i) the start-up, shut-down, scheduled energy, and up and down reserve capacity costs of generating units in the day-ahead market, (ii) the charged and discharged energy to EVs scheduled in the day-ahead market as well as the up and down reserve capacity to be provided, (iii) the expected cost of deploying up and down reserve by generating units and EVs in the real-time operation, and (iv) the expected cost of the involuntary unserved demand.

$$
\begin{aligned}
&\text{Minimize}_{\Phi} \\
&\sum_{t \in T} \sum_{g \in G} \left(c_{gt}^{\mathrm{G,SU}} + c_{gt}^{\mathrm{G,SD}} + C_{gt}^{\mathrm{G,D}} p_{gt}^{\mathrm{G,D}} + C_{gt}^{\mathrm{G,RCU}} r_{gt}^{\mathrm{G,CU}} + C_{gt}^{\mathrm{G,RCD}} r_{gt}^{\mathrm{G,CD}}\right) \\
&+ \sum_{t \in T} \sum_{b \in B} \sum_{k \in K_t} \Big(C_{kbt}^{\mathrm{EV,DD}} p_{kbt}^{\mathrm{EV,DD}} - C_{kbt}^{\mathrm{EV,CD}} p_{kbt}^{\mathrm{EV,CD}} \\
&\quad + C_{kbt}^{\mathrm{EV,RCU}} r_{kbt}^{\mathrm{EV,CU}} + C_{kbt}^{\mathrm{EV,RCD}} r_{kbt}^{\mathrm{EV,CD}} \Big) \\
&+ \sum_{t \in T} \sum_{\omega \in \Omega} \pi_{\omega} \Bigg(\sum_{g \in G^{\mathrm{D}}} \left(C_{gt\omega}^{\mathrm{G,RDU}} r_{gt\omega}^{\mathrm{G,DU}} - C_{gt\omega}^{\mathrm{G,RDD}} r_{gt\omega}^{\mathrm{G,DD}}\right) \\
&\quad + \sum_{b \in B} \sum_{k \in K_t} \left(C_{kbt\omega}^{\mathrm{EV,RDU}} r_{kbt\omega}^{\mathrm{EV,DU}} - C_{kbt\omega}^{\mathrm{EV,RDD}} r_{kbt\omega}^{\mathrm{EV,DD}} \right) + \sum_{b \in B} C^{\mathrm{UD}} p_{bt\omega}^{\mathrm{UD,B}} \Bigg)
\end{aligned} \tag{5.9}
$$

where $\Phi = \{c_{gt}^{\mathrm{G,SU}}, c_{gt}^{\mathrm{G,SD}}, p_{gt}^{\mathrm{G,D}}, r_{gt}^{\mathrm{G,CD}}, r_{gt}^{\mathrm{G,CU}}, \forall g \in G, \forall t \in T;\ p_{gt\omega}^{\mathrm{G,B}}, p_{gt\omega}^{\mathrm{G,BS}}, r_{gt\omega}^{\mathrm{G,DD}}, r_{gt\omega}^{\mathrm{G,DU}}, \forall g \in G, \forall t \in T, \forall \omega \in \Omega;\ p_{kbt}^{\mathrm{EV,CD}}, p_{kbt}^{\mathrm{EV,DD}}, r_{kbt}^{\mathrm{EV,CD}}, r_{kbt}^{\mathrm{EV,CU}}, \forall k \in K, \forall b \in B, \forall t \in T; e_{kbt\omega}^{\mathrm{EV}} p_{kbt\omega}^{\mathrm{EV,BC}}, p_{kbt\omega}^{\mathrm{EV,BD}} r_{kbt\omega}^{\mathrm{EV,DD}}, r_{kbt\omega}^{\mathrm{EV,DDC}}, r_{kbt\omega}^{\mathrm{EV,DDD}}, r_{kbt\omega}^{\mathrm{EV,DU}}, r_{kbt\omega}^{\mathrm{EV,DUC}}, r_{kbt\omega}^{\mathrm{EV,DUD}}, \forall k \in K, \forall b \in B, \forall t \in T, \forall \omega \in \Omega;\ p_{lt\omega}^{\mathrm{L,D}}, \forall l \in L, \forall t \in T;\ p_{lt\omega}^{\mathrm{L,B}}, \forall l \in L, \forall t \in T, \forall \omega \in \Omega;\ \theta_{bt}^{\mathrm{D}}, \forall b \in B, \forall t \in T;\ p_{bt\omega}^{\mathrm{UD,B}}, \theta_{bt\omega}^{\mathrm{B}}, \forall b \in B, \forall t \in T, \forall \omega \in \Omega\}$ is the set of optimization variables.

5.6.2 Constraints

The objective function (5.9) is subject to the following set of constraints:

- Day-ahead market constraints:

(Power balance)

$$
\begin{aligned}
&\sum_{g \in G_b} p_{gt}^{\mathrm{G,D}} + \sum_{k \in K_t} \left(p_{kbt}^{\mathrm{EV,DD}} - p_{kbt}^{\mathrm{EV,CD}} \right) - \sum_{\ell \in L_b^{\mathrm{O}}} p_{\ell t}^{\mathrm{L,D}} + \sum_{\ell \in L_b^{\mathrm{F}}} p_{\ell t}^{\mathrm{L,D}} = P_{bt}^{\mathrm{D,D}}, \\
&\forall b \in B, \forall t \in T,
\end{aligned} \tag{5.10}
$$

(Technical constraints of generating units)

$$u_{gt} P^{\mathrm{G}}_{\min,g} \leq p^{\mathrm{G,D}}_{gt} \leq u_{gt} P^{\mathrm{G}}_{\max,g}, \quad \forall g \in G^{\mathrm{D}}, \forall t \in T, \tag{5.11a}$$

$$p^{\mathrm{G,D}}_{gt} - p^{\mathrm{G,D}}_{gt-1} \leq P^{\mathrm{G}}_{\mathrm{UP},g}, \quad \forall g \in G^{\mathrm{D}}, \forall t \in T, \tag{5.11b}$$

$$p^{\mathrm{G,D}}_{gt-1} - p^{\mathrm{G,D}}_{gt} \leq P^{\mathrm{G}}_{\mathrm{DW},g}, \quad \forall g \in G^{\mathrm{D}}, \forall t \in T, \tag{5.11c}$$

$$0 \leq p^{\mathrm{G,D}}_{gt} \leq A^{\mathrm{D}}_{gdt} P^{\mathrm{G}}_{\max,g}, \quad \forall g \in G^{\mathrm{I}}, \forall t \in T, \tag{5.11d}$$

$$c^{\mathrm{G,SU}}_{gt} = C^{\mathrm{G,SU}}_{g}(u_{gt} - u_{g,t-1}), \forall g \in G^{\mathrm{D}}, \forall t \in T, \tag{5.11e}$$

$$c^{\mathrm{G,SD}}_{gt} = C^{\mathrm{G,SD}}_{g}(u_{g,t-1} - u_{gt}), \forall g \in G^{\mathrm{D}}, \forall t \in T, \tag{5.11f}$$

$$c^{\mathrm{G,SU}}_{gt}, c^{\mathrm{G,SD}}_{gt} \geq 0, \forall g \in G^{\mathrm{D}}, \forall t \in T, \tag{5.11g}$$

(Technical constraints of EVs)

$$0 \leq p^{\mathrm{EV,CD}}_{kbt} \leq N^{\mathrm{EV}}_{kb} P^{\mathrm{EV}}_{\max,k}, \quad \forall k \in K, \forall b \in B, \forall t \in T_k, \tag{5.12a}$$

$$0 \leq p^{\mathrm{EV,DD}}_{kbt} \leq N^{\mathrm{EV}}_{kb} P^{\mathrm{EV}}_{\max,k}, \quad \forall k \in K, \forall b \in B, \forall t \in T_k, \tag{5.12b}$$

(Reserve capacity limits)

$$0 \leq r^{\mathrm{G,CU}}_{gt} \leq u_{gt} P^{\mathrm{G}}_{\max,g} - p^{\mathrm{G,D}}_{g,t}, \quad \forall g \in G^{\mathrm{D}}, \forall t \in T, \tag{5.13a}$$

$$0 \leq r^{\mathrm{G,CD}}_{gt} \leq p^{\mathrm{G,D}}_{g,t}, \quad \forall g \in G^{\mathrm{D}}, \forall t \in T, \tag{5.13b}$$

$$r^{\mathrm{EV,DU}}_{kbt} \geq 0, \quad \forall k \in K, \forall b \in B, \forall t \in T_k, \tag{5.13c}$$

$$r^{\mathrm{EV,DD}}_{kbt} \geq 0, \quad \forall k \in K, \forall b \in B, \forall t \in T_k, \tag{5.13d}$$

(Network power flow constraints)

$$p^{\mathrm{L,D}}_{\ell t} = \frac{1}{X_\ell}\left(\theta^{\mathrm{D}}_{O(\ell)t} - \theta^{\mathrm{D}}_{F(\ell)t}\right), \quad \forall \ell \in L, \forall t \in T, \tag{5.14a}$$

$$-P^{\mathrm{L}}_{\max,\ell} \leq p^{\mathrm{L,D}}_{\ell t} \leq P^{\mathrm{L}}_{\max,\ell}, \quad \forall \ell \in L, \forall t \in T, \tag{5.14b}$$

$$-\pi \leq \theta^{\mathrm{D}}_{bt} \leq \pi, \quad \forall b \in B, \forall t \in T, \tag{5.14c}$$

- Real-time operation constraints
  (Power balance)

$$\Big\{ \sum_{g \in G_b} p^{\mathrm{G,B}}_{gt\omega} + \sum_{k \in K_t} \left(p^{\mathrm{EV,BD}}_{kbt\omega} - p^{\mathrm{EV,BC}}_{kbt\omega}\right) - \sum_{\ell \in L^{\mathrm{O}}_b} p^{\mathrm{L,B}}_{\ell t\omega} + \sum_{\ell \in L^{\mathrm{F}}_b} p^{\mathrm{L,B}}_{\ell t\omega} = P^{\mathrm{D,B}}_{bt\omega} - p^{\mathrm{UD,B}}_{bt\omega}, \quad \forall b \in B, \forall t \in T, \tag{5.15}$$

(Limit on the unserved demand)

$$0 \leq p^{\mathrm{UD,B}}_{dt\omega} \leq P^{\mathrm{D,B}}_{dt\omega}, \quad \forall d \in D, \forall t \in T, \tag{5.16}$$

(Technical constraints of generating units)

$$p_{gt\omega}^{\mathrm{G,B}} = p_{gt}^{\mathrm{G,D}} + r_{gt\omega}^{\mathrm{G,DU}} - r_{gt\omega}^{\mathrm{G,DD}}, \quad \forall g \in G^{\mathrm{D}}, \forall t \in T, \tag{5.17a}$$

$$u_{gt} P_{\min,g}^{\mathrm{G}} \le p_{gt\omega}^{\mathrm{G,B}} \le u_{gt} P_{\max,g}^{\mathrm{G}}, \ \forall g \in G^{\mathrm{D}}, \forall t \in T, \tag{5.17b}$$

$$p_{gt\omega}^{\mathrm{G,B}} - p_{gt-1,\omega}^{\mathrm{G,B}} \le P_{\mathrm{UP},g}^{\mathrm{G}}, \quad \forall g \in G^{\mathrm{D}}, \forall t \in T, \tag{5.17c}$$

$$p_{gt-1,\omega}^{\mathrm{G,B}} - p_{gt\omega}^{\mathrm{G,B}} \le P_{\mathrm{DW},g}^{\mathrm{G}}, \quad \forall g \in G^{\mathrm{D}}, \forall t \in T, \tag{5.17d}$$

$$p_{gt\omega}^{\mathrm{G,B}} + p_{gt\omega}^{\mathrm{G,BS}} = A_{gt\omega}^{\mathrm{B}} P_{\max,g}^{\mathrm{G}}, \quad \forall g \in G^{\mathrm{I}}, \forall t \in T, \tag{5.17e}$$

$$p_{gt\omega}^{\mathrm{G,B}} \ge 0, \quad \forall g \in G^{\mathrm{I}}, \forall t \in T \tag{5.17f}$$

$$p_{gt\omega}^{\mathrm{G,BS}} \ge 0, \quad \forall g \in G^{\mathrm{I}}, \forall t \in T, \tag{5.17g}$$

(Technical constraints of EVs)

$$e_{kbt\omega}^{\mathrm{EV}} = N_{kb}^{\mathrm{EV}} \alpha_{k\omega}^{\mathrm{EV,SI}} E_{\max,k}^{\mathrm{EV}}, \quad \forall k \in K, \forall b \in B, t = t_k^{\mathrm{I}} - 1, \tag{5.18a}$$

$$e_{kbt\omega}^{\mathrm{EV}} = e_{kbt-1\omega}^{\mathrm{EV}} + \eta^{\mathrm{EV}} p_{kbt\omega}^{\mathrm{EV,BC}} - \frac{1}{\eta^{\mathrm{EV}}} p_{kbt\omega}^{\mathrm{EV,BD}}, \forall k \in K, \forall b \in B, t \in T_k, \tag{5.18b}$$

$$e_{kbt\omega}^{\mathrm{EV}} \ge N_{kb}^{\mathrm{EV}} \alpha_{k\omega}^{\mathrm{EV,SE}} E_{\max,k}^{\mathrm{EV}}, \quad \forall k \in K, \forall b \in B, t = t_k^{\mathrm{F}} \tag{5.18c}$$

$$N_{kb}^{\mathrm{EV}} \alpha_k^{\mathrm{EV,SMIN}} E_{\max,k}^{\mathrm{EV}} \le e_{kbt\omega}^{\mathrm{EV}} \le N_{kb}^{\mathrm{EV}} E_{\max,k}^{\mathrm{EV}}, \forall k \in K, \forall b \in B, t \in T_k, \tag{5.18d}$$

$$p_{kbt\omega}^{\mathrm{EV,BC}} = p_{kbt}^{\mathrm{EV,CD}} - r_{kbt\omega}^{\mathrm{EV,DUC}} + r_{kbt\omega}^{\mathrm{EV,DDC}}, \quad \forall k \in K, \forall b \in B, \forall t \in T_k, \tag{5.18e}$$

$$p_{kbt\omega}^{\mathrm{EV,BD}} = p_{kbt}^{\mathrm{EV,DD}} + r_{kbt\omega}^{\mathrm{EV,DUD}} - r_{kbt\omega}^{\mathrm{EV,DDC}}, \quad \forall k \in K, \forall b \in B, \forall t \in T_k, \tag{5.18f}$$

$$0 \le p_{kbt\omega}^{\mathrm{EV,BC}} \le N_{kb}^{\mathrm{EV}} P_{\max,k}^{\mathrm{EV}}, \quad \forall k \in K, \forall b \in B, \forall t \in T_k, \tag{5.18g}$$

$$0 \le p_{kbt\omega}^{\mathrm{EV,BD}} \le N_{kb}^{\mathrm{EV}} P_{\max,k}^{\mathrm{EV}}, \quad \forall k \in K, \forall b \in B, \forall t \in T_k \tag{5.18h}$$

$$r_{kbt\omega}^{\mathrm{EV,DU}} = r_{kbt\omega}^{\mathrm{EV,DUC}} + r_{kbt\omega}^{\mathrm{EV,DUD}}, \quad \forall k \in K, \forall b \in B, \forall t \in T_k, \tag{5.18i}$$

$$0 \le r_{kbt\omega}^{\mathrm{EV,DUC}} \le p_{kbt}^{\mathrm{EV,CD}}, \quad \forall k \in K, \forall b \in B, \forall t \in T_k, \tag{5.18j}$$

$$0 \le r_{kbt\omega}^{\mathrm{EV,DUD}} \le N_{kb}^{\mathrm{EV}} P_{\max}^{\mathrm{EV}} - p_{kbt}^{\mathrm{EV,DD}}, \quad \forall k \in K, \forall b \in B, \forall t \in T_k, \tag{5.18k}$$

$$r_{kbt\omega}^{\mathrm{EV,DD}} = r_{kbt\omega}^{\mathrm{EV,DDC}} + r_{kbt\omega}^{\mathrm{EV,DDD}}, \quad \forall k \in K, \forall b \in B, \forall t \in T_k, \tag{5.18l}$$

$$0 \le r_{kbt\omega}^{\mathrm{EV,DDC}} \le N_{kb}^{\mathrm{EV}} P_{\max}^{\mathrm{EV}} - p_{kbt}^{\mathrm{EV,CD}}, \quad \forall k \in K, \forall b \in B, \forall t \in T_k, \tag{5.18m}$$

$$0 \le r_{kbt\omega}^{\mathrm{EV,DDD}} \le p_{kbt}^{\mathrm{EV,DD}}, \quad \forall k \in K, \forall b \in B, \forall t \in T_k, \tag{5.18n}$$

(Limits on the reserve deployment)

$$0 \le r_{gt\omega}^{\mathrm{G,DU}} \le r_{gt}^{\mathrm{G,CU}}, \quad \forall g \in G^{\mathrm{D}}, \forall t \in T, \tag{5.19a}$$

$$0 \le r_{gt\omega}^{\mathrm{G,DD}} \le r_{gt}^{\mathrm{G,CD}}, \quad \forall g \in G^{\mathrm{D}}, \forall t \in T, \tag{5.19b}$$

$$0 \le r_{kbt\omega}^{\mathrm{EV,DU}} \le r_{kbt}^{\mathrm{EV,CU}}, \quad \forall k \in K, \forall b \in B, \forall t \in T_k, \tag{5.19c}$$

$$0 \le r_{kbt\omega}^{\mathrm{EV,DD}} \le r_{kbt}^{\mathrm{EV,CD}}, \quad \forall k \in K, \forall b \in B, \forall t \in T_k, \tag{5.19d}$$

(Network power flow constraints)

$$\left.\begin{aligned} p^{\mathrm{L,B}}_{\ell t\omega} &= \frac{1}{X_\ell}\left(\theta^{\mathrm{B}}_{O(\ell)t\omega} - \theta^{\mathrm{B}}_{F(\ell)t\omega}\right), \quad \forall \ell \in L, \forall t \in T, && (5.20\mathrm{a})\\ -P^{\mathrm{L}}_{\max,\ell} &\le p^{\mathrm{L,B}}_{\ell t\omega} \le P^{\mathrm{L}}_{\max,\ell}, \quad \forall \ell \in L, \forall t \in T, && (5.20\mathrm{b})\\ -\pi &\le \theta^{\mathrm{B}}_{bt\omega} \le \pi, \quad \forall b \in B, \forall t \in T \end{aligned}\right\}, \forall \omega \in \Omega. \qquad (5.20\mathrm{c})$$

The first group of constraints are related to the day-ahead market. Constraints (5.10) enforce the energy balance in each bus and period. Constraints (5.11a)–(5.11g) describe the technical operation of generating units. Constraints (5.11a) limit the minimum and maximum power output of dispatchable units, which is determined according to the unit-commitment variable that establishes the on/off status of the unit. Constraints (5.11b) and (5.11c) impose the up and down ramping limits of dispatchable units. Constraints (5.11d) limit the power output of intermittent renewable units according to the availability factor of the renewable resource in each hour. Constraints (5.11e)–(5.11g) define the start-up and shut-down costs considering whether the generating unit changes its on/off status from one period to the next. Both costs must be higher than or equal to zero.

Constraints (5.12a) and (5.12b) limit the charged and discharged power scheduled in the day-ahead market for each group of k of EVs, of each bus and for each period.

Constraints (5.13a)–(5.13d) impose the limits on the reserve capacity scheduled in the day-ahead market. For the generating units, (5.13a) limit the up reserve capacity taking into account the capacity of the unit and the energy scheduled, while (5.13b) set the maximum down reserve capacity equal to the energy scheduled. For the EVs, (5.13c) and (5.13d) simply enforce the non-negativity of the up and down reserve capacity, respectively.

Constraints (5.14a) define the power flow through the transmission lines considering a DC representation of the network. Constraints (5.14b) limit the minimum and maximum power flow through each line, whereas (5.14c) establish that the voltage angles cannot take values higher (lower) than π ($-\pi$) rad.

The second set of constraints, i.e., (5.15)–(5.20), describe the real-time operation. Constraints (5.15) enforce the power balance in each bus, in each period and scenario. Constraints (5.16) impose the limit on the maximum unserved demand.

Constraints (5.17) refer to the technical operation of the generating units. Constraints (5.17a) define the power generated by dispatchable units in the real-time operation considering the energy scheduled in the day-ahead market and the up and down reserve deployed in each scenario. Constraints (5.17b) establish the capacity limits of the unit considering the on/off status determined in the day-ahead market, whereas (5.17c) and (5.17d) impose the ramping limits. Constraints (5.17e)–(5.17g) are related to the intermittent renewable units, such (5.17e) that establish that the power supplied plus the spilled power in the real time must be equal to the power available in the corresponding scenario, while (5.17f) and (5.17g) indicate the non-negativity of the supplied and spilled renewable power, respectively.

Constraints (5.18) describe the operation of EVs in the real time. The explanation of these constraints is provided in Sect. 5.5.

Constraints (5.19) limit the up and down reserve deployed by the generating units and the EVs to the reserve capacity scheduled in the day ahead. Finally, (5.20) define the limits on the power flow through the transmission lines in the real-time operation.

The problem described in this section is formulated as a two-stage stochastic programming problem with continuous and integer variables, whose size grows linearly with the number of scenarios.

5.7 Illustrative Case Study

This section presents an illustrative case study that allows to better understand the day-ahead scheduling model proposed above.

5.7.1 Input Data

The network considered in this illustrative example is depicted in Fig. 5.1. The power system comprises three buses with a demand and a generating unit in each bus. The three buses are connected by three transmission lines as represented in Fig. 5.1. Table 5.1 indicates the capacity, the minimum power output, the up/down ramp capability, and the operating cost of the generating units. Generating unit 1 and 2 are thermal power plants, while unit 3 represents an intermittent renewable power plant. The start-up costs correspond to the cost of working at full capacity during one day (19,200 and 28,800 € for generating units 1 and 2, respectively), while the shut-down costs are assumed to be half of the start-up costs (9,600 and 14,400 € for generating units 1 and 2, respectively).

It is assumed that the intermittent renewable unit cannot provide reserve services. The cost of scheduling up/down reserve by generating units 1 and 2 is 1 and 2 €/MW,

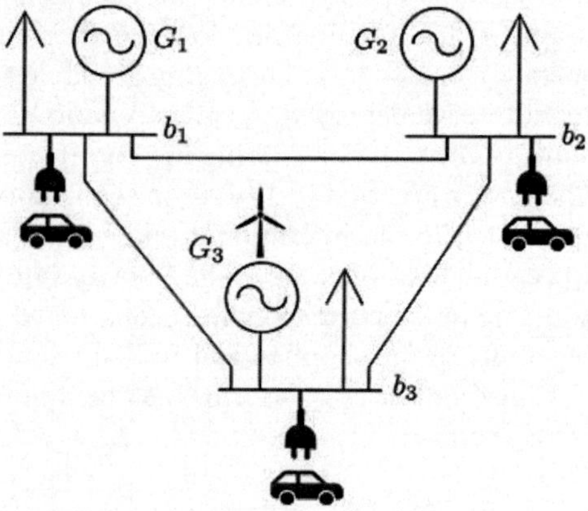

Fig. 5.1 Illustrative case study: schema of the network

Table 5.1 Illustrative case study: technical and economic data of generating units

Unit	Capacity (MW)	Min. power (MW)	Ramp up/down (MW)	Operating cost (€/MWh)
G_1	80	32	40	10
G_2	60	10	30	20
G_3	100	–	100	2

Table 5.2 Illustrative case study: demand power in each hour and bus

Hour/Bus	b_1	b_2	b_3
t_1	10.0	12.0	14.0
t_2	30.0	36.0	42.0
t_3	60.0	72.0	84.0

Table 5.3 Illustrative case study: number of EVs in each group and bus

Group	Bus	# EVs
k_1	b_1	120
	b_2	360
	b_3	720
k_2	b_1	80
	b_2	240
	b_3	480

respectively. The cost of deploying up reserve by generating units 1 and 2 is 10 and 20 €/MWh, respectively, while the cost of deploying down reserve is 8 and 16 €/MWh, respectively for each unit.

The planning period considers 3 h. The availability factor of the renewable source in each hour is 0.4, 0.6, and 0.5, respectively.

Table 5.2 gives the demand power in each hour and bus of the network. The cost of the involuntary unserved demand is 1000 €/MWh.

The capacity of the transmission lines is 150 MW and the reactance is 0.0005 Ω.

Two groups of EVs are considered depending on their charging period. Group 1 is connected in the second and third periods and Group 2 is connected just in the last period. Both groups of EVs are located in each bus of the network. The capacity of the battery is 40 kWh, while the maximum charging/discharging power is 11 kW. The efficiency rate of the charging/discharging of EVs is equal to 0.88. The minimum energy level in the battery is 8 kWh. The battery must be full at the end of the connecting period.

Table 5.3 provides the total number of EVs per group in each bus.

The initial SOC of the battery is the only uncertain parameter considered in this example and it is modelled as an uncertain parameter using two scenarios. Table 5.4

Table 5.4 Illustrative case study: initial state-of-charge of the EV batteries

Scenario	Group	Bus	Energy level
ω_1	k_1	b_1	0.7
		b_2	0.7
		b_3	0.7
	k_2	b_1	0.9
		b_2	0.9
		b_3	0.9
ω_2	k_1	b_1	0.6
		b_2	0.6
		b_3	0.6
	k_2	b_1	0.8
		b_2	0.8
		b_3	0.8

indicates the level of the battery at the beginning of the charging period for each group of each bus and for each scenario. The scenarios are equiprobable.

The cost of scheduling in the day-ahead market power to charge the EVs is null, while the cost of discharging the batteries of EV groups 1 and 2 is 8 and 10 €/MWh, respectively. The cost of scheduling up/down reserve capacity in the day-ahead market is 1 €/MWh. The cost of deploying down reserve in the real-time operation is 5 €/MWh, whereas the cost of deploying up reserve of EV groups 1 and 2 is 10 and 12 €/MWh, respectively.

5.7.2 Results

This section provides the results obtained from applying the input data described above to the formulation described in Sect. 5.6. The problem has been solved using CPLEX 12.6.1 under GAMS [8]. The number of variables is 452, from which 6 are binary variables, and the number of constraints is 715.

The total expected cost arises to 2943.5 €. The costs of the energy scheduled by generating units in the day-ahead market corresponds to 2928.0 € The cost of the reserve capacity scheduled in the day ahead is 47.3 €, while the expected cost of deploying reserves in the real-time operation is -31.8 €, which is due to the deploying of down reserve from EVs.

Figure 5.2 depicts the energy scheduled in the day-ahead market by generating units and the total demand. There is no power spilled by the renewable unit. Since generating unit 1 is the cheapest between the two dispatchable units, it supplies large part of the energy needs, arriving to its maximum capacity in the last hour. The

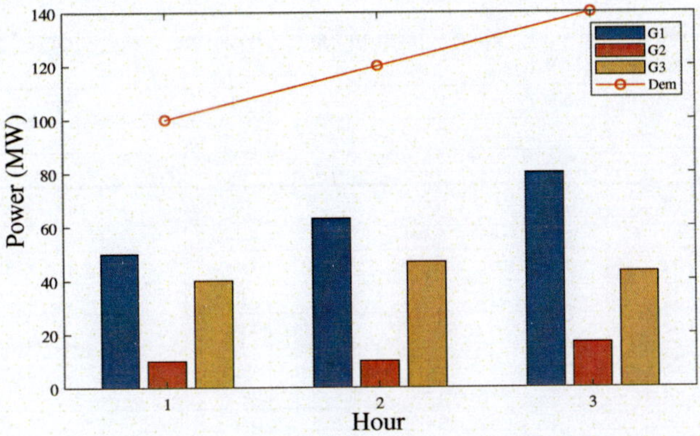

Fig. 5.2 Illustrative case study: day-ahead energy scheduling of generating units and total demand

Table 5.5 Illustrative case study: down reserve capacity scheduled by EVs in the day-ahead market (MW)

Group	Bus	Hour		
		t_1	t_2	t_3
k_1	b_1	0.0	1.3	0.9
	b_2	0.0	4.0	2.6
	b_3	0.0	7.9	5.2
k_2	b_1	0.0	0.0	0.7
	b_2	0.0	0.0	2.2
	b_3	0.0	0.0	4.4

reserve capacity scheduled by generating units is 0 in all periods, as well as, the power to be discharged/charged in the day-ahead market by EVs.

Table 5.5 indicates the down reserve capacity scheduled by EVs in the day-ahead market. Down reserve capacity is scheduled when the EVs are connected. The up reserve capacity scheduled is zero.

Table 5.6 describes the SOC of the EV batteries in each scenario of the real-time operation, namely the energy level, as well as the energy charged and discharged in the batteries. For instance, the energy level in the batteries of group 1 in bus 3 in hour 2 and scenario 1 is equal to 27.1 MWh, which results from charging 7.9 MWh to the energy level in hour 1 that is 20.2 MWh. Note that the efficiency rate of the charging is applying to the power charged.

Finally, Table 5.7 indicates the down reserve deployed by EVs in the real-time operation, which is due to charging the EV batteries.

Table 5.6 Illustrative case study: energy level in the EV batteries, power charged and discharged in the real-time operation (MWh)

			Hour								
			t_1			t_2			t_3		
Group	Bus	Scenario	$e^{EV}_{kbt\omega}$	$p^{EV,BC}_{kbt\omega}$	$p^{EV,BD}_{kbt\omega}$	$e^{EV}_{kbt\omega}$	$p^{EV,BC}_{kbt\omega}$	$p^{EV,BD}_{kbt\omega}$	$e^{EV}_{kbt\omega}$	$p^{EV,BC}_{kbt\omega}$	$p^{EV,BD}_{kbt\omega}$
k_1	b_1	ω_1	3.4	0.0	0.0	4.5	1.3	0.0	4.8	0.3	0.0
		ω_2	2.9	0.0	0.0	4.0	1.3	0.0	4.8	0.9	0.0
	b_2	ω_1	10.1	0.0	0.0	13.6	4.0	0.0	14.4	0.9	0.0
		ω_2	8.6	0.0	0.0	12.1	4.0	0.0	14.4	2.6	0.0
	b_3	ω_1	20.2	0.0	0.0	27.1	7.9	0.0	28.8	1.9	0.0
		ω_2	17.3	0.0	0.0	24.2	7.9	0.0	28.8	5.2	0.0
k_2	b_1	ω_1	–	–	–	2.9	0.0	0.0	3.2	0.4	0.0
		ω_2	–	–	–	2.6	0.0	0.0	3.2	0.7	0.0
	b_2	ω_1	-	-	-	8.6	0.0	0.0	9.6	1.1	0.0
		ω_2	–	–	–	7.7	0.0	0.0	9.6	2.2	0.0
	b_3	ω_1	–	–	–	17.3	0.0	0.0	19.2	2.2	0.0
		ω_2	-	-	-	15.4	0.0	0.0	19.2	4.4	0.0

Table 5.7 Illustrative case study: down reserve deployed by EVs in the real-time operation (MWh)

Group	Bus	Scenario	Hour		
			t_1	t_2	t_3
k_1	b_1	ω_1	0.0	1.3	0.3
		ω_2	0.0	1.3	0.9
	b_2	ω_1	0.0	4.0	0.9
		ω_2	0.0	4.0	2.6
	b_3	ω_1	0.0	7.9	1.9
		ω_2	0.0	7.9	5.2
k_2	b_1	ω_1	0.0	0.0	0.4
		ω_2	0.0	0.0	0.7
	b_2	ω_1	0.0	0.0	1.1
		ω_2	0.0	0.0	2.2
	b_3	ω_1	0.0	0.0	2.2
		ω_2	0.0	0.0	4.4

5.8 Realistic Case Study

This section analyzes a case study based on the isolated power system of Lanzarote-Fuerteventura (LZ-FV) in Spain. The objective of this case study is to test the formulation presented in Sect. 5.6 in a realistic case study.

All mathematical problems in this section have been solved by CPLEX 12.6.1 under GAMS [8] on a Linux-based server with four 3.0 GHz processors and 250 GB of RAM.

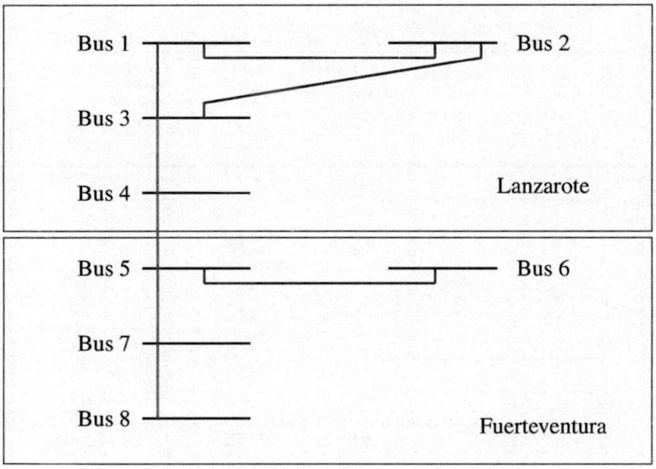

Fig. 5.3 Realistic case study: Lanzarote-Fuerteventura power system

Table 5.8 Realistic case study: transmission line parameters

Line bus	Origin bus	Destination	Length (km)	Capacity (MW)	Voltage (kV)	Reactance (Ω/km)
1	1	2	8.9	62	66	0.1276
2	2	3	10.8	62	66	0.3712
3	3	1	25	80	132	0.0613
4	3	4	18	160	132	0.3616
5	4	5	20	120	132	0.0588
6	5	6	25.5	62	66	0.3712
7	5	7	26	160	132	0.3616
8	7	8	23	160	132	0.3616

5.8.1 Input Data

The characterization of the LZ-FV power system is based on the representation of this system described in [1, 3]. Figure 5.3 shows the power system topology. Blue lines correspond to 66 kV transmission lines, whereas 132 kV lines are represented in red. For simplicity, the graphical representation of the transformers placed between 132 and 66 kV zones are omitted. The parameters that characterize the considered transmission system are included in Table 5.8.

This power system comprises eight buses, eight lines, and 38 generating units. The technologies of these candidate generating units are diesel, gas, wind, and solar PV. The technical characteristics of generating units are provided in Table 5.9. It is assumed that minimum power outputs, startup and operation costs of wind and solar PV units are negligible. The shutdown costs of thermal units are assumed to be 0.1

Table 5.9 Realistic case study: technical characteristics of dispatchable units

Techn.	Unit	Node	Installed capacity (MW)	Start-up cost (€)	Operating cost (€/MWh)	Minimum power output (MW)
Diesel	1–3	1	6.49	527	180.235	4.20
	4,5	1	12.85	1494	145.041	8.70
	6	1	20.51	2287	145.051	14.10
	7,8	1	17.20	1529	156.439	11.80
	9,10	1	17.60	1829	156.575	11.80
	11,12	6	3.82	270	174.018	2.40
	13	6	4.11	440	180.507	2.80
	14,15	6	6.21	227	180.805	4.20
	16	6	20.51	2287	145.510	14.09
	17	6	17.20	1529	156.439	11.80
	18,19	6	17.20	1529	156.439	11.80
Gas	20	1	19.60	2468	292.051	6.79
	21	1	32.34	2468	245.011	6.79
	22	6	29.40	2255	251.703	9.69
	23	6	11.74	2255	343.795	2.93
	24	6	21.85	2468	283.135	6.79
Wind	25	1	38.250	–	–	–
	26	2	5.625	–	–	–
	27	5	8.500	–	–	–
	28	8	5.625	–	–	–
	29	8	7.695	–	–	–
PV	30	1	7.115	–	–	–
	31	2	8.160	–	–	–
	32	3	1.530	–	–	–
	33	4	2.040	–	–	–
	35	5	1.555	–	–	–
	36	6	18.870	–	–	–
	37	7	7.225	–	–	–
	38	8	3.060	–	–	–

times their startup costs. The up and down reserve capacity costs are equal to 10% of the operation costs. The deployment of up reserves by thermal units is 1.1 times the operation costs, whereas the deployment of down reserves is 0.85 times the operation costs.

Three groups of EVs are considered. The description of the characteristics of each group is included in Table 5.10. Each EV is equipped with a 40 kWh battery and it is considered that the average daily distance driven by each vehicle is 35 km. The

Table 5.10 Realistic case study: electric vehicle groups

Type	Beginning hour	Ending hour	Duration (h)	Percentage of EVs (%)
1	17.00	8.00 (next day)	15	50
2	20.00	8.00 (next day)	12	35
3	8.00	17.00	9	15

Table 5.11 Realistic case study: number of vehicles

Bus								Total
1	2	3	4	5	6	7	8	
2177	9824	35182	7242	10055	16332	11302	9534	101648

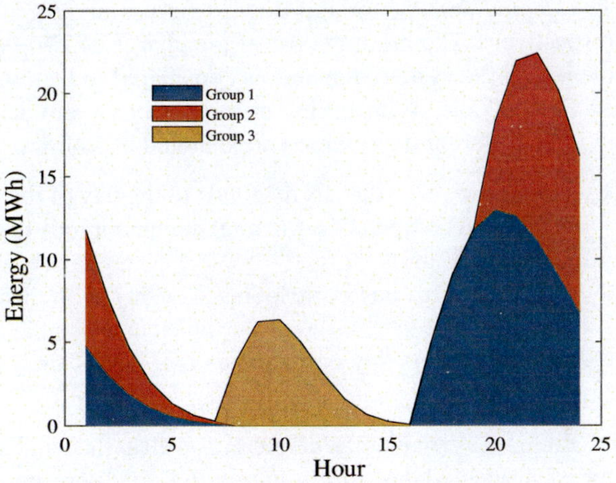

Fig. 5.4 Realistic case study: charging profiles of electric vehicle groups (30% of electric vehicles)

energy consumption is 0.2 kWh/km. The efficiency of charging and discharging the batteries is equal to 0.88. The batteries of EVs are charged under a maximum power transfer rate of 11 kW.

The total number of vehicles per bus, including electric and combustion-engine vehicles, is listed in Table 5.11. The percentage of EVs is specified afterwards for each case study.

The aggregated charging profiles of passive vehicles in each group considering a penetration of 30% of EVs is represented in Fig. 5.4. As stated above, we consider that passive EVs are those that do not participate actively in the different electricity markets. The charging profiles of passive EVs have been generated considering that the charge of vehicles belonging to groups 1, 2, and 3 are characterized by Weibull distributions with parameters {6, 2}, {4.8, 2}, and {3.5, 2}, respectively. Observe that most of the energy is charged by EVs at night hours.

Table 5.12 Realistic case study: day-ahead values of demand and wind and solar PV production in each selected day

Day	Lanzarote			Fuerteventura		
	Demand (MWh)	Wind power (MWh)	Solar PV power (MWh)	Demand (MWh)	Wind power (MWh)	Solar PV power (MWh)
AveNetDem	2320.9	91.4	56.7	1700.0	183.7	91.8
HighWind	2036.9	828.8	79.1	1742.2	465.8	128.5
HighSoPV	2325.1	173.4	145.3	1710.1	50.9	250.0

In this case study, the discharge cost of EVs is assumed to be equal to 250 €/MWh, whereas the up and down reserve capacity costs are equal to 15% of this value. The deployment of up and down reserves are equal to 262.5 and 12.5 €/MWh, respectively. The bidding cost of the charge of EVs, as well as the bidding cost of the rest of the system demand, is considered to be null.

The day-ahead scheduling of the power system described above has been tested in three different days, which are denoted as follows:

- *AveNetDem*: This day corresponds to the day of the year that achieves the median net demand computed as the total demand minus the available production of wind and solar PV units.
- *HighWind*: This day corresponds to a day of the year with high wind power production.
- *HighSoPV*: This day corresponds to a day of the year with high solar PV power production.

Table 5.12 provides the demand, as well as the wind and solar PV power production for each selected day. Furthermore, Fig. 5.5 represents the hourly value of demand, wind and solar PV power production in Lanzarote (LZ) and Fuerteventura (FV) for each day. Observe that the demand of Lanzarote is higher than in Fuerteventura.

In order to represent the balancing market, 12 scenarios are considered to model the uncertainty of demand and wind and solar PV power production. Table 5.13 lists the expected and the average standard deviation of the demand and wind and solar PV power scenarios. Observe that the expected values of these scenarios are identical to the day-ahead values provided in Table 5.12. As an example, the scenarios considered for *AveNetDem* day are graphically represented in Fig. 5.6.

5.8.2 Results

This section discuses the results obtained from solving the case described above. The number of constraints, continuous and binary variables are $137.3 \cdot 10^3$, $106.5 \cdot 10^3$, and 576, respectively. The solution time of each case solved in this section is less than 1 min.

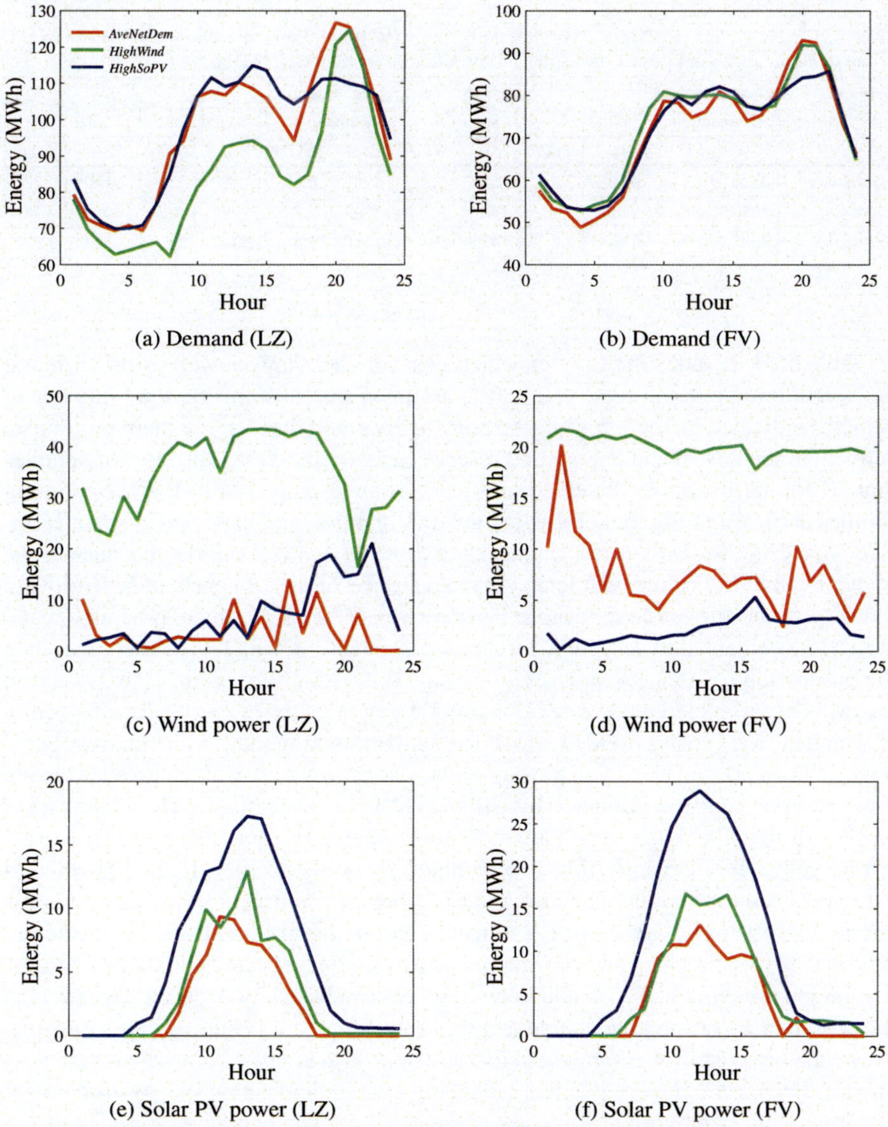

(a) Demand (LZ) (b) Demand (FV)

(c) Wind power (LZ) (d) Wind power (FV)

(e) Solar PV power (LZ) (f) Solar PV power (FV)

Fig. 5.5 Realistic case study: demand, wind, and solar PV values for selected days

Table 5.13 Realistic case study: expected daily value (Exp.) and hourly standard deviation (STD) of the demand and wind power and solar PV production in each day (MWh)

Day	Lanzarote						Fuerteventura					
	Demand		Wind		Solar PV		Demand		Wind		Solar PV	
	Exp.	STD	Exp.	STD	Exp.	STD	Exp.	STD	Exp.	STD	Exp.	STD
AveNetDem	2320.9	1.1	91.4	2.6	56.7	0.2	1700.0	0.8	183.7	2.6	91.8	0.2
HighWind	2036.9	2.5	828.8	6.5	79.1	0.3	1742.2	2.2	465.8	4.6	128.5	0.4
HighSoPV	2325.1	3.0	173.4	5.8	145.3	0.4	1710.1	1.8	50.9	3.3	250.0	0.5

Table 5.14 includes the total expected cost for each day, as well as the different items that comprise the total value. It is assumed that 30% of the total number of vehicles indicated in Table 5.11 correspond to EVs, and that 50% of them participate actively in the day-ahead energy and reserve capacity markets. The 50% of vehicles that do not participate in the electricity markets are assumed to follow the charging profiles depicted in Fig. 5.4. Table 5.14 shows that the total expected costs in *HighWind* and *HighSoPV* days are smaller than those in *AveNetDem* day because of the higher availability of renewable and inexpensive resources. In particular, *HighWind* day attains the lowest cost because, as shown in Table 5.12, this day is associated with a low demand and very high wind power production, which is much higher than the production of solar PV in *HighSoPV* day. It is also observed that the day-ahead energy cost associated with generation units (DA gen) is the highest value of all costs in which the total cost is divided into. It is also worth noting that, in the three analyzed days, EVs do not sell energy in the day-ahead energy market and they provide up and down reserve capacities, denoted by UR and DR EVs, respectively. The advantage of participating in the reserve capacity markets instead of the day-ahead energy market is that the reserve capacity that is scheduled is not necessarily deployed, whereas the energy sold in the day-ahead market is a firm commitment. Also, note that the energy that is discharged from EVs must be compulsorily purchased at a different time period to satisfy that EVs be charged at a desired level at the end of the charging period. On the other hand, up and down reserves can be easily deployed by reducing the charge or increasing the charge, respectively. Finally, it is interesting to note that the balancing cost in *AveNetDem* and *HighWind* days is positive, whereas it is negative in *HighSoPV* day. A positive balancing cost indicates that the cost increment resulting from deploying up reserves is greater than the cost decrease for reducing production when down reserves are deployed. In fact, it is observed that the down reserve scheduled in *HighSoPV* day is much higher than in the rest of analyzed days.

Table 5.15 provides the expected energy provided in the day-ahead energy market. The energy associated with EVs corresponds with the discharged energy that is sold in this market. As mentioned above, this value is equal to zero in the set of analyzed days. We observe that diesel units supply most of the electricity demand in each day. Note that the energy produced by all technologies is higher that the demand values listed in Table 5.12. The reason of this result is that the power demanded by EVs is not accounted for in the demand values provided in Table 5.12. It also worth noting

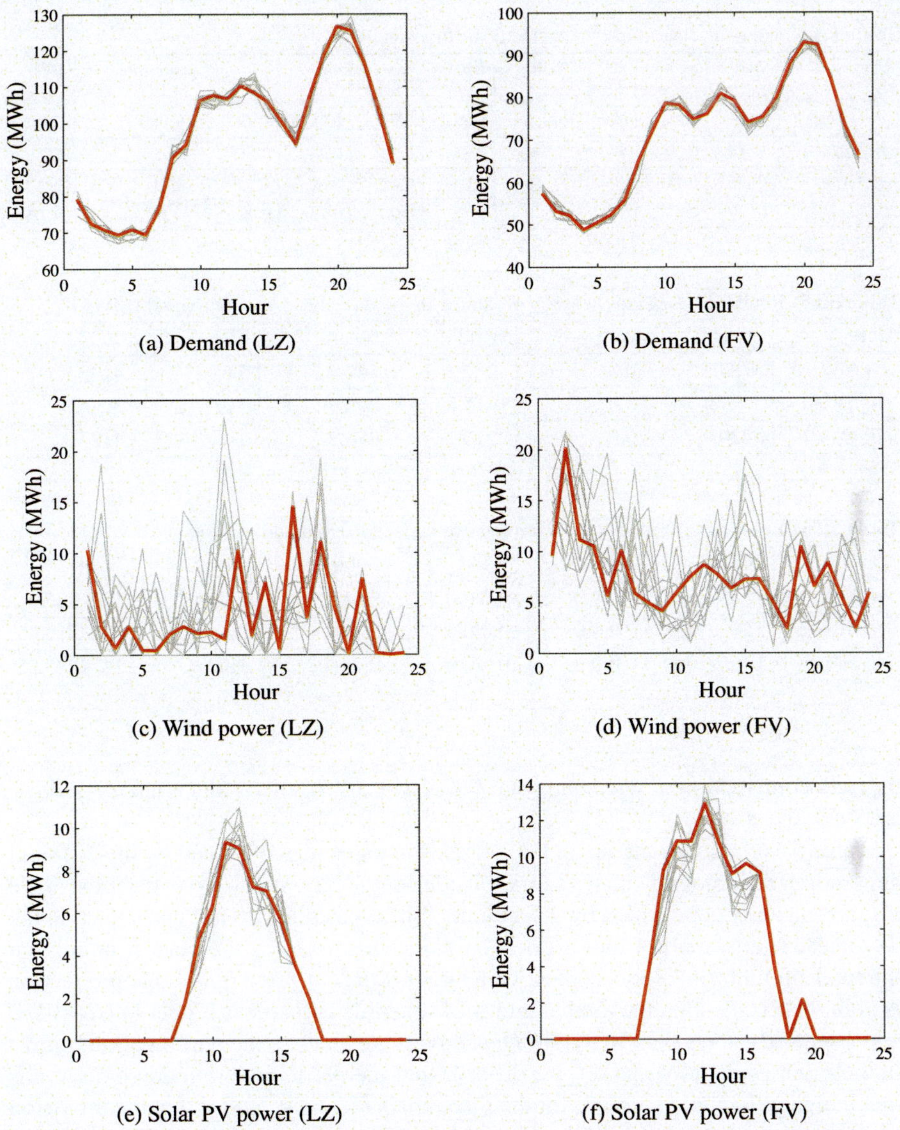

Fig. 5.6 Realistic case study: demand, wind, and solar PV scenarios for *AveNetDem* day

Table 5.14 Realistic case study: expected cost (thousand €)

Day	Start-up	Shut-down	DA	UR	DR	DA	UR	DR	Balancing	Total
			Gen	Gen	Gen	EVs	EVs	EVs		
AveNetDem	21.14	0.29	576.09	2.41	0.33	0.00	0.24	0.55	1.95	602.28
HighWind	19.26	0.41	364.02	7.88	1.08	0.00	0.62	1.25	4.74	398.71
HighSoPV	18.49	0.02	548.77	2.29	3.87	0.00	0.03	0.08	−3.06	567.56

Table 5.15 Realistic case study: energy produced in the day-ahead energy market (MWh)

Case	Diesel	Gas	Wind	Solar PV	EVs	Total
AveNetDem	3783.7	0.0	262.2	147.9	0.0	4193.8
HighWind	2421.4	0.0	1315.8	206.7	0.0	3944.0
HighSoPV	3620.9	0.0	210.7	356.3	0.0	4187.9

Table 5.16 Realistic case study: up/down reserve capacity scheduled (MW)

Case	Diesel	Gas	Wind	Solar PV	EVs	Total
AveNetDem	151.4/21.2	0.0/0.0	0.0/0.0	0.0/0.0	6.3/14.7	157.7/35.9
HighWind	520.2/72.5	0.0/0.0	0.0/0.0	0.0/0.0	16.4/33.3	536.6/105.8
HighSoPV	146.2/250.5	0.0/0.0	0.0/0.0	0.0/0.0	0.8/2.0	147.0/252.5

the high contribution of wind and solar PV power in *HighWind* and *HighSoPV* days, respectively.

Table 5.16 includes the up and down reserve capacities provided by the different generation technologies and EVs for each considered day. We observe that most of the reserve capacity is provided by diesel units. Since gas units do not participate in the day-ahead energy market, they are not able to provide reserves. We can also notice that the total amounts of reserve capacity are larger if the expected renewable production is high. In this manner, the total amounts of scheduled reserve is higher in *HighWind* and *HighSoPV* days than in *AveNetDem* day. If we analyze the participation of EVs in the reserve capacity market, we observe that the participation in down reserve is much higher than in up reserve for the three analyzed days. Notice that the provision of down reserve can be easily performed by EVs by increasing the charge of the vehicles with respect to the values scheduled in the day-ahead market. In this manner, the deployment of down reserve by EVs helps to maintain the energy balance in the system, as well as being useful to charge the batteries of EVs.

Figure 5.7 illustrates the scheduled day-ahead energy and reserve capacity, and the participation of EVs in the different markets. Figure 5.7a–c represent, using positive values, the energy that is sold in the day-ahead market by generating units and active EVs. The energy charged by EVs is represented using negative values, differentiating between those EVs that are active participants in the electricity markets (A), and those that are passive participants (P). As stated above, the energy charged from passive

EVs correspond to the charging profiles represented in Fig. 5.4. These figures show that, as indicated in Table 5.15, the whole demand is supplied by diesel, wind, and solar PV units for the three considered days. We also notice that the charging patterns of active EVs are different for each considered day and from those of passive EVs. In this manner, we observe that active EVs are not charged in peak hours (close to 21:00 h). On the contrary, active EVs are mostly charged during the night, between 1:00 and 7:00 hours. In the case of EV group 3 that must be charged during the day time, we observe that EVs belonging to this group are mostly charged when the renewable production is maximum.

Figure 5.7d–f represent the up and down reserve capacities scheduled for generating units and active EV using positive and negative values, respectively. From these figures we state that the capacity requirements are supplied by diesel generating units and EVs. We also notice that the total reserve requirements grow as the expected renewable production increases. In this maner, we observe that the reserve requirements in *HighWind* and *HighSoPV* days are significantly higher than those

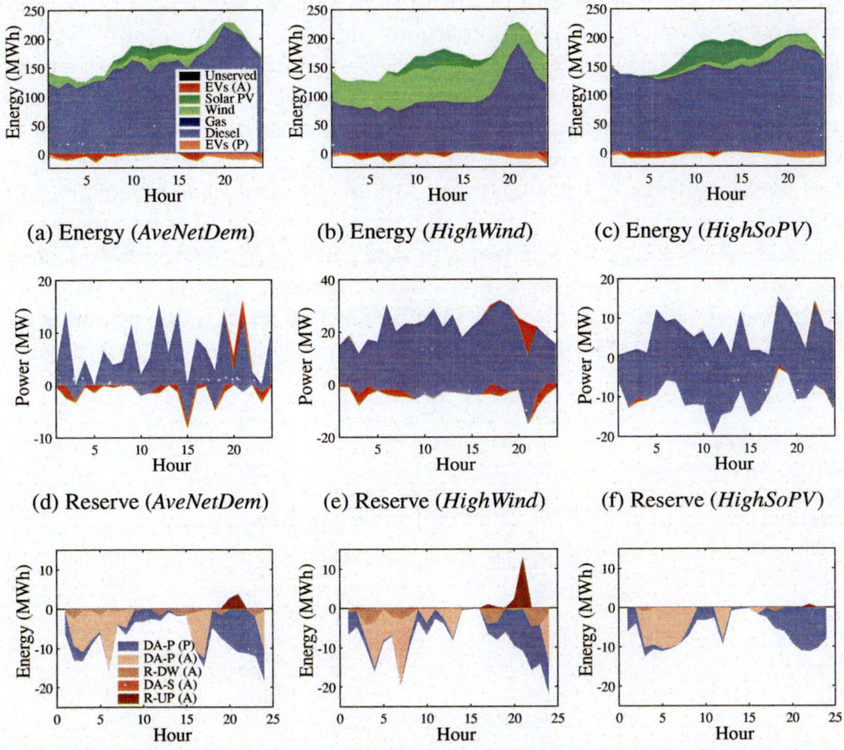

(a) Energy (*AveNetDem*) (b) Energy (*HighWind*) (c) Energy (*HighSoPV*)

(d) Reserve (*AveNetDem*) (e) Reserve (*HighWind*) (f) Reserve (*HighSoPV*)

(g) EVs (*AveNetDem*) (h) EVs (*HighWind*) (i) EVs (*HighSoPV*)

Fig. 5.7 Realistic case study: energy, reserve capacity, and EVs scheduling in *AveNetDem* day. Furthermore, we also observe that the contribution of EVs to the down-reserve capacity is much higher than to the up-reserve capacity.

Figure 5.7g–i show the participation of EVs in the different trading floors. As stated above, the participation of passive EVs corresponds to the purchase in the day-ahead energy market, DA-P (P), of the demand determined by the energy profiles depicted in Fig. 5.4. On the contrary, active EVs may either sell, DA-S (A), or purchase, DA-P (A), energy in the day-ahead energy market, and participate in the up, R-UP (A), and down, R-DW (A), reserve capacity markets. From Fig. 5.7g–i we observe that active EVs mainly participate in the day-ahead market to purchase energy. Then, it is observed that EVs prefer to purchase energy when the net demand is low. We also observe that up-reserve capacity is only scheduled for EVs when net demand is high, around 21:00 h. However, the total amount of up-reserve capacity that is scheduled is lower than the down-reserve capacity scheduled.

Finally, in order to analyze the impact of the number of EVs and the percentage of active EVs participating in the electricity markets, Fig. 5.8 represents the operation cost of *AveNetDem* day considering different numbers of EVs and percentages of active vehicles. The number of EVs is expressed in terms of the percentage of EVs over the total of vehicles, ranging between 0 and 100% with respect to the values included in Table 5.11. From this figure we observe that the total cost increases as the number EVs grows and the percentage of active EVs decreases. Therefore, the expected cost increases up to 20.8% if the number of EVs grows from 0 to 100% if EVs behave passively in the electricity markets. However, considering that the percentage of EVs is 100% and that 50% of the vehicles behave actively, the total cost decreases by 2.4 and 3.6% if all EVs are active participants. However, the dependency of the total cost with respect to the percentage of active EVs decreases as the number of EVs diminishes. For instance, if the percentage of EVs over the total number of vehicles is only 15%, the difference in terms of the operation costs of cases with 0 and 100% of active EVs is smaller than 0.4%. Therefore, we can

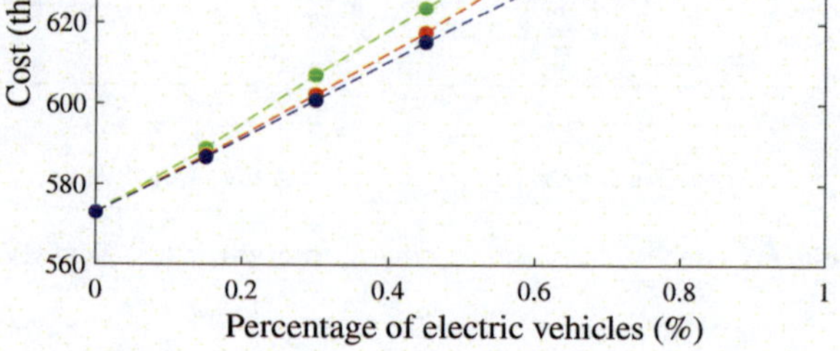

Fig. 5.8 Realistic case study: operation costs for different numbers of electric vehicles and percentages of controlled vehicles

conclude that the active participation of EVs in the electricity markets is of special interest in the analyzed case if the number of EVs is high enough.

5.9 Summary and Conclusions

This chapter analyzes the day-ahead scheduling problem of a power system with high presence of intermittent renewable generation and EVs. The possibility of controlling the charging and discharging batteries of EVs from the system operator provides high flexibility to counteract the lack of dispatchability of renewable sources.

The model is formulated through a mixed-integer two-stage stochastic programming problem. The first stage represents the day-ahead scheduling of energy and reserve capacity, while the second stage represents the variability of the real-time operation. Dispatchable generating units and EVs can provide reserve services. The uncertain parameters considered are the intermittent renewable and demand power, as well as the level of energy in the batteries of the EVs at the beginning of the charging period.

A case study based on the power system of Lanzarote-Fuerteventura is analyzed. Different scenarios of availability level of wind and solar PV production have been studied. The main conclusions obtained from this numerical study are:

(i) It has been observed that EVs do not sell energy in the day-ahead energy market, whereas they purchase energy in this market and they provide up and down reserve capacity in the three analyzed days; the participation of EVs in the reserve market can be explained by the fact that it is easier for EVs to modify the charging/discharging of their batteries to provide reserve services rather than make a commitment of selling energy in the day-ahead market that must be realized in the real-time operation for every scenario.

(ii) The cases analyzed result in a higher participation of EVs in providing down reserve than in up reserve; it is comparatively more beneficial to the system to charge the batteries of the EVs by deploying down reserve while maintaining the energy balance rather than scheduling the corresponding energy in the day-ahead market.

(iii) Considering the data describing the isolated power system under study, diesel units represent a cheaper alternative with respect to gas units; wind and solar PV generation could help to reduce the environmental impact of fossil fuel facilities.

(iv) Finally, it has been shown that the total operating cost increases as long as the number of EVs does; however, it is worth noting that the active participation of EVs helps in reducing the total costs.

5.10 Notation

The notation used in this chapter is included below for quick reference.

Acronyms

DC Direct current.
EV Electric vehicle.
GHG Greenhouse gas.
PV Photovoltaic.
SO System operator.
SOC State-of-charge of a battery.
V2G Vehicle to grid.

Sets and Indices

b Index of buses.
B Set of buses.
$F(\ell)$ Destination bus of line ℓ.
g Index of generating units.
G Set of generating units.
G_b Set of generating units located at bus b.
G^{D} Set of dispatchable generating units.
G^{I} Set of intermittent generating units.
k Index of electric vehicle groups.
K Set of electric vehicle groups.
ℓ Index of transmission lines.
L Set of transmission lines.
L_b^{F} Set of transmission lines whose destination bus is b.
L_b^{O} Set of transmission lines whose origin bus is b.
$O(\ell)$ Origin bus of line ℓ.
t Index of time periods.
T Set of time periods.
T_k Set of time periods in which electric vehicles of group k are connected to the network.
ω Index of scenarios.
Ω Set of scenarios.

Parameters

$A^{\mathrm{B}}_{gt\omega}$ Power availability of intermittent unit g in the real-time operation, in period t and scenario ω (pu).

A^{D}_{gt} Power availability of intermittent unit g in the day-ahead market in period t (pu).

$C^{\mathrm{G,D}}_{gt}$ Offering price in the day-ahead market of generating unit g in period t (€/MWh).

$C^{\mathrm{G,RCD}}_{gt}$ Offering price of down-reserve capacity of generating unit g in period t (€/MW).

$C^{\mathrm{G,RCU}}_{gt}$ Offering price of up-reserve capacity of generating unit g in period t (€/MW).

$C^{\mathrm{G,RDD}}_{gt}$ Offering price of deploying down-reserve of generating unit g in period t (€/MWh).

$C^{\mathrm{G,RDU}}_{gt}$ Offering price of deploying up-reserve of generating unit g in period t (€/MWh).

$C^{\mathrm{EV,CD}}_{kbt}$ Bidding price for charging energy in the day-ahead market by electric vehicle group k, at bus b in period t (€/MWh).

$C^{\mathrm{EV,DD}}_{kbt}$ Offering price for discharging energy scheduled in the day-ahead market by electric vehicle group k, at bus b in period t (€/MWh).

$C^{\mathrm{EV,RCD}}_{kbt}$ Offering price of down-reserve capacity of electric vehicle group k, at bus b in period t (€/MW).

$C^{\mathrm{EV,RCU}}_{kbt}$ Offering price of up-reserve capacity of electric vehicle group k, at bus b in period t (€/MW).

$C^{\mathrm{EV,RDD}}_{kbt}$ Offering price of deploying down-reserve of electric vehicle group k, at bus b in period t (€/MWh).

$C^{\mathrm{EV,RDU}}_{kbt}$ Offering price of deploying up-reserve of electric vehicle group k, at bus b in period t (€/MWh).

$C^{\mathrm{G,SU}}_{g}$ Startup cost parameter of generating unit g (€).

$C^{\mathrm{G,SD}}_{g}$ Shut-down cost parameter of generating unit g (€).

C^{UD} Cost of unserved demand (€/MWh).

$E^{\mathrm{EV}}_{\max,k}$ Maximum energy capacity of the batteries of electric vehicles belonging to group k (MWh).

N^{EV}_{kb} Number of electric vehicles in group k at bus b.

$P^{\mathrm{EV}}_{\max,k}$ Maximum charging/discharging power rate of EVs belonging to group k (MW).

$P^{\mathrm{D,B}}_{bt\omega}$ Power demand in the real-time operation at bus b, period t, and scenario ω (MW).

$P^{\mathrm{D,D}}_{bt}$ Power demand in the day-ahead market at bus b and period t (MW).

$P^{\mathrm{G}}_{\mathrm{DW},g}$ Down ramp factor of generating unit g (pu).

$P^{\mathrm{G}}_{\max,g}$ Upper limit of the power production of generators (MW).

$P^{\mathrm{G}}_{\min,g}$ Lower limit of the power production of generators (MW).

$P^{\mathrm{G}}_{\mathrm{UP},g}$ Upper ramp factor of generating unit g (MW).

$P^{\mathrm{L}}_{\max,\ell}$ Active power capacity of transmission line ℓ (MW).
X_ℓ Reactance of line ℓ (Ω).
$\alpha^{\mathrm{EV,SE}}_{kd\omega}$ Factor used to model the ending status of the batteries of the electric vehicles belonging to group k in scenario ω (pu).
$\alpha^{\mathrm{EV,SI}}_{kd\omega}$ Factor used to model the initial status of the batteries of the electric vehicles belonging to group k in scenario ω (pu).
$\alpha^{\mathrm{EV,SMIN}}_{kd}$ Factor used to model the minimum energy that must contain the batteries of the electric vehicles belonging to group k (pu).
η^{EV} Efficiency of charging and discharging electric vehicles (pu).
π_ω Probability of scenario ω (pu).

Variables

$c^{\mathrm{G,SU}}_{gt}$ Startup cost of generating unit g in period t (€).
$c^{\mathrm{G,SD}}_{gt}$ Shut-down cost of generating unit g in period t (€).
$e^{\mathrm{EV}}_{kbt\omega}$ Energy stored in the batteries of electric vehicles belonging to group k at bus b in period t and scenario ω (MW).
$p^{\mathrm{EV,CD}}_{kbt}$ Charging energy scheduled in the day-ahead market by electric vehicle group k, at bus b in period t (MWh).
$p^{\mathrm{EV,DD}}_{kbt}$ Discharging energy scheduled in the day-ahead market by electric vehicle group k, at bus b in period t (MWh).
$p^{\mathrm{EV,BC}}_{kbt\omega}$ Power charged by electric vehicle group k, at bus b in period t and scenario ω (MW).
$p^{\mathrm{EV,BD}}_{kbt\omega}$ Power discharged by electric vehicle group k, at bus b in period t and scenario ω (MW).
$p^{\mathrm{G,B}}_{gt\omega}$ Power generated in the real-time operation by generating unit g, in period t, and scenario ω (MW).
$p^{\mathrm{G,BS}}_{gt\omega}$ Power spillage of intermittent generating unit g, in period t, and scenario ω (MW).
$p^{\mathrm{G,D}}_{gt}$ Energy scheduled by generating unit g in the day-ahead market in period t (MWh).
$p^{\mathrm{L,B}}_{lt\omega}$ Power flow through line ℓ in the real-time operation, in period t and scenario ω (MW).
$p^{\mathrm{L,D}}_{lt\omega}$ Power flow through line ℓ in the day-ahead market in period t (MW).
$p^{\mathrm{UD,B}}_{bt\omega}$ Unserved demand at bus b in the real-time operation in period t and scenario ω (MWh).
$r^{\mathrm{EV,CD}}_{kbt}$ Down-reserve capacity scheduled by electric vehicle group k, at bus b in period t (MW).
$r^{\mathrm{EV,CU}}_{kbt}$ Up-reserve capacity scheduled by electric vehicle group k, at bus b in period t (MW).
$r^{\mathrm{EV,DD}}_{kbt\omega}$ Deployed down-reserve by electric vehicle group k, at bus b in period t and scenario ω (MWh).

$r^{\mathrm{EV,DDC}}_{kbt\omega}$ Down-reserve deployed from the charging of electric vehicle group k, at bus b in period t and scenario ω (MWh).

$r^{\mathrm{EV,DDD}}_{kbt\omega}$ Down-reserve deployed from the discharging of electric vehicle group k, at bus b in period t and scenario ω (MWh).

$r^{\mathrm{EV,DU}}_{kbt\omega}$ Deployed up-reserve electric vehicle group k, at bus b in period t and scenario ω (MWh).

$r^{\mathrm{EV,DUC}}_{kbt\omega}$ Up-reserve deployed from the charging of electric vehicle group k, at bus b in period t and scenario ω (MWh).

$r^{\mathrm{EV,DUD}}_{kbt\omega}$ Up-reserve deployed from the discharging of electric vehicle group k, at bus b in period t and scenario ω (MWh).

$r^{\mathrm{G,CD}}_{gt}$ Down-reserve capacity scheduled by generating unit g in period t (MW).

$r^{\mathrm{G,CU}}_{gt}$ Up-reserve capacity scheduled by generating unit g in period t (MW).

$r^{\mathrm{G,DD}}_{gt\omega}$ Deployed down-reserve by generating unit g, in period t, and scenario ω (MWh).

$r^{\mathrm{G,DU}}_{gt\omega}$ Deployed up-reserve by generating unit g, in period t, and scenario ω (MWh).

$\theta^{\mathrm{B}}_{bt\omega}$ Voltage angle at bus b in the real-time operation in period t and scenario ω (rad).

θ^{D}_{bt} Voltage angle at bus b in the day-ahead market in period t (rad).

5.11 GAMS Codes

This section includes the GAMS code of Illustrative Case Study solved in Sect. 5.7.

```
$title Scheduling model of renewable-dominated power systems with
    electric vehicles

***********************************************************************
* DATA
***********************************************************************

SETS

b    'Buses'            /b1*b3/
d    'Demands'          /d1*d3/
g    'Generators'       /g1*g3/
k    'PEV-types'        /k1*k2/
l    'Lines'            /l1*l3/
t    'Time periods'     /t1*t3/
w    'Scenarios'        /w1*w2/

dem2bus(d,b) 'Demands connected to buses'
/d1.b1,d2.b2,d3.b3/

gen2bus(g,b) 'Generators connceted to buses'
/g1.b1,g2.b2,g3.b3/

gen_con(g)  'Conventional (dispatchable) generators'
```

```
/g1,g2/

gen_ren(g) 'Renewable (non-dispatchable) generators'
/g3/

pev2bus(k,b) 'PEV-types connected to buses'
/(k1*k2).(b1*b3)/

pev_online(k,t) 'PEV-types online periods'
/k1.t2,k1.t3,k2.t3/

pev_ti(k,t) 'Initial period of online horizon of PEVs'
/k1.t2,k2.t3/

pev_tf(k,t) 'Final period of online horizon of PEVs'
/k1.t3,k2.t3/

busOline(b,l) 'Origin buses of lines'
/b1.l1,b2.l2,b3.l3/

lineObus(l,b) 'Lines with origing bus b'
/l1.b1,l2.b2,l3.b3/

busFline(b,l) 'Destination buses of lines'
/b2.l1,b3.l2,b1.l3/

lineFbus(l,b) 'Lines with destination bus b'
/l1.b2,l2.b3,l3.b1/

ref_bus(b) 'reference bus'
/b1/
;

SCALARS

etaev 'Efficiency of the charge/discharge of PEVs'
/0.88/

pevmax 'Maximum power charging and discharging rate of electric vehicles'
/0.011/

PARAMETERS

cud(d) 'Cost of unserved conventional demand'
/
d1   1000
d2   1000
d3   1000
/

pdd(d,t) 'Demand power'
/
d1.t1   10
d1.t2   12
d1.t3   14
d2.t1   30
d2.t2   36
d2.t3   42
```

```
d3.t1   60
d3.t2   72
d3.t3   84
/

pdb(d,t,w) 'Demand in real-time operation'
/
d1.t1.w1  10
d1.t2.w1  12
d1.t3.w1  14
d1.t1.w2  10
d1.t2.w2  12
d1.t3.w2  14
d2.t1.w1  30
d2.t2.w1  36
d2.t3.w1  42
d2.t1.w2  30
d2.t2.w2  36
d2.t3.w2  42
d3.t1.w1  60
d3.t2.w1  72
d3.t3.w1  84
d3.t1.w2  60
d3.t2.w2  72
d3.t3.w2  84
/

cosu(g) 'Start-up cost of conventional generators'
/
g1  19200
g2  28800
/

cosd(g) 'Shut-down cost of conventional generators'
/
g1  9600
g2  14400
/

cgd(g) 'Offering price of generators in the day-ahead market'
/
g1  10.0
g2  20.0
g3  2.0
/

cgrcd(g) 'Offering price of down-reserve capacity of conventional generators'
/
g1  1.0
g2  2.0
/

cgrcu(g) 'Offering price of up-reserve capacity of conventional generators'
/
g1  1.0
g2  2.0
/
```

```
cgrdd(g) 'Offering price of deploying down-reserve of conventional
     generators'
/
g1  8.0
g2  16.0
/

cgrdu(g) 'Offering price of deploying up-reserve of conventional
     generators'
/
g1  10.0
g2  20.0
/

pg0(g) 'Production of conventional generators prior to the first
     scheduling period'
/
g1  40
g2  10
/

pgmax(g) 'Maximum power output of generators'
/
g1  80
g2  60
g3  100
/

pgmin(g) 'Minimum power output of generators'
/
g1  32
g2  10
g3  0
/

pgup(g) 'Ramp-up limit of conventional generators'
/
g1  40
g2  30
/

pgdw(g) 'Ramp-down limit of conventional generators'
/
g1  40
g2  30
/

gammag(g,t) 'Predicted availability of renewable generators'
/
g3.t1  0.4
g3.t2  0.6
g3.t3  0.5
/

gammagw(g,t,w) 'Availability of renewable generators in real-time
     operation'
/
g3.t1.w1  0.4
g3.t2.w1  0.6
```

```
g3.t3.w1  0.5
g3.t1.w2  0.4
g3.t2.w2  0.6
g3.t3.w2  0.5
/

x(l) 'Series reactance of lines'
/
l1  5e-4
l2  5e-4
l3  5e-4
/

plmax(l) 'Maximum power flow through lines'
/
l1  150
l2  150
l3  150
/

cevcd(k) 'Bidding price of charging power of PEVs in the day-ahead market'
/
k1  0.0
k2  0.0
/

cevdd(k) 'Offering price of discharging power of PEVs in the day-ahead market'
/
k1  8.0
k2  10.0
/

cevrcd(k) 'Offering price of down-reserve capacity of PEVs'
/
k1  1.0
k2  1.0
/

cevrcu(k) 'Offering price of up-reserve capacity of PEVs'
/
k1  1.0
k2  1.0
/

cevrdd(k) 'Bidding price of deploying down-reserve of PEVs'
/
k1  5.0
k2  5.0
/

cevrdu(k) 'Offering price of deploying up-reserve of PEVs'
/
k1  10.0
k2  12.0
/
eevmax(k) 'Energy capacity of the battery of PEVs'
/
```

```
k1  0.04
k2  0.04
/
eevmin(k) 'Minimum energy that must remain in the battery of PEVs'
/
k1  0.008
k2  0.008
/

nev(k,b) 'Total number of k-type PEVs at buses'
/
k1.b1  120.0
k1.b2  360.0
k1.b3  720.0
k2.b1  80.0
k2.b2  240.0
k2.b3  480.0
/

socevf(k) 'Required final SOC of the battery of PEVs'
/
k1  1.0
k2  1.0
/

socev0(k,b,w) 'Initial SOC of the battery of PEVs'
/
k1.b1.w1  0.7
k1.b2.w1  0.7
k1.b3.w1  0.7
k2.b1.w1  0.9
k2.b2.w1  0.9
k2.b3.w1  0.9
k1.b1.w2  0.6
k1.b2.w2  0.6
k1.b3.w2  0.6
k2.b1.w2  0.8
k2.b2.w2  0.8
k2.b3.w2  0.8
/

piw(w) probability associated to scenarios
/
w1  0.5
w2  0.5
/
;

*****************************************************************
* DECLARATION OF VARIABLES AND MATHEMATICAL CHARACTERIZATION
*****************************************************************

VARIABLES

obj 'Value of the objective function'
cda 'Cost of the day-ahead market'
crc 'Cost of the reserve capacity market'
crd(w) 'Cost of the reserve deployment in real-time operation'
```

```
thetad(b,t) 'Voltage angle at buses for the day-ahead market
    solution'
thetab(b,t,w) 'Voltage angle at buses in the real-time operation'
pld(l,t) 'Power flow through lines for the day-ahead market
    solution'
plb(l,t,w) 'Power flow through lines in the real-time operation'
;

BINARY VARIABLES

u(g,t) 'Binary variable used to implement the minimum power output
     constraint of generators'
;

POSITIVE VARIABLES

cgsu(g,t)  'Start-up cost of conventional generators'
cgsd(g,t)  'Shut-down cost of conventional generators'

pud(d,t,w) 'Unserved power demand in the real-time operation'

pgd(g,t) 'Power scheduled of generators in the day-ahead market'
pgb(g,t,w) 'Power production of generators in the real-time
    operation'
pgbs(g,t,w) 'Power spillage of renewable generators in the real-
    time operation'

rgcd(g,t) 'Down-reserve capacity scheduled for conventional
    generators in the day-ahead market'
rgcu(g,t) 'Up-reserve capacity scheduled for conventional
    generators in the day-ahead market'

rgdd(g,t,w) 'Down-reserve deployed by conventional generators in
    the real-time operation'
rgdu(g,t,w) 'Up-reserve deployed by conventional generators in the
     real-time operation'

eev(k,b,t,w) 'Energy stored in the batteries of PEVs in the real-
    time operation'
pevbc(k,b,t,w) 'Power charged by PEVs in the real-time operation'
pevbd(k,b,t,w) 'Power discharged by PEVs in real-time operation'

pevcd(k,b,t) 'Charging power scheduled of PEVs in the day-ahead
    market'
pevdd(k,b,t) 'Discharging power scheduled of PEVs in the day-ahead
     market'

revcd(k,b,t) 'Down-reserve capacity scheduled for PEVs in the day-
    ahead market'
revcu(k,b,t) 'Up-reserve capacity scheduled for PEVs in the day-
    ahead market'
revdd(k,b,t,w) 'Down-reserve deployed by PEVs in the real-time
    operation'
revddc(k,b,t,w) 'Down-reserve deployed by PEVs in charging mode in
     the real-time operation'
revddd(k,b,t,w) 'Down-reserve deployed by PEVs in discharging mode
     in the real-time operation'
revdu(k,b,t,w) 'Up-reserve deployed by PEVs in the real-time
    operation'
```

```
revduc(k,b,t,w) 'Up-reserve deployed by PEVs in the charging mode
     in the real-time operation'
revdud(k,b,t,w) 'Up-reserve deployed by PEVs in the discharging
     mode in the real-time operation'
;

******************************************************************
*  EQUATIONS
******************************************************************

EQUATIONS

**        OBJECTIVE FUNCTION

obj_func  'Eq of the definition of the objective function value'
costda_eq 'Eq of the definition of the cost of the day-ahead
     market'
costrc_eq 'Eq of the definition of the cost of the reserve-
     capacity market'
costrd_eq 'Eq of the definition of the cost of the reserve
     deployment in real-time operation'

**        CONSTRAINTS

balanced_eq 'Eq of the power balance at buses in the day-ahead
     market'

pgdmin_con_eq 'Eq of minimum power output of conventional
     generators in the day-ahead market'
pgdmax_con_eq 'Eq of maximum power output of conventional
     generators in the day-ahead market'
pgdmax_ren_eq 'Eq of maximum power output of renewable generators
     in the day-ahead market'

rampdup_eq 'Eq of the ramp-up limit in the day-ahead market'
rampddw_eq 'Eq of the ramp-down limit in the day-ahead market'
rampdup_t1_eq 'Eq of the ramp-up limit for the first period in the
      day-ahead market'
rampddw_t1_eq 'Eq of the ramp-down limit for the first period in
     the day-ahead market'

costsu_eq  'Eq of the start-up cost of conventional generators'
costsd_eq  'Eq of the shut-down cost of conventional generators'

rgcu_limit_eq 'Eq of the upper limit of rgcu'
rgcd_limit_eq 'Eq of the upper limit of rgcd'

pld_eq 'Eq of the power flow through lines for the the day-ahead
     market solution'
thetadmax_eq 'Eq of the upper bound of thetad'
thetadmin_eq 'Eq of the lower bound of thetad'
thetadref_eq 'Eq to fix the value of thetad at the reference bus'
pldmax_eq 'Eq of the upper limit of pld'
pldmin_eq 'Eq of the upper limit of pld'

balanceb_eq 'Eq of the power balance at buses in real-time
     operation'

pudmax_eq 'Eq of the upper limit of pud'
```

```
pgb_con_eq 'Eq defining pgb for conventional generators'
pgb_ren_eq 'Eq defining pgb for renewable generators'

pgbmin_con_eq 'Eq of lower limit of the production of conventional
     generators in the real-time operation'
pgbmax_con_eq 'Eq of upper limit of the production of conventional
     generators in the real-time operation'
pgbmax_ren_eq 'Eq of maximum power output of renewable generators
    in the real-time operation'

rampbup_eq 'Eq of the ramp-up limit for the real-time operation'
rampbdw_eq 'Eq of the ramp-down limit for the real-time operation'
rampbup_t1_eq 'Eq of the ramp-up limit for the first period in the
     real-time operation'
rampbdw_t1_eq 'Eq of the ramp-down limit for the first period in
    the real-time operation'

pevcd_limit_eq 'Eq limiting the power to be purchased by PEVs in
    the day-ahead market'
pevdd_limit_eq 'Eq limiting the power to be sold by PEVs in the
    day-ahead market'

rgdu_limit_eq 'Eq of the upper limit of rgdu'
rgdd_limit_eq 'Eq of the upper limit of rgdd'

plb_eq 'Eq of the power flow through lines for the the real-time
    operation'
thetabmax_eq 'Eq of the upper bound of thetab'
thetabmin_eq 'Eq of the lower bound of thetab'
thetabref_eq 'Eq to fix the value of thetab at the reference bus'
plbmax_eq 'Eq of the upper limit of plb'
plbmin_eq 'Eq of the upper limit of plb'

eev_init_eq 'Eq for the definition of the energy stored in the
    batteries of PEVs - initial soc'
eev_online_eq 'Eq for the definition of the energy stored in the
    batteries of PEVs - online period'
eev_final_eq 'Eq for the definition of the energy stored in the
    batteries of PEVs - final soc'

pevbd_eev_eq 'Eq of the upper bound of the power discharged by
    PEVs based on the energy already stored in its battery'

pevbc_eq 'Eq defining the power charged by PEVs in real-time
    operation'
pevbd_eq 'Eq defining the power discharged by PEVs in real-time
    operation'

revdd_eq 'Eq defining the down-reserve deployed by PEVs in the
    balancing market'
revdu_eq 'Eq defining the up-reserve deployed by PEVs in the
    balancing market'

revdu_limit_eq 'Eq of the upper limit of revdu'
revdd_limit_eq 'Eq of the upper limit of revdd'

revduc_limit_eq 'Eq of upper limit of revduc'
revdud_limit_eq 'Eq of upper limit of revdud'
revddc_limit_eq 'Eq of upper limit of revddc'
```

```
revddd_limit_eq 'Eq of upper limit of revddd'

eevmin_eq 'Eq of the lower limit of eev'
eevmax_eq 'Eq of the upper limit of eev'
;

obj_func.. obj =e= cda + crc + sum(w,piw(w)*crd(w))
;
costda_eq.. cda =e=  sum(t,sum(g$gen_con(g),cgsu(g,t) + cgsd(g,t)))
                    + sum(t,sum(g,cgd(g)*pgd(g,t)))
                    + sum(t,sum(b,sum(k$pev_online(k,t),cevdd(k)*pevdd(k,b,t) - cevcd(k)*pevcd(k,b,t))))
;
costrc_eq.. crc =e= sum(t,sum(g,cgrcd(g)*rgcd(g,t) + cgrcu(g)*rgcu(g,t)))
                    + sum(t,sum(b,sum(k$pev_online(k,t),cevrcd(k)*revcd(k,b,t) + cevrcu(k)*revcu(k,b,t))))
;
costrd_eq(w).. crd(w) =e= sum(t,sum(g,cgrdu(g)*rgdu(g,t,w) - cgrdd(g)*rgdd(g,t,w)))
                          + sum(t,sum(b,sum(k$pev_online(k,t),cevrdu(k)*revdu(k,b,t,w) - cevrdd(k)*revdd(k,b,t,w))))
                          + sum(t,sum(d, cud(d)*pud(d,t,w)))
;

balanced_eq(b,t).. sum(g$gen2bus(g,b),pgd(g,t)) - sum(d$dem2bus(d,b), pdd(d,t))
                  + sum(k$(pev2bus(k,b) and pev_online(k,t)),pevdd(k,b,t) - pevcd(k,b,t))
                  =e= sum(l$lineObus(l,b), pld(l,t)) - sum(l$lineFbus(l,b), pld(l,t))
;

pgdmin_con_eq(gen_con(g),t).. pgd(g,t) =g= u(g,t)*pgmin(g)
;
pgdmax_con_eq(gen_con(g),t).. pgd(g,t) =l= u(g,t)*pgmax(g)
;
pgdmax_ren_eq(gen_ren(g),t).. pgd(g,t) =l= gammag(g,t)*pgmax(g)
;

rampdup_eq(gen_con(g),t)$(ord(t) gt 1).. pgd(g,t) - pgd(g,t-1) =l= pgup(g)
;
rampddw_eq(gen_con(g),t)$(ord(t) gt 1).. pgd(g,t-1) - pgd(g,t) =l= pgdw(g)
;
rampdup_t1_eq(gen_con(g),'t1').. pgd(g,'t1') - pg0(g) =l= pgup(g)
;
rampddw_t1_eq(gen_con(g),'t1').. pg0(g) - pgd(g,'t1') =l= pgdw(g)
;

costsu_eq(gen_con(g),t)$(ord(t) gt 1)..  cgsu(g,t) =e= cosu(g)*(u(g,t) - u(g,t-1))
;
costsd_eq(gen_con(g),t)$(ord(t) gt 1)..  cgsd(g,t) =e= cosd(g)*(u(g,t-1) - u(g,t))
;
```

```
pevcd_limit_eq(k,b,t)$pev_online(k,t)..  pevcd(k,b,t) =l= nev(k,b)
    *pevmax
;
pevdd_limit_eq(k,b,t)$pev_online(k,t)..  pevdd(k,b,t) =l= nev(k,b)
    *pevmax
;

rgcu_limit_eq(gen_con(g),t).. rgcu(g,t) =l= u(g,t)*pgmax(g) - pgd(
    g,t)
;
rgcd_limit_eq(gen_con(g),t).. rgcd(g,t) =l= pgd(g,t)
;

pld_eq(l,t)..    pld(l,t) =e= (1/x(l))*(sum(b$busOline(b,l),thetad
    (b,t)) -sum(b$busFline(b,l),thetad(b,t)))
;
thetadmax_eq(b,t).. thetad(b,t) =l= 3.141592
;
thetadmin_eq(b,t).. thetad(b,t) =g= -3.141592
;
thetadref_eq(ref_bus(b),t).. thetad(b,t) =e= 0
;
pldmax_eq(l,t).. pld(l,t) =l= plmax(l)
;
pldmin_eq(l,t).. pld(l,t) =g= -plmax(l)
;

balanceb_eq(b,t,w).. sum(g$gen2bus(g,b),pgb(g,t,w))
                    - sum(d$dem2bus(d,b), pdb(d,t,w) - pud(d,t,w))
                    + sum(k$(pev2bus(k,b) and pev_online(k,t)),
                        pevbd(k,b,t,w) - pevbc(k,b,t,w))
                     =e= sum(l$lineObus(l,b), plb(l,t,w)) - sum(
                        l$lineFbus(l,b), plb(l,t,w))
;

pudmax_eq(d,t,w).. pud(d,t,w) =l= pdb(d,t,w)
;

pgb_con_eq(gen_con(g),t,w).. pgb(g,t,w) =e= pgd(g,t) + rgdu(g,t,w)
      - rgdd(g,t,w)
;
pgb_ren_eq(gen_ren(g),t,w).. pgb(g,t,w) + pgbs(g,t,w) =e= gammagw(
    g,t,w)*pgmax(g)
;

pgbmin_con_eq(gen_con(g),t,w).. pgb(g,t,w) =g= u(g,t)*pgmin(g)
;
pgbmax_con_eq(gen_con(g),t,w).. pgb(g,t,w) =l= u(g,t)*pgmax(g)
;
pgbmax_ren_eq(gen_ren(g),t,w).. pgb(g,t,w) =l= gammagw(g,t,w)*
    pgmax(g)
;

rampbup_eq(gen_con(g),t,w)$(ord(t) gt 1).. pgb(g,t,w) - pgb(g,t-1,
    w) =l= pgup(g)
;
rampbdw_eq(gen_con(g),t,w)$(ord(t) gt 1).. pgb(g,t-1,w) - pgb(g,t,
    w) =l= pgdw(g)
;
```

```
rampbup_t1_eq(gen_con(g),'t1',w).. pgb(g,'t1',w) - pg0(g) =l= pgup(g)
;
rampbdw_t1_eq(gen_con(g),'t1',w).. pg0(g) - pgb(g,'t1',w) =l= pgdw(g)
;

rgdu_limit_eq(g,t,w).. rgdu(g,t,w) =l= rgcu(g,t)
;
rgdd_limit_eq(g,t,w).. rgdd(g,t,w) =l= rgcd(g,t)
;

plb_eq(l,t,w)..     plb(l,t,w) =e= (1/x(l))*(sum(b$busOline(b,l),thetab(b,t,w)) -sum(b$busFline(b,l),thetab(b,t,w)))
;
thetabmax_eq(b,t,w).. thetab(b,t,w) =l= 3.141592
;
thetabmin_eq(b,t,w).. thetab(b,t,w) =g= -3.141592
;
thetabref_eq(ref_bus(b),t,w).. thetab(b,t,w) =e= 0
;
plbmax_eq(l,t,w).. plb(l,t,w) =l= plmax(l)
;
plbmin_eq(l,t,w).. plb(l,t,w) =g= -plmax(l)
;

eev_init_eq(k,b,t,w)$pev_ti(k,t).. eev(k,b,t-1,w) =e= nev(k,b)*socev0(k,b,w)*eevmax(k)
;
eev_online_eq(k,b,t,w)$pev_online(k,t).. eev(k,b,t,w) =e= eev(k,b,t-1,w) + etaev*pevbc(k,b,t,w) - (1/etaev)*pevbd(k,b,t,w)
;
eev_final_eq(k,b,t,w)$pev_tf(k,t).. eev(k,b,t,w) =g= nev(k,b)*socevf(k)*eevmax(k)
;

pevbd_eev_eq(k,b,t,w)$pev_online(k,t).. (1/etaev)*pevbd(k,b,t,w) =l= eev(k,b,t,w)
;

pevbc_eq(k,b,t,w)$pev_online(k,t).. pevbc(k,b,t,w) =e= pevcd(k,b,t) - revduc(k,b,t,w) + revddc(k,b,t,w)
;
pevbd_eq(k,b,t,w)$pev_online(k,t).. pevbd(k,b,t,w) =e= pevdd(k,b,t) + revdud(k,b,t,w) - revddd(k,b,t,w)
;

revdd_eq(k,b,t,w)$pev_online(k,t).. revdd(k,b,t,w) =e= revddc(k,b,t,w) + revddd(k,b,t,w)
;
revdu_eq(k,b,t,w)$pev_online(k,t).. revdu(k,b,t,w) =e= revduc(k,b,t,w) + revdud(k,b,t,w)
;

revdu_limit_eq(k,b,t,w).. revdu(k,b,t,w) =l= revcu(k,b,t)
;
revdd_limit_eq(k,b,t,w).. revdd(k,b,t,w) =l= revcd(k,b,t)
;
```

```
revduc_limit_eq(k,b,t,w)$pev_online(k,t).. revduc(k,b,t,w) =l=
    pevcd(k,b,t)
;
revdud_limit_eq(k,b,t,w)$pev_online(k,t).. revdud(k,b,t,w) =l= nev
    (k,b)*pevmax - pevdd(k,b,t)
;
revddc_limit_eq(k,b,t,w)$pev_online(k,t).. revddc(k,b,t,w) =l= nev
    (k,b)*pevmax - pevcd(k,b,t)
;
revddd_limit_eq(k,b,t,w)$pev_online(k,t).. revddd(k,b,t,w) =l=
    pevdd(k,b,t)
;

eevmin_eq(k,b,t,w).. eev(k,b,t,w) =g= nev(k,b)*eevmin(k)
;
eevmax_eq(k,b,t,w).. eev(k,b,t,w) =l= nev(k,b)*eevmax(k)
;

*****************************************************************
*  MODEL AND SOLVING
*****************************************************************

Model EnergyAndReserve /all/;
EnergyAndReserve.optcr=0;
Option mip = cplex;

Solve EnergyAndReserve using mip minimizing obj;
```

References

1. Asensio, M., Contreras, J.: Stochastic unit commitment in isolated systems with renewable penetration under CVaR assessment. IEEE Trans. Smart Grid. **7**(3), 1356–1367 (2016)
2. Birge, J.R., Louveaux, F.: Introduction to Stochastic Programming, 2nd edn. Springer, New York, US (2011)
3. Carrión, M., Domínguez, R., Cañas-Carretón, M., Zárate-Miñano, R.: Scheduling isolated power systems considering electric vehicles and primary frequency response. Energy **168**, 1192–1207 (2019)
4. Carrión, M., Domínguez, R., Zárate-Miñano, R.: Influence of the controllability of electric vehicles on generation and storage capacity expansion decisions. Energy **189**, 116156 (2019)
5. Carrión, M., Zárate-Miñano, R.: Operation of intermittent-dominated power systems with a significant penetration of plug-in electric vehicles. Energy **90**, 827–835 (2015)
6. Conejo, A.J., Baringo, L.: Power System Operations, Springer, Cham (2018)
7. Donadee, J., Ilic, M.D.: Stochastic optimization of grid to vehicle frequency regulation capacity bids. IEEE Trans. Smart Grid. **5**(2), 1061–1069 (2014)
8. GAMS (2022). www.gams.com
9. International energy agency: World Energy Outlook 2019, IEA, Paris (2022). https://www.irena.org/Statistics
10. Kempton, W., Letendre, S.: Electric vehicles as a new power source for electric utilities. Transp. Res. Part D **2**, 157–175 (1997)
11. Kroposki, B., et al.: Achieving a 100% renewable grid: operating electric power systems with extremely high levels of variable renewable energy. IEEE Power Energy Mag. **15**(2), 61–73 (2017)

12. Lebrouhi, B.E., et al.: Key challenges for a large-scale development of battery electric vehicles: a comprehensive review. J. Energy Storage Part B **44**, 103273 (2021)
13. Liu, H., Hu, Z.C., Song, Y.H., Lin, J.: Decentralized vehicle-to-grid control for primary frequency regulation considering charging demands. IEEE Trans. Power Syst. **28**(3), 3480–3489 (2013)
14. Liu, H., Hu, Z., Song, Y., Wang, J., Xie, X.: Vehicle-to-grid control for supplementary frequency regulation considering charging demands. IEEE Trans. Power Syst. **30**(6), 3110–3119 (2015)
15. Mu, Y., Wu, J., Ekanayake, J., Jenkins, N., Jia, H.: Primary frequency response from electric vehicles in the Great Britain power system. IEEE Trans. Smart Grid **4**(2), 1142–1150 (2013)
16. Sarker, M.R., Dvorkin, Y., Ortega-Vazquez, M.A.: Optimal participation of an electric vehicle aggregator in day-ahead energy and reserve markets. IEEE Trans. Power Syst. **31**(5), 3506–3515 (2016)
17. Gonzáilez-Vayá, M., Andersson, G.: Optimal bidding strategy of a plug-in electric vehicle aggregator in day-ahead electricity markets. IEEE Trans. Power Syst. **24**(4), 1855–1866 (2013)
18. Vagropoulos, S.I., Kyriazidis, D.K., Bakirtzis, A.G.: Real-time charging management framework for electric vehicle aggregators in a market environment. IEEE Trans. Smart Grid **7**(2), 948–957 (2016)
19. Xu, B., et al.: Scalable planning for energy storage in energy and reserve markets. IEEE Trans. Power Syst. **32**(6), 4515–4527 (2017)
20. Yao, E., Wong, V.W.S., Schober, R.: Robust frequency regulation capacity scheduling algorithm for electric vehicles. IEEE Trans. Smart Grid. **8**(2), 984–997 (2017)

Chapter 6
Bidding Strategy of an Electric Vehicle Aggregator

This chapter analyzes the bidding strategy problem of an aggregator managing a set of electric vehicles. The electric vehicle aggregator determines the bidding decisions in the market that minimize the charging costs of electric vehicles and, at the same time, meet the driving requirements of electric vehicle users. The models developed in this chapter take into account the uncertainty in both market prices and driving needs using a set of scenarios, which allows formulating the problem using a stochastic programming approach. The impact of the bidding decisions of the electric vehicle aggregator on market prices is also addressed in the decision-making tool. A number of illustrative examples explain how to formulate and solve this type of problems.

6.1 Introduction

The number of Electric Vehicles (EVs) in electric energy systems is expected to increase in the near future. One of the most relevant problems is determining the charging strategy of EVs that minimizes their charging costs while meeting the driving requirements of EV users. In this aspect, it is important to note that the charging of EVs is generally flexible since they are parked most of the time and can be charged at the most suitable times.

However, determining the charging strategy of every individual EV has two main problems. Firstly, the limited flexibility of an individual EV. Secondly, the high computation effort that would be needed to manage a high number of EVs. To deal with these two issues, the figure of an aggregator, which can be seen as an energy management system in charge of an EV fleet, is considered. The aggregator manages the charging of these EVs with the aim of minimizing the charging costs and guaranteeing that the driving requirements of EV users are satisfied. The concept of EV aggregator has been widely considered in the technical literature [12, 19].

L. Baringo et al., *Electric Vehicles and Renewable Generation*,
Green Energy and Technology, https://doi.org/10.1007/978-3-031-09079-0_6

Under this context, this chapter focuses on the bidding strategy problem of an EV aggregator that participates in the Day-Ahead (DA) energy market. The aim of this problem is determining both the power quantities traded in this market, as well as the bid prices.

The bidding strategy problem is solved once a day, one day in advance, and on an hourly basis. Therefore, the EV aggregator faces a number of uncertain parameters at the time of deciding its bidding strategy, including market prices and driving requirements of EV users. In this chapter, these uncertainties are modeled using a set of scenarios so that the decision-making problem under uncertainty is formulated as a stochastic programming problem [5].

Other important aspect of EVs is that their batteries may be also used to inject energy into the network, i.e., charging can be bidirectional, also known as Vehicle-to-Grid (V2G), which allows the EV aggregator to participate as a power producer or as a power consumer in the market, depending on the market prices and the driving requirements. This is also considered in this chapter.

The remainder of this chapter is organized as follows. Section 6.2 describes the main characteristics of the bidding strategy problem faced by an EV aggregator. This problem is formulated in Sect. 6.3 using a deterministic and a stochastic approach. The problems formulated in Sect. 6.3 considers that the EV aggregator behaves as a price-taker, i.e., it is assumed that its bidding decisions do not alter market prices. Then, this problem is extended in Sect. 6.4 to consider a price-maker EV aggregator that is able to alter market prices to its own benefit. Sections 6.2–6.4 include a number of illustrative examples. Section 6.5 provides a brief summary and the main conclusions derived from the numerical experiments carried out. Section 6.6 defines the main notation used in the chapter. Finally, Sect. 6.7 provides the GAMS codes used to solve some of the illustrative examples included in Sects. 6.2–6.4.

6.2 Problem Description

This section describes the main characteristics of the bidding strategy problem faced by an EV aggregator.

6.2.1 Electric Vehicle Aggregator

As explained in Chap. 2, managing EVs individually may lead to computationally intractable problems if the number of EVs is large enough. This chapter considers the figure of an EV aggregator that manages a fleet of EVs. The EV aggregation is modeled as a *virtual battery* as explained in Sect. 2.4.3 of Chap. 2.

6.2.2 *Market Structure and Bidding Curves*

It is assumed that the EV aggregator participates in the DA energy market. This market is the one with the largest volume of energy traded and is cleared once a day, one day in advance, and on an hourly basis. Therefore, the EV aggregator must decide and submit its bidding curves for each hour of the following day one day in advance. Additional information about the DA energy market is provided in Sect. 2.2.1 of Chap. 2.

6.2.3 *Uncertainty Characterization*

The EV aggregator participates in the DA market. The bidding decisions in this market are made one day in advance. Therefore, the EV aggregator faces a number of uncertainties at the time of making these decisions, namely:

1. The market prices.
2. The driving requirements, which impact the upper bounds of charging and discharging power levels, as well as the minimum and maximum energy levels of the virtual battery representing the EV aggregation.

In this chapter, uncertainties are modeled using a finite set of scenarios [2]. Each scenario represents a plausible realization of the uncertain parameters and has associated a probability of occurrence.

Illustrative Example 6.1 Scenarios

We consider an EV aggregator deciding its self-scheduling strategy in the DA energy market. This problem is solved one day in advance, when the EV aggregator faces the uncertainty in the market prices, which are modeled using the two scenarios depicted in Fig. 6.1. Both scenarios have the same probability. In addition, two scenarios are also considered to model the uncertainty in the driving requirements, in particular, in the maximum charging and discharging power levels, as depicted in Fig. 6.2. Again, both scenarios are equiprobable. For the sake of simplicity, no uncertainty is considered in the energy bounds.

Scenarios for market prices and driving requirements are considered independent. Therefore, there is a total of four scenarios to represent the uncertainties in the problem, as summarized in Table 6.1. □

6.2.4 *Decision Framework*

Considering the uncertainty characterization described in Sect. 6.2.3, the bidding strategy problem is formulated as a two-stage stochastic programming problem [2]. The decision making process is as follows:

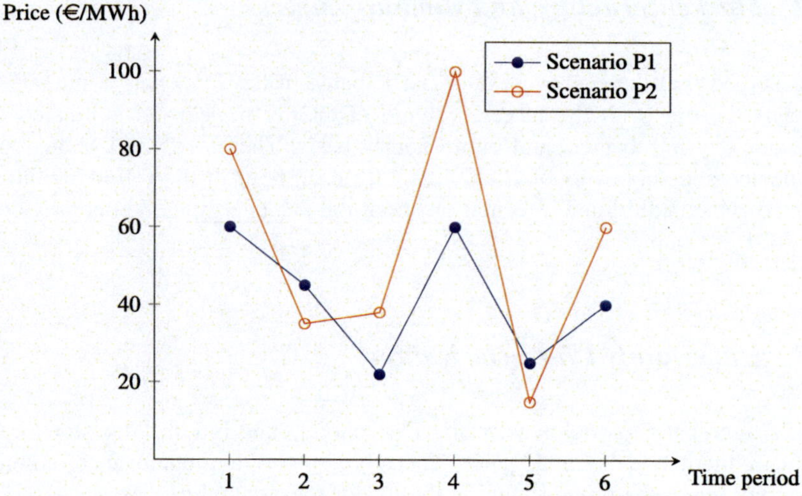

Fig. 6.1 Illustrative Example 6.1: scenarios for prices

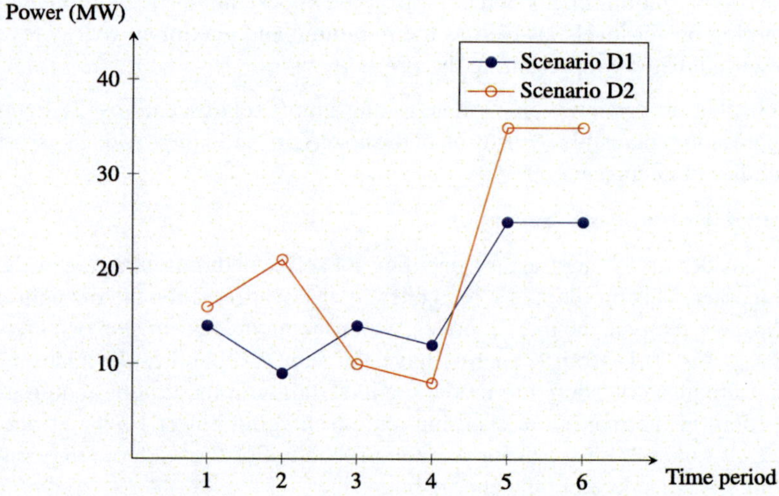

Fig. 6.2 Illustrative Example 6.1: scenarios for maximum charging and discharging power levels

Table 6.1 Illustrative Example 6.1: scenarios

Scenario	Price scenario	Driving scenario	Probability
1	P1	D1	$0.5 \cdot 0.5 = 0.25$
2	P1	D2	$0.5 \cdot 0.5 = 0.25$
3	P2	D1	$0.5 \cdot 0.5 = 0.25$
4	P2	D2	$0.5 \cdot 0.5 = 0.25$

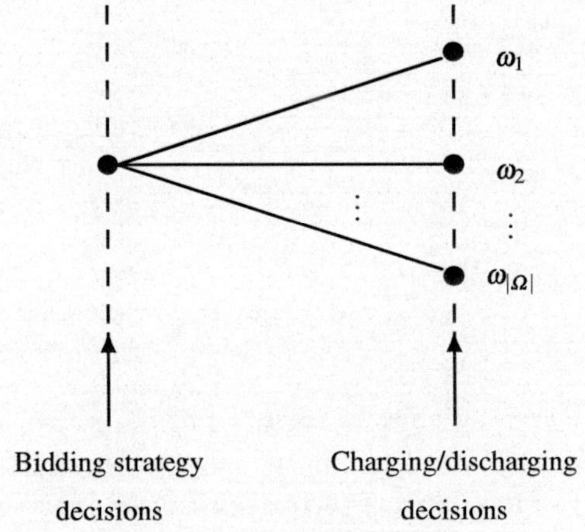

Fig. 6.3 Scenario tree for the two-stage stochastic bidding strategy problem

1. The EV aggregator determines its bidding strategy in the DA market. This is done one day in advance, i.e., before knowing the actual realization of the uncertain parameters. Therefore, these decisions are *here-and-now* since they are independent of the future scenarios.
2. A scenario ω is realized.
3. After knowing the actual scenario realization, the EV aggregator determines the charging and discharging power levels. These decisions depend on the scenario realization and therefore are *wait-and-see* decisions.

The above decision sequence corresponds with the scenario tree depicted in Fig. 6.3.

6.3 Problem Formulation

This section provides the formulation of the bidding strategy problem faced by an EV aggregator. Firstly, a deterministic problem in which uncertain parameters are represented using single-point forecasts is presented. Then, this deterministic problem is extended to a stochastic programming problem in which uncertain parameters are represented using a set of scenarios as described in Sect. 6.2.3.

6.3.1 Deterministic Problem

The bidding strategy problem of an EV aggregator considering a deterministic approach is formulated using the following model:

$$\text{Minimize}_{\Phi^{\text{D}}} \quad \sum_{t \in T} \tilde{\lambda}_t^{\text{D}} \left(p_t^{\text{EV,C}} - p_t^{\text{EV,D}} \right) \Delta_t \tag{6.1a}$$

subject to:

$$e_t^{\mathrm{EV}} = e_{t-1}^{\mathrm{EV}} + p_t^{\mathrm{EV,C}} \eta^{\mathrm{EV,C}} \Delta_t - \frac{p_t^{\mathrm{EV,D}} \Delta_t}{\eta^{\mathrm{EV,D}}} + E_t^{\mathrm{EV,IO}}, \quad \forall t \in T, \tag{6.1b}$$

$$0 \leq p_t^{\mathrm{EV,C}} \leq \overline{P}_t^{\mathrm{EV,C}}, \quad \forall t \in T, \tag{6.1c}$$

$$0 \leq p_t^{\mathrm{EV,D}} \leq \overline{P}_t^{\mathrm{EV,D}}, \quad \forall t \in T, \tag{6.1d}$$

$$\underline{E}_t^{\mathrm{EV}} \leq e_t^{\mathrm{EV}} \leq \overline{E}_t^{\mathrm{EV}}, \quad \forall t \in T, \tag{6.1e}$$

where variables in set $\Phi^{\mathrm{D}}=\left\{p_t^{\mathrm{EV,C}}, p_t^{\mathrm{EV,D}}, e_t^{\mathrm{EV}}, \forall t \in T\right\}$ are the optimization variables of the deterministic model (6.1a)–(6.1e).

Model (6.1a)–(6.1e) is a Linear Programming (LP) problem [17]. The charging costs minus discharging revenues are minimized in (6.1a). Constraints (6.1b) define the energy evolution in the virtual battery representing the EV aggregation. Constraints (6.1c) and (6.1d) impose bounds on the charging and discharging power levels, respectively. Finally, constraints (6.1e) impose limits on the energy levels.

The output of problem (6.1a)–(6.1e) comprises the charging and discharging power levels of the virtual battery representing the EV aggregation. However, note that the EV aggregator must submit an energy quantity-price pair to the DA market. In this sense, the EV aggregator has two options:

1. The optimal discharging power level is offered at zero price, while the bid price for the optimal charging power level is fixed to a large enough value. This way, the EV aggregator knows with certainty that its offer/bid will be accepted in the DA market. The disadvantage of this option is that if the price deviates from the forecast, the actual cost/revenue achieved by the EV aggregator may be much higher/lower.
2. The offer and bid prices submitted by the EV aggregator are fixed to the forecast market prices. The disadvantage of this option is that if the actual price deviates from this forecast, the offer/bid may be not accepted.

Illustrative Example 6.2 Deterministic bidding strategy

The bidding strategy throughout a 6-h planning horizon of an EV aggregator is analyzed below. The EV aggregation comprises 10,000 individual EVs whose driving requirements are represented using the maximum charging and discharging power levels depicted in Fig. 6.4, the minimum and maximum energy levels provided in Fig. 6.5, as well as the energy contributions/drops of EVs arriving to/departing from the virtual battery representing the EV aggregation depicted in Fig. 6.6. Charging and discharging efficiency rates are considered equal to 90%, while the initial energy level is 30 MWh.

The prices paid by/paid to the EV aggregator for charging/discharging the EVs are provided in Fig. 6.7.

The optimal charging and discharging profile of the virtual battery representing the EV aggregation is provided in Fig. 6.8. The solution of this problem is used to

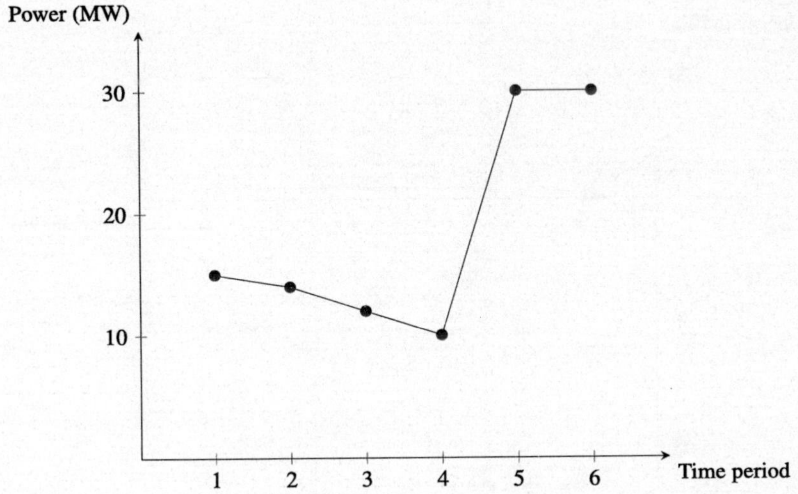

Fig. 6.4 Illustrative Example 6.2: maximum charging and discharging power levels

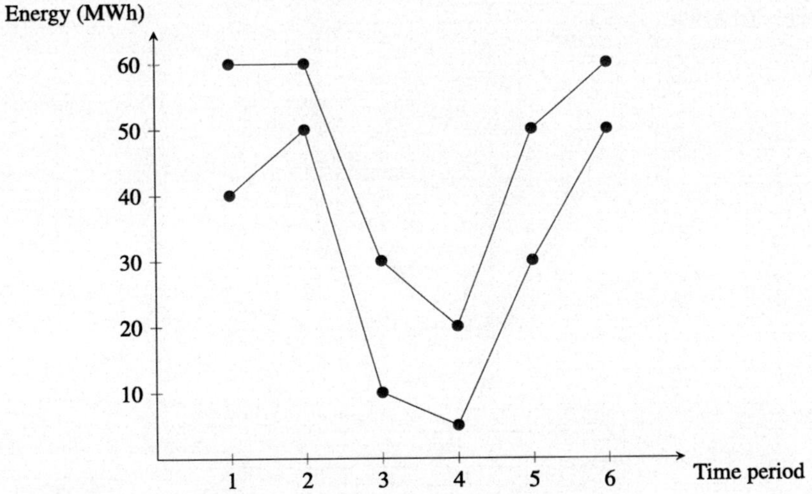

Fig. 6.5 Illustrative Example 6.2: minimum and maximum energy levels

determine the bidding strategy of the EV aggregator. Note that the EV aggregator charges the virtual battery representing the EV aggregation in time periods 1, 2, 3, and 5, while it discharges the battery in time periods 4 and 6. Two options are considered for the offer/bid prices, namely:

1. Option A. The EV aggregator considers a bid price of 180 €/MWh in time periods 1, 2, 3, and 5, and an offer price of 0 in time periods 4 and 6. This

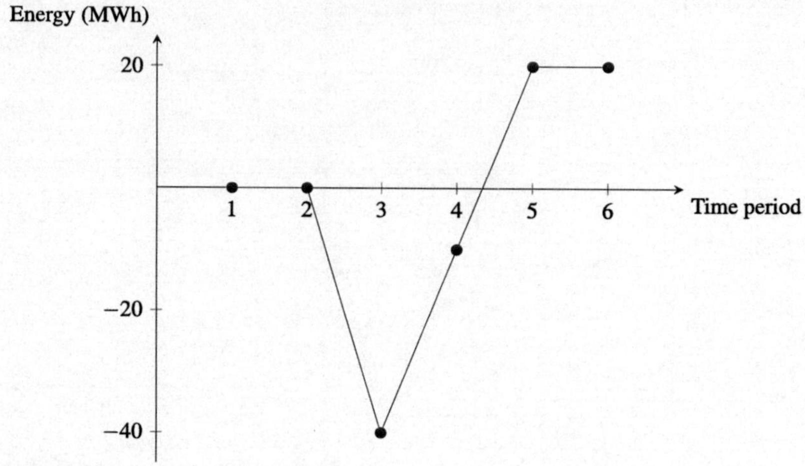

Fig. 6.6 Illustrative Example 6.2: energy contributions/drop of EVs arriving to/departing from the EV aggregation

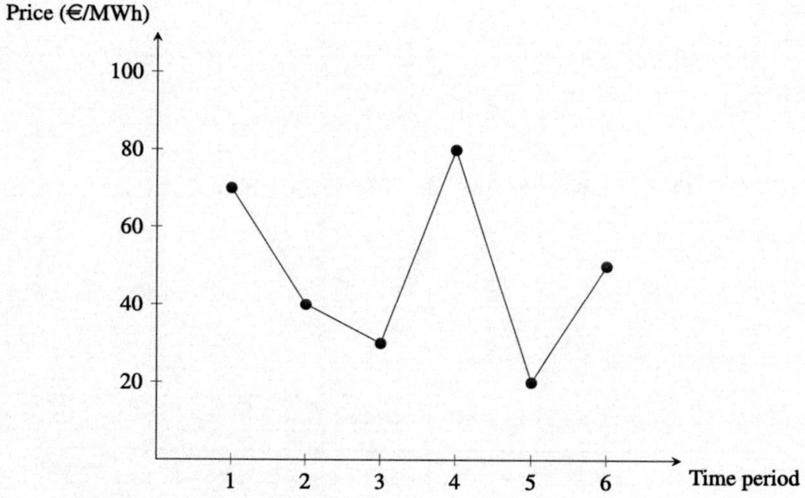

Fig. 6.7 Illustrative Example 6.2: prices

way, the EV aggregator knows that its offers/bids will be accepted by the market operator and it will be able to charge/discharge the EVs.

2. Option B. The EV aggregator considers bid/offer prices of 70, 40, 30, 80, 20, and 50 €/MWh in time periods 1, 2, 3, 4, 5, and 6, respectively, i.e., the forecast price in each time period.

The forecast charging cost minus discharging revenue of the EV aggregator is 748.53 €. Next, let us assume that the actual market prices are those depicted in

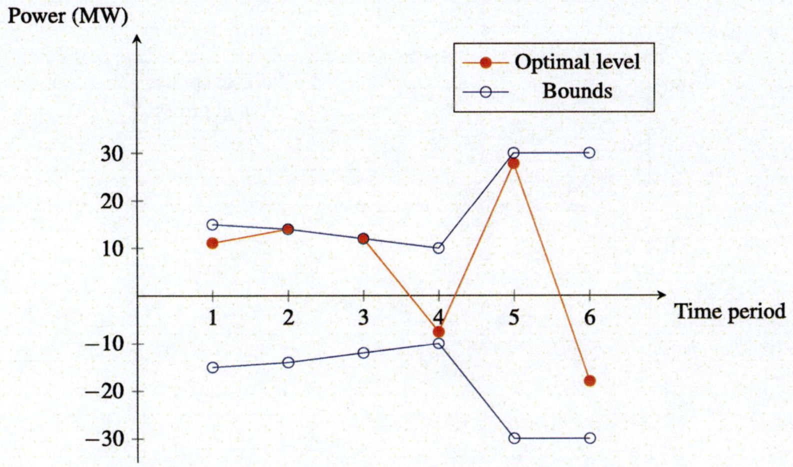

Fig. 6.8 Illustrative Example 6.2: charging and discharging power levels

Fig. 6.9. In such a case, the actual power levels scheduled in the DA market for the EV aggregator would be different depending on the option considered for offer/bid prices as explained below:

1. Option A. All the offers/bids of the EV aggregator would be accepted. However, the actual charging cost minus discharging revenue would be different. In such a case, this value can be computed as $75 \cdot 11.11 + 38 \cdot 14.00 + 42 \cdot 12.00 + 75 \cdot (-7.56) + 25 \cdot 27.78 + 52 \cdot (-18) = 1060.75$ €, i.e., the total cost faced by the EV aggregator increases by 41.71% in comparison with the forecast value.
2. Option B. In this case, the bid of the EV aggregator is only accepted in time period 2 because the bid prices are lower than the actual market prices in the remaining time periods. On the other hand, the offer of the EV aggregator in time period 6 is accepted since the offer price is lower than the market price; however, this is not the case of time period 4. This means that the EV would be only scheduled in the DA market in time periods 2 and 6, and would need to participate in other markets to charge the EVs in the remaining time periods, which may have a significant impact on the costs.

□

6.3.2 *Stochastic Programming Problem*

The formulation of the bidding strategy problem is extended in this section by considering a two-stage stochastic programming problem [2]. In this case, uncertain parameters are modeled using a set of scenarios as described in Sect. 6.2.3. For the sake of clarity, it is assumed that the EV aggregator cannot discharge the EVs. The resulting model is formulated below:

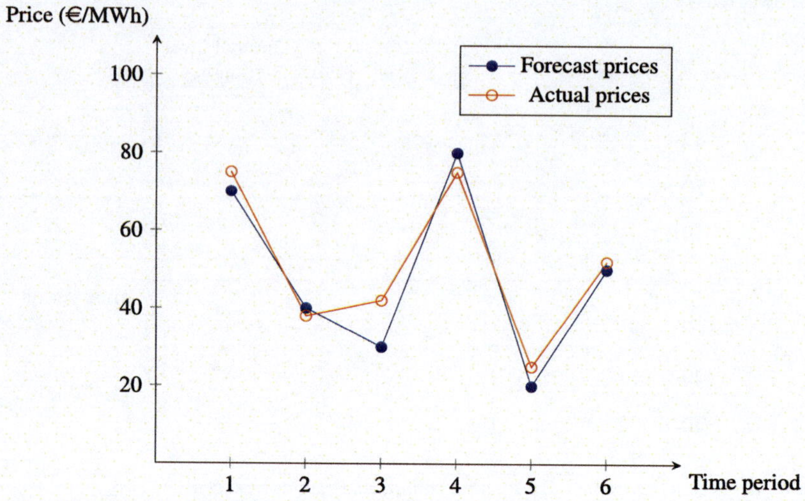

Fig. 6.9 Illustrative Example 6.2: actual DA market prices

$$\text{Minimize}_{\Phi^{\text{SP}}} \quad \sum_{\omega \in \Omega} \pi_\omega \sum_{t \in T} \tilde{\lambda}^{\text{D}}_{t\omega} p^{\text{EV,C}}_t \Delta_t \tag{6.2a}$$

subject to:

$$e^{\text{EV}}_{t\omega} = e^{\text{EV}}_{(t-1)\omega} + p^{\text{EV,C}}_t \eta^{\text{EV,C}} \Delta_t + E^{\text{EV,IO}}_{t\omega}, \quad \forall t \in T, \forall \omega \in \Omega, \tag{6.2b}$$

$$0 \leq p^{\text{EV,C}}_t \leq \overline{P}^{\text{EV,C}}_{t\omega}, \quad \forall t \in T, \forall \omega \in \Omega, \tag{6.2c}$$

$$\underline{E}^{\text{EV}}_{t\omega} \leq e^{\text{EV}}_{t\omega} \leq \overline{E}^{\text{EV}}_{t\omega}, \quad \forall t \in T, \forall \omega \in \Omega, \tag{6.2d}$$

where variables in set $\Phi^{\text{SP}} = \left\{ p^{\text{EV,C}}_t, \forall t \in T; e^{\text{EV}}_{t\omega}, \forall t \in T, \forall \omega \in \Omega \right\}$ are the optimization variables of the stochastic programming model (6.2a)–(6.2d).

Model (6.2a)–(6.2d) is an LP problem [17] and is very similar to the deterministic problem (6.1a)–(6.1e), but with three differences:

1. The objective function (6.2a) represents the expected charging costs over the considered scenarios.
2. Constraints (6.2b)–(6.2d) are imposed for all scenarios.
3. Model (6.2a)–(6.2d) does not consider the option of discharging the EVs.

Illustrative Example 6.3 Stochastic bidding strategy

The data of Illustrative Example 6.2 are considered again. However, in this case uncertain market prices and maximum charging power levels are modeled using the scenarios described in Illustrative Example 6.1, and it is assumed that EVs cannot discharge.

Considering these data, the problem is formulated below.

First, the objective function of the problem is the minimization of expected purchasing costs:

$$
\begin{aligned}
&0.25\Big[60p_1^{\mathrm{EV,C}} + 45p_2^{\mathrm{EV,C}} + 22p_3^{\mathrm{EV,C}} + 60p_4^{\mathrm{EV,C}} + 25p_5^{\mathrm{EV,C}} + 40p_6^{\mathrm{EV,C}}\Big] \\
&\quad + 0.25\Big[60p_1^{\mathrm{EV,C}} + 45p_2^{\mathrm{EV,C}} + 22p_3^{\mathrm{EV,C}} + 60p_4^{\mathrm{EV,C}} + 25p_5^{\mathrm{EV,C}} + 40p_6^{\mathrm{EV,C}}\Big] \\
&\quad + 0.25\Big[80p_1^{\mathrm{EV,C}} + 35p_2^{\mathrm{EV,C}} + 38p_3^{\mathrm{EV,C}} + 100p_4^{\mathrm{EV,C}} + 15p_5^{\mathrm{EV,C}} + 60p_6^{\mathrm{EV,C}}\Big] \\
&\quad + 0.25\Big[80p_1^{\mathrm{EV,C}} + 35p_2^{\mathrm{EV,C}} + 38p_3^{\mathrm{EV,C}} + 100p_4^{\mathrm{EV,C}} + 15p_5^{\mathrm{EV,C}} + 60p_6^{\mathrm{EV,C}}\Big].
\end{aligned}
$$

Next, the constraints of the problem are formulated:

- Energy evolution of the virtual battery representing the EV aggregation:

$$
\begin{aligned}
e_{1i}^{\mathrm{EV}} &= 30 + 0.9p_1^{\mathrm{EV,C}} + 0, \quad i = 1, \ldots, 4, \\
e_{2i}^{\mathrm{EV}} &= e_{1i}^{\mathrm{EV}} + 0.9p_2^{\mathrm{EV,C}} + 0, \quad i = 1, \ldots, 4, \\
e_{3i}^{\mathrm{EV}} &= e_{2i}^{\mathrm{EV}} + 0.9p_3^{\mathrm{EV,C}} - 40, \quad i = 1, \ldots, 4, \\
e_{4i}^{\mathrm{EV}} &= e_{3i}^{\mathrm{EV}} + 0.9p_4^{\mathrm{EV,C}} - 10, \quad i = 1, \ldots, 4, \\
e_{5i}^{\mathrm{EV}} &= e_{4i}^{\mathrm{EV}} + 0.9p_5^{\mathrm{EV,C}} + 20, \quad i = 1, \ldots, 4, \\
e_{6i}^{\mathrm{EV}} &= e_{5i}^{\mathrm{EV}} + 0.9p_6^{\mathrm{EV,C}} + 20, \quad i = 1, \ldots, 4.
\end{aligned}
$$

- Charging power limits:

$$
\begin{aligned}
0 &\le p_1^{\mathrm{EV,C}} \le 14, \\
0 &\le p_2^{\mathrm{EV,C}} \le 9, \\
0 &\le p_3^{\mathrm{EV,C}} \le 14, \\
0 &\le p_4^{\mathrm{EV,C}} \le 12, \\
0 &\le p_5^{\mathrm{EV,C}} \le 25, \\
0 &\le p_6^{\mathrm{EV,C}} \le 25, \\
0 &\le p_1^{\mathrm{EV,C}} \le 16, \\
0 &\le p_2^{\mathrm{EV,C}} \le 21, \\
0 &\le p_3^{\mathrm{EV,C}} \le 10, \\
0 &\le p_4^{\mathrm{EV,C}} \le 8, \\
0 &\le p_5^{\mathrm{EV,C}} \le 35, \\
0 &\le p_6^{\mathrm{EV,C}} \le 35, \\
0 &\le p_1^{\mathrm{EV,C}} \le 14, \\
0 &\le p_2^{\mathrm{EV,C}} \le 6,
\end{aligned}
$$

$$0 \le p_3^{\mathrm{EV,C}} \le 14,$$
$$0 \le p_4^{\mathrm{EV,C}} \le 12,$$
$$0 \le p_5^{\mathrm{EV,C}} \le 25,$$
$$0 \le p_6^{\mathrm{EV,C}} \le 25,$$
$$0 \le p_1^{\mathrm{EV,C}} \le 16,$$
$$0 \le p_2^{\mathrm{EV,C}} \le 21,$$
$$0 \le p_3^{\mathrm{EV,C}} \le 10,$$
$$0 \le p_4^{\mathrm{EV,C}} \le 8,$$
$$0 \le p_5^{\mathrm{EV,C}} \le 35,$$
$$0 \le p_6^{\mathrm{EV,C}} \le 35,$$

- Energy bounds:

$$40 \le e_{1i}^{\mathrm{EV}} \le 60, \quad i = 1, \dots, 4,$$
$$50 \le e_{2i}^{\mathrm{EV}} \le 60, \quad i = 1, \dots, 4,$$
$$10 \le e_{3i}^{\mathrm{EV}} \le 30, \quad i = 1, \dots, 4,$$
$$5 \le e_{4i}^{\mathrm{EV}} \le 20, \quad i = 1, \dots, 4,$$
$$30 \le e_{5i}^{\mathrm{EV}} \le 50, \quad i = 1, \dots, 4,$$
$$50 \le e_{6i}^{\mathrm{EV}} \le 60, \quad i = 1, \dots, 4.$$

Finally, the optimization variables of the problem are those included in set $\Phi^{\mathrm{IE6.3}} = \{p_j^{\mathrm{EV,C}}, j = 1, \dots, 6; e_{ji}^{\mathrm{EV}}, j = 1, \dots, 6, i = 1, \dots, 4\}$.

The above model is an LP problem [17] that is solved using CPLEX [18] under a GAMS environment [9].

The optimal charging profile of the virtual battery representing the EV aggregation is provided in Fig. 6.10. Note that the virtual battery representing the EV aggregation is preferably charged at those time periods with the comparatively lowest market prices and always complying with the power and energy bounds.

Finally, the optimal value of the objective function, i.e., the average total charging cost, is 1563.33 €. □

Illustrative Example 6.4 Impact of the stochastic solution

The data and results of Illustrative Example 6.3 are considered again. However, in this case uncertain parameters are modeled using their forecast values, i.e., considering a deterministic approach. Note that in this case, the problem formulation is identical to that used in Illustrative Example 6.2 but considering that the discharging power levels are 0, since the average values of the scenarios considered for market prices and maximum charging levels are equal to those considered in Illustrative Example 6.2.

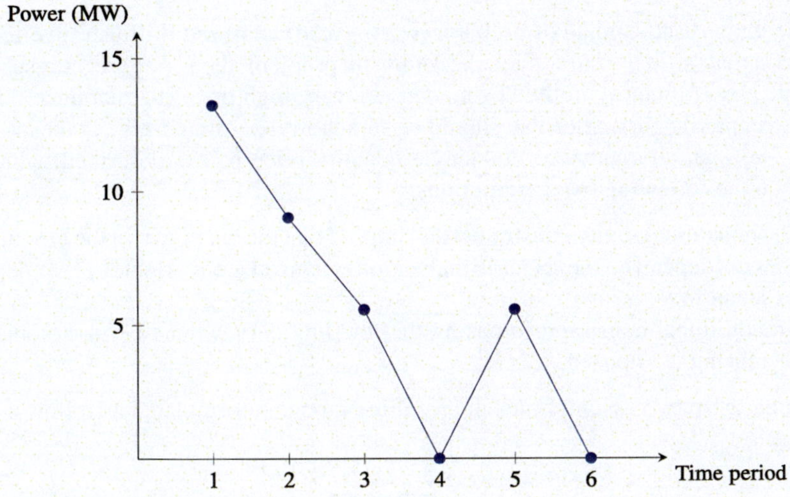

Fig. 6.10 Illustrative Example 6.3: charging power levels

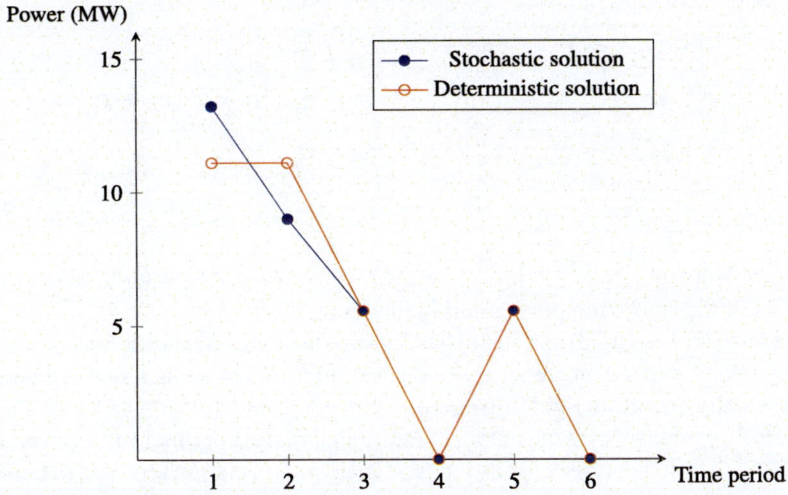

Fig. 6.11 Illustrative Example 6.4: charging power levels

Figure 6.11 depicts the optimal charging power levels for both the stochastic and deterministic approaches. Note that the solution is different in time periods 1 and 2. Moreover, implementing the solution of the deterministic approach in the second time period (charging 11.11 MW) would result in a violation of the constraints if scenarios 2 or 4 occur since the maximum charging power level in that time period for these scenarios is 9 MW. This result highlights the importance of an accurate uncertainty representation. □

The main disadvantage of problem (6.2a)–(6.2d) analyzed in Illustrative Example 6.3 is that, as in the case of the deterministic problem, the EV aggregator submits a single power quantity to the DA market at a very high price (to guarantee that its bid is accepted) or at a price that should be somehow determined, e.g., at the average price. Next, an alternative to this single quantity-price pair is described, which is based on the following two assumptions:

1. The nonanticipativity constraints that impose that the bid power in the DA market does not depend on the scenario realization is relaxed, i.e., variables $p_{t\omega}^{\mathrm{EV,C}}$ depend on scenario ω.
2. An additional constraint imposing that bidding curves must be monotonically decreasing is imposed.

Based on these two assumptions, the resulting model is formulated as follows:

$$\text{Minimize}_{\Phi^{\mathrm{SP2}}} \quad \sum_{\omega\in\Omega} \pi_\omega \sum_{t\in T} \tilde{\lambda}_{t\omega}^{\mathrm{D}} p_{t\omega}^{\mathrm{EV,C}} \Delta_t \tag{6.3a}$$

subject to:

$$p_{t\omega}^{\mathrm{EV,C}} \le p_{t\tilde{\omega}}^{\mathrm{EV,C}} \ \text{ if } \ \tilde{\lambda}_{t\omega}^{\mathrm{D}} \ge \tilde{\lambda}_{t\tilde{\omega}}^{\mathrm{D}}, \quad \forall t \in T, \forall \omega, \tilde{\omega} \in \Omega, \tag{6.3b}$$

$$e_{t\omega}^{\mathrm{EV}} = e_{(t-1)\omega}^{\mathrm{EV}} + p_{t\omega}^{\mathrm{EV,C}} \eta^{\mathrm{EV}} \Delta_t + E_{t\omega}^{\mathrm{EV,IO}}, \quad \forall t \in T, \forall \omega \in \Omega, \tag{6.3c}$$

$$0 \le p_{t\omega}^{\mathrm{EV,C}} \le \overline{P}_{t\omega}^{\mathrm{EV,C}}, \quad \forall t \in T, \forall \omega \in \Omega, \tag{6.3d}$$

$$\underline{E}_{t\omega}^{\mathrm{EV}} \le e_{t\omega}^{\mathrm{EV}} \le \overline{E}_{t\omega}^{\mathrm{EV}}, \quad \forall t \in T, \forall \omega \in \Omega, \tag{6.3e}$$

where variables in set $\Phi^{\mathrm{SP2}} = \left\{ p_{t\omega}^{\mathrm{EV,C}}, e_{t\omega}^{\mathrm{EV}}, \forall t \in T, \forall \omega \in \Omega \right\}$ are the optimization variables of the stochastic programming model (6.3a)–(6.3e).

Problems (6.2a)–(6.2d) and (6.3a)–(6.3e) have two important differences, namely, variables $p_{t\omega}^{\mathrm{EV,C}}$ depend on scenario ω and constraints (6.3b) are included to guarantee monotonically decreasing bidding curves.

The EV aggregator uses the output of the model, i.e., the optimal value of variables $p_{t\omega}^{\mathrm{EV,C}}$, and the market prices for different scenarios, i.e., parameters $\tilde{\lambda}_{t\omega}^{\mathrm{D}}$, to build the bid curves submitted to the DA market. This is illustrated in the following example.

Illustrative Example 6.5 Impact of the bidding strategy

The data of Illustrative Example 6.3 are considered again. However, in this case model (6.3a)–(6.3e) is implemented, obtaining a different charging profile for each scenario, as depicted in Fig. 6.12.

The EV aggregator can then use these charging power levels for different scenarios and the input market prices to build the bid curves. Two examples of these curves are provided in Figs. 6.13 and 6.14 that depict the bid curves in time periods 3 and 5, respectively. Note that the optimal power levels provided in Fig. 6.12 are combined with the market prices for different scenarios provided in Fig. 6.1 to build the bid curves. □

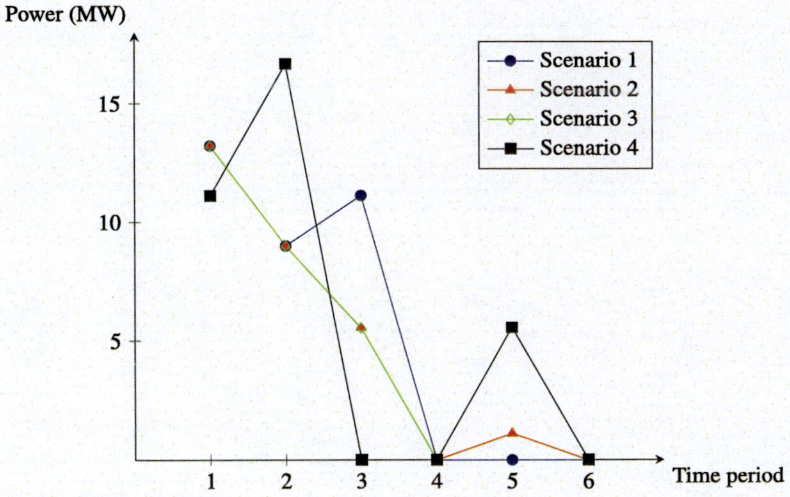

Fig. 6.12 Illustrative Example 6.5: charging power levels for different scenarios

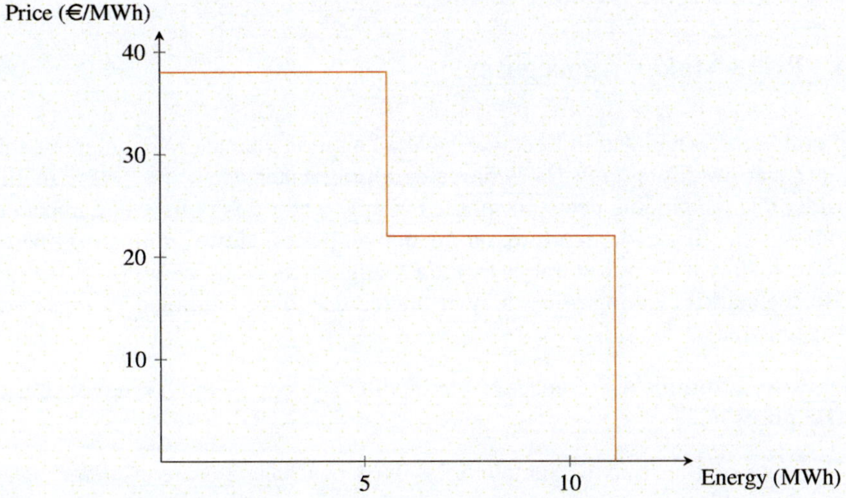

Fig. 6.13 Illustrative Example 6.5: bid curve for time period 3

Finally, note that problems (6.2a)–(6.2d) and (6.3a)–(6.3e) can be easily extended to consider EVs with V2G technologies, i.e., EVs that can also discharge their batteries and therefore participate in the market as a power producer.

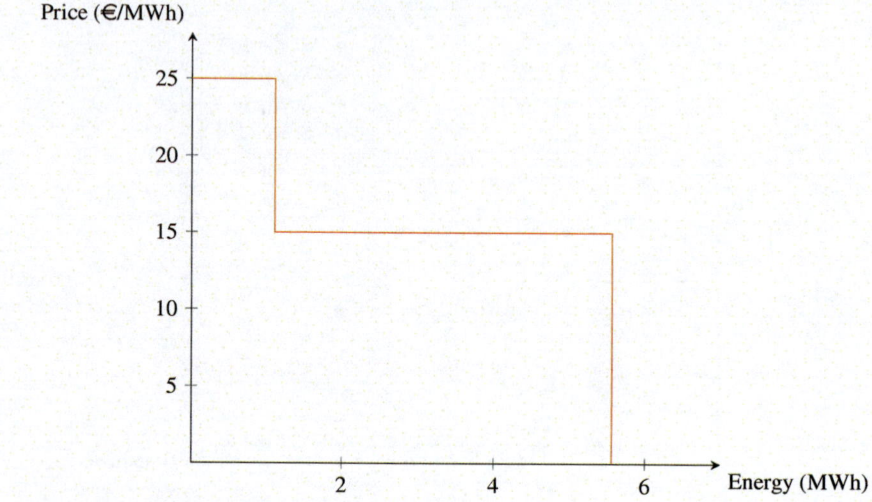

Fig. 6.14 Illustrative Example 6.5: bid curve for time period 5

6.4 Price-Maker Aggregator

The problems formulated in Sect. 6.3 have a common characteristic, namely, they assume that the EV aggregator behaves as a price-taker agent, i.e., these models consider that the bidding decisions of the EV aggregator do not have any impact on the market clearing and, therefore, on the market prices. However, this may be not the case, mainly if the power traded by the EV aggregator is sufficiently high. In such a case, the market clearing prices may be altered due to the bidding decisions of the EV aggregator.

Illustrative Example 6.6 Impact of the bidding strategy of the EV aggregator on market prices

The data of Illustrative Example 2.2 solved in Chap. 2 are considered again. However, in this case it is assumed that the EV aggregator participates also in the market by submitting its bid curve in time period 1, which comprises 13.22 MW at a high price, e.g., 100 €/MWh, as obtained in Illustrative Example 6.3. Figure 6.15 depicts both the resulting production offer (in blue) and consumption bid (in red) curves, considering the EV aggregator as an additional consumer. The crossing of these two curves determines the market clearing price, which is equal to 75 €/MWh. Note that due to the bidding decisions of the EV aggregator, the resulting market clearing price has increased from 70 €/MWh up to 75 €/MWh. □

The above illustrative example shows that the bidding strategy of the EV aggregator may influence the market prices. In order to account for this issue, it is necessary to explicitly represent the market clearing problem in the decision-making problem. To do so, the problem is formulated as a bi-level problem comprising two nested

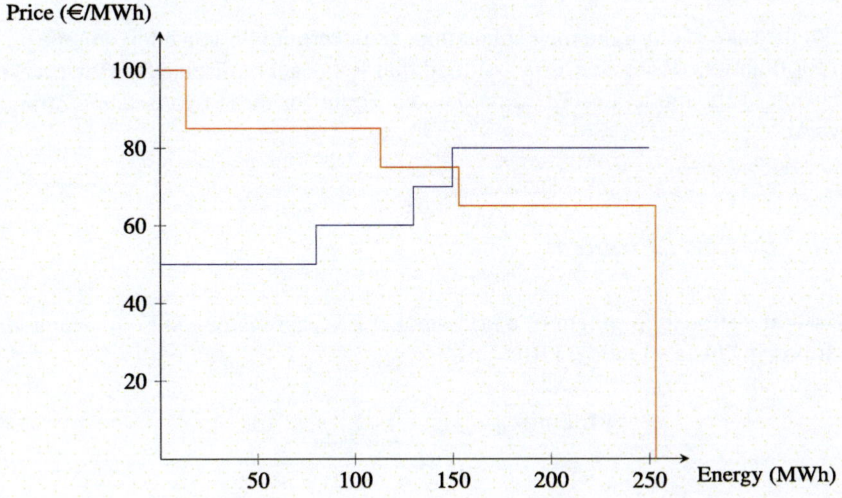

Fig. 6.15 Illustrative Example 6.6: accepted production/consumption levels and market clearing price

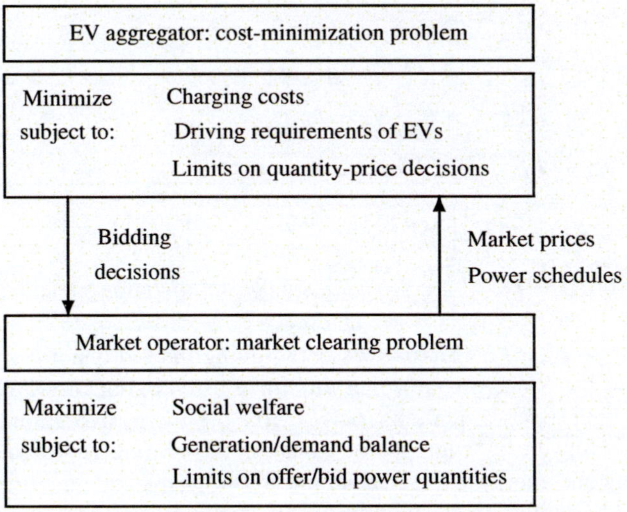

Fig. 6.16 Structure of the bi-level model for the bidding strategy of a price-maker EV aggregator

optimization problems. The upper-level problem of this bi-level model represents the bidding and operation decisions of the virtual battery representing the EV aggregation. These bidding decisions are then used in the lower-level problem, which represents the market clearing problem and provides the market prices. Figure 6.16 schematically depicts the structure of this bi-level problem.

For the sake of clarity in the explanation, no uncertainty is considered in the problem formulation. Moreover, it is assumed that EVs cannot discharge. However, the extension of the model described below to account for these two issues is straightforward.

6.4.1 Bi-Level Problem

The bidding strategy problem of a price-maker EV aggregator can be formulated as the following bi-level model [10]:

$$\text{Minimize}_{\Phi^{\text{UL}},\Phi^{\text{LL}}} \quad \sum_{t\in T} \lambda_t^{\text{D}} p_t^{\text{EV,C}} \Delta_t \tag{6.4a}$$

subject to:

$$e_t^{\text{EV}} = e_{(t-1)}^{\text{EV}} + p_t^{\text{EV,C}} \eta^{\text{EV,C}} \Delta_t + E_t^{\text{EV,IO}}, \quad \forall t \in T, \tag{6.4b}$$

$$0 \le p_t^{\text{EV,C}} \le \overline{P}_t^{\text{EV,C}}, \quad \forall t \in T, \tag{6.4c}$$

$$\underline{E}_t^{\text{EV}} \le e_t^{\text{EV}} \le \overline{E}_t^{\text{EV}}, \quad \forall t \in T, \tag{6.4d}$$

$$\overline{p}_t^{\text{EV,C}}, \xi_t^{\text{EV}} \ge 0, \quad \forall t \in T, \tag{6.4e}$$

$$\lambda_t^{\text{D}}, p_t^{\text{EV,C}} \in \Psi(\cdot), \quad \forall t \in T, \tag{6.4f}$$

where variables in sets $\Phi^{\text{UL}}=\left\{\overline{p}_t^{\text{EV,C}}, \xi_t^{\text{EV}}, e_t^{\text{EV}}, \forall t \in T\right\}$ and $\Phi^{\text{LL}}=\left\{p_{gt}^{\text{G}}, \forall g \in G, \forall t \in T;\ p_{dt}^{\text{D}}, \forall d \in D, \forall t \in T;\ p_t^{\text{EV,C}}, \forall t \in T\right\}$ are the optimization variables of problem (6.4a)–(6.4f).

The aim of problem (6.4a)–(6.4f) is determining the bid quantities ($\overline{p}_t^{\text{EV,C}}$) and prices ($\xi_t^{\text{EV}}$) of the EV aggregator that minimize its charging costs, represented in objective function (6.4a). Equations (6.4b) define the energy evolution in the virtual battery representing the EV aggregation. Constraints (6.4c) and (6.4d) impose bounds on the charging and energy levels. Constraints (6.4e) define bid quantities and prices as nonnegative variables. Finally, constraints (6.4f) impose that both the market prices and the charging power scheduled of the EV aggregator are obtained as the result of the market clearing problem represented by set $\Psi(\cdot)$, which will be formulated in Sect. 6.4.2. Note that the difference between bi-level problem (6.4a)–(6.4f) and the single-level problems described in Sect. 6.3 is that here the market prices are endogenously obtained, while in the single-level problems these market prices are input parameters. This way, the impact of the EV aggregator bidding decisions on the market clearing problem is taken into account.

Illustrative Example 6.7 Price-maker EV aggregator (bi-level problem)

The data of Illustrative Example 6.2 are considered again. However, here the EV aggregator behaves as a price-maker. Moreover, it is assumed that it cannot discharge the EVs. The resulting model is formulated as follows.

First, the objective function of the problem is the minimization of charging costs:

$$\lambda_1^{\mathrm{D}} p_1^{\mathrm{EV,C}} + \lambda_2^{\mathrm{D}} p_2^{\mathrm{EV,C}} + \lambda_3^{\mathrm{D}} p_3^{\mathrm{EV,C}} + \lambda_4^{\mathrm{D}} p_4^{\mathrm{EV,C}} + \lambda_5^{\mathrm{D}} p_5^{\mathrm{EV,C}} + \lambda_6^{\mathrm{D}} p_6^{\mathrm{EV,C}}.$$

Next, the constraints of the problem are formulated:

- Energy evolution of the virtual battery representing the EV aggregation:

$$\begin{aligned}
e_1^{\mathrm{EV}} &= 30 + 0.9 p_1^{\mathrm{EV,C}} + 0, \\
e_2^{\mathrm{EV}} &= e_1^{\mathrm{EV}} + 0.9 p_2^{\mathrm{EV,C}} + 0, \\
e_3^{\mathrm{EV}} &= e_2^{\mathrm{EV}} + 0.9 p_3^{\mathrm{EV,C}} - 40, \\
e_4^{\mathrm{EV}} &= e_3^{\mathrm{EV}} + 0.9 p_4^{\mathrm{EV,C}} - 10, \\
e_5^{\mathrm{EV}} &= e_4^{\mathrm{EV}} + 0.9 p_5^{\mathrm{EV,C}} + 20, \\
e_6^{\mathrm{EV}} &= e_5^{\mathrm{EV}} + 0.9 p_6^{\mathrm{EV,C}} + 20.
\end{aligned}$$

- Charging power limits:

$$\begin{aligned}
0 &\le p_1^{\mathrm{EV,C}} \le 15, \\
0 &\le p_2^{\mathrm{EV,C}} \le 14, \\
0 &\le p_3^{\mathrm{EV,C}} \le 12, \\
0 &\le p_4^{\mathrm{EV,C}} \le 10, \\
0 &\le p_5^{\mathrm{EV,C}} \le 30, \\
0 &\le p_6^{\mathrm{EV,C}} \le 30.
\end{aligned}$$

- Energy bounds:

$$\begin{aligned}
40 &\le e_1^{\mathrm{EV}} \le 60, \\
50 &\le e_2^{\mathrm{EV}} \le 60, \\
10 &\le e_3^{\mathrm{EV}} \le 30, \\
5 &\le e_4^{\mathrm{EV}} \le 20, \\
30 &\le e_5^{\mathrm{EV}} \le 50, \\
50 &\le e_6^{\mathrm{EV}} \le 60.
\end{aligned}$$

- Nonnegativity of bid power and price quantities:

$$\overline{p}_1^{\mathrm{EV,C}}, \overline{p}_2^{\mathrm{EV,C}}, \overline{p}_3^{\mathrm{EV,C}}, \overline{p}_4^{\mathrm{EV,C}}, \overline{p}_5^{\mathrm{EV,C}}, \overline{p}_6^{\mathrm{EV,C}}, \xi_1^{\mathrm{EV}}, \xi_2^{\mathrm{EV}}, \xi_3^{\mathrm{EV}}, \xi_4^{\mathrm{EV}}, \xi_5^{\mathrm{EV}}, \xi_6^{\mathrm{EV}} \geq 0.$$

- Market clearing conditions:

$$\lambda_1^{\mathrm{D}}, \lambda_2^{\mathrm{D}}, \lambda_3^{\mathrm{D}}, \lambda_4^{\mathrm{D}}, \lambda_5^{\mathrm{D}}, \lambda_6^{\mathrm{D}}, p_1^{\mathrm{EV,C}}, p_2^{\mathrm{EV,C}}, p_3^{\mathrm{EV,C}}, p_4^{\mathrm{EV,C}}, p_5^{\mathrm{EV,C}}, p_6^{\mathrm{EV,C}} \in \Psi(\cdot).$$

Note that the above problem is very similar to that considered in Illustrative Example 6.2. However, it considers that the market prices and charging power schedules are endogenously obtained from the market clearing problem and that EVs cannot discharge their batteries. □

6.4.2 Market Clearing Problem

As explained in Sect. 6.4.1, problem (6.4a)–(6.4f) includes constraints (6.4f) that indicate that both the market prices and power quantities scheduled in the DA market are obtained from the market clearing problem represented by set $\Psi(\cdot)$. This problem can be formulated as the following LP problem:

$\Psi(\cdot) = \Big\{$

$$\text{Minimize}_{\Phi^{\mathrm{LL}}} \sum_{t \in T} \left(\sum_{g \in G} \chi_{gt}^{\mathrm{G}} p_{gt}^{\mathrm{G}} - \sum_{d \in D} \chi_{dt}^{\mathrm{D}} p_{dt}^{\mathrm{D}} - \xi_t^{\mathrm{EV,C}} p_t^{\mathrm{EV,C}} \right) \tag{6.5a}$$

subject to:

$$\sum_{g \in G} p_{gt}^{\mathrm{G}} = \sum_{d \in D} p_{dt}^{\mathrm{D}} + p_t^{\mathrm{EV,C}} \ : \lambda_t^{\mathrm{D}}, \quad \forall t \in T, \tag{6.5b}$$

$$0 \leq p_{gt}^{\mathrm{G}} \leq \overline{P}_{gt}^{\mathrm{G}} \ : \underline{\mu}_{gt}^{\mathrm{G}}, \overline{\mu}_{gt}^{\mathrm{G}}, \quad \forall g \in G, \forall t \in T, \tag{6.5c}$$

$$0 \leq p_{dt}^{\mathrm{D}} \leq \overline{P}_{qt}^{\mathrm{D}} \ : \underline{\mu}_{dt}^{\mathrm{D}}, \overline{\mu}_{dt}^{\mathrm{D}}, \quad \forall d \in D, \forall t \in T, \tag{6.5d}$$

$$0 \leq p_t^{\mathrm{EV,C}} \leq \overline{p}_t^{\mathrm{EV,C}} \ : \underline{\mu}_t^{\mathrm{EV,C}}, \overline{\mu}_t^{\mathrm{EV,C}}, \quad \forall t \in T, \tag{6.5e}$$

$\Big\}$, where variables in set Φ^{LL} are the optimization variables of problem (6.5a)–(6.5e) and dual variables associated with the constraints are provided following a colon.

Problem (6.5a)–(6.5e) is very similar to the DA market clearing problem (2.1) described in Sect. 2.2.1 of Chap. 2. The only difference is that problem (6.5a)–(6.5e) includes also the bids of the EV aggregator, which participates in the market as an additional power consumer.

Illustrative Example 6.8 Market clearing problem (linear programming problem)

The data of Illustrative Example 6.6 are considered again. The market clearing problem is formulated as follows:

$$\text{Minimize}_{\Phi^{\text{IE6.8}}} \; 50p_1^{\text{G}} + 80p_2^{\text{G}} + 70p_3^{\text{G}} + 60p_4^{\text{G}} - 85p_1^{\text{D}} - 65p_2^{\text{D}} - 75p_3^{\text{D}} - 100p^{\text{EV,C}}$$

subject to:

$$\begin{aligned}
&p_1^{\text{G}} + p_2^{\text{G}} + p_3^{\text{G}} + p_4^{\text{G}} = p_1^{\text{D}} + p_2^{\text{D}} + p_3^{\text{D}} + p^{\text{EV,C}},\\
&0 \le p_1^{\text{G}} \le 80,\\
&0 \le p_2^{\text{G}} \le 100,\\
&0 \le p_3^{\text{G}} \le 20,\\
&0 \le p_4^{\text{G}} \le 50,\\
&0 \le p_1^{\text{D}} \le 40,\\
&0 \le p_2^{\text{D}} \le 100,\\
&0 \le p_3^{\text{D}} \le 100,\\
&0 \le p^{\text{EV,C}} \le 13.22,
\end{aligned}$$

where $\Phi^{\text{IE6.8}} = \left\{p_1^{\text{G}}, p_2^{\text{G}}, p_3^{\text{G}}, p_4^{\text{G}}, p_1^{\text{D}}, p_2^{\text{D}}, p_3^{\text{D}}, p^{\text{EV,C}}\right\}$. The above model is an LP problem [17] that is solved using CPLEX [18] under a GAMS environment [9].

The optimal solution of the problem is as follows:

$$\begin{aligned}
p_1^{\text{G}*} &= 80,\\
p_2^{\text{G}*} &= 0,\\
p_3^{\text{G}*} &= 20,\\
p_4^{\text{G}*} &= 50,\\
p_1^{\text{D}*} &= 40,\\
p_2^{\text{D}*} &= 0,\\
p_3^{\text{D}*} &= 96.78,\\
p^{\text{EV,C}*} &= 13.22.
\end{aligned}$$

Finally, the optimal value of the objective function, which represents the minus social welfare, is −3580 €. □

A GAMS [9] code for solving Illustrative Example 6.8 is provided in Sect. 6.7.

6.4.2.1 Market Clearing Problem: Dual Formulation

The market clearing problem (6.5a)–(6.5e) is an LP problem. Therefore, its dual optimization problem can be also formulated as an LP problem [17]:

$$\text{Maximize}_{\Phi^{\text{DP}}} \quad \sum_{t \in T} \left(-\sum_{g \in G} \overline{P}^{\text{G}}_{gt} \overline{\mu}^{\text{G}}_{gt} - \sum_{d \in D} \overline{P}^{\text{D}}_{dt} \overline{\mu}^{\text{D}}_{dt} - \overline{p}^{\text{EV,C}}_{t} \overline{\mu}^{\text{EV,C}}_{t} \right) \tag{6.6a}$$

subject to:

$$\chi^{\text{G}}_{gt} - \lambda^{\text{D}}_{t} + \overline{\mu}^{\text{G}}_{gt} - \underline{\mu}^{\text{G}}_{gt} = 0, \quad \forall g \in G, \forall t \in T, \tag{6.6b}$$

$$-\chi^{\text{D}}_{dt} + \lambda^{\text{D}}_{t} + \overline{\mu}^{\text{D}}_{dt} - \underline{\mu}^{\text{D}}_{dt} = 0, \quad \forall d \in D, \forall t \in T, \tag{6.6c}$$

$$-\xi^{\text{EV}}_{t} + \lambda^{\text{D}}_{t} + \overline{\mu}^{\text{EV,C}}_{t} - \underline{\mu}^{\text{EV,C}}_{t} = 0, \quad \forall t \in T, \tag{6.6d}$$

$$\overline{\mu}^{\text{G}}_{gt}, \underline{\mu}^{\text{G}}_{gt} \geq 0, \quad \forall g \in G, \forall t \in T, \tag{6.6e}$$

$$\overline{\mu}^{\text{D}}_{dt}, \underline{\mu}^{\text{D}}_{dt} \geq 0, \quad \forall d \in D, \forall t \in T, \tag{6.6f}$$

$$\overline{\mu}^{\text{EV,C}}_{t}, \underline{\mu}^{\text{EV,C}}_{t} \geq 0, \quad \forall t \in T, \tag{6.6g}$$

where dual variables in set $\Phi^{\text{DP}} = \left\{ \lambda^{\text{D}}_{t}, \forall t \in T; \underline{\mu}^{\text{G}}_{gt}, \overline{\mu}^{\text{G}}_{gt}, \forall g \in G, \forall t \in T; \underline{\mu}^{\text{D}}_{dt}, \overline{\mu}^{\text{D}}_{dt}, \forall d \in D, \forall t \in T; \underline{\mu}^{\text{EV,C}}_{t}, \overline{\mu}^{\text{EV,C}}_{t}, \forall t \in T \right\}$, are the optimization variables of dual problem (6.6a)–(6.6g). Details about how to derive the dual problem of an LP problem are provided in [17].

Illustrative Example 6.9 Market clearing problem (dual formulation)

The data of Illustrative Example 6.8 are considered again. The dual formulation of the market clearing problem is provided below:

$$\begin{aligned} \text{Maximize}_{\Phi^{\text{IE6.9}}} \quad & -80\overline{\mu}^{\text{G}}_{1} - 100\overline{\mu}^{\text{G}}_{2} - 20\overline{\mu}^{\text{G}}_{3} - 50\overline{\mu}^{\text{G}}_{4} - 40\overline{\mu}^{\text{D}}_{1} - 100\overline{\mu}^{\text{D}}_{2} - 100\overline{\mu}^{\text{D}}_{3} \\ & - 13.22\overline{\mu}^{\text{EV,C}} \end{aligned}$$

subject to:

$$\begin{aligned} & 50 - \lambda^{\text{D}} + \overline{\mu}^{\text{G}}_{1} - \underline{\mu}^{\text{G}}_{1} = 0, \\ & 80 - \lambda^{\text{D}} + \overline{\mu}^{\text{G}}_{2} - \underline{\mu}^{\text{G}}_{2} = 0, \\ & 70 - \lambda^{\text{D}} + \overline{\mu}^{\text{G}}_{3} - \underline{\mu}^{\text{G}}_{3} = 0, \\ & 60 - \lambda^{\text{D}} + \overline{\mu}^{\text{G}}_{4} - \underline{\mu}^{\text{G}}_{4} = 0, \\ & -85 + \lambda^{\text{D}} + \overline{\mu}^{\text{D}}_{1} - \underline{\mu}^{\text{D}}_{1} = 0, \\ & -65 + \lambda^{\text{D}} + \overline{\mu}^{\text{D}}_{2} - \underline{\mu}^{\text{D}}_{2} = 0, \end{aligned}$$

$$
\begin{aligned}
& -75 + \lambda^{D} + \overline{\mu}_3^{D} - \underline{\mu}_3^{D} = 0, \\
& -100 + \lambda^{D} + \overline{\mu}^{EV,C} - \underline{\mu}^{EV,C} = 0, \\
& \overline{\mu}_1^{G}, \underline{\mu}_1^{G}, \overline{\mu}_2^{G}, \underline{\mu}_2^{G}, \overline{\mu}_3^{G}, \underline{\mu}_3^{G}, \overline{\mu}_4^{G}, \underline{\mu}_4^{G} \geq 0, \\
& \overline{\mu}_1^{D}, \underline{\mu}_1^{D}, \overline{\mu}_2^{D}, \underline{\mu}_2^{D}, \overline{\mu}_3^{D}, \underline{\mu}_3^{D} \geq 0, \\
& \overline{\mu}^{EV,C}, \underline{\mu}^{EV,C} \geq 0,
\end{aligned}
$$

where $\Phi^{IE6.9} = \left\{\lambda^{D}, \underline{\mu}_1^{G}, \overline{\mu}_1^{G}, \underline{\mu}_2^{G}, \overline{\mu}_2^{G}, \underline{\mu}_3^{G}, \overline{\mu}_3^{G}, \underline{\mu}_4^{G}, \overline{\mu}_4^{G}, \underline{\mu}_1^{D}, \overline{\mu}_1^{D}, \underline{\mu}_2^{D}, \overline{\mu}_2^{D}, \underline{\mu}_3^{D}, \overline{\mu}_3^{D}, \underline{\mu}^{EV,C}, \overline{\mu}^{EV,C}\right\}$. The above model is an LP problem [17] that is solved using CPLEX [18] under a GAMS environment [9].

The optimal solution of the problem is as follows:

$$
\begin{aligned}
& \lambda^{D*} = 75, \\
& \underline{\mu}_1^{G*} = 0, \\
& \underline{\mu}_2^{G*} = 5, \\
& \underline{\mu}_3^{G*} = 0, \\
& \underline{\mu}_4^{G*} = 0, \\
& \overline{\mu}_1^{G*} = 25, \\
& \overline{\mu}_2^{G*} = 0, \\
& \overline{\mu}_3^{G*} = 5, \\
& \overline{\mu}_4^{G*} = 15, \\
& \underline{\mu}_1^{D*} = 0, \\
& \underline{\mu}_2^{D*} = 10, \\
& \underline{\mu}_3^{D*} = 0, \\
& \overline{\mu}_1^{D*} = 10, \\
& \overline{\mu}_2^{D*} = 0, \\
& \overline{\mu}_3^{D*} = 0, \\
& \underline{\mu}^{EV,C*} = 25, \\
& \overline{\mu}^{EV,C*} = 0.
\end{aligned}
$$

As explained in Sect. 6.2.2, the value of the dual variable associated with the power balance constraint, λ^{D}, is used as the market clearing price. Note that the optimal value of this variable is equal to the one obtained in Illustrative Example 6.6.

Finally, the optimal value of the objective function is −3580 €. □

A GAMS [9] code for solving Illustrative Example 6.9 is provided in Sect. 6.7.

6.4.2.2 Market Clearing Problem: Strong Duality Equality

The market clearing problem (6.5a)–(6.5e) is an LP problem. Therefore, the strong duality holds [3]. In other words, the duality gap is zero at the optimal solution, i.e., the optimal values of the objective functions of the primal optimization problem (6.5a)–(6.5e) and its corresponding dual optimization problem (6.6a)–(6.6g) are equal:

$$\sum_{t\in T}\left(\sum_{g\in G}\chi^{\mathrm{G}}_{gt}p^{\mathrm{G}}_{gt}-\sum_{d\in D}\chi^{\mathrm{D}}_{dt}p^{\mathrm{D}}_{dt}-\xi^{\mathrm{EV,C}}_{t}p^{\mathrm{EV,C}}_{t}\right)$$

$$=\sum_{t\in T}\left(-\sum_{g\in G}\overline{P}^{\mathrm{G}}_{gt}\overline{\mu}^{\mathrm{G}}_{gt}-\sum_{d\in D}\overline{P}^{\mathrm{D}}_{dt}\overline{\mu}^{\mathrm{D}}_{dt}-\overline{P}^{\mathrm{EV,C}}_{t}\overline{\mu}^{\mathrm{EV,C}}_{t}\right). \tag{6.7}$$

Illustrative Example 6.10 Market clearing problem (strong duality)

The data and results of Illustrative Examples 6.8 and 6.9 are considered again. The optimal value of the objective functions of both the primal and dual problems is the same, namely, −3580 €, which validates the strong duality equality.

As an alternative to obtain the solution of the market clearing problem, it is possible to replace the primal and dual optimization problems by a set of equations comprising:

1. The constraints of the primal optimization problem.
2. The constraints of the dual optimization problem.
3. The strong duality equality.

The above equations comprise a system of linear equations that can be easily solved and provides the solution of both primal and dual variables, whose optimal values are equal to those obtained in Illustrative Examples 6.8 and 6.9. □

6.4.2.3 Market Clearing Problem: Karush-Kuhn-Tucker Conditions

The market clearing problem (6.5a)–(6.5e) is a convex optimization problem as it is continuous and linear on its decision variables. Therefore, the Karush-Kuhn-Tucker (KKT) conditions are necessary and sufficient conditions for optimality [3]. Thus, the DA market clearing problem (6.5a)–(6.5e) can be replaced by the following KKT conditions:

$$\Psi(\cdot)=\Bigg\{$$

$$\sum_{g\in G}p^{\mathrm{G}}_{gt}=\sum_{d\in\Omega^{\mathrm{D}}}p^{\mathrm{D}}_{dt}+p^{\mathrm{EV,C}}_{t},\quad\forall t\in T, \tag{6.8a}$$

$$\chi^{\mathrm{G}}_{gt}-\lambda^{\mathrm{D}}_{t}+\overline{\mu}^{\mathrm{G}}_{gt}-\underline{\mu}^{\mathrm{G}}_{gt}=0,\quad\forall g\in G,\forall t\in T, \tag{6.8b}$$

$$- \chi_{dt}^{\mathrm{D}} + \lambda_t^{\mathrm{D}} + \overline{\mu}_{dt}^{\mathrm{D}} - \underline{\mu}_{dt}^{\mathrm{D}} = 0, \quad \forall d \in D, \forall t \in T, \tag{6.8c}$$

$$- \xi_t^{\mathrm{EV,C}} + \lambda_t^{\mathrm{D}} + \overline{\mu}_t^{\mathrm{EV,C}} - \underline{\mu}_t^{\mathrm{EV,C}} = 0, \quad \forall t \in T, \tag{6.8d}$$

$$0 \leq \underline{\mu}_{gt}^{\mathrm{G}} \perp p_{gt}^{\mathrm{G}} \geq 0, \quad \forall g \in G, \forall t \in T, \tag{6.8e}$$

$$0 \leq \overline{\mu}_{gt}^{\mathrm{G}} \perp \overline{P}_{gt}^{\mathrm{G}} - p_{gt}^{\mathrm{G}} \geq 0, \quad \forall g \in G, \forall t \in T, \tag{6.8f}$$

$$0 \leq \underline{\mu}_{dt}^{\mathrm{D}} \perp p_{dt}^{\mathrm{D}} \geq 0, \quad \forall d \in \Omega^{\mathrm{D}}, \forall t \in T, \tag{6.8g}$$

$$0 \leq \overline{\mu}_{dt}^{\mathrm{D}} \perp \overline{P}_{dt}^{\mathrm{D}} - p_{dt}^{\mathrm{D}} \geq 0, \quad \forall d \in D, \forall t \in T, \tag{6.8h}$$

$$0 \leq \underline{\mu}_t^{\mathrm{EV,C}} \perp p_t^{\mathrm{EV,C}} \geq 0, \quad \forall t \in T, \tag{6.8i}$$

$$0 \leq \overline{\mu}_t^{\mathrm{EV,C}} \perp \overline{p}_t^{\mathrm{EV,C}} - p_t^{\mathrm{EV,C}} \geq 0, \quad \forall t \in T \tag{6.8j}$$

$\}.$

Equations (6.8a)–(6.8d) are obtained by taking the derivatives of the Lagrange function of the market clearing problem (6.5a)–(6.5e) with respect to the dual variable λ_t^{D} and the primal variables p_{gt}^{G}, p_{dt}^{D}, and $p_t^{\mathrm{EV,C}}$, respectively. On the other hand, constraints (6.8e)–(6.8j) are the complementarity constraints regarding inequalities (6.5c)–(6.5e) of the market clearing problem (6.5a)–(6.5e). These complementarity constraints have the form $0 \leq y \perp z \geq 0$, which is equivalent to $y, z \geq 0$ and the slackness condition $y \cdot z = 0$.

Illustrative Example 6.11 Market clearing problem (KKT conditions)

The data of Illustrative Example 6.8 are considered again. The KKT conditions of the market clearing problem are provided below:

$$p_1^{\mathrm{G}} + p_2^{\mathrm{G}} + p_3^{\mathrm{G}} + p_4^{\mathrm{G}} = p_1^{\mathrm{D}} + p_2^{\mathrm{D}} + p_3^{\mathrm{D}} + p^{\mathrm{EV,C}},$$

$$50 - \lambda^{\mathrm{D}} + \overline{\mu}_1^{\mathrm{G}} - \underline{\mu}_1^{\mathrm{G}} = 0,$$

$$80 - \lambda^{\mathrm{D}} + \overline{\mu}_2^{\mathrm{G}} - \underline{\mu}_2^{\mathrm{G}} = 0,$$

$$70 - \lambda^{\mathrm{D}} + \overline{\mu}_3^{\mathrm{G}} - \underline{\mu}_3^{\mathrm{G}} = 0,$$

$$60 - \lambda^{\mathrm{D}} + \overline{\mu}_4^{\mathrm{G}} - \underline{\mu}_4^{\mathrm{G}} = 0,$$

$$- 85 + \lambda^{\mathrm{D}} + \overline{\mu}_1^{\mathrm{D}} - \underline{\mu}_1^{\mathrm{D}} = 0,$$

$$- 65 + \lambda^{\mathrm{D}} + \overline{\mu}_2^{\mathrm{D}} - \underline{\mu}_2^{\mathrm{D}} = 0,$$

$$- 75 + \lambda^{\mathrm{D}} + \overline{\mu}_3^{\mathrm{D}} - \underline{\mu}_3^{\mathrm{D}} = 0,$$

$$- 100 + \lambda^{\mathrm{D}} + \overline{\mu}^{\mathrm{EV,C}} - \underline{\mu}^{\mathrm{EV,C}} = 0,$$

$$0 \leq \underline{\mu}_1^{\mathrm{G}} \perp p_1^{\mathrm{G}} \geq 0,$$

$$0 \leq \underline{\mu}_2^{\mathrm{G}} \perp p_2^{\mathrm{G}} \geq 0,$$

$$0 \leq \underline{\mu}_3^{\mathrm{G}} \perp p_3^{\mathrm{G}} \geq 0,$$

$$0 \leq \underline{\mu}_4^{\mathrm{G}} \perp p_4^{\mathrm{G}} \geq 0,$$

$$
\begin{aligned}
&0 \leq \overline{\mu}_1^{\mathrm{G}} \perp 80 - p_1^{\mathrm{G}} \geq 0,\\
&0 \leq \overline{\mu}_2^{\mathrm{G}} \perp 100 - p_2^{\mathrm{G}} \geq 0,\\
&0 \leq \overline{\mu}_3^{\mathrm{G}} \perp 20 - p_3^{\mathrm{G}} \geq 0,\\
&0 \leq \overline{\mu}_4^{\mathrm{G}} \perp 50 - p_4^{\mathrm{G}} \geq 0,\\
&0 \leq \underline{\mu}_1^{\mathrm{D}} \perp p_1^{\mathrm{D}} \geq 0,\\
&0 \leq \underline{\mu}_2^{\mathrm{D}} \perp p_2^{\mathrm{D}} \geq 0,\\
&0 \leq \underline{\mu}_3^{\mathrm{D}} \perp p_3^{\mathrm{D}} \geq 0,\\
&0 \leq \overline{\mu}_1^{\mathrm{D}} \perp 40 - p_1^{\mathrm{D}} \geq 0,\\
&0 \leq \overline{\mu}_2^{\mathrm{D}} \perp 100 - p_2^{\mathrm{D}} \geq 0,\\
&0 \leq \overline{\mu}_3^{\mathrm{D}} \perp 100 - p_3^{\mathrm{D}} \geq 0,\\
&0 \leq \underline{\mu}^{\mathrm{EV,C}} \perp p^{\mathrm{EV,C}} \geq 0,\\
&0 \leq \overline{\mu}^{\mathrm{EV,C}} \perp 13.22 - p^{\mathrm{EV,C}} \geq 0.
\end{aligned}
$$

□

Note that the KKT conditions (6.8a)–(6.8j) comprise a system of nonlinear equations. The nonlinearities are due to the complementarity constraints (6.8e)–(6.8j). However, these nonlinear constraints have the form $y \cdot z = 0$ with $y, z \geq 0$ and can be linearized through the big-M method [7] using the following expressions:

$$0 \leq y \leq Mu, \tag{6.9a}$$

$$0 \leq z \leq M\,(1-u)\,, \tag{6.9b}$$

$$u \in \{0, 1\}, \tag{6.9c}$$

where M is a large enough positive constant.

Constraints (6.9a)–(6.9c) impose that either y or z (or both) must be equal to 0. If $u = 0$, then constraint (6.9a) imposes that $0 \leq y \leq 0$, i.e., $y = 0$, while (6.9b) imposes that $0 \leq z \leq M$. As M is a large enough positive constant, then this last constraint is not bounding the upper value of z. On the other hand, if $u = 1$, then constraint (6.9b) imposes that $0 \leq z \leq 0$, i.e., $z = 0$, while (6.9a) imposes that $0 \leq y \leq M$.

Therefore, using (6.9a)–(6.9c), nonlinear complementarity constraints (6.8e)–(6.8j) can be replaced by the following mixed-integer linear expressions:

$$0 \leq \underline{\mu}_{gt}^{\mathrm{G}} \leq \underline{N}_{gt}^{\mathrm{G}} \underline{u}_{gt}^{\mathrm{G}}, \quad \forall g \in G, \forall t \in T, \tag{6.10a}$$

$$0 \leq p_{gt}^{\mathrm{G}} \leq \underline{M}_{gt}^{\mathrm{G}}(1-\underline{u}_{gt}^{\mathrm{G}}), \quad \forall g \in G, \forall t \in T, \tag{6.10b}$$

$$0 \leq \overline{\mu}_{gt}^{\mathrm{G}} \leq \overline{N}_{gt}^{\mathrm{G}} \overline{u}_{gt}^{\mathrm{G}}, \quad \forall g \in G, \forall t \in T, \tag{6.10c}$$

$$0 \leq \overline{P}_{gt}^{\mathrm{G}} - p_{gt}^{\mathrm{G}} \leq \overline{M}_{gt}^{\mathrm{G}}(1-\overline{u}_{gt}^{\mathrm{G}}), \quad \forall g \in G, \forall t \in T, \tag{6.10d}$$

$$\underline{u}_{gt}^{\mathrm{G}}, \overline{u}_{gt}^{\mathrm{G}} \in \{0, 1\}, \quad \forall g \in G, \forall t \in T, \tag{6.10e}$$

$$0 \leq \underline{\mu}^{\mathrm{D}}_{dt} \leq \underline{N}^{\mathrm{D}}_{dt}\underline{u}^{\mathrm{D}}_{dt}, \quad \forall d \in D, \forall t \in T, \tag{6.10f}$$

$$0 \leq p^{\mathrm{D}}_{dt} \leq \underline{M}^{\mathrm{D}}_{dt}(1 - \underline{u}^{\mathrm{D}}_{dt}), \quad \forall d \in D, \forall t \in T, \tag{6.10g}$$

$$0 \leq \overline{\mu}^{\mathrm{D}}_{dt} \leq \overline{N}^{\mathrm{D}}_{dt}\overline{u}^{\mathrm{D}}_{dt}, \quad \forall d \in D, \forall t \in T, \tag{6.10h}$$

$$0 \leq \overline{P}^{\mathrm{D}}_{dt} - p^{\mathrm{D}}_{dt} \leq \overline{M}^{\mathrm{D}}_{dt}(1 - \overline{u}^{\mathrm{D}}_{dt}), \quad \forall d \in D, \forall t \in T, \tag{6.10i}$$

$$\underline{u}^{\mathrm{D}}_{dt}, \overline{u}^{\mathrm{D}}_{dt} \in \{0, 1\}, \quad \forall d \in D, \forall t \in T, \tag{6.10j}$$

$$0 \leq \underline{\mu}^{\mathrm{EV,C}}_{t} \leq \underline{N}^{\mathrm{EV,C}}_{t}\underline{u}^{\mathrm{EV,C}}_{t}, \quad \forall t \in T, \tag{6.10k}$$

$$0 \leq p^{\mathrm{EV,C}}_{t} \leq \underline{M}^{\mathrm{EV,C}}_{t}(1 - \underline{u}^{\mathrm{EV,C}}_{t}), \quad \forall t \in T, \tag{6.10l}$$

$$0 \leq \overline{\mu}^{\mathrm{EV,C}}_{t} \leq \overline{N}^{\mathrm{EV,C}}_{t}\overline{u}^{\mathrm{EV,C}}_{t}, \quad \forall t \in T, \tag{6.10m}$$

$$0 \leq \overline{P}^{\mathrm{EV,C}}_{t} - p^{\mathrm{EV,C}}_{t} \leq \overline{M}^{\mathrm{EV,C}}_{t}(1 - \overline{u}^{\mathrm{EV,C}}_{t}), \quad \forall t \in T, \tag{6.10n}$$

$$\underline{u}^{\mathrm{EV,C}}_{t}, \overline{u}^{\mathrm{EV,C}}_{t} \in \{0, 1\}, \quad \forall t \in T, \tag{6.10o}$$

where $\underline{N}^{\mathrm{G}}_{gt}, \underline{M}^{\mathrm{G}}_{gt}, \overline{N}^{\mathrm{G}}_{gt}, \overline{M}^{\mathrm{G}}_{gt}, \forall g \in G, \forall t \in T; \underline{N}^{\mathrm{D}}_{dt}, \underline{M}^{\mathrm{D}}_{dt}, \overline{N}^{\mathrm{D}}_{dt}, \overline{M}^{\mathrm{D}}_{dt}, \forall d \in D, \forall t \in T;$ $\underline{N}^{\mathrm{EV,C}}_{t}, \underline{M}^{\mathrm{EV,C}}_{t}, \overline{N}^{\mathrm{EV,C}}_{t}, \overline{M}^{\mathrm{EV,C}}_{t}, \forall t \in T$, are large enough positive constants.

Illustrative Example 6.12 Market clearing problem (linear KKT conditions)

The data of Illustrative Example 6.11 are considered again. The nonlinear complementarity constraints are replaced by the following set of linear constraints:

$$
\begin{aligned}
& p^{\mathrm{G}}_1 + p^{\mathrm{G}}_2 + p^{\mathrm{G}}_3 + p^{\mathrm{G}}_4 = p^{\mathrm{D}}_1 + p^{\mathrm{D}}_2 + p^{\mathrm{D}}_3 + p^{\mathrm{EV,C}}, \\
& 50 - \lambda^{\mathrm{D}} + \overline{\mu}^{\mathrm{G}}_1 - \underline{\mu}^{\mathrm{G}}_1 = 0, \\
& 80 - \lambda^{\mathrm{D}} + \overline{\mu}^{\mathrm{G}}_2 - \underline{\mu}^{\mathrm{G}}_2 = 0, \\
& 70 - \lambda^{\mathrm{D}} + \overline{\mu}^{\mathrm{G}}_3 - \underline{\mu}^{\mathrm{G}}_3 = 0, \\
& 60 - \lambda^{\mathrm{D}} + \overline{\mu}^{\mathrm{G}}_4 - \underline{\mu}^{\mathrm{G}}_4 = 0, \\
& -85 + \lambda^{\mathrm{D}} + \overline{\mu}^{\mathrm{D}}_1 - \underline{\mu}^{\mathrm{D}}_1 = 0, \\
& -65 + \lambda^{\mathrm{D}} + \overline{\mu}^{\mathrm{D}}_2 - \underline{\mu}^{\mathrm{D}}_2 = 0, \\
& -75 + \lambda^{\mathrm{D}} + \overline{\mu}^{\mathrm{D}}_3 - \underline{\mu}^{\mathrm{D}}_3 = 0, \\
& -100 + \lambda^{\mathrm{D}} + \overline{\mu}^{\mathrm{EV,C}} - \underline{\mu}^{\mathrm{EV,C}} = 0, \\
& 0 \leq \underline{\mu}^{\mathrm{G}}_1 \leq N\underline{u}^{\mathrm{G}}_1, \\
& 0 \leq p^{\mathrm{G}}_1 \leq M\left(1 - \underline{u}^{\mathrm{G}}_1\right), \\
& 0 \leq \underline{\mu}^{\mathrm{G}}_2 \leq N\underline{u}^{\mathrm{G}}_2, \\
& 0 \leq p^{\mathrm{G}}_2 \leq M\left(1 - \underline{u}^{\mathrm{G}}_2\right), \\
& 0 \leq \underline{\mu}^{\mathrm{G}}_3 \leq N\underline{u}^{\mathrm{G}}_3, \\
& 0 \leq p^{\mathrm{G}}_3 \leq M\left(1 - \underline{u}^{\mathrm{G}}_3\right), \\
& 0 \leq \underline{\mu}^{\mathrm{G}}_4 \leq N\underline{u}^{\mathrm{G}}_4,
\end{aligned}
$$

$$0 \le p_4^{\mathrm{G}} \le M\left(1 - \underline{u}_4^{\mathrm{G}}\right),$$
$$0 \le \overline{\mu}_1^{\mathrm{G}} \le N\overline{u}_1^{\mathrm{G}},$$
$$0 \le 80 - p_1^{\mathrm{G}} \le M\left(1 - \overline{u}_1^{\mathrm{G}}\right),$$
$$0 \le \overline{\mu}_2^{\mathrm{G}} \le N\overline{u}_2^{\mathrm{G}},$$
$$0 \le 100 - p_2^{\mathrm{G}} \le M\left(1 - \overline{u}_2^{\mathrm{G}}\right),$$
$$0 \le \overline{\mu}_3^{\mathrm{G}} \le N\overline{u}_3^{\mathrm{G}},$$
$$0 \le 20 - p_3^{\mathrm{G}} \le M\left(1 - \overline{u}_3^{\mathrm{G}}\right),$$
$$0 \le \overline{\mu}_4^{\mathrm{G}} \le N\overline{u}_4^{\mathrm{G}},$$
$$0 \le 50 - p_4^{\mathrm{G}} \le M\left(1 - \overline{u}_4^{\mathrm{G}}\right),$$
$$0 \le \underline{\mu}_1^{\mathrm{D}} \le N\underline{u}_1^{\mathrm{D}},$$
$$0 \le p_1^{\mathrm{D}} \le N\left(1 - \underline{u}_1^{\mathrm{D}}\right),$$
$$0 \le \underline{\mu}_2^{\mathrm{D}} \le N\underline{u}_2^{\mathrm{D}},$$
$$0 \le p_2^{\mathrm{D}} \le N\left(1 - \underline{u}_2^{\mathrm{D}}\right),$$
$$0 \le \underline{\mu}_3^{\mathrm{D}} \le N\underline{u}_3^{\mathrm{D}},$$
$$0 \le p_3^{\mathrm{D}} \le N\left(1 - \underline{u}_3^{\mathrm{D}}\right),$$
$$0 \le \overline{\mu}_1^{\mathrm{D}} \le N\overline{u}_1^{\mathrm{D}},$$
$$0 \le 40 - p_1^{\mathrm{D}} \le M\left(1 - \overline{u}_1^{\mathrm{D}}\right),$$
$$0 \le \overline{\mu}_2^{\mathrm{D}} \le N\overline{u}_2^{\mathrm{D}},$$
$$0 \le 100 - p_2^{\mathrm{D}} \le M\left(1 - \overline{u}_2^{\mathrm{D}}\right),$$
$$0 \le \overline{\mu}_3^{\mathrm{D}} \le N\overline{u}_3^{\mathrm{D}},$$
$$0 \le 100 - p_3^{\mathrm{D}} \le M\left(1 - \overline{u}_3^{\mathrm{D}}\right),$$
$$0 \le \underline{\mu}^{\mathrm{EV,C}} \le N\underline{u}^{\mathrm{EV,C}},$$
$$0 \le p^{\mathrm{EV,C}} \le M\left(1 - \underline{u}^{\mathrm{EV,C}}\right),$$
$$0 \le \overline{\mu}^{\mathrm{EV,C}} \le N\overline{u}^{\mathrm{EV,C}},$$
$$0 \le 13.22 - p^{\mathrm{EV,C}} \le M\left(1 - \overline{u}^{\mathrm{EV,C}}\right),$$
$$\underline{u}_1^{\mathrm{G}}, \underline{u}_2^{\mathrm{G}}, \underline{u}_3^{\mathrm{G}}, \underline{u}_4^{\mathrm{G}}, \overline{u}_1^{\mathrm{G}}, \overline{u}_2^{\mathrm{G}}, \overline{u}_3^{\mathrm{G}}, \overline{u}_4^{\mathrm{G}}, \underline{u}_1^{\mathrm{D}}, \underline{u}_2^{\mathrm{D}}, \underline{u}_3^{\mathrm{D}}, \overline{u}_1^{\mathrm{D}}, \overline{u}_2^{\mathrm{D}}, \overline{u}_3^{\mathrm{D}}, \underline{u}^{\mathrm{EV,C}}, \overline{u}^{\mathrm{EV,C}} \in \{0, 1\}.$$

The above constraints comprise a system of linear equations. It includes two parameters, N and M, that should be fixed to large enough values. Note that N and M impose upper bounds of dual and primal variables, respectively. Thus, the value of M can be fixed to the maximum value of the primal variables, in this particular case, 100. However, selecting the value of N is not straightforward since the maximum value of dual variables is unknown beforehand. Therefore, it has to be fixed to a large enough value, e.g., 1000, and then check that this value is not bounding the value of the dual variables. In case the chosen value bounds the dual variables, then a larger value should be selected.

The solution of the above system of equations is identical to that obtained in Illustrative Examples 6.8 and 6.9. □

A GAMS [9] code for solving Illustrative Example 6.12 is provided in Sect. 6.7.

6.4.3 MPEC Formulation

Recall that this section is devoted to the bidding strategy problem of a price-maker EV aggregator, which is formulated using the bi-level model (6.4a)–(6.4f). In order to obtain the solution, this bi-level model needs to be reformulated as a single-level problem. To do so, the lower-level problem, i.e., the market clearing problem (6.5a)–(6.5e), is replaced by its KKT conditions rendering the following single-level problem:

$$\text{Minimize}_{\Phi^{\text{UL}},\Phi^{\text{LL}},\Phi^{\text{DP}}} \quad \sum_{t\in T} \lambda_t^{\text{D}} p_t^{\text{EV,C}} \Delta_t \tag{6.11a}$$

subject to:

$$\text{Constraints (6.4b)}-\text{(6.4e)}, \tag{6.11b}$$

$$\text{Constraints (6.8)}. \tag{6.11c}$$

Problem (6.11a)–(6.11c) is generally known as a Mathematical Program with Equilibrium Constraints (MPEC) [8].

6.4.4 MILP Formulation

The resulting MPEC (6.11a)–(6.11c) is a single-level problem; however, it is nonlinear since it includes two types of nonlinearities:

1. Terms $\lambda_t^{\text{D}} p_t^{\text{EV,C}}$, $\forall t \in T$, in the objective function (6.11a).
2. Complementarity constraints (6.8e)–(6.8j).

As explained in Sect. 6.4.2.3, complementarity constraints (6.8e)–(6.8j) can be replaced by the equivalent mixed-integer linear constraints (6.10a)–(6.10o).

On the other hand, it is also possible to replace the nonlinear objective function (6.11a) by exact equivalent mixed-integer linear expressions as explained below [16].

Firstly, the KKT conditions (6.8d) are multiplied by $p_t^{\text{EV,C}}$:

$$-\xi_t^{\text{EV,C}} p_t^{\text{EV,C}} + \lambda_t^{\text{D}} p_t^{\text{EV,C}} + \overline{\mu}_t^{\text{EV,C}} p_t^{\text{EV,C}} - \underline{\mu}_t^{\text{EV,C}} p_t^{\text{EV,C}} = 0, \quad \forall t \in T. \tag{6.12}$$

In (6.12), term $\underline{\mu}_t^{\mathrm{EV,C}} p_t^{\mathrm{EV,C}}$ is equal to zero based on the complementarity constraints (6.8i). Moreover, term $\overline{\mu}_t^{\mathrm{EV,C}} p_t^{\mathrm{EV,C}}$ can be replaced by $\overline{\mu}_t^{\mathrm{EV,C}} \overline{p}_t^{\mathrm{EV,C}}$ based on the complementarity constraints (6.8j). Thus, Eq. (6.12) can be rewritten as:

$$\lambda_t^{\mathrm{D}} p_t^{\mathrm{EV,C}} = \xi_t^{\mathrm{EV,C}} p_t^{\mathrm{EV,C}} - \overline{\mu}_t^{\mathrm{EV,C}} \overline{p}_t^{\mathrm{EV,C}}, \quad \forall t \in T. \tag{6.13}$$

Note that the left-hand side of Eq. (6.13) is the nonlinear term in the objective function (6.11a) that needs to be replaced. However, the right-hand side of Eq. (6.13) is still nonlinear.

Then, the strong duality equality is imposed. As explained in Sect. 6.4.2.2, this equality states that, at the optimum, the values of the objective functions of the primal (6.5a)–(6.5e) and dual problem formulations (6.6a)–(6.6g) of the energy market clearing problem are equal, i.e.:

$$\begin{aligned} &\sum_{g \in G} \chi_{gt}^{\mathrm{G}} p_{gt}^{\mathrm{G}} - \sum_{d \in D} \chi_{dt}^{\mathrm{D}} p_{dt}^{\mathrm{D}} - \xi_t^{\mathrm{EV,C}} p_t^{\mathrm{EV,C}} \\ &\quad = -\sum_{g \in G} \overline{P}_{gt}^{\mathrm{G}} \overline{\mu}_{gt}^{\mathrm{G}} - \sum_{d \in D} \overline{P}_{dt}^{\mathrm{D}} \overline{\mu}_{dt}^{\mathrm{D}} - \overline{p}_t^{\mathrm{EV,C}} \overline{\mu}_t^{\mathrm{EV,C}}, \quad \forall t \in T. \end{aligned} \tag{6.14}$$

Using the strong duality (6.14), the right-hand side of Eq. (6.13) can be rewritten as follows:

$$\lambda_t^{\mathrm{D}} p_t^{\mathrm{EV,C}} = \sum_{g \in G} \chi_{gt}^{\mathrm{G}} p_{gt}^{\mathrm{G}} - \sum_{d \in D} \chi_{dt}^{\mathrm{D}} p_{dt}^{\mathrm{D}} + \sum_{g \in G} \overline{P}_{gt}^{\mathrm{G}} \overline{\mu}_{gt}^{\mathrm{G}} + \sum_{d \in D} \overline{P}_{dt}^{\mathrm{D}} \overline{\mu}_{dt}^{\mathrm{D}}, \quad \forall t \in T. \tag{6.15}$$

Note that the right-hand side of Eq. (6.15) is linear. Thus, nonlinear terms $\lambda_t^{\mathrm{D}} p_t^{\mathrm{EV,C}}$, $\forall t \in T$, in the objective function (6.11a) can be replaced by the right-hand side of Eq. (6.15) and the nonlinear MPEC (6.11a)–(6.11c) can be recast as the following Mixed-Integer Linear Programming (MILP) problem:

$$\begin{aligned} \text{Minimize}_{\Phi^{\mathrm{UL}}, \Phi^{\mathrm{LL}}, \Phi^{\mathrm{DP}}, \Phi^{\mathrm{L}}} \quad & \sum_{t \in T} \left(\sum_{g \in G} \chi_{gt}^{\mathrm{G}} p_{gt}^{\mathrm{G}} - \sum_{d \in D} \chi_{dt}^{\mathrm{D}} p_{dt}^{\mathrm{D}} \right. \\ & \left. + \sum_{g \in G} \overline{P}_{gt}^{\mathrm{G}} \overline{\mu}_{gt}^{\mathrm{G}} + \sum_{d \in D} \overline{P}_{dt}^{\mathrm{D}} \overline{\mu}_{dt}^{\mathrm{D}} \right) \Delta_t \end{aligned} \tag{6.16a}$$

subject to:

$$\text{Constraints (6.4b) – (6.4e),} \tag{6.16b}$$

$$\text{Constraints (6.8a) – (6.8e),} \tag{6.16c}$$

$$\text{Constraints (6.10),} \tag{6.16d}$$

Table 6.2 Illustrative Example 6.13: production offers

Time period	Energy (MWh)	Price (€/MWh)
1	80, 50, 20, 100	50, 60, 70, 80
2	100, 20, 10, 10	0, 20, 40, 60
3	100, 20, 10, 10	0, 10, 30, 50
4	50, 50, 50, 50	60, 70, 80, 90
5	100, 10, 10, 10	0, 10, 20, 30
6	100, 10, 10, 10	40, 50, 60, 70

Table 6.3 Illustrative Example 6.13: consumption bids

Time period	Energy (MWh)	Price (€/MWh)
1	40, 100, 100	85, 75, 55
2	100, 25, 30	65, 55, 45
3	100, 25, 30	65, 55, 45
4	60, 80, 40	95, 85, 75
5	90, 40, 20	25, 15, 5
6	80, 30, 20	65, 55, 45

where $\Phi^{\mathrm{L}}=\left\{\underline{u}^{\mathrm{G}}_{gt}, \overline{u}^{\mathrm{G}}_{gt}, \forall g \in G, \forall t \in T; \underline{u}^{\mathrm{D}}_{dt}, \overline{u}^{\mathrm{D}}_{dt}, \forall d \in D, \forall t \in T; \underline{u}^{\mathrm{EV,C}}_{t}, \overline{u}^{\mathrm{EV,C}}_{t}, \forall t \in T\right\}$.

Illustrative Example 6.13 Price-maker EV aggregator (MILP problem)

The data of Illustrative Example 6.7 are considered again. The price-maker EV aggregator participates in the DA energy market. The offer and bid data of the power producers and consumers (others than the EV aggregator) during the 6-h planning horizon are provided in Tables 6.2 and 6.3, respectively.

Using these data, model (6.16a)–(6.16d) is implemented. Figures 6.17 and 6.18 provide the optimal power traded by the EV aggregator and the resulting market clearing prices. These two figures compare the solution of both the price-taker approach, i.e., considering that the EV aggregator does not have any influence on market prices, whose solution corresponds to that obtained in Illustrative Example 6.2 without V2G, and the price-maker solution. Note that both the charging power and, mainly, the market prices, are different depending on the approach considered, which highlights the importance of explicitly representing the market clearing process in the bidding strategy problem. The total charging cost of the price-maker EV aggregator, i.e., the optimal value of the objective function, is 1869.44 €, while considering the price-taker approach this cost was 1500 €. This means that neglecting the impact of the EV aggregator bidding decisions on market prices results in an underestimation of charging costs. □

A GAMS [9] code for solving Illustrative Example 6.13 is provided in Sect. 6.7.

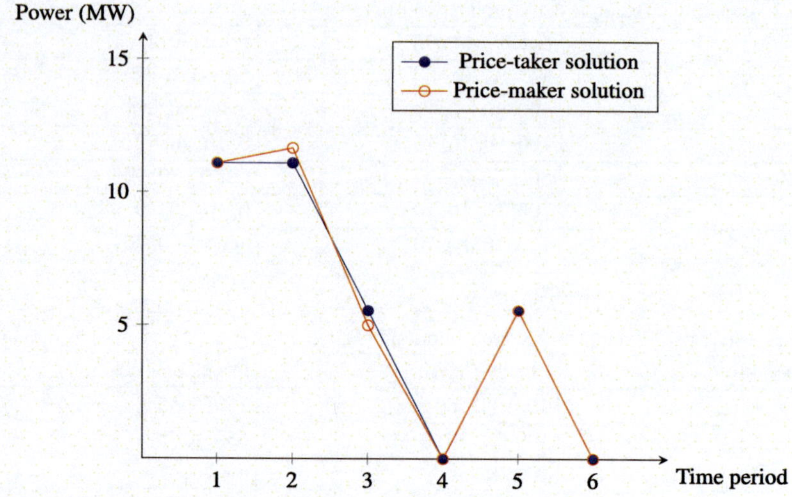

Fig. 6.17 Illustrative Example 6.13: charging power levels

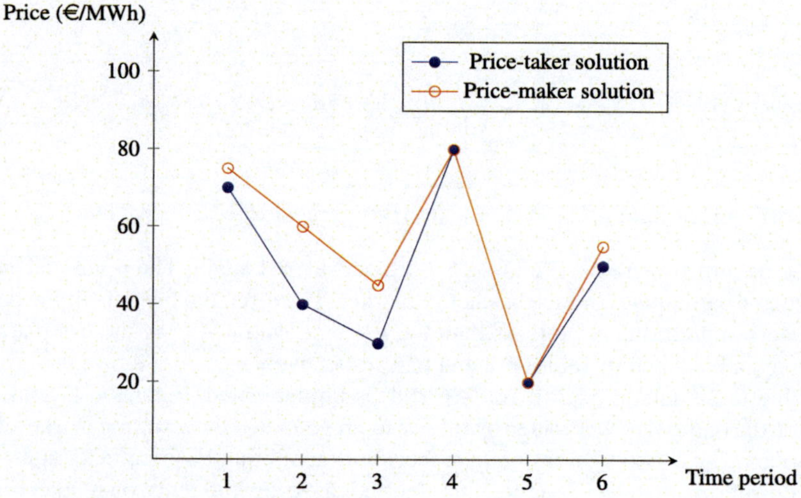

Fig. 6.18 Illustrative Example 6.13: market prices

6.5 Summary and Conclusions

This chapter analyzes the bidding strategy problem of an EV aggregator. Firstly, the main characteristics of the problem are described. Then, the problem is formulated using both a deterministic and a stochastic programming approach to account for the uncertainties in the driving requirements of EV users and market prices. Finally, the bidding strategy problem is extended to take into account the impact of the bidding decisions of the EV aggregator on the market clearing problem.

Considering the theoretical framework and the results of the illustrative examples provided in the chapter, the conclusions below are in order:

1. Managing the charging strategies of EVs individually may result in an intractable problem, mainly if the number of EVs is very large.
2. The bidding strategy of an EV aggregator in the DA market is highly influenced by the uncertainties in market prices and driving requirements of EV users. An accurate representation of these uncertainties is essential to obtain informed decisions.
3. The bidding strategy of the EV aggregator may impact the market clearing process and, therefore, the market prices. In such a case, it is important to explicitly represent the market clearing auction in the decision-making problem.

The interested reader is suggested to extend the models provided in this chapter to account for uncertainties in the bi-level model described in Sect. 6.4. Moreover, uncertainties in the bidding strategy problem may be also handled using robust optimization as done, for instance, in [1, 15].

6.6 Notation

The main notation used in this chapter is provided below for quick reference. Other symbols are defined as needed throughout the chapter.

Acronyms

DA	Day ahead.
EV	Electric vehicle.
KKT	Karush-Kuhn-Tucker.
LP	Linear programming.
MILP	Mixed-integer linear programming.
MPEC	Mathematical program with equilibrium constraints.

Indices and Sets

d	Power consumers participating in the market.
D	Set of power consumers participating in the market.
g	Power producers participating in the market.
G	Set of power producers participating in the market.

t Index of time periods.
T Set of time periods.
ω Index of scenarios.
Ω Set of scenarios.

Parameters

$\underline{E}^{\mathrm{EV}}_{t\omega}$ Minimum energy that must contain the virtual battery representing the EV aggregation in time period t and scenario ω (MWh).
$\overline{E}^{\mathrm{EV}}_{t\omega}$ Maximum energy that can contain the virtual battery representing the EV aggregation in time period t and scenario ω (MWh).
$E^{\mathrm{EV,IO}}_{t\omega}$ Energy drop/contribution of the EVs departing from/arriving to the virtual battery representing the EV aggregation in time period t and scenario ω (MWh).
$\overline{P}^{\mathrm{D}}_{dt}$ Power bid quantity of power consumer d in time period t (MW).
$\overline{P}^{\mathrm{G}}_{gt}$ Power offer quantity of power producer g in time period t (MW).
χ^{D}_{dt} Bid price of power consumer d in time period t (€/MWh).
χ^{G}_{gt} Offer price of power producer g in time period t (€/MWh).
Δ_t Duration of time period t (h).
$\eta^{\mathrm{EV,C}}$ Charging efficiency of the virtual battery representing the EV aggregation (pu).
$\eta^{\mathrm{EV,D}}$ Discharging efficiency of the virtual battery representing the EV aggregation (pu).
$\tilde{\lambda}^{\mathrm{D}}_{t\omega}$ DA energy price in time period t and scenario ω (€/MWh).
π_{ω} Probability of scenario ω (pu).

Variables

$e^{\mathrm{EV}}_{t\omega}$ Energy stored in the virtual battery representing the EV aggregation in time period t and scenario ω (MWh).
p^{D}_{dt} Consumption level of power consumer d in time period t (MW).
p^{G}_{dt} Generation level of power producer g in time period t (MW).
$p^{\mathrm{EV,C}}_{t\omega}$ Charging power level of the virtual battery representing the EV aggregation in time period t and scenario ω (MW).
$p^{\mathrm{EV,D}}_{t\omega}$ Discharging power level of the virtual battery representing the EV aggregation in time period t and scenario ω (MW).
$\lambda^{\mathrm{D}}_{t\omega}$ Dual variable used to compute the DA energy market price in time period t and scenario ω (€/MWh).

6.7 GAMS Codes

This section provides the GAMS [9] codes used to solve some of the illustrative examples of this chapter.

6.7.1 *Day-Ahead Market Clearing Problem: Primal Formulation*

A GAMS code for solving Illustrative Example 6.8 is provided below:

```
$title DAY-AHEAD MARKET CLEARING PROBLEM: PRIMAL FORMULATION

*******************************************************************
* DATA
*******************************************************************

SETS
d          'Consumers'                    /d1*d3/
g          'Producers'                    /g1*g4/
;

SCALARS
Deltat     'Duration of time steps' /1/
;

TABLE GDATA(g,*)
           PGmax    XiG
g1         80       50
g2         100      80
g3         20       70
g4         50       60;

TABLE DDATA(d,*)
           PDmax    XiD
d1         40       85
d2         100      65
d3         100      75;

SCALARS
XIEV       'Bid price EV aggregator' /100/
PEVmax     'Bid power EV aggregator' /13.22/;

*******************************************************************
* DECLARATION OF VARIABLES AND MATHEMATICAL CHARACTERIZATION
*******************************************************************

VARIABLES

**         OBJECTIVE FUNCTION
z              'Objective function'
;

POSITIVE VARIABLES
```

```
**          OPERATIONAL VARIABLES
pG(g)          'Production level'
pD(d)          'Consumption level'
pEV            'Charging power level'
;

*******************************************************************
*   EQUATIONS
*******************************************************************

EQUATIONS

**          OBJECTIVE FUNCTION
ObjectiveFunction     'Objective function'

**          CONSTRAINTS
Balance               'Power balance'
MaxPG                 'Maximum production levels'
MaxPD                 'Maximum consumption levels'
MaxEV                 'Maximum charging levels'
;

ObjectiveFunction..                 z=e=sum(g,GDATA(g,'XiG')*pG(g))-sum(d,DDATA(d,'XiD')*pD(d))-XIEV*pEV;

Balance..                           sum(g,pG(g))=e=sum(d,pD(d))+pEV;
MaxPG(g)..                          pG(g)=l=GDATA(g,'PGmax');
MaxPD(d)..                          pD(d)=l=DDATA(d,'PDmax');
MaxEV..                             pEV=l=PEVmax;

*******************************************************************
*   MODEL AND SOLVING
*******************************************************************

MODEL MarketClearing2 /all/;
MarketClearing2.optcr=0;
Option lp=cplex;

SOLVE MarketClearing2 USING lp MINIMIZING z;
```

6.7.2 Day-Ahead Market Clearing Problem: Dual Formulation

A GAMS code for solving Illustrative Example 6.9 is provided below:

```
$title DAY-AHEAD MARKET CLEARING PROBLEM: PRIMAL FORMULATION

*******************************************************************
* DATA
*******************************************************************
```

```
SETS
d          'Consumers'                         /d1*d3/
g          'Producers'                         /g1*g4/
;

SCALARS
Deltat    'Duration of time steps' /1/
;

TABLE GDATA(g,*)
          PGmax    XiG
g1        80       50
g2        100      80
g3        20       70
g4        50       60;

TABLE DDATA(d,*)
          PDmax    XiD
d1        40       85
d2        100      65
d3        100      75;

SCALARS
XIEV      'Bid price EV aggregator' /100/
PEVmax    'Bid power EV aggregator' /13.22/;

*****************************************************************
* DECLARATION OF VARIABLES AND MATHEMATICAL CHARACTERIZATION
*****************************************************************

VARIABLES

**         OBJECTIVE FUNCTION
z              'Objective function'
lambda         'Dual variable associated with the power balance'
;

POSITIVE VARIABLES

**         OPERATIONAL VARIABLES
MUGmax(g)    'Dual variable associated with the upper production
    level'
MUGmin(g)    'Dual variable associated with the lower production
    level'
MUDmax(d)    'Dual variable associated with the upper consumption
    level'
MUDmin(d)    'Dual variable associated with the lower consumption
    level'
MUEVmax      'Dual variable associated with the upper charging
    power level'
MUEVmin      'Dual variable associated with the lower charging
    power level'
;

*****************************************************************
*  EQUATIONS
*****************************************************************

EQUATIONS
```

```
**          OBJECTIVE FUNCTION
ObjectiveFunction    'Objective function'

**          CONSTRAINTS
EQ1
EQ2
EQ3
;

ObjectiveFunction..          z=e=-sum(g,GDATA(g,'PGmax')*MUGmax(g))-
     sum(d,DDATA(d,'PDmax')*MUDmax(D))-PEVmax*MUEVmax;

EQ1(g)..                     GDATA(g,'XiG')-lambda+MUGmax(g)-MUGmin(g
     )=e=0;
EQ2(d)..                     -DDATA(d,'XiD')+lambda+MUDmax(d)-MUDmin(
     d)=e=0;
EQ3..                        -XIEV+lambda+MUEVmax-MUEVmin=e=0;

*****************************************************************
*  MODEL AND SOLVING
*****************************************************************

MODEL DualMarketClearing /all/;
DualMarketClearing.optcr=0;
Option lp=cplex;

SOLVE DualMarketClearing USING lp MAXIMIZING z;
```

6.7.3 Day-Ahead Market Clearing Problem: Linear Karush-Kuhn-Tucker Conditions

A GAMS code for solving Illustrative Example 6.12 is provided below:

```
$title DAY-AHEAD MARKET CLEARING PROBLEM: KKT CONDITIONS

*****************************************************************
* DATA
*****************************************************************

SETS
d          'Consumers'                     /d1*d3/
g          'Producers'                     /g1*g4/
;

SCALARS
Deltat     'Duration of time steps' /1/
N          'Large enough constant'  /1000/
M          'Large enough constant'  /100/
;

TABLE GDATA(g,*)
           PGmax    XiG
```

```
g1        80        50
g2        100       80
g3        20        70
g4        50        60;

TABLE DDATA(d,*)
          PDmax     XiD
d1        40        85
d2        100       65
d3        100       75;

SCALARS
XIEV      'Bid price EV aggregator' /100/
PEVmax    'Bid power EV aggregator' /13.22/;

****************************************************************
* DECLARATION OF VARIABLES AND MATHEMATICAL CHARACTERIZATION
****************************************************************

VARIABLES

**        OBJECTIVE FUNCTION
z             'Objective function'
lambda        'Dual variable associated with the power balance'
;

POSITIVE VARIABLES

**        OPERATIONAL VARIABLES
pG(g)         'Production level'
pD(d)         'Consumption level'
pEV           'Charging power level'
MUGmax(g)     'Dual variable associated with the upper production level'
MUGmin(g)     'Dual variable associated with the lower production level'
MUDmax(d)     'Dual variable associated with the upper consumption level'
MUDmin(d)     'Dual variable associated with the lower consumption level'
MUEVmax       'Dual variable associated with the upper charging power level'
MUEVmin       'Dual variable associated with the lower charging power level'
;

BINARY VARIABLES

**        AUXILIARY VARIABLES TO REFORMULATE THE PROBLEM AS MILP
uGmin(g), uGmax(g), uDmin(d), uDmax(d), uEVmin, uEVmax;

****************************************************************
*  EQUATIONS
****************************************************************

EQUATIONS

**        OBJECTIVE FUNCTION
ObjectiveFunction   'Objective function'
```

```
**          CONSTRAINTS
Balance                     'Power balance'
MaxPG                       'Maximum production levels'
MaxPD                       'Maximum consumption levels'
MaxEV                       'Maximum charging levels'
EQ1,EQ2,EQ3
EQ4,EQ5,EQ6,EQ7
EQ8,EQ9,EQ10,EQ11,
EQ12,EQ13,EQ14,EQ15
;

ObjectiveFunction..         z=e=1;

Balance..                   sum(g,pG(g))=e=sum(d,pD(d))+pEV;
MaxPG(g)..                  pG(g)=l=GDATA(g,'PGmax');
MaxPD(d)..                  pD(d)=l=DDATA(d,'PDmax');
MaxEV..                     pEV=l=PEVmax;

EQ1(g)..                    GDATA(g,'XiG')-lambda+MUGmax(g)-MUGmin(g
    )=e=0;
EQ2(d)..                    -DDATA(d,'XiD')+lambda+MUDmax(d)-MUDmin(
    d)=e=0;
EQ3..                       -XIEV+lambda+MUEVmax-MUEVmin=e=0;

EQ4(g)..                    MUGmin(g)=l=N*uGmin(g);
EQ5(g)..                    pG(g)=l=M*(1-uGmin(g));
EQ6(g)..                    MUGmax(g)=l=N*uGmax(g);
EQ7(g)..                    GDATA(g,'PGmax')-pG(g)=l=M*(1-uGmax(g));

EQ8(d)..                    MUDmin(d)=l=N*uDmin(d);
EQ9(d)..                    pD(d)=l=M*(1-uDmin(d));
EQ10(d)..                   MUDmax(d)=l=N*uDmax(d);
EQ11(d)..                   DDATA(d,'PDmax')-pD(d)=l=M*(1-uDmax(d));

EQ12..                      MUEVmin=l=N*uEVmin;
EQ13..                      pEV=l=M*(1-uEVmin);
EQ14..                      MUEVmax=l=N*uEVmax;
EQ15..                      PEVmax-pEV=l=M*(1-uEVmax);

*******************************************************************
*  MODEL AND SOLVING
*******************************************************************

MODEL KKTConditions /all/;
KKTConditions.optcr=0;
Option lp=cplex;

SOLVE KKTConditions USING MIP MAXIMIZING z;
```

6.7.4 *Price-Maker Bidding Strategy: Mixed-Integer Linear Programming Problem*

A GAMS code for solving Illustrative Example 6.13 is provided below:

```
$title PRICE-MAKER BIDDING STRATEGY

************************************************************************
* DATA
************************************************************************

SETS
d         'Consumers'                          /d1*d3/
g         'Producers'                          /g1*g4/
t         'Time periods'                       /t1*t6/
;

SCALARS
Deltat    'Duration of time steps' /1/
E0        'Initial energy level'   /30/
etaC      'Charging efficiency'    /0.9/
N         'Large enough constant'  /1000/
M         'Large enough constant'  /100/
;

PARAMETER PCmax(t) 'maximum charging power level'
/t1       15
t2        14
t3        12
t4        10
t5        30
t6        30/
;

PARAMETER PDmax(t) 'maximum discharging power level'
/t1       15
t2        14
t3        12
t4        10
t5        30
t6        30/
;

PARAMETER Emin(t) 'minimum energy level'
/t1       40
t2        50
t3        10
t4        5
t5        30
t6        50/
;

PARAMETER Emax(t) 'maximum energy level'
/t1       60
t2        60
t3        30
t4        20
```

```
t5        50
t6        60/
;

PARAMETER EIO(t) 'energy input/output'
/t1       0
t2        0
t3        -40
t4        -10
t5        20
t6        20/
;

TABLE GDATA(g,t,*)
          PGmax    XiG
g1.t1     80       50
g2.t1     50       60
g3.t1     20       70
g4.t1     100      80
g1.t2     100      0
g2.t2     20       20
g3.t2     10       40
g4.t2     10       60
g1.t3     100      0
g2.t3     20       10
g3.t3     10       30
g4.t3     10       50
g1.t4     50       60
g2.t4     50       70
g3.t4     50       80
g4.t4     50       90
g1.t5     100      0
g2.t5     50       20
g3.t5     10       30
g4.t5     10       40
g1.t6     100      40
g2.t6     20       50
g3.t6     10       60
g4.t6     10       70;

TABLE DDATA(d,t,*)
          PDmax    XiD
d1.t1     40       85
d2.t1     100      75
d3.t1     100      55
d1.t2     100      65
d2.t2     25       65
d3.t2     30       45
d1.t3     100      65
d2.t3     25       55
d3.t3     30       45
d1.t4     60       95
d2.t4     80       85
d3.t4     40       75
d1.t5     90       35
d2.t5     40       25
d3.t5     20       15
d1.t6     80       65
d2.t6     30       55
```

```
d3.t6   20       45;

*******************************************************************
* DECLARATION OF VARIABLES AND MATHEMATICAL CHARACTERIZATION
*******************************************************************

VARIABLES

**        OBJECTIVE FUNCTION
z             'Objective function'
lambda(t)     'Dual variable associated with the power balance'
;

POSITIVE VARIABLES

**        OPERATIONAL VARIABLES
e(t)          'Energy level'
pev(t)        'Charging power'
pevmax(t)     'Bid power'
XIEV(t)       'Bid price'
pG(g,t)       'Production level'
pD(d,t)       'Consumption level'
MUGmax(g,t)  'Dual variable associated with the upper production
    level'
MUGmin(g,t)  'Dual variable associated with the lower production
    level'
MUDmax(d,t)  'Dual variable associated with the upper consumption
    level'
MUDmin(d,t)  'Dual variable associated with the lower consumption
    level'
MUEVmax(t)   'Dual variable associated with the upper charging
    power level'
MUEVmin(t)   'Dual variable associated with the lower charging
    power level'
;

BINARY VARIABLES

**        AUXILIARY VARIABLES TO REFORMULATE THE PROBLEM AS MILP
uGmin(g,t), uGmax(g,t), uDmin(d,t), uDmax(d,t), uEVmin(t), uEVmax
    (t);

*******************************************************************
*  EQUATIONS
*******************************************************************

EQUATIONS

**        OBJECTIVE FUNCTION
ObjectiveFunction    'Objective function'

**        CONSTRAINTS
EnergyEvolution      'Energy evolution in the battery'
EnergyEvolution0     'Energy evolution in the battery in the first
    period'
ChargingLimits       'Charging power limits'
EnergyLimitsL        'Lower energy limits'
EnergyLimitsU        'Upper energy limits'
```

```
Balance                  'Power balance'
MaxPG                    'Maximum production levels'
MaxPD                    'Maximum consumption levels'
MaxEV                    'Maximum charging levels'

EQ1,EQ2,EQ3,
EQ4,EQ5,EQ6,EQ7
EQ8,EQ9,EQ10,EQ11,
EQ12,EQ13,EQ14,EQ15
;

ObjectiveFunction..                          z=e=sum(t,sum(g,GDATA(g,t,'XiG')*pG(g,t))-sum(d,DDATA(d,t,'XiD')*pD(d,t))+sum(g,GDATA(g,t,'PGmax')*MUGmax(g,t))+sum(d,DDATA(d,t,'PDmax')*MUDmax(d,t)));

EnergyEvolution(t)$(ord(t) gt 1)..           e(t)=e=e(t-1)+pev(t)*etaC*Deltat+EIO(t);
EnergyEvolution0..                           e('t1')=e=E0+pev('t1')*etaC*Deltat+EIO('t1');
ChargingLimits(t)..                          pev(t)=l=PCmax(t);
EnergyLimitsL(t)..                           Emin(t)=l=e(t);
EnergyLimitsU(t)..                           e(t)=l=Emax(t);

Balance(t)..                                 sum(g,pG(g,t))=e=sum(d,pD(d,t))+pEV(t);
MaxPG(g,t)..                                 pG(g,t)=l=GDATA(g,t,'PGmax');
MaxPD(d,t)..                                 pD(d,t)=l=DDATA(d,t,'PDmax');
MaxEV(t)..                                   pEV(t)=l=PEVmax(t);

EQ1(g,t)..                                   GDATA(g,t,'XiG')-lambda(t)+MUGmax(g,t)-MUGmin(g,t)=e=0;
EQ2(d,t)..                                   -DDATA(d,t,'XiD')+lambda(t)+MUDmax(d,t)-MUDmin(d,t)=e=0;
EQ3(t)..                                     -XIEV(t)+lambda(t)+MUEVmax(t)-MUEVmin(t)=e=0;

EQ4(g,t)..                                   MUGmin(g,t)=l=N*uGmin(g,t);
EQ5(g,t)..                                   pG(g,t)=l=M*(1-uGmin(g,t));
EQ6(g,t)..                                   MUGmax(g,t)=l=N*uGmax(g,t);
EQ7(g,t)..                                   GDATA(g,t,'PGmax')-pG(g,t)=l=M*(1-uGmax(g,t));

EQ8(d,t)..                                   MUDmin(d,t)=l=N*uDmin(d,t);
EQ9(d,t)..                                   pD(d,t)=l=M*(1-uDmin(d,t));
EQ10(d,t)..                                  MUDmax(d,t)=l=N*uDmax(d,t);
EQ11(d,t)..                                  DDATA(d,t,'PDmax')-pD(d,t)=l=M*(1-uDmax(d,t));

EQ12(t)..                                    MUEVmin(t)=l=N*uEVmin(t);
EQ13(t)..                                    pEV(t)=l=M*(1-uEVmin(t));
EQ14(t)..                                    MUEVmax(t)=l=N*uEVmax(t);
EQ15(t)..                                    PEVmax(t)-pEV(t)=l=M*(1-uEVmax(t));

***********************************************************************
*  MODEL AND SOLVING
***********************************************************************
```

```
MODEL PriceMakerBidding /all/;
PriceMakerBidding.optcr=0;
Option lp=cplex;

SOLVE PriceMakerBidding USING MIP MINIMIZING z;
```

References

1. Baringo, L., Sánchez-Amaro, R.: A stochastic robust optimization approach for the bidding strategy of an electric vehicle aggregator. Electr. Pow. Syst. Res. **146**, 362–370 (2017)
2. Birge, J.R., Louveaux, F.: Introduction to Stochastic Programming, 2nd edn. Springer, New York (2011)
3. Boyd, S., Vandenberghe, L.: Convex Optimization. University Press, Cambridge (2009)
4. Chang, H.-J.: Economics. The User's Guide. Bloomsbury Press, New York (2014)
5. Conejo, A.J., Carrión, M., Morales, J.M.: Decision Making Under Uncertainty in Electricity Markets. Springer, New York (2010)
6. Conejo, A.J., Baringo, L., Kazempour, S.J., Siddiqui, A.S.: Investment in Electricity Generation and Transmission: Decision Making Under Uncertainty. Springer, Switzerland (2016)
7. Fortuny-Amat, J., McCarl, B.: A representation and economic interpretation of a two-level programming problem. J. Oper. Res. Soc. **32**(9), 783–792 (1981)
8. Gabriel, S.A., Conejo, A.J., Fuller, J.D., Hobbs, B.F., Ruiz, C.: Complementarity Modeling in Energy Markets. Springer, New York (2013)
9. GAMS (2021). www.gams.com
10. González-Vayá, M., Baringo, L., Andersson, G.: Integration of PEVs into power markets: a bidding strategy for a fleet aggregator. In: Plug In Electric Vehicles in Smart Grids. Springer, Singapore (2015)
11. González-Vayá, M., Baringo, L., Krause, T., Andersson, G., Almeida, P., Geth, F., Rapoport, S.: EV aggregation models for different charging scenarios. In: Proceedings of the 23rd International Conference on Electricity Distribution (CIRED) (2015)
12. Kristoffersen, T.K., Capion, K., Meibom, P.: Optimal charging of electric drive vehicles in a market environment. Appl. Energy **88**(5), 1940–1948 (2011)
13. Li, Z., Guo, Q., Sun, H., Wang, Y., Xin, S.: Emission-concerned wind-EV coordination on the transmission grid side with network constraints: concept and case study. IEEE Trans. Smart Grid. **4**(3), 1692–1704 (2013)
14. Pérez-Arriaga, J.I., Meseguer, C.: Wholesale marginal prices in competitive generation markets. IEEE Trans. Power Syst. **12**(2), 710–717 (1997)
15. Porras, Á., Fernández-Blanco, R., Morales, J.M., Pineda, S.: An efficient robust approach to the day-ahead operation of an aggregator of electric vehicles. IEEE Trans. Smart Grid **11**(6), 4960–4970 (2020)
16. Ruiz, C., Conejo, A.J.: Pool strategy of a producer with endogenous formation of locational marginal prices. IEEE Trans. Power Syst. **24**(4), 1855–1866 (2009)
17. Sioshansi, R., Conejo, A.J.: Optimization in Engineering. Models and Algorithms, Springer, New York (2017)
18. The ILOG CPLEX (2021). www.ilog.com/products/cplex/
19. Vandael, S., Claessens, B., Hommelberg, M., Holvoet, T., Deconinck, G.: A scalable three-step approach for demand side management of plug-in hybrid vehicles. IEEE Trans. Smart Grid. **4**(2), 720–728 (2013)

Chapter 7
Pricing Strategy of Electricity Suppliers for Electric Vehicles

This chapter describes and formulates the selling price determination problem faced by an electricity supplier. We assume that this supplier provides the power demand of a set of small consumers that do not participate in the electricity markets. We also consider that these consumers are electric vehicle users. In this problem, the electricity supplier has to handle a number of uncertainties: (i) the electricity price in the pool market, (ii) the consumption patterns of electric vehicles, and (iii) the selling prices offered by rival suppliers. Additionally, electricity suppliers must anticipate the response of their potential clients to the offered selling price. For these reasons, the selling price determination by electricity suppliers can be considered challenging from the modeling point of view.

7.1 Introduction

The liberalization of power systems in the late nineties of 20th century allowed the active participation of consumers in the electricity markets. However, only large consumers, most of them belonging to the industrial sector, currently participate in these markets. The rest of industrial consumers, and most of residential and commercial consumers, are represented by intermediate agents in the electricity markets. These intermediate agents, usually known as electricity suppliers or retailers, purchase the energy demanded by the consumers in the electricity markets to sell it afterwards to them. For doing that, consumers and electricity suppliers sign procurement contracts where the terms and conditions of the sale and purchase of electricity are established. Usually, the price paid by a consumer to its electricity supplier can be (i) a fixed price, (ii) a time-of-use rate, or (iii) a variable price associated with the price in the pool. In this chapter, it is assumed that the client portfolio of the analyzed electricity supplier consists exclusively of electric vehicle (EV) users and that the demand that has to be provided is that concerning the charge of EVs.

L. Baringo et al., *Electric Vehicles and Renewable Generation*,
Green Energy and Technology, https://doi.org/10.1007/978-3-031-09079-0_7

In this problem, it is necessary to model the response of the EV users to the price offered by the electricity supplier. If EV users behave as rational agents, they will prefer to sign a procurement contract with a supplier offering a low selling price. Therefore, if the electricity supplier offers a low selling price, it will expect to sign a high number of contracts. However, if the selling price is too low, the margin between the price of the electricity sold to their clients and the acquisition price of electricity in the pool can be quite narrow, even negative. Therefore, how to determinate an adequate selling price to offer to electricity consumers is not a trivial task at all.

Additionally to the modeling of the response to prices of consumers, the electricity supplier has also to handle different sources of uncertainty. On one hand, it has to face the uncertainty pertaining to the purchase price of electricity in the different electricity markets. For instance, the electricity price in the pool is unknown and it can be unexpectedly altered in the medium-term because of weather phenomena, changes in economic cycles or any other unforeseen episode. On the other hand, the hourly consumption of energy by EVs is also uncertain and it can be also affected by external factors as those described for the case of electricity prices. The last source of uncertainty are the selling prices offered by rival suppliers, which have a relevant influence on the response of consumers to the price offered by the considered supplier. As a result, an appropriate modeling of the uncertainty in this problem is mandatory. Besides, risk modeling tools should be also considered in order to tailor the pricing strategy to the risk aversion degree selected by the electricity supplier.

Another aspect that must be contemplated in this problem are the new capabilities that smart grids can offer. Using these capabilities, electricity suppliers may be able to exercise a direct control on the charging process of the EVs of its clients. From an economical point of view, the direct control of the charge of EVs can help the electricity supplier to reduce the electricity purchasing cost by charging the EVs in those hours with low electricity prices. Additionally, the vehicle-to-grid (V2G) capability can be also used to sell part of the energy stored in the batteries of the EVs if electricity prices are high enough.

Relevant references dealing with electricity suppliers of EVs include [1, 2, 5, 6, 8, 10, 13, 14, 19–27].

This chapter is organized as follows. Section 7.1 introduces this problem. The decision framework is described in Sect. 7.2. The response of the clients to the price offered by the electricity supplier is modeled in Sect. 7.3. The uncertainty characterization is described in Sect. 7.4. The formulation of the problem is provided in Sect. 7.5. In Sects. 7.6 and 7.7, two illustrative examples are solved and comprehensible discussed in order to highlight the main features of the proposed models. In Sect. 7.8, the performance of the proposed model is tested on a realistic case study. Section 7.9 provides a summary of the chapter together with some relevant conclusions. Section 7.10 defines the main notation used in this chapter. Finally, Sect. 7.11 provides the GAMS codes used to solve the illustrative examples analyzed in this chapter.

7.2 Decision Framework

Without loss of generality, let us to consider an electricity supplier that desires to provide the electricity demand of a set of EVs. In order to provide the electricity demand of its clients, the electricity supplier participates directly in the electricity markets. We also assume that the electricity supplier has no market-power capability in these markets.

Considering the above, the objective of the supplier is to determine the selling price offered to its clients to maximize the expected profit obtained by the sale of electricity. This is a medium-term decision-making problem where the selling price determination is made at the beginning of a given planning horizon, e.g., one year. In order to make informed selling price decisions, the purchasing and selling processes of electricity during the year must be considered in the decision-making process. It is assumed that the electricity supplier acquires the electricity in the pool. The pool comprises all those short-term markets in which electricity is traded. For instance, day-ahead, adjustment, and balancing markets are part of the pool. However, the energy traded in the pool is mostly negotiated in the day-ahead energy market, whereas the rest of markets are only used to make final energy adjustments. For this reason, we assume in this problem that the pool refers to the day-ahead energy market. Pool prices result from the offers and bids submitted by the pool agents and they are independent of the actions made by price-taker agents. Therefore, at the beginning of the year, the electricity supplier does not know the future electricity prices and consumption profiles of their clients. Additionally, it has not perfect knowledge about the response of its potential clients to the offered price and the pricing strategies of rival suppliers, either. Because of the presence of uncertainty, the usage of a stochastic programming approach for modeling this decision-making problem is advisable. In this manner, uncertainty parameters can be characterized using credible realizations or scenarios. For each scenario, the participation of the electricity supplier in the electricity market and the provision of the demand of its clients has to be modeled. Furthermore, if the electricity supplier has the capability of controlling the charge and discharge of EVs, these charging and discharging decisions have to be also considered for each scenario.

Illustrative Example 7.1 Electricity supplier

We consider the case represented in Fig. 7.1 in which an electricity supplier must provide the demand of their clients during two time periods. The clients of the supplier are grouped together in two different groups of EVs. The supplier participates in the pool to provide the electricity demand of EVs.

As shown in Fig. 7.1, the supplier purchases in the pool 5 and 20 MWh at prices 70 and 100 €/MWh in periods 1 and 2, respectively. Likewise, the supplier sells 5 and 20 MWh to EV groups 1 and 2 at price 120 €/MWh, respectively. EVs belonging to group 1 consume 5 MWh in the first period, whereas EVs belonging to group 2 consume 20 MWh in the second one. The purchasing cost paid by the supplier is equal to 5 MW $\cdot$ 70 €/MWh + 20 MW $\cdot$ 100 €/MWh = 2,350 €/h. Additionally,

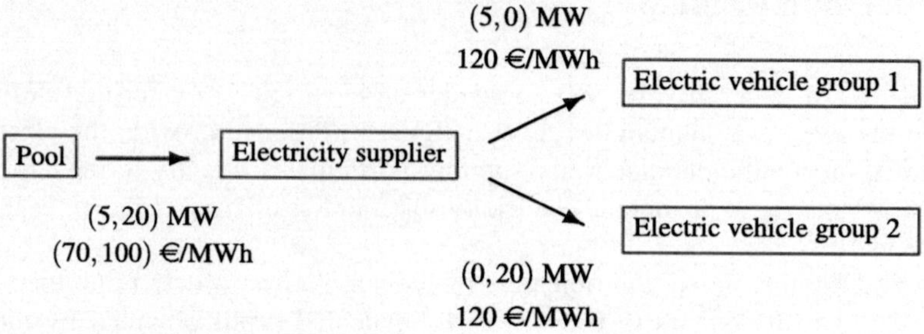

Fig. 7.1 Illustrative Example 7.1: electricity supplier

the revenue obtained from selling energy to their clients is equal to $(5+20)\cdot$ 120 €/MWh $=$ 3,000 €/h. Therefore, the profit achieved by the electricity supplier is equal to $(3,000 - 2,350) = 650$ €/h. □

Figure 7.2 represents graphically the decision-making process described above. The black point represents the node in which the offered selling price is decided. Note that this is a *here-and-now* decision that is made before knowing the actual realizations of the uncertain parameters. Therefore, this node represents a single decision that is made for each possible realization of the uncertain parameters. On the contrary, operation decisions, as those related to the purchase of energy in the pool, are made after knowing the actual realizations of the uncertain parameters. For this reason, we refer to these decisions as *wait-and-see* and they can be different for each scenario. Note that the scenario tree represented in Fig. 7.2 is the typical decision tree used in two-stage stochastic programming problems with recourse. In this case, the recourse decisions are those related to the short-term operation of the electricity supplier.

In order to model the risk-aversion degree of the electricity supplier, the Conditional Value-at-Risk (CVaR) of the profit has been considered. A detailed description of the CVaR can be consulted in Sect. 3.2.4. In this problem, the CVaR may be roughly defined as the expected value of the profit in worst scenarios. A risk-neutral supplier will prefer to have a high expected profit even if there is a certain probability of experiencing very low profits in some scenarios. On the contrary, a risk-averse supplier will prefer to increase the profit associated to worst scenarios at expenses of reducing the total expected profit. The formulation of the CVaR and its integration in the formulation of this problem will be discussed in Sect. 7.5.

7.3 Response of Electric Vehicle Users to Prices

The objective of this section is to model the response of the EV users to the prices offered by electricity suppliers. The modeling of the behaviour of the clients from the perspective of the electricity supplier is key for determining an adequate pricing

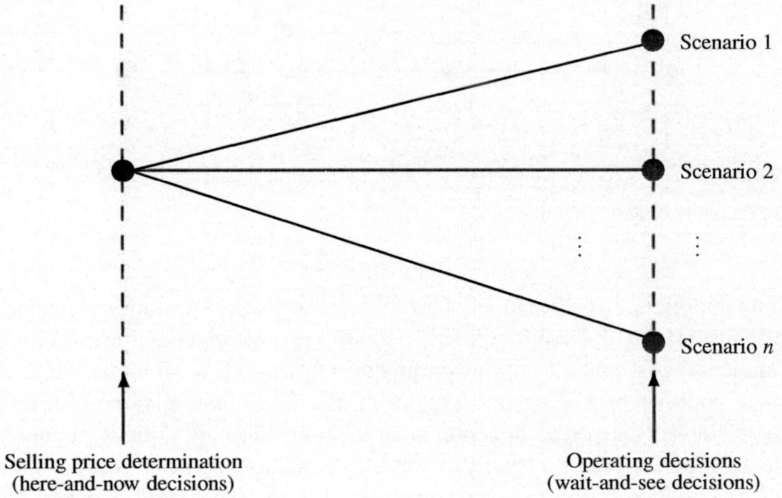

Fig. 7.2 Scenario tree

strategy. Observe that, if the selling price is too low or too high, electricity suppliers may experience low profits or even losses.

The preferences of EV users on electricity suppliers can be characterized using a Hotelling model. This model has been traditionally used to characterize the behaviour of consumers in markets where homogeneous goods are traded, such as electricity, [11]. Together with the selling price, this model allows to consider other elements in order to simulate the selection of electricity supplier. Even though the selling price is the most relevant factor in the selection of supplier, it is well known from the real world experience that not all consumers purchase electricity from the cheapest supplier. Other factors as corporate image or customer service are also important in the selection process of the supplier.

In order to describe the Hotelling model, let us assume an EV user that can purchase its electricity demand from two different suppliers. Based on the Hotelling model, we consider that both suppliers are placed at the two ends of a segment line of length 1, denoted as Hotelling line. The EV user is located in a given position within the segment line. This Hotelling line is graphically represented in Fig. 7.3. We define $\theta \in [0, 1]$ as the distance along the Hotelling line between the EV user and the first supplier. In this manner, the distance between the EV user and the second supplier is $(1 - \theta)$. The Hotelling model states that the preference of the EV user over a given supplier increases as the distance between the EV user and the supplier along the Hotelling line decreases. In this manner, if $\theta < (1 - \theta)$ the EV user will present a higher preference to select the first supplier. A value of $\theta = 0.5$ indicates that the EV user has a neutral position with respect to both suppliers.

Considering the Hotelling model described above, the transportation cost per unit, T^{C} (€/pu), is used to quantify economically the preferences of the potential clients for

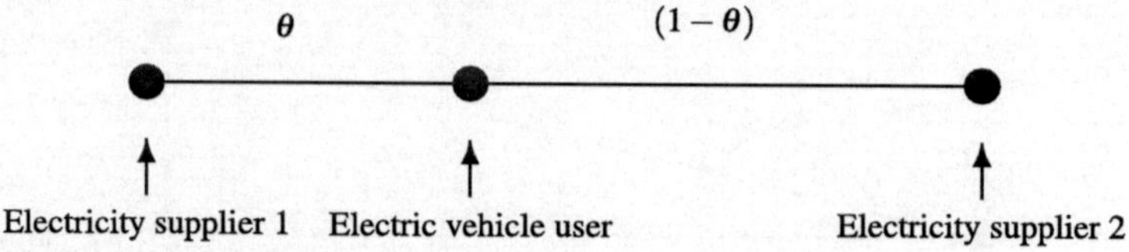

Fig. 7.3 Example of Hotelling line

electricity suppliers. In this manner, $T^{C}\theta$ represents the transportation costs incurred by the EV user if the first supplier is selected. Please, note that the transportation cost is not an actual cost, but a fictitious parameter that is used to simulate the selection process of supplier by EV users. In addition, EV users may respond in a different manner to the selling prices depending on whether they are current clients of the considered supplier or not. Usually, electricity consumers present a certain level of resistance to change suppliers. Therefore, a loyalty reward, L (€), is used to model the incentive of an EV user to purchase again from its current supplier. Similarly to the case of the transportation cost, the loyalty reward is a fictitious term considered only for modeling the decision process of electricity consumers.

Considering that the annual demand of a given EV is denoted by E, and that the selling prices offered by suppliers 1 and 2 are denoted by $\hat{\lambda}^{S,1}$ and $\hat{\lambda}^{S,2}$, respectively, the selection of electricity supplier by an EV user that is a current client of supplier 1 can be mathematically formulated as follows:

$$\text{Minimize}_{y\in\{0,1\}} \quad \left(\hat{\lambda}^{S,1}E + T^{C}\theta - L\right)y + \left(\hat{\lambda}^{S,2}E + T^{C}(1-\theta)\right)(1-y). \qquad (7.1)$$

The selection made by the EV user is denoted by binary variable y. If $y = 1$, the EV user will select the first supplier. In this case, the total expected cost incurred by the EV user is equal to the energy procurement cost plus the transportation costs minus the loyalty reward. Observe that the expected energy procurement cost is equal to the selling price offered by the considered supplier, $\hat{\lambda}^{S,1}$, times the expected energy supplied, E. The second term of (7.1) formulates the total expected costs in the case in which the EV user selects the second supplier, $y = 0$. Note that the loyalty reward is not part of the expected procurement cost of the EV user in this case.

The formulation of problem (7.2) included below is equivalent to that of problem (7.1) but assuming that the EV user is currently a client of the second supplier. In this case, the loyalty reward, L, is considered in the second term of (7.2), which corresponds with the expected cost incurred by the consumer when it procures its energy demand from the second supplier.

$$\text{Minimize}_{y\in\{0,1\}} \quad \left(\hat{\lambda}^{S,1}E + T^{C}\theta\right)y + \left(\hat{\lambda}^{S,2}E + T^{C}(1-\theta) - L\right)(1-y). \qquad (7.2)$$

Observe that problems (7.1) and (7.2) are formulated using a deterministic approach. In other words, problems (7.1) and (7.2) assume that parameters $\hat{\lambda}^{S,1}$,

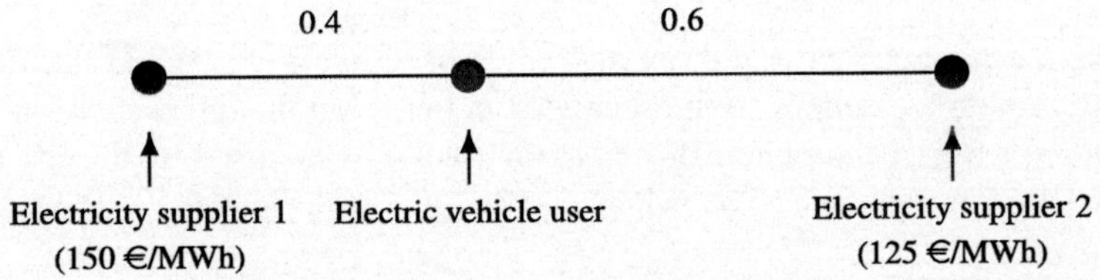

Fig. 7.4 Illustrative Example 7.2: Hotelling line model

$\hat{\lambda}^{S,2}$, and E are known. However, this assumption may not be realistic if the modeling of the response of the EV users is made from the perspective of electricity suppliers. For instance, the supplier cannot accurately forecast either the future consumed demand or the pricing strategies of rival retailers. The characterization of the uncertainty of this problem is described in the next section.

Illustrative Example 7.2 Hotelling line model

Let us consider the case of an EV user that desires to select its electricity supplier from two possible candidates. In order to determine the best supplier, a Hotelling line model is adopted for modeling this selection process. The selling price offered by the first supplier is 150 €/MWh, whereas the second supplier offers a price equal to 125 €/MWh. The expected annual demand of the EV user is 2 MWh, and the per-unit transportation cost and the loyalty reward are equal to 50 €/pu and 45 €, respectively. In this example, we assume that the EV user has a higher preference for the first supplier and it is placed at 0.4 distance units from this supplier in the Hotelling line. On the contrary, the EV user is located at 0.6 distance units from the second supplier. Figure 7.4 represents graphically the Hotelling line model associated with this example.

If the EV user is currently a client of the first supplier, the formulation of the electricity supplier selection problem based on the Hotelling line model is included below. According to this formulation, the first supplier is selected if $y = 1$, whereas the second supplier is selected when $y = 0$.

$$\text{Minimize}_{y \in \{0,1\}} \big(150 \cdot 2 + 50 \cdot 0.4 - 45\big)y + \big(125 \cdot 2 + 50 \cdot 0.6\big)(1 - y).$$

In this case the optimal value of variable y is 1 and the EV user selects the first supplier. Since the EV user is a current client of the first supplier and has also preference for this supplier, the EV prefers to pay a higher purchasing price ($150 > 125$ €/MWh) to continue being a client of the first supplier. Observe that if $y = 1$, the total cost incurred by the EV user, including the transportation cost and the loyalty reward, is $150 \cdot 2 + 50 \cdot 0.4 - 45 = 275$ €, which is smaller than that associated with $y = 0$, that is equal to $125 \cdot 2 + 50 \cdot 0.6 = 280$ €.

If the EV user is currently a client of the second supplier, the formulation of the electricity supplier selection problem is:

$$\text{Minimize}_{y \in \{0,1\}} \big(150 \cdot 2 + 50 \cdot 0.4\big)y + \big(125 \cdot 2 + 50 \cdot 0.6 - 45\big)(1 - y).$$

Unlike the previous case, the optimal value of variable y is 0 and the EV user prefers to remain as a client of the second supplier, even though it shows a higher preference for the first supplier. If $y = 0$, the total cost incurred by the EV user is $125 \cdot 2 + 50 \cdot 0.6 - 45 = 235$ €, whereas it is equal to $150 \cdot 2 + 50 \cdot 0.4 = 320$ € for $y = 1$.

Finally, if the EV user is not a client of any of the suppliers, the selection of electricity supplier is the following:

$$\text{Minimize}_{y \in \{0,1\}} \big(150 \cdot 2 + 50 \cdot 0.4\big)y + \big(125 \cdot 2 + 50 \cdot 0.6\big)(1 - y).$$

In this case, the EV user selects the second supplier $y = 0$. If $y = 0$, the total cost incurred by the EV user is $125 \cdot 2 + 50 \cdot 0.6 = 280$ €, whereas it is equal to $150 \cdot 2 + 50 \cdot 0.4 = 320$ € for $y = 1$. □

7.4 Uncertainty Characterization

An electricity supplier has to face two main sources of uncertainty in the medium-term problem described above: pool prices and the energy to supply to its clients. Considering that the electricity supplier is a price-taker agent in the pool, prices are independent on the supplier actions. However, the energy to provide by the supplier is strongly dependent on the selling price offered. Additionally, the energy supplied to the clients also depends on the prices offered by the rival suppliers, which are also uncertain.

7.4.1 Pool Prices

As stated above, it is assumed that electricity suppliers participate in the pool to purchase the electricity that they sell afterwards to their clients. Note that a price-taker agent is a participant with no capability of altering prices. Future pool prices are unknown for the electricity supplier when it has to determine the offered selling price to its clients.

In order to model the uncertainty of future pool prices, we define the pool price in period $t \in T$ as a random variable denoted by λ_t^{P}. As it is customary in stochastic programming, this random variable is represented using a finite set of scenarios indexed by $\omega \in \Omega$ [3]. As a result, pool prices are represented by the following set of scenarios: $\lambda_{t\omega}^{\mathrm{P}}, \forall t, \forall \omega$. As explained in Chap. 3, there exist different techniques to generate pool price scenarios. Traditionally, they are generated using historical data, time series models, or other statistical approaches.

Table 7.1 Illustrative Example 7.3: pool price scenarios

Scenario	Probability	Pool price (€/MWh)		
		Period		
		1	2	3
1	1/3	30	50	70
2	1/3	70	100	120
3	1/3	40	60	80

Illustrative Example 7.3 Pool price scenarios

Table 7.1 provides an example of pool price scenarios. The market horizon spans three time periods and the number of pool price scenarios is three. The probability of each scenario is 1/3. □

7.4.2 Demand Provided by the Electricity Supplier

The demand that the electricity supplier must provide during a given planning horizon is an uncertain parameter that depends on both endogenous and exogenous factors. In this regard, the future demand to procure will depend on the selling price offered by the supplier, but also on the particular preferences of the consumers and on the pricing strategies of rival suppliers.

In order to reduce the computational size of the problem, we assume that EVs with similar behavior patterns are clustered together in groups denoted by $k \in K$. In the same manner, the total number of hourly periods in the planning horizon are characterized using a set of representative hourly periods, $t \in T$, whose weight is denoted by W_t. Similarly to the case of pool prices, the future hourly demand of the consumers is a random variable that is characterized using a set of scenarios indexed by $\omega \in \Omega$. Considering this, the demand associated with group k, period t and scenario ω is denoted by $E^{\mathrm{D}}_{kt\omega}$.

Illustrative Example 7.4 Demand scenarios

Table 7.2 provides an example of the characterization of the demand of two EV groups. The first group charge their batteries in periods 2 and 3, whereas the second group charges the batteries in periods 1 and 2. Similarly to the case of pool prices characterized in Illustrative Example 7.3, three periods and three equiprobable scenarios are considered. Observe that the demand provided in Table 7.2 refers to the total demand of the EV groups, of which a part may be provided by the considered supplier. □

We would like to recall that $E^{\mathrm{D}}_{kt\omega}$ represents the total demand of the EVs, not the demand that has to be procured by the electricity supplier. This last demand is

Table 7.2 Illustrative Example 7.4: demand scenarios

Scenario	Probability	Group 1 (MWh)			Group 2 (MWh)		
		Period			Period		
		1	2	3	1	2	3
1	1/3	0	15	17	4	3	0
2	1/3	0	20	22	5	4	0
3	1/3	0	25	27	6	5	0

denoted by variable $e^{\mathrm{S}}_{kt\omega}$ and satisfies that $e^{\mathrm{S}}_{kt\omega} \leq E^{\mathrm{D}}_{kt\omega}$. The value of $e^{\mathrm{S}}_{kt\omega}$ can be computed using the Hotelling model described in Sect. 7.3. For doing that, let us consider that γ^{C}_{k} denotes the per-unit number of EVs belonging to group k that are current clients of the electricity supplier, whereas γ^{R}_{rk} refers to the per-unit number of EVs that are clients of rival supplier r. Without loss of generality, we denote by $e^{\mathrm{S,C}}_{kt\omega}$ and $e^{\mathrm{S,NC}}_{kt\omega}$ the energy to procure to current clients and non-clients of the supplier. In this manner, it is satisfied that $e^{\mathrm{S}}_{kt\omega} = e^{\mathrm{S,C}}_{kt\omega} + e^{\mathrm{S,NC}}_{kt\omega}$. The energy supplied to current clients of the supplier can be computed as $e^{\mathrm{S,C}}_{kt\omega} = \gamma^{\mathrm{S}}_{k} E^{\mathrm{D}}_{kt\omega} y^{\mathrm{S}}_{k\omega}$, where $y^{\mathrm{S}}_{k\omega}$ is a binary variable that is obtained from solving the following optimization problem based on the Hotelling model:

$$\Bigg\{ \text{Minimize}_{\{y^{\mathrm{S}}_{k\omega}, y^{\mathrm{R}}_{rk\omega}\} \in \{0,1\}}$$

$$\left(\sum_{t \in T} W_t \lambda^{\mathrm{S}}_{k} E^{\mathrm{D}}_{kt\omega} + T^{\mathrm{C}}_{k} \theta^{\mathrm{S}}_{k\omega} - L_k \right) y^{\mathrm{S}}_{k\omega}$$

$$+ \sum_{r=1}^{N_{\mathrm{R}}} \left(\sum_{t \in T} W_t \lambda^{\mathrm{R}}_{rkt\omega} E^{\mathrm{D}}_{kt\omega} + T^{\mathrm{C}}_{k} (1 - \theta^{\mathrm{R}}_{rk\omega}) \right) y^{\mathrm{R}}_{rk\omega} \tag{7.3a}$$

subject to:

$$y^{\mathrm{S}}_{k\omega} + \sum_{r=1}^{N_{\mathrm{R}}} y^{\mathrm{R}}_{rk\omega} = 1 \Bigg\}, \quad \forall k \in K, \forall \omega \in \Omega. \tag{7.3b}$$

Observe that problem (7.3) must be solved for each group k and scenario ω.

The energy supplied to non-current clients of the supplier is equal to $e^{\mathrm{S,NC}}_{kt\omega} = (1 - \gamma^{\mathrm{S}}_{k}) E^{\mathrm{D}}_{kt\omega} y^{\mathrm{S}}_{k\omega}$, where $y^{\mathrm{S}}_{k\omega}$ is obtained from:

$$\Bigg\{ \text{Minimize}_{\{y^{\mathrm{S}}_{k\omega}, y^{\mathrm{R}}_{rk\omega}\} \in \{0,1\}}$$

$$\left(\sum_{t \in T} W_t \lambda^{\mathrm{S}}_{k} E^{\mathrm{D}}_{kt\omega} + T^{\mathrm{C}}_{k} \theta^{\mathrm{S}}_{k\omega} \right) y^{\mathrm{S}}_{k\omega}$$

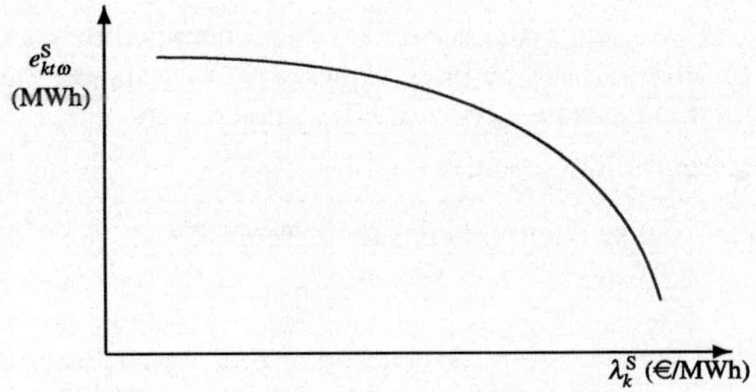

Fig. 7.5 Example of price-quota curve

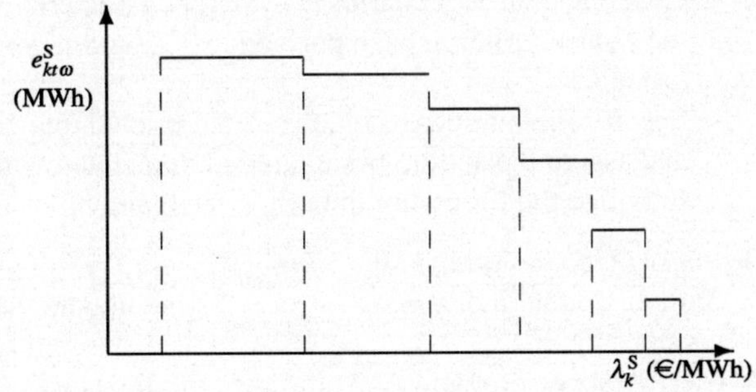

Fig. 7.6 Example of step-wise price-quota curve

$$+\sum_{r=1}^{N_{\mathrm{R}}}\left(\sum_{t\in T} W_t \lambda^{\mathrm{R}}_{rkt\omega} E^{\mathrm{D}}_{kt\omega} + T^{\mathrm{C}}_{k}(1-\theta^{\mathrm{R}}_{rk\omega}) - L^{\mathrm{R}}_{kr}\right) y^{\mathrm{R}}_{rk\omega} \tag{7.4a}$$

subject to:

$$\left. y^{\mathrm{S}}_{k\omega} + \sum_{r=1}^{N_{\mathrm{R}}} y^{\mathrm{R}}_{rk\omega} = 1 \right\}, \quad \forall k \in K, \forall \omega \in \Omega. \tag{7.4b}$$

The parameter L^{R}_{kr} is the loyalty reward of group k associated with rival supplier r. This parameter is greater than zero if the considered EV user is a current client of supplier r, being equal to zero otherwise.

From the formulations of problems (7.3) and (7.4) it is observed that the energy that the electricity supplier must provide is a function of the offered selling price, λ^{S}_{k}. Considering this, the relationship between the selling price and the supplied demand can be expressed explicitly using price-quota curves. A price-quota curve determines the quantity of electricity that the supplier must provide for each offered selling price. Figure 7.5 represents an example of price-quota curve. From this figure we observe that the supplied demand decreases as the selling price offered increases.

In order to preserve the linearity of the formulation, it is recommended to represent the price-quota curve using a step-wise approximation. For instance, Fig. 7.6 depicts a step-wise price-quota curve using six blocks.

Price-quota curves can be easily computed by solving problems (7.3) and (7.4) for different selling price values, λ_k^{S}. For instance, the following procedure can be used to generate step-wise price-quota curves:

- Step 0: Initialization.
  - Define the set of consumer groups $k \in K$.
  - Generate a set of scenarios $\omega \in \Omega$ for parameter $\theta_{k\omega}^{\mathrm{S}}$ for each consumer group k.
  - Generate a set of scenarios $\omega \in \Omega$ for characterizing the energy consumed by each consumer group k in each period t, $E_{kt\omega}^{\mathrm{D}}$.
  - Generate a set of scenarios $\omega \in \Omega$ for characterizing the selling price offered by the rival supplier in each period t, $\lambda_{rkt\omega}^{\mathrm{R}}$, and parameter $\theta_{rk\omega}^{\mathrm{R}}$ for each consumer group k.
  - Specify the number of blocks of the price quota curve $m \in M = \{1, \cdots, N_{\mathrm{M}}\}$ and assign the selling price interval associated with each block, $[\hat{\lambda}_{km}^{\mathrm{S}}, \hat{\lambda}_{km+1}^{\mathrm{S}}]$.
  - Initialize the following indices: $k \leftarrow 0$ and $m \leftarrow 0$.
- Step 1: Update index k, $k \leftarrow k + 1$.
- Step 2: Update index m, $m \leftarrow m + 1$, and update the selling price offered by the supplier $\lambda_k^{\mathrm{S}} = \hat{\lambda}_{km}^{\mathrm{S}}$.
- Step 3: Solve problem (7.3) for current clients of the supplier belonging to consumer group k and store the optimal decision, $\hat{Y}_{mk\omega}^{\mathrm{C}} = (y_{k\omega}^{\mathrm{S}})^*$.
- Step 4: Solve problem (7.4) for current non-clients of the supplier belonging to consumer group k and store the optimal decision, $\hat{Y}_{mk\omega}^{\mathrm{NC}} = (y_{k\omega}^{\mathrm{S}})^*$.
- Step 5:
  - If $m < N_{\mathrm{M}}$, go to Step 2.
  - Else, if $k < N_{\mathrm{K}}$, go to Step 1.
  - Else, the process is finished.

Illustrative Example 7.5 Price-quota curve

Figure 7.7 depicts an example of a price-quota curve for two different EV groups. The price-quota curve is expressed in per-unit magnitudes. In this manner, a value equal to 1 indicates that the whole demand of the considered EV group will be provided by the supplier. To build these two price-quota curves it is considered a single rival retailer that offers a selling price equal to the pool prices listed in Table 7.1 plus a fixed rate equal to 20 €/MWh. The per-unit transportation cost and the loyalty reward are equal to 50 €/pu and 100€, respectively. The GAMS codes used to generate these figures are provided in Sect. 7.11.1.

Figure 7.7 shows that the supplier will provide the complete demand of EV group 1 if the selling price offered is lower than 70 €/MWh. As the selling price offered increases, the energy sold to each EV group decreases. Observe that the demand supplied to EV group 1 is greater than or equal to the demand supplied to EV group 2. The reason of this result is that the EVs belonging to group 1 charge their batteries in periods 2 and 3, which are associated to pool prices that are more expensive than

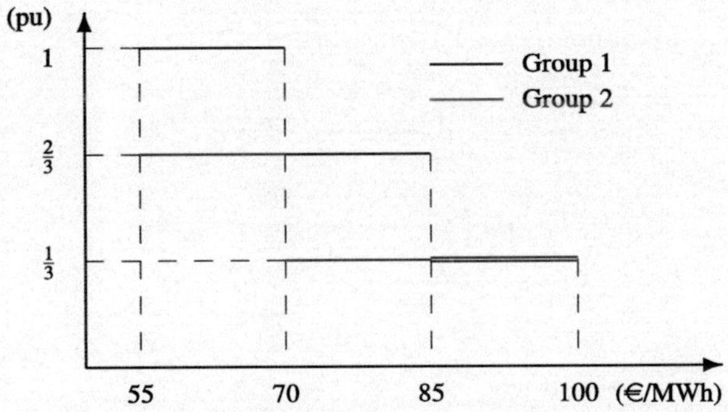

Fig. 7.7 Illustrative Example 7.5: three-block price-quota curve

those in periods 1 and 2 in which vehicles belonging to group 2 charge their batteries. Therefore, considering that the selling price offered by the rival supplier is indexed to pool prices, the selling price offered by the supplier is more competitive for client group 1 than for client group 2.

□

7.5 Formulation

In this section, we provide the complete formulation of the electricity supplier problem. For doing so, two different cases are considered. The first one assumes that the electricity supplier has not control over the charging processes of EVs. Then, EVs behave as inflexible demands that cannot be modified by the decisions of the electricity supplier. Additionally, we will consider a second case in which the electricity supplier makes use of the capabilities of smart grids. Therefore, in this case it is assumed that the electricity supplier can exercise a direct control over the charging processes of EVs. In this second case, it is also considered that EVs have V2G capabilities and that the electricity supplier can sell in the pool part of the electricity stored in the batteries of EVs. The description of notation used in the next sections is provided in Sect. 7.10 for quick reference.

7.5.1 Uncontrolled Charge of Electric Vehicles

In this case, it is assumed that, from the point of view of the supplier, EVs behave as typical inflexible loads. Then, the charging profiles of EVs cannot be modified by the supplier actions. The formulation of this problem is the following:

$$\text{Maximize}_{\Xi_1}$$

$$\beta \sum_{\omega \in \Omega} \sum_{t \in T} \pi_\omega W_t \left(\sum_{k \in K} \sum_{m \in M} \lambda^{\mathrm{S}}_{km} \left(\gamma^{\mathrm{C}}_k \hat{Y}^{\mathrm{C}}_{km\omega} + \gamma^{\mathrm{NC}}_k \hat{Y}^{\mathrm{NC}}_{km\omega} \right) E^{\mathrm{D}}_{kt\omega} - \lambda^{\mathrm{P}}_{t\omega} \sum_{k \in K} e^{\mathrm{S}}_{kt\omega} \right)$$

$$+ (1 - \beta) \left(\xi - \frac{1}{1-\alpha} \sum_{\omega \in \Omega} \pi_\omega \eta_\omega \right) \tag{7.5a}$$

subject to:

$$e^{\mathrm{S}}_{kt\omega} = \left(\gamma^{\mathrm{C}}_k y^{\mathrm{C}}_{k\omega} + \gamma^{\mathrm{NC}}_k y^{\mathrm{NC}}_{k\omega} \right) E^{\mathrm{D}}_{kt\omega}, \quad \forall k \in K, \forall t \in T, \forall \omega \in \Omega, \tag{7.5b}$$

$$\hat{\lambda}^{\mathrm{S}}_{km} \upsilon_{km} \leq \lambda^{\mathrm{S}}_{km} \leq \hat{\lambda}^{\mathrm{S}}_{km+1} \upsilon_{km}, \quad \forall k \in K, \forall m \in M, \tag{7.5c}$$

$$\sum_{m \in M} \upsilon_{km} = 1, \quad \forall k \in K, \tag{7.5d}$$

$$\sum_{m \in M} \lambda^{\mathrm{S}}_{km} = \lambda^{\mathrm{S}}_k, \quad \forall k \in K, \tag{7.5e}$$

$$\lambda^{\mathrm{S}}_{k'} = \lambda^{\mathrm{S}}_k, \quad \forall \{k, k'\} \in K, \tag{7.5f}$$

$$y^{\mathrm{C}}_{k\omega} = \sum_{m \in M} \hat{Y}^{\mathrm{C}}_{km\omega} \upsilon_{km}, \quad \forall k \in K, \forall \omega \in \Omega, \tag{7.5g}$$

$$y^{\mathrm{NC}}_{k\omega} = \sum_{m \in M} \hat{Y}^{\mathrm{NC}}_{km\omega} \upsilon_{km}, \quad \forall k \in K, \forall \omega \in \Omega, \tag{7.5h}$$

$$\upsilon_{km} \in \{0, 1\}, \quad \forall k \in K, \forall m \in M, \tag{7.5i}$$

$$\sum_{t \in T} W_t \left(- \left(\sum_{k \in K} \sum_{m \in M} \lambda^{\mathrm{S}}_{km} \left(\gamma^{\mathrm{C}}_k \hat{Y}^{\mathrm{C}}_{km\omega} + \gamma^{\mathrm{NC}}_k \hat{Y}^{\mathrm{NC}}_{km\omega} \right) E^{\mathrm{D}}_{kt\omega} - \lambda^{\mathrm{P}}_{t\omega} \sum_{k \in K} e^{\mathrm{S}}_{kt\omega} \right) \right.$$

$$\left. + \xi - \eta_\omega \right) \leq 0, \quad \forall \omega \in \Omega, \tag{7.5j}$$

$$\eta_\omega \geq 0, \quad \forall \omega \in \Omega, \tag{7.5k}$$

where $\Xi_1 = \{\lambda^{\mathrm{S}}_k, \forall k; \lambda^{\mathrm{S}}_{km}, \forall k, \forall m; e^{\mathrm{S}}_{kt\omega}, \forall k, \forall t, \forall \omega; y^{\mathrm{C}}_{k\omega}, \forall k, \forall \omega; y^{\mathrm{NC}}_{k\omega}, \forall k, \forall \omega; \upsilon_{km}, \forall k, \forall m; \xi; \eta_\omega, \forall \omega\}$.

The objective function (7.5a) represents the weighted sum of the expected profit achieved by the electricity supplier plus the CVaR. Parameter $\beta \in [0, 1]$ is used to model the tradeoff between expected profit and risk. The risk is represented by the CVaR, which is defined as the average profit in the worst $(1 - \alpha)$ scenarios. A conservative supplier will prefer maximizing the CVaR, so it chooses a low value of β to increase the weight of the risk measure in (7.5a). On the contrary, a risky supplier might be willing to assume higher risk in the hope of obtaining a higher profit, so its selected value for β would be close to 1. The expected profit is computed as the weighted average over all scenarios of the expected revenue obtained by the supplier minus the expected procurement cost from purchasing in the pool. The revenue in each period and scenario is computed using a linear expression that is equal to the product of the variable representing the selling price in block m of the price-quota curve, λ^{S}_{km}, and the energy associated with block m of the price-quota curve, $\left(\gamma^{C}_{k}\hat{Y}^{C}_{km\omega} + \gamma^{NC}_{k}\hat{Y}^{NC}_{km\omega}\right) E^{D}_{kt\omega}$. The expected procurement cost of the electricity supplier is computed, for each period and scenario, as the product of the electricity price in the pool market, $\lambda^{P}_{t\omega}$, and the total energy supplied, $\sum_{k\in K} e^{S}_{kt\omega}$.

Constraints (7.5b) calculate the total energy provided by the electricity supplier in each period and scenario. Constraints (7.5c)–(7.5i) formulate the price-quota curves. Binary variable υ_{km} is equal to 1 if the selling price offered by the electricity supplier to group k is within the interval $[\hat{\lambda}^{S}_{km}, \hat{\lambda}^{S}_{km+1}]$, being equal to 0 otherwise (constraints (7.5c)). Constraints (7.5d) enforce that the selling price must belong to one block of the price-quota curve. The relationship between the offered price λ^{S}_{k} and auxiliary variables λ^{S}_{km} is established by constraints (7.5e). Constraint (7.5f) can be enforced if it is desired that the same selling price must be offered to all groups k. The per-unit demand provided by the electricity supplier is equal to the level of energy of the price-quota curve indicated by (7.5g) and (7.5h). The binary nature of variable υ_{km} is established by constraints (7.5i). Finally, constraints (7.5j) and (7.5k) are used to formulate the CVaR. The optimal value of ξ represents the smallest value such that the probability that the profit be smaller or equal to ξ is less than or equal to $(1-\alpha)$. Also, ξ is known as the value-at-risk (VaR). Variable η_{ω} is the difference between the profit of scenario ω and the VaR.

Problem (7.5) is a mixed-integer linear programming problem that provides the optimal selling price that an electricity supplier may offer to EV users to maximize the expected profit during a given planning horizon.

7.5.2 Controlled Charge and Discharge of Electric Vehicles

Section 7.5.1 has described the formulation of the electricity supplier problem considering that the charging behaviour of EVs cannot be modified by the supplier actions. However, the new technologies associated with the smart grids allow to control remotely the charging processes of EVs. Considering that EVs are parked most of the time, an EV user is mainly interested in that the status of the battery of its vehicle be at the desired energy level at the moment of usage, but not so much in which specific hours the vehicle is being charged. Then, it is expected that EV users be willing to leave in hands of an external agent the charge of their vehicles during a specified interval of hours if (i) the battery is charged up to the desired level at the end of the charging period and (ii) it can profit from a lower charging costs.

Considering the above, the problem of the electricity supplier is next formulated considering that the charge of EVs is controlled by the electricity supplier. Additionally, it is also considered that EVs are equipped with the V2G capability and they can inject part of the energy stored in the batteries into the grid. Unlike to the formulation presented in Sect. 7.5.1, the charge and discharge of EVs are considered as decisions made by the electricity supplier. In this manner, $e^{\mathrm{C}}_{kt\omega}$ and $e^{\mathrm{D}}_{kt\omega}$ denote the charge and discharge of EVs belonging to group k, in period t and scenario ω. The proposed formulation is provided below:

$$\text{Maximize}_{\Xi_2}$$

$$\beta \sum_{\omega\in\Omega}\sum_{t\in T} \pi_\omega W_t \left(\sum_{k\in K}\sum_{m\in M} \lambda^{\mathrm{S}}_{km} \left(\gamma^{\mathrm{C}}_k \hat{Y}^{\mathrm{C}}_{km\omega} + \gamma^{\mathrm{NC}}_k \hat{Y}^{\mathrm{NC}}_{km\omega}\right) E^{\mathrm{D}}_{kt\omega} - \lambda^{\mathrm{P}}_{t\omega} \sum_{k\in K} e^{\mathrm{S}}_{kt\omega} \right)$$

$$+(1-\beta)\left(\xi - \frac{1}{1-\alpha}\sum_{\omega\in\Omega}\pi_\omega \eta_\omega\right) \quad (7.6a)$$

subject to:

$$\text{Price-quota curve constraints (7.5c)–(7.5i)} \quad (7.6b)$$

$$e^{\mathrm{S}}_{kt\omega} = (e^{\mathrm{C}}_{kt\omega} - e^{\mathrm{D}}_{kt\omega}), \quad \forall k\in K, \forall t \in T_{k\omega}, \forall \omega\in\Omega, \quad (7.6c)$$

$$soc_{kt\omega} = E^{\mathrm{O}}_{k\omega}\left(\gamma^{\mathrm{C}}_k y^{\mathrm{C}}_{k\omega} + \gamma^{\mathrm{NC}}_k y^{\mathrm{NC}}_{k\omega}\right), \quad \forall k\in K, \forall t\in T^{\mathrm{O}}_{k\omega}, \forall\omega\in\Omega, \quad (7.6d)$$

$$soc_{kt\omega} \geq E^{\mathrm{E}}_{k}\left(\gamma^{\mathrm{C}}_k y^{\mathrm{C}}_{k\omega} + \gamma^{\mathrm{NC}}_k y^{\mathrm{NC}}_{k\omega}\right), \quad \forall k\in K, \forall t\in T^{\mathrm{E}}, \forall\omega\in\Omega, \quad (7.6e)$$

$$soc_{kt\omega} = soc_{kt-1\omega} + \eta^{\mathrm{E}} e^{\mathrm{C}}_{kt\omega} - \frac{1}{\eta^{\mathrm{E}}} e^{\mathrm{D}}_{kt\omega}, \quad \forall k\in K, \forall t\in T_{k\omega}, \forall\omega\in\Omega, \quad (7.6f)$$

$$S^{\min}_k\left(\gamma^{\mathrm{C}}_k y^{\mathrm{C}}_{k\omega} + \gamma^{\mathrm{NC}}_k y^{\mathrm{NC}}_{k\omega}\right) \leq soc_{kt\omega} \leq S^{\max}_k\left(\gamma^{\mathrm{C}}_k y^{\mathrm{C}}_{k\omega} + \gamma^{\mathrm{NC}}_k y^{\mathrm{NC}}_{k\omega}\right),$$

$$\forall k\in K, \forall t\in T_{k\omega}, \forall\omega\in\Omega, \quad (7.6g)$$

$$0 \leq e^{\mathrm{C}}_{kt\omega} \leq C^{\mathrm{R}}_k N_k \left(\gamma^{\mathrm{C}}_k y^{\mathrm{C}}_{k\omega} + \gamma^{\mathrm{NC}}_k y^{\mathrm{NC}}_{k\omega}\right), \quad \forall k\in K, \forall t\in T_{k\omega}, \forall\omega\in\Omega, \quad (7.6h)$$

$$0 \leq e^{\mathrm{D}}_{kt\omega} \leq C^{\mathrm{R}}_k N_k \left(\gamma^{\mathrm{C}}_k y^{\mathrm{C}}_{k\omega} + \gamma^{\mathrm{NC}}_k y^{\mathrm{NC}}_{k\omega}\right), \quad \forall k\in K, \forall t\in T_{k\omega}, \forall\omega\in\Omega, \quad (7.6i)$$

$$\sum_{t\in T} W_t \left(-\left(\sum_{k\in K}\sum_{m\in M} \lambda^{\mathrm{S}}_{km}\left(\gamma^{\mathrm{C}}_k \hat{Y}^{\mathrm{C}}_{km\omega} + \gamma^{\mathrm{NC}}_k \hat{Y}^{\mathrm{NC}}_{km\omega}\right) E^{\mathrm{D}}_{kt\omega} - \lambda^{\mathrm{P}}_{t\omega}\sum_{k\in K} e^{\mathrm{S}}_{kt\omega} \right)\right.$$

$$\left. + \xi - \eta_\omega \right) \leq 0, \quad \forall\omega\in\Omega, \quad (7.6j)$$

$$\eta_\omega \geq 0, \quad \forall\omega\in\Omega, \quad (7.6k)$$

where $\Xi_2 = \{\lambda^{\mathrm{S}}_k; \lambda^{\mathrm{S}}_{km}, \forall m; e^{\mathrm{S}}_{t\omega}, \forall t, \forall\omega; e^{\mathrm{C}}_{kt\omega}, \forall k, \forall t, \forall\omega; e^{\mathrm{D}}_{kt\omega}, \forall k, \forall t, \forall\omega; soc_{kt\omega}, \forall k, \forall t, \forall\omega; v_{km}, \forall k, \forall m; \xi; \eta_\omega, \forall\omega\}$.

The objective function (7.6a) represents the weighted sum of the expected profit of the electricity supplier and the CVaR, and it is identical to the objective function (7.5a) included in the uncoordinated charge formulation. Note that the price-quota curve constraints (7.6b) are also included in the formulation of this problem. Constraints (7.6c) compute the energy traded in the pool, $e^{\mathrm{S}}_{kt\omega}$, as the difference between the energy charged, $e^{\mathrm{C}}_{kt\omega}$, and the energy discharged, $e^{\mathrm{D}}_{kt\omega}$, from EVs. Observe that set $T_{k\omega}$ comprises all time periods in which EVs belonging to group k are connected to their chargers in scenario ω. The initial and final status of the batteries of EVs, $E^{\mathrm{O}}_{k\omega}$ and E^{E}_k, are enforced by constraints (7.6d) and (7.6e), respectively. Observe that the final status of the batteries after the charge is a parameter that can be decided by the EV users. The energy stored in the batteries of each group k in each period t and scenario ω is computed by constraints (7.6f). The upper and lower limits of the energy stored in the batteries of the EVs are enforced by constraints (7.6g). The limits of the energy charged and discharged from EVs are formulated by constraints (7.6h) and (7.6i), respectively.

Note that if the value of variable $e^{\mathrm{D}}_{kt\omega}$ is fixed to 0, the model described above can be used to control exclusively the charge of the vehicles, avoiding the discharge of the batteries into the grid.

As formulated in (7.6a), the revenue obtained from selling in the pool the energy stored in the EVs is entirely destined to the supplier. However, since the charge and discharge may result in a degradation of the batteries, it is reasonable to consider that EV users have to be economically rewarded for a possible aging of their batteries. In order to estimate the degradation costs of the batteries of EVs, the linear regression model described in [16] is used. This model assumes that the capacity degradation of the batteries is a linear function of the number of round trip charging cycles. Assuming that batteries are usually replaced when their capacity retention is less than 80% of their initial capacity, a battery life of $N^{\mathrm{C}} = 4370$ round trip cycles can be considered as reasonable. Therefore, if B^{C} (€/MWh) denotes the battery cost, the expected battery degradation cost caused by the controlled discharge of energy stored in EVs belonging to group k, c^{Deg}_k, is:

$$c^{\mathrm{Deg}}_k = \frac{B^{\mathrm{C}}}{N^{\mathrm{C}}} \sum_{t \in T} \sum_{\omega \in \Omega} \pi_\omega W_t e^{\mathrm{D}}_{kt\omega}, \quad \forall k \in K. \tag{7.7}$$

In order to explicitly account for the battery degradation costs in the decision-making problem solved by the electricity supplier when it controls the charge and discharge of EVs, the objective function (7.6a) and the CVaR constraint (7.6j) may be respectively replaced by:

$$\beta \sum_{\omega \in \Omega} \sum_{t \in T} \pi_\omega W_t \Bigg(\sum_{k \in K} \sum_{m \in M} \lambda^{\mathrm{S}}_{km} \left(\gamma^{\mathrm{C}}_k \hat{Y}^{\mathrm{C}}_{km\omega} + \gamma^{\mathrm{NC}}_k \hat{Y}^{\mathrm{NC}}_{km\omega} \right) E^{\mathrm{D}}_{kt\omega} - \lambda^{\mathrm{P}}_{t\omega} \sum_{k \in K} e^{\mathrm{S}}_{kt\omega}$$
$$- \alpha^{\mathrm{D}} \sum_{k \in K} \frac{B^{\mathrm{C}}}{N^{\mathrm{C}}} e^{\mathrm{D}}_{kt\omega} \Bigg) + (1-\beta) \left(\xi - \frac{1}{1-\alpha} \sum_{\omega \in \Omega} \pi_\omega \eta_\omega \right) \tag{7.8}$$

and

$$\sum_{t \in T} W_t \Bigg(- \Bigg(\sum_{k \in K} \sum_{m \in M} \lambda^{\mathrm{S}}_{km} \left(\gamma^{\mathrm{C}}_k \hat{Y}^{\mathrm{C}}_{km\omega} + \gamma^{\mathrm{NC}}_k \hat{Y}^{\mathrm{NC}}_{km\omega} \right) E^{\mathrm{D}}_{kt\omega} - \lambda^{\mathrm{P}}_{t\omega} \sum_{k \in K} e^{\mathrm{S}}_{kt\omega}$$
$$- \alpha^{\mathrm{D}} \sum_{k \in K} \frac{B^{\mathrm{C}}}{N^{\mathrm{C}}} e^{\mathrm{D}}_{kt\omega} \Bigg) + \xi - \eta_\omega \Bigg) \le 0, \quad \forall \omega \in \Omega, \tag{7.9}$$

where α^{D} is a parameter used to model the portion of the degradation cost of the battery caused by the discharge of energy that is returned back to EV users. Observe that if $\alpha^{D} = 0$, expressions (7.8) and (7.9) are identical to (7.6a) and (7.6j), respectively.

Finally, note that the two mathematical problems included in this section are mixed-integer linear programming problems. Since the responsiveness of EV users to the selling price offered by the electricity supplier is modeled by using step-wise price-quota curves, the resulting computational size of these problems makes them solvable in real-world applications using commercially available software.

7.6 Illustrative Case Study: Uncontrolled Charge

We present an example of reduced size to illustrate the formulation of the supplier problem considering uncontrolled charge of EVs presented in Sect. 7.5.1.

We consider a planning horizon of three hourly periods and two groups of EVs. It is assumed that 20% and 50% of the vehicles belonging to the first and second groups are current clients of the supplier, respectively. The total electricity demand of EVs and pool prices are modeled using a set of three equiprobable scenarios. The values of these parameters are provided in Tables 7.1 and 7.2, respectively.

The response of the EV users to the price offered to current clients and non-clients of the supplier is characterized through the price-quota curves represented in Figs. 7.8 and 7.9, respectively. Observe that the energy procured by the supplier decreases as the selling price increases. According to Figs. 7.8 and 7.9, the per-unit energy procured by the supplier to current clients is always higher than or equal to that supplied to non-current clients of the supplier for each selling price.

In this example, parameter α used to compute the CVaR is equal to 0.95.

Problem (7.5) is characterized by 52 constraints, 49 continuous variables, and 6 binary variables and is solved using CPLEX under GAMS [7].

The expected profit and CVaR for values of β equal to 1 and 0 are presented in Table 7.3. As expected, the solution achieved in the risk-neutral case ($\beta = 0$) attains the highest expected profit and the highest risk in terms of the CVaR of the profit

Fig. 7.8 Illustrative case study—uncontrolled charge: price-quota curve for current clients of the supplier

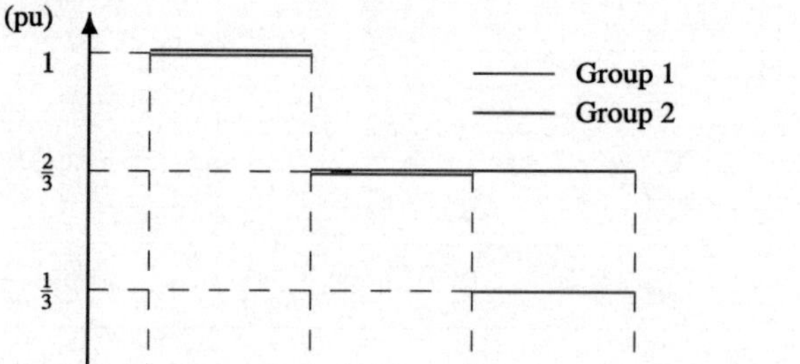

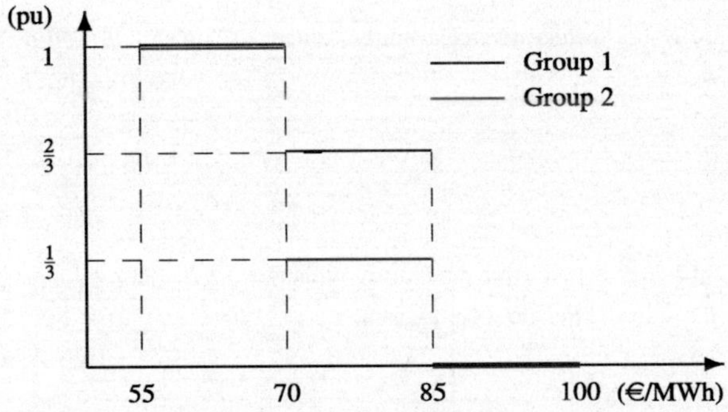

Fig. 7.9 Illustrative case study—uncontrolled charge: price-quota curve for current non-clients of the supplier

Table 7.3 Illustrative case study—uncontrolled charge: expected profit and CVaR (€)

β	Expected profit	CVaR
1	160.3	–536.1
0	145.1	18.7

Table 7.4 Illustrative case study—uncontrolled charge: profit per scenario (€)

β	Profit		
	Scenario 1	Scenario 2	Scenario 3
1	792.8	224.2	–536.1
0	257.7	159.0	18.7

distribution. If $\beta = 1$, a risk-averse behaviour is adopted and the CVaR of the profit increases from -536.1 € to 18.7 € at expenses of decreasing the expected profit by 9.5%.

Table 7.4 provides the profit associated with each scenario for β equal to 1 and 0. The profits in scenarios 1 and 2 are much higher in the risk-neutral case ($\beta = 1$) than in the risk-averse case ($\beta = 0$). This fact causes that the expected profit considering all scenarios be higher in the risk-neutral case. However, the profit associated with scenario 3 is much smaller in the risk-neutral case. In fact, the profit in this scenario is negative, which yields in a certain probability of experiencing losses. On the contrary, in the risk-averse case the profit in the worst scenario is positive and the probability of experiencing losses is null.

The optimal selling prices offered by the supplier are provided in Table 7.5. This table shows that the selling price increases as the supplier becomes more risk averse. Observe that as the selling price increases, the difference between the selling price and the purchasing price in the pool market also grows. In this manner, the probability of experiencing losses decreases. However, if the selling price increases, the energy provided by the supplier decreases, as shown in the price-quota curves represented in Figs. 7.8 and 7.9. This causes a reduction in the profit attained in most favourable scenarios (1 and 2).

Table 7.5 Illustrative case study—uncontrolled charge: selling price (€/MWh)

β	Selling price
1	85
0	100

Table 7.6 Illustrative case study—uncontrolled charge: energy supplied (MWh)

β	Period	Group 1			Group 2		
		Scenario			Scenario		
		1	2	3	1	2	3
1	1	0.0	0.0	0.0	2.0	2.5	3.0
	2	10.0	13.3	16.7	1.5	2.0	2.5
	3	11.3	14.7	18.0	0.0	0.0	0.0
0	1	0.0	0.0	0.0	0.7	0.8	1
	2	2.0	2.7	3.3	0.5	0.7	0.8
	3	2.3	2.9	3.6	0.0	0.0	0.0

The energy sold by the supplier for each EV group in each time period and scenario is listed in Table 7.6. In this table it is observed that group 1 is not available for charging in period 1, whereas group 2 is unavailable in period 3. Besides, it is also noted that the total amount of energy sold to the clients is much smaller when the supplier is risk averse ($\beta = 0$). The reason of this result is that the selling price offered in case $\beta = 0$ is 100 €/MWh, which is significantly higher than that offered for $\beta = 1$, which is equal to 85 €/MWh.

7.7 Illustrative Case Study: Controlled Charge

In this section, we analyze an illustrative example to test the formulation of the supplier problem considering controlled charge of EV described in Sect. 7.5.2.

We consider a planning horizon of three hourly periods and two groups of EVs. For the sake of comparison, the description of the EV groups, pool prices, and price-quota curves are identical to those described in the Illustrative Example analyzed in Sect. 7.6. It is assumed that vehicles belonging to groups 1 and 2 can charge their batteries in periods {2,3} and {1,2}, respectively. The number of vehicles in each group are 15,000 and 5,000, respectively. The rate power of EV chargers is 11 kW, the efficiency of the charge and discharge of the batteries is 0.9, and the size of the battery of EVs is 40 kWh. It is assumed that batteries are totally discharged when they are connected to the grid and that they require to charge a quantity of energy equal to the values provided in Table 7.2. In this case, $\alpha^{\mathrm{D}} = 0$, and all the revenues obtained from selling energy in the pool go to the considered supplier.

Table 7.7 Illustrative case study—controlled charge: expected profit and CVaR (€)

β	Expected profit	CVaR
1	1365.4	95.8
0	1169.0	486.6

Table 7.8 Illustrative case study—controlled charge: profit per scenario (€)

β	Profit		
	Scenario 1	Scenario 2	Scenario 3
1	2335.3	1665.0	95.8
0	1676.9	1343.6	486.6

Problem (7.6) is characterized by 124 constraints, 91 continuous variables, and 6 binary variables and is solved using CPLEX under GAMS [7].

The resulting expected profit and CVaR are provided in Table 7.7. The solution achieved by the risk-neutral case ($\beta = 0$) has the highest expected profit and lowest CVaR. Observe that the expected profits and the CVaR obtained in this case are higher than those obtained in the case in which the charge is not controlled (see Table 7.3). This result indicates that the earnings of the supplier can be higher if the charge of the EVs is decided by the electricity supplier.

Table 7.8 provides the profit associated with each scenario for β equal to 1 and 0. Likewise to the results obtained from the model in which the charge is not controlled by the supplier, the profits attained in scenarios 1 and 2 are higher in the risk-neutral case ($\beta = 1$) than in the risk-averse case ($\beta = 0$), but the profit in scenario 3 is smaller in the risk-neutral case. However, the profit in scenario 3 is positive in this case. Therefore, the probability of experiencing losses in this example is null when the charge is controlled by the supplier.

The optimal selling prices offered by the supplier are listed in Table 7.9. Once again, we observe that the selling price increases as the supplier becomes more risk averse. Comparing these prices with respect to those provided in Table 7.5 resulting from the case when the charge is not controlled by the supplier, we notice that the controlability of the charge of the EVs allows the supplier the possibility to offer lower prices. The reason of this result can be obtained from analyzing Table 7.10. This table shows the energy charged by each EV group in each period and scenario. From this table we observe that vehicles are not charged in period 2, which has associated the highest pool prices (see Table 7.1). Instead, the charge of vehicles belonging to group 1 is done in period 3, whereas the charge of vehicles belonging to group 2 is realized in period 1. In this manner, the supplier can profit from purchasing energy in periods with low prices.

Table 7.11 shows the energy discharged from EVs to be sold in the pool. As expected, the energy is sold in the pool during period 2, which is associated with a high price. From this table we also observe that only energy from EV group 2

Table 7.9 Illustrative case study—controlled charge: selling price (€/MWh)

β	Selling price
1	70
0	85

Table 7.10 Illustrative case study—controlled charge: energy charged (MWh)

β	Period	Group 1			Group 2		
		Scenario			Scenario		
		1	2	3	1	2	3
1	1	0.0	0.0	0.0	55.0	55.0	55.0
	2	0.0	0.0	0.0	0.0	0.0	0.0
	3	35.6	46.7	57.8	0.0	0.0	0.0
0	1	0.0	0.0	0.0	27.5	27.5	27.5
	2	0.0	0.0	0.0	0.0	0.0	0.0
	3	23.7	31.1	38.5	0.0	0.0	0.0

Table 7.11 Illustrative case study—controlled charge: energy discharged (MWh)

β	Period	Group 1			Group 2		
		Scenario			Scenario		
		1	2	3	1	2	3
1	1	0.0	0.0	0.0	0.0	0.0	0.0
	2	0.0	0.0	0.0	38.2	36.4	34.6
	3	0.0	0.0	0.0	0.0	0.0	0.0
0	1	0.0	0.0	0.0	0.0	0.0	0.0
	2	0.0	0.0	0.0	19.1	18.2	17.3
	3	0.0	0.0	0.0	0.0	0.0	0.0

is discharged. The reason of this fact is that EVs belonging to group 1 are initially discharged in period 2. However, EVs belonging to group 2 can charge their batteries in period 1 to discharge part of this energy in period 2. Observe that the sale of energy in the pool implies an additional revenue for the supplier.

7.8 Realistic Case Study

In this section, a realistic case study is solved to test the proposed procedure. A planning horizon of one year and a set of 348,000 potential clients of the electricity supplier are considered. All mathematical problems in this section have been solved

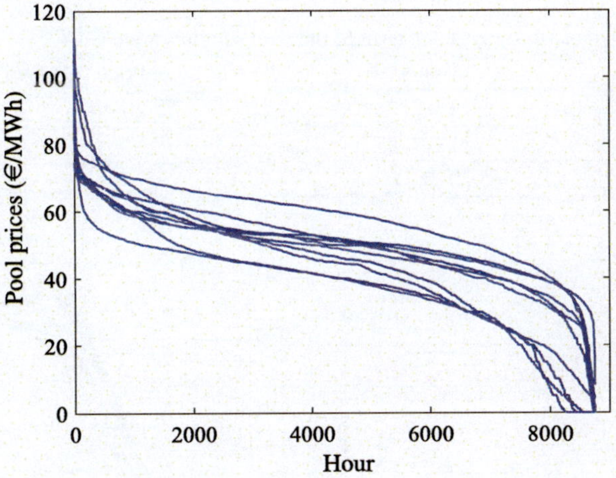

Fig. 7.10 Realistic case study: historical pool prices

by CPLEX 12.6.1 under GAMS [7] on a Linux-based server with four 3.0 GHz processors and 250 GB of RAM.

7.8.1 Input Data

The target year is represented by a set of five representative weeks divided into hourly periods. The selection of the weeks and their weights are determined by using the fast-forward scenario reduction algorithm described in [9].

Pool prices are characterized using ten equiprobable scenarios. These scenarios are generated from a ten-year historical price series data pertaining to the Spanish power system between 2010–2019, which are available at [17]. Figure 7.10 provides the price duration curve for each considered year. The pool price scenarios for the considered representative weeks are represented in Fig. 7.11.

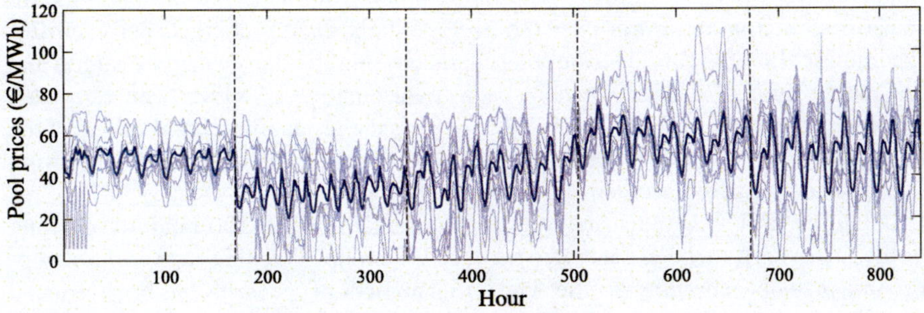

Fig. 7.11 Realistic case study: pool price scenarios

Table 7.12 Realistic case study: fixed term of the rival supplier price

Block	Hours	Price (€/MWh)
1	13–22	20
2	7–12, 23	5
3	0–6	1

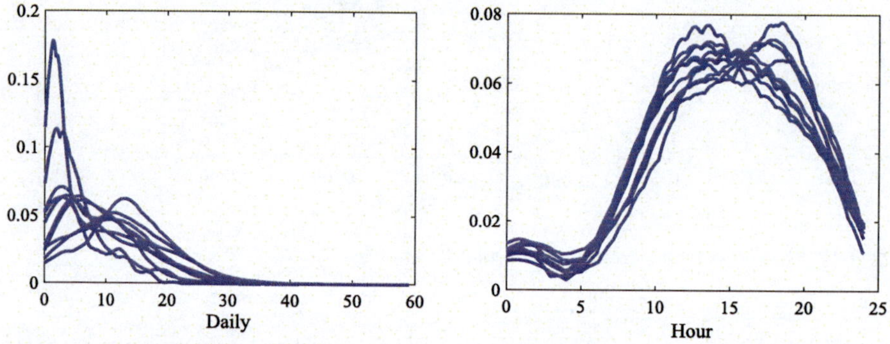

(a) Daily energy charged **(b) Starting hour of the charging**

Fig. 7.12 Realistic case study: probability distributions of the daily energy charged and the starting hour of the charging

Rival supplier prices are defined as a regulated time-of-use rate indexed to pool prices. The price payed by users of this tariff in each hour is equal to the pool price plus a fixed term. Table 7.12 provides the value of the fixed part of the price for each time interval. Observe that competitive prices are offered in off-peak hours.

EVs are equipped by a 40 kWh battery that is charged at home from the grid. The efficiency of charging and discharging processes is 0.874 and the charging rate is 1.92 kW.

A set of ten EV groups is considered. The behavior patterns of each EV group are generated from the charging profile data provided in [15] and denoted by L1. The resulting set of EV groups has been obtained from applying the fast-forward scenario reduction algorithm over the average daily energy charged per each EV. The empirical probability distributions characterizing the daily energy charged and the starting hour of the charging for each representative EV have been estimated and represented in Fig. 7.12. Figure 7.12a depicts the probability distributions of the daily energy charged, whereas the probability distribution characterizing the starting hour of charging is represented in Fig. 7.12b.

Using the probability distributions depicted in Fig. 7.12, 100 random charging profile scenarios have been generated for each EV group. Note that each scenario represents a plausible charging profile. The initial number of scenarios has been reduced

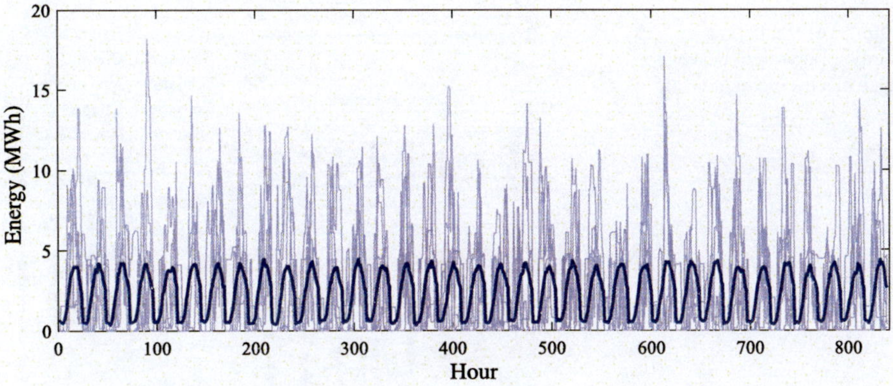

Fig. 7.13 Realistic case study: charging profile scenarios

to 5 using the fast-forward scenario reduction algorithm. The resulting charging scenarios are depicted in Fig. 7.13. In this case, it is considered that all the revenues obtained from selling energy in the pool are property of the considered supplier ($\alpha^{D} = 0$).

The initial status of the batteries is assigned for each scenario and each day. This value is equal to the capacity of the batteries minus the amount of energy that has to be charged during the day, which is randomly generated using the probability distribution represented in Fig. 7.12a.

We assume that 15% of all potential clients are currently clients of the electricity supplier. We consider that the annual per-unit transportation cost and loyalty reward per hour are equal to 20 €/pu/h and 10 €/h, respectively. These values are expressed per hour for the sake of clarity. Price-quota curves comprise 16 blocks and are computed using the procedure described in Sect. 7.4.2. The resulting price-quota curves for current clients and non-clients of each EV group are represented in Fig. 7.14. In this case, two scenarios are considered with probabilities 0.75 and 0.25, respectively. The values of the price-quota curve in the second scenario are equal to 0.6 times the values of the first scenario. Note that for identical selling prices, the percentage of demand supplied to current clients of the electricity supplier is significantly greater than that provided to non-current clients.

Finally, observe that considering the number of scenarios used to characterize: (i) pool prices, (ii) EV demand, and (iii) price-quota curves, a total number of $10 \cdot 5 \cdot 2 = 100$ scenarios are considered in this case study.

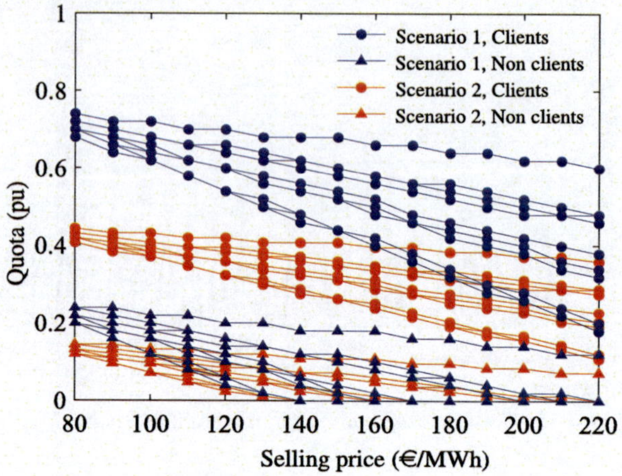

Fig. 7.14 Realistic case study: price-quota curves for each electric vehicle group

7.8.2 *Results*

This section provides the results of three different cases:

- *Case 1: Uncontrolled case*. In this case the charge of EVs is not controlled by the electricity supplier. Therefore, the charging profile of each group is characterized by the charging scenarios represented in Fig. 7.13.
- *Case 2: Controlled charge*. In this case it is assumed that the supplier decides in which periods EVs are charged. The charge of an EV group begins at a starting hour that is randomly generated. It is assumed that EVs have to be fully charged at 6:00 am of each day. In this case the sale of energy discharged from the batteries of EVs is not allowed.
- *Case 3: Controlled charge and discharge* This case is similar to *Case 2* but allowing the sale of energy discharged from the batteries of the EVs in the pool.

Table 7.13 lists the expected profit and the CVaR achieved by the electricity supplier, the selling price offered to EV users and the percentage of clients of the electricity supplier from the total number of potential clients. As expected the smallest expected profit is obtained in the uncontrolled case, which is significantly smaller than that in cases 2 and 3. Note also that, for this case and $\beta = 0$, the offered selling price, 150 €/MWh, is the highest of the three analyzed cases, yielding 14.52% of clients. On the contrary, cases 2 and 3 are able to offer smaller selling prices. It is worth noting that the effect of considering the CVaR in the formulation, $\beta = 0$, is more relevant in the uncontrolled case. In fact, in *Case 2* the solutions with $\beta = 1$ and $\beta = 0$ are identical. In cases 1 and 3, the incorporation of the CVaR in the problem formulation increases the risk-aversion level of the supplier, which yields in an increase of the selling price offered.

It is remarkable the increment of the expected profit obtained in *Case 3* with regard to that resulting from *Case 2*. This result indicates that the sale of energy

Table 7.13 Realistic case study: base case results

Case #	β (pu)	Expected profit (millions €)	CVaR (millions €)	Selling price (€/MWh)	Percentage of clients (%)
1	1	1.742	1.018	130	17.19
	0	1.719	1.023	150	14.52
2	1	2.093	1.253	130	17.19
	0	2.093	1.253	130	17.19
3	1	2.850	1.498	110	19.56
	0	2.815	1.509	120	17.96

Table 7.14 Realistic case study: computational size

Case #	Constraints (millions)	Con. variables (millions)	Binary variables	Solution time (s)
1	0.360	0.940	150	500
2	2.995	2.244	150	10000
3	3.115	2.364	150	85000

in the pool market increases significantly the expected revenues attained by the electricity supplier. However, observe that identical positions in the Hotelling line, transportation costs, and loyalty rewards are assumed in the three cases solved in this section. The main purpose of considering the same input data in all cases is only to facilitate the comparison of the performance of the proposed formulations. Note that assuming identical price-quota curves in all cases may not be realistic for analyzing the outcomes resulting from the different control strategies for EVs that can be adopted by the electricity supplier. In fact, it is reasonable to expect that EVs users will be reluctant to leave the control of the charging process of their vehicles in hands of an external agent.

Table 7.14 provides the computational information associated with each of the three analyzed cases. As expected, *Case 1* requires a significantly smaller number of constraints and variables, which results in the smallest solution time. On the contrary, *Case 3*, with the highest number of constraints and variables, needs the highest solution time. However, observe that all reported solution times are reasonable for solving a medium-term decision-making problem.

7.8.3 Sensitivity Analysis

In order to test the performance of the proposed formulations for different input data, four sensitivity analyses are discussed focusing on the influence of: (i) initial client portfolio, (ii) per-unit transportation cost, (iii) loyalty reward, and (iv) reimbursement of the degradation costs of batteries to EV users. For the sake of conciseness, only results of *Case 2* are reported.

7.8.3.1 Initial Client Portfolio of the Electricity Supplier

This section studies the effect of the number of initial clients of the electricity supplier. Figure 7.15 provides the selling prices, the percentage of clients, and the expected profits for different initial client portfolios expressed in percentage over the total number of potential clients. Figure 7.15a shows that the initial portfolio of the electricity supplier is highly relevant for deciding the selling price offered. In particular, it is noted that the higher is the initial number of clients of the supplier, the higher is the selling price offered. As the number of initial clients increases, and considering that the loyalty reward discourages the change of supplier of the current clients of the supplier, the number of clients increases even though the selling price increases. This fact causes a high increase in the expected profit and the CVaR, as shown in Fig. 7.15b.

7.8.3.2 Per-Unit Transportation Cost

The effect of the per-unit transportation cost, T_k^{C}, is analyzed in this section. For doing so, price-quota curves are again generated considering different values of T_k^{C}. As an example, Fig. 7.16 represents graphically the resulting price-quota curves for T_k^{C} equal to 15 and 25 €/pu/h. Comparing the resulting price-quota curves in both cases, we observe that if the per-unit transportation cost is low, the preferences of EVs over supplier is not relevant on the selection process and it is very difficult to sign new contracts with non-clients of the supplier. However, if the per-unit transportation

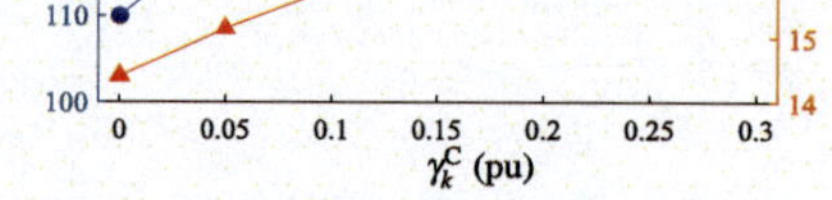

(a) Selling price (SP) and percentage of clients (PERC)

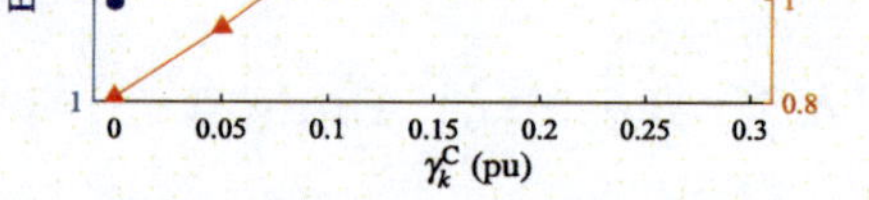

(b) Expected profit (EP) and CVaR

Fig. 7.15 Realistic case study: effect of the initial portfolio of clients

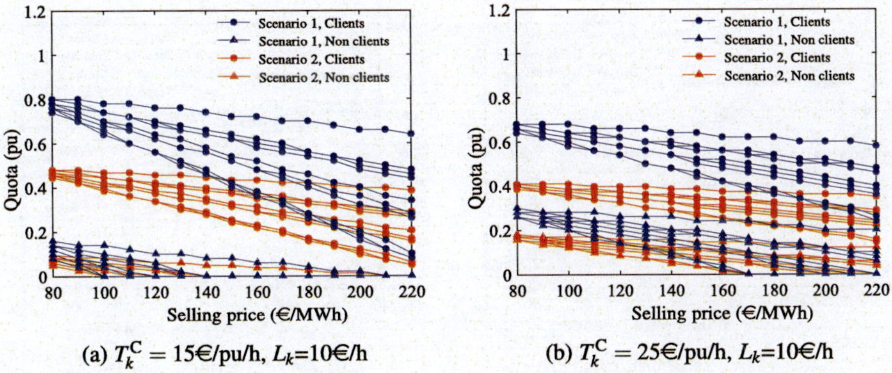

Fig. 7.16 Realistic case study: price-quota curves for different per-unit transportation costs

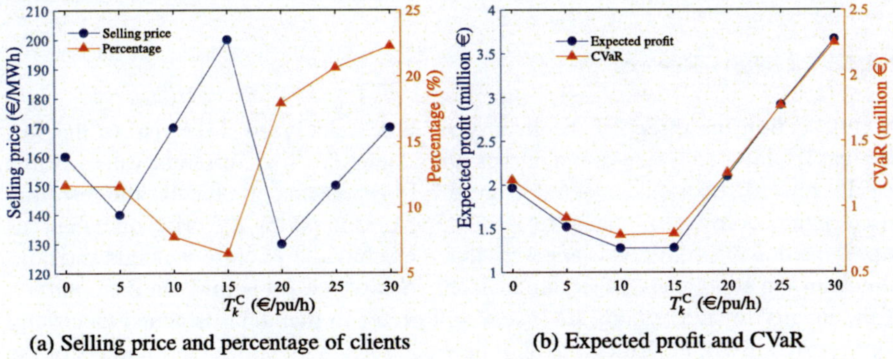

Fig. 7.17 Realistic case study: effect of the transportation cost per unit

cost is high, it will be possible to attract new clients belonging to the rival supplier if the offered price is competitive.

Figure 7.17 provides the resulting selling prices, the percentage of clients and the expected profit and CVaR for different transportation costs. As observed in Fig. 7.17, if the per-unit transportation cost is low, the position within the Hotelling of EV users is not relevant and the value of the offered selling price and the loyalty reward are the main drivers for selecting the supplier. Therefore, the supplier prefers to offer a high price assuming that, because of the loyalty reward, a significant part of their clients will remain as its clients regardless of the offered price. However, as the per-unit transportation cost increases, the preferences of EV users about the supplier become more important in the selection of electricity supplier and the supplier may profit from attracting previous non-clients. In this manner, if the per-unit transportation cost is high, the electricity supplier can afford to increase the selling price in the hope of attracting those EV users that have a high preference for the considered supplier.

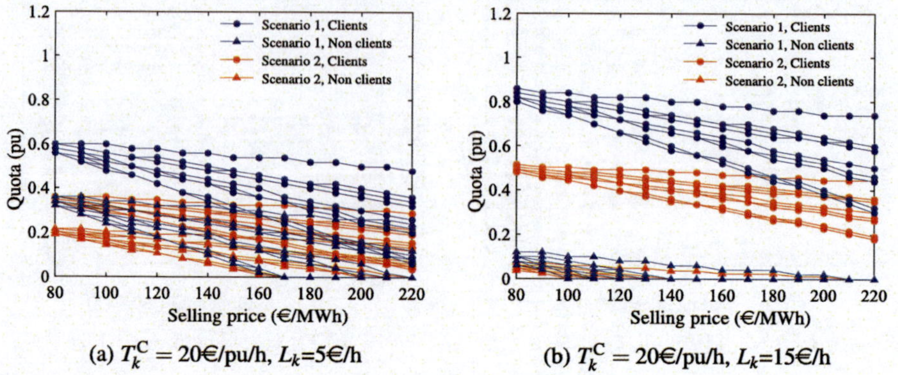

Fig. 7.18 Realistic case study: price-quota curves for different loyalty rewards

7.8.3.3 Loyalty Reward

In this section the effect of the loyalty reward is analyzed. Likewise to the former case, additional price-quota curves are generated. Now, different values of the loyalty reward, L_k are considered. Figure 7.18 represents graphically the resulting price-quota curves for L_k equal to 5 and 15 €/h. Comparing the price-quota curves represented in this figure, it is apparent that the behaviour of current clients and non-clients of the supplier is quite similar if the loyalty reward is low. On the contrary, if the loyalty reward is high, EV users will prefer to maintain its actual electricity supplier and it will be difficult to sign new contracts with current non-clients of the supplier.

The resulting selling prices, percentages of clients and expected profits and CVaR for different loyalty rewards are represented in Fig. 7.19. Note that low loyalty rewards reduce the importance of the initial client portfolio of the electricity supplier. In this manner, given that the initial client portfolio of the electricity supplier is only 15%, Fig. 7.19a shows that a low loyalty reward allows the electricity supplier to attract a large number of non-clients that have preferences over the considered supplier. This causes a high expected profit and CVaR for low loyalty rewards, as shown in Fig. 7.19b. As the loyalty reward increases, the number of non-clients that select the considered supplier decreases significantly and the expected profit is also reduced. Finally, if the loyalty reward is high enough, the electricity supplier can profit from increasing the offered selling price hopping that most of its initial clients will continue purchasing from it.

7.8.3.4 Influence of Battery Degradation Costs

This subsection analyzes the influence of considering the battery degradation costs in the case in which the electricity supplier is allowed to control the charge and discharge of the batteries of EVs.

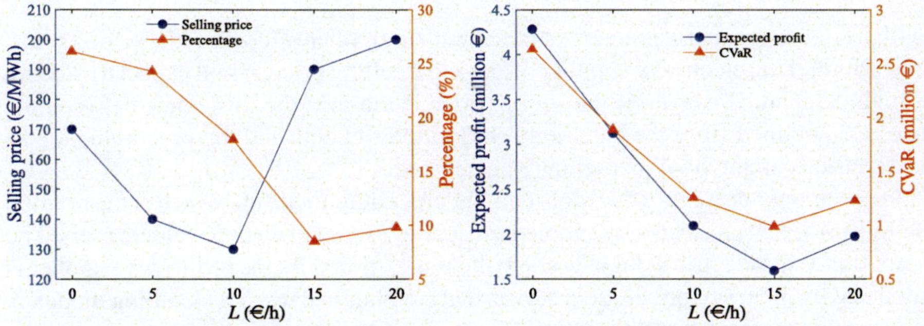

(a) Selling price and percentage of clients (b) Expected profit and CVaR

Fig. 7.19 Realistic case study: effect of the loyalty reward

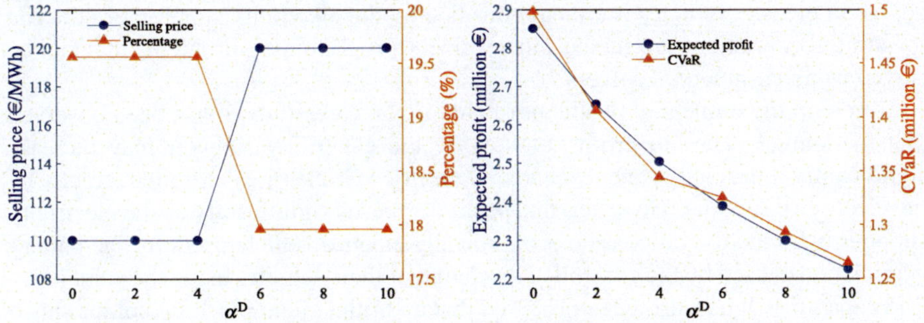

(a) Selling price and percentage of clients (b) Expected profit and CVaR

Fig. 7.20 Realistic case study: effect of the reimbursement of the degradation costs of electric vehicles

Figure 7.20 provides the results obtained for *Case 3* considering different values of parameter α^{D}. Figure 7.20a shows that the selling price offered by the electricity supplier remains equal to 110 €/MWh for values of α^{D} within the interval [0, 4]. For greater values of α^{D}, the electricity supplier decides to increase the selling price up to 120 €/MWh. As expected, Fig. 7.20b shows that both the profit attained by the electricity supplier and the CVaR decrease as the value of α^{D} increases.

7.9 Summary and Conclusions

The determination of the selling price offered to electric vehicle users constitutes a significant challenge for electricity suppliers due to the uncertainty inherent to: (i) the response of electric vehicle users to the offered selling price, (ii) pool prices, and (iii) pricing strategies of rival suppliers. If electric vehicle users behave rationally, they will prefer to sign a procurement contract with a supplier offering a low selling price. Therefore, if the electricity supplier offers a low selling price, it will expect to procure a high demand. However, if the selling price is too low, the difference between the revenue obtained from the electricity sold to their clients and the procurement cost of electricity in the pool market may be negative.

This chapter presents a decision-making procedure to decide which is the optimal selling price that an electricity supplier must offer to a set of electric vehicle users. The response of the electric vehicle users to the price offered by the electricity supplier is modeled by means of price-quota curves that are obtained using a Hotelling model. A price-quota curve is useful to determine the quantity of electricity that the electricity supplier must provide for each offered selling price. The Hotelling model allows to characterize the preferences of electric vehicle users with respect to different electricity suppliers. Two different formulations are derived considering that the charge of electric vehicles is either controlled by the electricity supplier or not. The resulting formulations are mixed-integer linear programming problems that can be solved using commercial solvers.

Based on the numerical results presented in the case study, it has been observed that the annual expected profit obtained by the electricity supplier may increase significantly if the charge and discharge of electric vehicles are controlled directly by the electricity supplier. However, this profit decreases significantly as the electricity supplier is forced to compensate economically electric vehicle users for the battery degradation caused by the controlled discharge of their vehicles using the vehicle-to-grid capability. It has also been observed that an initial client portfolio comprising a large number of clients allows the electricity supplier to increase the offered selling price and to achieve large expected profits.

7.10 Notation

The main notation used along this chapter is provided below.

Acronyms

CVaR Conditional value-at-risk.
EV Electric vehicles.
V2G Vehicle-to-grid.

Sets and Indices

k Index of groups of electric vehicles.
K Set of groups of electric vehicles.
m Index of price-quota curve blocks.
M Set of price-quota curve blocks.
t Index of time periods used for characterizing the planning horizon.
T Set of time periods used for characterizing the planning horizon.
T_{kw} Set of time periods in which electric vehicles belonging to group k can be charged in scenario ω.
T^{O}_{kw} Set of starting time periods of all charging intervals in which EVs belonging to group k and scenario ω can be charged.
T^{E}_{k} Set of ending time periods in which electric vehicles belonging to group k can be charged.
ω Index of scenarios.
Ω Set of scenarios.

Parameters

B^{C} Battery cost (€/MWh).
C^{R}_{k} Maximum power charging/discharging rate of electric vehicles in group k (MW).
$E^{\mathrm{D}}_{kt\omega}$ Energy charged by electric vehicle group k in period t and scenario ω (MWh).
E^{E}_{k} Final status required for the batteries of electric vehicles belonging to group k (MWh).
$E^{\mathrm{O}}_{k\omega}$ Initial status of the batteries of electric vehicles belonging to group k in scenario ω (MWh).
N_k Number of electric vehicles in group k.
N^{C} Number of round trip cycles of the batteries of electric vehicles.
L_k Loyalty reward of electric vehicle group k (€).
L^{R}_{kr} Loyalty reward of electric vehicles belonging to group k that are current clients of rival supplier r (€).
$S^{\max}_{k}$ Energy capacity of the batteries of electric vehicles in group k (MWh).
$S^{\min}_{k}$ Minimum value of the energy that must remain in the batteries of electric vehicles in group k (MWh).
T^{C}_{k} Per-unit transportation cost of electric vehicle group k (€/pu).
W_t Weight of period t (h).
$\hat{Y}^{\mathrm{C}}_{km\omega}$ Per unit value of the demand of electric vehicles that are clients of the supplier belonging to group k that is supplied in block m of the price-quota curve in scenario ω (pu).

$\hat{Y}^{\text{NC}}_{km\omega}$ Per-unit value of the demand of electric vehicles that are non clients of the supplier belonging to group k that is supplied in block m of the price-quota curve in scenario ω (pu).

α Confidence level used in the calculation of the CVaR (pu).

α^{D} Parameter used to model the portion of the battery degradation cost caused by the controlled discharge of energy that is returned back to EV users (pu).

β Weighting factor used for materialize the tradeoff between expected profit and the CVaR (pu).

γ^{C}_{k} Per-unit number of electric vehicles belonging to group k that are current clients of the supplier (pu).

γ^{NC}_{k} Per-unit number of electric vehicles belonging to group k that are current non-clients of the supplier (pu).

$\lambda^{\text{P}}_{t\omega}$ Pool price in period t and scenario ω (€/MWh).

$\lambda^{\text{R}}_{rkt\omega}$ Selling price offered by the rival supplier r to group k in period t and scenario ω (€/MWh).

$\hat{\lambda}^{\text{S}}_{mk}$ Lower bound of block m of the price-quota curve (€/MWh).

η^{E} Efficiency of charging/discharging processes of EVs

π_{ω} Probability of scenario ω (pu).

$\theta^{\text{R}}_{rk\omega}$ Per-unit distance between the clients of rival supplier r belonging to group k in scenario ω in the Hotelling line (pu).

$\theta^{\text{S}}_{k\omega}$ Per-unit distance between the clients of the supplier belonging to group k in scenario ω in the Hotelling line (pu).

Variables

c^{Deg}_{k} Expected battery degradation cost of electric vehicles belonging to group k (€).

$e^{\text{C}}_{kt\omega}$ Energy charged from the grid by electric vehicles belonging to group k in period t and scenario ω (MWh).

$e^{\text{D}}_{kt\omega}$ Energy discharged to the grid by electric vehicles belonging to group k in period t and scenario ω (MWh).

$e^{\text{S}}_{kt\omega}$ Total energy supplied by the electricity supplier to electric vehicles belonging to group k in period t and scenario ω (MWh).

$soc_{kt\omega}$ State-of-charge of the batteries of electric vehicles belonging to group k in period t and scenario ω (MWh).

$y^{\text{C}}_{k\omega}$ Per-unit demand provided to current clients of the supplier belonging to group k in scenario ω (pu).

$y^{\text{NC}}_{k\omega}$ Per-unit demand provided to current non-clients of the supplier belonging to group k in scenario ω (pu).

$y^{\text{R}}_{rk\omega}$ Binary variable that is equal to 1 if electric vehicle users belonging to group k decide to buy energy to rival supplier r in scenario ω, being 0 otherwise.

$y^{S}_{k\omega}$ Binary variable that is equal to 1 if electric vehicle users belonging to group k decide to buy energy to the considered electricity supplier in scenario ω, being 0 otherwise.

v_{km} Binary variable that is equal to 1 if the selling price offered by the electricity supplier to group k belongs to the interval $[\hat{\lambda}^{S}_{km}, \hat{\lambda}^{S}_{km+1}]$.

λ^{S}_{k} Selling price offered by the supplier to group k (€/MWh).

λ^{S}_{km} Selling price associated with block m of the price-quota curve of group k (€/MWh).

ξ Auxiliary variable used to calculate the CVaR (€).

η_{ω} Auxiliary variable related to scenario ω used to calculate the CVaR (€).

7.11 GAMS Codes

This section provides the GAMS [7] codes used to solve some of the illustrative examples of this chapter.

7.11.1 Price-Quota Curve

A GAMS code for solving Illustrative Example 7.5 is provided below:

```
$title ELECTRICITY SUPPLIER PRICE-QUOTA CURVE

*********************************************************************
* DATA
*********************************************************************


SETS
w          'Scenarios'                                      /w1*w3/,
t          'Time periods'                                   /t1*t3/,
k          'Electric vehicle groups'                        /k1*k2/,
m          'Set of blocks in price-quota curve'             /m1*m3/
r          'Rival suppliers'                                /r1/
SK(k)      'Set of active electric vehicle groups'
SM(m)      'Set of active price-quota curve blocks'
SW(w)      'Set of active scenarios'
;

alias(k,kk);alias(w,ww);alias(m,mm);

PARAMETERS
pi(w)               'Scenario probability'
Tc(k)               'Transportation cost per unit'
L(k)                'Loyalty reward'
Wg(t)               'Period weight'
lambdaR(r,t,w)      'Selling price by rival supplier'
thetaR(r,k,w)       'Posittion in Hotelling line'
```

```
YMC(k,m,w)          'Per-unit number of clients'
YMNC(k,m,w)         'Per-unit number of non-clients'
YMC_EXP(k,m)        'Expected per-unit number of clients'
YMNC_EXP(k,m)       'Expected per-unit number of non-clients';


TABLE thetaS(k,w) 'Position in Hotelling line'
          w1       w2       w3
k1        0.7      0.8      0.9
k2        0.2      0.3      0.4;

TABLE ED(k,t,w) 'Demand of electric vehicles'
          w1       w2       w3
k1.t1     0        0        0
k1.t2     15       20       25
k1.t3     17       22       27
k2.t1     4        5        6
k2.t2     3        4        5
k2.t3     0        0        0;

TABLE lambdaP(t,w) 'Pool price'
          w1       w2       w3
t1        30       50       70
t2        70       100      120
t3        40       60       80;

TABLE lambdaHatS(k,m) 'Selling price in price-quota curve'
          m1       m2       m3
k1        70       85       100
k2        70       85       100;

pi(w)=1/card(w);
Tc(k)=50;
L(k)=100;
Wg(t)=1;
lambdaR(r,t,w)=lambdaP(t,w)+20;
thetaR(r,k,w)=1-thetaS(k,w);


*****************************************************************
* DECLARATION OF VARIABLES AND MATHEMATICAL CHARACTERIZATION
*****************************************************************


VARIABLES

** OBJECTIVE FUNCTION
z           'Objective function'

;

BINARY VARIABLES     yS(k,w), yR(r,k,w);


*****************************************************************
*  EQUATIONS
*****************************************************************
```

```
EQUATIONS

** OBJECTIVE FUNCTIONS
ObjectiveFunction_C
ObjectiveFunction_NC

** RATIONALE SELECTION SUPPLIER

Logic_cons;


ObjectiveFunction_C(SK(k),SW(w),SM(m)).. z=e= (sum(t,Wg(t)*lambdaHatS(k,m)*ED(k,t,w)) + Tc(k)*thetaS(k,w)-L(k))*yS(k,w)+sum(r,(sum(t,Wg(t)*lambdaR(r,t,w)*ED(k,t,w)) + Tc(k)*thetaR(r,k,w))*yR(r,k,w));
ObjectiveFunction_NC(SK(k),SW(w),SM(m)).. z=e= (sum(t,Wg(t)*lambdaHatS(k,m)*ED(k,t,w)) + Tc(k)*thetaS(k,w))*yS(k,w)+sum(r,(sum(t,Wg(t)*lambdaR(r,t,w)*ED(k,t,w)) + Tc(k)*thetaR(r,k,w)-L(k))*yR(r,k,w));

Logic_cons(SK(k),SW(w)).. yS(k,w)+sum(r,yR(r,k,w))=e= 1;


****************************************************************
*  MODEL AND SOLVING
****************************************************************

MODEL Hotelling_C  'model for current clients' / ObjectiveFunction_C, Logic_cons /;
MODEL Hotelling_NC 'model for current non-clients'  / ObjectiveFunction_NC, Logic_cons /;
Hotelling_C.optcr=0;
Hotelling_NC.optcr=0;
Option mip=cplex;


loop((kk,ww,mm),
        SK(k)=no;SM(m)=no;SW(w)=no;
        SK(k)$(ord(k)=ord(kk))=yes;SM(m)$(ord(m)=ord(mm))=yes;SW(w)$(ord(w)=ord(ww))=yes;

        SOLVE Hotelling_C USING mip MINIMIZING z;
        YMC(kk,mm,ww)= yS.l(kk,ww);


        SOLVE Hotelling_NC USING mip MINIMIZING z;
        YMNC(kk,mm,ww)= yS.l(kk,ww);

);

YMC_EXP(kk,mm)= sum(ww,pi(ww)*YMC(kk,mm,ww));
YMNC_EXP(kk,mm)= sum(ww,pi(ww)*YMNC(kk,mm,ww));
```

7.11.2 Uncontrolled Charge

A GAMS code for solving Illustrative Case Study of Sect. 7.6 is provided below:

```
$title ELECTRICITY SUPPLIER UNCONTROLLED

*********************************************************************
* DATA
*********************************************************************
SETS
w          'Scenarios'                                   /w1*w3/,
t          'Time periods'                                /t1*t3/,
k          'Electric vehicle groups'                     /k1*k2/,
m          'Set of blocks in price-quota curve'          /m1*m3/
;


SCALARS
beta       'Risk weighting factor' /0.9999/,
alpha      'Confidence level'      /0.95/
;


PARAMETERS
pi(w)                'Scenario probability'
Wg(t)                'Period weight'
gammaC(k)            'Initial number of clients'
/k1 0.2
k2  0.5/,
gammaNC(k)           'Initial number of non-clients';

TABLE ED(k,t,w)   'Demand of electric vehicles'
          w1      w2      w3
k1.t1     0       0       0
k1.t2     15      20      25
k1.t3     17      22      27
k2.t1     4       5       6
k2.t2     3       4       5
k2.t3     0       0       0;

TABLE YMC(k,m,w) 'Per-unit number of clients'
          w1      w2      w3
k1.m1     1.000   1.000   1.000
k1.m2     0.667   0.667   0.667
k1.m3     0.667   0.667   0.667
k2.m1     1.000   1.000   1.000
k2.m2     0.667   0.667   0.667
k2.m3     0.333   0.333   0.333;


TABLE YMNC(k,m,w) 'Per-unit number of non-clients'
          w1      w2      w3
k1.m1     1.000   1.000   1.000
k1.m2     0.667   0.667   0.667
k1.m3     0.000   0.000   0.000
k2.m1     0.667   0.667   0.667
k2.m2     0.333   0.333   0.333
```

```
k2.m3     0.000    0.000    0.000;

TABLE lambdaP(t,w) 'Pool price'
         w1       w2       w3
t1       30       50       70
t2       70       100      120
t3       40       60       80;

TABLE lambdaHatS(k,m) 'Selling price in price-quota curve'
         m1       m2       m3
k1       70       85       100
k2       70       85       100;

pi(w)=1/card(w);
Wg(t)=1;
gammaNC(k)=1-gammaC(k);


************************************************************************
* DECLARATION OF VARIABLES AND MATHEMATICAL CHARACTERIZATION
************************************************************************
VARIABLES

** OBJECTIVE FUNCTION
z                'Objective function'

** FIRST-STAGE VARIABLES
lambdaS(k)       'Selling price offered to electric vehicles'
lambdaMS(k,m)    'Auxiliary variable used to compute the selling
     price'
v(k,m)           'Selection of block of the price-quota curve'

** SECOND-STAGE VARIABLES
yC(k,w)          'Per-unit number of current clients'
yNC(k,w)         'Per-unit number of non-clients'
eS(k,t,w)        'Demand provided by the supplier'

** VARIABLES ASSOCIATED WITH RISK MODELING
xi               'Auxiliary variable used to calculate the CVaR'
eta(w)           'Auxiliary variable used to calculate the CVaR'
;

POSITIVE VARIABLES  eta(w), lambdaMS(k,m);
BINARY VARIABLES    v(k,m);


************************************************************************
*  EQUATIONS
************************************************************************


EQUATIONS

**        OBJECTIVE FUNCTION
ObjectiveFunction

**        CALCULATION OF THE DEMAND TO BE SUPPLIED
Demand
```

```
**          PRICE-QUOTA CURVE CONSTRAINTS
PriceQuota_01
PriceQuota_02
PriceQuota_03
PriceQuota_04
PriceQuota_05
PriceQuota_06


**          SELLING PRICE CONSTRAINTS
PriceEqu_01
PriceEqu_02

**          CVAR CONSTRAINTS
CVaR;


ObjectiveFunction..                              z=e= beta*sum((w,t),pi(w)
     *Wg(t)*(sum((k,m),lambdaMS(k,m)*(gammaC(k)*YMC(k,m,w)+gammaNC(
     k)*YMNC(k,m,w))*ED(k,t,w))-lambdaP(t,w)*sum(k,eS(k,t,w))))+(1-
     beta)*(xi-1/(1-alpha)*sum(w,pi(w)*eta(w)));

Demand(k,t,w)..                                  eS(k,t,w)=e=(gammaC(k)*yC
     (k,w)+gammaNC(k)*yNC(k,w))*ED(k,t,w);

PriceQuota_01(k,m)$(ord(m) gt 1)..               lambdaHatS(k,m-1)*v(k,m)
     =l= lambdaMS(k,m);
PriceQuota_02(k,m)..                             lambdaHatS(k,m)*v(k,m) =g
     = lambdaMS(k,m);
PriceQuota_03(k)..                               sum(m,v(k,m)) =e= 1;
PriceQuota_04(k,w)..                             yC(k,w)=e=sum(m,YMC(k,m,w
     )*v(k,m));
PriceQuota_05(k,w)..                             yNC(k,w)=e=sum(m,YMNC(k,m
     ,w)*v(k,m));
PriceQuota_06(k)..                               sum(m,lambdaMS(k,m)) =e=
     lambdaS(k);

PriceEqu_01(k)$(ord(k) lt card(k))..             lambdaS(k) =e= lambdaS(k
     +1);
PriceEqu_02(k,m)$(ord(k) lt card(k))..           v(k,m) =e= v(k+1,m);

CVaR(w)..                                        -sum(t,Wg(t)*(sum((k,m),
     lambdaMS(k,m)*(gammaC(k)*YMC(k,m,w)+gammaNC(k)*YMNC(k,m,w))*ED
     (k,t,w))-lambdaP(t,w)*sum(k,eS(k,t,w))))+xi-eta(w) =l= 0;


*******************************************************************
*  MODEL AND SOLVING
*******************************************************************


MODEL Supplier /all/;
Supplier.optcr=0;
Option mip=cplex;

SOLVE Supplier USING mip MAXIMIZING z;
```

7.11.3 Controlled Charge

A GAMS code for solving Illustrative Case Study of Sect. 7.7 is provided below:

```
$title ELECTRICITY SUPPLIER CONTROLLED


***************************************************************
* DATA
***************************************************************


SETS
w          'Scenarios'                                   /w1*w3/,
t          'Time periods'                                /t1*t3/,
k          'Electric vehicle groups'                     /k1*k2/,
m          'Set of blocks in price-quota curve'          /m1*m3/,
BEG(k,t),
END(k,t),
CH(k,t)
;

BEG('k1','t2')=yes;
BEG('k2','t1')=yes;
END('k1','t3')=yes;
END('k2','t2')=yes;
CH('k1','t2')=yes;
CH('k1','t3')=yes;
CH('k2','t1')=yes;
CH('k2','t2')=yes;

SCALARS
beta      'Risk weighting factor' /0.9999/,
alpha     'Confidence level'      /0.95/,
eta_B     'Battery efficiency'   /0.9/
;


PARAMETERS
pi(w)               'Scenario probability'
Wg(t)               'Period weight'
gammaC(k)           'Inital number of clients'
/k1 0.2
k2  0.5/,
gammaNC(k)          'Initial number of non-clients'
E0(k,w)             'Initial status of the batteries'
EE(k,w)             'Ending status of the batteries'
SOC_max(k)          'Maximum status of the batteries'

SOC_min(k)          'Maximum status of the batteries'
/k1 0
k2 0/,

CR(k)               'Peak power rate'
/k1 0.011
k2 0.011/,
```

```
N(k)                'Number of electric vehicles'
/k1  15000
k2   5000/
;


SOC_max(k)= N(k)*0.04;


TABLE ED(k,t,w)   'Demand of electric vehicles'
           w1       w2       w3
k1.t1      0        0        0
k1.t2      15       20       25
k1.t3      17       22       27
k2.t1      4        5        6
k2.t2      3        4        5
k2.t3      0        0        0;

TABLE YMC(k,m,w) 'Per-unit number of clients'
           w1       w2       w3
k1.m1      1.000    1.000    1.000
k1.m2      0.667    0.667    0.667
k1.m3      0.667    0.667    0.667
k2.m1      1.000    1.000    1.000
k2.m2      0.667    0.667    0.667
k2.m3      0.333    0.333    0.333;


TABLE YMNC(k,m,w) 'Per-unit number of non-clients'
           w1       w2       w3
k1.m1      1.000    1.000    1.000
k1.m2      0.667    0.667    0.667
k1.m3      0.000    0.000    0.000
k2.m1      0.667    0.667    0.667
k2.m2      0.333    0.333    0.333
k2.m3      0.000    0.000    0.000;

TABLE lambdaP(t,w) 'Pool price'
           w1       w2       w3
t1         30       50       70
t2         70       100      120
t3         40       60       80;

TABLE lambdaHatS(k,m) 'Selling price in price-quota curve'
           m1       m2       m3
k1         70       85       100
k2         70       85       100;

pi(w)=1/card(w);
Wg(t)=1;
gammaNC(k)=1-gammaC(k);
E0(k,w)=0;
EE(k,w)=E0(k,w)+sum(t,ED(k,t,w)) ;


*********************************************************************
* DECLARATION OF VARIABLES AND MATHEMATICAL CHARACTERIZATION
*********************************************************************
```

```
VARIABLES

** OBJECTIVE FUNCTION
z                   Objective function

** FIRST-STAGE VARIABLES
lambdaS(k)          Selling price offered to electric vehicles
lambdaMS(k,m)       Auxiliary variable used to compute the selling price
v(k,m)              Selection of block of the price-quota curve

** SECOND-STAGE VARIABLES
yC(k,w)             Per-unit number of current clients of the supplier belonging to group k that desires to buy from the supplier
yNC(k,w)            Per-unit number of non-clients of the supplier belonging to group k that desires to buy from the supplier
eS(k,t,w)           Demand provided by the supplier
eCch(k,t,w)           Energy charged to the batteries of electric vehicles
eDch(k,t,w)           Energy discharged from the batteries of electric vehicles
soc(k,t,w)          Energy stored in the batteries of electric vehicles

** VARIABLES ASSOCIATED WITH RISK MODELING
xi                  Auxiliary variable used to calculate the CVaR
eta(w)              Auxiliary variable used to calculate the CVaR
;

POSITIVE VARIABLES  eta(w), lambdaMS(k,m),eCch(k,t,w),eDch(k,t,w), soc(k,t,w);
BINARY VARIABLES    v(k,m);


*****************************************************************
*  EQUATIONS
*****************************************************************


EQUATIONS

** OBJECTIVE FUNCTION
ObjectiveFunction

** CALCULATION OF THE DEMAND TO BE SUPPLIED
Demand

*. STATUS BATTERIES
StatusBattery_01
StatusBattery_02
StatusBattery_03
StatusBattery_04
StatusBattery_05

**  CHARGE AND DISCHARGE CONSTRAINTS
MaxCharge_01
MaxDisCharge_01
```

```
MaxCharge_02
MaxDisCharge_02

** PRICE-QUOTA CURVE CONSTRAINTS
PriceQuota_01
PriceQuota_02
PriceQuota_03
PriceQuota_04
PriceQuota_05
PriceQuota_06

** SELLING PRICE CONSTRAINTS
PriceEqu_01
PriceEqu_02

** CVAR CONSTRAINTS
CVaR;


ObjectiveFunction..                          z=e= beta*sum((w,t),pi(w)*Wg(t)*(sum((k,m),lambdaMS(k,m)*(gammaC(k)*YMC(k,m,w)+gammaNC(k)*YMNC(k,m,w))*ED(k,t,w)))-lambdaP(t,w)*sum(k$Ch(k,t),eS(k,t,w)))+(1-beta)*(xi-1/(1-alpha)*sum(w,pi(w)*eta(w)));
;
Demand(k,t,w)$CH(k,t)..                      eS(k,t,w)=e= eCch(k,t,w)-eDch(k,t,w);

StatusBattery_01(k,t,w)$BEG(k,t)..           soc(k,t,w)=e= E0(k,w)*(gammaC(k)*yC(k,w)+gammaNC(k)*yNC(k,w))+eta_B*eCch(k,t,w)-1/eta_B*eDch(k,t,w);
StatusBattery_02(k,t,w)$END(k,t)..           soc(k,t,w)=g= EE(k,w)*(gammaC(k)*yC(k,w)+gammaNC(k)*yNC(k,w));
StatusBattery_03(k,t,w)$(CH(k,t) and (not(BEG(k,t))))..          soc(k,t,w)=e=soc(k,t-1,w)+eta_B*eCch(k,t,w)-1/eta_B*eDch(k,t,w) ;
StatusBattery_04(k,t,w)$CH(k,t)..            soc(k,t,w)=l=SOC_max(k)*(gammaC(k)*yC(k,w)+gammaNC(k)*yNC(k,w)) ;
StatusBattery_05(k,t,w)$CH(k,t)..            soc(k,t,w)=g=SOC_min(k)*(gammaC(k)*yC(k,w)+gammaNC(k)*yNC(k,w)) ;

MaxCharge_01(k,t,w)$CH(k,t)..                      eCch(k,t,w)=l=N(k)*CR(k)*(gammaC(k)*yC(k,w)+gammaNC(k)*yNC(k,w)) ;
MaxDisCharge_01(k,t,w)$CH(k,t)..                   eDch(k,t,w)=l=N(k)*CR(k)*(gammaC(k)*yC(k,w)+gammaNC(k)*yNC(k,w)) ;
MaxCharge_02(k,t,w)$(not(CH(k,t)))..               eCch(k,t,w)=l=0;
MaxDisCharge_02(k,t,w)$(not(CH(k,t)))..            eDch(k,t,w)=l=0 ;

PriceQuota_01(k,m)$(ord(m) gt 1)..           lambdaHatS(k,m-1)*v(k,m)=l= lambdaMS(k,m);
PriceQuota_02(k,m)..                         lambdaHatS(k,m)*v(k,m) =g= lambdaMS(k,m);
PriceQuota_03(k)..                           sum(m,v(k,m)) =e= 1;
PriceQuota_04(k,w)..                         yC(k,w)=e=sum(m,YMC(k,m,w)*v(k,m));
PriceQuota_05(k,w)..                         yNC(k,w)=e=sum(m,YMNC(k,m,w)*v(k,m));
PriceQuota_06(k)..                           sum(m,lambdaMS(k,m)) =e= lambdaS(k);
```

```
PriceEqu_01(k)$(ord(k) lt card(k))..                lambdaS(k) =e= lambdaS
    (k+1);
PriceEqu_02(k,m)$(ord(k) lt card(k))..              v(k,m) =e= v(k+1,m);

CVaR(w)..                                    -sum(t,Wg(t)*(sum((k,m),
    lambdaMS(k,m)*(gammaC(k)*YMC(k,m,w)+gammaNC(k)*YMNC(k,m,w))*ED
    (k,t,w))-lambdaP(t,w)*sum(k$Ch(k,t),eS(k,t,w))))+xi-eta(w) =l=
     0;


*************************************************************************
*  MODEL AND SOLVING
*************************************************************************


MODEL Supplier /all/;

Supplier.optcr=0;
Option mip=cplex;

SOLVE Supplier USING mip MAXIMIZING z;
```

References

1. Alizadeh, M., Wai, H.T., Goldsmith, A., Scaglione, A.: Retail and wholesale electricity pricing considering electric vehicle mobility. IEEE Trans. Control Netw. Syst. **6**(1), 249–260 (2019)
2. Bae, M., Kim, H., Kim, E., Chung, A.Y., Kim, H., Roh, J.H.: Toward electricity retail competition: survey and case study on technical infrastructure for advanced electricity market system. Appl. Energy **133**, 252–273 (2014)
3. Birge, J.R., Louveaux, F.: Introduction to Stochastic Programming, 2nd edn. Springer, New York (2011)
4. Carrión, M., Zárate-Miñano, R.: Operation of intermittent-dominated power systems with a significant penetration of plug-in electric vehicles. Energy **90**, 827–835 (2015)
5. Carrión, M.: Determination of the selling price offered by electricity suppliers to electric vehicle users. IEEE Trans. Smart Grid **10**(6), 6655–6666 (2019)
6. Dave, S., Sooriyabandaraa, M., Yearworth, M.: System behaviour modelling for demand response provision in a smart grid. Energy Policy **61**, 172–181 (2013)
7. GAMS (2021). www.gams.com
8. Ghosh, A., Aggarwal, V.: Menu-based pricing for charging of electric vehicles with vehicle-to-grid service. IEEE Trans. Veh. Technol. **67**(11), 10268–10280 (2018)
9. Gröwe-Kuska, N., Heitsch, H., Römisch, W.: Scenario reduction and scenario tree construction for power management problems. In: Proceedings of the IEEE Bologna Power Technology Conference, Bologna, Italy (2003)
10. Hanif, S., Creutzburg, P., Gooi, H.B., Hamacher, T.: Pricing mechanism for flexible loads using distribution grid hedging rights. IEEE Trans. Power Syst. **34**(5), 4048–4059 (2019)
11. Hotelling, H.: Stability in competition. Econ. J. **39**(153), 41–57 (1929)
12. MATLAB. https://es.mathworks.com/products/matlab.html
13. Momber, I., Wogrin, S., Gómez San Roman, T.: Retail pricing: a bilevel program for PEV aggregator decisions using indirect load control. IEEE Trans. Power Syst. **31**(1), 464–473 (2018)
14. Muratori, M., Rizzoni, G.: Residential demand response: dynamic energy management and time-varying electricity pricing. IEEE Trans. Power Syst. **31**(2), 1108–1117 (2016)

15. Muratori, M.: Impact of uncoordinated plug-in electric vehicles charging on residential power demand. Nat. Energy **3** (2018)
16. Ortega-Vazquez, M.A.: Optimal scheduling of electric vehicle charging and vehicle-to-grid services at household level including battery degradation and price uncertainty. IET Gener. Transmiss. Distribut. **8**(6), 1007–1016 (2014)
17. Spanish Market Operator (OMIE) (2022). http://www.omie.es
18. Regulated Electricity Prices for 2017 (2022). https://www.boe.es/boe/dias/2016/12/29/pdfs/BOE-A-2016-12464.pdf
19. Shaukat, N., Khan, B., Ali, S.M., et al.: A survey on electric vehicle transportation within smart grid system. Renew. Sus. Energy Rev. **81**, 1329–1349 (2018)
20. Shuai, W., Maillé, P., Pelov, A.: Charging electric vehicles in the smart city: a survey of economy-driven approaches. IEEE Trans. Intell. Transp. Syst. **17**(8), 2089–2106 (2016)
21. Sioshansi, R.: OR forum-modeling the impacts of electricity tariffs on plug-in hybrid electric vehicle charging, costs, and emissions. Oper. Res. **60**(3), 506–516 (2012)
22. Xi, X., Sioshansi, R.: Using price-based signals to control plug-in electric vehicle fleet charging. IEEE Trans. Smart Grid **5**(3), 1451–1464 (2014)
23. Soltani, N.Y., Kim, S.J., Giannakis, G.B.: Real-time load elasticity tracking and pricing for electric vehicle charging. IEEE Trans. Smart Grid **6**(3), 1303–1313 (2015)
24. U.S. Department of Energy: Benefits of demand response in electricity markets and recommendations for achieving them. A report to the United States Congress pursuant to Section 1252 of the Energy Policy Policy Act of 2005 (2006)
25. Yang, J., Zhao, J., Luo, F., Wen, F., Dong, Z.Y.: Decision-making for electricity retailers: a brief survey. IEEE Trans. Smart Grid **9**(5), 4140–4153 (2017)
26. Yang, J., Zhao, J., Wen, F., Dong, Z.Y.: A framework of customizing electricity retail prices. IEEE Trans. Power Syst. **33**(3), 2415–2428 (2018)
27. Yuan, W., Huang, J., Zhang, Y.J.: Competitive charging station pricing for plug-in electric vehicles. IEEE Trans. Smart Grid **8**(2), 627–639 (2017)

Part III
Planning Models

Chapter 8
Generation Expansion Planning Considering Renewable Energies

This chapter analyzes the generation expansion planning problem of a power system with renewable energies. The problem is formulated under the perspective of a central planner that aims at determining the generation expansion plans that minimize both the investment and operation costs. Then, this central planner promotes the building of the most suitable generating units among private profit-oriented investors. Uncertain variables are modeled using both scenarios and confidence bounds, which allows formulating the generation expansion planning problem using stochastic programming and robust optimization approaches. A static approach is adopted, where investment decisions are made at the beginning of the planning horizon. This chapter includes a number of illustrative examples that easily explain how the proposed approaches work and a case study based on the European power system.

8.1 Introduction

The Generation Expansion Planning (GEP) problem [6] is one of the most important problems in power system planning analysis. The GEP problem determines the optimal investments in generation facilities. These decisions are made with different objectives such as maximizing profits, minimizing generation costs, minimizing load-shedding costs, or increasing the share of renewable energy resources. In this sense, the GEP problem can be tackled considering two approaches. Firstly, a central approach that determines the most convenient generation expansion plan for the power system as a whole. Secondly, a market approach in which the generation expansion plans are made by private investors that aim at maximizing their own profits.

L. Baringo et al., *Electric Vehicles and Renewable Generation*,
Green Energy and Technology, https://doi.org/10.1007/978-3-031-09079-0_8

In this chapter we solve the GEP problem considering the perspective of a central planner. This planner determines the investment in generation facilities with the aim of minimizing both the operation and investment costs. Then, it promotes the generation-capacity investments in the most suitable locations and technologies among private profit-oriented investors. However, encouraging the investment in the optimal generation facilities is not a trivial problem. For example, the central planner may design incentive polices to promote the investment in a given technology or in a particular location [22]. Nevertheless, this is out of the scope of this chapter.

A relevant issue at the time of determining the generation investment decisions is the modeling of the uncertainties. In this sense, two different approaches are generally considered in the technical literature. Firstly, uncertainties can be modeled using scenarios, i.e., using a scenario-based stochastic programming approach [5], as done in [1, 7]. However, one of the drawbacks of stochastic programming is that a very large number of scenarios is generally needed to obtain an accurate modeling of the uncertain variables, which may result in very computationally complex and even intractable problems. An alternative is modeling uncertain variables using confidence bounds, which allows using a robust optimization approach [3], as done in [19]. This chapter analyzes both approaches.

Specifically, the GEP problem is solved considering a static approach [6] for the sake of simplicity, in which all investment decisions are made at a single point in time, namely, at the beginning of the planning horizon. This model can be extended to a dynamic approach in which investment decisions can be made at different points in time. This is analyzed in Chap. 9.

The remainder of this chapter is organized as follows. Section 8.2 describes the main characteristics of the GEP problem. Section 8.3 formulates the GEP problem considering a stochastic programming approach. Then, Sect. 8.4 reformulates the GEP problem using an adaptive robust optimization model. Sections 8.2, 8.3, and 8.4 include a number of clarifying examples. Section 8.5 provides the results of a realistic case study. Section 8.6 concludes the chapter with a summary and the main conclusions of the models and results reported. Section 8.7 defines the main notation used in this chapter. Finally, Sect. 8.8 provides the GAMS codes used to solve some of the illustrative examples included in the chapter.

8.2 Problem Description

This section describes the main characteristics of the GEP problem.

8.2.1 Aim and Approach

The overall aim of the GEP problem is to determine the new generation facilities to be built in the power system under study given an investment budget. These decisions are made with the objective of minimizing the total costs, namely, the generation costs, the load-shedding costs, and the generation investment costs. Both conventional and renewable generation units are considered as candidate expansion options.

The problem is formulated under the perspective of a central planner. This central planner does not actually build the generation facilities; however, it may suggest and encourage private profit-oriented investors to build these facilities.

For the sake of simplicity, the problem is solved considering a static approach, i.e., expansion decisions are made at the beginning of the planning horizon and for a future target year [6].

8.2.2 Uncertainty Characterization

Investment decisions in generation facilities are made for a long-term planning horizon, e.g., 30 years. Therefore, at the time of making these decisions there is a number of uncertainties that may affect the investment plans and that must be taken into account in order to achieve informed expansion decisions. These uncertainties include the future peak demand and the future generation costs of generating units.

To model these uncertainties, two different approaches are considered in this chapter, namely, stochastic programming [5] and robust optimization [3]. On the one hand, considering a scenario-based stochastic programming approach, uncertainties are modeled using a set of scenarios modeling different realizations of the uncertain variables. On the other hand, confidence bounds are used to represent uncertainties within a robust optimization approach. In this chapter, both scenarios and confidence bounds are used to represent uncertainties.

Note that the models described in this chapter for the GEP problem are general and additional sources of uncertainty may be considered, e.g., the decommissioning of existing generating units, the availability of transmission lines, or the maintenance of generating units.

8.2.3 Short-Term Variability

In order to make adequate generation expansion decisions it is necessary to model the variability of demand and stochastic renewable production conditions during the planning horizon. These demand and production conditions are based on historical data in the system under study. However, working with a large historical data set may result in intractable problems.

Therefore, to reduce the computational burden of the problem, historical data can be reduced into a set of clusters or groups using the K-means method described in [2]. This method provides a set of operating conditions indexed by t that represent both the variability and the correlation between demand and production of intermittent renewable generating units at different locations.

This method groups historical data that are similar among them. Moreover, it assigns a weight (N_t^{H}) to each operating condition t depending on the number of historical demand and intermittent renewable production conditions that are grouped in that cluster. Additional information can be found in [2] and in Sect. 2.5 of Chap. 2.

Illustrative Example 8.1 Operating conditions

Historical data of demand and renewable production in a power system are used to obtain a set of operating conditions that model the short-term variability in this system. Particularly, historical data comprise hourly data for a whole year, i.e., there are 8760 historical operating points, each one comprising values of the demand consumption and renewable production in the system. These historical data are grouped into four clusters giving rise to four operating conditions with the following characteristics:

1. Operating condition 1: characterized by a demand capacity factor K_{d1}^{D} of 0.9 pu and a renewable capacity factor K_{g1}^{G} of 0.9 pu. This means that under such condition the demand consumption and renewable production of the system are 90% of the peak demand and installed renewable capacity, respectively. The weight of this operating condition (N_1^{H}) is 1000 h. This means that 1000 of the 8760 historical hourly data are grouped and represented by this operating condition.
2. Operating condition 2: characterized by a demand capacity factor K_{d2}^{D} of 0.8 pu and a renewable capacity factor K_{g2}^{G} of 0.6 pu. The weight of this operating condition (N_2^{H}) is 3000 h.
3. Operating condition 3: characterized by a demand capacity factor K_{d3}^{D} of 0.6 pu and a renewable capacity factor K_{g3}^{G} of 0.7 pu. The weight of this operating condition (N_3^{H}) is 3000 h.
4. Operating condition 4: characterized by a demand capacity factor K_{d4}^{D} of 0.5 pu and a renewable capacity factor K_{g4}^{G} of 0.2 pu. The weight of this operating condition (N_4^{H}) is 1760 h.

These data are summarized in Table 8.1. Note that the sum of the weights of all operating conditions is 8760 h, i.e., the number of historical conditions. □

Table 8.1 Illustrative Example 8.1: operating conditions

Operating condition (t)	K_{dt}^{D} (pu)	K_{gt}^{G} (pu)	N_t^{H} (h)
1	0.9	0.9	1000
2	0.8	0.6	3000
3	0.6	0.7	3000
4	0.5	0.2	1760

8.3 Stochastic Programming Approach

This section describes the GEP problem considering a stochastic programming approach. Further details on the fundamentals and the practical application of stochastic programming can be found in Sect. 3.2 of Chap. 3.

8.3.1 *Uncertainty Characterization*

As mentioned in Sect. 8.2.2, the GEP is solved considering a set of uncertain variables. Under a stochastic approach, the uncertainty is modeled using a set of scenarios. The uncertainty comprising the future peak demand and production costs of conventional generating units in the system under study is modeled using a set of scenarios indexed by ω. Each of these scenarios has associated a probability of occurrence denoted by π_ω.

Illustrative Example 8.2 Scenarios

Long-term uncertainties in a power system comprise the future production costs of conventional generating units and the future peak demands. Uncertainties in future production costs are modeled using two scenarios that consider that the future production costs can be 20% higher than or 10% lower than the actual production costs. Moreover, the probability of each scenario is 0.4 and 0.6, respectively. On the other hand, uncertainties in future peak demand are modeled using two equiprobable scenarios that consider that the peak demands can be 30% higher than or equal to the actual peak demands. Uncertainties in future production costs and peak demands are assumed to be independent so that there is a total of $2 \cdot 2 = 4$ scenarios. These scenario data are summarized in Table 8.2. □

The aim of the GEP problem is determining the capacity of candidate generating units to be built in the power system under study. Note that these decisions are made at the beginning of the planning horizon, i.e., before knowing the actual realization of uncertain variables. Therefore, these decisions are *here-and-now* and do not depend on future scenario realizations. Then, after the candidate units are built, the information about the actual uncertainty realizations becomes available and the system

Table 8.2 Illustrative Example 8.2: scenarios

Scenario (ω)	Future production cost (%)	Future peak demand (%)	π_ω
1	+20	+30	$0.4 \cdot 0.5 = 0.2$
2	−10	+30	$0.6 \cdot 0.5 = 0.3$
3	+20	+0	$0.4 \cdot 0.5 = 0.2$
4	−10	+0	$0.6 \cdot 0.5 = 0.3$

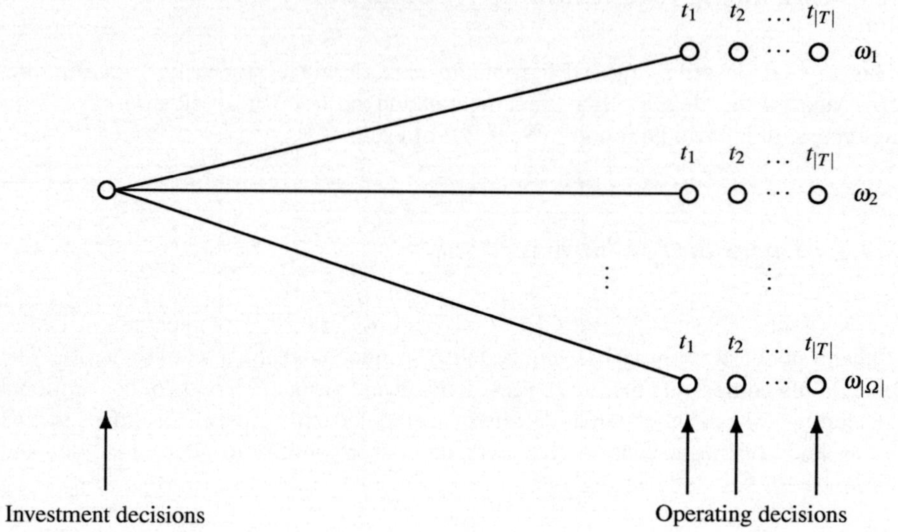

Fig. 8.1 Scenario tree used in the GEP problem considering a scenario-based stochastic programming approach

is operated. These operating decisions are *wait-and-see* since they depend on the scenario realizations. This decision sequence can be schematically represented by the scenario tree depicted in Fig. 8.1.

8.3.2 Formulation

The GEP problem considering a stochastic approach can be formulated using the following linear programming (LP) model:

Minimize$_{\Phi^{\mathrm{S}}}$

$$\sum_{g\in\tilde{G}} C_g^{\mathrm{IG}} p_g^{\mathrm{CG,max}} + \sum_{\omega\in\Omega} \pi_\omega \sum_{t\in T} N_t^{\mathrm{H}} \left(\sum_{g\in G} C_{g\omega}^{\mathrm{OG}} p_{gt\omega}^{\mathrm{G}} + \sum_{d\in D} C_d^{\mathrm{UD}} p_{dt\omega}^{\mathrm{UD}} \right) \tag{8.1a}$$

subject to:

$$\sum_{g\in\tilde{G}} C_g^{\mathrm{IG}} p_g^{\mathrm{CG,max}} \le \overline{C}^{\mathrm{IG}}, \tag{8.1b}$$

$$0 \le p_g^{\mathrm{CG,max}} \le \overline{P}_g^{\mathrm{CG,max}}, \quad \forall g \in \tilde{G}, \tag{8.1c}$$

$$\sum_{g \in G_b} p^{\mathrm{G}}_{gt\omega} - \sum_{\ell | O(\ell)=b} p^{\mathrm{L}}_{\ell t\omega} + \sum_{\ell | F(\ell)=b} p^{\mathrm{L}}_{\ell t\omega}$$

$$= \sum_{d \in D_b} \left(K^{\mathrm{D}}_{dt} \overline{P}^{\mathrm{D}}_{d\omega} - p^{\mathrm{UD}}_{dt\omega} \right), \quad \forall b \in B, \forall t \in T, \forall \omega \in \Omega, \tag{8.1d}$$

$$p^{\mathrm{L}}_{\ell t\omega} = \frac{1}{X_\ell} \left(\delta_{O(\ell)t\omega} - \delta_{F(\ell)t\omega} \right), \quad \forall \ell \in L, \forall t \in T, \forall \omega \in \Omega, \tag{8.1e}$$

$$-\overline{P}^{\mathrm{L}}_{\ell} \le p^{\mathrm{L}}_{\ell t\omega} \le \overline{P}^{\mathrm{L}}_{\ell}, \quad \forall \ell \in L, \forall t \in T, \forall \omega \in \Omega, \tag{8.1f}$$

$$0 \le p^{\mathrm{G}}_{gt\omega} \le K^{\mathrm{G}}_{gt} \overline{P}^{\mathrm{G}}_{g}, \quad \forall g \backslash g \in \tilde{G}, \forall t \in T, \forall \omega \in \Omega, \tag{8.1g}$$

$$0 \le p^{\mathrm{G}}_{gt\omega} \le K^{\mathrm{G}}_{gt} p^{\mathrm{CG,max}}_{g}, \quad \forall g \in \tilde{G}, \forall t \in T, \forall \omega \in \Omega, \tag{8.1h}$$

$$0 \le p^{\mathrm{UD}}_{dt\omega} \le K^{\mathrm{D}}_{dt} \overline{P}^{\mathrm{D}}_{d\omega}, \quad \forall d \in D, \forall t \in T, \forall \omega \in \Omega, \tag{8.1i}$$

$$\delta_{bt\omega} = 0, \quad b: \text{ref.}, \forall t \in T, \forall \omega \in \Omega, \tag{8.1j}$$

where variables in set $\Phi^{\mathrm{S}} = \Big\{ p^{\mathrm{CG,max}}_{g}, \forall g \in \tilde{G};\ p^{\mathrm{G}}_{gt\omega}, \forall g \in G, \forall t \in T, \forall \omega \in \Omega;$ $p^{\mathrm{UD}}_{dt\omega}, \forall d \in D, \forall t \in T, \forall \omega \in \Omega;\ p^{\mathrm{L}}_{\ell t\omega}, \forall \ell \in L, \forall t \in T, \forall \omega \in \Omega;\ \delta_{bt\omega}, \forall b \in B, \forall t \in$ $T, \forall \omega \in \Omega \Big\}$ are the optimization variables of problem (8.1).

The objective function (8.1) represents the expected annualized investment and operation costs and comprises the following terms:

1. Term $\sum_{g \in \tilde{G}} C^{\mathrm{IG}} p^{\mathrm{CG,max}}_{g}$ is the annualized investment cost of building new generating units. Investment cost parameters are annualized using the capital recovery factor, as explained in Sect. 2.5.1 of Chap. 2.
2. Terms $\sum_{g \in G} C^{\mathrm{OG}}_{g\omega} p^{\mathrm{G}}_{gt\omega}$ are the production costs of generating units.
3. Terms $\sum_{d \in D} C^{\mathrm{UD}}_{d} p^{\mathrm{UD}}_{dt\omega}$ are the load-shedding costs of unserved demands.

Note that the operation costs in terms 2 and 3 are multiplied by the probability of each scenario π_ω and by factors N^{H}_t in order to make both the investment and operation costs comparable.

Constraint (8.1b) imposes annualized investment budgets for building candidate generating units. Constraints (8.1c) define the maximum capacity that can be built of candidate generating units. Constraints (8.1d)–(8.1j) model the operation of the system for all time periods and scenarios. Constraints (8.1d) impose the generation-demand balance at each bus of the system. Constraints (8.1e) define the power flows through transmission lines, which are limited by the corresponding capacities by constraints (8.1f). Constraints (8.1g), (8.1h), and (8.1i) impose bounds on the power production levels of existing generating units, candidate generating units, and load shed of demands, respectively. Note that factors K^{G}_{gt} in constraints (8.1g) and (8.1h) model the available production of generating unit in each time period, being these factors equal to 1 for conventional generating units. For the sake of simplicity, no inter-temporal constraints for thermal power units are considered. Finally, constraints (8.1j) define the voltage angle at the reference node.

Table 8.3 Illustrative Example 8.3: data of existing generating units

Generating unit	Production cost (€/MWh)	Capacity (MW)
1	40	200
2	0	100

Table 8.4 Illustrative Example 8.3: data of demands

Demand	Unserved cost (€/MWh)	Peak demand (MW)
1	200	400

Table 8.5 Illustrative Example 8.3: data of candidate generating units

Generating unit	Production cost (€/MWh)	Capacity (MW)	Investment cost (€/MW)
3	25	150	150,000
4	0	100	200,000

Illustrative Example 8.3 Stochastic GEP problem

The GEP problem is considered for a single-node power system comprising a conventional generating unit (g_1), a wind-power unit (g_2), and a demand (d_1). The data of the conventional generating unit and wind-power unit are provided in Table 8.3, while the data of the demand are provided in Table 8.4.

The candidate generating units comprise a conventional generating unit (g_3) and a wind-power unit (g_4), whose data are provided in Table 8.5. The investment budget is 30 million €.

The short-term variability is modeled using four operating conditions, whereas the uncertainty is characterized by four scenarios, as provided in Illustrative Examples 8.1 and 8.2, respectively.

Considering the above data, the GEP problem considering a scenario-based stochastic programming approach is formulated.

Firstly, the objective function of the problem comprises the minimization of the expected investment and operation costs and is formulated as follows:

$$\begin{aligned}
&150000 p_3^{\mathrm{CG,max}} + 200000 p_4^{\mathrm{CG,max}} \\
&\quad + 0.2\Big[1000\Big(1.2 \cdot 40 p_{111}^{\mathrm{G}} + 1.2 \cdot 25 p_{311}^{\mathrm{G}} + 200 p_{111}^{\mathrm{UD}}\Big) \\
&\qquad + 3000\Big(1.2 \cdot 40 p_{121}^{\mathrm{G}} + 1.2 \cdot 25 p_{321}^{\mathrm{G}} + 200 p_{121}^{\mathrm{UD}}\Big) \\
&\qquad + 3000\Big(1.2 \cdot 40 p_{131}^{\mathrm{G}} + 1.2 \cdot 25 p_{331}^{\mathrm{G}} + 200 p_{131}^{\mathrm{UD}}\Big)
\end{aligned}$$

$$
\begin{aligned}
&+ 1760\Big(1.2 \cdot 40 p_{141}^{\mathrm{G}} + 1.2 \cdot 25 p_{341}^{\mathrm{G}} + 200 p_{141}^{\mathrm{UD}}\Big)\Big] \\
+ 0.3\Big[&1000\Big(0.9 \cdot 40 p_{112}^{\mathrm{G}} + 0.9 \cdot 25 p_{312}^{\mathrm{G}} + 200 p_{112}^{\mathrm{UD}}\Big) \\
&+ 3000\Big(0.9 \cdot 40 p_{122}^{\mathrm{G}} + 0.9 \cdot 25 p_{322}^{\mathrm{G}} + 200 p_{122}^{\mathrm{UD}}\Big) \\
&+ 3000\Big(0.9 \cdot 40 p_{132}^{\mathrm{G}} + 0.9 \cdot 25 p_{332}^{\mathrm{G}} + 200 p_{132}^{\mathrm{UD}}\Big) \\
&+ 1760\Big(0.9 \cdot 40 p_{142}^{\mathrm{G}} + 0.9 \cdot 25 p_{342}^{\mathrm{G}} + 200 p_{142}^{\mathrm{UD}}\Big)\Big] \\
+ 0.2\Big[&1000\Big(1.2 \cdot 40 p_{113}^{\mathrm{G}} + 1.2 \cdot 25 p_{313}^{\mathrm{G}} + 200 p_{113}^{\mathrm{UD}}\Big) \\
&+ 3000\Big(1.2 \cdot 40 p_{123}^{\mathrm{G}} + 1.2 \cdot 25 p_{323}^{\mathrm{G}} + 200 p_{123}^{\mathrm{UD}}\Big) \\
&+ 3000\Big(1.2 \cdot 40 p_{133}^{\mathrm{G}} + 1.2 \cdot 25 p_{333}^{\mathrm{G}} + 200 p_{133}^{\mathrm{UD}}\Big) \\
&+ 1760\Big(1.2 \cdot 40 p_{143}^{\mathrm{G}} + 1.2 \cdot 25 p_{343}^{\mathrm{G}} + 200 p_{143}^{\mathrm{UD}}\Big)\Big] \\
+ 0.3\Big[&1000\Big(0.9 \cdot 40 p_{114}^{\mathrm{G}} + 0.9 \cdot 25 p_{314}^{\mathrm{G}} + 200 p_{114}^{\mathrm{UD}}\Big) \\
&+ 3000\Big(0.9 \cdot 40 p_{124}^{\mathrm{G}} + 0.9 \cdot 25 p_{324}^{\mathrm{G}} + 200 p_{124}^{\mathrm{UD}}\Big) \\
&+ 3000\Big(0.9 \cdot 40 p_{134}^{\mathrm{G}} + 0.9 \cdot 25 p_{334}^{\mathrm{G}} + 200 p_{134}^{\mathrm{UD}}\Big) \\
&+ 1760\Big(0.9 \cdot 40 p_{144}^{\mathrm{G}} + 0.9 \cdot 25 p_{344}^{\mathrm{G}} + 200 p_{144}^{\mathrm{UD}}\Big)\Big].
\end{aligned}
$$

The first line represents the investment costs. The second, third, fourth, and fifth lines represent the operating cost related to the first, second, third, and fourth operating conditions, respectively, for the first scenario. Note that these terms are multiplied by the probability of the first scenario and the weight of each operating condition. Similarly, lines 6–9, 10–13, and 14–17 represent the operating costs in the second, third, and fourth scenarios, respectively.

Then, the constraints of the problem are formulated below:

- Investment budget:

$$
150000 p_3^{\mathrm{CG,max}} + 200000 p_4^{\mathrm{CG,max}} \leq 30000000.
$$

- Maximum capacity of candidate generating units:

$$0 \leq p_3^{\text{CG,max}} \leq 150,$$
$$0 \leq p_4^{\text{CG,max}} \leq 100.$$

- Power balance:

$$p_{111}^{\text{G}} + p_{211}^{\text{G}} + p_{311}^{\text{G}} + p_{411}^{\text{G}} = 0.9 \cdot 1.3 \cdot 400 - p_{111}^{\text{UD}},$$
$$p_{121}^{\text{G}} + p_{221}^{\text{G}} + p_{321}^{\text{G}} + p_{421}^{\text{G}} = 0.8 \cdot 1.3 \cdot 400 - p_{121}^{\text{UD}},$$
$$p_{131}^{\text{G}} + p_{231}^{\text{G}} + p_{331}^{\text{G}} + p_{431}^{\text{G}} = 0.6 \cdot 1.3 \cdot 400 - p_{131}^{\text{UD}},$$
$$p_{141}^{\text{G}} + p_{241}^{\text{G}} + p_{341}^{\text{G}} + p_{441}^{\text{G}} = 0.5 \cdot 1.3 \cdot 400 - p_{141}^{\text{UD}},$$
$$p_{112}^{\text{G}} + p_{212}^{\text{G}} + p_{312}^{\text{G}} + p_{412}^{\text{G}} = 0.9 \cdot 1.3 \cdot 400 - p_{112}^{\text{UD}},$$
$$p_{122}^{\text{G}} + p_{222}^{\text{G}} + p_{322}^{\text{G}} + p_{422}^{\text{G}} = 0.8 \cdot 1.3 \cdot 400 - p_{122}^{\text{UD}},$$
$$p_{132}^{\text{G}} + p_{232}^{\text{G}} + p_{332}^{\text{G}} + p_{432}^{\text{G}} = 0.6 \cdot 1.3 \cdot 400 - p_{132}^{\text{UD}},$$
$$p_{142}^{\text{G}} + p_{242}^{\text{G}} + p_{342}^{\text{G}} + p_{442}^{\text{G}} = 0.5 \cdot 1.3 \cdot 400 - p_{142}^{\text{UD}},$$
$$p_{113}^{\text{G}} + p_{213}^{\text{G}} + p_{313}^{\text{G}} + p_{413}^{\text{G}} = 0.9 \cdot 1.0 \cdot 400 - p_{113}^{\text{UD}},$$
$$p_{123}^{\text{G}} + p_{223}^{\text{G}} + p_{323}^{\text{G}} + p_{423}^{\text{G}} = 0.8 \cdot 1.0 \cdot 400 - p_{123}^{\text{UD}},$$
$$p_{133}^{\text{G}} + p_{233}^{\text{G}} + p_{333}^{\text{G}} + p_{433}^{\text{G}} = 0.6 \cdot 1.0 \cdot 400 - p_{133}^{\text{UD}},$$
$$p_{143}^{\text{G}} + p_{243}^{\text{G}} + p_{343}^{\text{G}} + p_{443}^{\text{G}} = 0.5 \cdot 1.0 \cdot 400 - p_{143}^{\text{UD}},$$
$$p_{114}^{\text{G}} + p_{214}^{\text{G}} + p_{314}^{\text{G}} + p_{414}^{\text{G}} = 0.9 \cdot 1.0 \cdot 400 - p_{114}^{\text{UD}},$$
$$p_{124}^{\text{G}} + p_{224}^{\text{G}} + p_{324}^{\text{G}} + p_{424}^{\text{G}} = 0.8 \cdot 1.0 \cdot 400 - p_{124}^{\text{UD}},$$
$$p_{134}^{\text{G}} + p_{234}^{\text{G}} + p_{334}^{\text{G}} + p_{434}^{\text{G}} = 0.6 \cdot 1.0 \cdot 400 - p_{134}^{\text{UD}},$$
$$p_{144}^{\text{G}} + p_{244}^{\text{G}} + p_{344}^{\text{G}} + p_{444}^{\text{G}} = 0.5 \cdot 1.0 \cdot 400 - p_{144}^{\text{UD}}.$$

- Production limits of existing generating units:

$$0 \leq p_{1t\omega}^{\text{G}} \leq 200, \quad t = 1, 2, 3, 4, \omega = 1, 2, 3, 4,$$
$$0 \leq p_{21\omega}^{\text{G}} \leq 0.9 \cdot 100, \quad \omega = 1, 2, 3, 4,$$
$$0 \leq p_{22\omega}^{\text{G}} \leq 0.6 \cdot 100, \quad \omega = 1, 2, 3, 4,$$
$$0 \leq p_{23\omega}^{\text{G}} \leq 0.7 \cdot 100, \quad \omega = 1, 2, 3, 4,$$
$$0 \leq p_{24\omega}^{\text{G}} \leq 0.2 \cdot 100, \quad \omega = 1, 2, 3, 4.$$

- Production limits of candidate generating units:

$$0 \leq p_{3t\omega}^{\text{G}} \leq p_3^{\text{CG,max}}, \quad t = 1, 2, 3, 4, \omega = 1, 2, 3, 4,$$
$$0 \leq p_{41\omega}^{\text{G}} \leq 0.9 \cdot p_4^{\text{CG,max}}, \quad \omega = 1, 2, 3, 4,$$

$$0 \le p^{\mathrm{G}}_{42\omega} \le 0.6 \cdot p^{\mathrm{CG,max}}_4, \quad \omega = 1, 2, 3, 4,$$
$$0 \le p^{\mathrm{G}}_{43\omega} \le 0.7 \cdot p^{\mathrm{CG,max}}_4, \quad \omega = 1, 2, 3, 4,$$
$$0 \le p^{\mathrm{G}}_{44\omega} \le 0.2 \cdot p^{\mathrm{CG,max}}_4, \quad \omega = 1, 2, 3, 4.$$

- Maximum unserved demand:

$$0 \le p^{\mathrm{UD}}_{111} \le 0.9 \cdot 1.3 \cdot 400,$$
$$0 \le p^{\mathrm{UD}}_{121} \le 0.8 \cdot 1.3 \cdot 400,$$
$$0 \le p^{\mathrm{UD}}_{131} \le 0.6 \cdot 1.3 \cdot 400,$$
$$0 \le p^{\mathrm{UD}}_{141} \le 0.5 \cdot 1.3 \cdot 400,$$
$$0 \le p^{\mathrm{UD}}_{112} \le 0.9 \cdot 1.3 \cdot 400,$$
$$0 \le p^{\mathrm{UD}}_{122} \le 0.8 \cdot 1.3 \cdot 400,$$
$$0 \le p^{\mathrm{UD}}_{132} \le 0.6 \cdot 1.3 \cdot 400,$$
$$0 \le p^{\mathrm{UD}}_{142} \le 0.5 \cdot 1.3 \cdot 400,$$
$$0 \le p^{\mathrm{UD}}_{113} \le 0.9 \cdot 1.0 \cdot 400,$$
$$0 \le p^{\mathrm{UD}}_{123} \le 0.8 \cdot 1.0 \cdot 400,$$
$$0 \le p^{\mathrm{UD}}_{133} \le 0.6 \cdot 1.0 \cdot 400,$$
$$0 \le p^{\mathrm{UD}}_{143} \le 0.5 \cdot 1.0 \cdot 400,$$
$$0 \le p^{\mathrm{UD}}_{114} \le 0.9 \cdot 1.0 \cdot 400,$$
$$0 \le p^{\mathrm{UD}}_{124} \le 0.8 \cdot 1.0 \cdot 400,$$
$$0 \le p^{\mathrm{UD}}_{134} \le 0.6 \cdot 1.0 \cdot 400,$$
$$0 \le p^{\mathrm{UD}}_{144} \le 0.5 \cdot 1.0 \cdot 400.$$

Finally, note that the optimization variables of the problem are those included in set $\Phi^{\mathrm{IE8.3}} = \left\{p^{\mathrm{CG,max}}_g,\ g = 3, 4;\ p^{\mathrm{G}}_{gt\omega},\ g = 1, 2, 3, 4,\ t = 1, 2, 3, 4,\ \omega = 1, 2, 3, 4;\ p^{\mathrm{UD}}_{1t\omega},\ t = 1, 2, 3, 4,\ \omega = 1, 2, 3, 4\right\}$.

The above model is a linear programming problem [20] that is solved using CPLEX [21] under a GAMS environment [11].

The optimal investment plan results in building 132.31 MW of the candidate conventional generating unit (g_3) and 55.77 MW of the candidate wind-power unit (g_4).

Regarding the operation of the power system once the new generating units are built, wind-power units g_2 and g_4 are used as first option, using the whole available production taking into account the capacity built and the capacity factor of each operating condition. Then, the newly built conventional generating unit (g_3) is used as second option due to its comparatively lower operation costs. Finally, the existing

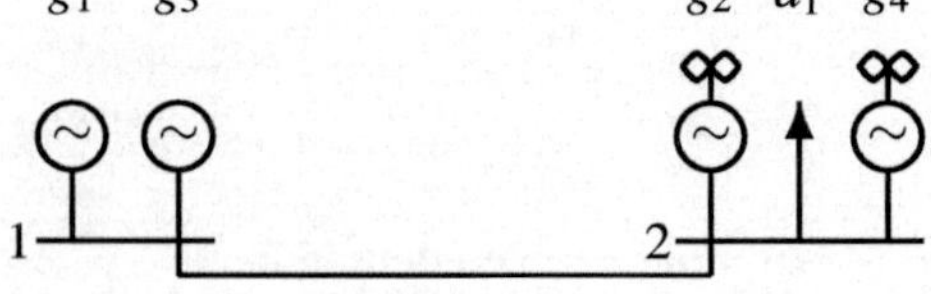

Fig. 8.2 Illustrative Example 8.4: two-bus system

conventional generating unit (g_1) is used to supply the rest of the demand, depending on the scenario and operating condition.

The optimal value of the objective function, i.e., the total cost, including both operation and investment costs, is 92.81 million €.

A GAMS [11] code for solving Illustrative Example 8.3 is provided in Sect. 8.8. □

Illustrative Example 8.4 Stochastic GEP problem: impact of the network constraints

The data of Illustrative Example 8.3 are considered again. However, next it is assumed that generating units and demands are interconnected within the two-bus network depicted in Fig. 8.2. Conventional generating units (g_1 and g_3) are connected in bus 1, while the wind-power units and the demand are connected in bus 2. Moreover, the transmission line connecting buses 1 and 2 has a capacity of 250 MW and a reactance of 1 Ω. The base power and voltage are 1 MW and 1 kV, respectively. Finally, bus 1 is considered as the reference bus.

The problem is solved considering the above network constraints and the optimal investment plan considers the building of 50 MW of the candidate conventional generating unit (g_3) and 100 MW of the candidate wind-power unit (g_4). The 250-MW capacity of the transmission line limits the generation level from bus 1 where the two conventional generating units are located. As a result, the capacity investment in the candidate conventional generating unit decreases as compared with Illustrative Example 8.3. The main consequence is that the optimal value of the objective function, i.e., the total cost, increases up to 105.00 million €. □

8.4 Adaptive Robust Optimization Approach

The stochastic GEP problem described in the previous section models the uncertainty using a set of scenarios. In order to have an accurate representation of the uncertainty it is essential:

1. To have information about the probability distribution functions of uncertain variables in order to generate realistic scenarios.
2. To generate a large enough number of scenarios to obtain a good representation of the uncertainty.

Knowing the probability distribution functions of uncertain variables is not an easy task since the GEP problem is solved for a long-term planning horizon. Moreover, even if this information is available, considering a very large number of scenarios may result in intractability issues for real-world power systems with hundreds of nodes and thousands of transmission lines.

As an alternative, this section analyzes the GEP problem considering an adaptive robust optimization approach. In this case, uncertainties are modeled using confidence bounds, which are generally easier to obtain than generating accurate scenarios. The interested reader is referred to Sect. 3.3 of Chap. 3 for additional information about the practical implementation of robust optimization approaches.

8.4.1 Problem Description

The GEP problem is solved for a long-term planning horizon following the decision sequence below:

1. The investment decisions are made for a long-term planning horizon. These decisions are made at the beginning of the planning horizon without knowing the future conditions in the power system under study.
2. Information about the system conditions becomes available. In other words, the actual uncertainty realizations are revealed.
3. The power system is operated after knowing both the investment decisions and uncertainty realizations.

Taking into account the above decision sequence it is possible to formulate the GEP problem using an adaptive robust optimization approach. First, the investment decisions are sought to minimize both investment and operating costs and by anticipating that, once they are made, the worst uncertainty will occur. Then, the worst-case uncertainty realization is determined as the one that maximizes the operating costs once the investment decisions are made and by anticipating that, after it occurs, the central entity in charge of the power system will take corrective actions to operate the system in such a way that operating costs are minimized.

Note that in this case the investment decisions are made with the aim of minimizing the investment and worst-case operating costs. This objective is different to that of the stochastic programming approach considered in Sect. 8.3 where investment decisions are sought to minimize the investment and average operating costs over scenarios.

This modeling framework allows formulating the GEP problem as the three-level programming problem described in the following section [4].

8.4.2 Three-Level Problem Formulation

The GEP problem considering an adaptive robust optimization approach is formulated below:

$$\text{Minimize}_{\Phi^{\text{L1}}} \sum_{g \in \tilde{G}} C^{\text{IG}} p_g^{\text{CG,max}}$$

$$+ \text{Maximize}_{\Phi^{\text{L2}} \in \Lambda} \text{Minimize}_{\Phi^{\text{L3}} \in \Xi(\cdot)} \sum_{t \in T} N_t^{\text{H}} \left(\sum_{g \in G} c_g^{\text{OG}} p_{gt}^{\text{G}} + \sum_{d \in D} C_d^{\text{UD}} p_{dt}^{\text{UD}} \right) \tag{8.2a}$$

subject to:

$$\sum_{g \in \tilde{G}} C^{\text{IG}} p_g^{\text{CG,max}} \le \overline{C}^{\text{IG}}, \tag{8.2b}$$

$$0 \le p_g^{\text{CG,max}} \le \overline{P}_g^{\text{CG,max}}, \quad \forall g \in \tilde{G}, \tag{8.2c}$$

where variables in sets $\Phi^{\text{L1}} = \left\{ p_g^{\text{CG,max}}, \forall g \in \tilde{G} \right\}$, $\Phi^{\text{L2}} = \left\{ \overline{p}_d^{\text{D}}, \forall d \in D; c_g^{\text{OG}}, \forall g \in G \right\}$, and $\Phi^{\text{L3}} = \left\{ p_{gt}^{\text{G}}, \forall g \in G, \forall t \in T;\ p_{dt}^{\text{UD}}, \forall d \in D, \forall t \in T;\ p_{\ell t}^{\text{L}}, \forall \ell \in L, \forall t \in T;\ \delta_{bt}, \forall b \in B, \forall t \in T \right\}$, are the optimization variables of the first-, second-, and third-level optimization problems, respectively, while sets Λ and Ξ represent the uncertainty and feasibility sets, respectively. These two sets are described in the following sections.

Problem (8.2) is driven by the minimization of the worst-case investment and operating costs. It comprises three nested optimization problems:

1. The first-level problem determines the investment decisions in new generating units. These decisions are made with the aim of minimizing the worst-case investment and operating costs (8.2a), considering the investment budget represented by constraint (8.2b) and the maximum capacity to be built of candidate generating units modeled by constraints (8.2c).
2. Given the first-level decisions, the second-level problem identifies the worst-case realization of uncertain variables, i.e., the values of these variables that maximize the operating cost. Note that the second-level decision variables in set Φ^{L2} must take values within the uncertainty set Λ. This uncertainty set is defined in Sect. 8.4.3.
3. Given first- and second-level decisions, the third-level problem models the operation of the power system with the aim of minimizing the operating cost. Note that the third-level decision variables in set Φ^{L3} must take values within the feasibility set Ξ. This feasibility set is related to the technical constraints of the power system under study and is defined in Sect. 8.4.4.

8.4.3 Definition of Uncertainty Set

Uncertain variables comprise the future peak demands and the future production costs of generating units, which are represented by variables $\overline{p}_d^{\text{D}}, \forall d \in D$, and $c_g^{\text{OG}}, \forall g \in G$, respectively. These uncertainties are modeled using the following confidence bounds [3]:

$$\overline{p}_d^{\text{D}} \in \left[\tilde{P}_d^{\text{D}} - \hat{P}_d^{\text{D}}, \tilde{P}_d^{\text{D}} + \hat{P}_d^{\text{D}}\right], \quad \forall d \in D, \tag{8.3a}$$

$$c_g^{\text{OG}} \in \left[\tilde{C}_g^{\text{OG}} - \hat{C}_g^{\text{OG}}, \tilde{C}_g^{\text{OG}} + \hat{C}_g^{\text{OG}}\right], \quad \forall g \in G. \tag{8.3b}$$

In addition, the so-called budgets Γ^{D} and Γ^{G} are defined to control the conservativeness of the model. These parameters respectively represent the maximum number of demands and generating units for which the future peak demand and future generation cost experience fluctuations with respect to their forecast values. The uncertainty budget Γ^{D} takes values between zero and $|D|$, i.e., the number of demands:

1. If $\Gamma^{\text{D}} = 0$, then the future peak demands of all loads in the system are considered to be equal to the corresponding forecast levels, i.e., $\overline{p}_d^{\text{D}} = \tilde{P}_d^{\text{D}}, \forall d \in D$.
2. If $\Gamma^{\text{D}} = |D|$, then the future peak demands of all loads can deviate from their forecast levels.
3. Intermediate values of Γ^{D} between 0 and $|D|$ indicate that the future peak demands of loads can deviate from their forecast levels in a maximum of Γ^{D} demands.

Similarly, Γ^{G} can take values between 0 and $|G|$, i.e., the total number of generating units. Note that by increasing the value of these uncertainty budgets it is possible to obtain more conservative investment plans.

As shown in [16], when polyhedral uncertainty sets as the one described above are considered, the worst-case uncertainty realization corresponds to an extreme or vertex of the polyhedron representing the uncertainty sets. Moreover, as shown in [18], the worst-case realizations of peak demand and generating cost will be never at the lower bound. Therefore, the polyhedral uncertainty set can be equivalently represented by a binary-variable-based equivalent uncertainty set that only models the finite set of extremes or vertexes of the polyhedron and is formulated as follows:

$$\Lambda = \Big\{ \overline{p}_d^{\text{D}} = \tilde{P}_d^{\text{D}} + u_d^{\text{D}} \hat{P}_d^{\text{D}}, \quad \forall d \in D, \tag{8.4a}$$

$$c_g^{\text{OG}} = \tilde{C}_g^{\text{OG}} + u_g^{\text{OG}} \hat{C}_g^{\text{OG}}, \quad \forall g \in G, \tag{8.4b}$$

$$\sum_{d \in D} u_d^{\text{D}} \le \Gamma^{\text{D}}, \tag{8.4c}$$

$$\sum_{g \in G} u_g^{\mathrm{OG}} \leq \Gamma^{\mathrm{G}}, \tag{8.4d}$$

$$u_d^{\mathrm{D}} \in \{0, 1\}, \quad \forall d \in D, \tag{8.4e}$$

$$\left. u_g^{\mathrm{OG}} \in \{0, 1\}, \quad \forall g \in G \right\}. \tag{8.4f}$$

Constraints (8.4a) and (8.4b) define uncertain variables $\overline{p}_d^{\mathrm{D}}$ and c_g^{OG}, respectively, as their forecast values plus their maximum possible increase. Constraints (8.4c) and (8.4d) impose bounds to the number of peak demands and operation costs that can deviate from their forecast values through the demand and generation uncertainty budgets, i.e., Γ^{D} and Γ^{G}, respectively. Finally, constraints (8.4e) and (8.4f) define the binary nature of variables u_d^{D} and u_g^{OG}, respectively.

Illustrative Example 8.5 Uncertainty set

This example defines the uncertainty set. To do so, data of Illustrative Example 8.3 are considered again, in which a single demand and four generating units (two existing and two candidate) are considered. On the one hand, it is assumed that the future peak demand has a forecast value of 400 MW that can experience a deviation of a maximum of 30% with respect to this forecast value, i.e., the future peak demand takes values within the following confidence bound:

$$\overline{p}_1^{\mathrm{D}} \in [400 - 120, 400 + 120] .$$

On the other hand, the future production costs of existing and candidate generating units take values within the confidence bounds below:

$$c_1^{\mathrm{OG}} \in [40 - 8, 40 + 8] ,$$
$$c_2^{\mathrm{OG}} \in [0, 0] ,$$
$$c_3^{\mathrm{OG}} \in [25 - 5, 25 + 5] ,$$
$$c_4^{\mathrm{OG}} \in [0, 0] .$$

Note that despite having four generating units, only two of them have uncertain operating costs (g_1 and g_3) since the other two (g_2 and g_4) are wind-power units that are assumed to have null operation costs.

As described in Eq. (8.4), these confidence bounds allow formulating the following uncertainty set:

$$\Lambda = \left\{ \overline{p}_1^{\mathrm{D}} = 400 + 120 u_1^{\mathrm{D}}, \right.$$
$$c_1^{\mathrm{OG}} = 40 + 8 u_1^{\mathrm{OG}},$$
$$c_3^{\mathrm{OG}} = 25 + 5 u_3^{\mathrm{OG}},$$

$$
\left.\begin{aligned}
&u_1^{\mathrm{D}} \leq \Gamma^{\mathrm{D}},\\
&u_1^{\mathrm{OG}} + u_3^{\mathrm{OG}} \leq \Gamma^{\mathrm{G}},\\
&u_1^{\mathrm{D}}, u_1^{\mathrm{OG}}, u_3^{\mathrm{OG}} \in \{0, 1\}
\end{aligned}\right\}.
$$

□

8.4.4 Definition of Feasibility Set

Once the generation investment decisions are made and the worst realization of the uncertain peak demands and operation costs of generating units has occurred, the system is operated in such a way that the operation costs are minimized, i.e., the available resources are used to adapt to this situation. This is modeled using the following feasibility set:

$$
\Xi\left(p_g^{\mathrm{CG,max}}, \forall g \in \tilde{G}; \overline{P}_d^{\mathrm{D}}, \forall d \in D; c_g^{\mathrm{OG}}, \forall g \in G\right) = \left\{\Phi^{\mathrm{L3}} :\right.
$$

$$
\sum_{g \in G_b} p_{gt}^{\mathrm{G}} - \sum_{\ell | O(\ell)=b} p_{\ell t}^{\mathrm{L}} + \sum_{\ell | F(\ell)=b} p_{\ell t}^{\mathrm{L}} = \sum_{d \in D_b} \left(K_{dt}^{\mathrm{D}} \overline{P}_d^{\mathrm{D}} - p_{dt}^{\mathrm{UD}}\right), \quad \forall b \in B, \forall t \in T, \tag{8.5a}
$$

$$
p_{\ell t}^{\mathrm{L}} = \frac{1}{X_\ell}\left(\delta_{O(\ell)t} - \delta_{F(\ell)t}\right), \quad \forall \ell \in L, \forall t \in T, \tag{8.5b}
$$

$$
-\overline{P}_\ell^{\mathrm{L}} \leq p_{\ell t}^{\mathrm{L}} \leq \overline{P}_\ell^{\mathrm{L}}, \quad \forall \ell \in L, \forall t \in T, \tag{8.5c}
$$

$$
0 \leq p_{gt}^{\mathrm{G}} \leq K_{gt}^{\mathrm{G}} \overline{P}_g^{\mathrm{G}}, \quad \forall g \backslash g \in \tilde{G}, \forall t \in T, \tag{8.5d}
$$

$$
0 \leq p_{gt}^{\mathrm{G}} \leq K_{gt}^{\mathrm{G}} p_g^{\mathrm{CG,max}}, \quad \forall g \in \tilde{G}, \forall t \in T, \tag{8.5e}
$$

$$
0 \leq p_{dt}^{\mathrm{UD}} \leq K_{dt}^{\mathrm{D}} \overline{P}_d^{\mathrm{D}}, \quad \forall d \in D, \forall t \in T, \tag{8.5f}
$$

$$
\left.\delta_{bt} = 0, \quad b: \text{ref.}, \forall t \in T\right\}. \tag{8.5g}
$$

Constraints (8.5a) impose the generation-demand balance at each bus of the system. Constraints (8.5b) define the power flows through transmission lines, which are limited by the corresponding capacities by constraints (8.5c). Constraints (8.5d)–(8.5f) impose bounds on the power quantities of existing generating units, candidate generating units, and load shed of demands, respectively. Finally, Eqs. (8.5g) define the voltage angle at the reference node. Note that constraints (8.5) are imposed for all operating conditions indexed by t and used to model the short-term variability. Moreover, observe that the feasibility set Ξ is parameterized in terms of investment and worst-case uncertainty decisions.

Illustrative Example 8.6 Feasibility set

The data of Illustrative Example 8.3 are considered again. However, in this case the uncertain peak demand and generation costs are modeled as uncertain variables as described in Illustrative Example 8.5. This allows formulating the following feasibility set:

$$\Xi\left(p_3^{\mathrm{CG,max}}, p_4^{\mathrm{CG,max}}, \overline{p}_1^{\mathrm{D}}, c_1^{\mathrm{OG}}, c_3^{\mathrm{OG}}\right) = \Bigg\{$$

$$\begin{aligned}
& p_{11}^{\mathrm{G}} + p_{21}^{\mathrm{G}} + p_{31}^{\mathrm{G}} + p_{41}^{\mathrm{G}} = 0.9\overline{p}_1^{\mathrm{D}} - p_{11}^{\mathrm{UD}},\\
& p_{12}^{\mathrm{G}} + p_{22}^{\mathrm{G}} + p_{32}^{\mathrm{G}} + p_{42}^{\mathrm{G}} = 0.8\overline{p}_1^{\mathrm{D}} - p_{12}^{\mathrm{UD}},\\
& p_{13}^{\mathrm{G}} + p_{23}^{\mathrm{G}} + p_{33}^{\mathrm{G}} + p_{43}^{\mathrm{G}} = 0.6\overline{p}_1^{\mathrm{D}} - p_{13}^{\mathrm{UD}},\\
& p_{14}^{\mathrm{G}} + p_{24}^{\mathrm{G}} + p_{34}^{\mathrm{G}} + p_{44}^{\mathrm{G}} = 0.5\overline{p}_1^{\mathrm{D}} - p_{14}^{\mathrm{UD}},\\
& 0 \le p_{1t}^{\mathrm{G}} \le 200, \quad t = 1, 2, 3, 4,\\
& 0 \le p_{21}^{\mathrm{G}} \le 0.9 \cdot 100,\\
& 0 \le p_{22}^{\mathrm{G}} \le 0.6 \cdot 100,\\
& 0 \le p_{23}^{\mathrm{G}} \le 0.7 \cdot 100,\\
& 0 \le p_{24}^{\mathrm{G}} \le 0.2 \cdot 100,\\
& 0 \le p_{3t}^{\mathrm{G}} \le p_3^{\mathrm{CG,max}}, \quad t = 1, 2, 3, 4,\\
& 0 \le p_{41}^{\mathrm{G}} \le 0.9 \cdot p_4^{\mathrm{CG,max}},\\
& 0 \le p_{42}^{\mathrm{G}} \le 0.6 \cdot p_4^{\mathrm{CG,max}},\\
& 0 \le p_{43}^{\mathrm{G}} \le 0.7 \cdot p_4^{\mathrm{CG,max}},\\
& 0 \le p_{44}^{\mathrm{G}} \le 0.2 \cdot p_4^{\mathrm{CG,max}},\\
& 0 \le p_{11}^{\mathrm{UD}} \le 0.9\overline{p}_1^{\mathrm{D}},\\
& 0 \le p_{12}^{\mathrm{UD}} \le 0.8\overline{p}_1^{\mathrm{D}},\\
& 0 \le p_{13}^{\mathrm{UD}} \le 0.6\overline{p}_1^{\mathrm{D}},\\
& 0 \le p_{14}^{\mathrm{UD}} \le 0.5\overline{p}_1^{\mathrm{D}}\Bigg\}.
\end{aligned}$$

The above feasibility set is parameterized in terms of the investment decisions $p_3^{\mathrm{CG,max}}$ and $p_4^{\mathrm{CG,max}}$, and long-term uncertain variables $\overline{p}_1^{\mathrm{D}}, c_1^{\mathrm{OG}}, c_3^{\mathrm{OG}}$. □

Illustrative Example 8.7 Generation expansion planning considering an adaptive robust optimization approach: detailed formulation

The data of Illustrative Examples 8.3 and 8.5 are considered again. The detailed formulation of the three-level optimization problem is as follows:

$$\text{Minimize}_{\Phi^{\text{L1}}} \; 150000 p_3^{\text{CG,max}} + 200000 p_4^{\text{CG,max}}$$

$$+ \text{Maximize}_{\Phi^{\text{L2}}} \; \text{Minimize}_{\Phi^{\text{L3}}} 1000\left(c_1^{\text{OG}} p_{11}^{\text{G}} + c_3^{\text{OG}} p_{31}^{\text{G}} + 200 p_{11}^{\text{UD}}\right)$$

$$+ 3000\left(c_1^{\text{OG}} p_{12}^{\text{G}} + c_3^{\text{OG}} p_{32}^{\text{G}} + 200 p_{12}^{\text{UD}}\right)$$

$$+ 3000\left(c_1^{\text{OG}} p_{13}^{\text{G}} + c_3^{\text{OG}} p_{33}^{\text{G}} + 200 p_{13}^{\text{UD}}\right)$$

$$+ 1760\left(c_1^{\text{OG}} p_{14}^{\text{G}} + c_3^{\text{OG}} p_{34}^{\text{G}} + 200 p_{14}^{\text{UD}}\right)$$

subject to

$$150000 p_3^{\text{CG,max}} + 200000 p_4^{\text{CG,max}} \le 30000000,$$

$$0 \le p_3^{\text{CG,max}} \le 150,$$

$$0 \le p_4^{\text{CG,max}} \le 100,$$

subject to

$$\overline{p}_1^{\text{D}} = 400 + 120 u_1^{\text{D}},$$

$$c_1^{\text{OG}} = 40 + 8 u_1^{\text{OG}}.$$

$$c_3^{\text{OG}} = 25 + 5 u_3^{\text{OG}}.$$

$$u_1^{\text{D}} \le \Gamma^{\text{D}},$$

$$u_1^{\text{OG}} + u_3^{\text{OG}} \le \Gamma^{\text{G}},$$

$$u_1^{\text{D}}, u_1^{\text{OG}}, u_3^{\text{OG}} \in \{0, 1\},$$

subject to

$$p_{11}^{\text{G}} + p_{21}^{\text{G}} + p_{31}^{\text{G}} + p_{41}^{\text{G}} = 0.9\overline{p}_1^{\text{D}} - p_{11}^{\text{UD}},$$

$$p_{12}^{\text{G}} + p_{22}^{\text{G}} + p_{32}^{\text{G}} + p_{42}^{\text{G}} = 0.8\overline{p}_1^{\text{D}} - p_{12}^{\text{UD}},$$

$$p_{13}^{\text{G}} + p_{23}^{\text{G}} + p_{33}^{\text{G}} + p_{43}^{\text{G}} = 0.6\overline{p}_1^{\text{D}} - p_{13}^{\text{UD}},$$

$$p_{14}^{\text{G}} + p_{24}^{\text{G}} + p_{34}^{\text{G}} + p_{44}^{\text{G}} = 0.5\overline{p}_1^{\text{D}} - p_{14}^{\text{UD}},$$

$$0 \le p_{1t}^{\text{G}} \le 200, \quad t = 1, 2, 3, 4,$$

$$0 \le p_{21}^{\text{G}} \le 0.9 \cdot 100,$$

$$0 \le p_{22}^{\text{G}} \le 0.6 \cdot 100,$$

$$0 \le p_{23}^{\text{G}} \le 0.7 \cdot 100,$$

$$0 \le p_{24}^{\text{G}} \le 0.2 \cdot 100,$$

$$0 \le p_{3t}^{\text{G}} \le p_3^{\text{CG,max}}, \quad t = 1, 2, 3, 4,$$

$$0 \le p_{41}^{\text{G}} \le 0.9 \cdot p_4^{\text{CG,max}},$$

$$0 \le p_{42}^{\text{G}} \le 0.6 \cdot p_4^{\text{CG,max}},$$

$$0 \leq p^{\text{G}}_{43} \leq 0.7 \cdot p^{\text{CG,max}}_{4},$$
$$0 \leq p^{\text{G}}_{44} \leq 0.2 \cdot p^{\text{CG,max}}_{4},$$
$$0 \leq p^{\text{UD}}_{11} \leq 0.9\overline{p}^{\text{D}}_{1},$$
$$0 \leq p^{\text{UD}}_{12} \leq 0.8\overline{p}^{\text{D}}_{1},$$
$$0 \leq p^{\text{UD}}_{13} \leq 0.6\overline{p}^{\text{D}}_{1},$$
$$0 \leq p^{\text{UD}}_{14} \leq 0.5\overline{p}^{\text{D}}_{1},$$

where variables in sets $\Phi^{\text{L1}} = \left\{p^{\text{CG,max}}_{3}, p^{\text{CG,max}}_{4}\right\}$, $\Phi^{\text{L2}} = \left\{\overline{p}^{\text{D}}_{1}, c^{\text{OG}}_{1}, c^{\text{OG}}_{3}\right\}$, and $\Phi^{\text{L3}} = \left\{p^{\text{G}}_{gt}, g = 1, 2, 3, 4, t = 1, 2, 3, 4, p^{\text{UD}}_{1t}, t = 1, 2, 3, 4\right\}$, are the optimization variables of the first-, second-, and third-level optimization problems, respectively. □

8.4.5 Solution Procedure

The three-level optimization problem (8.2) is solved using the Column-and-Constraint Generation Algorithm (CCGA) that decomposes problem (8.2) into a master problem and a subproblem that are iteratively solved until convergence [1, 15, 18]. The formulation of these problems, as well as the proposed solution procedure are described below.

8.4.5.1 Master problem

The formulation of the master problem is provided below:

Minimize$_{\Phi^{\text{M}}}$

$$\sum_{g \in \tilde{G}} C^{\text{IG}} p^{\text{CG,max}}_{g} + \theta \tag{8.6a}$$

subject to:

$$\text{Constraints (8.2}b\text{)}-\text{(8.2}c\text{)}, \tag{8.6b}$$

$$\theta \geq \sum_{t \in T} N^{\text{H}}_{t} \left(\sum_{g \in G} c^{\text{OG}(\nu')}_{g} p^{\text{G}}_{gt\nu'} + \sum_{d \in D} C^{\text{UD}}_{d} p^{\text{UD}}_{dt\nu'} \right), \quad \forall \nu' \leq \nu, \tag{8.6c}$$

$$\sum_{g \in G_b} p^{\mathrm{G}}_{gtv'} - \sum_{\ell | O(\ell)=b} p^{\mathrm{L}}_{\ell tv'} + \sum_{\ell | F(\ell)=b} p^{\mathrm{L}}_{\ell tv'}$$

$$= \sum_{d \in D_b} \left(K^{\mathrm{D}}_{dt} \overline{p}^{\mathrm{D}(v')}_{d} - p^{\mathrm{UD}}_{dtv'} \right), \quad \forall b \in B, \forall t \in T, \forall v' \leq v, \tag{8.6d}$$

$$p^{\mathrm{L}}_{\ell tv'} = \frac{1}{X_\ell} \left(\delta_{O(\ell)tv'} - \delta_{F(\ell)tv'} \right), \quad \forall \ell \in L, \forall t \in T, \forall v' \leq v, \tag{8.6e}$$

$$- \overline{P}^{\mathrm{L}}_{\ell} \leq p^{\mathrm{L}}_{\ell tv'} \leq \overline{P}^{\mathrm{L}}_{\ell}, \quad \forall \ell \in L, \forall t \in T, \forall v' \leq v, \tag{8.6f}$$

$$0 \leq p^{\mathrm{G}}_{gtv'} \leq K^{\mathrm{G}}_{gt} \overline{P}^{\mathrm{G}}_{g}, \quad \forall g \backslash g \in \tilde{G}, \forall t \in T, \forall v' \leq v, \tag{8.6g}$$

$$0 \leq p^{\mathrm{G}}_{gtv'} \leq K^{\mathrm{G}}_{gt} p^{\mathrm{CG,max}}_{g}, \quad \forall g \in \tilde{G}, \forall t \in T, \forall v' \leq v, \tag{8.6h}$$

$$0 \leq p^{\mathrm{UD}}_{dtv'} \leq K^{\mathrm{D}}_{dt} \overline{p}^{\mathrm{D}(v')}_{d}, \quad \forall d \in D, \forall t \in T, \forall v' \leq v, \tag{8.6i}$$

$$\delta_{btv'} = 0, \quad b : \text{ref.}, \forall t \in T, \forall v' \leq v, \tag{8.6j}$$

where variables in sets $\Phi^{\mathrm{M}} = \Big\{ \Phi^{\mathrm{L1}}; \theta; p^{\mathrm{G}}_{gtv'}, \forall g \in G, \forall t \in T, \forall v' \leq v; p^{\mathrm{UD}}_{dtv'}, \forall d \in D, \forall t \in T, \forall v' \leq v; p^{\mathrm{L}}_{\ell tv'}, \forall \ell \in L, \forall t \in T, \forall v' \leq v; \delta_{btv'}, \forall b \in B, \forall t \in T, \forall v' \leq v \Big\}$, are the optimization variables of problem (8.6) and v is the iteration counter.

Master problem (8.6) is a relaxed version of the original problem (8.2) that is progressively reconstructed at each iteration using auxiliary variable θ. Therefore, the size of problem (8.6) increases with the iteration counter.

Note that in master problem (8.6), uncertain variables $\overline{p}^{\mathrm{D}}_{d}$ and c^{OG}_{g} are considered as input data. These variables are obtained from the solution of the subproblem described in the following subsection.

8.4.5.2 Subproblem

The subproblem is provided below:

$\text{Maximize}_{\Phi^{\mathrm{L2}}}$ $\text{Minimize}_{\Phi^{\mathrm{L3}}}$

$$\sum_{t \in T} N^{\mathrm{H}}_{t} \left(\sum_{g \in G} c^{\mathrm{OG}}_{g} p^{\mathrm{G}}_{gt} + \sum_{d \in D} C^{\mathrm{LS}}_{d} p^{\mathrm{LS}}_{dt} \right) \tag{8.7a}$$

subject to:

$$\text{Constraints (8.4)}, \tag{8.7b}$$

subject to:

$$\sum_{g \in G_b} p^{\mathrm{G}}_{gt} - \sum_{\ell | O(\ell)=b} p^{\mathrm{L}}_{\ell t} + \sum_{\ell | F(\ell)=b} p^{\mathrm{L}}_{\ell t}$$
$$= \sum_{d \in D_b} \left(K^{\mathrm{D}}_{dt} \overline{P}^{\mathrm{D}}_{d} - p^{\mathrm{UD}}_{dt} \right) : \lambda_{bt}, \quad \forall b \in B, \forall t \in T, \tag{8.7c}$$

$$p^{\mathrm{L}}_{\ell t} = \frac{1}{X_\ell} \left(\delta_{O(\ell)t} - \delta_{F(\ell)t} \right) : \varphi^{\mathrm{L}}_{\ell t}, \quad \forall \ell \in L, \forall t \in T, \tag{8.7d}$$

$$-\overline{P}^{\mathrm{L}}_{\ell} \le p^{\mathrm{L}}_{\ell t} \le \overline{P}^{\mathrm{L}}_{\ell} : \underline{\mu}^{\mathrm{L}}_{\ell t}, \overline{\mu}^{\mathrm{L}}_{\ell t}, \quad \forall \ell \in L, \forall t \in T, \tag{8.7e}$$

$$0 \le p^{\mathrm{G}}_{gt} \le K^{\mathrm{G}}_{gt} \overline{P}^{\mathrm{G}}_{g} : \overline{\mu}^{\mathrm{G}}_{gt}, \quad \forall g \backslash g \in \tilde{G}, \forall t \in T, \tag{8.7f}$$

$$0 \le p^{\mathrm{G}}_{gt} \le K^{\mathrm{G}}_{gt} p^{\mathrm{CG,max}(\nu)}_{g} : \overline{\mu}^{\mathrm{G+}}_{gt}, \quad \forall g \in \tilde{G}, \forall t \in T, \tag{8.7g}$$

$$0 \le p^{\mathrm{UD}}_{dt} \le K^{\mathrm{D}}_{dt} \overline{P}^{\mathrm{D}}_{d} : \overline{\mu}^{\mathrm{D}}_{dt}, \quad \forall d \in D, \forall t \in T, \tag{8.7h}$$

$$\delta_{bt} = 0 : \xi_{bt}, \quad b : \text{ref.}, \forall t \in T, \tag{8.7i}$$

where dual variables associated with the constraints of the lower-level problem are provided following a colon.

Subproblem (8.7) is a bi-level problem. At each iteration ν, the upper-level problem determines the worst-case realization of uncertain variables for a given investment plan provided by the master problem, while the lower-level problem identifies the operation of the system.

Note that the lower-level problem in (8.7) is a linear programming problem on its decision variables. Therefore, it can be replaced by its dual formulation. On the other hand, the strong duality theorem states that, at the optimum, the values of the primal and dual objective functions are the same [20]. Since both the upper- and lower-level objective functions in (8.7) are the same, it is possible to replace the primal objective function by the dual objective function, resulting in the following single-level optimization problem:

Maximize$_{\Phi^{\mathrm{S}}}$

$$\sum_{t \in T} \Bigg[\sum_{b \in B} \lambda_{bt} \sum_{d \in D_b} K^{\mathrm{D}}_{dt} \overline{P}^{\mathrm{D}}_{d} - \sum_{\ell \in L} \overline{P}^{\mathrm{L}}_{\ell} \left(\underline{\mu}^{\mathrm{L}}_{\ell t} + \overline{\mu}^{\mathrm{L}}_{\ell t} \right) - \sum_{g \backslash g \in \tilde{G}} K^{\mathrm{G}}_{gt} \overline{P}^{\mathrm{G}}_{g} \overline{\mu}^{\mathrm{G}}_{gt}$$
$$- \sum_{g \in \tilde{G}} K^{\mathrm{G}}_{gt} p^{\mathrm{CG,max}(\nu)}_{g} \overline{\mu}^{\mathrm{G+}}_{gt} - \sum_{d \in D} K^{\mathrm{D}}_{dt} \overline{P}^{\mathrm{D}}_{d} \overline{\mu}^{\mathrm{D}}_{dt} \Bigg] \tag{8.8a}$$

subject to:

$$\text{Constraints (8.4)} \tag{8.8b}$$

$$\lambda_{b(g)t} - \overline{\mu}^{\mathrm{G}}_{gt} \le N^{\mathrm{H}}_{t} c^{\mathrm{OG}}_{g}, \quad \forall g \backslash g \in \tilde{G}, \forall t \in T, \tag{8.8c}$$

$$\lambda_{b(g)t} - \overline{\mu}^{\mathrm{G+}}_{gt} \le N^{\mathrm{H}}_{t} c^{\mathrm{OG}}_{g}, \quad \forall g \in \tilde{G}, \forall t \in T, \tag{8.8d}$$

$$\lambda_{n(d)t} - \overline{\mu}^{\mathrm{D}}_{dt} \le N^{\mathrm{H}}_{t} C^{\mathrm{UD}}_{d}, \quad \forall d \in D, \forall t \in T, \tag{8.8e}$$

$$-\lambda_{O(\ell)t} + \lambda_{F(\ell)t} + \varphi^{\mathrm{L}}_{\ell t} - \overline{\mu}^{\mathrm{L}}_{\ell t} + \underline{\mu}^{\mathrm{L}}_{\ell t} = 0, \quad \forall \ell \in L, \forall t \in T, \tag{8.8f}$$

$$-\sum_{\ell|O(\ell)=b} \frac{1}{X_\ell}\varphi^{\mathrm{L}}_{\ell t} + \sum_{\ell|F(\ell)=b} \frac{1}{X_\ell}\varphi^{\mathrm{L}}_{\ell t} = 0, \quad \forall b \backslash b : \text{ref.}, \forall t \in T, \tag{8.8g}$$

$$-\sum_{\ell|O(\ell)=b} \frac{1}{X_\ell}\varphi^{\mathrm{L}}_{\ell t} + \sum_{\ell|F(\ell)=b} \frac{1}{X_\ell}\varphi^{\mathrm{L}}_{\ell t} + \xi_{bt} = 0, \quad b : \text{ref.}, \forall t \in T, \tag{8.8h}$$

$$\underline{\mu}^{\mathrm{L}}_{\ell t}, \overline{\mu}^{\mathrm{L}}_{\ell t} \geq 0, \quad \forall \ell \in L, \forall t \in T, \tag{8.8i}$$

$$\overline{\mu}^{\mathrm{G}}_{gt} \geq 0, \quad \forall g \backslash g \in \tilde{G}, \forall t \in T, \tag{8.8j}$$

$$\overline{\mu}^{\mathrm{G+}}_{gt} \geq 0, \quad \forall g \in \tilde{G}, \forall t \in T, \tag{8.8k}$$

$$\overline{\mu}^{\mathrm{D}}_{dt} \geq 0, \quad \forall d \in D, \forall t \in T, \tag{8.8l}$$

where variables in set $\Phi^{\mathrm{S}} = \Big\{\Phi^{\mathrm{L2}}; \underline{\mu}^{\mathrm{L}}_{\ell t}, \overline{\mu}^{\mathrm{L}}_{\ell t}, \forall \ell \in L, \forall t \in T; \overline{\mu}^{\mathrm{G}}_{gt}, \forall g \backslash g \in \tilde{G}, \forall t \in T;$ $\overline{\mu}^{\mathrm{G+}}_{gt}, \forall g \in \tilde{G}, \forall t \in T; \overline{\mu}^{\mathrm{D}}_{dt}, \forall d \in D, \forall t \in T; \xi_{bt}, b : \text{ref.}, \forall t \in T\Big\}$, are the optimization variables of problem (8.8).

Objective function (8.8a) is the dual objective function of lower-level problem in (8.7). Constraints (8.8b) define the uncertainty sets and constraints (8.8c)–(8.8l) are the dual constraints of lower-level problem in (8.7).

Problem (8.8) is a mixed-integer nonlinear programming (MINLP) problem since it includes products of continuous variables in objective function (8.8a). Particularly, these nonlinear terms are $\lambda_{bt} K^{\mathrm{D}}_{dt} \overline{P}^{\mathrm{D}}_{d}$ and $\overline{P}^{\mathrm{D}}_{d} \overline{\mu}_{dt}$. However, these nonlinear can be replaced by exact equivalent mixed-integer linear expressions as explained below.

Firstly, using equations (8.4a):

$$\lambda_{bt} \sum_{d \in D_b} K^{\mathrm{D}}_{dt} \overline{P}^{\mathrm{D}}_{d} = \lambda_{bt} \sum_{d \in D_b} K^{\mathrm{D}}_{dt} \left(\tilde{P}^{\mathrm{D}}_{d} + u^{\mathrm{D}}_{d} \hat{P}^{\mathrm{D}}_{d}\right), \quad \forall b \in B, \forall t \in T. \tag{8.9}$$

Note that terms $\lambda_{bt} K^{\mathrm{D}}_{dt} \tilde{P}^{\mathrm{D}}_{d}$ are linear, while terms $\lambda_{bt} K^{\mathrm{D}}_{dt} u^{\mathrm{D}}_{d} \hat{P}^{\mathrm{D}}_{d}$ comprise the product of a continuous variable, a binary variable, and a positive constant, which can be replaced by the following exact mixed-integer linear expressions [10]:

$$\lambda_{bt} \sum_{d \in D_b} K^{\mathrm{D}}_{dt} \overline{P}^{\mathrm{D}}_{d} = \lambda_{bt} \sum_{d \in D_b} K^{\mathrm{D}}_{dt} \tilde{P}^{\mathrm{D}}_{d} + \sum_{d \in D_b} z^{\mathrm{B}}_{dbt}, \quad \forall b \in B, \forall t \in T, \tag{8.10a}$$

$$-Mu^{\mathrm{D}}_{d} \leq z^{\mathrm{B}}_{dbt} \leq Mu^{\mathrm{D}}_{d}, \quad \forall d \in D, \forall b \in B, \forall t \in T, \tag{8.10b}$$

$$-M\left(1-u^{\mathrm{D}}_{d}\right) \leq \lambda_{bt} K^{\mathrm{D}}_{dt} \hat{P}^{\mathrm{D}}_{d} - z^{\mathrm{B}}_{dbt} \leq M\left(1-u^{\mathrm{D}}_{d}\right), \quad \forall d \in D, \forall b \in B, \forall t \in T, \tag{8.10c}$$

where M is a large enough positive constant.

Equations (8.10) work as follows:

1. If $u^{\mathrm{D}}_{d} = 0$, then constraints (8.10b) impose that $z^{\mathrm{B}}_{dbt} = 0$, while constraints (8.10c) do not impose bounds provided that M is a large enough positive constant.

2. If $u^{\mathrm{D}}_{d} = 1$, then constraints (8.10c) impose that $z^{\mathrm{B}}_{dbt} = \lambda_{bt} K^{\mathrm{D}}_{dt} \hat{P}^{\mathrm{D}}_{d}$, while constraints (8.10b) do not impose bounds provided that M is a large enough positive constant.

Similarly, nonlinear terms $\overline{p}^{\mathrm{D}}_{d} \overline{\mu}_{dt}$ can be rewritten as:

$$\overline{\mu}_{dt} \overline{P}^{\mathrm{D}}_{d} = \overline{\mu}_{dt} \left(\tilde{P}^{\mathrm{D}}_{d} + u^{\mathrm{D}}_{d} \hat{P}^{\mathrm{D}}_{d} \right) = \overline{\mu}_{dt} \tilde{P}^{\mathrm{D}}_{d} + z^{\mathrm{D}}_{dt}, \quad \forall d \in D, \forall t \in T, \tag{8.11a}$$

$$0 \le z^{\mathrm{D}}_{dt} \le M u^{\mathrm{D}}_{d}, \quad \forall d \in D, \forall t \in T, \tag{8.11b}$$

$$0 \le \overline{\mu}^{\mathrm{D}}_{dt} \hat{P}^{\mathrm{D}}_{d} - z^{\mathrm{D}}_{dt} \le M \left(1 - u^{\mathrm{D}}_{d}\right), \quad \forall d \in D, \forall t \in T. \tag{8.11c}$$

Thus, subproblem (8.8) is finally formulated as the following mixed-integer linear programming (MILP) problem:

$$\text{Maximize}_{\Phi^{\mathrm{SL}}}$$

$$\sum_{t \in T} \Bigg[\sum_{b \in B} \Bigg(\lambda_{bt} \sum_{d \in D_b} K^{\mathrm{D}}_{dt} \tilde{P}^{\mathrm{D}}_{d} + \sum_{d \in D_b} z^{\mathrm{B}}_{dbt} \Bigg) - \sum_{\ell \in L} \overline{P}^{\mathrm{L}}_{\ell} \left(\underline{\mu}^{\mathrm{L}}_{\ell t} + \overline{\mu}^{\mathrm{L}}_{\ell t} \right) - \sum_{g \backslash g \in \tilde{G}} K^{\mathrm{G}}_{gt} \overline{P}^{\mathrm{G}}_{g} \overline{\mu}^{\mathrm{G}}_{gt}$$

$$- \sum_{g \in \tilde{G}} K^{\mathrm{G}}_{gt} p^{\mathrm{CG,max}(\nu)}_{g} \overline{\mu}^{\mathrm{G+}}_{gt} - \sum_{d \in D} K^{\mathrm{D}}_{dt} \left(\overline{\mu}_{dt} \tilde{P}^{\mathrm{D}}_{d} + z^{\mathrm{D}}_{dt} \right) \Bigg] \tag{8.12a}$$

subject to:

$$\text{Constraints (8.8b)–(8.81)}, \tag{8.12b}$$

$$\text{Constraints (8.10b)–(8.10c)}, \tag{8.12c}$$

$$\text{Constraints (8.11b)–(8.11c)}, \tag{8.12d}$$

where variables in set $\Phi^{\mathrm{SL}} = \left\{ \Phi^{\mathrm{S}};\ z^{\mathrm{B}}_{dbt}, \forall d \in D, \forall b \in B, \forall t \in T;\ z^{\mathrm{D}}_{dt}, \forall d \in D, \forall t \in T \right\}$, are the optimization variables of problem (8.12).

8.4.5.3 Algorithm

The solution of the master problem provides the generation investment decisions. These decisions are considered as input data of the subproblem, which provides the worst-case realizations of the long-term uncertain variables, namely, future peak demands and generation costs. Then, these worst realizations are considered as input data of the master problem. Finally, both the master problem and the subproblem are iteratively solved until convergence following the CCGA below:

1. Initialization:
   - Initialize lower and upper bounds: $LB = -\infty$ and $UB = \infty$, respectively.
   - Initialize the iteration counter: $\nu = 0$.
   - Set the convergence tolerance ε.
2. Master problem:
   - Set $c_g^{\mathrm{OG}(\nu)} = c_g^{\mathrm{OG}*}$ and $\overline{p}_d^{\mathrm{D}(\nu)} = \overline{p}_d^{\mathrm{D}*}$, where $c_g^{\mathrm{OG}*}$ and $\overline{p}_d^{\mathrm{D}*}$ are the optimal values of variables c_g^{OG} and $\overline{p}_d^{\mathrm{D}}$, respectively, obtained from the subproblem at iteration ν.
   - Solve master problem (8.6).
   - Update of lower-bound:

$$\mathrm{LB} = \sum_{g \in \tilde{G}} C_g^{\mathrm{IG}} p_g^{\mathrm{CG,max}*} + \theta^*, \tag{8.13}$$

   where $p_g^{\mathrm{CG,max}*}$ and θ^* are the optimal values of of variables $p_g^{\mathrm{CG,max}}$ and θ, respectively, obtained from the master problem.
3. Subproblem:
   - Set $p_g^{\mathrm{CG,max}(\nu)} = p_g^{\mathrm{CG,max}*}$.
   - Solve problem (8.12).
   - Update upper-bound:

$$\mathrm{UB} = \sum_{g \in \tilde{G}} C_g^{\mathrm{IG}} p_g^{\mathrm{CG,max}*} + z^{\mathrm{S}*}, \tag{8.14}$$

   where $z^{\mathrm{S}*}$ is the optimal value of the objective function of the subproblem (8.12a).
4. Convergence checking:
   - If $\mathrm{UB} - \mathrm{LB} < \varepsilon$, the algorithm stops.
   - If not, update the iteration counter $\nu \leftarrow \nu + 1$ and continue with step 2.

Figure 8.3 depicts the flowchart of the above algorithm.

Illustrative Example 8.8 Adaptive robust optimization problem

The GEP problem formulated in Illustrative Example 8.7 using an adaptive robust optimization approach is solved using the CCGA described in Sect. 8.4.5.3. Uncertainty budgets are fixed to $\Gamma^{\mathrm{D}} = 1$ and $\Gamma^{\mathrm{G}} = 1$. Constant M is set to 10^8. The step-by-step solution is provided below.

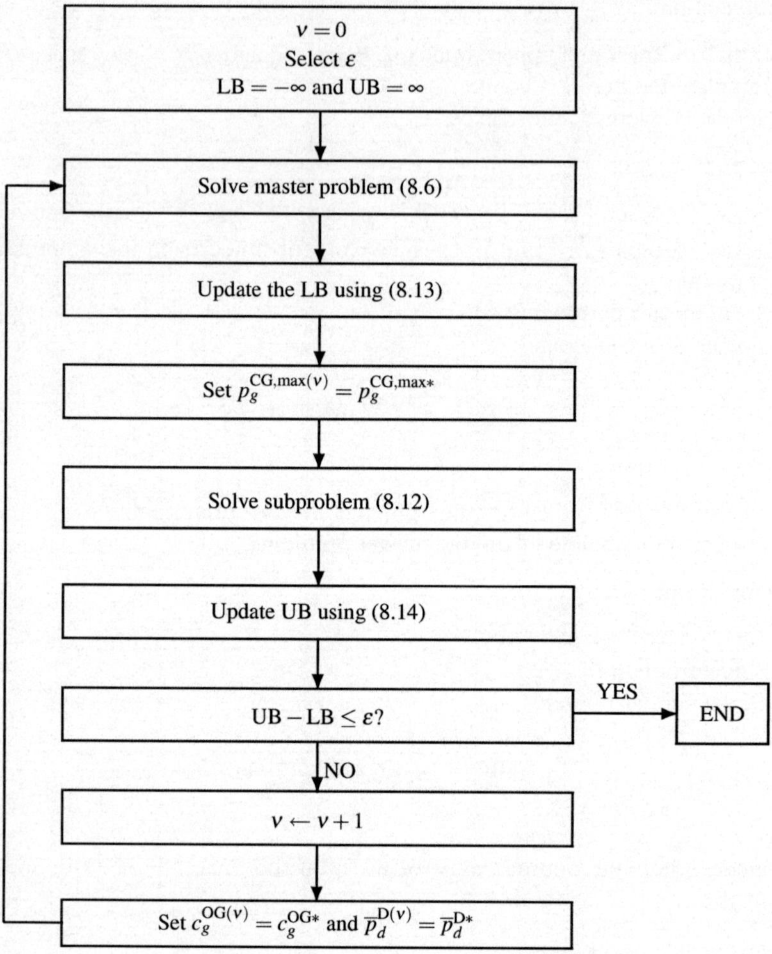

Fig. 8.3 Flowchart for the column-and-constraint generation algorithm

Step 1. Initialization:

- The lower and upper bounds are initialized to $LB = -\infty$ and $UB = \infty$, respectively.
- The iteration counter is initialized $\nu = 0$.
- The convergence tolerance is set to $\varepsilon = 0$.

Step 2. Master problem:

- The following problem is solved:
  Minimize$_{\Phi^{\mathrm{M}}}$

$$150000 p_3^{\mathrm{CG,max}} + 200000 p_4^{\mathrm{CG,max}} + \theta$$

subject to:

$$150000 p_3^{\mathrm{CG,max}} + 200000 p_4^{\mathrm{CG,max}} \le 30000000,$$
$$0 \le p_3^{\mathrm{CG,max}} \le 150,$$
$$0 \le p_4^{\mathrm{CG,max}} \le 100,$$

where $\Phi^{\mathrm{M}} = \left\{ p_3^{\mathrm{CG,max}}, p_4^{\mathrm{CG,max}}, \theta \right\}$.

The optimal solution of this problem is trivial, namely, $p_3^{\mathrm{CG,max}} = p_4^{\mathrm{CG,max}} = 0$ and $\theta = -\infty$.

- The lower bound is updated:

$$\mathrm{LB} = 150000 \cdot 0 + 200000 \cdot 0 + (-\infty) = -\infty.$$

Step 3. Subproblem:

- The investment decisions are set to $p_3^{\mathrm{CG,max}(0)} = p_4^{\mathrm{CG,max}(0)} = 0$.
- The following problem is solved:
  Maximize$_{\Phi^{\mathrm{SL}}}$

$$\begin{aligned}
&400 \cdot 0.9\lambda_1 + z_{11}^{\mathrm{B}} + 400 \cdot 0.8\lambda_2 + z_{12}^{\mathrm{B}} + 400 \cdot 0.6\lambda_3 + z_{13}^{\mathrm{B}} + 400 \cdot 0.5\lambda_4 + z_{14}^{\mathrm{B}} \\
&+ 200\overline{\mu}_{11}^{\mathrm{G}} + 200\overline{\mu}_{12}^{\mathrm{G}} + 200\overline{\mu}_{13}^{\mathrm{G}} + 200\overline{\mu}_{14}^{\mathrm{G}} + 0.9 \cdot 100\overline{\mu}_{21}^{\mathrm{G}} \\
&+ 0.6 \cdot 100\overline{\mu}_{22}^{\mathrm{G}} + 0.7 \cdot 100\overline{\mu}_{23}^{\mathrm{G}} + 0.2 \cdot 100\overline{\mu}_{24}^{\mathrm{G}} + 0 \cdot \overline{\mu}_{31}^{\mathrm{G+}} + 0 \cdot \overline{\mu}_{32}^{\mathrm{G+}} \\
&+ 0 \cdot \overline{\mu}_{33}^{\mathrm{G+}} + 0 \cdot \overline{\mu}_{34}^{\mathrm{G+}} + 0.9 \cdot 0\overline{\mu}_{41}^{\mathrm{G}} + 0.6 \cdot 0\overline{\mu}_{42}^{\mathrm{G+}} + 0.7 \cdot 0\overline{\mu}_{43}^{\mathrm{G+}} \\
&+ 0 \cdot 100\overline{\mu}_{44}^{\mathrm{G+}} + 0.9\left(400\overline{\mu}_{11}^{\mathrm{D}} + z_{11}^{\mathrm{D}}\right) + 0.8\left(400\overline{\mu}_{12}^{\mathrm{D}} + z_{12}^{\mathrm{D}}\right) \\
&+ 0.6\left(400\overline{\mu}_{13}^{\mathrm{D}} + z_{13}^{\mathrm{D}}\right) + 0.5\left(400\overline{\mu}_{14}^{\mathrm{D}} + z_{14}^{\mathrm{D}}\right)
\end{aligned}$$

subject to:

$$\overline{p}_1^{\mathrm{D}} = 400 + 120u_1^{\mathrm{D}},$$
$$c_1^{\mathrm{OG}} = 40 + 8u_1^{\mathrm{OG}},$$
$$c_3^{\mathrm{OG}} = 25 + 5u_3^{\mathrm{OG}},$$
$$u_1^{\mathrm{D}} \le 1,$$
$$u_1^{\mathrm{OG}} + u_3^{\mathrm{OG}} \le 1,$$
$$u_1^{\mathrm{D}}, u_1^{\mathrm{OG}}, u_3^{\mathrm{OG}} \in \{0, 1\},$$
$$\lambda_1 - \overline{\mu}_{11}^{\mathrm{G}} \le 1000c_1^{\mathrm{OG}},$$
$$\lambda_2 - \overline{\mu}_{12}^{\mathrm{G}} \le 3000c_1^{\mathrm{OG}},$$
$$\lambda_3 - \overline{\mu}_{13}^{\mathrm{G}} \le 3000c_1^{\mathrm{OG}},$$
$$\lambda_4 - \overline{\mu}_{14}^{\mathrm{G}} \le 1760c_1^{\mathrm{OG}},$$

$$
\begin{aligned}
&\lambda_t - \overline{\mu}_{2t}^{\mathrm{G}} \leq 0, \quad t = 1, 2, 3, 4, \\
&\lambda_1 - \overline{\mu}_{31}^{\mathrm{G}+} \leq 1000 c_3^{\mathrm{OG}}, \\
&\lambda_2 - \overline{\mu}_{32}^{\mathrm{G}+} \leq 3000 c_3^{\mathrm{OG}}, \\
&\lambda_3 - \overline{\mu}_{33}^{\mathrm{G}+} \leq 3000 c_3^{\mathrm{OG}}, \\
&\lambda_4 - \overline{\mu}_{34}^{\mathrm{G}+} \leq 1760 c_3^{\mathrm{OG}}, \\
&\lambda_t - \overline{\mu}_{4t}^{\mathrm{G}+} \leq 0, \quad t = 1, 2, 3, 4, \\
&\lambda_1 - \overline{\mu}_{11}^{\mathrm{D}} \leq 1000 \cdot 200, \\
&\lambda_2 - \overline{\mu}_{12}^{\mathrm{D}} \leq 3000 \cdot 200, \\
&\lambda_3 - \overline{\mu}_{13}^{\mathrm{D}} \leq 3000 \cdot 200, \\
&\lambda_4 - \overline{\mu}_{14}^{\mathrm{D}} \leq 1760 \cdot 200, \\
&\overline{\mu}_{gt}^{\mathrm{G}} \geq 0, \quad g = 1, 2; t = 1, 2, 3, 4, \\
&\overline{\mu}_{gt}^{\mathrm{G}+} \geq 0, \quad g = 3, 4; t = 1, 2, 3, 4, \\
&\overline{\mu}_{1t}^{\mathrm{D}} \geq 0, \quad t = 1, 2, 3, 4, \\
&-1 \cdot 10^8 u_1^{\mathrm{D}} \leq z_t^{\mathrm{B}} \leq 1000 u_1^{\mathrm{D}}, \quad t = 1, 2, 3, 4, \\
&-1 \cdot 10^8 \left(1 - u_1^{\mathrm{D}}\right) \leq 0.9 \cdot 120 \lambda_1 - z_{11}^{\mathrm{B}} \leq 1 \cdot 10^8 \left(1 - u_1^{\mathrm{D}}\right), \\
&-1 \cdot 10^8 \left(1 - u_1^{\mathrm{D}}\right) \leq 0.8 \cdot 120 \lambda_2 - z_{12}^{\mathrm{B}} \leq 1 \cdot 10^8 \left(1 - u_1^{\mathrm{D}}\right), \\
&-1 \cdot 10^8 \left(1 - u_1^{\mathrm{D}}\right) \leq 0.6 \cdot 120 \lambda_3 - z_{13}^{\mathrm{B}} \leq 1 \cdot 10^8 \left(1 - u_1^{\mathrm{D}}\right), \\
&-1 \cdot 10^8 \left(1 - u_1^{\mathrm{D}}\right) \leq 0.5 \cdot 120 \lambda_4 - z_{14}^{\mathrm{B}} \leq 1 \cdot 10^8 \left(1 - u_1^{\mathrm{D}}\right), \\
&0 \leq z_{1t}^{\mathrm{D}} \leq 1 \cdot 10^8 u_1^{\mathrm{D}}, \quad t = 1, 2, 3, 4, \\
&0 \leq 120 \overline{\mu}_{1t}^{\mathrm{D}} - z_{1t}^{\mathrm{D}} \leq 1 \cdot 10^8 \left(1 - u_1^{\mathrm{D}}\right), \quad t = 1, 2, 3, 4. \\
&\overline{\mu}_{1t}^{\mathrm{G}} \geq 0, \quad g = 1, 2, t = 1, 2, 3, 4, \\
&\overline{\mu}_{gt}^{\mathrm{G}+} \geq 0, \quad g = 3, 4, t = 1, 2, 3, 4, \\
&\overline{\mu}_{1t}^{\mathrm{D}} \geq 0, \quad t = 1, 2, 3, 4,
\end{aligned}
$$

where $\Phi^{\mathrm{SL}} = \left\{\overline{p}_1^{\mathrm{D}}, c_1^{\mathrm{OG}}, c_3^{\mathrm{OG}}, u_1^{\mathrm{D}}, u_1^{\mathrm{OG}}, u_3^{\mathrm{OG}}, \overline{\mu}_{gt}^{\mathrm{G}}, g = 1, 2, t = 1, 2, 3, 4, \overline{\mu}_{gt}^{\mathrm{G}+}, g = 3, 4, t = 1, 2, 3, 4, \overline{\mu}_{1t}^{\mathrm{D}}, t = 1, 2, 3, 4; z_{1t}^{\mathrm{B}}, t = 1, 2, 3, 4, z_{1t}^{\mathrm{D}}, t = 1, 2, 3, 4\right\}$. The optimal solution of the above problem provides the worst-case uncertainty realizations for the given investment plan identified by the master problem:

$$
\begin{aligned}
\overline{p}_1^{\mathrm{D}*} &= 520, \\
c_1^{\mathrm{OG}*} &= 48, \\
c_3^{\mathrm{OG}*} &= 25.
\end{aligned}
$$

Note that the uncertain peak demand is at its upper bound, while only one of the generation cost is at the upper bound since $\Gamma^{\mathrm{G}} = 1$. Particularly, the worst case corresponds to the case in which the generation cost of the first generating unit is at its upper bound since the subproblem was solved considering that the third generating unit is not built so that the generation cost of this unit has no impact on the solution in this case. On the other hand, the optimal value of the objective function is $2.53 \cdot 10^8$.

- Update upper-bound:

$$\mathrm{UB} = 150000 \cdot 0 + 200000 \cdot 0 + 2.53 \cdot 10^8 = 2.53 \cdot 10^8.$$

Step 4. Convergence checking:

- $\mathrm{UB} - \mathrm{LB} = 2.53 \cdot 10^8 - (-\infty) = \infty$.
- As $\mathrm{UB} - \mathrm{LB} \geq 0$, the iteration counter is updated $\nu \leftarrow 1$ and the algorithm continues with step 2.

Step 2. Master problem:

- The uncertain variables are set to $\overline{p}_1^{\mathrm{D}(1)} = 520$, $c_1^{\mathrm{OG}(1)} = 48$, and $c_3^{\mathrm{OG}(1)} = 25$.
- The following problem is solved:

Minimize$_{\Phi^{\mathrm{M}}}$

$$150000 p_3^{\mathrm{CG,max}} + 200000 p_4^{\mathrm{CG,max}} + \theta$$

subject to:

$$150000 p_3^{\mathrm{CG,max}} + 200000 p_4^{\mathrm{CG,max}} \leq 30000000,$$

$$0 \leq p_3^{\mathrm{CG,max}} \leq 150,$$

$$0 \leq p_4^{\mathrm{CG,max}} \leq 100,$$

$$\theta \geq 1000\left(48 p_{111}^{\mathrm{G}} + 25 p_{211}^{\mathrm{G}} + 200 p_{111}^{\mathrm{UD}}\right) + 3000\left(48 p_{121}^{\mathrm{G}} + 25 p_{221}^{\mathrm{G}} + 200 p_{121}^{\mathrm{UD}}\right)$$
$$+ 3000\left(48 p_{131}^{\mathrm{G}} + 25 p_{231}^{\mathrm{G}} + 200 p_{131}^{\mathrm{UD}}\right) + 1760\left(48 p_{141}^{\mathrm{G}} + 25 p_{241}^{\mathrm{G}} + 200 p_{141}^{\mathrm{UD}}\right),$$

$$p_{111}^{\mathrm{G}} + p_{211}^{\mathrm{G}} + p_{311}^{\mathrm{G}} + p_{411}^{\mathrm{G}} = 0.9 \cdot 520 - p_{111}^{\mathrm{UD}},$$

$$p_{121}^{\mathrm{G}} + p_{221}^{\mathrm{G}} + p_{321}^{\mathrm{G}} + p_{421}^{\mathrm{G}} = 0.8 \cdot 520 - p_{121}^{\mathrm{UD}},$$

$$p_{131}^{\mathrm{G}} + p_{231}^{\mathrm{G}} + p_{331}^{\mathrm{G}} + p_{431}^{\mathrm{G}} = 0.6 \cdot 520 - p_{131}^{\mathrm{UD}},$$

$$p_{141}^{\mathrm{G}} + p_{241}^{\mathrm{G}} + p_{341}^{\mathrm{G}} + p_{441}^{\mathrm{G}} = 0.5 \cdot 520 - p_{141}^{\mathrm{UD}},$$

$$0 \leq p_{1t1}^{\mathrm{G}} \leq 200, \quad t = 1, 2, 3, 4,$$

$$0 \leq p_{211}^{\mathrm{G}} \leq 0.9 \cdot 100,$$

$$0 \leq p_{221}^{\mathrm{G}} \leq 0.6 \cdot 100,$$

$$0 \leq p_{231}^{\mathrm{G}} \leq 0.7 \cdot 100,$$

$$0 \leq p_{241}^{\mathrm{G}} \leq 0.2 \cdot 100,$$

$$0 \leq p^{\mathrm{G}}_{3t1} \leq p^{\mathrm{CG,max}}_{3}, \quad t = 1, 2, 3, 4,$$
$$0 \leq p^{\mathrm{G}}_{411} \leq 0.9 \cdot p^{\mathrm{CG,max}}_{4},$$
$$0 \leq p^{\mathrm{G}}_{421} \leq 0.6 \cdot p^{\mathrm{CG,max}}_{4},$$
$$0 \leq p^{\mathrm{G}}_{431} \leq 0.7 \cdot p^{\mathrm{CG,max}}_{4},$$
$$0 \leq p^{\mathrm{G}}_{441} \leq 0.2 \cdot p^{\mathrm{CG,max}}_{4},$$
$$0 \leq p^{\mathrm{UD}}_{111} \leq 0.9 \cdot 520,$$
$$0 \leq p^{\mathrm{UD}}_{121} \leq 0.8 \cdot 520,$$
$$0 \leq p^{\mathrm{UD}}_{131} \leq 0.6 \cdot 520,$$
$$0 \leq p^{\mathrm{UD}}_{141} \leq 0.5 \cdot 520,$$

where $\Phi^{\mathrm{M}} = \left\{p^{\mathrm{CG,max}}_{3}, p^{\mathrm{CG,max}}_{4}, \theta, p^{\mathrm{G}}_{gt1}, g = 1, 2, 3, 4, t = 1, 2, 3, 4, p^{\mathrm{UD}}_{1t1}, t = 1, 2, 3, 4\right\}$.

The optimal solution of this problem is:

$$p^{\mathrm{CG,max}}_{3} = 150,$$
$$p^{\mathrm{CG,max}}_{4} = 37.5,$$
$$\theta = 8.50 \cdot 10^{7}.$$

- The lower-bound is updated:

$$\mathrm{LB} = 150000 \cdot 150 + 200000 \cdot 37.5 + 8.50 \cdot 10^{7} = 1.15 \cdot 10^{8}.$$

Step 3. Subproblem:

- The investment decisions are set to $p^{\mathrm{CG,max(1)}}_{3} = 150$ and $p^{\mathrm{CG,max(1)}}_{4} = 37.5$.
- The following problem is solved:
  Maximize$_{\Phi^{\mathrm{SL}}}$

$$\begin{aligned}
&400 \cdot 0.9\lambda_1 + z^{\mathrm{B}}_{11} + 400 \cdot 0.8\lambda_2 + z^{\mathrm{B}}_{12} + 400 \cdot 0.6\lambda_3 + z^{\mathrm{B}}_{13} + 400 \cdot 0.5\lambda_4 + z^{\mathrm{B}}_{14}\\
&\quad + 200\overline{\mu}^{\mathrm{G}}_{11} + 200\overline{\mu}^{\mathrm{G}}_{12} + 200\overline{\mu}^{\mathrm{G}}_{13} + 200\overline{\mu}^{\mathrm{G}}_{14} + 0.9 \cdot 100\overline{\mu}^{\mathrm{G}}_{21}\\
&\quad + 0.6 \cdot 100\overline{\mu}^{\mathrm{G}}_{22} + 0.7 \cdot 100\overline{\mu}^{\mathrm{G}}_{23} + 0.2 \cdot 100\overline{\mu}^{\mathrm{G}}_{24} + 150 \cdot \overline{\mu}^{\mathrm{G+}}_{31} + 150 \cdot \overline{\mu}^{\mathrm{G+}}_{32}\\
&\quad + 150 \cdot \overline{\mu}^{\mathrm{G+}}_{33} + 150 \cdot \overline{\mu}^{\mathrm{G+}}_{34} + 0.9 \cdot 37.5\overline{\mu}^{\mathrm{G}}_{41} + 0.6 \cdot 37.5\overline{\mu}^{\mathrm{G+}}_{42} + 0.7 \cdot 37.5\overline{\mu}^{\mathrm{G+}}_{43}\\
&\quad + 37.5 \cdot 100\overline{\mu}^{\mathrm{G+}}_{44} + 0.9\left(400\overline{\mu}^{\mathrm{D}}_{11} + z^{\mathrm{D}}_{11}\right) + 0.8\left(400\overline{\mu}^{\mathrm{D}}_{12} + z^{\mathrm{D}}_{12}\right)\\
&\quad + 0.6\left(400\overline{\mu}^{\mathrm{D}}_{13} + z^{\mathrm{D}}_{13}\right) + 0.5\left(400\overline{\mu}^{\mathrm{D}}_{14} + z^{\mathrm{D}}_{14}\right)
\end{aligned}$$

subject to:

$$\overline{P}^{\mathrm{D}}_{1} = 400 + 120u^{\mathrm{D}}_{1},$$

$$
\begin{aligned}
& c_1^{\mathrm{OG}} = 40 + 8u_1^{\mathrm{OG}}, \\
& c_3^{\mathrm{OG}} = 25 + 5u_3^{\mathrm{OG}}, \\
& u_1^{\mathrm{D}} \le 1, \\
& u_1^{\mathrm{OG}} + u_3^{\mathrm{OG}} \le 1, \\
& u_1^{\mathrm{D}}, u_1^{\mathrm{OG}}, u_3^{\mathrm{OG}} \in \{0, 1\}, \\
& \lambda_1 - \overline{\mu}_{11}^{\mathrm{G}} \le 1000 c_1^{\mathrm{OG}}, \\
& \lambda_2 - \overline{\mu}_{12}^{\mathrm{G}} \le 3000 c_1^{\mathrm{OG}}, \\
& \lambda_3 - \overline{\mu}_{13}^{\mathrm{G}} \le 3000 c_1^{\mathrm{OG}}, \\
& \lambda_4 - \overline{\mu}_{14}^{\mathrm{G}} \le 1760 c_1^{\mathrm{OG}}, \\
& \lambda_t - \overline{\mu}_{2t}^{\mathrm{G}} \le 0, \quad t = 1, 2, 3, 4, \\
& \lambda_1 - \overline{\mu}_{31}^{\mathrm{G+}} \le 1000 c_3^{\mathrm{OG}}, \\
& \lambda_2 - \overline{\mu}_{32}^{\mathrm{G+}} \le 3000 c_3^{\mathrm{OG}}, \\
& \lambda_3 - \overline{\mu}_{33}^{\mathrm{G+}} \le 3000 c_3^{\mathrm{OG}}, \\
& \lambda_4 - \overline{\mu}_{34}^{\mathrm{G+}} \le 1760 c_3^{\mathrm{OG}}, \\
& \lambda_t - \overline{\mu}_{4t}^{\mathrm{G+}} \le 0, \quad t = 1, 2, 3, 4, \\
& \lambda_1 - \overline{\mu}_{11}^{\mathrm{D}} \le 1000 \cdot 200, \\
& \lambda_2 - \overline{\mu}_{12}^{\mathrm{D}} \le 3000 \cdot 200, \\
& \lambda_3 - \overline{\mu}_{13}^{\mathrm{D}} \le 3000 \cdot 200, \\
& \lambda_4 - \overline{\mu}_{14}^{\mathrm{D}} \le 1760 \cdot 200, \\
& \overline{\mu}_{gt}^{\mathrm{G}} \ge 0, \quad g = 1, 2; t = 1, 2, 3, 4, \\
& \overline{\mu}_{gt}^{\mathrm{G+}} \ge 0, \quad g = 3, 4; t = 1, 2, 3, 4, \\
& \overline{\mu}_{t}^{\mathrm{D}} \ge 0, \quad t = 1, 2, 3, 4, \\
& \quad -1 \cdot 10^8 u_1^{\mathrm{D}} \le z_{1t}^{\mathrm{B}} \le 1000 u_1^{\mathrm{D}}, \quad t = 1, 2, 3, 4, \\
& \quad -1 \cdot 10^8 \left(1 - u_1^{\mathrm{D}}\right) \le 0.9 \cdot 120\lambda_1 - z_{11}^{\mathrm{B}} \le 1 \cdot 10^8 \left(1 - u_1^{\mathrm{D}}\right), \\
& \quad -1 \cdot 10^8 \left(1 - u_1^{\mathrm{D}}\right) \le 0.8 \cdot 120\lambda_2 - z_{12}^{\mathrm{B}} \le 1 \cdot 10^8 \left(1 - u_1^{\mathrm{D}}\right), \\
& \quad -1 \cdot 10^8 \left(1 - u_1^{\mathrm{D}}\right) \le 0.6 \cdot 120\lambda_3 - z_{13}^{\mathrm{B}} \le 1 \cdot 10^8 \left(1 - u_1^{\mathrm{D}}\right), \\
& \quad -1 \cdot 10^8 \left(1 - u_1^{\mathrm{D}}\right) \le 0.5 \cdot 120\lambda_4 - z_{14}^{\mathrm{B}} \le 1 \cdot 10^8 \left(1 - u_1^{\mathrm{D}}\right), \\
& \quad 0 \le z_{1t}^{\mathrm{D}} \le 1 \cdot 10^8 u_1^{\mathrm{D}}, \quad t = 1, 2, 3, 4, \\
& \quad 0 \le 120\overline{\mu}_{1t}^{\mathrm{D}} - z_{1t}^{\mathrm{D}} \le 1 \cdot 10^8 \left(1 - u_1^{\mathrm{D}}\right), \quad t = 1, 2, 3, 4. \\
& \overline{\mu}_{1t}^{\mathrm{G}} \ge 0, \quad g = 1, 2, t = 1, 2, 3, 4, \\
& \overline{\mu}_{gt}^{\mathrm{G+}} \ge 0, \quad g = 3, 4, t = 1, 2, 3, 4, \\
& \overline{\mu}_{1t}^{\mathrm{D}} \ge 0, \quad t = 1, 2, 3, 4,
\end{aligned}
$$

where $\Phi^{\mathrm{SL}} = \Big\{\overline{p}_1^{\mathrm{D}}, c_1^{\mathrm{OG}}, c_3^{\mathrm{OG}}, u_1^{\mathrm{D}}, u_1^{\mathrm{OG}}, u_3^{\mathrm{OG}}, \overline{\mu}_{1t}^{\mathrm{G}}, g = 1, 2, t = 1, 2, 3, 4, \overline{\mu}_{gt}^{\mathrm{G}+}, g = 3, 4, t = 1, 2, 3, 4, \overline{\mu}_{1t}^{\mathrm{D}}, t = 1, 2, 3, 4; z_{1t}^{\mathrm{B}}, t = 1, 2, 3, 4, z_{1t}^{\mathrm{D}}, t = 1, 2, 3, 4\Big\}$.

The optimal solution of the above problem provides the worst-case uncertainty realizations for the given investment plan identified by the master problem:

$$\overline{p}^{\mathrm{D}*} = 520,$$
$$c_1^{\mathrm{OG}*} = 48,$$
$$c_3^{\mathrm{OG}*} = 25.$$

On the other hand, the optimal value of the objective function is $8.50 \cdot 10^7$.

- Update upper-bound:

$$\mathrm{UB} = 150000 \cdot 150 + 200000 \cdot 37.5 + 8.50 \cdot 10^7 = 1.15 \cdot 10^8.$$

Step 4. Convergence checking:

- $\mathrm{UB} - \mathrm{LB} = 1.15 \cdot 10^8 - 1.15 \cdot 10^8 = 0$.
- As $\mathrm{UB} - \mathrm{LB} = 0$, the algorithm stops. The optimal generation investment plan is $p_3^{\mathrm{CG,max}} = 150$ MW and $p_4^{\mathrm{CG,max}} = 37.5$ MW.

A GAMS [11] code for solving Illustrative Example 8.8 is provided in Sect. 8.8.

□

Illustrative Example 8.9 Adaptive robust optimization problem: impact of the network constraints

Data of Illustrative Example 8.8 are considered again. However, in this case it is assumed that generating units and demands are connected within the two-node network depicted in Fig. 8.2 and described in Illustrative Example 8.4.

The optimal investment plan in this case is $p_3^{\mathrm{CG,max}} = 66.67$ MW and $p_4^{\mathrm{CG,max}} = 100$ MW, while the worst-case investment and operation cost increases up to 143.12 million € as a result of considering the network constraints. □

Illustrative Example 8.10 Impact of the uncertainty budgets

The data of Illustrative Example 8.8 are considered again. Next it is analyzed the impact of the uncertainty budgets on the expansion decisions. To do so, the problem is solved again for different values of the uncertainty budgets. The optimal expansion plans are provided in Table 8.6.

The investment decisions change as the uncertainty budgets are modified. If no uncertainty is considered ($\Gamma^{\mathrm{G}} = \Gamma^{\mathrm{D}} = 0$), then only the candidate wind-power unit is built. If the demand uncertainty budget increases, it becomes necessary to build

Table 8.6 Illustrative Example 8.10: generation investment plans for different uncertainty budgets (MW)

	$\Gamma^{\text{D}} = 0$	$\Gamma^{\text{D}} = 1$
$\Gamma^{\text{G}} = 0$	$p_3^{\text{CG,max}} = 0$, $p_4^{\text{CG,max}} = 100$	$p_3^{\text{CG,max}} = 132.3$, $p_4^{\text{CG,max}} = 50.8$
$\Gamma^{\text{G}} = 1$	$p_3^{\text{CG,max}} = 106.5$, $p_4^{\text{CG,max}} = 70.2$	$p_3^{\text{CG,max}} = 150$, $p_4^{\text{CG,max}} = 37.5$
$\Gamma^{\text{G}} = 2$	$p_3^{\text{CG,max}} = 66.7$, $p_4^{\text{CG,max}} = 100$	$p_3^{\text{CG,max}} = 132.3$, $p_4^{\text{CG,max}} = 50.8$

Table 8.7 Illustrative Example 8.10: worst-case investment and operating costs for different uncertainty budgets (€)

	$\Gamma^{\text{D}} = 0$	$\Gamma^{\text{D}} = 1$
$\Gamma^{\text{G}} = 0$	$7.45 \cdot 10^7$	$1.06 \cdot 10^8$
$\Gamma^{\text{G}} = 1$	$8.13 \cdot 10^7$	$1.15 \cdot 10^8$
$\Gamma^{\text{G}} = 2$	$8.48 \cdot 10^7$	$1.21 \cdot 10^8$

the candidate conventional generating unit since the worst-case uncertainty realization considers the increase in the peak demand. Similarly, if the generation cost uncertainty budget increases, then it becomes also optimal to build the candidate conventional generating unit. In this case, the reason behind this decision is the fact that the generation cost of the existing conventional generating unit is allowed to deviate from its forecast value.

The worst-case investment and operating costs for different uncertainty budgets are provided in Table 8.7. The worst-case costs increase as the uncertainty budgets increase. However, note that investment decisions are comparatively more robust. This means that if the selected investment plan is that related to the case without uncertainty ($\Gamma^{\text{G}} = \Gamma^{\text{D}} = 0$) and then the actual uncertainty realization corresponds to an increase in the peak demand and the generation costs, the actual costs may be much higher than those that would be obtained if the most conservative investment plan would have been selected. □

8.5 Realistic Case Study

This section analyzes the generation expansion planning problem in the European power system [8].

Table 8.8 Realistic case study: data of investment and operating costs of different technologies

Technology	Investment cost (k€/MW)	Operating cost (€/MWh)
Nucl	5400	10.54
Coal	1800	48.39
GasF	900	45.28
Fuel	1500	35.00
Wons	1802	3.50
Woff	4000	10.50
SoPV	800	2.70
SoTH	4590	4.60
Biom	2115	24.00

8.5.1 Data

The European power system is modeled considering the publicly available data provided by ENTSO-e [8]. This model comprises 24 nodes, each one representing an European country. Data of the transmission system are provided in Sect. 4.6 of Chap. 4.

The generating technologies considered in this case study are nuclear (Nucl), coal, gas-fired (GasF), fuel, onshore (Wons) and offshore (Woff) wind, solar PV (SoPV), solar thermal (SoTH), and biomass (Biom). Table 8.8 provides the investment and operating costs. These data are based on reports [13, 14, 17].

The planning horizon of this problem goes from 2016 to 2050. Table 8.9 indicates the peak demand in each country in 2015. Table 8.10 provides the existing generating capacity in each country in 2015. It is assumed that 50% of this capacity will be decommissioned at the end of the planning horizon. Finally, the limits on the generating capacity that can be built per technology and country are provided in Table 8.11 and based on the projection made by the European Commission in 2016 [9]. Note that no hydro nor storage technologies are considered in this case study since they require a specific modeling. The investment in storage technologies is considered in Chap. 10.

The demand and renewable production variability are represented using three representative days of 2015, selected using the method proposed in [12]. Since each representative day comprises 24 hourly data, $24 \cdot 3 = 72$ operating conditions are finally considered. These demand and renewable production conditions are represented in Figs. 8.4, 8.5, and 8.6.

Table 8.9 Realistic case study: data of peak demand (GW)

Country	Peak demand
AT	10.40
BE	12.33
CH	8.39
CZ	10.40
DE	71.98
DK	5.71
EE	1.39
ES	37.53
FI	14.41
FR	84.83
HU	6.16
IT	47.69
LT	1.72
LU	0.63
LV	1.23
NL	16.21
NR	24.19
PL	25.28
PT	7.72
RO	9.42
SE	26.09
SI	2.02
SK	4.26
UK	57.33

8.5.2 Results: Stochastic Programming

Considering the data described in the previous section, the GEP problem is firstly solved considering the stochastic approach described in Sect. 8.3. Uncertainties in the peak demand and production costs are modeled using a set of scenarios. Particularly, five scenarios are used to represent the uncertainties in the peak demand, whose data are provided in Table 8.12, while four scenarios are considered to represent the uncertainty in the operating costs of generating units based on coal, gas-fired, and fuel technologies, whose data are provided in Table 8.13. For the sake of simplicity, scenarios for peak demand and operating costs are independent and equiprobable, which results in a total of 20 scenarios with a probability of $1/20 = 0.05$.

Using these data, problem (8.1) is solved, rendering the generation expansion plan provided in Table 8.14. As observed, there is a high investment in renewable gener-

Table 8.10 Realistic case study: generating capacity per technology and country (GW)

Country	Nucl	Coal	GasF	Fuel	Wons	Woff	SoPV	SoTH	Biom
AT	0.00	0.87	4.07	0.97	2.41	0.00	0.88	0.00	0.64
BE	3.91	0.83	6.80	0.65	2.23	0.71	3.21	0.00	0.78
CH	3.38	0.00	0.00	0.00	0.00	0.00	0.00	0.00	0.00
CZ	4.01	9.66	1.22	0.13	0.28	0.00	2.27	0.00	0.17
DE	12.19	52.82	25.18	5.03	44.95	3.26	39.76	0.00	3.52
DK	0.00	4.23	2.27	1.02	5.06	1.27	0.84	0.00	1.78
EE	0.00	1.87	0.36	0.00	0.30	0.00	0.01	0.00	0.14
ES	7.40	10.32	31.33	7.50	23.03	0.00	7.13	2.30	1.19
FI	2.73	4.34	2.70	1.53	1.00	0.00	0.01	0.00	2.59
FR	63.25	5.38	9.65	7.69	10.36	0.01	6.10	0.00	1.25
HU	1.96	1.14	3.50	0.09	0.33	0.00	0.05	0.00	0.38
IT	0.00	9.51	52.04	13.93	8.96	0.00	18.90	0.00	5.58
LT	0.00	0.00	1.99	0.77	0.42	0.00	0.07	0.00	0.07
LU	0.00	0.00	0.47	0.00	0.06	0.00	0.12	0.00	0.02
LV	0.00	0.02	1.10	0.02	0.06	0.00	0.00	0.00	0.05
NL	0.49	6.98	17.36	0.20	3.43	0.65	1.24	0.00	1.14
NR	0.00	0.00	1.41	0.00	0.87	0.05	0.05	0.00	0.00
PL	0.00	28.54	1.66	0.40	5.10	0.00	0.04	0.00	1.57
PT	0.00	1.73	5.06	1.14	5.01	0.00	0.47	0.00	0.55
RO	1.41	6.44	4.17	1.36	2.98	0.00	1.79	0.00	0.10
SE	9.53	0.36	1.17	2.96	6.02	0.00	0.08	0.00	3.16
SI	0.70	0.79	0.47	0.09	0.00	0.00	0.26	0.00	0.05
SK	1.94	1.27	1.74	0.08	0.00	0.00	0.61	0.00	0.24
UK	9.37	18.73	33.95	2.23	13.60	5.20	9.72	0.00	3.63

ating units, mainly based on solar PV capacity. Regarding conventional generating units, most of the investment is made in gas-fired units.

Note that within a stochastic programming approach, the uncertainties in the peak demand and operating costs are modeled using a set of scenarios modeling different realizations of these uncertain variables. An alternative is to use a deterministic approach in which these parameters are modeled using their expected value. This is equivalent to solve problem (8.1) again with a single scenario, which comprises the expected peak demand and operating costs. In particular, considering the data provided in Tables 8.12 and 8.13, the expected peak demand increase is 20%, while the expected operating cost increase is 62.5% in generating units based on coal, gas-fired, and fuel technologies.

Table 8.11 Realistic case study: maximum capacity that can be built of each technology (GW)

Country	Nucl	Coal	GasF	Fuel	Wons	Woff	SoPV	SoTH	Biom
AT	0.00	0.00	0.00	0.00	8.50	0.00	5.01	0.00	0.85
BE	0.00	0.00	8.01	0.00	11.66	1.25	5.90	0.00	1.00
CH	0.00	0.00	0.00	0.00	0.63	0.00	0.03	0.00	0.20
CZ	2.84	0.00	0.61	0.00	1.05	0.00	3.86	0.00	0.64
DE	0.00	0.00	8.33	0.00	108.19	6.25	107.68	0.00	6.59
DK	0.00	0.00	2.90	0.00	9.05	3.13	1.05	0.00	2.60
EE	0.00	0.00	0.69	0.00	2.14	0.00	0.00	0.00	0.37
ES	0.00	0.00	0.00	0.00	58.93	0.00	61.70	2.80	2.15
FI	2.22	0.00	1.37	0.00	3.93	0.25	0.03	0.00	3.13
FR	0.00	0.00	26.18	0.00	71.96	0.63	56.50	0.00	3.64
HU	1.73	0.00	0.95	0.00	2.02	0.00	0.74	0.00	0.39
IT	0.00	0.00	3.32	0.00	32.45	0.00	70.96	0.00	6.81
LT	1.12	0.00	0.00	0.00	1.43	0.00	0.09	0.00	0.17
LU	0.00	0.00	0.78	0.00	0.61	0.00	0.20	0.00	0.04
LV	0.00	0.00	0.00	0.00	0.85	0.00	0.00	0.00	0.13
NL	0.00	0.00	5.50	0.00	16.01	1.38	7.34	0.00	2.64
NR	0.00	0.00	0.00	0.00	3.23	1.00	0.25	0.00	0.00
PL	8.25	0.00	2.53	0.00	23.60	0.00	0.44	0.00	3.02
PT	0.00	0.00	0.00	0.00	8.88	0.00	4.58	0.30	0.60
RO	1.41	0.00	0.00	0.00	9.31	0.00	5.18	0.00	0.31
SE	2.07	0.00	3.57	0.00	14.02	0.00	0.12	0.00	3.41
SI	0.42	0.00	0.41	0.00	0.35	0.00	1.08	0.00	0.16
SK	1.08	0.00	0.00	0.00	0.21	0.00	1.40	0.00	0.46
UK	7.93	0.00	11.70	0.00	51.84	8.75	14.07	0.00	18.03

Table 8.15 provides a summary of the investment decisions depending on whether a stochastic programming approach (Stoch) or a deterministic approach (Det) is considered. Note that there are some differences on the capacity built of different technologies, which highlights the importance of an accurate modeling of the uncertainties in the generation expansion planning problem. This uncertainty modeling may impact not only the total costs but also the feasibility of the problem, e.g., in the case that one of the scenarios with a high peak demand increase is finally realized, since this possibility is not considered when a deterministic problem is solved.

Finally, Table 8.16 provides the number of variables and constraints, as well as the computation time required to solve both the stochastic and deterministic generation expansion planning problems. Note that the computation burden of the stochastic programming problem is significantly higher than that of the deterministic approach since second-stage variables as well as the constraints modeling the operation of the power system are considered for all scenarios.

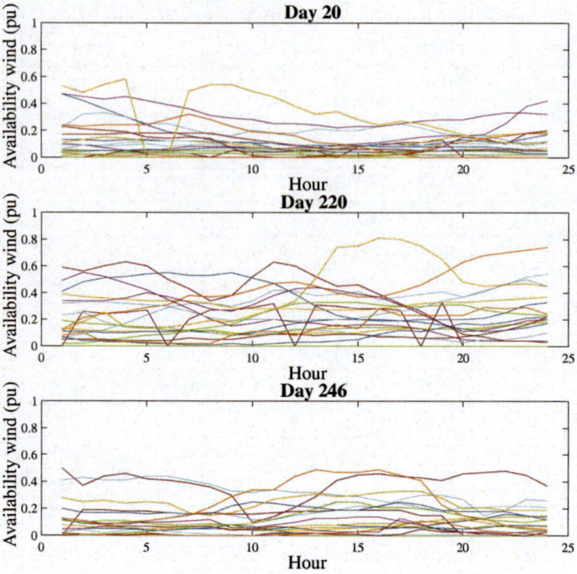

Fig. 8.4 Realistic case study: wind power availability in three representative days

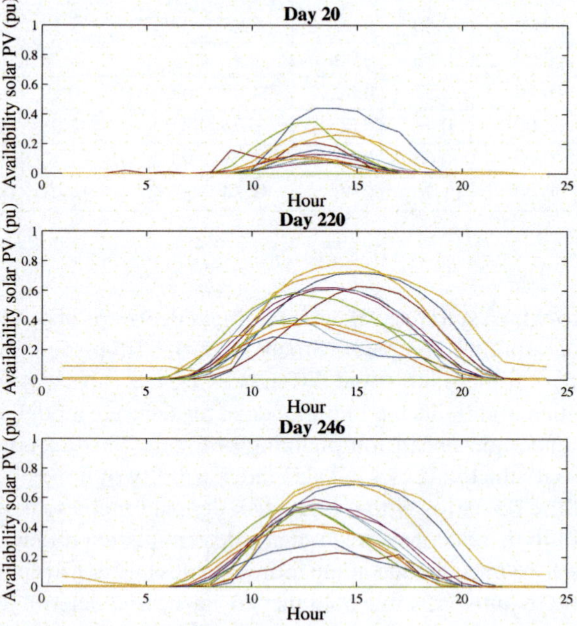

Fig. 8.5 Realistic case study: solar power availability in three representative days

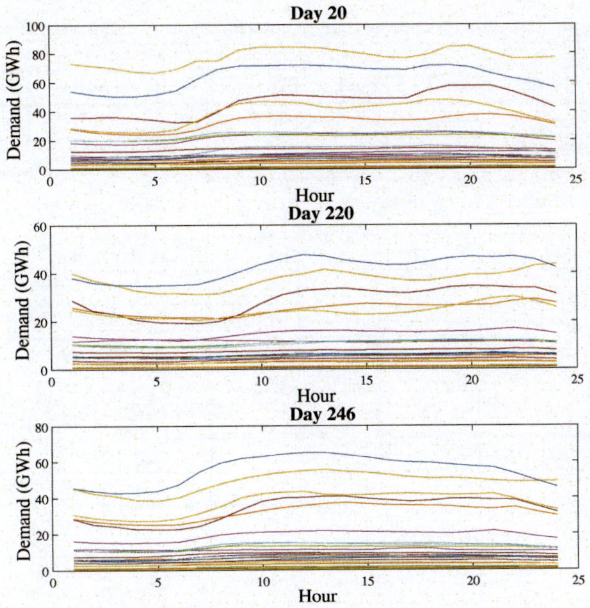

Fig. 8.6 Realistic case study: demand in three representative days

Table 8.12 Realistic case study: scenarios for peak demand

Scenario	Peak demand increase (%)
ω_1^D	+0
ω_2^D	+10
ω_3^D	+20
ω_4^D	+30
ω_5^D	+40

Table 8.13 Realistic case study: scenarios for operating costs

Scenario	Operating cost increase (%)
ω_1^C	+20
ω_2^C	+50
ω_3^C	+80
ω_4^C	+100

Table 8.14 Realistic case study: capacity built of each technology using a stochastic programming approach (GW)

Country	Nucl	Coal	GasF	Fuel	Wons	Woff	SoPV	SoTH	Biom
AT	–	–	–	–	5.53	–	5.01	–	0.85
BE	–	–	8.01	–	–	1.01	5.90	–	1.00
CH	–	–	–	–	0.18	–	0.03	–	0.20
CZ	2.84	–	0.61	–	1.05	–	3.46	–	0.64
DE	–	–	8.33	–	–	6.25	28.42	–	6.59
DK	–	–	1.15	–	–	–	–	–	2.60
EE	–	–	0.69	–	–	–	–	–	0.37
ES	–	–	–	–	2.44	–	36.97	–	2.15
FI	2.22	–	1.37	–	3.93	–	–	–	3.13
FR	–	–	26.18	–	–	–	41.91	–	–
HU	1.73	–	0.95	–	–	–	–	–	0.39
IT	–	–	3.32	–	5.89	–	31.07	–	6.81
LT	0.22	–	–	–	1.43	–	0.09	–	0.17
LU	–	–	0.78	–	–	–	–	–	0.04
LV	–	–	–	–	–	–	–	–	0.13
NL	–	–	5.50	-	2.16	1.38	7.34	–	2.64
NR	–	–	–	–	3.23	–	–	–	–
PL	–	–	2.53	–	23.60	–	–	–	3.02
PT	–	–	–	–	3.95	–	4.58	–	0.60
RO	1.41	–	–	–	–	–	1.4	–	0.31
SE	2.07	–	3.57	–	10.94	–	–	–	3.41
SI	0.42	–	0.41	–	–	–	1.08	–	0.16
SK	1.08	–	–	–	–	–	1.40	–	0.46
UK	–	–	10.50	–	–	–	14.07	–	18.03
Total	11.99	–	73.90	–	64.16	8.64	186.51	–	57.34

Table 8.15 Realistic case study: comparison between the capacity built of each technology considering a stochastic programming and a deterministic approach (GW)

Approach	Nucl	Coal	GasF	Fuel	Wons	Woff	SoPV	SoTH	Biom
Stoch	11.99	–	73.90	–	64.16	8.64	186.51	–	57.34
Det	12.08	–	67.87	–	67.74	7.96	177.29	–	57.34

Table 8.16 Realistic case study: computation burden of the stochastic and deterministic generation expansion planning problems

Approach	Number of variables	Number of constraints	Computation time (s)
Stochastic	508,585	626,666	171.13
Deterministic	25,681	31,586	0.75

Table 8.17 Realistic case study: capacity built of each technology considering an adaptive robust optimization approach (GW)

Approach	Nucl	Coal	GasF	Fuel	Wons	Woff	SoPV	SoTH	Biom
Rob	21.60	–	71.07	–	94.65	8.88	219.37	–	57.34
Stoch	11.99	–	73.90	–	64.16	8.64	186.51	–	57.34
Det	12.08	–	67.87	–	67.74	7.96	177.29	–	57.34

8.5.3 Results: Adaptive Robust Optimization

Next, the GEP problem is solved considering the adaptive robust optimization approach described in Sect. 8.4. In this case, uncertainties in the peak demand and production costs are modeled using confidence bounds. Particularly, the peak demand in each country is assumed to increase between 0 and 40% in the target year in comparison with its actual value, while the operating costs of generating units based on coal, gas-fired, and fuel technologies, are assumed to increase between 20 and 100% in the target year in comparison with its actual value.

Regarding the uncertainty budgets, Γ^{D} is assumed to be equal to 10, which means that the peak demand may increase in up to 10 countries, while Γ^{G} is fixed to 1, which means that only one of technologies may achieve its upper operating cost.

Considering these data, the adaptive robust optimization problem described in Sect. 8.4 is solved, rendering the generation expansion plan provided in Table 8.17. This table includes also the results of the stochastic and deterministic approaches. Note that considering an adaptive robust optimization approach, the investment in generation capacity increases, mainly on nuclear, onshore wind, and solar PV. This way the solution is comparatively more conservative to deal with the possible worst-case uncertainty realizations.

Next, we analyze the impact of the uncertainty budgets on the generation expansion plans. Tables 8.18 and 8.19 provide the investment in different technologies as we vary the demand and operating cost uncertainty budgets, respectively. As observed, as the demand uncertainty budget is increased, the total generation capacity installed in the power system increases since it is considered that the peak demand increases in a higher number of countries. In other words, the generation expansion plan is in this case more conservative to guarantee the supply of the demand. Regarding the variation of the operating cost uncertainty budget, as it increases, the investment in capacity based on nuclear and renewable generating units also increases since their operating costs are not subject to uncertainty.

The average computation time to achieve the solution of the adaptive robust optimization problem for different values of the uncertainty budgets is 185 s.

Table 8.18 Realistic case study: capacity built of each technology considering an adaptive robust optimization approach and different demand uncertainty budgets (GW)

Γ^{D}	Nucl	Coal	GasF	Fuel	Wons	Woff	SoPV	SoTH	Biom
0	11.51	–	40.52	–	27.86	2.53	114.94	–	57.34
5	18.47	–	54.50	–	53.66	8.88	203.00	–	57.34
10	21.60	–	71.07	–	94.65	8.88	219.37	–	57.34
15	22.63	–	71.11	–	107.48	8.88	227.60	–	57.34
20	22.76	–	72.99	–	114.01	8.88	228.13	–	57.34
25	23.59	–	72.99	–	114.64	8.88	228.49	–	57.34

Table 8.19 Realistic case study: capacity built of each technology considering an adaptive robust optimization approach and different operating cost uncertainty budgets (GW)

Γ^{G}	Nucl	Coal	GasF	Fuel	Wons	Woff	SoPV	SoTH	Biom
0	12.13	–	72.08	–	66.18	6.81	205.73	–	57.34
1	21.60	–	71.07	–	94.65	8.88	219.37	–	57.34
2	25.32	–	67.73	–	106.28	8.88	227.95	–	57.34
3	25.32	–	67.74	–	109.83	8.88	230.71	–	57.34

8.6 Summary and Conclusions

This chapter analyzes the generation expansion planning problem in a power system with renewable energy resources. The problem is tackled using the perspective of a central planner that determines the investment decisions that minimize the investment and operation costs. This central planner does not actually build the generation facilities; however, it encourages the building of these units among profit-oriented investors.

The main conclusions from the models and results provided in this chapter are summarized below:

1. Generation investment decisions are made for a long-term planning horizon. Therefore, it is important an accurate modeling of uncertainties in the decision-making problem to obtain informed expansion decisions.
2. Short-term variability refers to the variation of demand and renewable production conditions during the planning horizon. It can be modeled using a set of operating conditions based on historical data in the power system under study.
3. Uncertain variables comprise the future peak demand, generation costs, and available capacity in the system. They can be modeled using scenarios or confidence bounds, which allow formulating the generation expansion planning problem using a stochastic and an adaptive robust optimization approach, respectively.
4. The main disadvantage of using a stochastic approach is that it needs the probability distribution functions of uncertain variables to generate accurate scenarios,

which is not always available. Moreover, if a very large number of scenarios are considered, the problem may be computationally intractable. These two issues are solved using an adaptive robust optimization approach.

5. The adaptive robust optimization problem incorporates the so-called uncertainty budgets to control the level of robustness in the solution. The values of these control parameters have a great impact on the investment plans.
6. The main drawback of the adaptive robust optimization approach in comparison with the stochastic programming approach is its comparatively higher mathematical complexity.
7. Network constraints have also a great impact on the optimal generation investment plans.

The interested reader is referred to the monograph by Conejo et al. [6] for further information about investment in generation facilities in power systems.

8.7 Notation

The main notation used in this chapter is provided below for quick reference. Other symbols are defined as needed throughout the chapter. Note that if the symbols below include a subscript or superscript ω/ν, it denotes the value of the corresponding symbol in scenario ω/iteration ν.

Acronyms

DC	Direct current.
CCGA	Column-and-constraint generation algorithm.
GEP	Generation expansion planning.
MILP	Mixed-integer linear programming.
MINLP	Mixed-integer non-linear programming.
PV	Photovoltaic.

Sets and Indices

b	Index of buses.
B	Set of buses.
d	Index of demands.
D	Set of demands.
D_b	Set of demands located at bus b.
$F(\ell)$	Destination bus of line ℓ.
g	Index of generating units.

G Set of generating units.
$\tilde{G}$ Set of candidate generating units.
G_b Set of generating units located at bus b.
ℓ Index of transmission lines.
L Set of transmission lines.
$O(\ell)$ Origin bus of line ℓ.
t Index of time periods.
T Set of time periods.
ν Index of iterations of the column-and-constraint generation algorithm.
ω Index of scenarios.
Ω Set of scenarios.

Parameters

C_g^{OG} Operation cost of generating unit g (€/MWh).
$\tilde{C}_g^{\mathrm{OG}}$ Forecast value of the operation cost of generating unit g (€/MWh).
$\hat{C}_g^{\mathrm{OG}}$ Maximum increase of the forecast operation cost of generating unit g (€/MWh).
$C_g^{\mathrm{I,G}}$ Annualized investment cost of candidate generating unit g (€/MW).
$\overline{C}^{\mathrm{IG}}$ Annualized investment budget for building candidate generating units (€).
C^{UD} Cost of unserved demand (€/MWh).
K_{dt}^{D} Demand factor of demand d in time period t (pu).
K_{gt}^{G} Capacity factor of generating unit g in time period t (pu).
N_t^{H} Number of hours grouped in time period t (h).
$\overline{P}_d^{\mathrm{D}}$ Peak power consumption of demand d (MW).
$\tilde{P}_d^{\mathrm{D}}$ Forecast value of the peak power consumption of demand d (MW).
$\tilde{P}_d^{\mathrm{D}}/\hat{P}_d^{\mathrm{D}}$ Maximum increase of the forecast peak power consumption of demand d (MW).
$P_g^{\mathrm{CG,max}}$ Capacity of generation unit g (MW).
$\overline{P}_g^{\mathrm{CG,max}}$ Maximum capacity that can be built of candidate generating unit g (MW).
$\overline{P}_\ell^{\mathrm{L}}$ Power capacity of transmission line ℓ (MW).
X_ℓ Reactance of line ℓ [Ω].
Γ^{D} Uncertainty budget in the peak demand.
Γ^{G} Uncertainty budget in the operation cost of generating units.
π_ω Probability of scenario ω (pu).

Variables

c_g^{OG} Operation cost of generating unit g (€/MWh).
$p_g^{\mathrm{CG,max}}$ Capacity to be built of candidate generating unit g (MW).
$\overline{p}_d^{\mathrm{D}}$ Peak power consumption of demand d (MW).
p_{gt}^{G} Power produced by generating unit g in time period t (MW).
$p_{\ell t}^{\mathrm{L}}$ Power flow through transmission line ℓ in time period t (MW).
p_{dt}^{UD} Unserved demand of demand d in time period t (MW).
δ_{bt} Voltage angle at bus b in time period t (rad).

8.8 GAMS Codes

This section provides the GAMS [11] codes used to solve some of the illustrative examples of this chapter.

8.8.1 Stochastic Generation Expansion Planning Problem

A GAMS code for solving Illustrative Example 8.3 is provided below:

```
$title STOCHASTIC GEP PROBLEM

***************************************************************
* DATA
***************************************************************

SETS
d         'Demands'                      /d1*d1/
g         'Generating units'             /g1*g4/
t         'Operating conditions'         /t1*t4/
w         'Scenarios'                    /w1*w4/
can(g)    'Candidate generating units'   /g3,g4/
;

SCALARS
Cmax      'Investment budget'       /30e6/
;

TABLE GDATA(g,*)
          PGmax    CG       IC
g1        200      40       0
g2        100      0        0
g3        150      25       150000
g4        100      0        200000
;

TABLE DDATA(d,*)
```

```
          PDmax    CUD
d1        400      200
;

TABLE KG(g,t)      'Generation capacity factor'
      t1  t2  t3  t4
g1    1   1   1   1
g2    0.9 0.6 0.7 0.2
g3    1   1   1   1
g4    0.9 0.6 0.7 0.2
;

TABLE KD(d,t)      'Demand factor'
      t1  t2  t3  t4
d1    0.9 0.8 0.6 0.5
;

PARAMETER NH(t) 'Weight of operating conditions'
/t1 1000
t2  3000
t3  3000
t4  1760/;

PARAMETER pi(w) 'Weight of scenarios'
/w1 0.2
w2  0.3
w3  0.2
w4  0.3/;

PARAMETER CD(w) 'Generation cost per scenario'
/w1 1.2
w2  0.9
w3  1.2
w4  0.9/;

PARAMETER DINC(w) 'Demand increase per scenario'
/w1 1.3
w2  1.3
w3  1
w4  1/;


*****************************************************************
* DECLARATION OF VARIABLES AND MATHEMATICAL CHARACTERIZATION
*****************************************************************

VARIABLES

**       OBJECTIVE FUNCTION
z             'Objective function'
;

POSITIVE VARIABLES

**       OPERATIONAL VARIABLES
pGmax(g)      'Capacity built'
pG(g,t,w)     'Production'
pUD(d,t,w)    'Unserved demand'
;
```

```
********************************************************************
*   EQUATIONS
********************************************************************

EQUATIONS

**        OBJECTIVE FUNCTION
ObjectiveFunction   'Objective function'

**        CONSTRAINTS
Balance             'Power balance'
IBudget             'Investment budget'
MaxCap              'Maximum capacity of candidate units'
MaxPG               'Maximum production levels of existing units'
MaxPG2              'Maximum production levels of candidate units'
MaxPD               'Maximum unserved demand'
;

ObjectiveFunction..          z=e=sum(g$can(g),GDATA(g,'IC')*pGmax(g
    ))+sum(w,pi(w)*sum(t,NH(t)*(sum(g,CD(w)*GDATA(g,'CG')*pG(g,t,w
    ))+sum(d,DDATA(d,'CUD')*pUD(d,t,w)))));
IBudget..                    sum(g$can(g),GDATA(g,'IC')*pGmax(g))=l
    =Cmax;
MaxCap(g)$(can(g))..         pGmax(g)=l=GDATA(g,'PGmax');
Balance(t,w)..               sum(g,pG(g,t,w))=e=sum(d,DINC(w)*KD(d,
    t)*DDATA(d,'PDmax')-pUD(d,t,w));
MaxPG(g,t,w)$(not can(g)).. pG(g,t,w)=l=KG(g,t)*GDATA(g,'PGmax');
MaxPG2(g,t,w)$(can(g))..     pG(g,t,w)=l=KG(g,t)*pGmax(g);
MaxPD(d,t,w)..               pUD(d,t,w)=l=DINC(w)*KD(d,t)*DDATA(d,'
    PDmax');

********************************************************************
*   MODEL AND SOLVING
********************************************************************

MODEL StochasticGEP /all/;
StochasticGEP.optcr=0;
Option lp=cplex;

SOLVE StochasticGEP USING lp MINIMIZING z;
```

8.8.2 Adaptive Robust Optimization Generation Expansion Planning Problem

A GAMS code for solving Illustrative Example 8.8 is provided below:

```
$title ARO GEP PROBLEM

********************************************************************
* DATA
********************************************************************

SETS
```

```
d          'Demands'                         /d1*d1/
g          'Generating units'                /g1*g4/
t          'Operating conditions'            /t1*t4/
can(g)     'Candidate generating units'      /g3,g4/
it         'iteration counter'               /it1*it5/;
ALIAS(it,itt);

SCALARS
Cmax       'Investment budget'       /30e6/
;

TABLE GDATA(g,*)
        PGmax   CG      IC
g1      200     40      0
g2      100     0       0
g3      150     25      150000
g4      100     0       200000
;

PARAMETER CGvar(g)
/g1      8
g2       0
g3       5
g4       0/;

TABLE DDATA(d,*)
        PDmax   CUD
d1      400     200
;

PARAMETER PDmaxvar(d)
/d1      120/;

TABLE KG(g,t)    'generation capacity factor'
     t1  t2  t3  t4
g1   1   1   1   1
g2   0.9 0.6 0.7 0.2
g3   1   1   1   1
g4   0.9 0.6 0.7 0.2
;

TABLE KD(d,t)    'demand factor'
     t1  t2  t3  t4
d1   0.9 0.8 0.6 0.5
;

PARAMETER NH(t) 'Weigth of operating conditions'
/t1  1000
t2   3000
t3   3000
t4   1760/;

SCALAR GammaG    'generation cost uncertainty budget' /1/;

SCALAR GammaD    'demand uncertainty budget' /1/;

SCALAR M         'large positive constant' /1e8/;

SCALAR ITAUX     'auxiliary parameter' /0/;
```

```
PARAMETERS
PGMAX_AUX(g)     'Capacity built'
CG_AUX(g,it)     'Operation cost'
PDMAX_AUX(d,it) 'Peak demand';
CG_AUX(g,it)=0;
PDMAX_AUX(d,it)=0;

*****************************************************************
* DECLARATION OF VARIABLES AND MATHEMATICAL CHARACTERIZATION
*****************************************************************

VARIABLES

Z_MP          'objective function of master problem'
Z_SP          'objective function of subproblem'
theta         'auxiliary variable'
lambda(t)     'dual variable';
theta.lo=-1e10;

POSITIVE VARIABLES

pGmax(g)      'Capacity built'
pG(g,t,it)    'Production'
pUD(d,t,it)   'Unserved demand'
CG(g)         'Variable cost of generating units'
PDmax(d)      'Peak demand'
muG(g,t)      'Dual variable'
muGm(g,t)     'Dual variable'
muD(d,t)      'Dual variable'
zD(d,t)       'Auxiliary variable'
zB(d,t)       'Auxiliary variable'
;

BINARY VARIABLES
UD(d)          'Peak demand at upper bound'
UG(g)          'Generation cost at upper bound';

*****************************************************************
*  EQUATIONS
*****************************************************************

EQUATIONS
MP_of     'Master problem'
MP_1      'Master problem: investment budget'
MP_2      'Master problem: maximum capacity of candidate units'
MP_3      'Master problem: bounds for theta'
MP_4      'Master problem: power balance'
MP_5      'Master problem: maximum production levels of existing units'
MP_6      'Master problem: maximum production levels of candidate units'
MP_7      'Master problem: Maximum unserved demand'
SP_of     'Subproblem: objective function'
SP_1      'Subproblem: peak demand'
SP_2      'Subproblem: generation cost'
SP_3      'Subproblem: demand uncertainty budget'
SP_4      'Subproblem: generation cost uncertainty budget'
SP_5      'Subproblem: auxiliary constraint'
```

```
SP_6        'Subproblem: auxiliary constraint'
SP_7        'Subproblem: auxiliary constraint'
SP_8        'Subproblem: auxiliary constraint'
SP_9        'Subproblem: auxiliary constraint'
SP_10       'Subproblem: auxiliary constraint'
SP_11       'Subproblem: dual constraint'
SP_12       'Subproblem: dual constraint'
SP_13       'Subproblem: dual constraint'
;

MP_of..                                  Z_MP=e=sum(g$can(g),GDATA
    (g,'IC')*pGmax(g))+theta;
MP_1..                                   sum(g$can(g),GDATA(g,'IC'
    )*pGmax(g))=l=Cmax;
MP_2(g)$(can(g))..                       pGmax(g)=l=GDATA(g,'PGmax
    ');
MP_3(itt)$(ord(itt) LT (ITAUX+1))..      theta=G=sum(t,NH(t)*(sum(
    g,CG_AUX(g,itt)*pG(g,t,itt))+sum(d,DDATA(d,'CUD')*pUD(d,t,itt)
    )));
MP_4(t,itt)$(ord(itt) LT (ITAUX+1))..    sum(g,pG(g,t,itt))=e=sum(
    d,KD(d,t)*PDMAX_AUX(d,itt)-pUD(d,t,itt));
MP_5(g,t,itt)$((ord(itt) LT (ITAUX+1)) AND not can(g))..  pG(g,t,
    itt)=l=KG(g,t)*GDATA(g,'PGmax');
MP_6(g,t,itt)$((ord(itt) LT (ITAUX+1)) AND can(g))..      pG(g,t,
    itt)=l=KG(g,t)*pGmax(g);
MP_7(d,t,itt)$(ord(itt) LT (ITAUX+1))..  pUD(d,t,itt)=l=KD(d,t)*
    PDMAX_AUX(d,itt);

SP_of..                            Z_SP=e=SUM(t,lambda(t)*sum(d,KD(d
    ,t)*DDATA(d,'PDmax'))+sum(d,zB(d,t))-sum(G$(not can(g)),KG(g,t
    )*GDATA(g,'PGmax')*muG(g,t))-sum(G$(can(g)),KG(g,t)*PGMAX_AUX(
    g)*muGm(g,t))-SUM(d,KD(d,t)*(DDATA(d,'PDmax')*muD(d,t)+zD(d,t)
    )));
SP_1(d)..                          PDmax(d)=e=DDATA(d,'PDmax')+UD(d)
    *PDmaxvar(d);
SP_2(g)..                          CG(g)=e=GDATA(g,'CG')+UG(g)*CGvar
    (g);
SP_3..                             sum(d,UD(d))=l=GammaD;
SP_4..                             sum(g,UG(g))=l=GammaG;
SP_5(d,t)..                        -M*UD(d)=l=ZB(d,t);
SP_6(d,t)..                        ZB(d,t)=l=M*UD(d);
SP_7(d,t)..                        -M*(1-UD(d))=l=lambda(t)*KD(d,t)*
    PDmaxvar(d)-ZB(d,t);
SP_8(d,t)..                        lambda(t)*KD(d,t)*PDmaxvar(d)-ZB(
    d,t)=l=M*(1-UD(d));
SP_9(d,t)..                        ZD(d,t)=l=M*UD(d);
SP_10(d,t)..                       muD(d,t)*PDmaxvar(d)-ZD(d,t)=l=M
    *(1-UD(d));
SP_11(g,t)$(not can(g))..          lambda(t)-muG(g,t)=l=NH(t)*CG(g);
SP_12(g,t)$(can(g))..              lambda(t)-muGm(g,t)=l=NH(t)*CG(g)
    ;
SP_13(d,t)..                       lambda(t)-muD(d,t)=l=NH(t)*DDATA(
    d,'CUD');

*************************************************************
*  MODELS
*************************************************************

MODEL Master /MP_of,MP_1,MP_2,MP_3,MP_4,MP_5,MP_6,MP_7/;
```

```
Master.optcr=0;
MODEL Subproblem /SP_of,SP_1,SP_2,SP_3,SP_4,SP_5,SP_6,SP_7,SP_8,
    SP_9,SP_10,SP_11,SP_12,SP_13/;
Subproblem.optcr=0;
*Option mip=cplex,lp=cplex;

****************************************************************
*  ALGORITHM
****************************************************************

************
***Step 1***
************
*Iteration counter it=0
SCALAR TOL        tolerance                         /0/;
SCALAR LB         lower bound                       /-1e10/;
SCALAR UB         upper bound                       /1e10/;
************
***Step 2***
************
SOLVE Master using lp minimizing Z_MP;
DISPLAY pGmax.l, Z_MP.l, theta.l;
LB=Z_MP.l;
DISPLAY LB;
************
***Step 3***
************
PGMAX_AUX(g)=pGmax.l(g);
SOLVE Subproblem using mip maximizing Z_SP;
DISPLAY PDmax.l, CG.l, Z_SP.l;
UB=min(sum(g$can(g),GDATA(g,'IC')*pGmax.l(g))+Z_SP.l,UB);
DISPLAY UB;
************
***Step 4***
************
LOOP(it$((UB-LB) GT TOL),
************
***Step 2***
************
CG_AUX(g,it)=CG.l(g);
PDMAX_AUX(d,it)=PDmax.l(d);
ITAUX=ORD(IT);
DISPLAY ITAUX, CG_AUX, PDMAX_AUX;
SOLVE Master using lp minimizing Z_MP;
DISPLAY pGmax.l, Z_MP.l, theta.l;
LB=Z_MP.l;
DISPLAY LB;
************
***Step 3***
************
PGMAX_AUX(g)=pGmax.l(g);
SOLVE Subproblem using mip maximizing Z_SP;
DISPLAY PDmax.l, CG.l, Z_SP.l;
UB=min(sum(g$can(g),GDATA(g,'IC')*pGmax.l(g))+Z_SP.l,UB);
DISPLAY UB,LB;
);
DISPLAY pGmax.l;
```

References

1. Baringo, L., Conejo, A.J.: Transmission and wind power investment. IEEE Trans. Power Syst. **27**(2), 885–893 (2012)
2. Baringo, L., Conejo, A.J.: Correlated wind-power production and electric load scenarios for investment decisions. App. Energy **101**, 475–482 (2013)
3. Bertsimas, D., Sim, M.: The price of robustness. Oper. Res. **52**, 35–53 (2004)
4. Bialas, W.F., Karwan, M.H.: Multilevel linear programming. State University of New York at Buffalo, Buffalo, NY, USA, Technical Report No. 78-1 (1978)
5. Birge, J.R., Louveaux, F.: Introduction to Stochastic Programming, 2nd edn. Springer, New York, NY (2011)
6. Conejo, A.J., Baringo, L., Kazempour, S.J., Siddiqui, A.S.: Investment in Electricity Generation and Transmission: Decision Making Under Uncertainty. Springer, Switzerland (2016)
7. Domínguez, R., Conejo, A.J., Carrión, M.: Toward fully renewable electric energy systems. IEEE Trans. Power Syst. **30**(1), 316–326 (2015)
8. ENTSO-E transparency platform (2022). https://transparency.entsoe.eu
9. European Commission. EU Reference Scenario (2016) Energy, transport and GHG emissions. Trends to 2050 (2022). https://ec.europa.eu/energy/node/186
10. Floudas, C.A.: Nonlinear and Mixed-Integer Optimization, Fundamentals and Applications. Oxford University Press, New York (1995)
11. GAMS (2022). www.gams.com
12. Gröwe-Kuska, N., Heitsch, H., Römisch, W.: Scenario reduction and scenario tree construction for power management problems In: Proceedings IEEE Bologna Power Tech Conference, Bologna, Italy (2003)
13. International Energy Agency (IEA). World Energy Outlook 2019 (2022). https://www.iea.org/weo/
14. International Renewable Energy Agency (IRENA). Renewable power generation costs in 2019 (2022). https://www.irena.org/publications/2020/Jun/Renewable-Power-Costs-in-2019
15. Jabr, R.A.: Robust transmission network expansion planning with uncertain renewable generation and loads. IEEE Trans. Power Syst. **28**(4), 4558–4567 (2013)
16. Jiang, R., Zhang, M., Li, G., Guan, Y.: Two-stage network constrained robust unit commitment problem. Eur. J. Oper. Res. **234**(3), 751–762 (2014)
17. Kost, C.: Levelized cost of electricity renewable energy technlogies. Fraunhofer Institute for Solar Energy Systems ISE (2021)
18. Mínguez, R., García-Bertrand, R.: Robust transmission network expansion planning in energy systems: improving computational performance. Eur. J. Oper. Res. **248**(1), 21–32 (2016)
19. Moreira, A., Pozo, D., Street, A., Sauma, E.: Reliable renewable generation and transmission expansion planning: co-optimizing system's resources for meeting renewable targets. IEEE Trans. Power Syst. **32**(4), 3246–3257 (2017)
20. Sioshansi, R., Conejo, A.J.: Optimization in Engineering, Models and Algorithms. Springer, New York (2017)
21. The ILOG CPLEX (2021). www.ilog.com/products/cplex/
22. Zhou, Y., Wang, L., McCalley, J.D.: Designing effective and efficient incentive policies for renewable energy in generation expansion planning. App. Energy D **88**, 2201–2209 (2011)

Chapter 9
Multi-stage Modeling of the Generation Expansion Planning

This chapter describes a generation expansion problem in which investment decisions can be made at different stages along the planning horizon. This planning problem usually consists of a long-term decision process that is subject to multiple uncertain sources. Therefore, a precise representation of the evolution of the uncertain parameters helps the decision maker to adopt the optimal solutions considering its final targets. In this chapter, a multi-stage investment model in generation capacity considering long-term uncertainties is presented. Specifically, the demand growth per year and the investment costs of the units are characterized as random variables. The proposed model is formulated using two different approaches, namely stochastic programming and linear decision rules. These approaches differ in how the uncertain parameters are represented. Illustrative examples are provided to ease the understanding of the proposed formulations. A realistic case study based on the European power system is solved. The outcomes obtained from the stochastic and linear decision rules approaches are compared for different number of decision stages.

9.1 Introduction

Power systems around the world are immersed in a transformation towards sustainability [10]. Burning massively fossil fuels to get energy has led the planet to a climate change whose consequences can be only mitigated by totally changing the energy system. Moreover, the continuous development of renewable energy technologies has driven wind and solar power plants up to a competitive situation from both the technical and economic points of view with respect to conventional thermal power plants. For these reasons, investors and social planners are making their capacity expansion plans for the next decades to achieve renewable-dominated power systems.

L. Baringo et al., *Electric Vehicles and Renewable Generation*,
Green Energy and Technology, https://doi.org/10.1007/978-3-031-09079-0_9

A capacity expansion plan usually involves a long-term exercise and, consequently, multiple uncertainty sources arise, such as the evolution of the electricity consumption, the investment cost of the generating and storage units, the cost of fossil fuels, the policies about the nuclear power, etc. The decision planner makes investment decisions in different facilities as long as there is a better knowledge of those uncertainties. The planner may take those decisions seeking to minimize the total cost, maximize the system reliability, or achieve a specific goal as, for instance, a renewable-dominated system, or a net-zero CO_2 emissions, etc. In this chapter, the decision planner is considered to be a central planner that aims at minimizing the total cost while increasing the renewable capacity.

From the modeling point of view, a multi-stage capacity expansion model allows to represent the decision-making process involved in this type of problems in a more precise way than single-stage approaches, as the one presented in the Chap. 8 in which it is considered that all investment decisions are made at a single time period. However, increasing the number of decision stages and representing the uncertainty sources in a precise way lead to a large problem that often results in computational intractability. Hence, the planner must find a trade-off between modeling accuracy and computational tractability.

This problem has been widely studied in the technical literature and different approaches to efficiently solve the problem have been proposed. For instance, in [1], the multi-stage investment problem of a wind producer is tackled considering the risk in the objective function. In [13], a two-stage capacity expansion problem is formulated and solved under a rolling window framework as many times as decision stages considered in the planning horizon. In [16], the progressive-hedging algorithm is applied to solve a multi-stage capacity expansion problem in generating, transmission and storage capacity. In [17], a heuristic approach is proposed to iteratively solve a two-stage deterministic problem such that the multi-stage decision process is simulated. In [3], four different approaches based on stochastic programming and linear decision rules (LDRs) are presented to solve a multi-stage generating capacity expansion model and the pros and cons of each approach are evaluated.

In this chapter, a multi-stage generating capacity expansion model is formulated using stochastic programming and LDRs. Capacity expansion decisions can be made at different stages along the planning horizon considering long-term uncertainties, such as the demand growth and the investment costs of the generating units. The stochastic approach implies to represent the random variables through a set of scenarios, while the LDR approach considers the intervals where the random variables are defined and establishes a linear dependency between the random and the decision variables. Additionally, the variability that can be found in the daily demand and intermittent renewable production is represented in this model by including a set of representative days of the year. Even though no capacity expansion in transmission or storage units is considered in this model, the operating constraints describing those facilities are incorporated in the model.

This chapter is structured as follows. In Sect. 9.2, the modeling of the planning horizon is explained. Section 9.3 describes the uncertainty characterization of the random variables. In Sect. 9.4, the decision framework considered to formulate the multi-stage capacity expansion model is presented. Section 9.5 provides the stochastic formulation of the multi-stage capacity expansion model. In Sect. 9.6, an illustrative case study is solved to clarify the stochastic formulation. In Sect. 9.7, the LDR approach is described and the corresponding formulation of the multi-stage capacity expansion model is provided. In Sect. 9.8, the illustrative case study solved using the stochastic formulation is now solved using the LDR-based formulation. Section 9.9 presents the realistic case study solved to compare the outcomes of both approaches and the influence of the number of decision stages. Section 9.10 concludes the chapter with a summary and the main conclusions obtained. Finally, the list of the main notation used along this chapter is included in Sect. 9.11 and the GAMS [8] codes corresponding to the illustrative case studies are provided in Sect. 9.12.

9.2 Modeling of the Planning Horizon

We consider that the planning horizon comprises several years and investment decisions can be made at different time points, which are called stages. The stages divide the planning horizon into periods, which can comprise more than one year. A year is represented through a set of representative days, comprising 24 hourly operating periods each one.

The uncertainty in the long-term horizon is modeled, including the demand growth and the investment costs of the generating units, which are assumed to be unknown and, hence, their values evolve randomly along the planning horizon.

Two types of decisions are considered, namely investment and operating decisions. Investment decisions determine the generating capacity to be built and are made at each stage. Operating decisions establish how to supply the demand in each time period of a representative day considering the renewable power availability and the technical constraints characterizing the system operation. The optimal investment decisions are determined with the information of the investment cost at the current stage and the realization of the demand growth of the previous investment period (*here-and-now* decisions). The operating decisions are made knowing the realization of the demand growth in the current period and the capacity to be built (*wait-and-see* decisions).

Figure 9.1 depicts an example of the decision-making process involved in this problem. As it can be seen, the planning horizon is divided into three periods and investment decisions can be made at the beginning of each period, i.e., at each stage. In this way, at the beginning of the planning horizon (Stage 1) investment decisions are made considering as known the investment costs of the units. Once the capacity to be built is decided, the system is operated for each year comprising *Period 1*, which means that the demand is supplied for each operating period considering the availability of the renewable power and the possible values of the demand growth. Then, at the beginning of *Period 2* (Stage 2) new investment decisions can be made taking into account the realization of the uncertain parameters, i.e., the demand growth along *Period 1* and the investment costs at Stage 2. As previously, the system

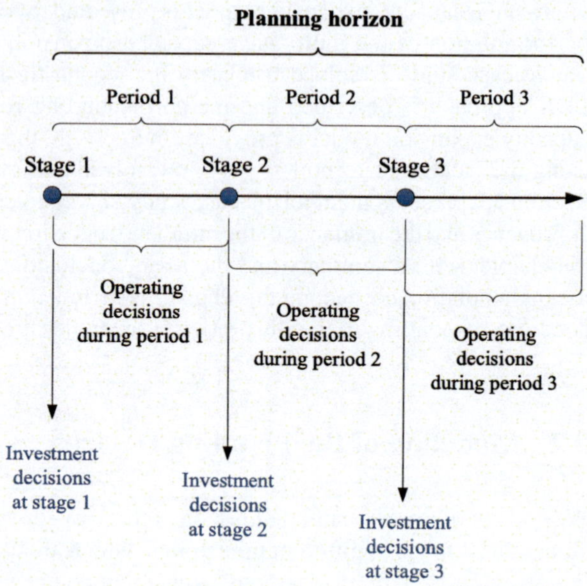

Fig. 9.1 Multi-stage modeling: planning horizon

is operated knowing the available generating capacity and the demand is supplied for each operating condition. This process follows until the end of the planning horizon.

9.3 Uncertainty Characterization

In this model, we consider that the random variables take values within specific intervals, which may be defined according to historical data and experts opinion. Let us assume that $\tilde{c}^{\mathrm{I}}_{gn}$ represents the investment cost of generating technology $g \in G$ in stage $n \in N$ and $\tilde{d}^{\mathrm{D}}_{bn}$ the demand growth attained in bus $b \in B$ and stage $n \in N$. The intervals comprising the values of those variables are defined as follows:

$$\bar{c}^{\mathrm{I}}_{gn}(1-\theta^{\mathrm{I}}_{gn}) \leq \tilde{c}^{\mathrm{I}}_{gn} \leq \bar{c}^{\mathrm{I}}_{gn}(1+\theta^{\mathrm{I}}_{gn}), \quad \forall g \in G, \forall n \in N, \tag{9.1a}$$

$$\bar{d}^{\mathrm{D}}_{bn}(1-\theta^{\mathrm{D}}_{bn}) \leq \tilde{d}^{\mathrm{D}}_{bn} \leq \bar{d}^{\mathrm{D}}_{bn}(1+\theta^{\mathrm{D}}_{bn}), \quad \forall b \in B, \forall n \in N, \tag{9.1b}$$

where $\bar{c}^{\mathrm{I}}_{gn}$ and $\bar{d}^{\mathrm{D}}_{bn}$ represent the expected value of the investment cost and the demand growth in each stage, respectively; and θ^{I}_{gn} and θ^{D}_{bn} fix the size of the interval.

In addition, we consider a temporal correlation among the values of a random variable between consecutive stages. This can be modeled as follows:

$$\tilde{c}^{\mathrm{I}}_{gn} = a^{\mathrm{I}}_{g} + b^{\mathrm{I}}_{g} \cdot \tilde{c}^{\mathrm{I}}_{g,n-1}, \quad \forall g \in G, \forall n > 1, \tag{9.2a}$$

$$\tilde{d}^{\mathrm{D}}_{bn} = a^{\mathrm{D}}_{b} + b^{\mathrm{D}}_{b} \cdot \tilde{d}^{\mathrm{D}}_{b,n-1}, \quad \forall b \in B, \forall n > 1, \tag{9.2b}$$

where a_g^{I}, b_g^{I}, a_b^{D}, and b_b^{D} are parameters characterizing the temporal correlation.

Consequently, the definition of the intervals containing the values of the random variables results as follows:

$$(a_g^{\mathrm{I}} + b_g^{\mathrm{I}} \cdot \bar{c}_{g,n-1}^{\mathrm{I}})(1 - \theta_{gn}^{\mathrm{I}}) \le \tilde{c}_{gn}^{\mathrm{I}} \le (a_g^{\mathrm{I}} + b_g^{\mathrm{I}} \cdot \bar{c}_{g,n-1}^{\mathrm{I}})(1 + \theta_{gn}^{\mathrm{I}}), \quad \forall g \in G, \forall n > 1, \tag{9.3a}$$

$$(a_b^{\mathrm{D}} + b_b^{\mathrm{D}} \cdot \bar{d}_{b,n-1}^{\mathrm{D}})(1 - \theta_{bn}^{\mathrm{D}}) \le \tilde{d}_{bn}^{\mathrm{D}} \le (a_b^{\mathrm{D}} + b_b^{\mathrm{D}} \cdot \bar{d}_{b,n-1}^{\mathrm{D}})(1 + \theta_{bn}^{\mathrm{D}}), \quad \forall b \in B, \forall n > 1, \tag{9.3b}$$

$$(a_g^{\mathrm{I}} + b_g^{\mathrm{I}} \cdot \bar{c}_g^{\mathrm{I,ini}})(1 - \theta_{gn}^{\mathrm{I}}) \le \tilde{c}_{gn}^{\mathrm{I}} \le (a_g^{\mathrm{I}} + b_g^{\mathrm{I}} \cdot \bar{c}_g^{\mathrm{I,ini}})(1 + \theta_{gn}^{\mathrm{I}}), \quad \forall g \in G, n = 1, \tag{9.3c}$$

$$(a_b^{\mathrm{D}} + b_b^{\mathrm{D}} \cdot \bar{d}_b^{\mathrm{D,ini}})(1 - \theta_{bn}^{\mathrm{D}}) \le \tilde{d}_{bn}^{\mathrm{D}} \le (a_b^{\mathrm{D}} + b_b^{\mathrm{D}} \cdot \bar{d}_b^{\mathrm{D,ini}})(1 + \theta_{bn}^{\mathrm{D}}), \quad \forall b \in B, n = 1. \tag{9.3d}$$

In the first stage, the interval is defined considering the investment cost and demand growth occurred before the starting of the planning horizon, represented by $\bar{c}_g^{\mathrm{I,ini}}$ and $\bar{d}_b^{\mathrm{D,ini}}$.

Illustrative Example 9.1 Uncertainty intervals

This example considers a capacity expansion problem in generating capacity of a power system with N_B buses connected by N_L transmission lines. The planning horizon is divided into two periods, i.e., investment decisions can be made at two stages. The demand growth and the investment costs of generating units (indexed by g) are considered as random variables that take values within specific intervals according to (9.3). The parameters establishing a temporal correlation between consecutive stages of the random variables are indicated in Table 9.1.

The size of the intervals is determined by parameters θ_{gn}^{I} and θ_{bn}^{D}, whose values are provided in Table 9.2. Note that the investment costs are known in the first stage, i.e., the uncertainty interval is null.

Considering the values above, the investment costs at stage 2 take values according to the following interval:

Table 9.1 Illustrative Example 9.1: temporal correlation (pu)

$a_g^{\mathrm{I}}, \forall g \in G$	0.01	$a_b^{\mathrm{D}}, \forall b \in B$	0.20
$b_g^{\mathrm{I}}, \forall g \in G$	0.85	$b_b^{\mathrm{D}}, \forall b \in B$	1.00

Table 9.2 Illustrative Example 9.1: size of the interval (pu)

	Stage	
	n_1	n_2
$\theta_{gn}^{\mathrm{I}}, \forall g \in G$	0.00	0.15
$\theta_{bn}^{\mathrm{D}}, \forall b \in B$	0.05	0.10

$$(0.01 + 0.85 \cdot \tilde{c}^{\text{I,G}}_{g1}) \cdot (1 - 0.15) \leq \tilde{c}^{\text{I,G}}_{g2} \leq (0.01 + 0.85 \cdot \tilde{c}^{\text{I,G}}_{g1}) \cdot (1 + 0.15), \quad \forall g \in G,$$

whereas the intervals where the demand growth takes values in stages 1 and 2 are:

$$(0.2 + 1.0 \cdot \tilde{d}^{\text{D}}_{b0}) \cdot (1 - 0.05) \leq \tilde{d}^{\text{D}}_{b1} \leq (0.2 + 1.0 \cdot \tilde{d}^{\text{D}}_{b0}) \cdot (1 + 0.05), \quad \forall b \in B$$
$$(0.2 + 1.0 \cdot \tilde{d}^{\text{D}}_{b1}) \cdot (1 - 0.1) \leq \tilde{d}^{\text{D}}_{b2} \leq (0.2 + 1.0 \cdot \tilde{d}^{\text{D}}_{b1}) \cdot (1 + 0.1), \quad \forall b \in B.$$

Observe that the investment cost at stage 1 is assumed to be known. □

9.4 Decision Framework

The decision framework considered in this model is described as follows:

- We adopt the perspective of a central planner, such as a system operator (SO), which seeks to define the expansion plan towards achieving a renewable-dominated power system minimizing the total cost.
- Investment decisions can be made in generation capacity. The existing generation capacity may be sequentially decommissioned along the planning horizon. The generation system comprises dispatchable and intermittent renewable units. Each generating unit is located in a specific bus.
- The demand growth and the investment costs are assumed to be unknown and considered as random variables. How these random variables are represented in the decision-making process depends on the approach implemented. For the sake of simplicity, the evolution of the investment costs and the demand growth is assumed to be the same for all generating technologies and for all buses, respectively. Other long-term random variables such as the cost of fossil fuels or CO_2 emissions could be easily introduced in the model.
- The variability of the renewable power and the demand level along the year is represented by considering a set of representative days. The representative days are selected using the scenario reduction technique presented in [9] that is described in Chap. 2. Each representative day receives a different weight so that the total number of hours of the initial input set are considered in the objective function.
- In this model, no unit-commitment variables of thermal units are considered, which is a common assumption in long-term expansion models [4, 16]. However, ramping constraints of thermal units are imposed.
- The transmission system is represented using a DC model. No expansion in transmission capacity is considered.
- Storage units such as hydro-pump power plants or batteries are assumed as existing elements in the current power system and are not decommissioned. The inter-temporal constraints characterizing the operation of such units are included in the model. It is assumed that storage units belong to the SO to ensure the flexibility in the system operation and, their operating cost is considered to be null.

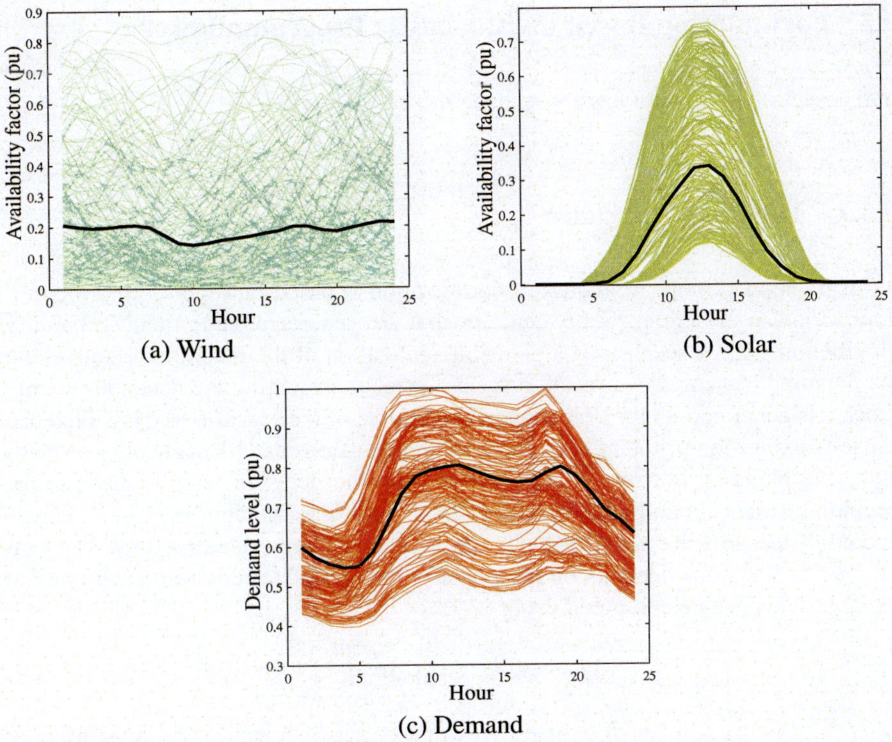

Fig. 9.2 Illustrative Example 9.2: example of a representative day

Illustrative Example 9.2 Representative days

This example illustrates how to select a representative day of the year. Figure 9.2 depicts in light-blue, yellow, and red lines historical data of the wind, solar, and demand power level, respectively, in Spain in each hour of the 366 days of 2016 [5]. With a black line, the wind, solar, and demand power level corresponding to the representative day are depicted. In this case, we used the scenario reduction technique presented in [9] to select the representative day of the year, which is applied to the net demand of the system, i.e., the electricity consumption minus the wind and solar production. The day selected as the most representative of 2016 in terms of net demand corresponds to day 246 of the year. Hence, the weight of this day is 366.

□

9.5 Formulation Based on Stochastic Programming

In this section, the multi-stage stochastic model is described.

9.5.1 *Modeling Approach*

As explained in Chap. 3, stochastic optimization is based on representing the random variables through a set of scenarios that are generated using their probability distributions [2]. A scenario is a plausible realization of the random variable along the planning horizon. As a result, a scenario tree representing the decision-making process is generated. Figure 9.3 depicts an example of the scenario tree involved in a capacity expansion problem where investment decisions can be made at two stages along the planning horizon. In this case, investment decisions can be made at the beginning (Stage 1) and in the middle (Stage 2) of the planning horizon, whereas operating decisions are made for each time period of the representative days. The total number of scenarios depends on the scenarios generated to represent each random variable. This can be computed using (9.4).

$$\Omega = (\Omega_n^{\mathrm{D}})^{N_N} \cdot (\Omega_n^{\mathrm{I}})^{(N_N-1)}, \tag{9.4}$$

where Ω_n^{D} is the number of demand growth scenarios per stage, N_N is the number of stages, and Ω_n^{I} is the number of investment cost scenarios per stage.

Illustrative Example 9.3 Scenarios in a multi-stage capacity expansion model

The data presented in Illustrative Example 9.1 are used in this example. The investment costs of the generating units and the demand growth are considered as random variables. The evolution of their values until the end of the planning horizon is represented through a set of scenarios. In particular, two scenarios are generated for each parameter at each stage, which results in a total number of eight scenarios according to (9.4). The scenarios are specified randomly taking values from the intervals defined in Illustrative Example 9.1. Table 9.3 indicates the values of the random variables in each scenario. In the case of the investment cost, at the beginning of the planning horizon those costs are assumed to be known and that is why the value of the scenarios in column 2 of Table 9.3 is equal to 1. Then, at the beginning of Period 2 the investment costs are unknown and their possible reductions are indicated in column 4. In the case of the demand growth, its possible growth by the end of Periods 1 and 2 is indicated in columns 3 and 5, respectively. The scenarios are equiprobable with a probability of $\pi_\omega = 1 \cdot 0.5 \cdot 0.5 \cdot 0.5 = 0.125$. □

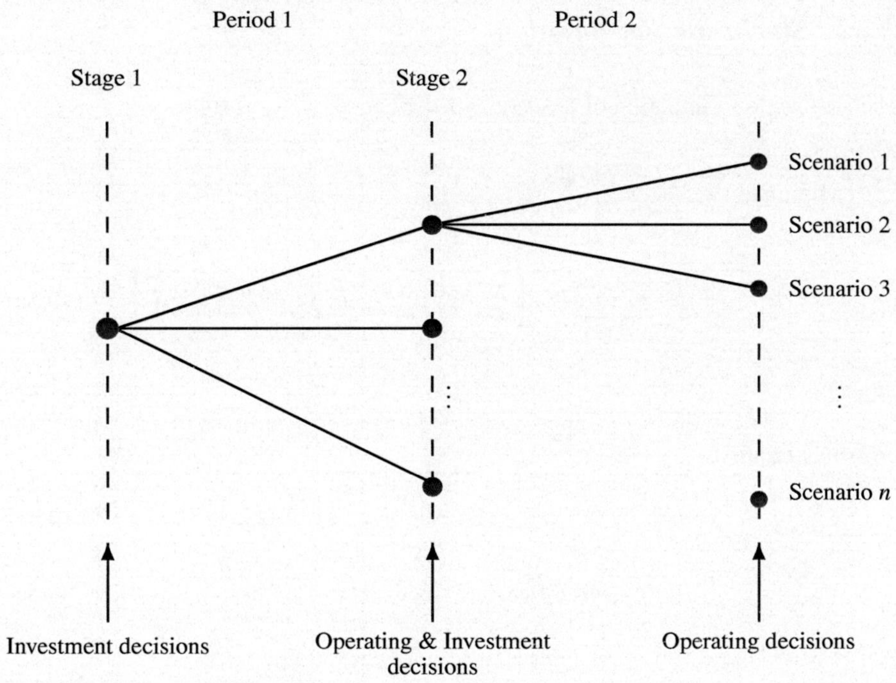

Fig. 9.3 Multi-stage modeling: scenario tree

Table 9.3 Illustrative Example 9.3: investment cost and demand growth scenarios (p.u.)

Scenario	Period 1		Period 2	
	Invest. cost	Dem. growth	Invest. cost	Dem. growth
ω_1	1	1.24	0.90	1.51
ω_2	1	1.24	0.90	1.44
ω_3	1	1.24	0.80	1.51
ω_4	1	1.24	0.80	1.44
ω_5	1	1.18	0.90	1.37
ω_6	1	1.18	0.90	1.29
ω_7	1	1.18	0.80	1.37
ω_8	1	1.18	0.80	1.29

9.5.2 Stochastic Formulation

The stochastic formulation of the model described above is as follows:

$$\underset{\Theta^{\mathrm{St}}}{\text{Minimize}} \sum_{n\in N}\sum_{\omega\in\Omega}\pi_{\omega}\Bigg[\sum_{g\in G^{\mathrm{C}}}\tilde{c}^{\mathrm{I}}_{gn\omega}\, p^{\mathrm{I,G}}_{gn\omega} + \sum_{d\in D}\sum_{t\in T} W_d\left(\sum_{g\in G} C^{\mathrm{O}}_g\, p^{\mathrm{G,D}}_{gdtn\omega} + \sum_{b\in B} C^{\mathrm{UD}}\, p^{\mathrm{UD,D}}_{bdtn\omega}\right)\Bigg] \quad (9.5a)$$

subject to:

Investment limits:

$$P^{\mathrm{C,G}}_{\max,gn} - \sum_{j\le n} p^{\mathrm{I,G}}_{gj\omega} \ge 0, \quad \forall g \in G^{\mathrm{C}}, \forall n \in N, \forall \omega \in \Omega, \quad (9.5b)$$

$$p^{\mathrm{I,G}}_{gn\omega} = p^{\mathrm{I,G}}_{gn\omega'}, \quad \forall g \in G^{\mathrm{C}}, \forall n \in N, \forall \omega, \omega' \in \Omega : \tilde{\xi}_n(\omega) = \tilde{\xi}_n(\omega'). \quad (9.5c)$$

Power balance:

$$\sum_{g\in G_b} p^{\mathrm{G,D}}_{gdtn\omega} + \sum_{s\in S_b}(p^{\mathrm{S,D}}_{sdtn\omega} - p^{\mathrm{S,C}}_{sdtn\omega}) - \sum_{l|O(\ell)=b} p^{\mathrm{L,D}}_{\ell dtn\omega} + \sum_{l|F(\ell)=b} p^{\mathrm{L,D}}_{\ell dtn\omega}$$
$$= \tilde{p}^{\mathrm{D}}_{bdtn\omega} - p^{\mathrm{UD,D}}_{bdtn\omega}, \quad \forall b \in B, \forall d \in D, \forall t \in T, \forall n \in N, \forall \omega \in \Omega. \quad (9.5d)$$

Generating units operation:

$$A^{\mathrm{D}}_{gdt} \sum_{j\le n} p^{\mathrm{I,G}}_{gj\omega} - p^{\mathrm{G,D}}_{gdtn\omega} \ge 0, \quad \forall g \in G^{\mathrm{C}}, \forall d \in D, \forall t \in T, \forall n \in N, \forall \omega \in \Omega, \quad (9.5e)$$

$$A^{\mathrm{D}}_{gdt} P^{\mathrm{E,G}}_{\max,gn} - p^{\mathrm{G,D}}_{gdtn\omega} \ge 0, \quad \forall g \in G^{\mathrm{E}}, \forall d \in D, \forall t \in T, \forall n \notin N^{\mathrm{off}}_g, \forall \omega \in \Omega, \quad (9.5f)$$

$$p^{\mathrm{G,D}}_{gdtn\omega} = 0, \quad \forall g \in G^{\mathrm{E}}, \forall d \in D, \forall t \in T, \forall n \in N^{\mathrm{off}}_g, \forall \omega \in \Omega, \quad (9.5g)$$

$$p^{\mathrm{G,D}}_{gdtn\omega} - p^{\mathrm{G,D}}_{gdt-1,n\omega} \le R^{\mathrm{UP}}_g P^{\mathrm{E,G}}_{\max,gn}, \quad \forall g \in G^{\mathrm{D}} \cup G^{\mathrm{E}}, \forall d \in D, \forall t \in T, \forall n \in N, \forall \omega \in \Omega, \quad (9.5h)$$

$$p^{\mathrm{G,D}}_{gdt-1,n\omega} - p^{\mathrm{G,D}}_{gdtn\omega} \le R^{\mathrm{DW}}_g P^{\mathrm{E,G}}_{\max,gn}, \quad \forall g \in G^{\mathrm{D}} \cup G^{\mathrm{E}}, \forall d \in D, \forall t \in T, \forall n \in N, \forall \omega \in \Omega, \quad (9.5i)$$

$$p^{\mathrm{G,D}}_{gdtn\omega} - p^{\mathrm{G,D}}_{gdt-1,n\omega} \le R^{\mathrm{UP}}_g \sum_{j\le n} p^{\mathrm{I,G}}_{gj\omega}, \quad \forall g \in G^{\mathrm{D}} \cup G^{\mathrm{C}}, \forall d \in D, \forall t \in T, \forall n \in N, \forall \omega \in \Omega, \quad (9.5j)$$

$$p^{\mathrm{G,D}}_{gdt-1,n\omega} - p^{\mathrm{G,D}}_{gdtn\omega} \le R^{\mathrm{DW}}_g \sum_{j\le n} p^{\mathrm{I,G}}_{gj\omega}, \quad \forall g \in G^{\mathrm{D}} \cup G^{\mathrm{C}}, \forall d \in D, \forall t \in T, \forall n \in N, \forall \omega \in \Omega, \quad (9.5k)$$

$$p^{\mathrm{G,D}}_{gdtn\omega}, p^{\mathrm{I,G}}_{gn\omega} \ge 0, \quad \forall g \in G, \forall d \in D, \forall t \in T, \forall n \in N, \forall \omega \in \Omega. \quad (9.5l)$$

Demand and transmission lines limits:

$$0 \leq p^{\mathrm{UD,D}}_{bdtn\omega} \leq \tilde{p}^{\mathrm{D}}_{bdtn\omega}, \quad \forall b \in B, \forall d \in D, \forall t \in T, \forall n \in N, \forall \omega \in \Omega, \tag{9.5m}$$

$$p^{\mathrm{L,D}}_{\ell dtn\omega} = \frac{1}{X_\ell}(\delta_{O(\ell)dtn\omega} - \delta_{F(\ell)dtn\omega}), \quad \forall \ell \in L, \forall d \in D, \forall t \in T, \forall n \in N, \forall \omega \in \Omega, \tag{9.5n}$$

$$-P^{\mathrm{L}}_{\max,\ell} \leq p^{\mathrm{L,D}}_{\ell dtn\omega} \leq P^{\mathrm{L}}_{\max,\ell}, \quad \forall \ell \in L, \forall d \in D, \forall t \in T, \forall n \in N, \forall \omega \in \Omega, \tag{9.5o}$$

$$-\pi \leq \delta_{bdtn\omega} \leq \pi, \quad \forall b \in B, \forall d \in D, \forall t \in T, \forall n \in N, \forall \omega \in \Omega. \tag{9.5p}$$

Storage units operation:

$$0 \leq p^{\mathrm{S,D}}_{sdtn\omega} \leq P^{\mathrm{C,SP}}_{\max,s}, \quad \forall s \in S, \forall d \in D, \forall t \in T, \forall n \in N, \forall \omega \in \Omega, \tag{9.5q}$$

$$0 \leq p^{\mathrm{S,C}}_{sdtn\omega} \leq P^{\mathrm{C,SP}}_{\max,s}, \quad \forall s \in S, \forall d \in D, \forall t \in T, \forall n \in N, \forall \omega \in \Omega, \tag{9.5r}$$

$$e^{\mathrm{SD}}_{sdtn\omega} = e^{\mathrm{SD}}_{sdt-1,n\omega} - (1/\eta^{\mathrm{S}}_s) p^{\mathrm{S,D}}_{sdtn\omega} \cdot 1 + \eta^{\mathrm{S}}_s \cdot p^{\mathrm{S,C}}_{sdtn\omega} \cdot 1, \quad \forall s \in S, \forall d \in D, \forall t > 1, \forall n \in N, \forall \omega \in \Omega, \tag{9.5s}$$

$$e^{\mathrm{SD}}_{sdtn\omega} = E^{\mathrm{C,SE}}_{\mathrm{ini},s} - (1/\eta^{\mathrm{S}}_s) p^{\mathrm{S,D}}_{sdtn\omega} \cdot 1 + \eta^{\mathrm{S}}_s \cdot p^{\mathrm{S,C}}_{sdtn\omega} \cdot 1, \quad \forall s \in S, \forall d \in D, t = 1, \forall n \in N, \forall \omega \in \Omega, \tag{9.5t}$$

$$0 \leq e^{\mathrm{SD}}_{sdtn\omega} \leq H^{\mathrm{S}}_s P^{\mathrm{C,SP}}_{\max,s}, \quad \forall s \in S, \forall d \in D, \forall t \in T, \forall n \in N, \forall \omega \in \Omega, \tag{9.5u}$$

$$e^{\mathrm{SD}}_{sdtn\omega} \geq E^{\mathrm{C,SE}}_{\mathrm{ini},s}, \quad \forall s \in S, \forall d \in D, t = 24, \forall n \in N, \forall \omega \in \Omega, \tag{9.5v}$$

where $\Theta^{\mathrm{St}} = \{p^{\mathrm{I,G}}_{gn\omega}, \forall g \in G, \forall n \in N, \forall \omega \in \Omega;\ p^{\mathrm{G,D}}_{gdtn\omega}, \forall g \in G, \forall d \in D, \forall t \in T, \forall n \in N, \forall \omega \in \Omega;\ p^{\mathrm{L,D}}_{\ell dtn\omega}, \forall \ell \in L, \forall d \in D, \forall t \in T, \forall n \in N, \forall \omega \in \Omega;\ e^{\mathrm{SD}}_{sdtn\omega}, p^{\mathrm{S,D}}_{sdtn\omega}, p^{\mathrm{S,C}}_{sdtn\omega}, \forall s \in S, \forall d \in D, \forall t \in T, \forall n \in N, \forall \omega \in \Omega\}$ is the set of optimization variables.

The objective function (9.5a) comprises the total expected investment and operating costs. Specifically, the first term of the equation represents the annualized investment cost of the generating capacity to be built considering as a random variable the investment cost of the technologies. The second and third terms of the objective function represent the operating cost of the generating units to supply the demand and the penalization cost of the unserved demand, respectively. Note that the latter terms are multiplied by the weight of the selected representative days.

The objective function is subject to several constraints. Constraints (9.5b) impose limits on the capacity that can be built of the candidate generating units to the potential capacity of each unit. For generalization purposes, those limits are imposed at each decision stage. In an explicit stochastic formulation (or scenario-stage formulation), non-anticipativity constraints as (9.5c) must be enforced so that the same investment decisions are taken for equal values of the random variables at the current stage. Set $\tilde{\xi}_n(\omega)$ comprises the random variables between stages 1 and n in scenario ω.

The remaining constraints describe the system operation. Constraints (9.5d) enforce the power balance at each bus, characteristic day, hour, decision stage, and scenario, considering the power generated by the generating units, the power discharged and charged by the storage units, the power flow through the transmission lines, and the hourly demand and the possible unserved demand. The hourly demand $\tilde{p}^{\mathrm{D}}_{bdtn\omega}$ is characterized as a random variable that takes into account the demand growth attained in each scenario ω of decision stage n, but also the demand level at each bus b in each hour t of representative day d.

Constraints (9.5e)–(9.5l) define the generating units operation. Constraints (9.5e) limit the power generated by the new units considering the capacity to be built up to stage n and the availability factor of the energy source. Similarly, (9.5f) impose equivalent limits to the existing units as long as they are not decommissioned. As stated by (9.5g), when the unit is decommissioned the power generated is null. Constraints (9.5h)–(9.5k) enforce ramping limits to dispatchable units of existing and new generation units, respectively. Constraints (9.5l) define as positive the investment and operating variables of the generation units.

Constraints (9.5m) limit the unserved demand to the actual demand in the corresponding bus. Constraints (9.5n)–(9.5p) model the DC power flows in transmission lines.

Constraints (9.5q)–(9.5v) characterize the operation of storage units. In particular, (9.5q) and (9.5r) limit the power discharged and charged by each storage unit, respectively. Constraints (9.5s) and (9.5t) define the energy balance in the storage unit, considering the state of charge at the previous period (or the beginning of the day) and the power charged and discharged in the current hour. Note that we assume that each period t corresponds to one hour. Constraints (9.5u) enforce the capacity limits of the stored energy, while (9.5v) impose as minimum energy level at the end of the day the energy level at the beginning of the day.

Alternatively, constraints limiting the investment budget or the GHG emissions, or imposing minimum renewable generation or capacity can be easily added to this type of capacity expansion models. We leave these modeling exercises for the interested reader.

Finally, note that problem (9.5) is a linear-programming (LP) model.

9.6 Illustrative Case Study: Multi-Stage Stochastic Capacity Expansion Problem

A multi-stage capacity expansion model formulated using stochastic optimization is solved. In this example, we consider one bus and a coal ('Coal') unit as existing unit. Along the planning horizon, it is possible to build a wind ('Wind') and a solar photovoltaic ('SoPV') unit. Figure 9.4 depicts the schema of the power system. The planning horizon is divided into two periods. Only the demand growth is assumed to be a random variable (the investment costs are known) and two scenarios per stage

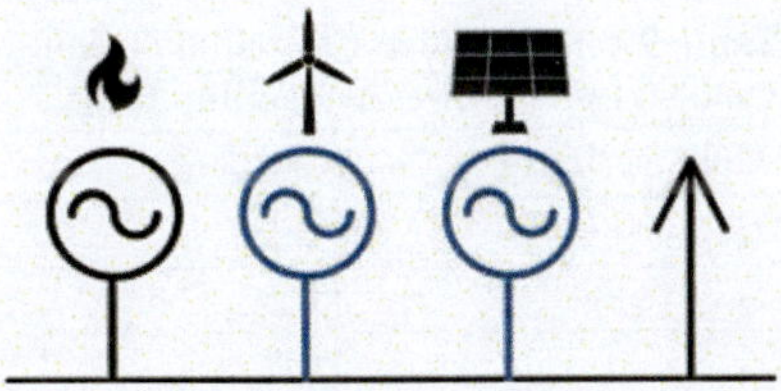

Fig. 9.4 Illustrative Case Study—Multi-stage stochastic capacity expansion: schema of the power system

are considered, which results in a total of four scenarios. The value of the demand growth per scenario is that presented in Illustrative Example 9.3. All scenarios are equiprobable, with a probability equal to 0.25.

The investment cost of the wind and solar PV unit is 1000 and 600 €/kW, respectively. The capital recovery factor applied to annualize the investment cost is equal to 0.07, which results from considering a life time of 20 years and an interest rate of 3%. The operating cost of the coal, wind, and solar PV units is 40, 3, and 1 €/MWh, respectively. The capacity of the coal unit is 150 MW, while the potential capacities to be built of wind and solar PV units are 150 and 100 MW, respectively. The peak demand at the beginning of the planning horizon is 110 MW. The cost of unserved demand is 1000 €/MWh.

The operating variability in this example is considered using one representative day that comprises only 8 operating periods instead of 24. Hence, the weight of the representative day is 365 and the weight of each operating period is 3 (24/8). Therefore, the weight of each period results in $365 \cdot 3 = 1095$. As a result, parameter W_d is equal to 1095 in the objective function (9.5a).

The demand level and wind and solar power availability for each operating period are indicated in Table 9.4. The power availability of the coal unit is equal to one.

The formulation of this problem is as follows. Note that since there is just one representative day, we can avoid to include index d representing the days.

$$\text{Minimize} \sum_{n=1}^{2}\sum_{\omega=1}^{4} 0.25\left(70 \cdot p^{\text{I,G}}_{\text{Wind},n\omega} + 42 \cdot p^{\text{I,G}}_{\text{SoPV},n\omega}\right)$$

$$+\sum_{n=1}^{2}\sum_{\omega=1}^{4} 0.25 \cdot \sum_{t=1}^{8} 1095\left(0.04 \cdot p^{\text{G,D}}_{\text{Coal},tn\omega} + 0.003 \cdot p^{\text{G,D}}_{\text{Wind},tn\omega}\right.$$

$$\left. +0.001 \cdot p^{\text{G,D}}_{\text{SoPV},tn\omega} + 1 \cdot p^{\text{UD,D}}_{dtn\omega}\right)$$

subject to:

$$p^{\text{C,G}}_{\max,gn} - \sum_{j=1}^{n} p^{\text{I,G}}_{gj\omega} \geq 0, \quad \forall g \in \{\text{Wind, SoPV}\}, n = 1, 2; \omega = 1, ..., 4,$$

$$p^{\text{I,G}}_{gn\omega} = p^{\text{I,G}}_{gn\omega'}, \quad \forall g \in \{\text{Wind, SoPV}\}, n = 1, 2; \forall \omega, \omega' : \tilde{\xi}_n(\omega) = \tilde{\xi}_n(\omega')$$

Table 9.4 Illustrative Case Study—Multi-stage stochastic capacity expansion: demand level and wind and solar power availability (pu)

Oper. period	Demand	Wind	SoPV
t_1	0.6	0.7	0.1
t_2	0.5	0.7	0.3
t_3	0.7	0.5	0.7
t_4	0.8	0.3	1.0
t_5	0.9	0.4	1.0
t_6	0.8	0.4	0.8
t_7	0.7	0.6	0.3
t_8	0.5	0.7	0.0

$$\sum_{g \in G} p^{\mathrm{G,D}}_{gtn\omega} = \tilde{p}^{\mathrm{D}}_{tn\omega} - p^{\mathrm{UD,D}}_{tn\omega}, \quad t = 1, ..., 8; n = 1, 2; \omega = 1, ..., 4,$$

$$A^{\mathrm{D}}_{gt} \sum_{j=1}^{n} p^{\mathrm{I,G}}_{gj\omega} - p^{\mathrm{G,D}}_{gtn\omega} \geq 0, \quad \forall g \in \{\text{Wind, SoPV}\}, t = 1, ..., 8; n = 1, 2; \omega = 1, ..., 4,$$

$$1 \cdot P^{\mathrm{E,G}}_{\max,\mathrm{Coal},1} - p^{\mathrm{G,D}}_{\mathrm{Coal},t1\omega} \geq 0, \quad t = 1, ..., 8; \omega = 1, ..., 4,$$

$$p^{\mathrm{G,D}}_{\mathrm{Coal},t2\omega} = 0, \quad t = 1, ..., 8; \omega = 1, ..., 4,$$

$$p^{\mathrm{I,G}}_{gn\omega} \geq 0, \forall g \in \{\text{Wind, SoPV}\}, \quad n = 1, 2; \omega = 1, ..., 4,$$

$$p^{\mathrm{G,D}}_{gtn\omega} \geq 0, \forall g \in \{\text{Coal, Wind, SoPV}\}, \quad t = 1, ..., 8; n = 1, 2; \omega = 1, ..., 4,$$

$$p^{\mathrm{UD,D}}_{tn\omega} \geq 0, \quad t = 1, ..., 8; n = 1, 2; \omega = 1, ..., 4.$$

This LP problem comprises 275 variables and 289 equations. It can be solved with the educational license of GAMS [8]. The code is provided in Sect. 9.12.1.

Considering the previous input data and the fact that the coal unit is not decommissioned, the optimal values of the capacity to be built per technology in each stage and scenario are those provided in Table 9.5. As it can be observed, high wind and solar capacity is built at the first decision stage. This means that a lower cost is attained by building new renewable capacity even though the existing coal unit could supply the

Table 9.5 Illustrative Case Study—Multi-stage stochastic capacity expansion: capacity to be built (MW)

Unit	Stage	Scenario			
		ω_1	ω_2	ω_3	ω_4
Wind	n_1	107.9	107.9	107.9	107.9
	n_2	10.4	10.4	0.0	0.0
	Total	118.3	118.3	107.9	107.9
SoPV	n_1	100.0	100.0	100.0	100.0
	n_2	0.0	0.0	0.0	0.0
	Total	100.0	100.0	100.0	100.0

Table 9.6 Illustrative Case Study—Multi-stage stochastic capacity expansion: energy generated by each unit in each operating period, stage and scenario (MWh)

Oper. period	Coal							
	Stage n_1				Stage n_2			
	ω_1	ω_2	ω_3	ω_4	ω_1	ω_2	ω_3	ω_4
t_1	0.0	0.0	0.0	0.0	6.6	1.9	4.6	0.0
t_2	0.0	0.0	0.0	0.0	0.0	0.0	0.0	0.0
t_3	0.0	0.0	0.0	0.0	0.0	0.0	0.0	0.0
t_4	0.0	0.0	0.0	0.0	0.0	0.0	0.0	0.0
t_5	0.0	0.0	0.0	0.0	1.8	0.0	0.0	0.0
t_6	0.0	0.0	0.0	0.0	5.2	0.0	0.0	0.0
t_7	0.0	0.0	0.7	0.7	15.0	9.6	10.4	5.0
t_8	0.0	0.0	0.0	0.0	0.0	0.0	0.0	0.0
Oper. period	Wind							
	Stage n_1				Stage n_2			
	ω_1	ω_2	ω_3	ω_4	ω_1	ω_2	ω_3	ω_4
t_1	67.8	67.8	71.8	71.8	82.8	82.8	75.5	75.5
t_2	34.8	34.8	38.1	38.1	52.8	49.0	45.1	41.3
t_3	20.8	20.8	25.4	25.4	46.0	40.6	35.2	29.8
t_4	3.8	3.8	9.0	9.0	32.5	26.4	20.2	14.0
t_5	16.7	16.7	22.7	22.7	47.3	42.2	35.2	28.3
t_6	23.8	23.8	29.0	29.0	47.3	46.4	40.2	34.0
t_7	60.8	60.8	64.7	64.7	71.0	71.0	64.7	64.7
t_8	64.8	64.8	68.1	68.1	82.8	79.0	75.1	71.3
Oper. period	SoPV							
	Stage n_1				Stage n_2			
	ω_1	ω_2	ω_3	ω_4	ω_1	ω_2	ω_3	ω_4
t_1	10.0	10.0	10.0	10.0	10.0	10.0	10.0	10.0
t_2	30.0	30.0	30.0	30.0	30.0	30.0	30.0	30.0
t_3	70.0	70.0	70.0	70.0	70.0	70.0	70.0	70.0
t_4	100.0	100.0	100.0	100.0	100.0	100.0	100.0	100.0
t_5	100.0	100.0	100.0	100.0	100.0	100.0	100.0	100.0
t_6	80.0	80.0	80.0	80.0	80.0	80.0	80.0	80.0
t_7	30.0	30.0	30.0	30.0	30.0	30.0	30.0	30.0
t_8	0.0	0.0	0.0	0.0	0.0	0.0	0.0	0.0

total demand at least in the first stage. Moreover, note that in scenarios 3 and 4 when the demand growth is comparatively lower, no wind capacity is built at the second stage.

The total expected cost arises to 16.088 M€, being 12.118 M€ and 3.970 M€ the investment and operating costs, respectively.

Table 9.6 gives the power generated by each unit in each operating period, decision stage and scenario. As expected from the previous results, the coal unit does not

Table 9.7 Illustrative Case Study—Multi-stage stochastic capacity expansion: capacity to be built with the decommissioning of the coal unit (MW)

Unit	Stage	Scenario			
		ω_1	ω_2	ω_3	ω_4
Wind	n_1	109.0	109.0	109.0	109.0
	n_2	34.3	34.3	16.3	16.3
	Total	143.3	143.3	125.3	125.3
SoPV	n_1	100.0	100.0	100.0	100.0
	n_2	0.0	0.0	0.0	0.0
	Total	100.0	100.0	100.0	100.0

generate much energy, especially in the first stage where the demand growth is not high. However, the solar PV generates the maximum power available in each hour, whereas there is wind power spillage in some hours.

Then, we analyse how the total expected cost would be if no wind and solar units can be built, i.e., if the coal unit must supply all the demand. In this case, assuming a coal capacity of 250 MW, the total expected cost, which is equal to the total operating cost in this case, would be 69.162 M€, which is around 4 times the investment and operating cost of the previous case.

Finally, let us investigate how the solution changes if the coal unit is decommissioned at stage 2. The capacity to be built in each stage is provided in Table 9.7. As it can be observed, the wind capacity built increases in all scenarios with respect to the initial case. The total expected cost in this case is 16.947 M€.

9.7 Formulation Based on Linear Decision Rules

This section describes the formulation of the multi-stage capacity expansion model using LDRs. This formulation is based on the approach presented in [3, 15].

9.7.1 *Modeling Approach*

As explained in Chap. 2, the LDR approach assumes a linear dependency between the random and the decision variables. This implies that the space of optimal decisions is limited to those fulfilling such linear dependency. As shown in [3], this approach leads to problems with lower size than the corresponding stochastic formulation when the number of stages or scenarios is high.

Following the description of Sect. 9.3, we define $\tilde{\boldsymbol{\xi}}$ as a vector comprising the random variables involved in the model. For modeling purposes and without loss of generality, the first element of this vector is equal to 1. Therefore, the vector is defined as follows:

$$\tilde{\boldsymbol{\xi}} = [1, \tilde{c}^{\mathrm{I}}_{11}, ..., \tilde{c}^{\mathrm{I}}_{\mathrm{N^G}1}, \tilde{d}^{\mathrm{D}}_{11}, ..., \tilde{d}^{\mathrm{D}}_{\mathrm{N^B}1}, ..., \tilde{c}^{\mathrm{I}}_{1\mathrm{N^N}}, ..., \tilde{c}^{\mathrm{I}}_{\mathrm{N^G}\mathrm{N^N}}, \tilde{d}^{\mathrm{D}}_{1\mathrm{N^N}}, ..., \tilde{d}^{\mathrm{D}}_{\mathrm{N^B}\mathrm{N^N}}],$$

being $\mathrm{N^G}$, $\mathrm{N^B}$, and $\mathrm{N^N}$ the number of generating units, buses, and decision stages, respectively.

In addition, $\tilde{\boldsymbol{\xi}}_m$ is defined as the m element of the vector comprising the random variables, $\tilde{\boldsymbol{\xi}}$. Thus, we can establish the relationship between the random variables and the investment and operating decisions as follows:

$$p^{\mathrm{I,G}}_{gn} = \sum_{m \in M^I_n} \hat{p}^{\mathrm{I,G}}_{gnm} \tilde{\xi}_m, \quad \forall g \in G, \forall n \in N, \tag{9.6a}$$

$$p^{\mathrm{G,D}}_{gdtn} = \sum_{m \in M^O_n} \hat{p}^{\mathrm{G,D}}_{gdtnm} \tilde{\xi}_m, \quad \forall g \in G, \forall d \in D, \forall t \in T, \forall n \in N, \tag{9.6b}$$

where $\hat{p}^{\mathrm{I,G}}_{gnm}$ and $\hat{p}^{\mathrm{G,D}}_{gdtnm}$ are the new decision variables; set M^I_n comprises the positions in array $\tilde{\boldsymbol{\xi}}$ of those random variables known when investment decisions at stage n are made; and set M^O_n comprises the positions in array $\tilde{\boldsymbol{\xi}}$ of random variables known when operating decisions at stage n are made. The definition of sets M^I_n and M^O_n, including only those random variables known up to stage n, implies the non-anticipativity on the decisions defined through (9.6).

Following the same reasoning, it is possible to establish the linear dependency of all variables and deterministic parameters with respect to vector $\tilde{\xi}$. The following examples illustrate how to define the LDRs.

Illustrative Example 9.4 Decision rules of the variables

This example considers the planning horizon and the random variables of Illustrative Example 9.3. It is assumed that the investment cost of all generating units evolves in the same way along the planning horizon. Similarly, the demand growth is the same in all buses. Therefore, for the sake of clarity, we can eliminate index g and b from the symbols representing those random variables and just include index n representing the decision stage. Thus, vector $\tilde{\xi}$ is defined as follows:

$$\tilde{\boldsymbol{\xi}} = \left[1, \tilde{c}^{\mathrm{I}}_1, \tilde{d}^{\mathrm{D}}_1, \tilde{c}^{\mathrm{I}}_2, \tilde{d}^{\mathrm{D}}_2\right]. \tag{9.7}$$

Then, the linear decision rules applied to the investment decisions in stages 1 and 2 are formulated as follows:

First stage: $n = 1;\ M_1^I = \{1, 2\};\ \tilde{\boldsymbol{\xi}}_{M_1^I} = \{1, \tilde{c}_1^{\mathrm{I}}\}$

$$p_1^{\mathrm{I,G}} = \sum_{m=1}^{2} \hat{p}_{g1m}^{\mathrm{I,G}} \tilde{\xi}_m = \hat{p}_{g11}^{\mathrm{I,G}} \tilde{\xi}_1 + \hat{p}_{g12}^{\mathrm{I,G}} \tilde{\xi}_2 = \hat{p}_{g11}^{\mathrm{I,G}} + \hat{p}_{g12}^{\mathrm{I,G}} \tilde{c}_1^{\mathrm{I}}, \quad \forall g \in G.$$

Second stage: $n = 2;\ M_2^I = \{1, 2, 3, 4\};\ \tilde{\boldsymbol{\xi}}_{M_2^I} = \{1, \tilde{c}_1^{\mathrm{I}}, \tilde{d}_1^{\mathrm{D}}, \tilde{c}_2^{\mathrm{I}}\}$

$$\begin{aligned} p_{g2}^{\mathrm{I,G}} &= \sum_{m=1}^{4} \hat{p}_{g2m}^{\mathrm{I,G}} \tilde{\xi}_m = \hat{p}_{g21}^{\mathrm{I,G}} \tilde{\xi}_1 + \hat{p}_{g22}^{\mathrm{I,G}} \tilde{\xi}_2 + \hat{p}_{g23}^{\mathrm{I,G}} \tilde{\xi}_3 + \hat{p}_{g24}^{\mathrm{I,G}} \tilde{\xi}_4 \\ &= \hat{p}_{g21}^{\mathrm{I,G}} + \hat{p}_{g22}^{\mathrm{I,G}} \tilde{c}_1^{\mathrm{I}} + \hat{p}_{g23}^{\mathrm{I,G}} \tilde{d}_1^{\mathrm{D}} + \hat{p}_{g24}^{\mathrm{I,G}} \tilde{c}_2^{\mathrm{I}}, \quad \forall g \in G. \end{aligned}$$

Note that the investment decisions at the first stage do not depend on the demand growth since the realization of this random variable is known at the end of the first investment period. However, in the second stage the demand growth in the first period has been realized and the investment decisions are made considering those values and those of the investment costs at stage 2.

Similarly, the LDRs representing an operating variable such as the power generated by each unit can be written as follows:

First stage: $n = 1;\ M_1^O = \{1, 2, 3\};\ \tilde{\boldsymbol{\xi}}_{M_1^O} = \{1, \tilde{c}_1^{\mathrm{I}}, \tilde{d}_1^{\mathrm{D}}\}$

$$\begin{aligned} p_{gdt1}^{\mathrm{G,D}} &= \sum_{m=1}^{3} \hat{p}_{gdt1m}^{\mathrm{G,D}} \tilde{\xi}_m = \hat{p}_{gdt11}^{\mathrm{G,D}} \tilde{\xi}_1 + \hat{p}_{gdt12}^{\mathrm{G,D}} \tilde{\xi}_2 + \hat{p}_{gdt13}^{\mathrm{G,D}} \tilde{\xi}_3 \\ &= \hat{p}_{gdt11}^{\mathrm{G,D}} + \hat{p}_{gdt12}^{\mathrm{G,D}} \tilde{c}_1^{\mathrm{I}} + \hat{p}_{gdt12}^{\mathrm{G,D}} \tilde{d}_1^{\mathrm{D}}, \quad \forall g \in G, \forall d \in D, \forall t \in T. \end{aligned}$$

Second stage: $n = 2;\ M_2^O = \{1, 2, 3, 4, 5\};\ \tilde{\boldsymbol{\xi}}_{M_2^O} = \{1, \tilde{c}_1^{\mathrm{I}}, \tilde{d}_1^{\mathrm{D}}, \tilde{c}_2^{\mathrm{I}}, \tilde{d}_2^{\mathrm{D}}\}$

$$\begin{aligned} p_{gdt2}^{\mathrm{G,D}} &= \sum_{m=1}^{5} \hat{p}_{gdt2m}^{\mathrm{G,D}} \tilde{\xi}_m \\ &= \hat{p}_{gdt21}^{\mathrm{G,D}} \tilde{\xi}_1 + \hat{p}_{gdt22}^{\mathrm{G,D}} \tilde{\xi}_2 + \hat{p}_{gdt23}^{\mathrm{G,D}} \tilde{\xi}_3 + \hat{p}_{gdt24}^{\mathrm{G,D}} \tilde{\xi}_4 + \hat{p}_{gdt25}^{\mathrm{G,D}} \tilde{\xi}_5 \\ &= \hat{p}_{gdt21}^{\mathrm{G,D}} + \hat{p}_{gdt22}^{\mathrm{G,D}} \tilde{c}_1^{\mathrm{I}} + \hat{p}_{gdt23}^{\mathrm{G,D}} \tilde{d}_1^{\mathrm{D}} + \hat{p}_{gdt24}^{\mathrm{G,D}} \tilde{c}_2^{\mathrm{I}} + \hat{p}_{gdt25}^{\mathrm{G,D}} \tilde{d}_2^{\mathrm{D}}, \quad \forall g \in G, \forall d \in D, \forall t \in T. \end{aligned}$$

Observe that the operating decisions at stage n are made considering the realization of the demand growth up to such stage.

□

Illustrative Example 9.5 Decision rule of a random variable

The same framework as in Illustrative Example 9.4 is considered. This example shows how to define a random variable through vector $\tilde{\xi}$. The LDR of the investment cost and the demand growth are formulated as follows:

$$\tilde{c}^{\mathrm{I}}_{gn} = \sum_{m \in M^I_n} \hat{C}^{\mathrm{I}}_{gnm}\, \tilde{\xi}_m, \quad \forall g \in G^{\mathrm{C}}, \forall n \in N,$$

$$\tilde{d}^{\mathrm{D}}_{bn} = \sum_{m \in M^O_n} \hat{D}^{\mathrm{D}}_{bnm}\, \tilde{\xi}_m, \quad \forall b \in B, \forall n \in N.$$

Since the investment cost is applied to the investment variables, set M^I_n is considered to define its corresponding LDR. However, the demand growth is also known when operating decisions are made and, hence, set M^O_n applies.

Considering the definition (9.7) of vector $\tilde{\xi}$, the values of the deterministic parameters $\hat{C}^{\mathrm{I}}_{gnm}$ and $\hat{D}^{\mathrm{D}}_{nm}$ are computed as follows:

$$\left\{ \begin{array}{l} \hat{C}^{\mathrm{I}}_{gnm} = 1, \text{ if } \tilde{\xi}_m = \tilde{c}^{\mathrm{I}}_{gn} \\ \hat{C}^{\mathrm{I}}_{gnm} = 0, \text{ if } \tilde{\xi}_m \neq \tilde{c}^{\mathrm{I}}_{gn} \end{array} \right\},$$

$$\left\{ \begin{array}{l} \hat{D}^{\mathrm{I}}_{nm} = 1, \text{ if } \tilde{\xi}_m = \tilde{d}^{\mathrm{D}}_{n} \\ \hat{D}^{\mathrm{I}}_{nm} = 0, \text{ if } \tilde{\xi}_m \neq \tilde{d}^{\mathrm{D}}_{n} \end{array} \right\}.$$

For the sake of clarity, the values of parameters $\hat{C}^{\mathrm{I}}_{gnm}$ and $\hat{D}^{\mathrm{D}}_{nm}$ can be organized in matrices $\hat{C}^{\mathrm{I}}_{g}$ and $\hat{D}^{\mathrm{D}}$ as follows:

$$\hat{C}^{\mathrm{I}}_{g}, \ \forall g = \{1, 2\}, \ \in R^{2\times 5} = \begin{bmatrix} \hat{C}^{\mathrm{I}}_{g11} & \hat{C}^{\mathrm{I}}_{g12} & \hat{C}^{\mathrm{I}}_{g13} & \hat{C}^{\mathrm{I}}_{g14} & \hat{C}^{\mathrm{I}}_{g15} \\ \hat{C}^{\mathrm{I}}_{g21} & \hat{C}^{\mathrm{I}}_{g22} & \hat{C}^{\mathrm{I}}_{g23} & \hat{C}^{\mathrm{I}}_{g24} & \hat{C}^{\mathrm{I}}_{g25} \end{bmatrix} = \begin{bmatrix} 0 & 1 & 0 & 0 & 0 \\ 0 & 0 & 0 & 1 & 0 \end{bmatrix},$$

$$\hat{D}^{\mathrm{D}} \in R^{2\times 5} = \begin{bmatrix} \hat{D}^{\mathrm{D}}_{11} & \hat{D}^{\mathrm{D}}_{12} & \hat{D}^{\mathrm{D}}_{13} & \hat{D}^{\mathrm{D}}_{14} & \hat{D}^{\mathrm{D}}_{15} \\ \hat{D}^{\mathrm{D}}_{21} & \hat{D}^{\mathrm{D}}_{22} & \hat{D}^{\mathrm{D}}_{23} & \hat{D}^{\mathrm{D}}_{24} & \hat{D}^{\mathrm{D}}_{25} \end{bmatrix} = \begin{bmatrix} 0 & 0 & 1 & 0 & 0 \\ 0 & 0 & 0 & 0 & 1 \end{bmatrix}.$$

For example, $\hat{C}^{\mathrm{I}}_{112}$ is equal to 1 because the investment cost of the wind unit at the first stage, $\tilde{c}^{\mathrm{I}}_{11}$, is in the second position of array $\tilde{\boldsymbol{\xi}}$.

□

Illustrative Example 9.6 Decision rule of a deterministic parameter

The same framework as in Illustrative Example 9.4 is considered. In this example, the LDR of a deterministic parameter is defined. For instance, let us consider the parameter representing the potential capacity that can be built of a new generating unit, i.e., $P^{\mathrm{C,G}}_{\max,gn}$. Since this parameter is related to the investment decisions, set M^I_n is considered. The LDR is formulated as follows:

$$P^{\mathrm{C,G}}_{\max,gn} = \sum_{m \in M^I_n} \hat{P}^{\mathrm{C,G}}_{\max,gnm} \, \tilde{\xi}_m, \quad \forall g \in G, \forall n \in N.$$

Since $\tilde{\xi}_1 = 1$, the previous assignment holds if:

$$\left\{ \begin{array}{lll} \hat{P}^{\mathrm{C,G}}_{\max,gnm} = P^{\mathrm{C,G}}_{\max,gn}, & \text{if } m = 1 \\ \hat{P}^{\mathrm{C,G}}_{\max,gnm} = 0, & \text{if } m > 1 \end{array} \right\}.$$

Let us consider the power system of Illustrative Example 9.6. Then, matrices $\hat{P}^{\mathrm{C,G}}_{\max,\mathrm{Wind}}$ and $\hat{P}^{\mathrm{C,G}}_{\max,\mathrm{SoPV}}$ can be obtained using the values of parameters $\hat{P}^{\mathrm{C,G}}_{\max,gnm}$ as follows:

$$\hat{P}^{\mathrm{C,G}}_{\max,\mathrm{Wind}} \in R^{2\times 5} = \begin{bmatrix} P^{\mathrm{C,G}}_{\max,\mathrm{Wind}} \; 0\,0\,0\,0 \\ P^{\mathrm{C,G}}_{\max,\mathrm{Wind}} \; 0\,0\,0\,0 \end{bmatrix} = \begin{bmatrix} 150\;0\;0\;0\;0 \\ 150\;0\;0\;0\;0 \end{bmatrix} (\mathrm{MW}),$$

$$\hat{P}^{\mathrm{C,G}}_{\max,\mathrm{SoPV}} \in R^{2\times 5} = \begin{bmatrix} P^{\mathrm{C,G}}_{\max,\mathrm{SoPV}} \; 0\,0\,0\,0 \\ P^{\mathrm{C,G}}_{\max,\mathrm{SoPV}} \; 0\,0\,0\,0 \end{bmatrix} = \begin{bmatrix} 100\;0\;0\;0\;0 \\ 100\;0\;0\;0\;0 \end{bmatrix} (\mathrm{MW}).$$

The same reasoning applies for all deterministic parameters. □

If the LDRs previously defined are applied to the multi-stage capacity expansion model presented in Sect. 9.5.2, the resulting formulation is an LP optimization model whose space of optimal decisions is limited to those variables holding a linear relationship with the random variables. Thus, the objective function (9.5a) would be rewritten as follows:

$$\begin{aligned} \mathbb{E}_{\tilde{\xi}} \Bigg\{ & \sum_{n\in N} \sum_{g \in G^{\mathrm{C}}} \Bigg[\sum_{m \in M^I_n} \hat{C}^{\mathrm{I}}_{gnm} \tilde{\xi}_m \Bigg(\sum_{r \in M^I_n} \hat{p}^{\mathrm{I,G}}_{gnr} \tilde{\xi}_r \Bigg) \Bigg] \\ & + \sum_{n\in N} \sum_{d\in D} \sum_{t\in T} W_d \sum_{g\in G} \sum_{m \in M^O_n} \hat{C}^{\mathrm{O}}_{gnm} \tilde{\xi}_m \Bigg(\sum_{r\in M^O_n} \hat{p}^{\mathrm{G,D}}_{gdtnr} \tilde{\xi}_r \Bigg) \\ & + \sum_{n\in N} \sum_{d\in D} \sum_{t\in T} W_d \sum_{b\in B} \sum_{m \in M^O_n} \hat{C}^{\mathrm{UD}}_{nm} \tilde{\xi}_m \Bigg(\sum_{r\in M^O_n} \hat{p}^{\mathrm{UD,D}}_{bdtnr} \tilde{\xi}_r \Bigg) \Bigg\}, \end{aligned} \quad (9.8)$$

where $\hat{C}^{\mathrm{I}}_{gnm}$, $\hat{C}^{\mathrm{O}}_{gnm}$, and $\hat{C}^{\mathrm{UD}}_{nm}$ are deterministic parameters defining the values of the investment, operating, and unserved demand costs in the LDRs, respectively. Variable $\hat{p}^{\mathrm{UD,D}}_{bdtnm}$ is the new variable representing the unserved demand. The operator $\mathbb{E}_{\tilde{\xi}}$ represents the expected value of the objective function with respect to the random variables in $\tilde{\xi}$.

As it will be explained below, all the variables and deterministic parameters depend on the random variables through vector $\tilde{\xi}$.

All constraints of the mathematical optimization problem can be reformulated using a similar approach to that used to obtain the expression (9.8) for the objective function. For instance, constraints limiting the capacity to be built (9.5b) are reformulated in the following way:

$$\sum_{m\in M^I_n} \hat{p}^{\mathrm{C,G}}_{\max,gnm}\, \tilde{\xi}_m - \sum_{j\le n}\left(\sum_{m\in M^I_j} \hat{p}^{\mathrm{I,G}}_{gjm}\tilde{\xi}_m\right) \ge 0, \quad \forall g \in G^{\mathrm{C}}, \forall n \in N. \tag{9.9}$$

The rest of constraints would be equivalently reformulated.

In addition, considering the uncertainty characterization described in Sect. 9.3, the intervals containing the random variable values can be defined through vector $\tilde{\xi}$ and parameters J and h as follows:

$$\sum_{m\in M^O_n} J_{km}\tilde{\xi}_m \ge h_k, \quad \forall k \in K, \tag{9.10}$$

being $K = \{1, ..., N_K\}$ and $N_K = 2 \cdot N_{M^O_n}$.

Illustrative Example 9.7 Uncertainty intervals using parameters J and h

Considering the data used in Illustrative Example 9.1 and the definition (9.7) of vector $\tilde{\xi}$, parameters J and h describing the intervals of the random variables take the following values:

$$J = \begin{bmatrix} 1 & 0 & 0 & 0 & 0 \\ -1 & 0 & 0 & 0 & 0 \\ -(a^{\mathrm{I}}_g + b^{\mathrm{I}}_g \cdot \bar{c}^{\mathrm{I}}_{g,0})(1-\theta^{\mathrm{I}}_{g1}) & 1 & 0 & 0 & 0 \\ (a^{\mathrm{I}}_g + b^{\mathrm{I}}_g \cdot \bar{c}^{\mathrm{I}}_{g,0})(1+\theta^{\mathrm{I}}_{g1}) & -1 & 0 & 0 & 0 \\ -(a^{\mathrm{D}}_b + b^{\mathrm{D}}_b \cdot \bar{d}^{\mathrm{D}}_{b,0})(1-\theta^{\mathrm{D}}_{b1}) & 0 & 1 & 0 & 0 \\ (a^{\mathrm{D}}_b + b^{\mathrm{D}}_b \cdot \bar{d}^{\mathrm{D}}_{b,0})(1+\theta^{\mathrm{D}}_{b1}) & 0 & -1 & 0 & 0 \\ -a^{\mathrm{I}}_g(1-\theta^{\mathrm{I}}_{g2}) & -b^{\mathrm{I}}_g(1-\theta^{\mathrm{I}}_{g2}) & 0 & 1 & 0 \\ a^{\mathrm{I}}_g(1+\theta^{\mathrm{I}}_{g2}) & b^{\mathrm{I}}_g(1+\theta^{\mathrm{I}}_{g2}) & 0 & -1 & 0 \\ -a^{\mathrm{D}}_b(1-\theta^{\mathrm{D}}_{b2}) & 0 & -b^{\mathrm{D}}_b(1-\theta^{\mathrm{D}}_{b2}) & 0 & 1 \\ a^{\mathrm{D}}_b(1+\theta^{\mathrm{D}}_{b2}) & 0 & b^{\mathrm{D}}_b(1+\theta^{\mathrm{D}}_{b2}) & 0 & -1 \end{bmatrix}$$

$$
= \begin{bmatrix} 1 & 0 & 0 & 0 & 0 \\ -1 & 0 & 0 & 0 & 0 \\ -1 & 1 & 0 & 0 & 0 \\ 1 & -1 & 0 & 0 & 0 \\ -1.149 & 0 & 1 & 0 & 0 \\ 1.270 & 0 & -1 & 0 & 0 \\ -0.008 & -0.722 & 0 & 1 & 0 \\ 0.011 & 0.977 & 0 & -1 & 0 \\ -0.180 & 0 & -0.900 & 0 & 1 \\ 0.220 & 0 & 1.100 & 0 & -1 \end{bmatrix},
$$

$$
h = \begin{bmatrix} 1 & -1 & 0 & 0 & 0 & 0 & 0 & 0 & 0 & 0 & 0 \end{bmatrix}. \qquad \square
$$

Still, the presence of vector $\tilde{\xi}$ in the formulation makes the resulting problem difficult to solve. Therefore, the next step is to remove that vector from all mathematical expressions of the optimization model. To do this, basic algebraic applies to the objective function, whereas the intervals definition (9.10) allows to reformulate the constraints.

Firstly, observe that in objective function (9.8) all the parameters are deterministic except for vector $\tilde{\xi}$, which implies that it is the only parameter affected by the expectation operator. Therefore, the following equivalences can be stated. For the sake of clarity, only one of the three cost terms in (9.8) is included.

$$
\begin{aligned}
&\mathbb{E}_{\tilde{\xi}}\left\{\sum_{n\in N}\sum_{g\in G^{\mathrm{C}}}\left[\sum_{m\in M_n^I}\hat{C}^{\mathrm{I}}_{gnm}\tilde{\xi}_m\left(\sum_{r\in M_n^I}\hat{p}^{\mathrm{I,G}}_{gnr}\tilde{\xi}_r\right)\right]\right\}\\
&=\mathbb{E}_{\tilde{\xi}}\left\{\sum_{n\in N}\sum_{g\in G^{\mathrm{C}}}\sum_{m\in M_n^I}\sum_{r\in M_n^I}\hat{p}^{\mathrm{I,G}}_{gnr}\hat{C}^{\mathrm{I}}_{gnm}\tilde{\xi}_m\tilde{\xi}_r\right\}\\
&=\sum_{n\in N}\sum_{g\in G^{\mathrm{C}}}\sum_{m\in M_n^I}\sum_{r\in M_n^I}\hat{p}^{\mathrm{I,G}}_{gnr}\hat{C}^{\mathrm{I}}_{gnm}\mathbb{E}_{\tilde{\xi}}\left\{\tilde{\xi}_m\tilde{\xi}_r\right\}\\
&=\sum_{n\in N}\sum_{g\in G^{\mathrm{C}}}\sum_{m\in M_n^I}\sum_{r\in M_n^I}\hat{p}^{\mathrm{I,G}}_{gnr}\hat{C}^{\mathrm{I}}_{gnm}M^{\mathrm{M}}_{mr},
\end{aligned}
\tag{9.11}
$$

where M^{M}_{mr} is the second order moment matrix and is computed as: $M^{\mathrm{M}}_{mr} = \mathbb{E}_{\tilde{\xi}}\left\{\tilde{\xi}_m\tilde{\xi}_r\right\} = \bar{\xi}_m\bar{\xi}_r + var\{\tilde{\xi}_m, \tilde{\xi}_r\}$.

As a consequence, all the parameters in (9.11) are deterministic.

Illustrative Example 9.8 Second order moment matrix

This example considers the data of Illustrative Example 9.1. In this case, the size of the second order moment matrix of vector $\tilde{\xi}$, M^{M}, is 5×5 and its components are computed as follows:

$$M^{\mathrm{M}} =$$

$$\begin{bmatrix} 1 & \bar{c}^{\mathrm{I}}_1 & \bar{d}^{\mathrm{D}}_1 & \bar{c}^{\mathrm{I}}_2 & \bar{d}^{\mathrm{D}}_2 \\ \bar{c}^{\mathrm{I}}_1 & (\bar{c}^{\mathrm{I}}_1)^2 & (\bar{c}^{\mathrm{I}}_1 \bar{d}^{\mathrm{D}}_1) & (\bar{c}^{\mathrm{I}}_1 \cdot \bar{c}^{\mathrm{I}}_2) & (\bar{c}^{\mathrm{I}}_1 \cdot \bar{d}^{\mathrm{D}}_2) \\ \bar{d}^{\mathrm{D}}_1 & (\bar{d}^{\mathrm{D}}_1 \cdot \bar{c}^{\mathrm{I}}_1) & ((\bar{d}^{\mathrm{D}}_1)^2 + \mathrm{var}\{\tilde{d}^{\mathrm{D}}_1\}) & (\bar{d}^{\mathrm{D}}_1 \cdot \bar{c}^{\mathrm{I}}_2) & (\bar{d}^{\mathrm{D}}_1 \cdot \bar{d}^{\mathrm{D}}_2 + \mathrm{cov}\{\tilde{d}^{\mathrm{D}}_1 \tilde{d}^{\mathrm{D}}_2\}) \\ \bar{c}^{\mathrm{I}}_2 & (\bar{c}^{\mathrm{I}}_2 \cdot \bar{c}^{\mathrm{I}}_1) & (\bar{c}^{\mathrm{I}}_2 \cdot \bar{d}^{\mathrm{D}}_1) & ((\bar{c}^{\mathrm{I}}_2)^2 + \mathrm{var}\{\tilde{c}^{\mathrm{I}}_2\}) & (\bar{c}^{\mathrm{I}}_2 \cdot \bar{d}^{\mathrm{D}}_2) \\ \bar{d}^{\mathrm{D}}_2 & (\bar{d}^{\mathrm{D}}_2 \cdot \bar{c}^{\mathrm{I}}_1) & (\bar{d}^{\mathrm{D}}_2 \cdot \bar{d}^{\mathrm{D}}_1 + \mathrm{cov}\{\tilde{d}^{\mathrm{D}}_1, \tilde{d}^{\mathrm{D}}_2\}) & (\bar{d}^{\mathrm{D}}_2 \cdot \bar{c}^{\mathrm{I}}_2) & ((\bar{d}^{\mathrm{D}}_2)^2 + \mathrm{var}\{\tilde{d}^{\mathrm{D}}_2\}) \end{bmatrix}$$

$$= \begin{bmatrix} 1.000 & 1.000 & 1.209 & 0.843 & 1.401 \\ 1.000 & 1.000 & 1.209 & 0.843 & 1.401 \\ 1.209 & 1.209 & 1.463 & 1.017 & 1.686 \\ 0.843 & 0.843 & 1.017 & 0.716 & 1.178 \\ 1.401 & 1.401 & 1.686 & 1.178 & 1.969 \end{bmatrix}.$$

Note that the covariance between a constant and a random variable is 0, as well as the covariance between the demand and the costs. □

Secondly, Proposition 1 in [15] is applied to the inequality constraints of optimization problem (9.5). This proposition states that being Z_{nm} a variable holding

$$\sum_{m \in M_n} Z_{nm} \tilde{\xi}_m \geq 0, \quad \forall n \in N, \tag{9.12}$$

and $\tilde{\xi}$ taking values according to (9.10), there exists a variable Λ_{nk} that allows to rewrite (9.12) as:

$$Z_{nm} = \sum_{k \in K} J_{km} \lambda_{nk}, \quad \forall n \in N, \forall m \in M_n, \tag{9.13a}$$

$$\sum_{k \in K} h_k \Lambda_{nk} \geq 0, \quad \forall n \in N, \tag{9.13b}$$

$$\Lambda_{nk} \geq 0, \quad \forall n \in N, \forall k \in K. \tag{9.13c}$$

Finally, the elimination of $\tilde{\xi}$ in the case of equality constraints is straightforward. Let us consider the equality constraint (9.14):

$$\sum_{m \in M_n} Z_{nm} \tilde{\xi}_m = 0, \quad \forall n \in N. \tag{9.14}$$

Then, considering expression (9.14) and that all elements in $\tilde{\xi}$ are, in general, different from zero, a feasible solution of (9.14) is:

$$Z_{nm} = 0, \quad \forall n \in N, \forall m \in M_n. \tag{9.15}$$

The resulting formulation is an objective function and a set of deterministic constraints that do not depend on array $\tilde{\xi}$.

9.7.2 LDR Formulation

In this section, the formulation of the multi-stage capacity expansion problem using LDRs is provided. The problem below is deterministic and can be solved with standard linear programming techniques [18]. Note that the symbols with ^ are equivalent to those enumerated in the notation list of Sect. 9.11 but now referring to the corresponding LDR variable/parameter.

$$\begin{aligned}\underset{\Theta^{\mathrm{LDR}}}{\text{Minimize}} \; & \sum_{n\in N}\sum_{g\in G^{\mathrm{C}}}\sum_{m\in M_n^I}\sum_{r\in M_n^I} \hat{C}^{\mathrm{I}}_{gnm}\hat{p}^{\mathrm{I,G}}_{gnr}M^{\mathrm{M}}_{mr} \\ & + \sum_{n\in N}\sum_{d\in D}\sum_{t\in T} W_d \Bigg[\sum_{g\in G}\sum_{m\in M_n^O}\sum_{r\in M_n^O} \hat{C}^{\mathrm{O}}_{gnm}\hat{p}^{\mathrm{G,D}}_{gdtnr}M^{\mathrm{M}}_{mr} \\ & + \sum_{b\in B}\sum_{m\in M_n^O}\sum_{r\in M_n^O} \hat{C}^{\mathrm{UD}}_{nm}\hat{p}^{\mathrm{UD,D}}_{bdtnr}M^{\mathrm{M}}_{mr} \Bigg]\end{aligned} \tag{9.16}$$

subject to:

Investment limits:

$$\hat{P}^{\mathrm{C,G}}_{\max,gnm} - \sum_{\substack{j\le n \\ m\in M_j^I}} \hat{p}^{\mathrm{I,G}}_{gjm} = \sum_{k\in K} J_{km}\Lambda_{gnk}, \quad \forall g \in G^{\mathrm{C}}, \forall n \in N, \forall m \in M_n^I, \tag{9.17a}$$

$$\sum_{k\in K} h_k \Lambda^1_{gnk} \ge 0, \quad \forall g \in G^{\mathrm{C}}, \forall n \in N. \tag{9.17b}$$

Power balance:

$$\begin{aligned}&\sum_{g\in G_b} \hat{p}^{\mathrm{G,D}}_{gdtnm} + \sum_{s\in S_b}(\hat{p}^{\mathrm{S,D}}_{sdtnm} - \hat{p}^{\mathrm{S,C}}_{sdtnm}) - \sum_{l|O(\ell)=b} \hat{p}^{\mathrm{L,D}}_{\ell dtnm} + \sum_{l|F(\ell)=b} \hat{p}^{\mathrm{L,D}}_{\ell dtnm} \\ &\quad = \hat{P}^{\mathrm{D}}_{bdtnm} - \hat{p}^{\mathrm{UD,D}}_{bdtnm}, \quad \forall b \in B, \forall d \in D, \forall t \in T, \forall n \in N, \forall m \in M_n^O.\end{aligned} \tag{9.18}$$

Generating units operation:

$$A^{\mathrm{D}}_{gdt} \sum_{\substack{j\le n\\ \mathrm{RV}\cdot j\ge m}} \hat{p}^{\mathrm{I,G}}_{gjm} - \hat{p}^{\mathrm{G,D}}_{gdtnm}$$
$$= \sum_{k\in K} J_{km}\Lambda^{2}_{gdtnk}, \forall g \in G^{\mathrm{C}}, \forall d \in D, \forall t \in T, \forall n \in N, \forall m \in M^{O}_{n}, \tag{9.19a}$$

$$A^{\mathrm{D}}_{gdt}\hat{P}^{\mathrm{E,G}}_{\max,gnm} - \hat{p}^{\mathrm{G,D}}_{gdtnm}$$
$$= \sum_{k\in K} J_{km}\Lambda^{3}_{gdtnk}, \quad \forall g \in G^{\mathrm{E}}, \forall d \in D, \forall t \in T, \forall n \notin N^{\mathrm{off}}_{g}, \forall m \in M^{O}_{n}, \tag{9.19b}$$

$$\hat{p}^{\mathrm{G,D}}_{gdtnm} = 0, \quad \forall g \in G^{\mathrm{E}}, \forall d \in D, \forall t \in T, \forall n \in N^{\mathrm{off}}_{g}, \forall m \in M^{O}_{n}, \tag{9.19c}$$

$$R^{\mathrm{UP}}_{g}\hat{P}^{\mathrm{E,G}}_{\max,gnm} - \hat{p}^{\mathrm{G,D}}_{gdtnm} + \hat{p}^{\mathrm{G,D}}_{gdt-1,nm}$$
$$= \sum_{k\in K} J_{km}\Lambda^{4}_{gdtnk}, \quad \forall g \in G^{\mathrm{D}} \cup G^{\mathrm{E}}, \forall d \in D, \forall t \in T, \forall n \in N, \forall m \in M^{O}_{n}, \tag{9.19d}$$

$$R^{\mathrm{DW}}_{g}\hat{P}^{\mathrm{E,G}}_{\max,gnm} - \hat{p}^{\mathrm{G,D}}_{gdt-1,nm} + \hat{p}^{\mathrm{G,D}}_{gdtnm}$$
$$= \sum_{k\in K} J_{km}\Lambda^{5}_{gdtnk}, \quad \forall g \in G^{\mathrm{D}} \cup G^{\mathrm{E}}, \forall d \in D, \forall t \in T, \forall n \in N, \forall m \in M^{O}_{n}, \tag{9.19e}$$

$$R^{\mathrm{UP}}_{g} \sum_{\substack{j\le n\\ \mathrm{RV}\cdot j\ge m}} \hat{p}^{\mathrm{I,G}}_{gjm} - \hat{p}^{\mathrm{G,D}}_{gdtnm} + \hat{p}^{\mathrm{G,D}}_{gdt-1,nm}$$
$$= \sum_{k\in K} J_{km}\Lambda^{6}_{gdtnk}, \forall g \in G^{\mathrm{D}} \cup G^{\mathrm{C}}, \quad \forall d \in D, \forall t \in T, \forall n \in N, \forall m \in M^{O}_{n}, \tag{9.19f}$$

$$R^{\mathrm{DW}}_{g} \sum_{\substack{j\le n\\ \mathrm{RV}\cdot j\ge m}} \hat{p}^{\mathrm{I,G}}_{gjm} - \hat{p}^{\mathrm{G,D}}_{gdt-1,nm} + \hat{p}^{\mathrm{G,D}}_{gdtnm}$$
$$= \sum_{k\in K} J_{km}\Lambda^{7}_{gdtnk}, \quad \forall g \in G^{\mathrm{D}} \cup G^{\mathrm{C}}, \forall d \in D, \forall t \in T, \forall n \in N, \forall m \in M^{O}_{n}, \tag{9.19g}$$

$$\hat{p}^{\mathrm{I,G}}_{gnm} = \sum_{k\in K} J_{km}\Lambda^{8}_{gnk}, \quad \forall g \in G^{\mathrm{C}}, \forall n \in N, \forall m \in M^{I}_{n}, \tag{9.19h}$$

$$\hat{p}^{\mathrm{G,D}}_{gdtnm} = \sum_{k\in K} J_{km}\Lambda^{9}_{gdtnk}, \quad \forall g \in G, \forall d \in D, \forall t \in T, \forall n \in N, \forall m \in M^{O}_{n}, \tag{9.19i}$$

$$\sum_{k\in K} h_k\Lambda^{2}_{gdtnk} \ge 0, \quad \forall g \in G^{\mathrm{C}}, \forall d \in D, \forall t \in T, \forall n \in N, \tag{9.19j}$$

$$\sum_{k\in K} h_k\Lambda^{3}_{gdtnk} \ge 0, \quad \forall g \in G^{\mathrm{E}}, \forall d \in D, \forall t \in T, \forall n, \le N^{\mathrm{off}}_{g}, \tag{9.19k}$$

$$\sum_{k \in K} h_k \Lambda^q_{gdtnk} \geq 0, \quad q = \{4, 5\}, \forall g \in G^{\mathrm{D}} \cup G^{\mathrm{E}}, \forall d \in D, \forall t \in T, \forall n \in N, \tag{9.19l}$$

$$\sum_{k \in K} h_k \Lambda^q_{gdtnk} \geq 0, \quad q = \{6, 7\}, \forall g \in G^{\mathrm{D}} \cup G^{\mathrm{C}}, \forall d \in D, \forall t \in T, \forall n \in N, \tag{9.19m}$$

$$\sum_{k \in K} h_k \Lambda^8_{gnk} \geq 0, \quad \forall g \in G^{\mathrm{C}}, \forall n \in N, \tag{9.19n}$$

$$\sum_{k \in K} h_k \Lambda^9_{gdtnk} \geq 0, \quad \forall g \in G, \forall d \in D, \forall t \in T, \forall n \in N. \tag{9.19o}$$

Demand and transmission lines limits:

$$\hat{P}^{\mathrm{D}}_{bdtnm} - \hat{p}^{\mathrm{UD,D}}_{bdtnm} = \sum_{k \in K} J_{km} \Lambda^{10}_{bdtnk}, \quad \forall b \in B, \forall d \in D, \forall t \in T, \forall n \in N, \forall m \in M^O_n, \tag{9.20a}$$

$$\hat{p}^{\mathrm{UD,D}}_{bdtnm} = \sum_{k \in K} J_{km} \Lambda^{11}_{bdtnk}, \quad \forall b \in B, \forall d \in D, \forall t \in T, \forall n \in N, \forall m \in M^O_n, \tag{9.20b}$$

$$\hat{p}^{\mathrm{L,D}}_{\ell dtnm} + \hat{P}^{\mathrm{L}}_{\max,\ell} = \sum_{k \in K} J_{km} \Lambda^{12}_{\ell dtnk}, \quad \forall \ell \in L, \forall d \in D, \forall t \in T, \forall n \in N, \forall m \in M^O_n, \tag{9.20c}$$

$$\hat{P}^{\mathrm{L}}_{\max,\ell} - \hat{p}^{\mathrm{L,D}}_{\ell dtnm} = \sum_{k \in K} J_{km} \Lambda^{13}_{\ell dtnk}, \quad \forall \ell \in L, \forall d \in D, \forall t \in T, \forall n \in N, \forall m \in M^O_n, \tag{9.20d}$$

$$\sum_{k \in K} h_k \Lambda^q_{bdtnk} \geq 0, \quad q = \{10, 11\}, \forall b \in B, \forall d \in D, \forall t \in T, \forall n \in N, \tag{9.20e}$$

$$\sum_{k \in K} h_k \Lambda^q_{\ell dtnk} \geq 0, \quad q = \{12, 13\}, \forall \ell \in L, \forall d \in D, \forall t \in T, \forall n \in N. \tag{9.20f}$$

Storage units operation:

$$\hat{P}^{\mathrm{C,SP}}_{\max,snm} - \hat{p}^{\mathrm{S,D}}_{sdtnm} = \sum_{k \in K} J_{km} \Lambda^{14}_{sdtnk}, \quad \forall s \in S, \forall d \in D, \forall t \in T, \forall n \in N, \forall m \in M^O_n, \tag{9.21a}$$

$$\hat{p}^{\mathrm{S,D}}_{sdtnm} = \sum_{k \in K} J_{km} \Lambda^{15}_{sdtnk}, \quad \forall s \in S, \forall d \in D, \forall t \in T, \forall n \in N, \forall m \in M^O_n, \tag{9.21b}$$

$$\hat{P}^{\mathrm{C,SP}}_{\max,snm} - \hat{p}^{\mathrm{S,C}}_{sdtnm} = \sum_{k \in K} J_{km} \Lambda^{16}_{sdtnk}, \quad \forall s \in S, \forall d \in D, \forall t \in T, \forall n \in N, \forall m \in M^O_n, \tag{9.21c}$$

$$\hat{p}^{\mathrm{S,C}}_{sdtnm} = \sum_{k \in K} J_{km} \Lambda^{17}_{sdtnk}, \quad \forall s \in S, \forall d \in D, \forall t \in T, \forall n \in N, \forall m \in M^O_n, \tag{9.21d}$$

$$\hat{e}^{\mathrm{SD}}_{sdtnm} = \hat{e}^{\mathrm{SD}}_{sdt-1,nm} - \hat{p}^{\mathrm{S,D}}_{sdtnm} + \eta^{\mathrm{S}}_s \, \hat{p}^{\mathrm{S,C}}_{sdtnm}, \quad \forall s \in S, \forall d \in D, \forall t > 1, \forall n \in N, \forall m \in M^O_n, \tag{9.21e}$$

$$\hat{e}^{\text{SD}}_{sdtnm} = \hat{P}^{\text{C,SE}}_{\text{ini},snm} - \hat{p}^{\text{S,D}}_{sdtnm} + \eta^{\text{S}}_{s}\,\hat{p}^{\text{S,C}}_{sdtnm}, \quad \forall s \in S, \forall d \in D, t = 1, \forall n \in N, \forall m \in M^{O}_{n}, \tag{9.21f}$$

$$H^{\text{S}}_{s}\,\hat{P}^{\text{C,SP}}_{\text{max},snm}\hat{e}^{\text{SD}}_{sdtnm} = \sum_{k\in K} J_{km}\,\Lambda^{18}_{sdtnk}, \quad \forall s \in S, \forall d \in D, \forall t \in T, \forall n \in N, \forall m \in M^{O}_{n}, \tag{9.21g}$$

$$\hat{e}^{\text{SD}}_{sdtnm} = \sum_{k\in K} J_{km}\,\Lambda^{19}_{sdtnk}, \quad \forall s \in S, \forall d \in D, \forall t \in T, \forall n \in N, \forall m \in M^{O}_{n}, \tag{9.21h}$$

$$\hat{P}^{\text{C,SE}}_{\text{ini},snm} - \hat{e}^{\text{SD}}_{sdtnm} = \sum_{k\in K} J_{km}\,\Lambda^{20}_{sdtnk}, \quad \forall s \in S, \forall d \in D, t = N^{T}, \forall n \in N, \forall m \in M^{O}_{n}, \tag{9.21i}$$

$$\sum_{k\in K} h_k\,\Lambda^{q}_{sdtnk} \geq 0, q = \{14, ..., 19\}, \quad \forall s \in S, \forall d \in D, \forall t \in T, \forall n \in N, \tag{9.21j}$$

$$\sum_{k\in K} h_k\,\Lambda^{20}_{sdtnk} \geq 0, \quad \forall s \in S, \forall d \in D, t = N^{T}, \forall n \in N. \tag{9.21k}$$

Non-negative variables:

$$\Lambda^{1}_{gnk}, \Lambda^{8}_{gnk} \geq 0, \quad \forall g \in G^{\text{C}}, \forall n \in N, \forall k \in K, \tag{9.22a}$$

$$\Lambda^{q}_{gdtnk} \geq 0, \quad \forall g \in G, \forall d \in D, \forall t \in T, \forall n \in N, \forall k \in K, q = \{2, ..., 7, 9\}, \tag{9.22b}$$

$$\Lambda^{q}_{bdtnk} \geq 0, \quad \forall b \in B, \forall d \in D, \forall t \in T, \forall n \in N, \forall k \in K, q = \{10, 11\}, \tag{9.22c}$$

$$\Lambda^{q}_{\ell dtnk} \geq 0, \quad \forall \ell \in L, \forall d \in D, \forall t \in T, \forall n \in N, \forall k \in K, q = \{12, 13\}, \tag{9.22d}$$

$$\Lambda^{q}_{sdtnk} \geq 0, \quad \forall s \in S, \forall d \in D, \forall t \in T, \forall n \in N, \forall k \in K, q = \{14, ..., 20\}, \tag{9.22e}$$

where $\Theta^{\text{LDR}} = \{\hat{p}^{\text{I,G}}_{gnm}, \forall g \in G^{\text{C}}, \forall n \in N, \forall m \in M^{I}_{n};\ \hat{p}^{\text{G,D}}_{gdtnm}, \forall g \in G, \forall d \in D, \forall t \in T, \forall n \in N, \forall m \in M^{O}_{n};\ \hat{p}^{\text{S,D}}_{sdtnm}, \hat{p}^{\text{S,C}}_{sdtnm}, \hat{e}^{\text{SD}}_{sdtnm}, \forall s \in S, \forall d \in D, \forall t \in T, \forall n \in N, \forall m \in M^{O}_{n};\ \hat{p}^{\text{UD,D}}_{bdtnm}, \forall b \in B, \forall d \in D, \forall t \in T, \forall n \in N, \forall m \in M^{O}_{n};\ \hat{p}^{\text{L,D}}_{\ell dtnm}, \forall \ell \in L, \forall d \in D, \forall t \in T, \forall n \in N, \forall m \in M^{O}_{n};\ \Lambda^{1}_{gnk}, \Lambda^{8}_{gnk}, \forall g \in G, \forall n \in N, \forall k \in K;\ \Lambda^{q}_{gdtnk}, \forall g \in G, \forall d \in D, \forall t \in T, \forall n \in N, \forall k \in K, q = \{2, ..., 7, 9\};\ \Lambda^{q}_{bdtnk}, \forall b \in B, \forall d \in D, \forall t \in T, \forall n \in N, \forall k \in K, q = \{10, 11\};\ \Lambda^{q}_{\ell dtnk}, \forall \ell \in L, \forall d \in D, \forall t \in T, \forall n \in N, \forall k \in K, q = \{12, 13\};\ \Lambda^{q}_{sdtnk}, \forall s \in S, \forall d \in D, \forall t \in T, \forall n \in N, \forall k \in K, q = \{14, ..., 20\}\}$ is the set of the optimization variables.

The equations comprising problem (9.16)–(9.22) are equivalent to those belonging to problem (9.5), but considering the reformulation established by the proposition (9.13). For instance, (9.5e) is reformulated using an LDR approach through (9.19a), (9.19j), and (9.22a) for $q = 2$. As it can be observed, there is a variable Λ for each constraint included in problem (9.5). Recall that this formulation implicitly considers the non-anticipativity between the investment decisions. Note that there are constraints such as (9.18) that are equalities, and so Proposition (9.13) does not apply. Finally, observe that problem (9.16)–(9.22) is an LP model.

Illustrative Example 9.9 Comparison of the number of variables

This example compares the size of a multi-stage capacity expansion problem formulated using stochastic programming or an LDR approach.

In problem (9.5), the number of variables is computed as:

$$\mathrm{VAR}^{\mathrm{sto}} = \mathrm{N}^{\mathrm{N}} \cdot \mathrm{N}^{\Omega} \left(\mathrm{N}^{\mathrm{G}} + (\mathrm{N}^{\mathrm{G}} + \mathrm{N}^{\mathrm{L}} + \mathrm{N}^{\mathrm{B}} + 3 \cdot \mathrm{N}^{\mathrm{S}})(\mathrm{N}^{\mathrm{D}} \cdot \mathrm{N}^{\mathrm{T}})\right) = \mathrm{N}^{\mathrm{N}} \cdot \mathrm{N}^{\Omega} \cdot \mathrm{X},$$

being $\mathrm{X} = \mathrm{N}^{\mathrm{G}} + (\mathrm{N}^{\mathrm{G}} + \mathrm{N}^{\mathrm{L}} + \mathrm{N}^{\mathrm{B}} + 3 \cdot \mathrm{N}^{\mathrm{S}})(\mathrm{N}^{\mathrm{D}} \cdot \mathrm{N}^{\mathrm{T}})$. Therefore, the number of variables is proportional to the number of scenarios.

In problem (9.16)–(9.22) the variables do not depend on the scenarios, but the introduction of auxiliary variables $\lambda^{q}_{\mathrm{nk}}$ makes the number of variables to be computed as:

$$\begin{aligned}\mathrm{VAR}^{\mathrm{LDR}} &= \mathrm{N}^{\mathrm{N}} \cdot \mathrm{N}^{\mathrm{M}} \cdot \mathrm{X} \\ &+ \mathrm{N}^{\mathrm{N}} \cdot \mathrm{N}^{\mathrm{K}} \left(2 \cdot \mathrm{N}^{\mathrm{G}} + (5 \cdot \mathrm{N}^{\mathrm{G}} + 2 \cdot \mathrm{N}^{\mathrm{B}} + 2 \cdot \mathrm{N}^{\mathrm{L}} + 5 \cdot \mathrm{N}^{\mathrm{S}})(\mathrm{N}^{\mathrm{D}} \cdot \mathrm{N}^{\mathrm{T}})\right),\end{aligned}$$

where N^{M} is equal to the number of elements comprising vector $\tilde{\xi}$. Therefore, the size of an LDR problem depends on the number of stages and random variables since N^{M} and N^{K} are defined according to how many random variables there are per decision stage (N^{K} depends on the size of matrices J and h used to establish the uncertainty bounds).

This implies that for low number of stages and/or scenarios, the LDR problem is more computationally expensive than the stochastic problem in terms of number of variables. However, as long as the number of stages and/or scenarios grows, the size of the stochastic problem increases significantly much faster than the LDR one and, hence, becomes more computationally expensive.

To numerically test the previous statements, let us consider a power system comprising 10 candidate generating units, 10 existing generating units, 10 buses and one storage unit per bus, and 15 transmission lines. The representative days considered are 25 comprising 24 h each. Moreover, we consider three demand growth and two investment cost scenarios per stage. Then, four planning schemes are considered. In this first case, investment decisions can be done only at the beginning of the planning horizon (1 period). In the other cases, the planning horizon is divided into 2, 3, or 4 periods so that investment decisions can be made at 2, 3, or 4 stages, respectively. Thus, the approximate number of variables comprising the stochastic and LDR formulations, respectively, is presented in Table 9.8.

Table 9.8 Illustrative Example 9.9: number of variables in each problem

Stages	1	2	3	4
Stochastic	153,030	1,836,360	16,527,240	132,217,920
LDR	582,100	2,910,500	6,112,050	10,477,800

As it can be observed, the number of variables grows exponentially as long as the number of decision stages increases in the stochastic problem. It is easy to check that something similar happens with the number of constraints. □

9.8 Illustrative Case Study: Multi-Stage Capacity Expansion Model Using LDRs

The same data as in Illustrative Case Study of Sect. 9.6 are considered. Therefore, there is only one random variable in this problem and array $\tilde{\xi}$ takes the values $[1, \tilde{d}_1^{\mathrm{D}}, \tilde{d}_2^{\mathrm{D}}]$. Thus, $M_1^I = \{1\}$, $M_2^I = \{1, 2\}$, and $M_1^O = \{1, 2\}$, $M_2^O = \{1, 2, 3\}$.

The deterministic parameters are redefined as follows. Three generating units are considered, Coal, Wind, and SoPV. However, only Wind and SoPV can be installed. As a result, the number of rows and columns in matrices defining investment costs, $\hat{C}^{\mathrm{I}}_{\mathrm{Wind}}$ and $\hat{C}^{\mathrm{I}}_{\mathrm{SoPV}}$, are respectively two and three because there are two decision stages and three elements in array $\tilde{\xi}$. The rest of matrices including input data that are expressed above has the same dimension.

Investment costs:

$$\hat{C}^{\mathrm{I}}_{\mathrm{Wind}} = \begin{bmatrix} 70 \; 0 \; 0 \\ 70 \; 0 \; 0 \end{bmatrix} (\text{k€/MW}),$$

$$\hat{C}^{\mathrm{I}}_{\mathrm{SoPV}} = \begin{bmatrix} 42 \; 0 \; 0 \\ 42 \; 0 \; 0 \end{bmatrix} (\text{k€/MW}).$$

Operating costs:

$$\hat{C}^{\mathrm{O}}_{\mathrm{Coal}} = \begin{bmatrix} 0.04 \; 0 \; 0 \\ 0.04 \; 0 \; 0 \end{bmatrix} (\text{k€/MWh}),$$

$$\hat{C}^{\mathrm{O}}_{\mathrm{Wind}} = \begin{bmatrix} 0.003 \; 0 \; 0 \\ 0.003 \; 0 \; 0 \end{bmatrix} (\text{k€/MWh}),$$

$$\hat{C}^{\mathrm{O}}_{\mathrm{SoPV}} = \begin{bmatrix} 0.001 \; 0 \; 0 \\ 0.001 \; 0 \; 0 \end{bmatrix} (\text{k€/MWh}).$$

Potential capacity to be built:

$$\hat{P}^{\mathrm{C,G}}_{\mathrm{max,Wind}} = \begin{bmatrix} 150 \; 0 \; 0 \\ 150 \; 0 \; 0 \end{bmatrix} (\text{MW}),$$

$$\hat{P}^{\mathrm{C,G}}_{\mathrm{max,SoPV}} = \begin{bmatrix} 100\ 0\ 0 \\ 100\ 0\ 0 \end{bmatrix} \text{(MW)}.$$

Capacity of the coal unit:

$$\hat{P}^{\mathrm{E,G}}_{\mathrm{max,Coal}} = \begin{bmatrix} 150\ 0\ 0 \\ 150\ 0\ 0 \end{bmatrix} \text{(MW)}.$$

The values of the demand $\hat{P}^{\mathrm{D}}_{dtnm}$ are defined considering (1) that the demand grows along the planning horizon, which is stated by matrix $\hat{D}^{\mathrm{D}}$, (2) the demand level in each period indicated in Table 9.4, and (3) the peak demand of 110 MW. Note that $\hat{D}^{\mathrm{D}}$ is defined according to the LDR $\tilde{d}^{\mathrm{D}}_{n} = \sum_{m \in M^{O}_{n}} \hat{D}^{\mathrm{D}}_{nm}\, \tilde{\xi}_{m}, \forall n,$

$$\hat{D}^{\mathrm{D}} = \begin{bmatrix} 0\ 1\ 0 \\ 0\ 0\ 1 \end{bmatrix}.$$

The second order moment matrix has a dimension 3×3 (as the dimension of vector $\tilde{\xi}$) and takes the following values:

$$M^{\mathrm{M}} = \begin{bmatrix} 1.000 & 1.209 & 1.401 \\ 1.209 & 1.463 & 1.694 \\ 1.401 & 1.694 & 1.969 \end{bmatrix}.$$

Parameters J and h are defined as:

$$J = \begin{bmatrix} 1 & 0 & 0 \\ -1 & 0 & 0 \\ -(a^{\mathrm{D}} + b^{\mathrm{D}} \cdot \bar{d}^{\mathrm{D}}_{0})(1-\theta^{\mathrm{D}}_{1}) & 1 & 0 \\ (a^{\mathrm{D}} + b^{\mathrm{D}} \cdot \bar{d}^{\mathrm{D}}_{0})(1+\theta^{\mathrm{D}}_{1}) & -1 & 0 \\ -a^{\mathrm{D}}(1-\theta^{\mathrm{D}}_{2}) & -b^{\mathrm{D}}(1-\theta^{\mathrm{D}}_{2}) & 1 \\ a^{\mathrm{D}}(1+\theta^{\mathrm{D}}_{2}) & b^{\mathrm{D}}(1+\theta^{\mathrm{D}}_{2}) & -1 \end{bmatrix} = \begin{bmatrix} 1 & 0 & 0 \\ -1 & 0 & 0 \\ -1.149 & 1 & 0 \\ 1.270 & -1 & 0 \\ -0.180 & -0.900 & 1 \\ 0.220 & 1.100 & -1 \end{bmatrix},$$

$$h = \begin{bmatrix} 1\ -1\ 0\ 0\ 0\ 0 \end{bmatrix}.$$

Therefore, the LDR formulation of this problem is as follows. Note that since there is only one representative day, we avoid to include index d in the symbols.

$$\text{Minimize} \sum_{g \in G^{\mathrm{C}}} \hat{C}^{\mathrm{I}}_{g11} \hat{p}^{\mathrm{I,G}}_{g11} M^{\mathrm{M}}_{11}$$

$$+ \sum_{t=1}^{8} 1095 \left[\sum_{g \in G} \sum_{m=1}^{2} \sum_{r=1}^{2} \hat{C}^{\mathrm{O}}_{g1m} \hat{p}^{\mathrm{G,D}}_{gt1r} M^{\mathrm{M}}_{mr} + \sum_{m=1}^{2} \sum_{r=1}^{2} 1 \cdot \hat{p}^{\mathrm{UD,D}}_{t1r} M^{\mathrm{M}}_{mr} \right]$$

$$+\sum_{g\in G^{\mathrm{C}}}\sum_{m=1}^{2}\sum_{r=1}^{2}\hat{C}^{\mathrm{I}}_{g2m}\hat{p}^{\mathrm{I,G}}_{g2r}M^{\mathrm{M}}_{mr}$$

$$+\sum_{t=1}^{8}1095\left[\sum_{g\in G}\sum_{m=1}^{3}\sum_{r=1}^{3}\hat{C}^{\mathrm{O}}_{g2m}\hat{p}^{\mathrm{G,D}}_{gt2r}M^{\mathrm{M}}_{mr}+\sum_{m=1}^{3}\sum_{r=1}^{3}1\cdot\hat{p}^{\mathrm{UD,D}}_{t2r}M^{\mathrm{M}}_{mr}\right]$$

subject to:

$$\hat{P}^{\mathrm{C,G}}_{\max,\mathrm{Wind},11}-\hat{p}^{\mathrm{I,G}}_{\mathrm{Wind},11}=\sum_{k=1}^{6}J_{k1}\Lambda^{1}_{\mathrm{Wind},1k},$$

$$\hat{P}^{\mathrm{C,G}}_{\max,\mathrm{SoPV},11}-\hat{p}^{\mathrm{I,G}}_{\mathrm{SoPV},11}=\sum_{k=1}^{6}J_{k1}\Lambda^{1}_{\mathrm{SoPV},1k},$$

$$\hat{P}^{\mathrm{C,G}}_{\max,\mathrm{Wind},21}-(\hat{p}^{\mathrm{I,G}}_{\mathrm{Wind},11}+\hat{p}^{\mathrm{I,G}}_{\mathrm{Wind},21})=\sum_{k=1}^{6}J_{k1}\Lambda^{1}_{\mathrm{Wind},2k},$$

$$\hat{P}^{\mathrm{C,G}}_{\max,\mathrm{Wind},22}-(\hat{p}^{\mathrm{I,G}}_{\mathrm{Wind},11}+\hat{p}^{\mathrm{I,G}}_{\mathrm{Wind},21}+\hat{p}^{\mathrm{I,G}}_{\mathrm{Wind},22})=\sum_{k=1}^{6}J_{k2}\Lambda^{1}_{\mathrm{Wind},2k},$$

$$\hat{P}^{\mathrm{C,G}}_{\max,\mathrm{SoPV},21}-(\hat{p}^{\mathrm{I,G}}_{\mathrm{SoPV},11}+\hat{p}^{\mathrm{I,G}}_{\mathrm{SoPV},21})=\sum_{k=1}^{6}J_{k1}\Lambda^{1}_{\mathrm{SoPV},2k},$$

$$\hat{P}^{\mathrm{C,G}}_{\max,\mathrm{SoPV},22}-(\hat{p}^{\mathrm{I,G}}_{\mathrm{SoPV},11}+\hat{p}^{\mathrm{I,G}}_{\mathrm{SoPV},21}+\hat{p}^{\mathrm{I,G}}_{\mathrm{SoPV},22})=\sum_{k=1}^{6}J_{k2}\Lambda^{1}_{\mathrm{SoPV},2k},$$

$$\sum_{k=1}^{6}h_k\Lambda^{1}_{\mathrm{Wind},nk}\geq 0,\ n=1,2,$$

$$\sum_{k=1}^{6}h_k\Lambda^{1}_{\mathrm{SoPV},nk}\geq 0,\ n=1,2,$$

$$\hat{p}^{\mathrm{G,D}}_{\mathrm{Coal},tnm}+\hat{p}^{\mathrm{G,D}}_{\mathrm{Wind},tnm}+\hat{p}^{\mathrm{G,D}}_{\mathrm{SoPV},tnm}=\hat{P}^{\mathrm{D}}_{dtnm}-\hat{p}^{\mathrm{UD,D}}_{dtnm},\ t=1,\ldots,8;\ n=1,2;\ m=1,2,3,$$

$$A^{\mathrm{D}}_{\mathrm{Wind},t}\sum_{\substack{j\leq n\\ m\in M^{I}_{j}}}\hat{p}^{\mathrm{I,G}}_{\mathrm{Wind},jm}-\hat{p}^{\mathrm{G,D}}_{\mathrm{Wind},tnm}=\sum_{k=1}^{6}J_{km}\Lambda^{2}_{\mathrm{Wind},tnk},\ t=1,\ldots,8;\ n=1,2;\ m=1,2,3,$$

$$A^{\mathrm{D}}_{\mathrm{SoPV},t}\sum_{\substack{j\leq n\\ m\in M^{I}_{j}}}\hat{p}^{\mathrm{I,G}}_{\mathrm{SoPV},jm}-\hat{p}^{\mathrm{G,D}}_{\mathrm{SoPV},tnm}=\sum_{k=1}^{6}J_{km}\Lambda^{2}_{\mathrm{SoPV},tnk},\ t=1,\ldots,8;\ n=1,2;\ m=1,2,3,$$

$$1\cdot\hat{P}^{\mathrm{E,G}}_{\max,\mathrm{Coal},1m}-\hat{p}^{\mathrm{G,D}}_{\mathrm{Coal},t1m}=\sum_{k=1}^{6}J_{km}\Lambda^{3}_{\mathrm{Coal},t1k},\ t=1,\ldots,8;\ m=1,2,$$

$$\hat{p}^{\mathrm{G,D}}_{\mathrm{Coal},t2m}=0,\ t=1,\ldots,8;\ m=1,2,3,$$

$$\hat{p}^{\mathrm{I,G}}_{\mathrm{Wind},11}=\sum_{k=1}^{6}J_{k1}\Lambda^{4}_{\mathrm{Wind},1k},$$

$$\hat{p}^{\mathrm{I,G}}_{\mathrm{SoPV},11} = \sum_{k=1}^{6} J_{k1} \Lambda^{4}_{\mathrm{SoPV},1k},$$

$$\hat{p}^{\mathrm{I,G}}_{\mathrm{Wind},21} = \sum_{k=1}^{6} J_{k1} \Lambda^{4}_{\mathrm{Wind},2k},$$

$$\hat{p}^{\mathrm{I,G}}_{\mathrm{Wind},22} = \sum_{k=1}^{6} J_{k2} \Lambda^{4}_{\mathrm{Wind},2k},$$

$$\hat{p}^{\mathrm{I,G}}_{\mathrm{SoPV},21} = \sum_{k=1}^{6} J_{k1} \Lambda^{4}_{\mathrm{SoPV},2k},$$

$$\hat{p}^{\mathrm{I,G}}_{\mathrm{SoPV},22} = \sum_{k=1}^{6} J_{k2} \Lambda^{4}_{\mathrm{SoPV},2k},$$

$$\hat{p}^{\mathrm{G,D}}_{gtnm} = \sum_{k=1}^{6} J_{km} \Lambda^{5}_{gtnk}, \forall g \in \{\mathrm{Coal, Wind, SoPV}\}, t = 1, ..., 8; n = 1, 2; m = 1, 2, 3,$$

$$\hat{p}^{\mathrm{UD,D}}_{tnm} = \sum_{k=1}^{6} J_{km} \Lambda^{6}_{tnk}, t = 1, ..., 8; n = 1, 2; m = 1, 2, 3,$$

$$\sum_{k=1}^{6} h_k \Lambda^{2}_{gtnk} \geq 0, \forall g \in \{\mathrm{Wind, SoPV}\}, t = 1, ..., 8; n = 1, 2; m = 1, 2, 3,$$

$$\sum_{k=1}^{6} h_k \Lambda^{3}_{\mathrm{Coal},t1k} \geq 0, t = 1, ..., 8; 2; m = 1, 2, 3,$$

$$\sum_{k=1}^{6} h_k \Lambda^{4}_{gnk} \geq 0, \forall g \in \{\mathrm{Wind, SoPV}\}, n = 1, 2,$$

$$\sum_{k=1}^{6} h_k \Lambda^{5}_{gtnk} \geq 0, \forall g \in \{\mathrm{Coal, Wind, SoPV}\}, t = 1, ..., 8; n = 1, 2,$$

$$\sum_{k=1}^{6} h_k \Lambda^{6}_{tnk} \geq 0, t = 1, ..., 8; n = 1, 2,$$

$$\Lambda^{1}_{gnk} \geq 0, \forall g \in \{\mathrm{Wind, SoPV}\}, n = 1, 2; k = 1, ..., 6,$$

$$\Lambda^{2}_{gtnk} \geq 0, \forall g \in \{\mathrm{Coal, Wind, SoPV}\}, t = 1, ..., 8; n = 1, 2; k = 1, ..., 6,$$

$$\Lambda^{3}_{\mathrm{Coal},t1k} \geq 0, t = 1, ..., 8; k = 1, ..., 6,$$

$$\Lambda^{4}_{gnk} \geq 0, \forall g \in \{\mathrm{Wind, SoPV}\}, n = 1, 2; k = 1, ..., 6,$$

$$\Lambda^{5}_{gtnk} \geq 0, \forall g \in \{\mathrm{Coal, Wind, SoPV}\}, t = 1, ..., 8; n = 1, 2; k = 1, ..., 6,$$

$$\Lambda^{6}_{tnk} \geq 0, t = 1, ..., 8; n = 1, 2; k = 1, ..., 6.$$

This problem comprises 887 variables and 521 equations. It can be solved with GAMS [8] using the code provided in Sect. 9.12.2.

Before providing the results of this example, recall that variables $\hat{p}^{\mathrm{I,G}}_{gnr}$ or $\hat{p}^{\mathrm{G,D}}_{gtnm}$ are not the original variables since those result from the LDR transformation, which

Table 9.9 Illustrative Case Study—Multi-stage capacity expansion with LDRs: capacity to be built (MW)

Unit	Stage	LDR				Stochastic			
		Scenario				Scenario			
		ω_1	ω_2	ω_3	ω_4	ω_1	ω_2	ω_3	ω_4
Wind	n_1	116.6	116.6	116.6	116.6	107.9	107.9	107.9	107.9
	n_2	7.9	7.9	2.6	2.6	10.4	10.4	0.0	0.0
	Total	124.5	124.5	119.2	119.2	118.3	118.3	107.9	107.9
SoPV	n_1	92.8	92.8	92.8	92.8	100.0	100.0	100.0	100.0
	n_2	5.4	5.4	1.8	1.8	0.0	0.0	0.0	0.0
	Total	98.2	98.2	94.6	94.6	100.0	100.0	100.0	100.0

implies the linear dependency from the random variables. Hence, in order to compare the results provided from this approach with those obtained in Illustrative Case Study of Sect. 9.6, we need to give specific values to the components of $\tilde{\xi}$. This is done by considering the scenarios used in Illustrative Case Study of Sect. 9.6, such that (the superscript refers to the order of scenarios):

$$\tilde{\xi}^1 = [1,\ 1.179,\ 1.296],$$
$$\tilde{\xi}^2 = [1,\ 1.179,\ 1.366],$$
$$\tilde{\xi}^3 = [1,\ 1.239,\ 1.436],$$
$$\tilde{\xi}^4 = [1,\ 1.239,\ 1.506].$$

Therefore, considering the input data described above and the fact that the coal unit is not decommissioned, the total expected cost is 16.954 M€ (16.088 M€ from the stochastic problem). The capacity to be built of wind and solar PV is provided in Table 9.9. For comparison reasons, the results obtained from the stochastic problem are also indicated.

From these results we observe that the LDR approach leads to a higher wind capacity than the stochastic formulation. The opposite happens with the solar PV capacity. Moreover, the total new capacity is comparatively higher in the LDR case, and hence, the total expected cost. Note that the LDR takes into account the whole interval where the demand growth takes values, including the maximum value. However, the stochastic scenarios were defined as explained in Illustrative Example 9.3, i.e., the maximum demand growth is not considered. This leads to a more conservative solution coming from the LDR approach.

The power generated by each unit in each period, stage and scenario is provided in Table 9.10.

Finally, in the case that the coal unit is decommissioned by period 2, the capacity to be built is that given in Table 9.11. As previously, the results obtained from the stochastic problem are also indicated. The total new capacity is larger in the LDR

Table 9.10 Illustrative Case Study—Multi-stage capacity expansion with LDRs: energy generated by each unit in each operating period, stage and scenario (MWh)

Oper. period	Coal							
	Stage n_1				Stage n_2			
	ω_1	ω_2	ω_3	ω_4	ω_1	ω_2	ω_3	ω_4
t_1	0.0	0.0	0.0	0.0	1.4	3.3	3.7	5.5
t_2	0.0	0.0	0.0	0.0	0.0	0.0	0.0	0.0
t_3	0.0	0.0	0.0	0.0	0.0	0.0	0.0	0.0
t_4	0.0	0.0	0.0	0.0	0.8	1.8	2.0	3.0
t_5	0.0	0.0	0.0	0.0	1.7	3.9	4.4	6.6
t_6	0.0	0.0	0.0	0.0	2.1	4.8	5.5	8.2
t_7	0.0	0.0	0.7	0.7	3.4	7.7	8.7	13.0
t_8	0.0	0.0	0.0	0.0	0.0	0.0	0.0	0.0
Oper. period	Wind							
	Stage n_1				Stage n_2			
	ω_1	ω_2	ω_3	ω_4	ω_1	ω_2	ω_3	ω_4
t_1	68.5	68.5	72.5	72.5	74.6	77.4	81.3	84.1
t_2	37.0	37.0	40.3	40.3	42.9	46.8	49.5	53.4
t_3	25.8	25.8	30.5	30.5	33.6	39.0	41.9	47.3
t_4	11.0	11.0	16.2	16.2	18.7	23.9	26.2	31.4
t_5	23.9	23.9	29.9	29.9	32.0	36.8	39.6	44.3
t_6	29.5	29.5	34.8	34.8	36.3	39.7	42.4	45.8
t_7	62.9	62.9	67.6	67.6	68.0	69.1	72.4	73.5
t_8	64.8	64.8	68.1	68.1	71.3	75.1	79.0	82.8
Oper. period	SoPV							
	Stage n_1				Stage n_2			
	ω_1	ω_2	ω_3	ω_4	ω_1	ω_2	ω_3	ω_4
t_1	9.3	9.3	9.3	9.3	9.5	9.5	9.8	9.8
t_2	27.8	27.8	27.8	27.8	28.4	28.4	29.4	29.4
t_3	64.9	64.9	64.9	64.9	66.2	66.2	68.7	68.7
t_4	92.8	92.8	92.8	92.8	94.6	94.6	98.2	98.2
t_5	92.8	92.8	92.8	92.8	94.6	94.6	98.2	98.2
t_6	74.2	74.2	74.2	74.2	75.7	75.7	78.5	78.5
t_7	27.8	27.8	27.8	27.8	28.4	28.4	29.4	29.4
t_8	0.0	0.0	0.0	0.0	0.0	0.0	0.0	0.0

Table 9.11 Illustrative Case Study—Multi-stage capacity expansion with LDRs: capacity to be built decommissioned the coal unit (MW)

Unit	Stage	LDR				Stochastic			
		Scenario				Scenario			
		ω_1	ω_2	ω_3	ω_4	ω_1	ω_2	ω_3	ω_4
Wind	n_1	133.5	133.5	133.5	133.5	109.0	109.0	109.0	109.0
	n_2	12.2	12.2	4.1	4.1	34.3	34.3	16.3	16.3
	Total	145.8	145.8	137.6	137.6	143.3	143.3	125.3	125.3
SoPV	n_1	100.0	100.0	100.0	100.0	100.0	100.0	100.0	100.0
	n_2	0.0	0.0	0.0	0.0	0.0	0.0	0.0	0.0
	Total	100.0	100.0	100.0	100.0	100.0	100.0	100.0	100.0

approach than in the stochastic one, specially the wind capacity to be built in the first stage. The total expected cost in this case is equal to 20.862 M€.

9.9 Realistic Case Study

This section presents a description of the input data and the results of the case study analyzed in this chapter. The case study is based on the European power system. The multi-stage stochastic and LDR capacity expansion approaches are applied to the system described below. The results obtained from each approach are compared. The problems are solved using CPLEX [19] under GAMS [8] on a Linux-based server with four 3.0 GHz processors and 250 GB of RAM.

9.9.1 Input Data

The power system considered in this case study is equivalent to that presented in Chap. 4, which is based on the publicly available data provided by ENTSO-e [5]. Thus, 24 European countries are considered and each country is represented by a single bus. Figure 9.5 depicts the schema of the European network that comprises the buses and transmission lines connecting the countries. There are 24 countries in total and 42 transmission lines. Table 9.12 provides the technical characteristics of the transmission lines. It is defined the origin and destiny country, the power flow capacity, and the reactance of each line.

The generating technologies considered in this case study are nuclear (Nucl), coal, gas-fired (GasF), fuel, hydroelectric (Hydr), onshore (Wons) and offshore (Woff) wind, solar photovoltaic (SoPV), solar thermal (SoTH), biomass (Biom), and coal with carbon capture and storage (CCCS) units. Table 9.13 provides the investment

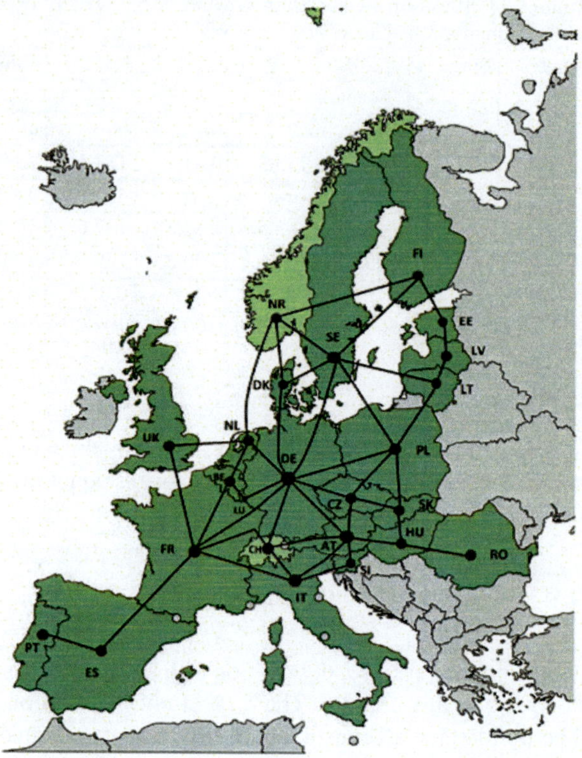

Fig. 9.5 Realistic case study: schema of the European network (created with mapchart.net)

and operating costs, the ramp rate and the initial power output level (prior to hour 1) of each generating technology. These data are based in reports [11, 12, 14].

Table 9.14 indicates the peak demand in each country in 2015 and the storage energy capacity considered. No investment decisions in storage capacity are considered in this chapter but an existing capacity, which is equal to 20% of the average daily consumption in each country in 2015. The ratio energy and power capacity of the storage units is 10.

The planning horizon of this problem goes from 2016 to 2050. At the end of 2015, the total existing generating capacity summed up to 988.2 GW. Figure 9.6 shows the existing capacity in each country (left subplot) and in total (right subplot) classified by renewable, nuclear and fossil-fuel technologies. As it can be observed, the renewable and fossil generating capacity in the whole system were very similar in 2015. Moreover, the limit on the generating capacity that can be built is based on the projection made by the European Commission in 2016 [7]. This projection considers a business-as-usual case where no specific expansion strategies are imposed in the countries. However, the goal of the European Commission is to attained a decarbonized electric energy system by 2050 [6].

Table 9.12 Realistic case study: characteristics of the transmission lines connecting the considered European countries

Line (#)	Origin	Destiny	Capacity (MW)	Reactance (Ω)	Line (#)	Origin	Destiny	Capacity (MW)	Reactance (Ω)
1	AT	CZ	2526	1.5	22	FR	ES	5567	1.6
2	AT	DE	4708	1.2	23	FR	CH	3769	1.9
3	AT	HU	1456	1.3	24	FR	UK	2024	1.5
4	AT	IT	300	1.4	25	DE	LU	819	1.4
5	AT	SI	1335	1.5	26	DE	NL	5134	1.3
6	AT	CH	1791	1.6	27	DE	PL	2139	1.7
7	BE	FR	4167	1.3	28	DE	SE	612	1.2
8	BE	LU	205	1.8	29	DE	CH	5352	1.3
9	BE	NL	3165	1.4	30	HU	RO	566	1.1
10	CZ	DE	3032	1.6	31	HU	SK	1807	1.7
11	CZ	PL	1749	1.2	32	IT	SI	1730	1.5
12	CZ	SK	2225	1.1	33	IT	CH	4905	1.6
13	DK	DE	2213	1.7	34	LV	LT	1315	1.8
14	DK	NR	1618	1.5	35	LT	PL	496	1.9
15	DK	SE	2269	1.4	36	LT	SE	734	1.4
16	EE	FI	1022	1.3	37	NL	NR	732	1.2
17	EE	LV	825	1.2	38	NL	UK	1103	1.3
18	FI	NR	120	1.3	39	NR	SE	4211	1.5
19	FI	SE	2770	1.6	40	PL	SK	1054	1.6
20	FR	DE	4569	1.7	41	PL	SE	601	1.7
21	FR	IT	3343	1.1	42	PT	ES	3978	1.8

Table 9.13 Realistic case study: characteristics of the generating technologies

Technology	Investment cost (k€/MW)	Operating cost (€/MWh)	Ramp rate (pu)	Initial state (pu)
Nucl	5400	10.54	0.2	0.9
Coal	1800	48.39	0.6	0.2
GasF	900	45.28	0.7	0.3
Fuel	1500	35.00	0.5	0.3
Hydr	2385	20.60	1.0	1.0
Wons	1802	3.50	–	–
Woff	4000	10.50	–	–
SoPV	800	2.70	–	–
SoTH	4590	4.60	0.8	0.3
Biom	2115	24.00	0.7	0.3
CCCS	5500	72.58	0.7	0.3

Table 9.14 Realistic case study: peak demand in 2015 and energy storage capacity in each country

Country	Peak demand (GW)	Energy storage capacity (GWh)
AT	10.40	35.30
BE	12.33	53.04
CH	8.39	32.46
CZ	10.40	34.30
DE	71.98	265.08
DK	5.71	16.82
EE	1.39	4.05
ES	37.53	152.79
FI	14.41	41.46
FR	84.83	233.17
HU	6.16	23.00
IT	47.69	177.96
LT	1.72	5.92
LU	0.63	2.68
LV	1.23	3.60
NL	16.21	62.84
NR	24.19	58.26
PL	25.28	91.34
PT	7.72	27.49
RO	9.42	30.60
SE	26.09	62.52
SI	2.02	7.41
SK	4.26	14.78
UK	57.33	161.64

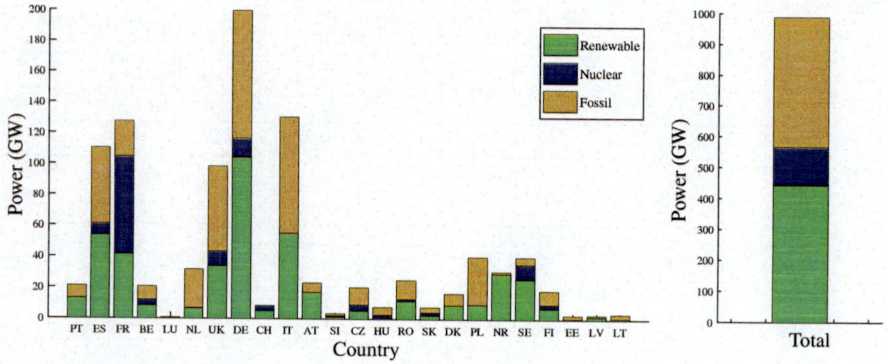

Fig. 9.6 Realistic case study: renewable, nuclear and fossil generating capacity in each country and in total in 2015

Table 9.15 Realistic case study: generating capacity that can be built up to 2050 (GW)

Country	Nucl	Coal	GasF	Fuel	Hydr	Wons	Woff	SoPV	SoTH	Biom	CCCS	Total
AT	0.00	0.00	0.00	0.00	14.04	8.50	0.00	5.01	0.00	0.85	0.00	28.4
BE	0.00	0.00	8.01	0.00	1.49	11.66	1.25	5.90	0.00	1.00	0.00	29.3
CH	0.00	0.00	0.00	0.00	4.99	0.63	0.00	0.03	0.00	0.20	0.00	5.8
CZ	2.84	0.00	0.61	0.00	2.78	1.05	0.00	3.86	0.00	0.64	1.32	13.1
DE	0.00	0.00	8.33	0.00	14.29	108.19	6.25	107.68	0.00	6.59	7.92	259.2
DK	0.00	0.00	2.90	0.00	0.01	9.05	3.13	1.05	0.00	2.60	0.40	19.1
EE	0.00	0.00	0.69	0.00	0.02	2.14	0.00	0.00	0.00	0.37	0.00	3.2
ES	0.00	0.00	0.00	0.00	20.59	58.93	0.00	61.70	2.80	2.15	0.00	146.2
FI	2.22	0.00	1.37	0.00	3.86	3.93	0.25	0.03	0.00	3.13	0.00	14.8
FR	0.00	0.00	26.18	0.00	26.56	71.96	0.63	56.50	0.00	3.64	0.40	185.9
HU	1.73	0.00	0.95	0.00	0.27	2.02	0.00	0.74	0.00	0.39	0.00	6.1
IT	0.00	0.00	3.32	0.00	22.39	32.45	0.00	70.96	0.00	6.81	0.00	135.9
LT	1.12	0.00	0.00	0.00	0.29	1.43	0.00	0.09	0.00	0.17	0.00	3.1
LU	0.00	0.00	0.78	0.00	0.05	0.61	0.00	0.20	0.00	0.04	0.00	1.7
LV	0.00	0.00	0.00	0.00	1.67	0.85	0.00	0.00	0.00	0.13	0.00	2.7
NL	0.00	0.00	5.50	0.00	0.04	16.01	1.38	7.34	0.00	2.64	0.25	33.2
NR	0.00	0.00	0.00	0.00	28.26	3.23	1.00	0.25	0.00	0.00	0.00	32.7
PL	8.25	0.00	2.53	0.00	3.23	23.60	0.00	0.44	0.00	3.02	4.95	46.0
PT	0.00	0.00	0.00	0.00	9.97	8.88	0.00	4.58	0.30	0.60	0.00	24.3
RO	1.41	0.00	0.00	0.00	6.69	9.31	0.00	5.18	0.00	0.31	1.86	24.8
SE	2.07	0.00	3.57	0.00	17.91	14.02	0.00	0.12	0.00	3.41	0.00	41.1
SI	0.42	0.00	0.41	0.00	1.47	0.35	0.00	1.08	0.00	0.16	0.00	3.9
SK	1.08	0.00	0.00	0.00	1.89	0.21	0.00	1.40	0.00	0.46	0.33	5.4
UK	7.93	0.00	11.70	0.00	1.82	51.84	8.75	14.07	0.00	18.03	0.45	114.6
EU	29.1	0.0	76.9	0.0	184.6	440.8	22.6	348.2	3.1	57.3	17.9	1180.5

Therefore, in this case study we assume that the wind and solar capacity that can be built up to 2050 is 25% higher than in [7] in each country. The limits on the capacity that can be built per technology and country are reported in Table 9.15. Note that the expansion in nuclear or fossil-fuel technologies is very low, even null in many countries. In fact, in most of the countries a large part of the nuclear, coal and fuel capacity will be decommissioned by 2050.

Since most of the hydroelectric resources are already exploited, we assume no expansion in this technology but no decommissioning. In addition, we assume that the onshore wind and solar photovoltaic units existing in 2015 are decommissioned along the planning horizon, but all that capacity can be restored.

The auxiliary parameters needed for the LDR approach are defined using the previous data and following the same reasoning as in Illustrative Case Study of Sect. 9.8.

To represent the operating variability along the year, three characteristic days are selected considering the hourly demand and wind and solar power of 2015 in each country [5]. The method used to select the representative days in that proposed

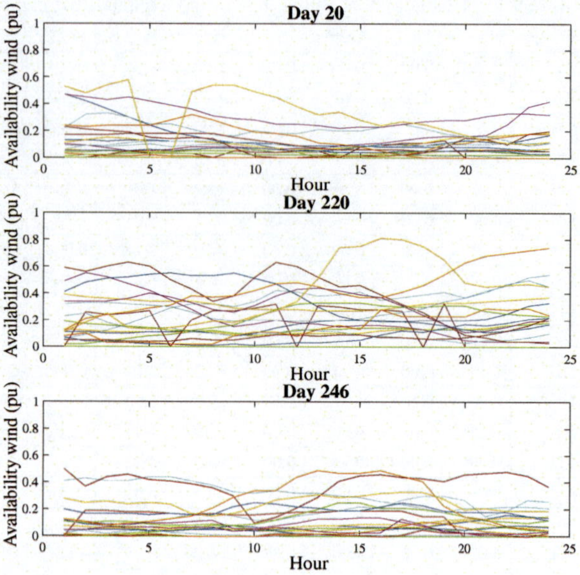

Fig. 9.7 Realistic case study: wind power availability in three representative days

in [9]. Figures 9.7 and 9.8 represent the wind and solar power availability, respectively, in each country in the selected representative days, while Fig. 9.9 represents the demand.

9.9.2 Uncertainty Characterization

As previously mentioned, the planning horizon of this case study goes from 2016 to 2050. The demand growth and the investment cost of the generating technologies are considered as random variables. In order to compare how the results change with the number of decision stages and the modeling approach, four case studies are solved, i.e., two modeling approaches (stochastic and LDR) for two planning schemas. The planning schema corresponding to Case N2 considers that the planning horizon is divided into two periods, leading to the possibility of making investment decisions at the beginning and in the middle of the planning horizon. In Case N3, the planning horizon is divided into three periods and, hence, investment decisions can be made around every 10 years. Figure 9.10 depicts schemas of the decision-making process of cases N2 and N3.

The intervals where the random variables take values are defined according to expressions (9.3) and considering the following aspects:

- the further from the beginning of the planning horizon, the higher the uncertainty and the higher the values of parameters θ^{I}_{gn} and θ^{D}_{bn};
- the expected annual demand growth is 1%;

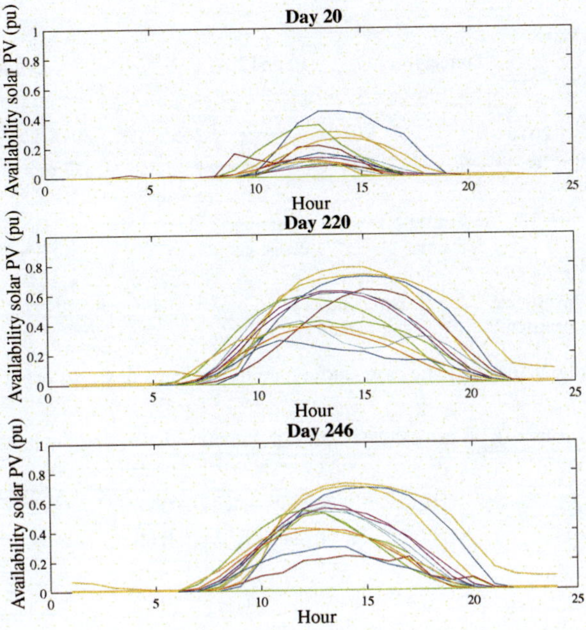

Fig. 9.8 Realistic case study: solar power availability in three representative days

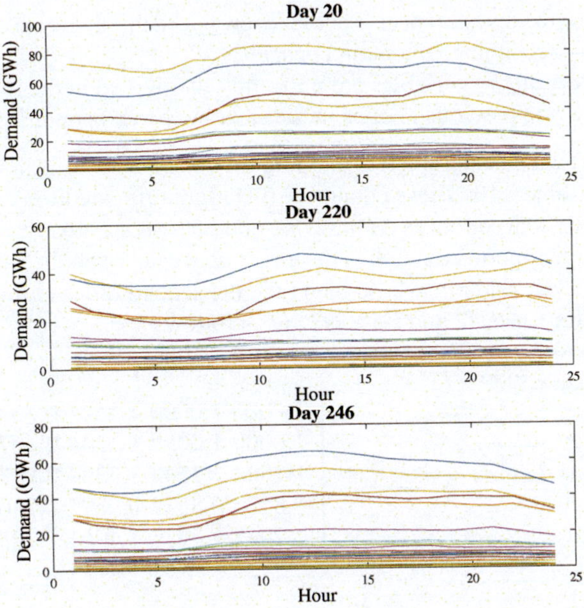

Fig. 9.9 Realistic case study: demand in three representative days

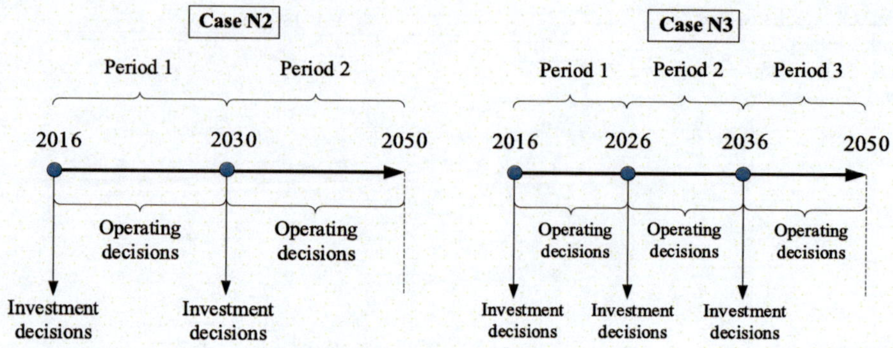

Fig. 9.10 Realistic case study: planning schema of Cases N2 and N3

Table 9.16 Realistic case study: parameters describing the intervals of the random variables (pu)

Case	a^{I}	b^{I}	θ_n^{I}	a^{D}	b^{D}	θ_n^{D}
N2	0.01	0.85	[0.00 0.15]	0.12	1.02	[0.08 0.12]
N3	0.005	0.90	[0 0.05 0.15]	0.09	1.00	[0.05 0.10 0.12]

- the investment cost of the renewable technologies decreases, while those of thermal technologies remain constant;
- random variables are uniformly distributed in the intervals;
- the previous applies in the same way to all countries.

Thus, the parameters defining the intervals of the random variables are indicated in Table 9.16. These data drive to a minimum, medium and maximum demand growth in 2050 of about 17%, 27%, and 45%, respectively, and a minimum, medium and maximum reduction in the investment costs of about 5%, 14%, and 22%, respectively.

Considering these intervals, the second order moment matrix M^{M} and matrix J for Case N2 and Case N3 are as follows:

Case N2:

$$M^{\mathrm{M}} = \begin{bmatrix} 1.0000 & 1.0000 & 1.1544 & 0.8571 & 1.2935 \\ 1.0000 & 1.0000 & 1.1544 & 0.8571 & 1.2935 \\ 1.1544 & 1.1544 & 1.3354 & 0.9895 & 1.4961 \\ 0.8571 & 0.8571 & 0.9895 & 0.7401 & 1.1087 \\ 1.2935 & 1.2935 & 1.4961 & 1.1087 & 1.6841 \end{bmatrix},$$

$$J = \begin{bmatrix} 1.000 & 0.000 & 0.000 & 0.000 & 0.000 \\ -1.000 & 0.000 & 0.000 & 0.000 & 0.000 \\ -1.000 & 1.000 & 0.000 & 0.000 & 0.000 \\ 1.000 & -1.000 & 0.000 & 0.000 & 0.000 \\ -1.062 & 0.000 & 1.000 & 0.000 & 0.000 \\ 1.246 & 0.000 & -1.000 & 0.000 & 0.000 \\ -0.008 & -0.722 & 0.000 & 1.000 & 0.000 \\ 0.012 & 0.977 & 0.000 & -1.000 & 0.000 \\ -0.105 & 0.000 & -0.897 & 0.000 & 1.000 \\ 0.134 & 0.000 & 1.142 & 0.000 & -1.000 \end{bmatrix}.$$

Case N3:

$$M^{\mathrm{M}} = \begin{bmatrix} 1.0000 & 1.0000 & 1.0977 & 0.9051 & 1.1950 & 0.8196 & 1.2930 \\ 1.0000 & 1.0000 & 1.0977 & 0.9051 & 1.1950 & 0.8196 & 1.2930 \\ 1.0977 & 1.0977 & 1.2059 & 0.9935 & 1.3127 & 0.8997 & 1.4193 \\ 0.9051 & 0.9051 & 0.9935 & 0.8199 & 1.0816 & 0.7424 & 1.1703 \\ 1.1950 & 1.1950 & 1.3127 & 1.0816 & 1.4338 & 0.9794 & 1.5509 \\ 0.8196 & 0.8196 & 0.8997 & 0.7424 & 0.9794 & 0.6773 & 1.0598 \\ 1.2930 & 1.2930 & 1.4193 & 1.1703 & 1.5509 & 1.0598 & 1.6857 \end{bmatrix},$$

$$J = \begin{bmatrix} 1.000 & 0.000 & 0.000 & 0.000 & 0.000 & 0.000 & 0.000 \\ -1.000 & 0.000 & 0.000 & 0.000 & 0.000 & 0.000 & 0.000 \\ -1.000 & 1.000 & 0.000 & 0.000 & 0.000 & 0.000 & 0.000 \\ 1.000 & -1.000 & 0.000 & 0.000 & 0.000 & 0.000 & 0.000 \\ -1.043 & 0.000 & 1.000 & 0.000 & 0.000 & 0.000 & 0.000 \\ 1.152 & 0.000 & -1.000 & 0.000 & 0.000 & 0.000 & 0.000 \\ -0.005 & -0.855 & 0.000 & 1.000 & 0.000 & 0.000 & 0.000 \\ 0.005 & 0.945 & 0.000 & -1.000 & 0.000 & 0.000 & 0.000 \\ -0.088 & 0.000 & -0.900 & 0.000 & 1.000 & 0.000 & 0.000 \\ 0.107 & 0.000 & 1.100 & 0.000 & -1.000 & 0.000 & 0.000 \\ -0.004 & 0.000 & 0.000 & -0.765 & 0.000 & 1.000 & 0.000 \\ 0.006 & 0.000 & 0.000 & 1.035 & 0.000 & -1.000 & 0.000 \\ -0.086 & 0.000 & 0.000 & 0.000 & -0.880 & 0.000 & 1.000 \\ 0.120 & 0.000 & 0.000 & 0.000 & 1.150 & 0.000 & -1.000 \end{bmatrix}.$$

Then, the scenarios are randomly generated considering the previous intervals. Three scenarios per random variable and per stage are generated, resulting in 27 scenarios for Case N2 and 243 scenarios for Case N3, respectively. The values of the demand growth and investment cost for each scenario of Case N2 are indicated in Table 9.17. For the sake of conciseness, the scenarios of Case N3 are not provided. The scenarios are equiprobable.

Table 9.17 Realistic case study: demand growth and investment cost scenarios for Case N2

Scenario	Period 1		Period 2	
	Cost	Demand	Cost	Demand
1	1	1.1082	0.7928	1.1693
2	1	1.1082	0.7928	1.2003
3	1	1.1082	0.7928	1.2314
4	1	1.1544	0.7928	1.2624
5	1	1.1544	0.7928	1.2935
6	1	1.1544	0.7928	1.3245
7	1	1.2006	0.7928	1.3555
8	1	1.2006	0.7928	1.3866
9	1	1.2006	0.7928	1.4176
10	1	1.1082	0.8571	1.1693
11	1	1.1082	0.8571	1.2003
12	1	1.1082	0.8571	1.2314
13	1	1.1544	0.8571	1.2624
14	1	1.1544	0.8571	1.2935
15	1	1.1544	0.8571	1.3245
16	1	1.2006	0.8571	1.3555
17	1	1.2006	0.8571	1.3866
18	1	1.2006	0.8571	1.4176
19	1	1.1082	0.9214	1.1693
20	1	1.1082	0.9214	1.2003
21	1	1.1082	0.9214	1.2314
22	1	1.1544	0.9214	1.2624
23	1	1.1544	0.9214	1.2935
24	1	1.1544	0.9214	1.3245
25	1	1.2006	0.9214	1.3555
26	1	1.2006	0.9214	1.3866
27	1	1.2006	0.9214	1.4176

9.9.3 Results

This section describes the outcomes obtained from applying the above input data to the multi-stage capacity expansion models previously explained.

Table 9.18 provides the minimum (Min.), expected (Exp.), and maximum (Max.) values of the new capacity to be built in each stage and in total for Case N2 and Case N3 for each approach. Comparing the approaches, (i) higher capacity is built in the first stage in the stochastic approach with respect to the LDR approach, while the opposite happens in the next stages; (ii) in the end of the planning horizon, the new capacity to be built is higher in the LDR than in the stochastic approach. The latter

Table 9.18 Realistic case study: new capacity built per stage and in total for each case

Approach	Case	Scenario	Stage 1	Stage 2	Stage 3	Total
Stochastic	N2	Min.	367.79	3.44	–	371.23
		Exp.	367.79	17.01	–	384.80
		Max.	367.79	30.70	–	398.49
	N3	Min.	363.45	52.15	3.44	419.04
		Exp.	363.45	60.90	14.35	438.70
		Max.	363.45	68.20	21.64	453.29
LDR	N2	Min.	349.67	25.04	–	374.71
		Exp.	349.67	46.63	–	396.30
		Max.	349.67	68.23	–	417.90
	N3	Min.	354.31	62.73	15.59	432.63
		Exp.	354.31	74.50	37.42	466.23
		Max.	354.31	86.26	59.27	499.84

is related to the fact that the LDR approach considers the whole interval where the demand growth takes values.

In addition, considering three decision stages (N3) leads to a higher value of the new capacity to be built with respect to the case with two decision stages (N2). This may be due to the fact that in the third stage the weight of scenarios with low investment cost is more relevant and, hence, it is more convenient to build generating capacity as with respect to the case with two decision stages.

Table 9.19 gives the expected new capacity built of nuclear, fossil-fuel and renewable technologies, respectively, for each stage, approach and planning schema. From these results we observe that most of the new capacity built corresponds to renewable technologies. The nuclear capacity to be built is relatively close to the maximum allowed, as indicated in Table 9.15. Even if available, not much fossil-fuel capacity is built in general.

In order to visualize what happens in each country and compare the evolution with respect to 2015, Figs. 9.11 and 9.12 depict the total expected generating capacity in 2050 for Cases N2 and N3 and for each approach, respectively, differentiating among renewable, nuclear and fossil-fuel technologies. The subplot on the right shows the total expected generating capacity in the whole system. In these figures it can be observed that the conclusions highlighted above apply to most of the countries. With respect to Fig. 9.6, countries like Germany, Spain, or Italy significantly increase the renewable capacity in detriment of the fossil-fuel capacity. In most of the countries the tendency is to also reduce the nuclear capacity, except for Poland.

Figures 9.13 and 9.14 depict the expected energy mix in the whole system by the end of the planning horizon (2050) resulting from Case N2 and Case N3 and for each approach. The electricity production is given in TWh. Comparing the differences derived from the number of stages, in Case N3 it can be observed a comparatively higher use of the nuclear generation, lower use of fossil fuels, and slightly lower use

Table 9.19 Realistic case study: new capacity built in expectation per technology and stage in each case (GW)

Approach	Case	Stage	Nuclear	Fossil-fuel	Renewable
Stochastic	N2	1	18.695	0.002	349.096
		2	0	0.723	16.287
	N3	1	27.821	1.9	335.626
		2	0	0	60.902
		3	0	0	14.348
LDR	N2	1	21.541	1.9	328.130
		2	0	13.919	32.713
	N3	1	29.077	0	325.236
		2	0	2.856	71.646
		3	0	17.618	19.804

of renewable sources. Moreover, comparing the outcomes from the stochastic and LDR approaches, the LDR approach leads to lower use of renewable sources, but higher use of fossil fuels, nuclear power and the storage units.

To analyze the influence of the storage capacity availability, Case N2 is solved for each approach considering that there is no storage capacity available in the system. The results are compared with those considering the storage capacity. Table 9.20 provides (i) the expected capacity built per stage and in total, (ii) the total expected capacity built per technology, (iii) the total expected cost excluding the penalization cost of the unserved demand, and (iv) the unserved demand. These results are indicated for the stochastic and LDR approaches and for the cases with and without storage capacity.

From these results it can be highlighted: (i) the lack of storage capacity implies to increase the generating capacity in about 13% and 26% in the stochastic and LDR approches, respectively, (ii) the latter implies an increase of the total cost, and iii) the LDR approach leads to unserved demand if no storage capacity is available in the studied system.

Finally, Table 9.21 provides the size of each problem. It can be observed the size of the stochastic problem increases exponentially with the number of scenarios, while the size of the LDR problem just doubles. However, this does not necessarily correlates with the execution time required to solve the problem because Case N3 with the stochastic formulation was solved in about 15.5 h, whereas the LDR formulation took 39.5 h. The models were solved in GAMS and CPLEX 12.6.111 on a Linux-based server of 10 3.0 GHz processors and 250 GB of RAM. The problems solved in [3] using the LDR approach did imply a reduction of the execution time. Therefore, the time issue may be related with the intertemporal constraints describing the storage unit operation.

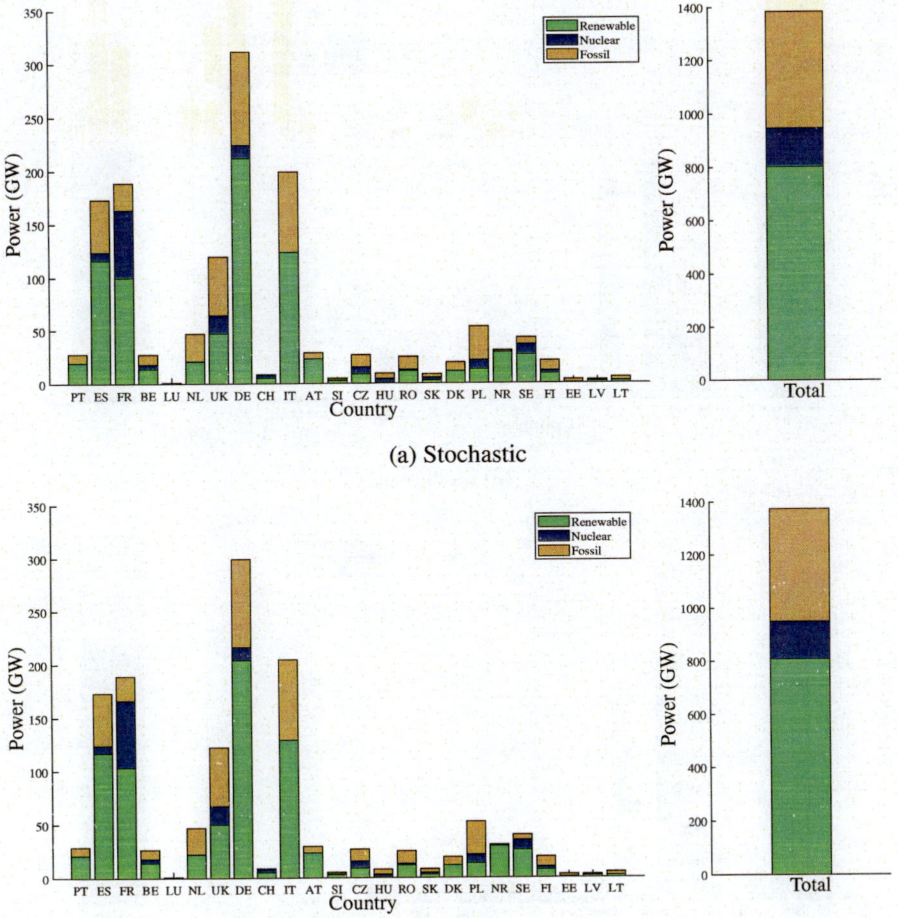

(a) Stochastic

(b) LDR

Fig. 9.11 Realistic case study: nuclear, fossil-fuel, renewable, and total generating capacity in each country in 2050 in Case N2

9.10 Summary and Conclusions

This chapter studies how to solve capacity expansion problems considering multiple decision stages and long-term uncertainties. Two approaches are proposed to formulate a multi-stage capacity expansion problem in generating and storage capacity under uncertainty. The first approach is based on stochastic optimization and, hence, the random variables are represented by a set of scenarios defined according to the probability distribution of those variables. The second approach consists in defining linear decision rules between the decision and the random variables, such that a lin-

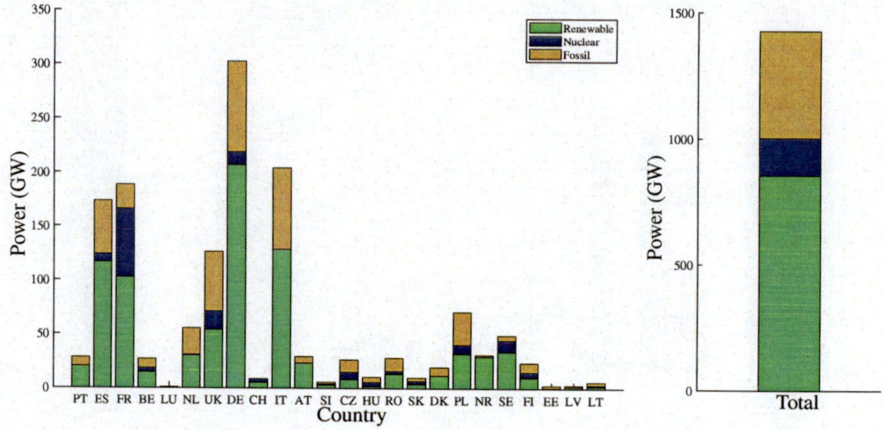

(a) Stochastic

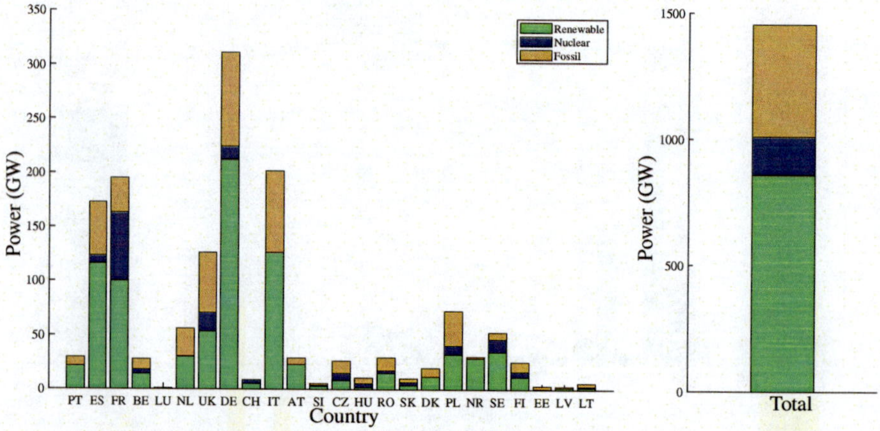

(b) LDR

Fig. 9.12 Realistic case study: nuclear, fossil-fuel, renewable, and total generating capacity in each country in 2050 in Case N3

ear dependency is established between them. Moreover, the short-term variability is account for through a set of representative days. In addition to the illustrative examples described along the chapter, a case study based on the European power system is solved to compare the outcomes derived from each approach and from different planning schemes.

The main conclusions obtained from this chapter are listed below:

- A planning exercise is a long-term process that involves high uncertainty and, consequently, investment decisions are usually made at different stages along the planning horizon.

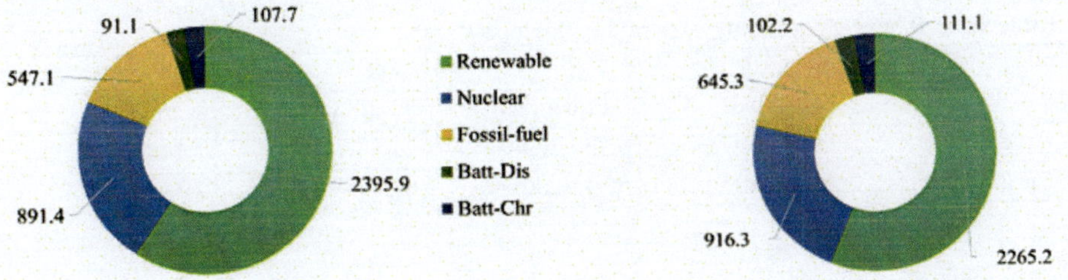

(a) Stochastic (b) LDR

Fig. 9.13 Realistic case study: expected energy mix in the whole system in Case N2 derived from both approaches (TWh)

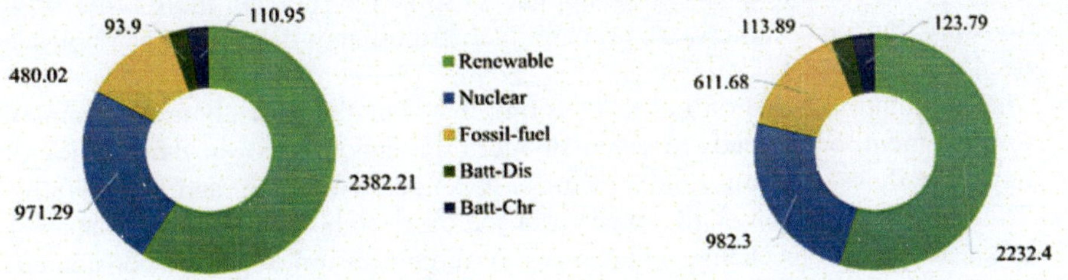

(a) Stochastic (b) LDR

Fig. 9.14 Realistic case study: expected energy mix in the whole system in Case N3 derived from both approaches (TWh)

Table 9.20 Realistic case study: outcomes of Case N2 with and without storage capacity

		Stochastic		LDR	
		With	Without	With	Without
Exp. cap. built (GW)	Stage 1	367.79	393.59	349.67	415.35
	Stage 2	17.01	39.76	46.63	86.29
	Total	384.80	433.35	396.30	501.64
Exp. total cap. (GW)	Nuclear	101.75	101.74	104.60	112.11
	Fossil-fuel	290.85	309.47	304.04	335.82
	Renew.	549.25	579.18	544.70	610.74
Total exp. cost (M€)		159,549.1	168,141.7	162,203.4	172,775.7
Unserved demand (MWh)		0	0	0	401.7

Table 9.21 Realistic case study: size of the problems

	Stochastic		LDR	
	# Variables	# Equations	# Variables	# Equations
Case N2	2,963,209	6,732,289	2,083,153	1,148,545
Case N3	21,240,409	50,446,801	4,047,913	2,115,649

- To accurately model the decision-making process of a multi-stage capacity expansion problem is crucial to obtain a realistic picture of how the future may evolve. However, considering a large number of decision stages with different sources of uncertainty usually leads to computationally intractable problems.
- Stochastic optimization is commonly used to model this kind of problems and decomposition techniques can be applied to solve the problem more efficiently. To apply this methodology, the probability distribution of the random variables is needed.
- The formulation based on linear decision rules is mathematically more complex to implement, but it leads to much smaller problems. Moreover, the solution is limited to those decisions holding a linear dependency with the random variables.
- The numerical outcomes obtained from the case study show how the capacity expansion decisions change whether two or three decision stages are considered. The case with three decision stages results in comparatively higher expected capacity built, which mainly corresponds to renewable capacity.
- The differences between the results of the stochastic and linear decision rule approaches are mainly due to the fact that the linear decision rule accounts for the whole range where the random variables take values and, hence, considering the maximum value of the demand growth leads to higher capacity built than in the stochastic approach.
- Finally, it was shown how much the lack of storage capacity affects the generating capacity to be built.

9.11 Notation

This section provides the list of symbols used to formulate the multi-stage capacity expansion models. Note that the symbols with ^ in this chapter are used in the LDR formulation and they are used to represent the same parameters and variables to those enumerated in this section. Other symbols used to derive the final LDR formulation are defined through the chapter.

Acronyms

CSP Concentrating solar power.
DC Direct current.
GHG Greenhouse gas.

LDR Linear decision rule.
LP Linear programming.
PV Photovoltaic.
SO System operator.
TSO Transmission system operator.

Sets and Indices

b Index of buses.
B Set of buses.
d Index of representative days.
D Set of representative days.
$F(\ell)$ Destination bus of line ℓ.
g Index of generating units.
G Set of generating units.
G_b Set of generating units located at bus b.
G^{C} Set of candidate generating units.
G^{D} Set of dispatchable generating units.
G^{E} Set of existing generating units.
ℓ Index of transmission lines.
L Set of transmission lines.
L_b^{F} Set of transmission lines whose destination bus is b.
L_b^{O} Set of transmission lines whose origin bus is b.
n Index of investment decision stage.
N Set of investment decision stages.
N_g^{off} Set of stages where generating unit g is decommissioned.
s Index of storage units.
S Set of storage units.
S_b Set of storage units located at bus b.
$O(\ell)$ Origin bus of line ℓ.
t Index of time periods.
T Set of time periods.
ω Index of scenarios.
Ω Set of scenarios.

Parameters

A_{gdt}^{D} Availability of generating unit g in representative day d and period t (pu).
C_g^{O} Operating cost of unit g (€/MWh).
C^{UD} Cost of unserved demand (€/MWh).
$E_{\mathrm{ini},s}^{\mathrm{C,SE}}$ Energy content at the beginning of the planning horizon in storage unit s (MWh).
H_s^{S} Hours of autonomy of storage unit s (h).
$P_{\mathrm{max},g}^{\mathrm{C,G}}$ Potential capacity of generation unit g (MW).
$P_{\mathrm{max},gn}^{\mathrm{E,G}}$ Existing power capacity of generation unit g at stage n (MW).
$P_{\mathrm{max},s}^{\mathrm{C,SP}}$ Peak power of storage unit s (MW).

$P^{\mathrm{G}}_{\mathrm{DW},g}$ Down ramp factor of generating unit g (pu).
$P^{\mathrm{G}}_{\mathrm{UP},g}$ Up ramp factor of generating unit g (pu).
$P^{\mathrm{L}}_{\max,\ell}$ Active power capacity of transmission line ℓ (MW).
W_d Weight of representative day d (day).
η^{S}_s Efficiency of charging/discharging storage unit s.
π_ω Probability of scenario ω (pu).

Random Variables

$\tilde{c}^{\mathrm{I}}_{gn\omega}$ Annualized investment cost of generating unit g at stage n in scenario ω (€/MWh).
$\tilde{d}^{\mathrm{Gr}}_{n\omega}$ Demand growth at stage n and scenario ω (pu).
$\tilde{P}^{\mathrm{D}}_{bdtn\omega}$ Power demand in bus b, representative day d, period t, stage n, and scenario ω (MW).

Variables

$e^{\mathrm{SB}}_{st\omega}$ Energy stored in storage unit s, period t, and scenario ω (MWh).
$p^{\mathrm{G,D}}_{gdtn\omega}$ Power generated by generating unit g in representative day d, period t, stage n, and scenario ω (MW).
$p^{\mathrm{I,G}}_{gn\omega}$ Capacity built of generating unit g at stage n in scenario ω (MW).
$p^{\mathrm{L,D}}_{\ell dtn\omega}$ Power flow through line ℓ in representative day d, period t, stage n, and scenario ω (MW).
$p^{\mathrm{S,C}}_{sdtn\omega}$ Power charged by storage s in representative day d, period t, stage n, and scenario ω (MW).
$p^{\mathrm{S,D}}_{sdtn\omega}$ Power discharged by storage s in representative day d, period t, stage n, and scenario ω (MW).
$p^{\mathrm{UD,D}}_{bdtn\omega}$ Unserved demand at bus b in representative day d, period t, stage n, and scenario ω (MW).
$\delta_{bdtn\omega}$ Voltage angle at bus b in representative day d, period t, stage n and scenario ω (rad).
$\Delta^q_{(\cdot)}$ Auxiliary variables used in the LDR formulation.

9.12 GAMS Codes

This section provides the GAMS [8] codes used to solve the Illustrative Case Studies of this chapter.

9.12.1 Multi-Stage Stochastic Capacity Expansion Problem

This section includes the GAMS code of Illustrative Case Study of Sect. 9.6.

```
$title Multi-stage Modeling of the Generation Expansion Planning -
    Stochastic

***********************************************************************
* DATA
***********************************************************************

SETS
g          'Generating units'         /Coal,Wind,SoPV/
b          'Buses'                    /b1/
n          'Decision stages'          /n1,n2/
d          'Representative days'      /d1/
t          'Hours'                    /t1*t8/
w          'Scenarios'                /w1*w4/

GExi(g)   'Existing units '      /Coal/
GCan(g)   'Candidate units'      /Wind,SoPV/
GDis(g)   'Dispatchable units'   /Coal/
;

ALIAS(w,ww);
ALIAS(n,j);

SCALARS
CDU        'Unserved demand cost (euro/MWh)' /1000/
;

SCALAR DemP  'Peak demand (MW)' /110/
;

PARAMETERS
CO(g) 'Operating cost (euro-MWh)' /
Coal     40
Wind     3
SoPV     1
/
CI(g) 'Investment cost (keuro-MW)' /
Wind     1000
SoPV     600
/
Noff(g) 'Decommissioning stage' /
Coal     1
/
PGmax(g) 'Capacity of existing generating units (MW)' /
Coal     150
/
PImax(g) 'Potentical capacity of candidate units (MW)' /
Wind     150
SoPV     100
/
;
* (keuro/MWh)
CO(g)=CO(g)/1000;


** REPRESENTATIVE DAYS
```

```
PARAMETER weight(t) 'Weight of each period (hour)';
weight(t)=1095;

TABLE DemH(t,d) 'Hourly demand level (p.u.)'
         d1
t1       0.6
t2       0.5
t3       0.7
t4       0.8
t5       0.9
t6       0.8
t7       0.7
t8       0.5
;

TABLE WindF(t,d) 'Wind availability (p.u.)'
         d1
t1       0.7
t2       0.7
t3       0.5
t4       0.3
t5       0.4
t6       0.4
t7       0.6
t8       0.7
;

TABLE SoPVF(t,d) 'Solar availability (p.u.)'
         d1
t1       0.1
t2       0.3
t3       0.7
t4       1.0
t5       1.0
t6       0.8
t7       0.3
t8       0.0
;
PARAMETER AF(g,d,t);
AF('Wind',d,t)=WindF(t,d);
AF('SoPV',d,t)=SoPVF(t,d);
AF(g,d,t)$GDis(g)=1;

** SCENARIOS

TABLE Scenarios(w,*) 'Data of scenarios'
              Dg1          Dg2          Prob
w1          1.179        1.506        0.25
w2          1.179        1.436        0.25
w3          1.239        1.366        0.25
w4          1.239        1.296        0.25
;
PARAMETER
Dem(d,t,n,w);
Dem(d,t,'n1',w)=DemP*DemH(t,d)*Scenarios(w,'Dg1');
Dem(d,t,'n2',w)=DemP*DemH(t,d)*Scenarios(w,'Dg2');
```

```
*****************************************************************
* DECLARATION OF VARIABLES AND MATHEMATICAL CHARACTERIZATION
*****************************************************************

VARIABLES
**        OBJECTIVE FUNCTION
z         'Objective function'
;

POSITIVE VARIABLES
**        INVESTMENT VARIABLES
pIG(g,n,w)          'Power to be built'
pIG1(g)

**        OPERATIONAL VARIABLES
pGD(g,d,t,n,w)      'Power output of generating units'
dUSD(d,t,n,w)       'Unserved demand'
;


*****************************************************************
*  EQUATIONS
*****************************************************************

EQUATIONS

**        OBJECTIVE FUNCTION
ObjectiveFunction    'Objective function value'

**        INVESTMENT CONSTRAINTS
MaxInvestCap           'Maximum new capacity built'
NonAnticipativityN1 'Non-anticipativity constraints stage 1'
NonAnticipativityN2 'Non-anticipativity constraints stage 2'

**        OPERATING CONSTRAINTS
BalanceD               'Power balance'
MaxPowGenCA            'Max generated power candidate'
MaxPowGenEX            'Max generated power existing'
DecommGenEX            'Decommiossioned'
;

ObjectiveFunction..      z =e= sum((g,n,w)$GCan(g),Scenarios(w,'prob
    ')*CI(g)*0.07*pIG(g,n,w)) +
                                sum((d,t,n,w),Scenarios(w,'prob')*
                                    weight(t)*(sum(g,CO(g)*pGD(g,d,t,
                                    n,w))
                                +CDU*dUSD(d,t,n,w)) )
;
**     INVESTMENT CONSTRAINTS
MaxInvestCap(g,n,w)$GCan(g)..  PImax(g) - sum(j$(ord(j) le ord(n))
    ,pIG(g,j,w)) =g= 0;
NonAnticipativityN1(g,w)$GCan(g).. pIG(g,'n1',w) =e= pIG1(g);
NonAnticipativityN2(g,w,ww)$(GCan(g) and (Scenarios(w,'Dg1') eq
    Scenarios(ww,'Dg1')) ).. pIG(g,'n2',w) =e= pIG(g,'n2',ww);

**     OPERATING CONSTRAINTS
BalanceD(d,t,n,w)..  sum(g,pGD(g,d,t,n,w)) =e= Dem(d,t,n,w) - dUSD
    (d,t,n,w);
```

```
MaxPowGenCA(g,d,t,n,w)$GCan(g)..    AF(g,d,t)*sum(j$(ord(j) le ord
     (n)),pIG(g,j,w)) - pGD(g,d,t,n,w) =g= 0;
MaxPowGenEX(g,d,t,n,w)$(GExi(g) and (ord(n) le Noff(g))).. AF(g,d
     ,t)*PGmax(g) - pGD(g,d,t,n,w) =g= 0;
DecommGenEX(g,d,t,n,w)$(GExi(g) and (ord(n) gt Noff(g))).. pGD(g,
     d,t,n,w) =e= 0;


*****************************************************************
*  MODEL AND SOLVING
*****************************************************************

MODEL ExampleStochastic /all/;

SOLVE ExampleStochastic USING lp MINIMIZING z;
```

9.12.2 *Multi-Stage Capacity Expansion Problem Using LDRs*

This section includes the GAMS code of Illustrative Case Study of Sect. 9.8.

```
$title Multi-Stage Modeling of the Generation Expansion Planning -
     LDR

*****************************************************************
* DATA
*****************************************************************

SETS
g          'Generating units'        /Coal,Wind,SoPV/
b          'Buses'                   /b1/
n          'Decision stages'         /n1,n2/
d          'Representative days'     /d1/
t          'Hours'                   /t1*t8/
w          'Scenarios'               /w1*w4/

GExi(g)    'Existing units '    /Coal/
GCan(g)    'Candidate units'    /Wind,SoPV/
GDis(g)    'Dispatchable units' /Coal/

m          'Elements vector \xi' /1*3/
k          'Uncertainty bounds'  /1*6/
set_NMI(n,m)
set_NMI2(n,m)
set_NMO(n,m)
set_NMO2(n,m)
;

SCALAR num_RV number random variables /1/;

ALIAS(n,j);
ALIAS(m,r);

set_NMI(n,m)=no;
set_NMI(n,m)$(ord(m) le (num_RV*ord(n)))=yes;
set_NMI2(n,m)=no;
```

```
set_NMI2(n,m)$(ord(m) gt (num_RV*ord(n)))=yes;

set_NMO(n,m)=no;
set_NMO(n,m)$(ord(m) le (num_RV*ord(n) + 1))=yes;
set_NMO2(n,m)=no;
set_NMO2(n,m)$(ord(m) gt (num_RV*ord(n) + 1))=yes;

SCALARS
CDU      'Unserved demand cost (keuro/MWh)' /1/
;

SCALAR DemP  'Peak demand (MW)' /110/
;

PARAMETERS
CO(g) 'Operating cost (euro-MWh)' /
Coal     40
Wind     3
SoPV     1
/
CI(g) 'Investment cost (keuro-MW)' /
Wind     1000
SoPV     600
/
Noff(g) 'Decommissioning stage' /
Coal     1
/
PGmax(g) 'Capacity of existing generating units (MW)' /
Coal     150
/
PImax(g) 'Potential capacity of candidate units (MW)' /
Wind     150
SoPV     100
/
;
* (keuro/MWh)
CO(g)=CO(g)/1000;


** REPRESENTATIVE DAYS

PARAMETER weight(t) 'Weight of each period (hour)';
weight(t)=1095;

TABLE DemH(t,d) 'Hourly demand level (p.u.)'
         d1
t1       0.6
t2       0.5
t3       0.7
t4       0.8
t5       0.9
t6       0.8
t7       0.7
t8       0.5
;

TABLE WindF(t,d) 'Wind availability (p.u.)'
         d1
t1       0.7
```

```
t2      0.7
t3      0.5
t4      0.3
t5      0.4
t6      0.4
t7      0.6
t8      0.7
;

TABLE SoPVF(t,d) 'Solar availability (p.u.)'
        d1
t1      0.1
t2      0.3
t3      0.7
t4      1.0
t5      1.0
t6      0.8
t7      0.3
t8      0.0
;
PARAMETER AF(g,d,t);
AF('Wind',d,t)=WindF(t,d);
AF('SoPV',d,t)=SoPVF(t,d);
AF(g,d,t)$GDis(g)=1;

** LDR PARAMETERS

TABLE FIXED(n,m) 'used for deterministic parameters'
     1    2    3
n1   1    0    0
n2   1    0    0
;

PARAMETER 'LDR parameters'
CO2(g,n,m), CI2(g,n,m), CDU2(n,m), PGmax2(g,n,m), PImax2(g,n,m);
CO2(g,n,m)=CO(g)*FIXED(n,m);
CI2(g,n,m)=CI(g)*FIXED(n,m);
CDU2(n,m)=CDU*FIXED(n,m);
PGmax2(g,n,m)=PGmax(g)*FIXED(n,m);
PImax2(g,n,m)=PImax(g)*FIXED(n,m);

TABLE RAND(n,m)  'used for random variables'
      1    2    3
n1   0    1    0
n2   0    0    1
;
PARAMETER
Dem(d,t,n,m) 'Hourly demand';
Dem(d,t,n,m)=DemP*DemH(t,d)*RAND(n,m);

TABLE MatM(m,m) 'Second order moment matrix'
        1        2        3
1       1        1.2095   1.4011
2       1.2095   1.4634   1.6942
3       1.4011   1.6942   1.9692
;

TABLE mat_W(k,m) 'Matrix defining the confidence intervals'
        1       2       3
```

```
1          1         0         0
2         -1         0         0
3         -1.149     1         0
4          1.270    -1         0
5         -0.180    -0.900     1
6          0.220     1.100    -1
;

PARAMETER h(k) 'Auxiliar parameter';
h('1')=1;
h('2')=-1;
h(k)$(ord(k) gt 2)=0;

*************************************************************
* DECLARATION OF VARIABLES AND MATHEMATICAL CHARACTERIZATION
*************************************************************

VARIABLES
**         OBJECTIVE FUNCTION
z          'Objective function'

**         INVESTMENT VARIABLES
pIG(g,n,m)          'Power to be built'

**         OPERATIONAL VARIABLES
pGD(g,d,t,n,m)      'Power output of generating units'
dUSD(d,t,n,m)       'Unserved demand'
;

POSITIVE VARIABLES
Lambda1(g,n,k)         'Auxiliary variables'
Lambda2(g,n,k)
Lambda3(g,d,t,n,k)
Lambda4(g,d,t,n,k)
Lambda5(g,d,t,n,k)
Lambda6(d,t,n,k)
;

*************************************************************
*  EQUATIONS
*************************************************************

EQUATIONS

**         OBJECTIVE FUNCTION
ObjectiveFunction   'Objective function value'

**         INVESTMENT CONSTRAINTS
MaxInvestCap           'Maximum new capacity built'
InvestPos              'Power to be built positive'
CLambda1               'Auxiliary constraint'
CLambda2               'Auxiliary constraint'

**         OPERATING CONSTRAINTS
BalanceD               'Power balance'
MaxPowGenCA            'Max generated power candidate'
MaxPowGenEX            'Max generated power existing'
DecommGenEX            'Decommiossioned'
GenPos                 'Power generated positive'
```

```
DUSpos                    'Unserved demand positive'

**      AUXILIARY CONSTRAINTS
CLambda3
CLambda4
CLambda5
CLambda6
Aux1
Aux2
Aux3
Aux4
Aux5
Aux6
;

ObjectiveFunction..   z =e= sum((g,n)$GCan(g),sum(m$set_NMI(n,m),(sum(r$set_NMI(n,r),0.07*CI2(g,n,m)*pIG(g,n,r)*MatM(m,r))))) +
 sum((g,d,t,n),weight(t)*sum(m$set_NMO(n,m),(sum(r$set_NMO(n,r),CO2(g,n,m)*pGD(g,d,t,n,r)*MatM(m,r)))))
+ sum((d,t,n),weight(t)*sum(m$set_NMO(n,m),(sum(r$set_NMO(n,r),CDU2(n,m)*dUSD(d,t,n,r)*MatM(m,r)))))
;

**      INVESTMENT CONSTRAINTS
MaxInvestCap(g,n,m)$(GCan(g) and set_NMI(n,m))..  PImax2(g,n,m) - sum(j$(ord(j) le ord(n) and set_NMI(j,m)),pIG(g,j,m)) =e= sum(k,Lambda1(g,n,k)*mat_W(k,m));
InvestPos(g,n,m)$(GCan(g) and set_NMI(n,m))..      pIG(g,n,m) =e= sum(k,Lambda2(g,n,k)*mat_W(k,m));
CLambda1(g,n)$GCan(g)..   sum(k,h(k)*Lambda1(g,n,k)) =g= 0;
CLambda2(g,n)$GCan(g)..   sum(k,h(k)*Lambda2(g,n,k)) =g= 0;

**      OPERATING CONSTRAINTS
BalanceD(d,t,n,m)$set_NMO(n,m)..  sum(g,pGD(g,d,t,n,m)) =e= Dem(d,t,n,m) - dUSD(d,t,n,m);

MaxPowGenCA(g,d,t,n,m)$(GCan(g) and set_NMO(n,m))..    AF(g,d,t)*sum(j$(ord(j) le ord(n) and set_NMI(j,m)),pIG(g,j,m)) - pGD(g,d,t,n,m) =e= sum(k,Lambda3(g,d,t,n,k)*mat_W(k,m));

MaxPowGenEX(g,d,t,n,m)$(GExi(g) and (ord(n) le Noff(g)) and set_NMO(n,m))..  AF(g,d,t)*PGmax2(g,n,m) - pGD(g,d,t,n,m) =e= sum(k,Lambda4(g,d,t,n,k)*mat_W(k,m));

DecommGenEX(g,d,t,n,m)$(GExi(g) and (ord(n) gt Noff(g)))..  pGD(g,d,t,n,m) =e= 0;

GenPos(g,d,t,n,m)$set_NMO(n,m)..  pGD(g,d,t,n,m) =e= sum(k,Lambda5(g,d,t,n,k)*mat_W(k,m));
DUSPos(d,t,n,m)$set_NMO(n,m)..      dUSD(d,t,n,m) =e= sum(k,Lambda6(d,t,n,k)*mat_W(k,m));

CLambda3(g,d,t,n)$GCan(g)..  sum(k,h(k)*Lambda3(g,d,t,n,k)) =g= 0;
CLambda4(g,d,t,n)$(GExi(g) and (ord(n) le Noff(g)))..  sum(k,h(k)*Lambda4(g,d,t,n,k)) =g= 0;
CLambda5(g,d,t,n)..                sum(k,h(k)*Lambda5(g,d,t,n,k)) =g= 0;
CLambda6(d,t,n)..                  sum(k,h(k)*Lambda6(d,t,n,k)) =g= 0;
```

```
Aux1(g,n,m)$(GCan(g) and set_NMI2(n,m)).. sum(k,Lambda1(g,n,k)*mat_W(k,m)) =e= 0;
Aux2(g,n,m)$(GCan(g) and set_NMI2(n,m)).. sum(k,Lambda2(g,n,k)*mat_W(k,m)) =e= 0;
Aux3(g,d,t,n,m)$(GCan(g) and set_NMO2(n,m)).. sum(k,Lambda3(g,d,t,n,k)*mat_W(k,m)) =e= 0;
Aux4(g,d,t,n,m)$(GExi(g) and (ord(n) le Noff(g)) and set_NMO2(n,m)).. sum(k,Lambda4(g,d,t,n,k)*mat_W(k,m)) =e= 0;
Aux5(g,d,t,n,m)$(set_NMO2(n,m)).. sum(k,Lambda5(g,d,t,n,k)*mat_W(k,m)) =e= 0;
Aux6(d,t,n,m)$(set_NMO2(n,m)).. sum(k,Lambda6(d,t,n,k)*mat_W(k,m)) =e= 0;

*********************************************************************
*  MODEL AND SOLVING
*********************************************************************

MODEL ExampleLDR /all/;

SOLVE ExampleLDR USING lp MINIMIZING z;

** OUTCOMES
** Stochastic scenarios

TABLE vector_psi(w,m)
              1         2         3
w1          1.0       1.179     1.296
w2          1.0       1.179     1.366
w3          1.0       1.239     1.436
w4          1.0       1.239     1.506
;

PARAMETERS
pIG_S(g,n,w)
pGD_S(g,d,t,n,w)
dUSD_S(d,t,n,w)
Dem_S(d,t,n,w)
;
pIG_S(g,n,w)$GCan(g)=sum(m$set_NMI(n,m),pIG.l(g,n,m)*vector_psi(w,m));
pGD_S(g,d,t,n,w)=sum(m$set_NMO(n,m),pGD.l(g,d,t,n,m)*vector_psi(w,m));
dUSD_S(d,t,n,w)=sum(m$set_NMO(n,m),dUSD.l(d,t,n,m)*vector_psi(w,m));
Dem_S(d,t,n,w)=sum(m$set_NMO(n,m), Dem(d,t,n,m)*vector_psi(w,m));
```

References

1. Baringo, L., Conejo, A.J.: Risk-constrained multi-stage wind power investment. IEEE Trans. Power Syst. **28**(1), 401–411 (2013)
2. Birge, J.R., Louveaux, F.: Introduction to Stochastic Programming, 2nd edn. Springer, New York, NY (2011)

3. Domínguez, R., Carrión, M., Conejo, A.J.: Influence of the number of decision stages on multi-stage renewable generation expansion models. Int. J. Electr. Power Energy Syst. **126**(A), 06588 (2021)
4. Domínguez, R., Carrión, M., Oggioni, G.: Planning and operating a renewable-dominated European power system under uncertainty. Appl. Energy **258**, 113989 (2020)
5. ENTSO-E transparency platform. https://transparency.entsoe.eu/
6. European Commission. Energy Roadmap 2050 (2011). https://ec.europa.eu/energy/sites/ener/files/documents/2012_energy_roadmap_2050_en_0.pdf
7. European Commission. EU Reference Scenario 2016. Energy, transport and GHG emissions. Trends to 2050 (2015). https://ec.europa.eu/energy/node/186
8. GAMS Development Corp. https://www.gams.com/download/
9. Gröwe-Kuska, N., Heitsch, H., Römisch, W.: Scenario reduction and scenario tree construction for power management problems. In: Proceedings of the IEEE Bologna Power Technology Conferences Bologna, Italy (2003)
10. International Energy Agency (IEA): Power systems in transition (2020). https://www.iea.org/reports/power-systems-in-transition
11. International Energy Agency (IEA). World Energy Outlook 2019. https://www.iea.org/weo/
12. International Renewable Energy Agency (IRENA). Renewable Power Generation Costs in 2019 (2019). https://www.irena.org/publications/2020/Jun/Renewable-Power-Costs-in-2019
13. Kaut, M., Midthun, K.T., Werner, A.S., Tomasgard, A., Hellemo, L., Fodstad, M.: Multi-horizon stochastic programming. Comput. Manag. Sci. **11**(1–2), 179–193 (2013)
14. Kost, C., et al.: Levelized cost of electricity renewable energy technologies, p 144. Fraunhofer Institute for Solar Energy Systems ISE (2013)
15. Kuhn, D., Wiesemann, W., Georghiou, A.: Primal and dual linear decision rules in stochastic and robust optimization. Math. Program. **130**(1), 199–209 (2011)
16. Liu, Y., Sioshansi, R., Conejo, A.J.: Multistage stochastic investment planning with multiscale representation of uncertainties and decisions. IEEE Trans. Pow. Syst. **33**(1), 781–791 (2018)
17. Min, D., Ryu, J., Choi, D.G.: A long-term capacity expansion planning model for an electric power system integrating large-size renewable energy technologies. Comp. Oper. Res. **96**, 244–255 (2018)
18. Sioshansi, R., Conejo, A.J.: Optimization in Engineering. Models and Algorithms. Springer Optimization (2017)
19. The IBM CPLEX. http://www-01.ibm.com/software/commerce/optimization/cplex-optimizer/

Chapter 10
Generation Expansion Planning Considering Electric Vehicles

This chapter studies the generation expansion problem considering explicitly the presence of electric vehicles. The demand associated with the charge of electric vehicles may change substantially the shape of the system demand curve and can influence the determination of the generation mix of future power systems. The model described in this chapter considers the possibility of using financial incentives to modify the consumption patterns of electric vehicle users. The aim of this financial tool is to place the charge of vehicles in convenient periods from the perspective of the power system operator. Therefore, it is assumed that electric vehicle users will be willing to leave the charge of their vehicles in hands of the power system operator if a high-enough financial incentive is offered to them. Price-quota curves are used to model the willingness of electric vehicle users to give the control of the charge/discharge of their vehicles to the system operator. The formulation of these curves implies the use of binary variables. The model analyzed in this chapter considers uncertainties such as the annual demand growth, the capital costs of generating and storage units and the number of electric vehicles. As a result, a two-stage stochastic programming problem is formulated. In this model, generation and storage capacity investments and financial incentives for electric vehicle users are decided in the first stage, whereas operating decisions are made in the second stage. The Benders' decomposition technique is applied to solve the resulting mathematical problem in reasonable computational times. A number of illustrative examples and a realistic case study are included in this chapter to show the performance of the described formulation. The realistic case study included at the end of the chapter is based on the isolated power system comprising Lanzarote and Fuerteventura islands in Spain.

L. Baringo et al., *Electric Vehicles and Renewable Generation*,
Green Energy and Technology, https://doi.org/10.1007/978-3-031-09079-0_10

10.1 Introduction

The generation capacity expansion problem is a decision-making problem consisting in determining the optimal generating mix for a given power system in a future planning horizon. Observe that the modeling of generation expansion problems considering, first, the presence of renewable units and, second, the possibility of making investment decisions at different time periods has been also analyzed in this book in Chaps. 8 and 9, respectively. Unlike the models presented in the aforementioned chapters, the influence of the presence of a large number of electric vehicles (EVs) in the optimal generation investments is analyzed in detail in this chapter. Considering that the energy demanded by an EV may double the daily energy consumption and triple the installed power requirement in a typical home, a large number of EVs may modify substantially the demand curve of a power system. Therefore, the generation expansion problem should consider the presence of EVs in order to decide properly an adequate generating mix.

Considering the above, in this chapter we assume the role of a power system planner that seeks to determine the optimal generation investment decisions to be implemented in an existing power system with a large number of EVs. In order to increase the flexibility of the power systems operation, we consider that the system operator may exercise a direct control over the charging process of EVs in exchange for a financial reimbursement that is paid to EV users. In this manner, the number of controlled vehicles is a function of the value of the offered incentives. These financial incentives represent additional costs for the power system operation that must be taken into account in the capacity expansion problem. The optimal values of these incentives are outputs of the proposed model. The control over the charging processes of EVs could be implemented in practice through financial intermediaries as aggregators or retailers, as described in Chap. 7. In order to avoid an overestimation of the effects of controlling EVs, investments in storage facilities are also considered in this problem. Note that storage units have been proven as efficient tools to increase the flexibility in the operation of power systems.

Finally, uncertainties as the annual demand growth, capital costs of generating and storage units, the number of EVs, and the degree of willingness of EV users to delegate the charge of their vehicles to the system operator are also characterized in this problem.

Relevant references dealing with the generation capacity expansion problem considering renewable power units are [4, 8, 12, 16, 17, 19, 20, 22–24, 26–28]. The participation of EVs in the generation capacity expansion problem has been analyzed in [7, 18, 25].

Illustrative Example 10.1 Generation expansion problem with electric vehicles

In this example, we describe the generation expansion problem faced by a power system planner that desires to determine the optimal generation mix for a future target year. The decisions that the generation expansion planner desires to determine are the following:

- The generation capacity installed from a set of candidate generating units.
- The energy and power storage capacity to install from a set of candidate storage units.
- The financial incentives offered to EV users.

The most relevant input data that the power system planner must either know or estimate are:

- The transmission system topology, including the number of buses and transmission lines parameters.
- The temporal characterization of the target period, for instance, one year represented using 10 characteristic days.
- The power capacity to install from a set of candidate generating units. Let us assume that a maximum of 20, 50, 30, and 10 GW of nuclear, gas, wind, and solar photovoltaic (PV) technologies can be installed, respectively. Nuclear and gas units are considered dispatchable units, whereas wind and solar PV units are intermittent units.
- The energy and power capacity to install from a set of candidate storage units. Let us assume that a maximum of 10 GW/30 GWh of power and energy capacity storage can be installed.
- The estimation of the availability of wind and solar PV power units for each hour of each characteristic day.
- The capital and operating costs of generating and storage units. Nuclear, wind, solar PV, and storage facilities have associated high capital costs, but low operation costs. On the contrary, gas units have low capital costs, but high operation costs.
- The estimation of the number of controlled vehicles as a function of financial incentives offered to EV users. Let us assume that the financial incentives can range between 0–200 € per vehicle and year, and that the number of controlled EVs increases linearly with respect to the offered incentive, being equal to 0 and 50% for incentives equal to 0 and 200 €, respectively.
- The estimation of the hourly demand for each hour of each characteristic day. The peak demand in the set of characteristic days is 50 GW.
- The estimation of the number of EVs and their charging patterns. Let us assume a penetration of 20% of EVs, which corresponds to 6 million of EVS and a maximum charging demand equal to 4 GW.

Observe that the input data described above are subject to different sources of uncertainty: demand and number of EVs in the target year, investment costs of generating and storage units, response of EV users to the offered incentives, etc.

Considering the above, the power system planner is interested in determining the optimal decisions that are able to satisfy the expected demand resulting in the minimum expected cost. This cost must comprise the investment cost associated with new generating and storage units, the incentive costs offered to EV users, and the power system operation costs. A possible solution for this problem could be the following:

- To install 20, 35, 15, and 5 GW of nuclear, gas, wind, and solar PV technologies, respectively.
- To install 5 GW/15 GWh of power and energy capacity storage.
- To offer 100 € per vehicle and year to control 25% of all EVs.

Note that the proposed solution is able to supply the peak demand, corresponding to $50 + 5 = 55$ GW, using dispatchable units. Controlled vehicles and storage units are used to enhance the exploitation of intermittent units, wind and solar PV units, which have associated the lowest operation costs.

□

The reminder of this chapter is organized as follows. The decision framework is described in Sect. 10.2. The uncertainty characterization is described in Sect. 10.3. The formulation of the problem is provided in Sect. 10.4. The proposed solution algorithm is described in Sect. 10.5. In Sect. 10.6, an illustrative case study is solved to show the performance of the analyzed model. In Sect. 10.7, a realistic case study is solved and comprehensively discussed. Section 10.8 provides a summary of the chapter and includes the most relevant conclusions. Section 10.9 defines the main notation used in this chapter. Finally, Sect. 10.10 provides the GAMS [13] code used to solve the illustrative case study described in Sect. 10.6.

10.2 Decision Framework

As stated above, in this chapter we assume the point of view of a power system planner that desires to decide the optimal generation mix of a given power system for a future planning horizon. The power system planner is interested in determining the generating capacity to install from a set of candidate technologies in each bus of the system. Note that the power system planner may install conventional and intermittent generating units. Conventional units refer to dispatchable units as thermal and hydro units, whereas intermittent units are those whose power output is based on uncontrollable physical phenomena, as wind speed and solar irradiation. Additionally, in order to cope with the variability of the production of intermittent units, it is considered that the power system planner may also install storage units.

In order to make appropriate investment decisions, the cost associated with the operation of the resulting power system must be also considered in the generation expansion problem. It is assumed that the resulting power system is operated by a power system operator aiming at satisfying the demand at minimum cost. The operation of the system is characterized by a day-ahead market in which energy and reserve capacities are simultaneously optimized. Considering this, the objective of the generation capacity expansion problem is to decide the optimal generation mix that minimizes the total cost.

The basic premise of the proposed model is that there will exist a high penetration of EVs in the target year. The demand associated with the charge of EVs has to be regarded in order to compute properly generation capacity investments. Furthermore, the uncoordinated charge of EVs can have a strong effect in the solution of the generation capacity expansion problem, resulting in a large generation capacity and high investment costs. With the goal of reducing possible demand peaks corresponding to the uncoordinated charge of EVs, here it is assumed that the power system operator is able to control the charging process of EVs. It is considered that some EV users will be willing to leave the control of the charge of their vehicles in hands of the power system operator in exchange for a financial incentive. For the convenience of EV users, it has to be satisfied that the batteries of all controlled vehicles must be charged up to a desired level at an agreed time period. Therefore, EV users must communicate to the system operator the following information: i) the time in which the charge of the batteries can start, ii) the stop time of the charge, and iii) the minimum value of the energy that must be stored in the batteries at the end of the charging period. The number of EVs that are controlled by the system operator is a function of the offered financial incentive. If a large financial incentive is offered, it is expected that a high number of EV users will be willing to allow the power system operator to control their vehicles. On the contrary, if the financial incentive is low, a small number of EVs will be controlled by the system operator. The financial incentives offered to EV users must be considered in the computation of the total cost associated with the generation capacity expansion problem.

Figure 10.1 represents graphically the decision-making process described above. The black point represents the time in which the generation expansion and the financial incentives for EV users are decided. Note that these are *here-and-now* decisions that are made before knowing the actual realizations of the uncertain parameters. The operation of the resulting power system is simulated in the second stage after knowing the actual realizations of the uncertain parameters. For this reason, operation decisions are considered as *wait-and-see* decisions that can be different for each scenario.

10.3 Uncertainty Characterization

The generation capacity expansion problem is subject to numerous uncertain parameters concerning the whole planning horizon. In this chapter, the capital costs of intermittent and storage units, the annual demand growth, the number of EVs, and the willingness of EV users to give the control of the charge of their vehicles to the power system operator are characterized as uncertain parameters. Note that intermittent generating units and energy storage systems can be considered as immature technologies that can evolve in forthcoming years. For this reason, the capital costs associated with these technologies are modeled as uncertain parameters.

Uncertain parameters are characterized as random variables and they are represented by the set of scenarios Ω. Because of the very nature of the kind of uncertain parameters mentioned above, they can be statistically characterized by expert judgment using the available information provided in technical reports.

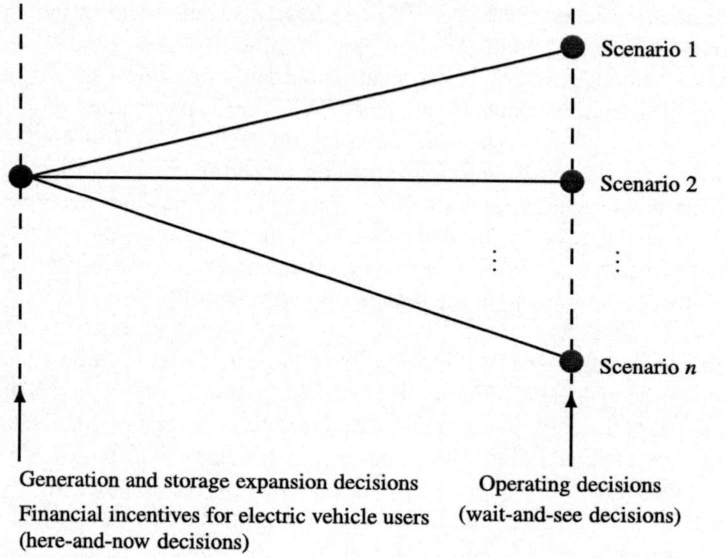

Fig. 10.1 Scenario tree

Table 10.1 Illustrative Example 10.2: uncertainty

Parameters	Low		Medium		High	
	Value	Prob.	Value	Prob.	Value	Prob.
Capital cost of wind power (€/kW)	1400	0.25	1600	0.50	1800	0.25
Demand growth per year (%)	0.0	0.25	1.0	0.50	3.0	0.25
Penetration of electric vehicles (%)	15.0	0.25	30.0	0.50	50.0	0.25

Illustrative Example 10.2 Uncertainty

Let us consider three uncertain parameters: the capital costs of wind power units, the annual demand growth, and the number of EVs. We consider three values (low, medium, and high) for each parameter in order to characterize mathematically these uncertain parameters. These values are provided in Table 10.1.

Considering that these uncertain parameters are not correlated, it is possible to generate the $3^3 = 27$ scenarios included in Table 10.2. The probability of each scenario corresponds to the product of the probabilities of all parameters considered in that scenario. For instance, scenario 1 with capital cost of wind power equal to

Table 10.2 Illustrative Example 10.2: scenarios

Scenarios	Capital cost of wind power (€/kW)	Demand growth per year (%)	Penetration of electric vehicles (%)	Probability
1	1400	0.0	15.0	0.015625
2	1400	0.0	30.0	0.031250
3	1400	0.0	50.0	0.015625
4	1400	1.0	15.0	0.031250
5	1400	1.0	30.0	0.062500
6	1400	1.0	50.0	0.031250
7	1400	3.0	15.0	0.015625
8	1400	3.0	30.0	0.031250
9	1400	3.0	50.0	0.015625
10	1600	0.0	15.0	0.031250
11	1600	0.0	30.0	0.062500
12	1600	0.0	50.0	0.031250
13	1600	1.0	15.0	0.062500
14	1600	1.0	30.0	0.125000
15	1600	1.0	50.0	0.062500
16	1600	3.0	15.0	0.031250
17	1600	3.0	30.0	0.062500
18	1600	3.0	50.0	0.031250
19	1800	0.0	15.0	0.015625
20	1800	0.0	30.0	0.031250
21	1800	0.0	50.0	0.015625
22	1800	1.0	15.0	0.031250
23	1800	1.0	30.0	0.062500
24	1800	1.0	50.0	0.031250
25	1800	3.0	15.0	0.015625
26	1800	3.0	30.0	0.031250
27	1800	3.0	50.0	0.015625

1400 €/kW (prob.$=0.25$), demand growth per year equal to 0.0% (prob.$=0.25$), and a 15% penetration of electric vehicles (prob.$=0.25$) has associated a probability equal to $0.25 \cdot 0.25 \cdot 0.25 = 0.015625$. The sum of the probabilities of all scenarios, $(0.015625 \cdot 8 + 0.03125 \cdot 12 + 0.0625 \cdot 6 + 0.125 \cdot 1)$, is equal to 1.

□

10.4 Formulation

This section presents the mathematical formulation of the problem described above. For the sake of clarity in the formulation, it is assumed that all generating units that are currently at work in the considered power system will be decommissioned prior to the considered target year. The list of notation used along this section is provided in Sect. 10.9.

10.4.1 Capacity Expansion Decisions

The capacity expansion decisions that the power system planner has to make are the following:

- Investments in generation units. It is considered that generation capacity investments can be made for a set of candidate units $g \in G$. The generation capacity that is installed, $p_g^{\mathrm{I,G}}$, has to be non-negative and lower than a maximum value denoted by $P_{\mathrm{max},g}^{\mathrm{I,G}}$. Considering that the planning horizon is one year, the investment costs are annualized as described in Chap. 2 using the Capital Recovery Factor (CRF), where $\mathrm{CRF} = \dfrac{r(1+r)^x}{(1+r)^x - 1}$, x is the plant life of the unit, and r is the interest rate. The annualized costs are represented as $C_{g\omega}^{\mathrm{I,G}}$. Note that investment costs of immature generation technologies are characterized as random variables and they are represented using a set of scenarios indexed by $\omega \in \Omega$.
- Investments in storage units. Similarly to the case of the installation of generation units, storage investments are made over a set of candidate units $s \in S$. Investments in storage units are made in terms of power and energy, $p_s^{\mathrm{I,SP}}$ and $e_s^{\mathrm{I,SE}}$. That is, the term of power refers to the maximum power that can be charged or discharged from the storage, whereas the energy term represents the maximum amount of energy that can be stored in the facility. The maximum power and energy capacities that can be installed are positive values and they are bounded by the maximum capacities $P_{\mathrm{max},s}^{\mathrm{I,SP}}$ and $E_{\mathrm{max},s}^{\mathrm{I,SE}}$, respectively. The power and energy terms have associated different annualized costs, $C_{s\omega}^{\mathrm{I,SP}}$ and $C_{s\omega}^{\mathrm{I,SE}}$. Considering that several utility-scale storage technologies are actually considered to be in a development stage, their investment costs are characterized as random variables. Finally, it should be noted that the values of power and energy capacities are usually dependent of each other in real-world applications. For instance, it is typical that the relationship between power and energy capacities in storage units used for participating in ancillary services be equal to 3, whereas larger values are common in applications related to the energy arbitrage in energy markets.

$$0 \le p_g^{\mathrm{I,G}} \le P_{\mathrm{max},g}^{\mathrm{I,G}}, \quad \forall g \in G, \tag{10.1a}$$

$$0 \le e_s^{\mathrm{I,SE}} \le E_{\mathrm{max},s}^{\mathrm{I,SE}}, \quad \forall s \in S, \tag{10.1b}$$

$$0 \le p_s^{\mathrm{I,SP}} \le P_{\mathrm{max},s}^{\mathrm{I,SP}}, \quad \forall s \in S, \tag{10.1c}$$

$$e_s^{\mathrm{I,SE}} = \gamma_s^{\mathrm{S,EP}} p_s^{\mathrm{I,SP}}, \quad \forall s \in S. \tag{10.1d}$$

Constraints (10.1a) formulate generation investment decisions, whereas constraints (10.1b)–(10.1d) are used to formulate storage investment decisions. Observe that constraints (10.1d) establish the relationship between energy and power capacities as a function of parameter $\gamma_s^{\mathrm{S,EP}}$.

10.4.2 *Financial Incentives Offered to Electric Vehicle Users*

As stated previously, here it is assumed that the power system operator is able to control the charge of a set of EVs. For doing that, a financial incentive is offered to EV users to compensate the lack of control of the charging process of their vehicles. Considering this, two different incentives are considered. The first one is offered to those EV users that only allow the control of the charge of their vehicles by the system operator. Therefore, the system operator decides in which time periods vehicles are charged. The second incentive is focused on those users that allow the vehicle-to-grid operation of their vehicles. In other words, these users are also willing to let the power system operator to discharge part of the energy stored in their vehicles into the grid if it is convenient for the power system operation. Of course, it has to be guaranteed that all EVs controlled by the power system operator have to be charged up to the desired level at the end of the charging period.

The financial incentive offered to EV users is mathematically characterized by variable $\lambda_{kc}^{\mathrm{EV}}$ and represents the amount annually received by each EV user belonging to group k and control mode c. Two different control modes are considered according to the two types of incentives described above. The first control mode, denoted by $c = 1$, represents only charging control, whereas $c = 2$ characterizes the discharging control mode. In this formulation, it is assumed that a single incentive is offered to all vehicles belonging to the same group. The per-unit number of EVs that are controlled by the power system operator, $x_{kc\omega}^{\mathrm{EV,C}}$, is an uncertain parameter that is a function of the offered incentive, $\lambda_{kc}^{\mathrm{EV}}$. If this incentive is high, a large number of controlled vehicles is expected. On the contrary, a low number of EVs will be controlled by the power system operator if the offered incentive is not competitive enough. Step-wise price-quota curves are used to model the number of EV users willing to leave the control of the charge and discharge of their vehicles to the power system operator as a function of the financial incentive. As an example, Fig. 10.2 represents a 6-block step-wise price-quota curve. Observe that the per-unit number of controlled vehicles depicted in this curve, $x_{kc\omega}^{\mathrm{EV,C}}$, grows as the offered incentive price $\lambda_{kc}^{\mathrm{EV}}$ increases. Note that the methodology behind the estimation of these curves

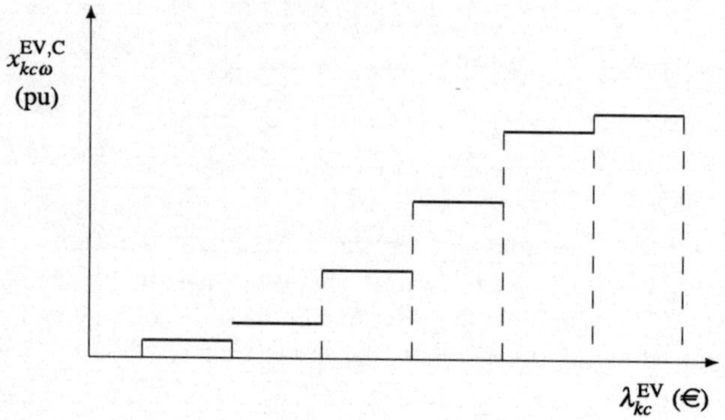

Fig. 10.2 Example of step-wise price-quota curve

is a complex econometric problem that can be addressed in practice through market research studies or econometric models. For the interested reader, Chap. 7 proposes a practical procedure based on the Hotelling model to derive price-quota curves in the case of electricity suppliers.

In order to establish mathematically the dependence of the number of controlled vehicles with regard to the financial incentive, binary variables are used. In this manner, v_{kcm}^{EV} is equal to 1 if the offered price belongs to block m of the piece-quota curve for EV group k and control mode c, being 0 otherwise. Constraints (10.2a)–(10.2f) mathematically formulate the step-wise price-quota curves used to compute the number of EVs controlled by the power system operator. Parameters $\hat{\lambda}_{kcm}^{\text{EV}}$ and $\hat{X}_{kcm\omega}^{\text{EV}}$ are used to characterize the step-wise curve, as shown in Fig. 10.3. The auxiliary variable $\lambda_{kcm}^{\text{EV,A}}$ is equal to λ_{kc}^{EV} if the incentive is in block m of the step-wise curve, being equal to 0 otherwise. Therefore, if $\lambda_{kc}^{\text{EV}} \in \{\hat{\lambda}_{kcm-1}^{\text{EV}}, \hat{\lambda}_{kcm}^{\text{EV}}\}$, then $\lambda_{kcm}^{\text{EV,A}} = \lambda_{kc}^{\text{EV}}$; if $\lambda_{kc}^{\text{EV}} \notin \{\hat{\lambda}_{kcm-1}^{\text{EV}}, \hat{\lambda}_{kcm}^{\text{EV}}\}$, then $\lambda_{kcm}^{\text{EV,A}} = 0$.

Constraints (10.2a) are used to establish the value of $\lambda_{kcm}^{\text{EV,A}}$ as a function of binary variable v_{kcm}^{EV}. Constraints (10.2b) enforce that the financial incentive can belong to only one block of the curve. The financial incentive λ_{kc}^{EV} is computed as a function of the auxiliary variables $\lambda_{kcm}^{\text{EV,A}}$ in constraints (10.2c). The per-unit number of vehicles controlled under each control mode c is calculated through constraints (10.2d). Constraints (10.2e) impose that the per-unit number of controlled vehicles has to be less than 1. Finally, (10.2f) establish that the number of controlled vehicles able to discharge energy in control mode $c = 2$ has to be less than or equal to the number of vehicles that under control mode $c = 1$. In other words, if the discharge of energy is controlled by the system operator, it is mandatory that the charge of the vehicle must be also controlled.

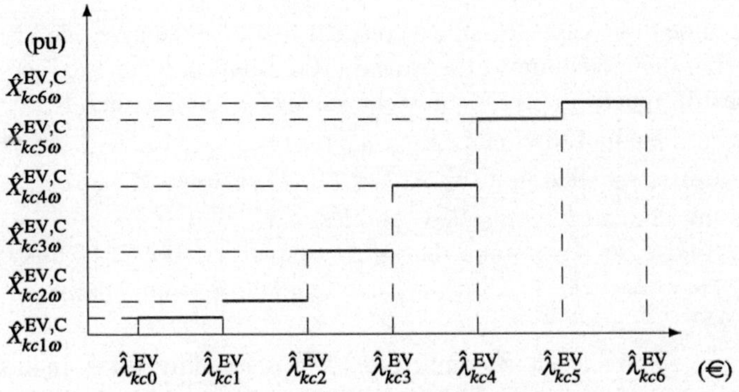

Fig. 10.3 Parameters used to model step-wise price-quota curves

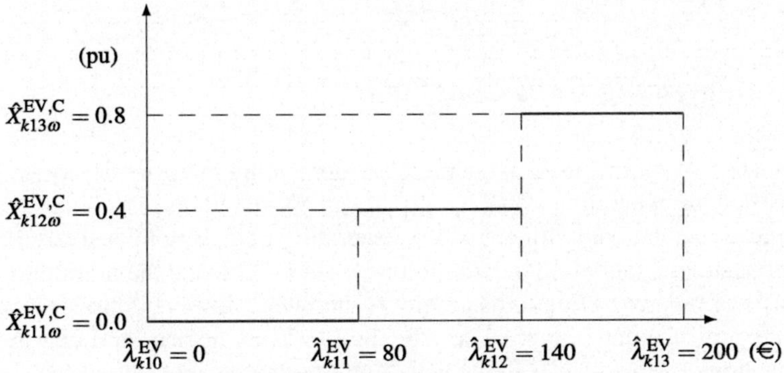

Fig. 10.4 Illustrative Example 10.3: example of step-wise price-quota curve

$$\hat{\lambda}^{\text{EV}}_{kcm-1} v^{\text{EV}}_{kcm} \le \lambda^{\text{EV,A}}_{kcm} \le \hat{\lambda}^{\text{EV}}_{kcm} v^{\text{EV}}_{kcm}, \quad \forall k \in K, \forall c \in C, \forall m \in M, \tag{10.2a}$$

$$\sum_{m \in M} v^{\text{EV}}_{kcm} = 1, \quad \forall k \in K, \forall c \in C, \tag{10.2b}$$

$$\sum_{m \in M} \lambda^{\text{EV,A}}_{kcm} = \lambda^{\text{EV}}_{kc}, \quad \forall k \in K, \forall c \in C, \tag{10.2c}$$

$$x^{\text{EV,C}}_{kc\omega} = \sum_{m \in M} \hat{X}^{\text{EV}}_{kcm\omega} v^{\text{EV}}_{kcm}, \quad \forall k \in K, \forall c \in C, \forall \omega \in \Omega, \tag{10.2d}$$

$$x^{\text{EV,C}}_{k1\omega} \le 1, \quad \forall k \in K, \forall \omega \in \Omega, \tag{10.2e}$$

$$x^{\text{EV,C}}_{k2\omega} \le x^{\text{EV,C}}_{k1\omega}, \quad \forall k \in K, \forall \omega \in \Omega. \tag{10.2f}$$

Illustrative Example 10.3 Incentives for electric vehicle users

Let us consider the 3-step price-quota curve for control mode $c = 1$ represented in Fig. 10.4. This curve is assumed to be identical for each EV group k and scenario ω.

Let us assume three cases. First, we consider that the incentive, λ^{EV}_{k1}, is equal to 0 €. Considering that constraints (10.2a) and (10.2b) enforce that auxiliary variables $\lambda^{\text{EV,A}}_{k1m}$ can be different from zero for only one value of m, constraints (10.2c) formulate that λ^{EV}_{k1} has to be equal to either $\lambda^{\text{EV,A}}_{k11}$, $\lambda^{\text{EV,A}}_{k12}$, or $\lambda^{\text{EV,A}}_{k13}$. In this case, considering that $\lambda^{\text{EV}}_{k1} = 0$, it is satisfied that $0 \leq \lambda^{\text{EV}}_{k1} \leq 80$. Therefore, $\lambda^{\text{EV}}_{k1} = \lambda^{\text{EV,A}}_{k11}$, $v^{\text{EV}}_{k11} = 1$, and the per-unit number of controlled vehicles is $\hat{X}^{\text{EV,C}}_{k11\omega} = 0$.

In the second case we assume that λ^{EV}_{k1} is equal to 100 €. In this case, $80 \leq \lambda^{\text{EV}}_{k1} \leq 140$. Therefore, $\lambda^{\text{EV}}_{k1} = \lambda^{\text{EV,A}}_{k12}$, $v^{\text{EV}}_{k12} = 1$, and the per-unit number of controlled vehicles is $\hat{X}^{\text{EV,C}}_{k12\omega} = 0.4$.

Finally, in the third case it is assumed that λ^{EV}_{k1} is equal to 180 €. In this case, it is satisfied that $140 \leq \lambda^{\text{EV}}_{k1} \leq 200$. Therefore, $\lambda^{\text{EV}}_{k1} = \lambda^{\text{EV,A}}_{k13}$, $v^{\text{EV}}_{k13} = 1$, and the per-unit number of controlled vehicles is $\hat{X}^{\text{EV,C}}_{k13\omega} = 0.8$. □

10.4.3 Generating Units Constraints

The technical constraints used to characterize generating units are described below. As it is usual in generation capacity expansion formulations, the binary variable that characterizes the commitment of the generating units is not considered. Therefore, the constraints modeling the minimum power output, and the minimum up and down times of the generating units are not formulated. However, observe that these constraints are inherent to typical low-flexible units as nuclear and coal units. In renewable dominated power systems with high presence of intermittent power units, it is expected the participation of high-flexible generators, mainly based on gas-fired units, which are characterized by low minimum power outputs and the capability of starting-up and shutting-down very often.

$$p^{\text{G,D}}_{gdt\omega} + r^{\text{G,U}}_{gdt\omega} \leq p^{\text{I,G}}_{g}, \quad \forall g \in G, \forall t \in T, \forall d \in D, \forall \omega \in \Omega, \tag{10.3a}$$

$$0 \leq p^{\text{G,D}}_{gdt\omega} - r^{\text{G,D}}_{gdt\omega}, \quad \forall g \in G, \forall t \in T, \forall d \in D, \forall \omega \in \Omega, \tag{10.3b}$$

$$p^{\text{G,D}}_{gdt\omega} + p^{\text{G,DS}}_{gdt\omega} = A^{\text{D}}_{gdt} p^{\text{I,G}}_{g}, \quad \forall g \in G^{\text{I}}, \forall t \in T, \forall d \in D, \forall \omega \in \Omega, \tag{10.3c}$$

$$\left(p^{\text{G,D}}_{gdt\omega} + r^{\text{G,U}}_{gdt\omega}\right) - \left(p^{\text{G,D}}_{gdt-1,\omega} - r^{\text{G,D}}_{gdt-1,\omega}\right) \leq P^{\text{G}}_{\text{up},g} p^{\text{I,G}}_{g},$$
$$\forall g \in G^{\text{D}}, \forall t \in T, \forall d \in D, \forall \omega \in \Omega, \tag{10.3d}$$

$$\left(p^{\text{G,D}}_{gdt-1,\omega} + r^{\text{G,U}}_{gdt-1,\omega}\right) - \left(p^{\text{G,D}}_{gdt\omega} - r^{\text{G,D}}_{gdt\omega}\right) \leq P^{\text{G}}_{\text{dw},g} p^{\text{I,G}}_{g},$$
$$\forall g \in G^{\text{D}}, \forall t \in T, \forall d \in D, \forall \omega \in \Omega, \tag{10.3e}$$

$$\{p^{\text{G,D}}_{gdt\omega}, r^{\text{G,U}}_{gdt\omega}, r^{\text{G,D}}_{gdt\omega}, p^{\text{G,DS}}_{gdt\omega}\} \geq 0, \quad \forall g \in G, \forall t \in T, \forall d \in D, \forall \omega \in \Omega, \tag{10.3f}$$

$$\{r^{\text{G,U}}_{gdt\omega}, r^{\text{G,D}}_{gdt\omega}\} = 0, \quad \forall g \in G^{\text{I}}, \forall t \in T, \forall d \in D, \forall \omega \in \Omega. \tag{10.3g}$$

Constraints (10.3a) establish for each generating unit that the power scheduled in the day-ahead energy market plus the up reserve capacity has to be smaller than the installed capacity. In the same manner, constraints (10.3b) ensure that the power scheduled in the day-ahead market minus the down reserve capacity is a non-negative value. Constraints (10.3c) state that the power scheduled by intermittent units in the day-ahead market plus the power spillage has to be equal to the available power in each period. The power spillage is the amount of power that is available to produce by an intermittent power unit but is not injected in the grid because of technical or economical reasons. For instance, the congestion of some specific lines could cause that the total production of a wind farm cannot be evacuated into the grid. Power ramp limits are characterized through constraints (10.3d) and (10.3e). In these constraints, up and down reserve capacities are considered in order to model the maximum variation of power production between two consecutive periods. Power ramps limits are expressed in per unit over the capacity installed of each unit, $p_g^{\mathrm{I,G}}$. Constraints (10.3f) define the non-negative nature of the variables characterizing the power scheduled in the day-ahead energy and reserve capacity markets and the power spillage. Finally, constraints (10.3g) state that intermittent power units are not allowed to participate in the reserve capacity market. In other words, it is assumed that only those units with dispatchable power output are authorized to participate in this market. However, note that the dispatchability of some generating technologies has significantly improved during the last years. For instance, this is the case of wind power units that, in some power systems, are allowed to provide primary and secondary frequency regulation and tertiary reserve. Therefore, the constraints that limit the participation of some generation units in the reserve capacity market have to be only applied to those power units that are not able to modify their production according to an external control signal given by the power system operator.

10.4.4 Storage Units

The flexibility of energy storage units is key to facilitate the integration of intermittent power units in future power systems. In this sense, the recent developments of electrochemical batteries have put on the table new options for increasing the potential of electricity energy storage. Note that the electricity storage based on batteries has significant advantages over the traditional utility-scale systems based on pumping hydro units. For instance, batteries have a compact size and a modular structure that make them suitable to be installed at any desired location. The high capital cost is currently its major drawback, but it has decreased during the last years and a significant cost drop is expected in the near future. Note that additionally to applications related to the temporal arbitrage in the day-ahead energy market (buying energy at low price and selling it afterwards when price is higher), energy storage units can also provide ancillary services as, for example, those related to the participation in the reserve capacity markets.

The mathematical formulation of the participation of energy storage systems in the day-ahead energy and reserve capacity markets in the generation capacity expansion problem is provided below. In this formulation, it is considered that new energy storage systems can be installed in different locations of the power grid. As stated in Sect. 10.4.1, the investment decisions for a candidate storage unit s have two components: energy (maximum amount of energy that can be stored) and power (maximum power that can be charged or discharged from the energy storage unit). These two components are denoted by $e_s^{\mathrm{I,SE}}$ and $p_s^{\mathrm{I,SP}}$, respectively. As stated in Sect. 10.2, hourly periods are considered and energy and power units are equivalent in this formulation. Considering this, the energies purchased and sold to the day-ahead energy market are characterized by $p_{sdt\omega}^{\mathrm{S,C}}$ and $p_{sdt\omega}^{\mathrm{S,D}}$, respectively. The energy purchased is used to charge the battery, whereas the energy sold in the day-ahead energy market is obtained from discharging the battery. The up and down reserve capacities that must be provided by the battery are denoted by $r_{sdt\omega}^{\mathrm{S,U}}$ and $r_{sdt\omega}^{\mathrm{S,D}}$, respectively.

Constraints (10.4a) state that the up reserve capacity can be provided either by decreasing the charge of the battery, $r_{sdt\omega}^{\mathrm{S,UC}}$, or by increasing the discharge of the battery, $r_{sdt\omega}^{\mathrm{S,UD}}$. In the same manner, constraints (10.4b) formulate the down reserve capacity can be supplied either by increasing the charge of the battery, $r_{sdt\omega}^{\mathrm{S,DC}}$, or by decreasing the discharge, $r_{sdt\omega}^{\mathrm{S,DD}}$. Constraints (10.4c) and (10.4d) limit the amount of power that can be discharged in each period. The maximum power that can be discharged considering the day-ahead energy and reserve capacity markets, $\left(p_{sdt\omega}^{\mathrm{S,D}} + r_{sdt\omega}^{\mathrm{S,UD}}\right)$, has to be less than the power capacity installed, $p_s^{\mathrm{I,SP}}$. Additionally, the minimum power that can be discharged, $\left(p_{sdt\omega}^{\mathrm{S,D}} - r_{sdt\omega}^{\mathrm{S,DD}}\right)$ has to be a non-negative value. In the same vein, the power that can be charged is limited by constraints (10.4e) and (10.4f). The maximum power that can be charged, $\left(p_{sdt\omega}^{\mathrm{S,C}} + r_{sdt\omega}^{\mathrm{S,DC}}\right)$, has to be less than the power capacity installed. The minimum power charged, $\left(p_{sdt\omega}^{\mathrm{S,C}} - r_{sdt\omega}^{\mathrm{S,UC}}\right)$, must be greater than or equal to 0. Finally, constraints (10.4g) and (10.4h) define the non-negative nature of the variables modeling the energy and power quantities traded in the day-ahead energy and reserve capacity markets.

$$r_{sdt\omega}^{\mathrm{S,U}} = r_{sdt\omega}^{\mathrm{S,UC}} + r_{sdt\omega}^{\mathrm{S,UD}}, \quad \forall s \in S, \forall t \in T, \forall d \in D, \forall \omega \in \Omega, \tag{10.4a}$$

$$r_{sdt\omega}^{\mathrm{S,D}} = r_{sdt\omega}^{\mathrm{S,DC}} + r_{sdt\omega}^{\mathrm{S,DD}}, \quad \forall s \in S, \forall t \in T, \forall d \in D, \forall \omega \in \Omega, \tag{10.4b}$$

$$p_{sdt\omega}^{\mathrm{S,D}} + r_{sdt\omega}^{\mathrm{S,UD}} \le p_s^{\mathrm{I,SP}}, \quad \forall s \in S, \forall t \in T, \forall d \in D, \forall \omega \in \Omega, \tag{10.4c}$$

$$0 \le p_{sdt\omega}^{\mathrm{S,D}} - r_{sdt\omega}^{\mathrm{S,DD}}, \quad \forall s \in S, \forall t \in T, \forall d \in D, \forall \omega \in \Omega, \tag{10.4d}$$

$$p_{sdt\omega}^{\mathrm{S,C}} + r_{sdt\omega}^{\mathrm{S,DC}} \le p_s^{\mathrm{I,SP}}, \quad \forall s \in S, \forall t \in T, \forall d \in D, \forall \omega \in \Omega, \tag{10.4e}$$

$$0 \le p_{sdt\omega}^{\mathrm{S,C}} - r_{sdt\omega}^{\mathrm{S,UC}}, \quad \forall s \in S, \forall t \in T, \forall d \in D, \forall \omega \in \Omega, \tag{10.4f}$$

$$p_{sdt\omega}^{\mathrm{S,C}}, p_{sdt\omega}^{\mathrm{S,D}}, r_{sdt\omega}^{\mathrm{S,U}}, r_{sdt\omega}^{\mathrm{S,D}} \ge 0, \quad \forall s \in S, \forall t \in T, \forall d \in D, \forall \omega \in \Omega, \tag{10.4g}$$

$$r_{sdt\omega}^{\mathrm{S,UD}}, r_{sdt\omega}^{\mathrm{S,UC}}, r_{sdt\omega}^{\mathrm{S,DD}}, r_{sdt\omega}^{\mathrm{S,DC}} \ge 0, \quad \forall s \in S, \forall t \in T, \forall d \in D, \forall \omega \in \Omega. \tag{10.4h}$$

Constraints (10.5a) and (10.5b) are used to compute the maximum and minimum levels of energy in the energy storage units in each period. Observe that it is needed to consider both cases because of the participation of storages in the day-ahead energy and reserve capacity markets. In this manner, the minimum energy stored is computed considering that (i) the power charged is equal to the energy purchased in the day-ahead energy market minus the scheduled reduction of the power charged used to provide up reserve capacity, and (ii) the power discharged is equal to the energy sold in the day-ahead market plus the scheduled increase of power discharged used to provide up reserve capacity. Similarly, the maximum energy stored is computed considering that (i) the power charged is equal to the energy purchased in the day-ahead energy market plus the scheduled increase of the power charged used to provide down reserve capacity, and (ii) the power discharged is equal to the energy sold in the day-ahead market minus the scheduled reduction of power discharged that is used to provide down reserve capacity. Finally, constraints (10.5c) and (10.5d) define the upper and lower limits of the energy stored according to the installed energy capacity, $e_s^{\text{I,SE}}$. In these constraints, parameter $\gamma_s^{\text{S,min}}$ is used to model the minimum per unit energy capacity that must remain in the energy storage unit at any time. This value depends on the considered technology and it typically ranges between 0 and 0.15 in electrochemical batteries. Constraints (10.5e) and (10.5f) define the initial status of the batteries for each day. The energy stored at the beginning of the first period, $t = 0$, depends on parameter $\gamma_{sbd}^{\text{S,0}}$, that indicates the per-unit energy stored at the energy storage unit. Besides, constraints (10.5g) and (10.5h) impose that the energy stored in the batteries at the end of the day, $t = 24$, has to be at least greater than a desired value, $\gamma_{sd}^{\text{S,F}}$.

$$e_{sdt\omega}^{\text{S,min}} = e_{sdt-1,\omega}^{\text{S,min}} + \eta^{\text{S}} \left(p_{sdt\omega}^{\text{S,C}} - r_{sdt\omega}^{\text{S,UC}} \right) - \frac{1}{\eta^{\text{S}}} \left(p_{sdt\omega}^{\text{S,D}} + r_{sdt\omega}^{\text{S,UD}} \right),$$
$$\forall s \in S, \forall t \in T, \forall d \in D, \forall \omega \in \Omega, \quad (10.5a)$$

$$e_{sdt\omega}^{\text{S,max}} = e_{sdt-1,\omega}^{\text{S,max}} + \eta^{\text{S}} \left(p_{sdt\omega}^{\text{S,C}} + r_{sdt\omega}^{\text{S,DC}} \right) - \frac{1}{\eta^{\text{S}}} \left(p_{sdt\omega}^{\text{S,D}} - r_{sdt\omega}^{\text{S,DD}} \right),$$
$$\forall s \in S, \forall t \in T, \forall d \in D, \forall \omega \in \Omega, \quad (10.5b)$$

$$\gamma_s^{\text{S,min}} e_s^{\text{I,SE}} \le e_{sdt\omega}^{\text{S,max}} \le e_s^{\text{I,SE}}, \quad \forall s \in S, \forall t \in T, \forall d \in D, \forall \omega \in \Omega, \quad (10.5c)$$

$$\gamma_s^{\text{S,min}} e_s^{\text{I,SE}} \le e_{sdt\omega}^{\text{S,min}} \le e_s^{\text{I,SE}}, \quad \forall s \in S, \forall t \in T, \forall d \in D, \forall \omega \in \Omega, \quad (10.5d)$$

$$e_{sdt\omega}^{\text{S,min}} = \gamma_{sd}^{\text{S,0}} e_s^{\text{I,SE}}, \quad \forall s \in S, t = 0, \forall d \in D, \forall \omega \in \Omega, \quad (10.5e)$$

$$e_{sdt\omega}^{\text{S,max}} = \gamma_{sd}^{\text{S,0}} e_s^{\text{I,SE}}, \quad \forall s \in S, t = 0, \forall d \in D, \forall \omega \in \Omega, \quad (10.5f)$$

$$e_{sdt\omega}^{\text{S,min}} \ge \gamma_{sd}^{\text{S,F}} e_s^{\text{I,SE}}, \quad \forall s \in S, t = 0, \forall d \in D, \forall \omega \in \Omega, \quad (10.5g)$$

$$e_{sdt\omega}^{\text{S,max}} \ge \gamma_{sd}^{\text{S,F}} e_s^{\text{I,SE}}, \quad \forall s \in S, t = 0, \forall d \in D, \forall \omega \in \Omega. \quad (10.5h)$$

Illustrative Example 10.4 Maximum and minimum energy stored

Let us consider a 1 MW/3 MWh storage with efficiency η^{S} equal to 1 that participates for a given day d and scenario ω in the day-ahead energy and reserve capacity markets in a planning horizon comprising 5 hourly time periods. The initial energy status of the storage is 1 MWh. A possible participation of this storage unit in these markets that satisfies constraints (10.4a)–(10.4h) is provided in Table 10.3.

According to the participation of the storage in the day-ahead energy and reserve capacity markets described above and to constraints (10.5a)–(10.5h), Table 10.4

Table 10.3 Illustrative Example 10.4: participation of a storage unit in the day-ahead and reserve capacity markets

	Periods				
	1	2	3	4	5
$p^{S,C}_{sdt\omega}$	0.8	0.5	0.0	0.0	0.0
$p^{S,D}_{sdt\omega}$	0.0	0.0	0.0	0.5	0.5
$r^{S,U}_{sdt\omega}$	0.1	0.0	0.0	0.2	0.0
$r^{S,D}_{sdt\omega}$	0.0	0.3	0.0	0.0	0.3
$r^{S,UC}_{sdt\omega}$	0.1	0.0	0.0	0.0	0.0
$r^{S,UD}_{sdt\omega}$	0.0	0.0	0.0	0.2	0.0
$r^{S,DC}_{sdt\omega}$	0.0	0.3	0.0	0.0	0.0
$r^{S,DD}_{sdt\omega}$	0.0	0.0	0.0	0.0	0.3

Table 10.4 Illustrative Example 10.4: minimum and maximum amounts of energy stored (MWh)

	Periods					
	0	1	2	3	4	5
$e^{S,min}_{sdt\omega}$	1.0	1.7	2.2	2.2	1.5	1.0
$e^{S,max}_{sdt\omega}$	1.0	1.8	2.6	2.6	2.1	1.9

provides the resulting minimum and maximum amounts of energy that can be stored in each period. Observe that the energy stored in each period is always between 0 and 3 MWh, and that the energy stored at the end of the 5-h planning horizon is greater than or equal to the energy stored in the initial period, 1 MWh.

The values shown in Table 10.4 are obtained as follows:

- Period 1

$$\begin{aligned} e^{S,min}_{sd1\omega} &= e^{S,min}_{sdt0\omega} + \eta^{S}\left(p^{S,C}_{sd1\omega} - r^{S,UC}_{sd1\omega}\right) - \frac{1}{\eta^{S}}\left(p^{S,D}_{sd1\omega} + r^{S,UD}_{sd1\omega}\right) \\ &= 1.0 + 1\cdot(0.8-0.1) - 0 = 1.7\,\text{MWh}, \\ e^{S,max}_{sd1\omega} &= e^{S,max}_{sdt0\omega} + \eta^{S}\left(p^{S,C}_{sd1\omega} + r^{S,DC}_{sd1\omega}\right) - \frac{1}{\eta^{S}}\left(p^{S,D}_{sd1\omega} - r^{S,DD}_{sd1\omega}\right) \\ &= 1.0 + 1\cdot(0.8) - 0 = 1.8\,\text{MWh}. \end{aligned}$$

- Period 2

$$\begin{aligned} e^{S,min}_{sd2\omega} &= e^{S,min}_{sdt1\omega} + \eta^{S}\left(p^{S,C}_{sd2\omega} - r^{S,UC}_{sd2\omega}\right) - \frac{1}{\eta^{S}}\left(p^{S,D}_{sd2\omega} + r^{S,UD}_{sd2\omega}\right) \\ &= 1.7 + 1\cdot(0.5) - 0 = 2.2\,\text{MWh}, \end{aligned}$$

$$e^{\text{S,max}}_{sd2\omega} = e^{\text{S,max}}_{sdt1\omega} + \eta^{\text{S}} \left(p^{\text{S,C}}_{sd2\omega} + r^{\text{S,DC}}_{sd2\omega} \right) - \frac{1}{\eta^{\text{S}}} \left(p^{\text{S,D}}_{sd2\omega} - r^{\text{S,DD}}_{sd2\omega} \right)$$
$$= 1.8 + 1 \cdot (0.5 + 0.3) - 0 = 2.6\,\text{MWh}.$$

- Period 3

$$e^{\text{S,min}}_{sd3\omega} = e^{\text{S,min}}_{sdt2\omega} + \eta^{\text{S}} \left(p^{\text{S,C}}_{sd3\omega} - r^{\text{S,UC}}_{sd3\omega} \right) - \frac{1}{\eta^{\text{S}}} \left(p^{\text{S,D}}_{sd3\omega} + r^{\text{S,UD}}_{sd3\omega} \right)$$
$$= 2.2 + 0 - 0 = 2.2\,\text{MWh},$$
$$e^{\text{S,max}}_{sd3\omega} = e^{\text{S,max}}_{sdt2\omega} + \eta^{\text{S}} \left(p^{\text{S,C}}_{sd3\omega} + r^{\text{S,DC}}_{sd3\omega} \right) - \frac{1}{\eta^{\text{S}}} \left(p^{\text{S,D}}_{sd3\omega} - r^{\text{S,DD}}_{sd3\omega} \right)$$
$$= 2.6 + 0 - 0 = 2.6\,\text{MWh}.$$

- Period 4

$$e^{\text{S,min}}_{sd4\omega} = e^{\text{S,min}}_{sdt3\omega} + \eta^{\text{S}} \left(p^{\text{S,C}}_{sd4\omega} - r^{\text{S,UC}}_{sd4\omega} \right) - \frac{1}{\eta^{\text{S}}} \left(p^{\text{S,D}}_{sd4\omega} + r^{\text{S,UD}}_{sd4\omega} \right)$$
$$= 2.2 + 0 + 0 - 1 \cdot (0.5 + 0.2) = 1.5\,\text{MWh},$$
$$e^{\text{S,max}}_{sd1\omega} = e^{\text{S,max}}_{sdt3\omega} + \eta^{\text{S}} \left(p^{\text{S,C}}_{sd1\omega} + r^{\text{S,DC}}_{sd1\omega} \right) - \frac{1}{\eta^{\text{S}}} \left(p^{\text{S,D}}_{sd1\omega} - r^{\text{S,DD}}_{sd1\omega} \right)$$
$$= 2.6 + 0 - 1 \cdot (0.5) = 2.1\,\text{MWh}.$$

- Period 5

$$e^{\text{S,min}}_{sd5\omega} = e^{\text{S,min}}_{sdt4\omega} + \eta^{\text{S}} \left(p^{\text{S,C}}_{sd5\omega} - r^{\text{S,UC}}_{sd5\omega} \right) - \frac{1}{\eta^{\text{S}}} \left(p^{\text{S,D}}_{sd5\omega} + r^{\text{S,UD}}_{sd5\omega} \right)$$
$$= 1.5 + 0 - 1 \cdot (0.5) = 1.0\,\text{MWh},$$
$$e^{\text{S,max}}_{sd5\omega} = e^{\text{S,max}}_{sdt4\omega} + \eta^{\text{S}} \left(p^{\text{S,C}}_{sd5\omega} + r^{\text{S,DC}}_{sd5\omega} \right) - \frac{1}{\eta^{\text{S}}} \left(p^{\text{S,D}}_{sd5\omega} - r^{\text{S,DD}}_{sd5\omega} \right)$$
$$= 2.1 + 0 - 1 \cdot (0.5 - 0.3) = 1.9\,\text{MWh}.$$

□

10.4.5 Electric Vehicles Modeling

This section describes the mathematical modeling of those EVs that are controlled by the power system operator. From a modeling point of view, EVs can be considered as energy storage units that are available to be charged and discharged only during certain time periods per day. The initial statuses of the batteries of EVs will depend on the driving patterns of their users, whereas the amount of energy stored in the batteries at the end of the charging period has to be equal to or greater than a prefixed value decided by electric vehicle users. Additionally, the number of EVs that are controlled by the power system operator is not a known parameter that depends on the financial incentives offered to EV users. In this sense, as described in Sect. 10.4.2, two different control modes are distinguished, which are denoted by $c \in C = \{1, 2\}$. The first control mode is assigned to all vehicles in which the discharge of energy into the grid is not allowed, $c = 1$. Therefore, the power system operator can only control the charging process of these vehicles. The vehicles belonging to users that also agree to the discharge of their vehicles are under the second control mode, $c = 2$. Therefore, the number of vehicles under the first control mode from group k at bus b in scenario ω can be computed as $N^{\mathrm{EV}}_{kb\omega} x^{\mathrm{EV,C}}_{k1\omega}$, where $N^{\mathrm{EV}}_{kb\omega}$ is the total number of EVs belonging to group k at bus b in scenario ω. As indicated in Sect. 10.4.2, $x^{\mathrm{EV,C}}_{k1\omega}$ is a variable ranging in interval [0, 1] that represents the per-unit number of vehicles controlled in mode $c = 1$. Considering that the charge of those vehicles belonging to second control mode is also controlled by the power system operator, the total number of vehicles whose charge is controlled is equal to $N^{\mathrm{EV}}_{kb\omega} x^{\mathrm{EV,C}}_{k1\omega}$. Considering the above, the number of vehicles that will be able to inject the energy discharged from their batteries into the grid is equal to $N^{\mathrm{EV}}_{kb\omega} x^{\mathrm{EV}}_{k2\omega}$.

Similarly to the case of energy storage units, it is assumed that the energy associated with the charge and discharge of EVs can be traded in the day-ahead energy and reserve capacity markets.

Constraints (10.6a) state that the up reserve capacity can be provided either by decreasing the charge of the EVs, $r^{\mathrm{EV,UC}}_{kbdt\omega}$, or by increasing the discharge, $r^{\mathrm{EV,UD}}_{kbdt\omega}$. In the same manner, constraints (10.6b) state the down reserve capacity can be supplied either by increasing the charge, $r^{\mathrm{EV,DC}}_{kbdt\omega}$, or by decreasing the discharge, $r^{\mathrm{EV,DD}}_{kbdt\omega}$. Constraints (10.6c) and (10.6d) limit the total quantity of power that can be discharged in each period by EVs belonging to group k at bus b. The maximum power that can be discharged considering the day-ahead energy and reserve capacity markets, $\left(p^{\mathrm{EV,D}}_{kbdt\omega} + r^{\mathrm{EV,UD}}_{kbdt\omega}\right)$, has to be less than the number of EVs that are controlled by the power system operator at bus b under mode $c = 2$, $N^{\mathrm{EV}}_{kb\omega} x^{\mathrm{EV,C}}_{k2\omega}$, times the average power rate of the charger, $P^{\mathrm{EV}}_{\max}$. Additionally, the minimum power that can be discharged, $\left(p^{\mathrm{EV,D}}_{kbdt\omega} - r^{\mathrm{EV,DD}}_{kbdt\omega}\right)$ has to be a non-negative value. In the same vein, the power that can be charged is limited by constraints (10.6e) and (10.6f). Again, the maximum power that can be charged, $\left(p^{\mathrm{EV,C}}_{kbdt\omega} + r^{\mathrm{EV,DC}}_{kbdt\omega}\right)$, has to be less than the number of controlled vehicles under mode $c = 1$, $N^{\mathrm{EV}}_{kb\omega} x^{\mathrm{EV,C}}_{k1\omega}$, times the average power rate of the charger. The minimum power charged, $\left(p^{\mathrm{EV,C}}_{kbdt\omega} - r^{\mathrm{EV,UC}}_{kbdt\omega}\right)$, must be higher than 0. Finally, constraints (10.6g) and (10.6h) define the positive nature of the variables modeling the energy and power quantities traded in the day-ahead energy and reserve capacity markets.

$$r^{\mathrm{EV,U}}_{kbdt\omega} = r^{\mathrm{EV,UC}}_{kbdt\omega} + r^{\mathrm{EV,UD}}_{kbdt\omega},$$

$$\forall k \in K, \forall b \in B, \forall t \in T_{kd}, \forall d \in D, \forall \omega \in \Omega, \tag{10.6a}$$

$$r^{\mathrm{EV,D}}_{kbdt\omega} = r^{\mathrm{EV,DC}}_{kbdt\omega} + r^{\mathrm{EV,DD}}_{kbdt\omega},$$

$$\forall k \in K, \forall b \in B, \forall t \in T_{kd}, \forall d \in D, \forall \omega \in \Omega, \quad (10.6b)$$

$$p^{\text{EV,D}}_{kbdt\omega} + r^{\text{EV,UD}}_{kbdt\omega} \leq N^{\text{EV}}_{kb\omega} x^{\text{EV,C}}_{k2\omega} P^{\text{EV}}_{\max},$$
$$\forall k \in K, \forall b \in B, \forall t \in T_{kd}, \forall d \in D, \forall \omega \in \Omega, \quad (10.6c)$$

$$0 \leq p^{\text{EV,D}}_{kbdt\omega} - r^{\text{EV,DD}}_{kbdt\omega}, \quad \forall k \in K, \forall b \in B, \forall t \in T_{kd}, \forall d \in D, \forall \omega \in \Omega, \quad (10.6d)$$

$$p^{\text{EV,C}}_{kbdt\omega} + r^{\text{EV,DC}}_{kbdt\omega} \leq N^{\text{EV}}_{kb\omega} x^{\text{EV,C}}_{k1\omega} P^{\text{EV}}_{\max},$$
$$\forall k \in K, \forall b \in B, \forall t \in T_{kd}, \forall d \in D, \forall \omega \in \Omega, \quad (10.6e)$$

$$0 \leq p^{\text{EV,C}}_{kbdt\omega} - r^{\text{EV,UC}}_{kbdt\omega}, \quad \forall k \in K, \forall b \in B, \forall t \in T_{kd}, \forall d \in D, \forall \omega \in \Omega, \quad (10.6f)$$

$$p^{\text{EV,C}}_{kbdt\omega}, p^{\text{EV,D}}_{kbdt\omega}, r^{\text{EV,U}}_{kbdt\omega}, r^{\text{EV,D}}_{kbdt\omega} \geq 0,$$
$$\forall k \in K, \forall b \in B, \forall t \in T_{kd}, \forall d \in D, \forall \omega \in \Omega, \quad (10.6g)$$

$$r^{\text{EV,UD}}_{kbdt\omega}, r^{\text{EV,UC}}_{kbdt\omega}, r^{\text{EV,DD}}_{kbdt\omega}, r^{\text{EV,DC}}_{kbdt\omega} \geq 0,$$
$$\forall k \in K, \forall b \in B, \forall t \in T_{kd}, \forall d \in D, \forall \omega \in \Omega. \quad (10.6h)$$

Constraints (10.7a) and (10.7b) are used to compute the maximum and minimum levels of energy stored in the set of controlled vehicles resulting in each set of EVs, bus, and period according to the participation in the day-ahead energy and reserve capacity markets. As stated in Chap. 5, if one hour periods are considered, power (W) and energy (Wh) units are equivalent in the modeling of the state-of-charge of the batteries. Then, similarly to the modeling of energy storage units, the minimum energy stored is computed considering that (i) the power charged is equal to the energy purchased in the day-ahead energy market minus the scheduled reduction of the power charged used to provide up reserve capacity, and (ii) the power discharged is equal to the energy sold in the day-ahead market plus the scheduled increase of power discharged used to provide up reserve capacity. Similarly, the maximum energy stored is computed considering that (i) the power charged is equal to the energy purchased in the day-ahead energy market plus the scheduled increase of the power charged used to provide down reserve capacity, and (ii) the power discharged is equal to the energy sold in the day-ahead market minus the scheduled reduction of power discharged that is used to provide down reserve capacity. Constraints (10.7c) and (10.7d) define the upper and lower limits of the energy stored according to the number of controlled vehicles. In these constraints, parameters $\gamma^{\text{EV,min}}$ and $P^{\text{EV}}_{\max}$ are the minimum per-unit energy capacity that must remain in the EV batteries and the average energy capacity of the battery of an EV, respectively. Constraints (10.7e) and (10.7f) define the initial status of the batteries for each day. The energy stored in the period previous to the first charging period, T^0_{kd}, depends on parameter $\gamma^{\text{EV},0}_{kbd}$, that indicates the per-unit energy stored at the battery. This value depends on the daily distance driven by the vehicle. In the same manner, constraints (10.7g) and (10.7h) enforce that the energy stored in the batteries at the end of the charging period, T^{F}_{kd}, has to be at least greater than a desired value, $\gamma^{\text{EV,F}}_{kbd}$.

$$e^{\text{EV,min}}_{kbdt\omega} = e^{\text{EV,min}}_{kbdt-1,\omega} + \eta^{\text{EV}} \left(p^{\text{EV,C}}_{kbdt\omega} - r^{\text{EV,UC}}_{kbdt\omega} \right) - \frac{1}{\eta^{\text{EV}}} \left(p^{\text{EV,D}}_{kbdt\omega} + r^{\text{EV,SUD}}_{kbdt\omega} \right),$$
$$\forall k \in K, \forall b \in B, \forall t \in T_{kd}, \forall d \in D, \forall \omega \in \Omega, \quad (10.7a)$$

$$e^{\text{EV,max}}_{kbdt\omega} = e^{\text{EV,max}}_{kbdt-1,\omega} + \eta^{\text{EV}}\left(p^{\text{EV,C}}_{kbdt\omega} + r^{\text{EV,DC}}_{kbdt\omega}\right) - \frac{1}{\eta^{\text{EV}}}\left(p^{\text{EV,D}}_{kbdt\omega} - r^{\text{EV,DD}}_{kbdt\omega}\right),$$
$$\forall k \in K, \forall b \in B, \forall t \in T_{kd}, \forall d \in D, \forall \omega \in \Omega, \quad (10.7b)$$
$$\gamma^{\text{EV,min}} N^{\text{EV}}_{kb\omega} x^{\text{EV}}_{k\omega} E^{\text{EV,B}}_{\text{max}} \leq e^{\text{EV,max}}_{kbdt\omega} \leq N^{\text{EV}}_{kb\omega} x^{\text{EV,C}}_{k1\omega} E^{\text{EV,B}}_{\text{max}},$$
$$\forall k \in K, \forall b \in B, \forall t \in T_{kd}, \forall d \in D, \forall \omega \in \Omega, \quad (10.7c)$$
$$\gamma^{\text{EV,min}} N^{\text{EV}}_{kb\omega} x^{\text{EV}}_{k\omega} E^{\text{EV,B}}_{\text{max}} \leq e^{\text{EV,min}}_{kbdt\omega} \leq N^{\text{EV}}_{kb\omega} x^{\text{EV,C}}_{k1\omega} E^{\text{EV,B}}_{\text{max}},$$
$$\forall k \in K, \forall b \in B, \forall t \in T_{kd}, \forall d \in D, \forall \omega \in \Omega, \quad (10.7d)$$
$$e^{\text{EV,min}}_{kbdt\omega} = \gamma^{\text{EV,0}}_{kbd} N^{\text{EV}}_{kb\omega} x^{\text{EV,C}}_{k1\omega} E^{\text{EV,B}}_{\text{max}},$$
$$\forall k \in K, \forall b \in B, \forall t \in T^{0}_{kd}, \forall d \in D, \forall \omega \in \Omega, \quad (10.7e)$$
$$e^{\text{EV,max}}_{kbdt\omega} = \gamma^{\text{EV,0}}_{kbd} N^{\text{EV}}_{kb\omega} x^{\text{EV,C}}_{k1\omega} E^{\text{EV,B}}_{\text{max}},$$
$$\forall k \in K, \forall b \in B, \forall t \in T^{0}_{kd}, \forall d \in D, \forall \omega \in \Omega, \quad (10.7f)$$
$$e^{\text{EV,min}}_{kbdt\omega} \geq \gamma^{\text{EV,F}}_{kbd} N^{\text{EV}}_{kb\omega} x^{\text{EV,C}}_{k1\omega} E^{\text{EV,B}}_{\text{max}},$$
$$\forall k \in K, \forall b \in B, \forall t \in T^{\text{F}}_{kd}, \forall d \in D, \forall \omega \in \Omega, \quad (10.7g)$$
$$e^{\text{EV,max}}_{kbdt\omega} \geq \gamma^{\text{EV,F}}_{kbd} N^{\text{EV}}_{kb\omega} x^{\text{C,EV}}_{kb1\omega} E^{\text{EV,B}}_{\text{max}},$$
$$\forall k \in K, \forall b \in B, \forall t \in T^{\text{F}}_{kd}, \forall d \in D, \forall \omega \in \Omega. \quad (10.7h)$$

Illustrative Example 10.5 Operation of controlled electric vehicles

Let us consider a set of 100 EVs belonging to group k at bus b that are controlled by the power system operator in charge/discharge mode ($c = 2$). The average capacity of the batteries of these vehicles is equal to 50 kWh, and they can charge at a maximum of 10 kW with charge efficiency η^{EV} equal to 1. We assume that a planning horizon comprising five hourly time periods in day d and scenario ω. The charging period starts in period 2 and ends at period 4. The initial energy status of the batteries of EVs is 25 kWh and the final status of the batteries at the end of the charging period has to be greater than or equal to 40 kWh. Table 10.5 provides an example of participation of this group of EVs in the day-ahead energy and reserve capacity markets that satisfies constraints (10.6a)–(10.6h). Note that EVs discharge energy into the grid in period 2, and charges during periods 3 and 4. In period 1 and 5, vehicles are not connected to the grid. The charge of the battery results from consuming energy in the day-ahead market, $p^{\text{EV,C}}_{kbdt\omega}$, whereas the energy discharged is delivered in this market, $p^{\text{EV,D}}_{kbdt\omega}$. In all periods, the power charged and discharged from the participation in the day-ahead markets is smaller than the maximum power that can be injected or extracted from the grid, that is $100 \cdot 0.01$MW=1 MW.

EVs provide up reserve $r^{\text{EV,U}}_{kbdt\omega}$ in period 4, and down reserve $r^{\text{EV,D}}_{kbdt\omega}$ in period 2. Since vehicles are charging in period 4, ($p^{\text{EV,C}}_{kbd4\omega} = 1.0$ MW), the provision of up reserve is made by decreasing the charge of power, $r^{\text{EV,UC}}_{sd4\omega} = 0.5$ MW. Observe that $1.0 - 0.5 \geq 0$. On the contrary, vehicles are discharging in period 2, ($p^{\text{EV,D}}_{sd2\omega} = 0.5$ MW). Therefore, the provision of down reserve can only be performed by decreasing the discharge of power, $r^{\text{EV,DD}}_{sd2\omega} = 0.5$ MW. Observe that the total amount of power that can be discharged is a non-negative value, $0.5 - 0.5 \geq 0$.

Table 10.5 Illustrative Example 10.5: participation of electric vehicles in day-ahead and reserve capacity markets

	Periods				
	1	2	3	4	5
$p^{\text{EV,C}}_{kbdt\omega}$	–	0.0	1.5	1.0	–
$p^{\text{EV,D}}_{kbdt\omega}$	–	0.5	0.0	0.0	–
$r^{\text{EV,U}}_{kbdt\omega}$	–	0.0	0.0	0.5	–
$r^{\text{EV,D}}_{kbdt\omega}$	–	0.5	0.0	0.0	–
$r^{\text{EV,UC}}_{kbdt\omega}$	–	0.0	0.0	0.5	–
$r^{\text{EV,UD}}_{kbdt\omega}$	–	0.0	0.0	0.0	–
$r^{\text{EV,DC}}_{kbdt\omega}$	–	0.0	0.0	0.0	–
$r^{\text{EV,DD}}_{kbdt\omega}$	–	0.5	0.0	0.0	–

Table 10.6 Illustrative Example 10.5: minimum and maximum amounts of energy stored in electric vehicles (MWh)

	Periods				
	1	2	3	4	5
$e^{\text{EV,min}}_{kbdt\omega}$	2.5	2.0	3.5	4.0	–
$e^{\text{EV,max}}_{kbdt\omega}$	2.5	2.5	4.0	5.0	–

Table 10.6 provides the resulting minimum and maximum amounts of energy stored by EVs in each period according to constraints (10.7a)–(10.7h). Observe that the energy stored in each period is always between 0 and 5 MWh, where 5 MWh is equal to the product of the number of EVs and the average battery capacity per vehicle, $100 \cdot 0.05$ MWh. Observe also that the energy stored at the end of the charging period, is greater than the minimum value $100 \cdot 0.04 = 4$ MWh.

The values shown in Table 10.6 are obtained as follows:

- Period 2

$$e^{\text{EV,min}}_{kbd2\omega} = e^{\text{EV,min}}_{kbdt1\omega} + \eta^{\text{EV}}\left(p^{\text{EV,C}}_{kbd2\omega} - r^{\text{EV,UC}}_{kbd2\omega}\right) - \frac{1}{\eta^{\text{EV}}}\left(p^{\text{EV,D}}_{kbd2\omega} + r^{\text{EV,UD}}_{kbd2\omega}\right)$$

$$= 2.5 + 0 - 1 \cdot (0.5) = 2.0\,\text{MWh},$$

$$e^{\text{EV,max}}_{kbd2\omega} = e^{\text{EV,max}}_{kbdt1\omega} + \eta^{\text{EV}}\left(p^{\text{EV,C}}_{kbd2\omega} + r^{\text{EV,DC}}_{kbd2\omega}\right) - \frac{1}{\eta^{\text{EV}}}\left(p^{\text{EV,D}}_{kbd2\omega} - r^{\text{EV,DD}}_{kbd2\omega}\right)$$

$$= 2.5 + 0 - 1 \cdot (0.5 - 0.5) = 2.5\,\text{MWh}.$$

- Period 3

$$e^{\text{EV,min}}_{kbd3\omega} = e^{\text{EV,min}}_{kbdt2\omega} + \eta^{\text{EV}} \left(p^{\text{EV,C}}_{kbd3\omega} - r^{\text{EV,UC}}_{kbd3\omega} \right) - \frac{1}{\eta^{\text{EV}}} \left(p^{\text{EV,D}}_{kbd3\omega} + r^{\text{EV,UD}}_{kbd3\omega} \right)$$
$$= 2.0 + 1 \cdot (1.5) - 0 = 3.5\,\text{MWh},$$
$$e^{\text{EV,max}}_{kbd3\omega} = e^{\text{EV,max}}_{kbdt2\omega} + \eta^{\text{EV}} \left(p^{\text{EV,C}}_{kbd3\omega} + r^{\text{EV,DC}}_{kbd3\omega} \right) - \frac{1}{\eta^{\text{EV}}} \left(p^{\text{EV,D}}_{kbd3\omega} - r^{\text{EV,DD}}_{kbd3\omega} \right)$$
$$= 2.5 + 1 \cdot (1.5) - 0 = 4.0\,\text{MWh}.$$

- Period 4

$$e^{\text{EV,min}}_{kbd4\omega} = e^{\text{EV,min}}_{kbdt3\omega} + \eta^{\text{EV}} \left(p^{\text{EV,C}}_{kbd4\omega} - r^{\text{EV,UC}}_{kbd4\omega} \right) - \frac{1}{\eta^{\text{EV}}} \left(p^{\text{EV,D}}_{kbd4\omega} + r^{\text{EV,UD}}_{kbd4\omega} \right)$$
$$= 3.5 + 1 \cdot (1.0 - 0.5) - 0 = 4.0\,\text{MWh},$$
$$e^{\text{EV,max}}_{kbd4\omega} = e^{\text{EV,max}}_{kbdt3\omega} + \eta^{\text{EV}} \left(p^{\text{EV,C}}_{kbd4\omega} + r^{\text{EV,DC}}_{kbd4\omega} \right) - \frac{1}{\eta^{\text{EV}}} \left(p^{\text{EV,D}}_{kbd4\omega} - r^{\text{EV,DD}}_{kbd4\omega} \right)$$
$$= 4.0 + 1 \cdot (1.0) - 0 = 5.0\,\text{MWh}.$$

□

10.4.6 Day-Ahead Market Constraints

In this model it is assumed that energy and up and down reserve capacities are simultaneously optimized in the day-ahead market. This type of market structure is specially convenient for scheduling renewable dominated power systems where the appropriate determination of reserves is key to ensure the maximum integration of intermittent power sources. The DC approximation of the transmission network is used to account for the structure of the transmission system in the operation of the power system. Considering that a long-term investment model is being solved, this approximation is reasonable. Therefore, once the day-ahead market is cleared by the market operator, the power system operator checks the feasibility of the obtained scheduling and modifies it if necessary according to some prefixed rules.

Constraints (10.8a) formulate the day-ahead energy balance for each bus, period, and scenario. The different terms in the energy balance constraint comprise the power output of generating units, the power discharged and charged from energy storage units and controlled EVs, outgoing and incoming transmission power flows, nodal demand, the power charged from non-controlled vehicles, and the unserved demand. Considering that the total expected power charged by EVs belonging to group k in bus b, day d, period t, and scenario ω is characterized by $P^{\text{EV}}_{kbdt\omega}$ and that the per-unit number of controlled vehicles is $x^{\text{EV,C}}_{k1\omega}$, the power charged by non-controlled vehicles can be computed as $(1 - x^{\text{EV,C}}_{k1\omega})P^{\text{EV}}_{kbdt\omega}$. The power charged to and discharged from controlled vehicles is represented by $p^{\text{EV,DC}}_{kbdt\omega}$ and $p^{\text{EV,DD}}_{kbdt\omega}$, respectively. Constraints (10.8b) state that the unserved demand has to be non-negative and lower than the inflexible demand. The power flows in the transmission lines are formulated through constraints (10.8c)–(10.8e). The power flow in line ℓ depends on the reactance of the line, X_ℓ , and on the difference of voltage angles between the origin bus of the line, $O(\ell)$, and the destination bus, $F(\ell)$.

$$\sum_{g \in G_b} p^{\text{G,D}}_{gdt\omega} + \sum_{s \in S_b} \left(p^{\text{S,D}}_{sdt\omega} - p^{\text{S,C}}_{sdt\omega} \right) + \sum_{k \in K_t} \left(p^{\text{EV,D}}_{kbdt\omega} - p^{\text{EV,C}}_{kbdt\omega} \right)$$
$$- \sum_{\ell \in L^{\text{O}}_b} p^{\text{L,D}}_{\ell dt\omega} + \sum_{\ell \in L^{\text{F}}_b} p^{\text{L,D}}_{\ell dt\omega} = P^{\text{D}}_{bdt\omega} + \sum_{k \in K_t} (1 - x^{\text{EV,C}}_{k1\omega}) P^{\text{EV}}_{kbdt\omega} - \sum_{b \in B} p^{\text{UD}}_{bdt\omega},$$
$$\forall b \in B, \forall t \in T, \forall d \in D, \forall \omega \in \Omega, \tag{10.8a}$$

$$0 \le p^{\text{UD}}_{bdt\omega} \le P^{\text{D}}_{bdt\omega} + \sum_{k \in K_t} (1 - x^{\text{EV,C}}_{k1\omega}) P^{\text{EV}}_{kbdt\omega},$$
$$\forall b \in B, \forall t \in T, \forall d \in D, \forall \omega \in \Omega, \tag{10.8b}$$

$$p^{\text{L,D}}_{\ell dt\omega} = \frac{1}{X_\ell} \left(\Theta^{\text{D}}_{O(\ell)dt\omega} - \Theta^{\text{D}}_{F(\ell)dt\omega} \right), \quad \forall \ell \in L, \forall t \in T, \forall d \in D, \forall \omega \in \Omega, \tag{10.8c}$$

$$-P^{\text{L}}_{\max,\ell} \le p^{\text{L,D}}_{\ell dt\omega} \le P^{\text{L}}_{\max,\ell}, \quad \forall \ell \in L, \forall t \in T, \forall d \in D, \forall \omega \in \Omega, \tag{10.8d}$$

$$-\pi \le \Theta^{\text{D}}_{bdt\omega} \le \pi, \quad \forall b \in B, \forall t \in T, \forall d \in D, \forall \omega \in \Omega. \tag{10.8e}$$

The up and down reserve capacity requirements are established by constraints (10.9a) and (10.9b), respectively. It is assumed that only generating units, storage units, and controlled vehicles are allowed to provide reserve capacity. The up reserve capacity requirement is formulated in (10.9a) as a function on the expected demand and the scheduled power of intermittent units. Parameters $\gamma^{\text{CU,D}}$ and $\gamma^{\text{CU,I}}$ are used to weight the reserve requirement amounts. Typical values of $\gamma^{\text{CU,D}}$ and $\gamma^{\text{CU,I}}$ are 0.03 and 0.08, respectively. The down reserve requirement is stated in (10.9b) and is analogous to the up reserve capacity requirement formulation.

$$\sum_{g \in G^{\text{D}}} r^{\text{G,U}}_{gdt\omega} + \sum_{s \in S} r^{\text{S,U}}_{sdt\omega} + \sum_{k \in K_t} \sum_{b \in B} r^{\text{EV,U}}_{kbdt\omega}$$
$$\ge \gamma^{\text{CU,D}} \sum_{b \in B} \left(P^{\text{D}}_{bdt\omega} + \sum_{k \in K_t} (1 - x^{\text{EV,C}}_{k1\omega}) P^{\text{EV}}_{kbdt\omega} \right) + \gamma^{\text{CU,I}} \sum_{g \in G^{\text{I}}} p^{\text{G,D}}_{gdt\omega},$$
$$\forall t \in T, \forall d \in D, \forall \omega \in \Omega, \tag{10.9a}$$

$$\sum_{g \in G^{\text{D}}} r^{\text{G,D}}_{gdt\omega} + \sum_{s \in S} r^{\text{S,D}}_{sdt\omega} + \sum_{k \in K_t} \sum_{b \in B} r^{\text{EV,D}}_{kbdt\omega}$$
$$\ge \gamma^{\text{CD,D}} \sum_{b \in B} \left(P^{\text{D}}_{bdt\omega} + \sum_{k \in K_t} (1 - x^{\text{EV,C}}_{k1\omega}) P^{\text{EV}}_{kbdt\omega} \right) + \gamma^{\text{CD,I}} \sum_{g \in G^{\text{I}}} p^{\text{G,D}}_{gdt\omega},$$
$$\forall t \in T, \forall d \in D, \forall \omega \in \Omega. \tag{10.9b}$$

10.4.7 Objective Function

In this section the objective of the problem is described. This objective function is formulated in (10.10) as the expected annual cost. This cost comprises the annualized investment costs plus the operating cost during a year. The annualized cost includes the financial incentives offered to EV users. Parameter π_ω refers to the probability of scenario ω, whereas W_{dt} is a parameter used to assign a weight to each day d and hourly period t. Therefore, considering 24 hourly periods per day, it should be satisfied that $\sum_{\omega\in\Omega} \pi_\omega = 1$ and $\sum_{d\in D}\sum_{t=1}^{24} W_{dt} = 8760$. The operation costs consider the participation of generating units and storage units in the day-ahead energy and reserve capacity markets.

$$
\begin{aligned}
&\sum_{\omega\in\Omega}\pi_\omega\Bigg[\sum_{g\in G} C^{\mathrm{G,C}}_{g\omega} p^{\mathrm{G,C}}_g + \sum_{s\in S}\left(C^{\mathrm{SE,C}}_{s\omega} e^{\mathrm{SE,C}}_s + C^{\mathrm{SP,C}}_{s\omega} p^{\mathrm{SP,C}}_s\right)\\
&+\sum_{k\in K}\sum_{b\in B}\sum_{c\in C}\sum_{m\in M} N^{\mathrm{EV}}_{kb\omega}\hat{X}^{\mathrm{EV}}_{kcm\omega}\lambda^{\mathrm{EV,A}}_{kcm}\\
&+\sum_{d\in D}\sum_{t\in T} W_{dt}\Bigg(\sum_{g\in G^{\mathrm{D}}}\left(C^{\mathrm{G,D}}_{gdt\omega} p^{\mathrm{G,D}}_{gdt\omega} + C^{\mathrm{G,RU}}_{gdt\omega} r^{\mathrm{G,U}}_{gdt\omega} + C^{\mathrm{G,RD}}_{gdt\omega} r^{\mathrm{G,D}}_{gdt\omega}\right)\\
&\quad+\sum_{s\in S}\left(C^{\mathrm{S,D}}_{sdt\omega} p^{\mathrm{S,D}}_{sdt\omega} - C^{\mathrm{S,C}}_{sdt\omega} p^{\mathrm{S,C}}_{sdt\omega} + C^{\mathrm{S,RU}}_{sdt\omega} r^{\mathrm{S,U}}_{sdt\omega} + C^{\mathrm{S,RD}}_{sdt\omega} r^{\mathrm{S,D}}_{sdt\omega}\right)\\
&\quad+\sum_{b\in B} C^{\mathrm{UD}} p^{\mathrm{UD}}_{bdt\omega}\Bigg)\Bigg].
\end{aligned}
\tag{10.10}
$$

10.4.8 Problem Formulation

The generation expansion planning problem is finally formulated as the following mixed-integer linear programming problem:

$$\text{Minimize}_{\Theta} \quad (10.10) \tag{10.11a}$$

subject to:

$$\text{Constraints (10.1a)–(10.9b)}, \tag{10.11b}$$

where set Θ comprises the decision variables.

10.5 Solution Algorithm: Benders' Decomposition

The problem formulated in the previous section is a mixed-integer linear programming problem that may become intractable if a large number of characteristic days and scenarios is considered. Therefore, the usage of a decomposition technique may be advisable to solve the proposed formulation in reasonable computational times. Observe that if first-stage variables are fixed, i.e., those corresponding to investment and incentive decisions, the analyzed problem can be decomposed by characteristic day and scenario. Considering this, the Benders' decomposition approach may be adequate for solving this problem. Each of the resulting decoupled problems may constitute a subproblem in the Benders' decomposition algorithm emulating the day-ahead market operation for a given characteristic day and scenario. The investment and incentive decisions are obtained by solving a master problem which constitutes a relaxation of the original problem, where the Benders' cuts that are obtained from solving each subproblem are added in an iterative manner. A general description of the Benders' decomposition approach and some illustrative examples of its practical application can be consulted in Sect. 3.5 of Chap. 3.

The practical implementation of Benders' decomposition to solve the proposed problem is described in the rest of this section. In the following, superscript (ν) is used to denote the iteration counter.

10.5.1 Subproblems

As stated above, each subproblem simulates the day-ahead market operation for a given representative day assuming that the values of first-stage variables $p_g^{\mathrm{I,G}}$, $e_s^{\mathrm{I,SE}}$, $p_s^{\mathrm{I,SP}}$, and $\lambda_{kc}^{\mathrm{EV}}$ are known. These variables are obtained from the previous execution of the master problem. Note that, if $\lambda_{kc}^{\mathrm{EV}}$ is known, the value of variable $x_{kc\omega}^{\mathrm{EV,C}}$ can be straightforwardly obtained through constraints (10.2a)–(10.2d). The solution of the subproblems provides the sensitivities of the first-stage decisions. Then, the following problem is solved for each representative day d and each scenario ω:

$$
\begin{aligned}
&\text{Minimize}_{\Theta_{d\omega}^{\mathrm{SP}}}\\
&\pi_\omega \sum_{t \in T} W_{dt} \Bigg(\sum_{g \in G} \Big(C_{gdt\omega}^{\mathrm{G,D}} p_{gdt\omega}^{\mathrm{G,D}} + C_{gdt\omega}^{\mathrm{G,RU}} r_{gdt\omega}^{\mathrm{G,U}} + C_{gdt\omega}^{\mathrm{G,RD}} r_{gdt\omega}^{\mathrm{G,D}} \Big)\\
&\quad + \sum_{s \in S} \Big(C_{gdt\omega}^{\mathrm{S,D}} p_{sdt\omega}^{\mathrm{S,D}} - C_{sdt\omega}^{\mathrm{S,C}} p_{sdt\omega}^{\mathrm{S,C}} + C_{sdt\omega}^{\mathrm{S,U}} r_{sdt\omega}^{\mathrm{S,RU}} + C_{sdt\omega}^{\mathrm{G,RD}} r_{sdt\omega}^{\mathrm{S,D}} \Big)\\
&\quad + \sum_{b \in B} C^{\mathrm{UD}} p_{bdt\omega}^{\mathrm{UD}} \Bigg)
\end{aligned} \tag{10.12a}
$$

subject to:

$$\text{Constraints (10.3a)–(10.9b)}, \tag{10.12b}$$

$$p_g^{\text{I,G}} = P_g^{\text{I,G},(\nu)} \; : \; (\alpha_{gd\omega}^{\text{I,G},(\nu)}), \quad \forall g \in G, \tag{10.12c}$$

$$e_s^{\text{I,SE}} = E_s^{\text{I,SE},(\nu)} \; : \; (\alpha_{sd\omega}^{\text{I,SE},(\nu)}), \quad \forall s \in S, \tag{10.12d}$$

$$p_s^{\text{I,S}} = P_s^{\text{I,SP},(\nu)} \; : \; (\alpha_{sd\omega}^{\text{I,SP},(\nu)}), \quad \forall i \in S, \tag{10.12e}$$

$$x_{kc\omega}^{\text{EV,C}} = X_{kc\omega}^{\text{EV,C},(\nu)} \; : \; (\alpha_{kcd\omega}^{\text{EV,C},(\nu)}), \quad \forall k \in K, \forall c \in C, \tag{10.12f}$$

where $\Theta_{d\omega}^{\text{SP}}$ is the set of optimization variables of the subproblem associated with representative day d and scenario ω. The optimal objective function of the subproblem in iteration (ν) is denoted by $SP_{d\omega}^{(\nu)}$, which represents the weighted operation cost in representative day d and scenario ω. Variables $\alpha_{gd\omega}^{\text{I,G},(\nu)}$, $\alpha_{sd\omega}^{\text{I,SE},(\nu)}$, $\alpha_{sd\omega}^{\text{I,SP},(\nu)}$, and $\alpha_{kcd\omega}^{\text{EV,C},(\nu)}$ denote the dual variables corresponding with constraints, (10.12c), (10.12d), (10.12e), and (10.12f), respectively. These dual variables will be used in the master problem to characterize mathematically the sensitivity of the first-stage decisions on the operation costs in each characteristic day and scenario.

10.5.2 Master Problem

The master problem is a relaxed approximation of the original generation capacity expansion problem. The objective of the master problem is to determine the first-stage variables, i.e., capacity investments in generation and storage units and the financial incentives offered to EV users. The formulation of the master problem is:

$$\text{Minimize}_{\Theta^{\text{MP}}}$$

$$\sum_{\omega\in\Omega} \pi_\omega \Big[\sum_{g\in G} C_{g\omega}^{\text{I,G}} p_g^{\text{I,G}} + \sum_{s\in S} (C_{s\omega}^{\text{I,SE}} e_s^{\text{I,SE}} + C_{s\omega}^{\text{I,SP}} p_s^{\text{I,SP}})$$

$$+ \sum_{k\in K} \sum_{b\in B} \sum_{c\in C} \sum_{m\in M} N_{kb\omega}^{\text{EV}} \hat{X}_{kcm\omega}^{\text{EV}} \lambda_{kcm}^{\text{EV,A}} \Big] + \beta \tag{10.13a}$$

subject to:

$$\text{Constraints (10.1a)–(10.2e)}, \tag{10.13b}$$

$$\beta \geq \sum_{\omega\in\Omega} \sum_{d\in D} \Big[SP_{d\omega}^{(\nu)} + \sum_{g\in G} \alpha_{gd\omega}^{\text{I,G},(\nu)} \left(p_g^{\text{I,G}} - P_g^{\text{I,G},(\nu)} \right)$$

$$+ \sum_{s\in S} \left(\alpha_{sd\omega}^{\text{I,SE},(\nu)} \left(e_s^{\text{I,SE}} - E_s^{\text{I,SE},(\nu)} \right) + \alpha_{sd\omega}^{\text{I,SP},(\nu)} \left(p_s^{\text{I,SP}} - P_s^{\text{I,SP},(\nu)} \right) \right)$$

$$+\sum_{k\in K,c\in C}\alpha_{kcd\omega}^{\mathrm{EV,C,}(\nu)}\left(x_{kc\omega}^{\mathrm{EV,C}}-X_{kc\omega}^{\mathrm{EV,C,}(\nu)}\right),\quad\forall\nu,\tag{10.13c}$$

where Θ^{MP} is the set of optimization variables of the master problem. The Benders' cuts formulated by constraints (10.13b) are used in (10.13a) to approximate the objective function (10.10). After solving the master problem, the optimal values of variables $p_g^{\mathrm{I,G}}$, $e_s^{\mathrm{I,SE}}$, $p_s^{\mathrm{I,SP}}$, $x_{kcm}^{\mathrm{EV,C}}$, and β obtained in iteration (ν) are stored in parameters $P_g^{\mathrm{I,G,}(\nu)}$, $E_s^{\mathrm{I,SE,}(\nu)}$, $P_s^{\mathrm{I,SP,}(\nu)}$, $X_{kcm}^{\mathrm{EV,C,}(\nu)}$, and $\beta^{(\nu)}$, respectively. Observe that an additional Benders' cut is added to the master problem in each iteration (ν).

10.5.3 Upper and Lower Bounds

Considering that each subproblem is solved for fixed values of the first-stage variables, an upper bound of the original problem can be computed as follows:

$$z_{\mathrm{up}}^{(\nu)}=\sum_{\omega\in\Omega}\pi_\omega\left[\sum_{g\in G}C_{g\omega}^{\mathrm{I,G}}p_g^{\mathrm{I,G}}+\sum_{s\in S}\left(C_{s\omega}^{\mathrm{I,SE}}e_s^{\mathrm{I,SE}}+C_{s\omega}^{\mathrm{I,SP}}p_s^{\mathrm{I,SP}}\right)\right.$$
$$\left.+\sum_{k\in K}\sum_{b\in B}\sum_{c\in C}\sum_{m\in M}N_{kb\omega}^{\mathrm{EV}}\hat{X}_{kcm\omega}^{\mathrm{EV}}\lambda_{kcm}^{\mathrm{EV,A}}\right]+\sum_{\omega\in\Omega}\sum_{d\in D}SP_{d\omega}^{(\nu)}.\tag{10.14}$$

Since the master problem is a relaxed version of the original problem, the optimal value of its objective function (10.13a) at each iteration (ν) constitutes a lower bound of the original problem:

$$z_{\mathrm{lo}}^{(\nu)}=\sum_{\omega\in\Omega}\pi_\omega\left[\sum_{g\in G}C_{g\omega}^{\mathrm{I,G}}p_g^{\mathrm{I,G}}+\sum_{s\in S}\left(C_{s\omega}^{\mathrm{I,SE}}e_s^{\mathrm{I,SE}}+C_{s\omega}^{\mathrm{I,SP}}p_s^{\mathrm{I,SP}}\right)\right.$$
$$\left.+\sum_{k\in K}\sum_{b\in B}\sum_{c\in C}\sum_{m\in M}N_{kb\omega}^{\mathrm{EV}}\hat{X}_{kcm\omega}^{\mathrm{EV}}\lambda_{kcm}^{\mathrm{EV,A}}\right]+\beta^{(\nu)}.\tag{10.15}$$

If the difference between the upper and lower bounds is smaller than a pre-specified tolerance ϵ, the algorithm concludes and the optimal decisions are $P_g^{\mathrm{I,G,}(\nu)}$, $E_s^{\mathrm{I,SE,}(\nu)}$, $P_s^{\mathrm{I,SP,}(\nu)}$, and $X_{kc\omega}^{\mathrm{EV,C,}(\nu)}$. Otherwise, the algorithm continues.

Finally, Fig. 10.5 represents graphically the proposed solution algorithm.

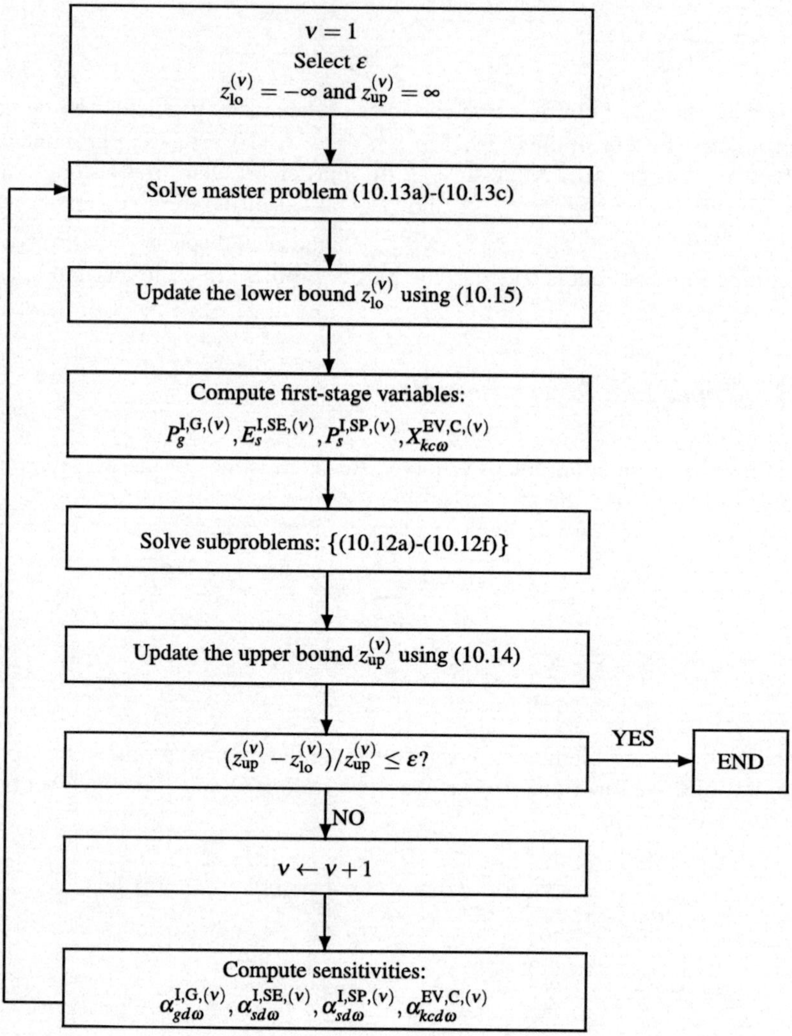

Fig. 10.5 Flowchart for the Benders' decomposition algorithm

10.6 Illustrative Case Study

We present an example to illustrate the formulation of the generation capacity expansion problem considering controlled charge of EVs presented in Sect. 10.4. To that end, we consider a planning horizon of one year characterized by a single day comprising three hourly periods. The weight of each hourly period, W_{dt}, is equal to $365 \cdot 8 = 2920$. The power system is composed by three buses and three lines, as represented in Fig. 10.6. A generating unit and a storage facility can be installed in

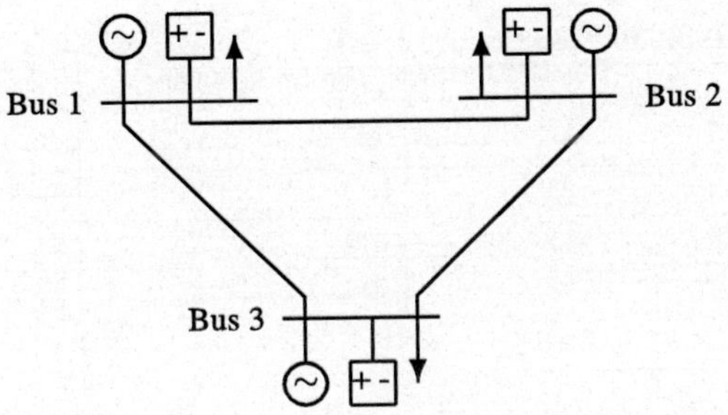

Fig. 10.6 Illustrative case study: three-node system

Table 10.7 Illustrative case study: demand (MWh)

Bus	Period 1	Period 2	Period 3
1	80	100	90
2	220	250	230
3	35	50	40

Table 10.8 Illustrative case study: generating units data

Unit	Bus	$P^{\text{I,G}}_{\text{max},g}$ (MW)	$P^{\text{G}}_{\text{up},g}$ (pu)	$C^{\text{I,G}}_{g\omega}$ (€/MW)	$C^{\text{G,D}}_{gdt\omega}$ (€/MWh)	$C^{\text{G,RU}}_{gdt\omega}$ (€/MW)
1	1	400	0.25	2000	0.40	0.25
2	2	400	0.25	6000	0.25	0.30
3	3	600	–	3200	0.00	–

each bus of the system. The number of EVs is the only uncertain parameter considered in this problem.

The hourly demand associated to each bus and period, with the exception of the electricity demanded by EVs, is provided in Table 10.7.

The parameters describing candidate generating units are listed in Table 10.8. Units 1 and 2 are dispatchable units, whereas unit 3 is an intermittent unit. Therefore, unit 3 is not allowed to participate in the reserve capacity market. The down power ramp, $P^{\text{G}}_{\text{dw},g}$ is assumed to be equal to the upper power ramp, $P^{\text{G}}_{\text{up},g}$. In the same manner, up and down reserve capacity costs, $C^{\text{G,RU}}_{gdt\omega}$ and $C^{\text{G,RD}}_{gdt\omega}$, are considered to be identical. The availability of unit 3 in each time period, A^{D}_{gdt}, is 0.4, 0.9, and 0.6, respectively.

The parameters describing the candidate storage facilities are included in Table 10.9. Note that different relationships between energy and power capacities, γ^{SEP}_{s}, are considered for each candidate unit. We assume that the cost of charging and discharging energy in the day-ahead energy market is null. Similarly to the case of generating units, up and down reserve capacity costs, $C^{\text{S,RU}}_{sdt\omega}$ and $C^{\text{S,RD}}_{sdt\omega}$, are assumed to be equal. The initial status of the storage units is 50% of the full capacity, and the minimum quantity of energy that must remain in the storage units in every period is

Table 10.9 Illustrative case study: storage units data

Storage	Bus	$E^{\mathrm{I,SE}}_{\max,s}$ (MWh)	$P^{\mathrm{I,SP}}_{\max,s}$ (MW)	$\gamma^{\mathrm{SEP}}_{s}$ (h)	$C^{\mathrm{I,SE}}_{s\omega}$ (€/MWh)	$C^{\mathrm{I,SP}}_{s\omega}$ (€/MW)	$C^{\mathrm{S,RU}}_{sdt\omega}$ (€/MW)
1	1	100	100	3	500	3000	0.2
2	2	100	100	6	500	3000	0.2
3	3	100	100	9	500	3000	0.2

Table 10.10 Illustrative case study: number of electric vehicles

Group	Bus	Scenario 1	Scenario 2	Scenario 3
1	1	550	750	950
	2	800	1000	1200
	3	400	600	800
2	1	300	500	700
	2	400	600	800
	3	150	350	550

Table 10.11 Illustrative case study: electric vehicle demand (kWh)

Group	Period 1	Period 2	Period 3
1	0	15	5
2	15	5	0

10% of the capacity. The energy stored at the end of the last period has to be greater than or equal to the energy stored at the beginning of the planning horizon.

Two groups of EVs are considered for each bus. In order to characterize the uncertainty associated with the number of EVs, three scenarios are considered representing low, average, and high number of EVs. The probability of each scenario is 0.25, 0.5, and 0.25, respectively. The number of vehicles in each bus and scenario is listed in Table 10.10. The expected charging demand of a single vehicle belonging to each group is included in Table 10.11. The power rate of EV chargers is 15 kW. It is assumed that vehicles belonging to group 1 can charge their batteries during periods 2 and 3, whereas vehicles in group 2 are charged in periods 1 and 2. The capacity of the battery of each single vehicle is equal to 40 kWh, the initial status of the batteries is 20 kWh, and they have to be fully charged at the end of the charging period.

The response of EV users to the incentive offered by the power system operator under control modes 1 and 2 is represented in Figs. 10.7 and 10.8, respectively. It is assumed that the response of the vehicles belonging to groups 1 and 2 is the same. Observe that the number of vehicles controlled by the power system operator grows as the incentive increases.

Considering the input data described above, problem (10.11) has been solved considering three different cases:

Fig. 10.7 Illustrative case study: price-quota curve of incentives for vehicles under control mode 1 (charge)

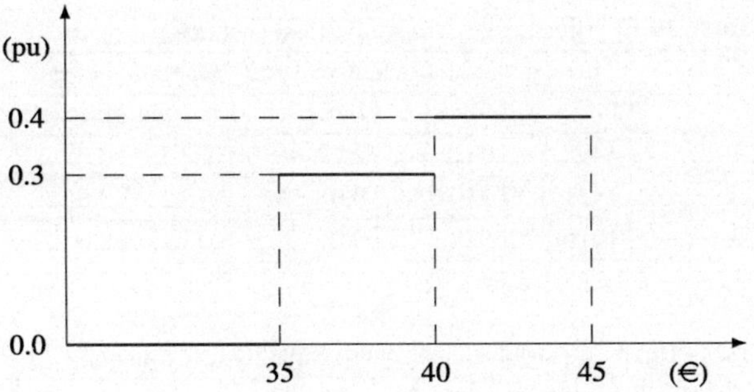

Fig. 10.8 Illustrative case study: price-quota curve of incentives for vehicles under control mode 2 (discharge)

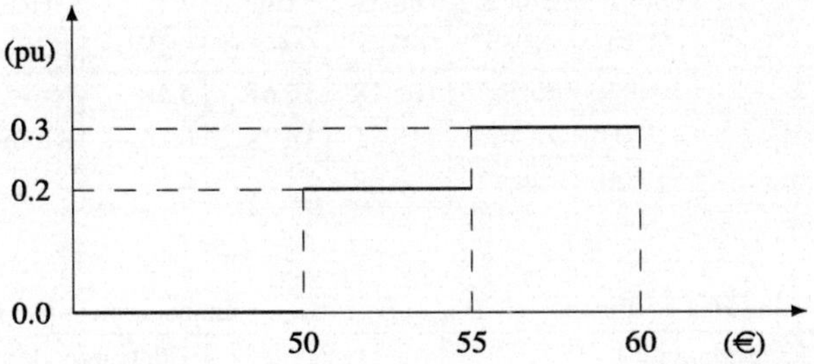

- *Case 1*: Problem (10.11) is solved considering investment in storage units and the direct control of the charging process of EVs.
- *Case 2*: Problem (10.11) is solved considering that EVs cannot be controlled ($\lambda_{kc}^{\mathrm{EV}} = 0$, $x_{kc}^{\mathrm{EV}} = 0$).
- *Case 3*: Problem (10.11) is solved considering that EVs cannot be controlled ($\lambda_{kc}^{\mathrm{EV}} = 0$, $x_{kc}^{\mathrm{EV}} = 0$), and that storage units cannot be installed ($e_s^{\mathrm{I,SE}} = 0$, $p_s^{\mathrm{I,SP}} = 0$).

Problem (10.11) is characterized by 1,142 constraints, 857 continuous variables, and 12 binary variables and is solved using CPLEX under GAMS [11].

The resulting expected cost is provided in Table 10.12 for each case. It is observed that the total cost is reduced by 0.5% if the possibility of controlling the charge of EVs is considered. Comparing *Case 1* and *Case 2*, it can be noticed that the controlled charge of vehicles allows to reduce the investment costs associated with the installation of new generation and storage units. If storage units are not considered (*Case 3*), the resulting investment cost of generating units increases by 6.2% from that obtained in *Case 2*. A similar conclusion can be derived for the expected operation cost, which increases 5.2% from *Case 2* to *Case 3*.

The capacity installed from generating and storage units is provided in Table 10.13. From this table we observe that unit 1 achieves the greatest installed generation capacity in all cases. The reason for this result is that this unit has a significant lower investment cost than the rest of candidate units. The results contained in this table also reveal that considering the possibility of controlling EVs, *Case 1*, reduces the installed capacity of both generating and storage units. Finally, we observe that storage unit

Table 10.12 Illustrative case study: total cost (k€)

Case	Investment and incentive costs			Operation costs				Total
	Inv. gen.	Inv. sto.	Inc. EV	Ene. gen.	Ene. sto.	Res. gen.	Res. sto.	
1	1498.5	94.1	24.7	1022.3	0.0	216.7	15.3	2871.6
2	1521.6	110.6	0.0	1018.2	0.0	216.3	18.1	2884.9
3	1616.7	0.0	0.0	1071.1	0.0	222.4	0.0	2910.2

Table 10.13 Illustrative case study: capacity investments

Case	Generating units			Storage units					
	Unit 1	Unit 2	Unit 3	Unit 1		Unit 2		Unit 3	
	(MW)	(MW)	(MW)	(MWh)	(MW)	(MWh)	(MW)	(MWh)	(MW)
1	306.89	91.36	105.18	16.67	5.56	69.12	11.52	0.00	0.00
2	310.03	91.46	110.25	18.78	6.26	82.46	13.74	0.00	0.00
3	317.20	106.72	106.88	–	–	–	–	–	–

Table 10.14 Illustrative case study: *Case 1*—financial incentive (€)

Control mode 1		Control mode 2	
Group 1	Group 2	Group 1	Group 2
35.0	0.0	0.0	0.0

Table 10.15 Illustrative case study: energy charged by electric vehicles (MWh)

Case	Period 1	Period 2	Period 3	Total
1	21.75	36.33	19.47	77.55
2–3	21.75	43.09	12.71	77.55

1, which has the greatest energy and power relationship parameter, $\gamma_s^{\text{SEP}} = 9$ h, is not installed in any case.

Table 10.14 provides the financial incentive offered to each EV group for each control mode. It is observed that this incentive is different from 0 only in the case of the first group of vehicles under control mode 1. In this manner, the discharge of the batteries into the grid is not allowed. Considering the price-quota curve represented in Fig. 10.7, an incentive equal to 35 € for control mode 1 (only charge), implies the control of 30% of the vehicles belonging to the first group. In this case, vehicles do not participate in the reserve capacity market, and the whole energy charged by the batteries of these vehicles comes from the day-ahead energy market. Table 10.15 includes the total expected amount of power that is charged from the grid by EVs considering simultaneously controlled and non-controlled vehicles. From this table, it is observed that the control of the charge of the vehicles allows to reduce by 15.7% the peak EV demand in period 2.

Table 10.16 Illustrative case study: *Case 1*–Expected day-ahead energy scheduling (MW)

Period	Generating units			Storage units					
	Unit 1	Unit 2	Unit 3	Unit 1		Unit 2		Unit 3	
				(Ch.)	(Disch.)	(Ch.)	(Disch.)	(Ch.)	(Disch.)
1	237.82	91.36	40.94	4.86	0.11	8.63	0.00	–	–
2	233.23	91.36	94.66	0.00	5.56	0.00	11.52	–	–
3	232.87	91.36	59.65	1.04	0.00	3.37	0.00	–	–

Table 10.17 Illustrative case study: *Case 1*—expected up reserve scheduling (MW)

Period	Generating units			Storage units		
	Unit 1	Unit 2	Unit 3	Unit 1	Unit 2	Unit 3
1	43.86	0.00	–	0.00	0.00	–
2	62.18	0.00	–	0.00	0.00	–
3	48.82	0.00	–	0.00	0.00	–

Tables 10.16, 10.17 and 10.18 provide the expected participation of generating and storage units in the day-ahead energy and reserve capacity markets in each period. From Table 10.16 it is observed that generating unit 2, with the lowest operation cost, is always operating at its maximum capacity in the day-ahead energy market, 93.66 MW. In the same manner, generating unit 3 produces its maximum available power, which is bounded by the availability factor A^{D}_{gdt}. Generating unit 1, which has the most expensive operation cost, is a ramping unit that adapts its production depending on the values of demand and intermittent power production in each period. Storage units 1 and 2 show a similar behaviour, and they charge energy in low-demand periods, 1 and 3, and discharge energy in high-demand period 2.

Tables 10.17 and 10.18 show the up and down reserve capacities scheduling of each unit. As commented above, EVs do not participate in this market. From Table 10.17 it is noticed that only generating unit 1 provides up reserve capacity. This result is a consequence of the fact that this unit is the most expensive one in terms of operation cost. Therefore, generating units 2 and 3 are mainly used to supply energy in the day-ahead market. Storage units do not provide up reserve capacity, and they prefer to supply down reserve capacity, as shown in Table 10.18. This result is a consequence of the establishment of a minimum amount of energy to be stored in the last period of the planning horizon, constraints (10.5g) and (10.5h), in a planning horizon comprising only 3 periods.

Finally, Table 10.19 includes the maximum and minimum expected energy stored in the storage units. For the sake of clarity, the status of the storage units in the period prior to the start of the planning horizon is also included. It can be observed that the minimum energy stored in the last period coincides with the energy stored at the beginning of the planning horizon.

Table 10.18 Illustrative case study: *Case 1*–expected down reserve scheduling (MW)

Period	Generating units			Storage units		
	Unit 1	Unit 2	Unit 3	Unit 1	Unit 2	Unit 3
1	19.13	22.45	–	0.00	2.28	–
2	28.26	21.75	–	1.45	10.72	–
3	14.17	22.84	–	3.66	8.15	–

Table 10.19 Illustrative case study: *Case 1*—expected energy stored (MWh)

Period	Minimum energy stored			Maximum energy stored		
	Unit 1	Unit 2	Unit 3	Unit 1	Unit 2	Unit 3
0	11.67	48.38	–	11.67	48.38	–
1	16.32	56.84	–	16.32	59.07	–
2	10.65	45.08	–	12.07	57.83	–
3	11.67	48.38	–	16.67	69.12	–

10.7 Realistic Case Study

This section analyzes a case study based on the isolated power system of Lanzarote-Fuerteventura (LZ-FV) in Spain. Lanzarote and Fuerteventura are part of the Canary Islands, which are located at the northwest coast of Africa.

The objective of this case study is to determine the optimal generation and storage capacity mix at year 2030. Additionally, the financial incentives offered to EV users are also decided in this problem. It is assumed that all generating units that are currently at work will be decommissioned prior to the considered target year. A one-year horizon planning has been considered, which has been represented by a set of nine characteristic days and nine scenarios. All mathematical problems in this section have been solved by CPLEX 12.6.1 under GAMS [11] on a Linux-based server with four 3.0 GHz processors and 250 GB of RAM.

10.7.1 Input Data

The characterization of the LZ-FV power system is based on the stylized representation of this system described in [1, 5]. This power system comprises eight buses, eight lines, and candidates sets of 45 generating units and 16 storage facilities. The technologies of these candidate generating units are diesel, gas, wind, and solar PV. Figure 10.9 shows the power system topology. Blue lines correspond to 66 kV transmission lines, whereas 132 kV lines are represented in red. For simplicity, the graphical representation of the transformers placed between 132 and 66 kV zones

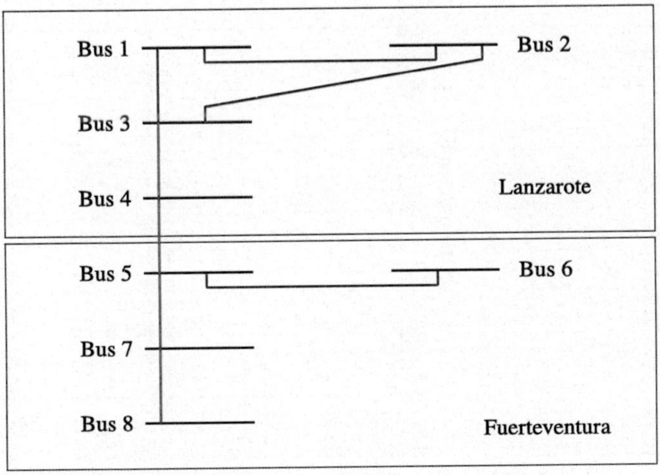

Fig. 10.9 Realistic case study: Lanzarote-Fuerteventura power system

Table 10.20 Realistic case study: transmission line parameters

Line	Origin bus	Destination bus	Length (km)	Capacity (MW)	Voltage (kV)	Reactance (Ω/km)
1	1	2	8.9	62	66	0.1276
2	2	3	10.8	62	66	0.3712
3	3	1	25	80	132	0.0613
4	3	4	18	160	132	0.3616
5	4	5	20	120	132	0.0588
6	5	6	25.5	62	66	0.3712
7	5	7	26	160	132	0.3616
8	7	8	23	160	132	0.3616

are omitted. The parameters that characterize the considered transmission system are included in Table 10.20.

The target year is represented by a set of characteristic days. These days have been selected using the procedure proposed in [21], which consists in solving the full optimization problem a number of instances equal to the total number of days considering only one day at a time. For solving each single-day case, the expected values of the uncertain parameters are considered. Afterwards, days are selected using the fast-forward algorithm described in [15] over the cost associated with each single day. In this case study, nine characteristic days have been selected. Figure 10.10 represents the total cost associated with the solution of the problem for each single day. For the sake of clarity, costs are increasingly ordered. Circles indicate the costs associated with the selected days. The expected (ED) and maximum (MD) demands, as well as the expected wind (EWA) and solar PV (ESA) availability per

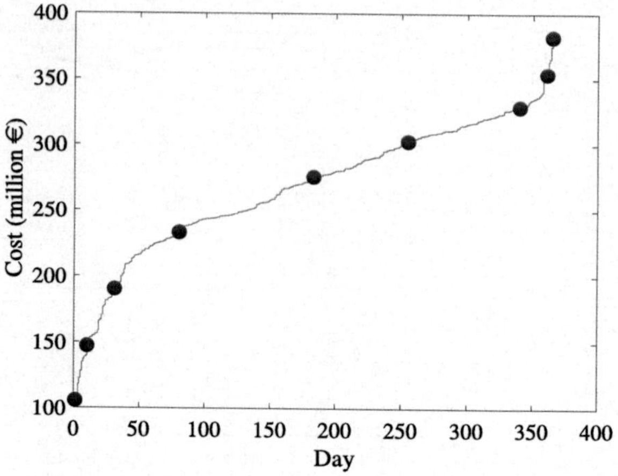

Fig. 10.10 Realistic case study: selection of characteristic days

Table 10.21 Realistic case study: characteristic days

Day	Lanzarote				Fuerteventura				Weight
	ED (MWh)	MD (MWh)	EWA (pu)	ESA (pu)	ED (MWh)	MD (MWh)	EWA (pu)	ESA (pu)	
1	96.9	125.5	0.17	0.12	70.1	91.8	0.07	0.12	58
2	95.4	124.2	0.22	0.12	70.9	93.3	0.21	0.15	75
3	92.1	120.0	0.02	0.08	67.4	88.2	0.13	0.09	6
4	95.0	124.2	0.72	0.16	70.8	92.8	0.83	0.14	16
5	93.1	119.8	0.69	0.24	72.0	93.3	0.19	0.16	23
6	96.7	113.5	0.49	0.28	73.2	88.3	0.17	0.31	95
7	104.1	123.3	0.97	0.29	79.6	96.7	0.28	0.32	4
8	99.7	123.5	0.22	0.08	75.2	96.8	0.10	0.07	2
9	99.1	127.1	0.26	0.18	77.0	102.3	0.05	0.20	86

island for each selected day, together with its weights, $\sum_{t=1}^{24} W_{dt}$, are provided in Table 10.21.

Figure 10.11 represents the demand per hour in each island for each characteristic day. Observe that the demand in Lanzarote is higher than in Fuerteventura in all days.

The allocation of the total demand per bus in each island is performed using the share factors provided in Table 10.22.

Table 10.23 provides the considered potentials of wind and solar PV technologies per bus. These potentials represent 60% of the maximum potentials reported in [14]. Observe that the sum of the potentials of these renewable technologies is much higher than the maximum peak demand observed in Table 10.29, which is equal to 192.6 MW.

The wind and solar PV availabilities for each hour for the power units located in each island are depicted in Figs. 10.12 and 10.13 for each characteristic day.

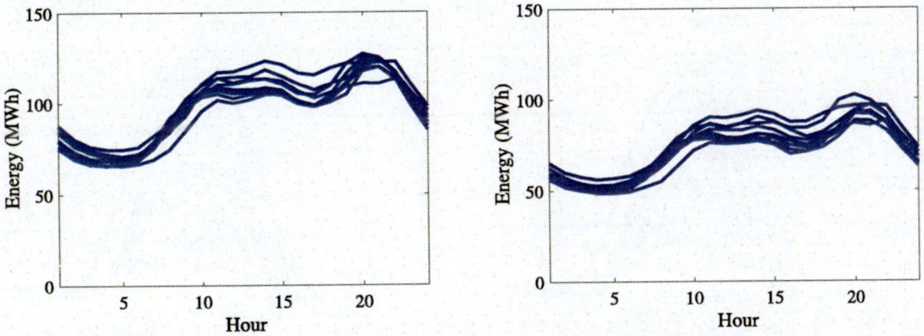

(a) Lanzarote (b) Fuerteventura

Fig. 10.11 Realistic case study: energy demand per island

Table 10.22 Realistic case study: demand share factor (%)

Bus							
Lanzarote				Fuerteventura			
1	2	3	4	5	6	7	8
4.0	18.1	64.6	13.3	21.3	34.6	23.9	20.2

Table 10.23 Realistic case study: potentials of renewable capacity

Node	Capacity (MW)		Node	Capacity (MW)	
	Wind	PV		Wind	PV
1	267.6	131.8	5	426.2	14.5
2	39.4	151.1	6	667.8	175.7
3	0.0	28.3	7	0.0	67.3
4	0.0	37.8	8	0.0	28.5

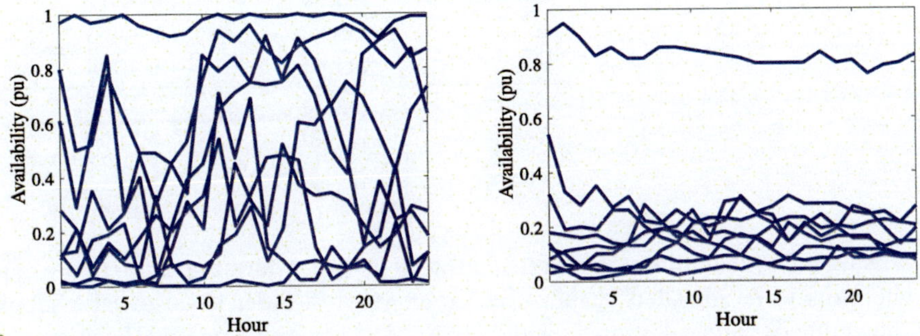

(a) Lanzarote (b) Fuerteventura

Fig. 10.12 Realistic case study: wind availability per island

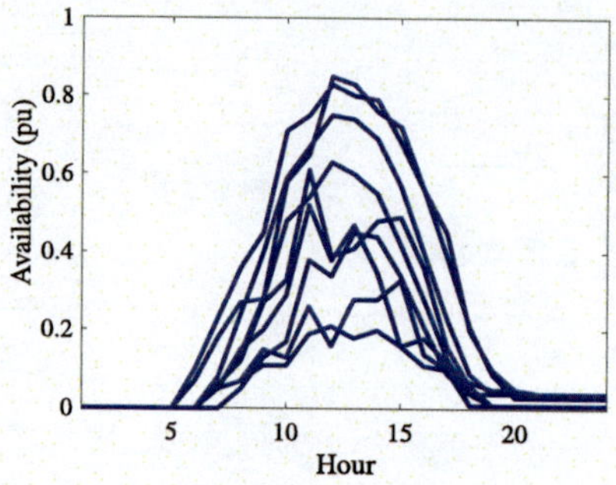

(a) Lanzarote

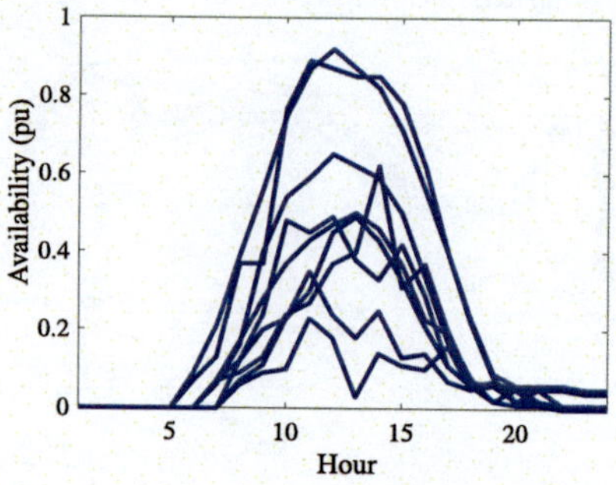

(b) Fuerteventura

Fig. 10.13 Realistic case study: solar PV availability per island

Table 10.24 Realistic case study: technical characteristics of candidate generating units

Techn.	Unit	Node	Potential capacity (MW)	Operating cost (€/MWh)
Diesel	1–3	1	6.49	180.235
	4, 5	1	12.85	145.041
	6	1	20.51	145.051
	7, 8	1	17.20	156.439
	9, 10	1	17.60	156.575
	11, 12	6	3.82	174.018
	13	6	4.11	180.507
	14, 15	6	6.21	180.805
	16	6	20.51	145.510
	17	6	17.20	156.439
	18, 19	6	17.20	156.439
Gas	20	1	19.60	292.051
	21	1	32.34	245.011
	22	6	29.40	251.703
	23	6	11.74	343.795
	24	6	21.85	283.135
	25–33	1–8	30.00	295.000

It is assumed that all thermal units that were at usage during year 2017 are candidate units to be installed again in the target year. In order to ensure the supply of the demand, eight additional candidate 30-MW gas units, one per bus, are also considered. Table 10.24 provides the capacity, location, and operating cost of each candidate thermal unit. The cost of up and down reserve capacities is equal to 0.15 times the operation costs. The up and down power ramps are assumed to be identical and they are equal to 0.4 times the installed capacity for diesel units, and 0.75 times the installed capacity for gas units.

Table 10.25 Realistic case study: capital costs

Technology	Capital cost (€/kW)	Technology	Capital cost (€/kW)
Diesel	900	PV	950
Gas	850	Storage (energy)	60$^{(*)}$
Wind	1450	Storage (power)	1500

$^{(*)}$: (€/kWh)

Table 10.26 Realistic case study: electric vehicle groups

Type	Beginning hour	Ending hour	Duration (h)	Percentage of EVs (%)
1	19.00	6.00 (next day)	12	50
2	22.00	6.00 (next day)	9	35
3	9.00	17.00	9	15

Table 10.27 Realistic case study: number of vehicles

Bus								Total
1	2	3	4	5	6	7	8	
2853	12872	46100	9489	13176	21400	14810	12493	120017

For each bus, two candidate storage facilities with relationships between energy and power (γ_s^{SP}) equal to 3 and 6 are considered. Therefore, a total number of 16 candidate storage facilities is accounted for. The maximum energy capacity that can be installed per bus is 100 MWh. The cost of charge and discharge energy in the day-ahead market is null. The cost of up and down reserve capacities is equal to 2 €/MW.

The capital costs of generating and storage units are listed in Table 10.25. These costs are afterwards annualized using the capital recovery factor described in Sect. 10.4.1. Considering a plant life parameter equal to 20 years and the interest rate is 9%, the capital recovery factor is equal to 10.95%.

Three groups of EVs are considered. The description of the characteristics of each group is included in Table 10.26. Each EV is equipped with a 40 kWh battery and it is considered that the average daily distance driven by each vehicle is 40 km. The energy consumption is 0.2 kWh/km. The efficiency of charging and discharging the batteries is equal to 0.88. EVs are charged under a maximum power transfer rate of 11 kW.

The total number of vehicles per bus, including electric and combustion-engine vehicles, is listed in Table 10.27. The percentage of these vehicles that are plug-in EVs is an uncertain parameter ranging between 20 and 60% that is specified below for each considered scenario.

The aggregated charging profiles of uncontrolled vehicles in each group considering a penetration of 40% of EVs is represented in Fig. 10.14. From the observation

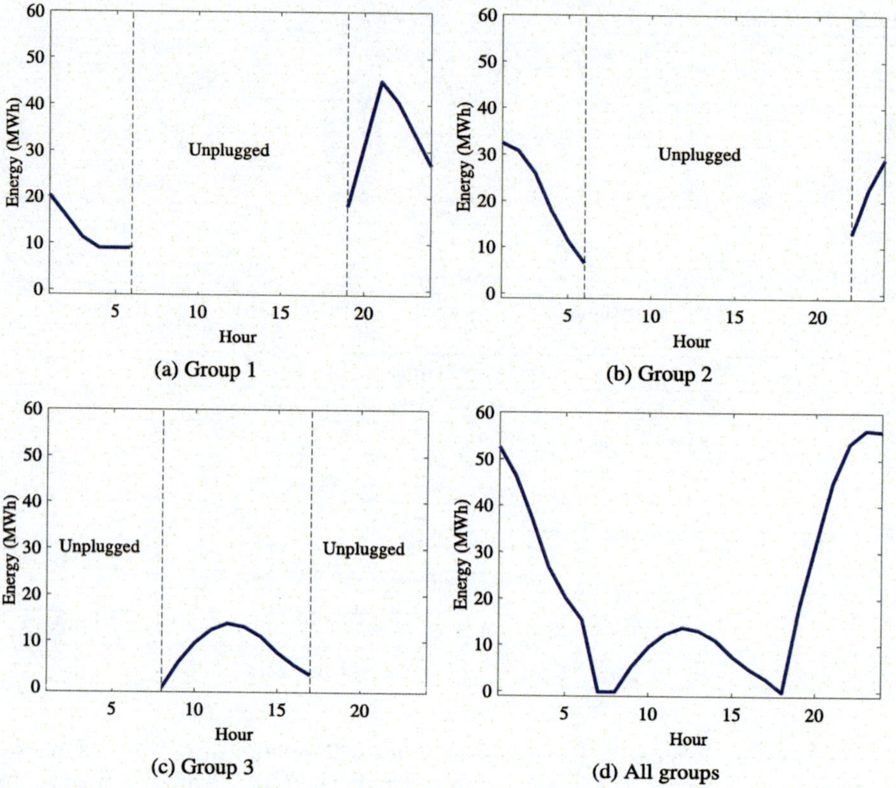

Fig. 10.14 Realistic case study: charging profiles of electric vehicles

of the aggregated charging profile considering all groups, it can be concluded that vehicles are mainly charged at night.

The willingness of EV users to leave the control of the charging processes of their vehicles to the power system operator is modeled with 10-block step-wise price-quota curves. These curves are generated for cases $c = \{1, 2\}$ by discretizing two normal distributions with means 450 and 600 €/year, respectively, and standard deviations equal to 50 €/year. These curves are scaled to establish a limit of a maximum of 40% of EVs in control mode 1, and 25% in control mode 2. The resulting price-quota curves for both control modes are depicted in Fig. 10.15.

The capital costs of storage units and renewable units, the annual demand growth, the number of EVs, and the controllability of EVs are characterized as uncertain parameters. For this purpose, we consider three values (low, medium, and high) for each parameter according to the values provided in Table 10.28. The variation of the capital costs of storage and renewable units provided in Table 10.28 are computed over the values listed in Table 10.25. The demand growth included in Table 10.28 is applied yearly to the actual demand in LZ-FV in year 2017. The penetration of

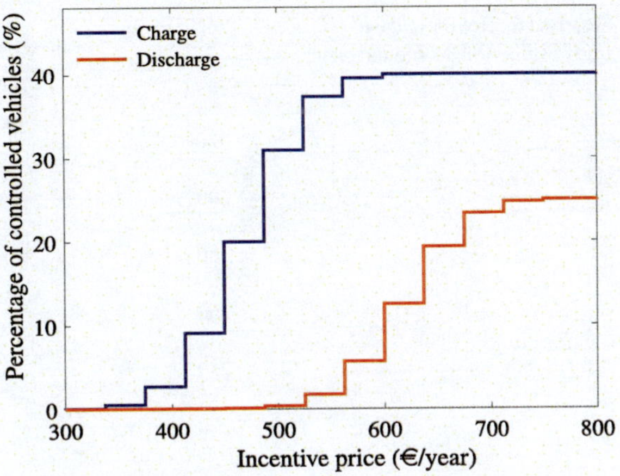

Fig. 10.15 Realistic case study: incentive price-quota curves

Table 10.28 Realistic case study: parameters modeled through scenarios

Parameters (%)	Low		Medium		High	
	Value	Prob.	Value	Prob.	Value	Prob.
(1) Capital cost of storages	75.0	0.25	100.0	0.50	125.0	0.25
(2) Capital cost of renewable units	90.0	0.25	100.0	0.50	110.0	0.25
(3) Demand growth per year	0.0	0.25	1.0	0.50	3.0	0.25
(4) Penetration of electric vehicles	20.0	0.25	40.0	0.50	60.0	0.25
(5) Horizontal shift of price-quota curves	75.0	0.25	100.0	0.50	125.0	0.25

EVs is computed over the number of vehicles listed in Table 10.27. The uncertainty associated with the willingness of EV users to delegate the control of the charge of their vehicles to the power system operator is modeled shifting horizontally the price-quota curves represented in Fig. 10.15 using the values indicated in Table 10.28.

Considering that three possible values are defined for each of the five uncertain parameters described above, a total number of $3^5 = 243$ scenarios are generated. For

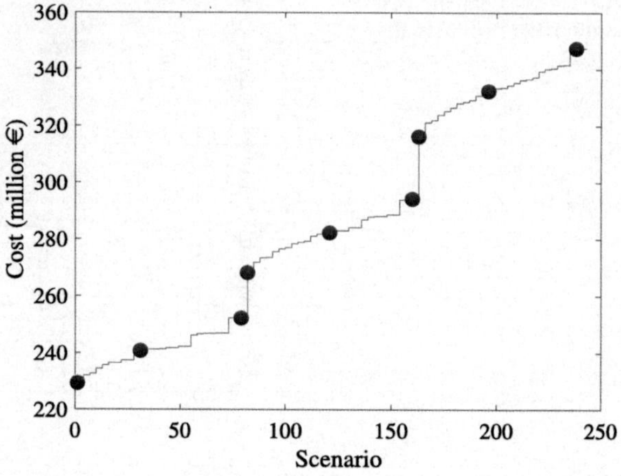

Fig. 10.16 Realistic case study: selection of scenarios

Table 10.29 Realistic case study: scenarios

Scenario	Parameters					Probability
	(1)	(2)	(3)	(4)	(5)	
1	75.0	90.0	1.00	20.0	75.0	0.0494
2	75.0	90.0	1.17	20.0	125.0	0.0494
3	75.0	90.0	1.37	20.0	75.0	0.0494
4	75.0	100.0	1.00	40.0	75.0	0.1852
5	75.0	110.0	1.17	20.0	75.0	0.2222
6	100.0	90.0	1.37	60.0	125.0	0.1975
7	100.0	110.0	1.17	60.0	75.0	0.0617
8	100.0	110.0	1.37	60.0	100.0	0.0864
9	125.0	110.0	1.00	60.0	75.0	0.0988

the sake of tractability, this set of scenarios has been reduced to 9 by using the scenario reduction technique described in [21]. In this case, the full optimization problem is solved as many times as scenarios are initially generated, 243, considering only one scenario at a time. Afterwards, scenarios are selected according to the cost associated with each single scenario. The performance of the scenario reduction algorithm is depicted in Fig. 10.16. The cost associated with each single scenario is represented by blue lines, whereas circles are used to denote the scenarios that are finally selected by the scenario reduction algorithm.

The resulting scenarios are included in Table 10.29. Note that each parameter is denoted by the number indicated in Table 10.28.

Table 10.30 Realistic case study: size of the problems

Problem	Full	Master	Subproblem
Constraints (#)	12309	10763	177
Continuous variables (#)	10763	1791	111
Binary variables (#)	480	480	0

10.7.2 Results

This section discuses the results obtained from solving three different cases:

- *Case 0*: The problem is solved without considering the presence of EVs and assuming that storage units cannot be installed ($e_s^{\mathrm{I,SE}} = 0,\ p_s^{\mathrm{I,SP}} = 0$).
- *Case 1*: The problem is solved considering investments in storage units and direct control over the charging and discharging processes of EVs.
- *Case 2*: The problem is solved considering that EVs cannot be controlled ($\lambda_{kc}^{\mathrm{EV}} = 0$, $x_{kc\omega}^{\mathrm{EV,C}} = 0$) and that storage unis cannot be installed ($e_s^{\mathrm{I,SE}} = 0,\ p_s^{\mathrm{I,SP}} = 0$).

The resulting problems have been solved using Benders' decomposition considering a tolerance equal to 0.1%. The number of constraints, continuous and binary variables in *Case 1* are $2.2{\cdot}10^6$, $1.5{\cdot}10^6$, and 480, respectively.

Table 10.30 provides the computational size of the full problem formulated in Sect. 10.4, and that pertaining to the master problem and subproblems resulting from applying Benders' decomposition in *Case 0*. The number of constraints in the master problem corresponds to the first iteration. The average solution time per iteration is 303 s. In total, 255 iterations have been needed to reach an optimality gap equal to 0.1% in *Case 1*, which results in a total computational time of 21.5 h. Note that this computational time is compatible with the practical application of an expansion planning study such as the one considered in this chapter.

Figure 10.17 depicts the evolution of the upper and lower bounds for each iteration.

Table 10.31 includes the expected costs of each case. As expected, the smallest cost is obtained if EVs are not considered. The presence of EVs increases total costs between 9.0 and 11.6% in *Case 1* and *Case 2* with respect to those reported in *Case 0*, respectively. It is also relevant to note that if the possibility of controlling EVs and installing storage unis is not considered in *Case 2*, the total expected cost increases by 2.4% with respect to *Case 1*. The investment cost in *Case 1* is 12.8% greater than that in *Case 2*, but the operation cost decreases by 32.5%.

Table 10.32 provides the generation and storage capacity to be installed in each case for each technology. It is worth noting that if EVs are not controlled by the power system operator (*Case 2*), the investments in wind power units decrease by 31.1% with respect to *Case 1*.

Table 10.33 provides the expected energy provided in the day-ahead energy market. The energy associated with storage units and EVs corresponds with the dis-

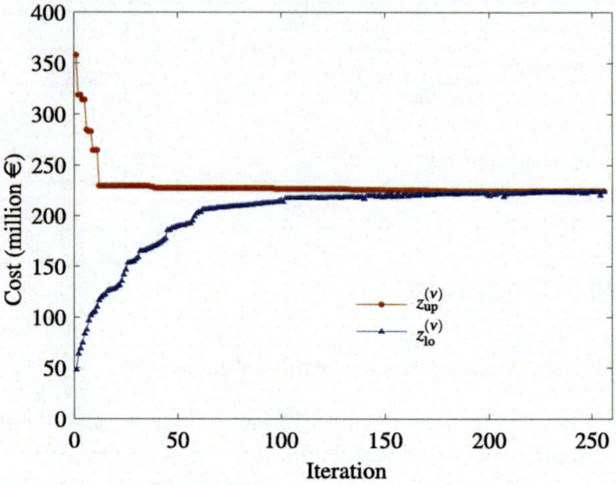

Fig. 10.17 Realistic case study: upper and lower bounds in the Benders' decomposition procedure

Table 10.31 Realistic case study: expected cost (million €)

Case	Annual investment	Financial incentive	Day-ahead energy	Day-ahead reserve	Unserved demand	Total
Case 0	94.2	0.00	109.4	6.1	1.0	210.7
Case 1	114.0	12.6	84.9	4,0	0.1	229.7
Case 2	101.1	0.00	125.7	6.6	1.7	235.2

Table 10.32 Realistic case study: capacity to be installed

Case	Diesel (MW)	Gas (MW)	Wind (MW)	Solar PV (MW)	Storage pow. (MW)	Storage ene. (MWh)
Case 0	231.6	35.2	317.6	188.8	–	–
Case 1	211.9	26.4	493.1	158.8	72.1	394.6
Case 2	231.6	72.5	348.7	176.3	–	–

charged energy that is sold in this market. Observe that the total energy produced in *Case 1* is the greatest one. The reason of this result is that the efficiency of charging and discharging storage units and EVs is less than 1 and all the energy that is discharged has to be charged in a different time period. As a result, the discharge of a certain amount of energy from storage units and EVs carries a higher amount of energy that must be charged. Note also that the percentage of energy coming from thermal units is much smaller in *Case 1*, 28.5%, than in *Case 0* and *Case 2*, 42.1% and 43.7%, respectively.

Figure 10.18 represents the expected day-ahead energy capacity scheduled for each technology in each hour for scenario 4, which has associated average values of demand and number of EVs. The fixed load in this figure refers to the electricity demand without considering EVs and storage units, whereas the total load is equal

Table 10.33 Realistic case study: energy produced in the day-ahead energy market (GWh)

Case	Diesel	Gas	Wind	Solar PV	Storage	Vehicles	Total
Case 0	713.5	6.3	781.5	206.6	0.0	0.0	1707.9
Case 1	563.7	0.7	1133.8	198.0	79.5	1.84	1977.5
Case 2	792.9	20.6	838.8	209.3	0.0	0.0	1861.6

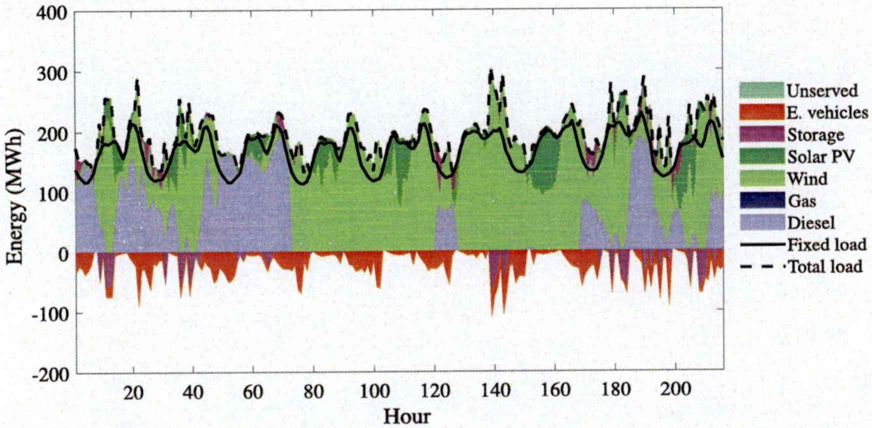

Fig. 10.18 Realistic case study: day-ahead energy market scheduling in scenario 4

to the fixed load plus the energy charged by EVs and storage units. It is observed that most of the energy scheduled corresponds to renewable units, specially wind units. Diesel units play the role of peak units that supply energy if renewable units are unavailable.

Tables 10.34 and 10.35 list the expected up and down reserve capacities per technology and case. Given that the expected renewable generation in *Case 1* is the highest one, it can be observed that the total amounts of up and down reserve capacities are the greatest for this case. In the three analyzed cases, it can be noticed that diesel units are mainly scheduled for providing up reserve. Diesel units are preferred to provide up reserve because these units have a smaller operation cost than gas units, and most of the time they are operating below their capacities. The down reserve capacity is also provided in a major extent by diesel units in *Case 0* and *Case 2*, but it is mainly scheduled for storage units and EVs in *Case 1*. The reason of this result is that the down reserve capacity cost is cheaper for storage units and EVs. Additionally, unlike to the provision of up reserve, the scheduling of down reserve does not involve a decrease of the energy stored in storage units and EVs.

Figure 10.19 represents graphically the expected reserve capacity scheduled for each technology in each hour in scenario 4. Positive values indicate the scheduling of up reserve capacity, whereas the scheduling of down reserve is represented using negative values. As observed in Tables 10.34 and 10.35, most of the up reserve

Table 10.34 Realistic case study: up reserve capacity (GW)

Case	Diesel	Gas	Wind	Solar PV	Storage	Vehicles	Total
Case 0	118.2	12.1	0.0	0.0	0.0	0.0	130.3
Case 1	150.2	3.8	0.0	0.0	0.2	6.0	160.2
Case 2	121.7	18.0	0.0	0.0	0.0	0.0	139.7

Table 10.35 Realistic case study: down reserve capacity (GW)

Case	Diesel	Gas	Wind	Solar PV	Storage	Vehicles	Total
Case 0	130.0	0.2	0.0	0.0	0.0	0.0	130.3
Case 1	13.7	0.0	0.0	0.0	68.3	78.2	160.2
Case 2	139.1	0.6	0.0	0.0	0.0	0.0	139.7

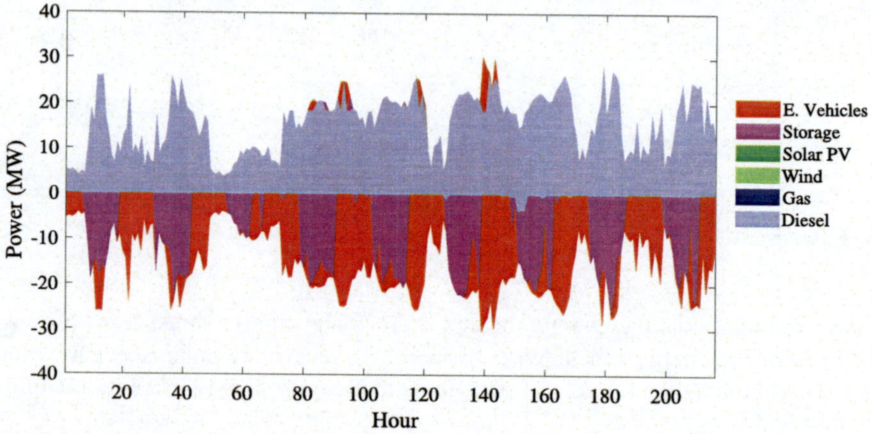

Fig. 10.19 Realistic case study: day-ahead reserve capacity market scheduling in scenario 4

capacity is provided by thermal units. The scheduling of down reserve capacity is mainly provided by storage units and EVs.

In order to illustrate the operation of storage units, Fig. 10.20 represents the aggregated values of energy charged, discharged, and stored considering all storage units in day 1 and scenario 8. Note that the maximum and minimum power charged and discharged are depicted, considering the participation in the day-ahead energy and reserve capacity markets.

Table 10.36 lists the resulting financial incentives offered by the power system operator to controllable vehicles in *Case 1*. Observe that the financial incentives offered to all groups in charging control mode are quite similar, resulting in 35.4% of controlled vehicles of groups 1 and 2, and 35.8% of group 1 according to the price-quota curves represented in Fig. 10.15. The financial incentives offered for discharge control are different from zero for groups 1 and 3, resulting in the control

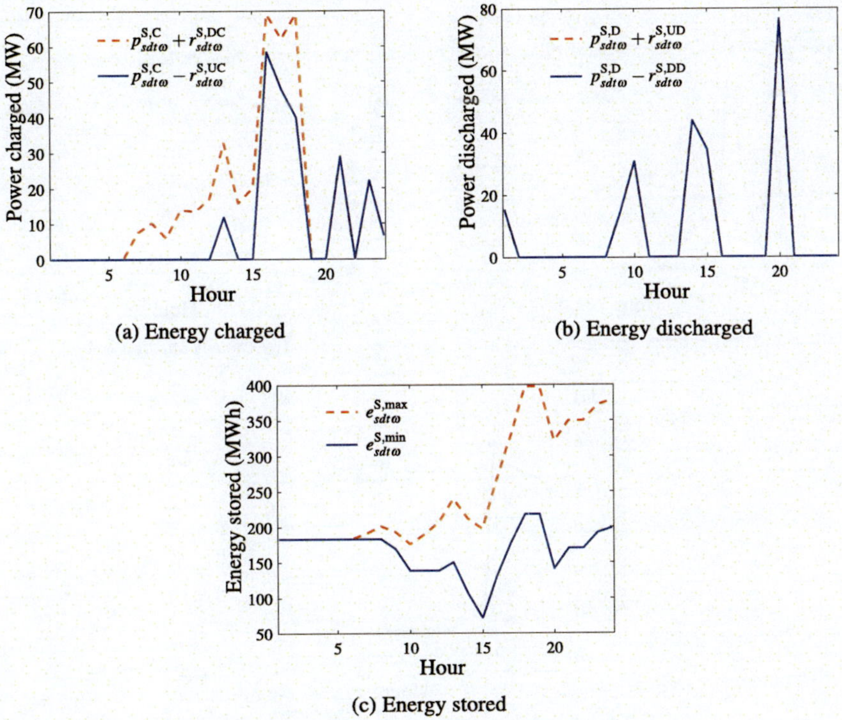

Fig. 10.20 Realistic case study: operation of storage units in day 1 and scenario 8

Table 10.36 Realistic case study: financial incentives for each electric vehicle group (€/year)

Group	Charging	Charging/discharging
1	525.0	637.5
2	525.0	0.0
3	562.5	562.5

of 17.3 and 11.2% of vehicles, respectively. The reason for this result is that the power system operator may have a higher interest in discharging vehicles during peak hours. However, this is not possible for vehicles in group 2, which can only charge or discharge during the night, between 22:00 and 6:00.

Figure 10.21 illustrates the aggregated values of energy charged, discharged, and stored considering all controlled EVs in day 1 and scenario 8. Observe that the power discharged in much smaller than the power charged.

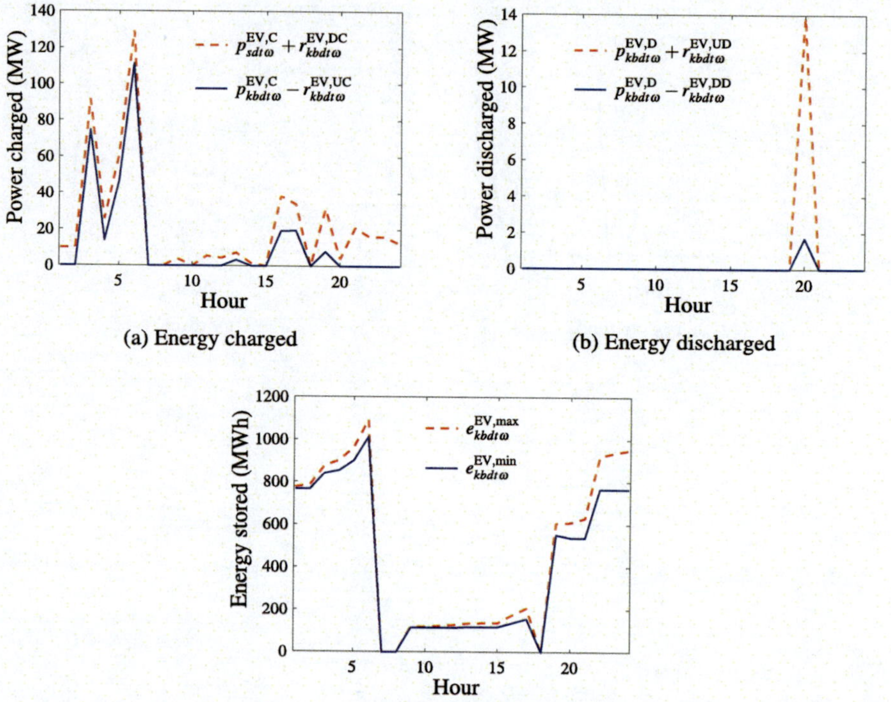

(c) Energy stored

Fig. 10.21 Realistic case study: operation of controlled vehicles in day 1 and scenario 8

10.8 Summary and Conclusions

This chapter presents the formulation of a two-stage stochastic programming problem to model the generation capacity expansion of a power system considering the presence of storage units and a large number of electric vehicles. The proposed model considers that electric vehicles have the vehicle-to-grid capability and that they can be discharged into the grid. Furthermore, the charge and discharge of the vehicles may be controlled by the power system operator, if electric vehicle users agree. The willingness of electric vehicle users to leave the charge and discharge of their vehicles in hands of the power system operator has been modeled by means of price-quota curves. The mathematical formulation of these curves require the usage of binary variables. In order to make appropriate generation and storage capacity decisions, the operation of the resulting power system is modeled considering the day-ahead energy and reserve capacity markets. The final mathematical problem is a large-scale mixed-integer linear programming problem that can be solved using decomposition techniques for large-scale problems.

The numerical results obtained from a realistic case study indicate that investment decisions are affected by the presence of storage units and the controllability of the charge and discharge of electric vehicles. It has been also observed that the economical incentives offered to electric vehicle users depend significantly on the interval of hours in which the vehicles are available for being controlled.

10.9 Notation

This section defines the main notation used in this chapter.

Acronyms

CRF	Cost recovery factor.
DC	Direct current.
EV	Electric vehicle.

Sets and Indices

b	Index of buses.
B	Set of buses.
c	Index of controlling modes of electric vehicles.
C	Set of controlling modes of electric vehicles.
d	Index of characteristic days.
D	Set of characteristic days.
g	Index of generating units.
G	Set of generating units.
G_b	Set of generating units located at bus b.
G^{D}	Set of dispatchable generating units.
G^{I}	Set of intermittent generating units.
G_b^{D}	Set of dispatchable generating units located at bus b.
G_b^{I}	Set of intermittent generating units located at bus b.
k	Index of plug-in electric vehicle groups.
K	Set of plug-in electric vehicle groups.
K_t	Set of plug-in electric vehicle groups that are connected to the grid in period t.
ℓ	Index of transmission lines.
L	Set of transmission lines.
m	Index of blocks in piece-wise price-quota curves.
M	Set of blocks in piece-wise price-quota curves.

s Index of storage units.
S Set of storage units.
S_b Set of storage units located at bus b.
t Index of time periods.
T Set of time periods.
T_k Set of time periods in which electric vehicle group k can be charged from or discharged to the grid.
ν Index used the denote the iteration counter in the Benders' decomposition algorithm.
ω Index of scenarios.
Ω Set of scenarios.

Parameters

A^{D}_{gdt} Availability of intermittent unit g in the day-ahead market in characteristic day d and period t (pu).
$C^{\mathrm{G,D}}_{gdt\omega}$ Operation cost of generating unit g in the day-ahead energy market in characteristic day d, period t, and scenario ω (€/MWh).
$C^{\mathrm{G,RD}}_{gdt\omega}$ Offering cost of down reserve capacity of generating unit g in characteristic day d, period t, and scenario ω (€/MW).
$C^{\mathrm{G,RU}}_{gdt\omega}$ Offering cost of up reserve capacity of generating unit g in characteristic day d, period t, and scenario ω (€/MW).
$C^{\mathrm{I,G}}_{g\omega}$ Annualized capital cost of generating unit g in scenario ω (€/MW).
$C^{\mathrm{I,SE}}_{s\omega}$ Annualized capital cost of the energy component of storage unit s in scenario ω (€/MW).
$C^{\mathrm{I,SP}}_{s\omega}$ Annualized capital cost of the power component of storage unit s in scenario ω (€/MW).
$C^{\mathrm{S,C}}_{sdt\omega}$ Consumption bid of storage unit s in the day-ahead energy market in characteristic day d, period t, and scenario ω (€/MWh).
$C^{\mathrm{S,D}}_{sdt\omega}$ Offering cost of storage unit s in the day-ahead energy market in characteristic day d, period t, and scenario ω (€/MWh).
$C^{\mathrm{S,RD}}_{sdt\omega}$ Offering cost of down reserve capacity of storage unit g in characteristic day d, period t, and scenario ω (€/MW).
$C^{\mathrm{S,RU}}_{sdt\omega}$ Offering cost of up reserve capacity of storage unit s in characteristic day d, period t, and scenario ω (€/MW).
C^{UD} Cost of unserved demand (€/MWh).
$E^{\mathrm{EV,B}}_{\max}$ Capacity of the battery of an electric vehicle (MWh).
$E^{\mathrm{I,SE}}_{\max,s}$ Maximum energy capacity that can be installed of storage unit s (MWh).
$N^{\mathrm{EV}}_{kn\omega}$ Number of electric vehicles pertaining to group k and bus b in scenario ω.
$P^{\mathrm{D}}_{bdt\omega}$ Demand in the day-ahead market at bus b and characteristic day d, period t, and scenario ω (MWh).

$P^{\mathrm{EV}}_{kbdt\omega}$ Demand in the day-ahead market of electric vehicle group k at bus b and characteristic day d, period t, and scenario ω (MWh).

$P^{\mathrm{EV}}_{\max}$ Maximum charged/discharged power in the battery of electric vehicles (MW).

$P^{\mathrm{EV,S}}_{\min,k}$ Minimum energy level stored in EV group k (MWh).

$P^{\mathrm{G}}_{\mathrm{up},g}$ Upper ramp factor of generating unit g (pu).

$P^{\mathrm{G}}_{\mathrm{dw},g}$ Down ramp factor of generating unit g (pu).

$P^{\mathrm{I,G}}_{\max,g}$ Maximum power capacity that can be installed of generating unit g (MW).

$P^{\mathrm{I,SP}}_{\max,s}$ Maximum power capacity that can be installed of storage unit s (MW).

$P^{\mathrm{L}}_{\max,\ell}$ Capacity of transmission line ℓ (MW).

$SP^{(\nu)}_{\omega d}$ Parameter that is equal to the optimal objective function of the subproblem associated with scenario ω and characteristic day d in iteration (ν) of the Benders' decomposition algorithm (€).

t^{F}_{kd} Final charging time of the battery of electric vehicles belonging to group k in characteristic day d.

t^{O}_{kd} Initial charging time of the battery of electric vehicles belonging to group k in characteristic day d.

W_{dt} Weight of period t and characteristic day d (h).

X_ℓ Reactance of line ℓ (Ω).

$\hat{X}^{\mathrm{EV}}_{kcm\omega}$ Auxiliary parameter used to compute the per unit number of electric vehicles in group k pertaining to charging mode c, block m, and scenario ω (pu).

$z^{(\nu)}_{\mathrm{lo}}$ Lower bound computed in iteration (ν) of the Benders' decomposition algorithm (€).

$z^{(\nu)}_{\mathrm{up}}$ Upper bound computed in iteration (ν) of the Benders' decomposition algorithm (€).

$\gamma^{\mathrm{CD,D}}$ Parameter used to compute the down-reserve capacity needs of the system associated with the demand (pu).

$\gamma^{\mathrm{CD,I}}$ Parameter used to compute the down-reserve capacity needs of the system associated with the intermittent production (pu).

$\gamma^{\mathrm{CU,D}}$ Parameter used to compute the up-reserve capacity needs of the system associated with the demand (pu).

$\gamma^{\mathrm{CU,I}}$ Parameter used to compute the up-reserve capacity needs of the system associated with the intermittent production (pu).

$\gamma^{\mathrm{EV,O}}_{kbd}$ Initial status of the battery of vehicles belonging to group k, at bus b in characteristic day d (pu).

$\gamma^{\mathrm{EV,F}}_{kbd}$ Final status of the battery of vehicles belonging to group k, at bus b in characteristic day d (pu).

$\gamma^{\mathrm{S,O}}_s$ Factor used to model the initial status of the storage unit s (pu).

$\gamma^{\mathrm{S,F}}_s$ Factor used to model the final status of the storage unit s (pu).

$\gamma^{\mathrm{S,EP}}_s$ Relationship between energy and power capacities in storage unit s (h).

$\gamma^{\mathrm{S,min}}_s$ Factor used to model the minimum energy that must contain storage unit s (pu).

η^{EV} Efficiency of charging/discharging electric vehicles (pu).

η^{S} Efficiency of charging/discharging storage units (pu).
$\hat{\lambda}^{\text{EV}}_{kcm\omega}$ Auxiliary parameter used to compute the incentive offered to electric vehicle group k in charging mode c, block m, and scenario ω (€).
π_{ω} Probability of scenario ω (pu).

Variables

$e^{\text{EV,max}}_{kbdt\omega}$ Maximum energy level in the batteries of electric vehicle group k at bus b and characteristic day d, period t, and scenario ω (MWh).
$e^{\text{EV,min}}_{kbdt\omega}$ Minimum energy level in the batteries of electric vehicle group k at bus b and characteristic day d, period t, and scenario ω (MWh).
$e^{\text{S,max}}_{sdt\omega}$ Maximum energy level in the battery of storage unit s, characteristic day d, period t, and scenario ω (MWh).
$e^{\text{S,min}}_{sdt\omega}$ Minimum energy level in the battery of storage unit s, characteristic day d, period t, and scenario ω (MWh).
$p^{\text{EV,C}}_{kbdt\omega}$ Consumption power scheduled by electric vehicles belonging to group k at bus b in the day-ahead market, in characteristic day d, period t, and scenario ω (MW).
$p^{\text{EV,D}}_{kbdt\omega}$ Discharged power scheduled by electric vehicles belonging to group k at bus b in the day-ahead market, in characteristic day d, period t, and scenario ω (MW).
$p^{\text{G,D}}_{gdt\omega}$ Energy scheduled by generating unit g in the day-ahead market, in characteristic day d, period t, and scenario ω (MWh).
$p^{\text{G,DS}}_{gdt\omega}$ Power spillage of intermittent generating unit g in the day-ahead market, in characteristic day d, period t, and scenario ω (MWh).
$p^{\text{I,G}}_{g}$ Capacity built of generating unit g (MW).
$p^{\text{I,SE}}_{s}$ Energy capacity built of storage unit s (MWh).
$p^{\text{I,SP}}_{s}$ Peak power of storage unit s (MW).
$p^{\text{L,D}}_{\ell dt\omega}$ Power flow through line ℓ in the day-ahead market, characteristic day d, period t, and scenario ω (MW).
$p^{\text{S,C}}_{sdt\omega}$ Consumption power scheduled by storage unit s in the day-ahead market, in characteristic day d, period t, and scenario ω (MW).
$p^{\text{S,D}}_{sdt\omega}$ Discharged power scheduled by storage unit s in the day-ahead market, in characteristic day d, period t, and scenario ω (MW).
$p^{\text{UD}}_{bdt\omega}$ Unserved demand at bus b in characteristic day d, period t, and scenario ω (MWh).
$r^{\text{EV,D}}_{kbdt\omega}$ Down reserve capacity scheduled by by electric vehicles belonging to group k at bus b and characteristic day d, period t, and scenario ω (MW).
$r^{\text{EV,U}}_{kbdt\omega}$ Up reserve capacity scheduled by by electric vehicles belonging to group k at bus b and characteristic day d, period t, and scenario ω (MW).

$r^{\text{G,D}}_{gdt\omega}$ Down reserve capacity scheduled by generating unit g, characteristic day d, period t, and scenario ω (MW).

$r^{\text{G,U}}_{gdt\omega}$ Up reserve capacity scheduled by generating unit g, characteristic day d, period t, and scenario ω (MW).

$r^{\text{S,D}}_{sdt\omega}$ Down reserve capacity scheduled by storage unit s, characteristic day d, period t, and scenario ω (MW).

$r^{\text{S,U}}_{sdt\omega}$ Up reserve capacity scheduled by storage unit s, characteristic day d, period t, and scenario ω (MW).

$r^{\text{EV,DC}}_{kbdt\omega}$ Down reserve capacity scheduled by electric vehicle group k at bus b in charging mode, characteristic day d, period t, and scenario ω (MW).

$r^{\text{EV,DD}}_{kbdt\omega}$ Down reserve capacity scheduled by electric vehicle group k at bus b in discharging mode, characteristic day d, period t, and scenario ω (MW).

$r^{\text{EV,UC}}_{kbdt\omega}$ Up reserve capacity scheduled by electric vehicle group k at bus b in charging mode, characteristic day d, period t, and scenario ω (MW).

$r^{\text{EV,UD}}_{kbdt\omega}$ Up reserve capacity scheduled by electric vehicle group k at bus b in discharging mode, characteristic day d, period t, and scenario ω (MW).

$r^{\text{G,D}}_{gdt\omega}$ Down reserve capacity scheduled by generating unit g, characteristic day d, period t, and scenario ω (MW).

$r^{\text{G,U}}_{gdt\omega}$ Up reserve capacity scheduled by generating unit g, characteristic day d, period t, and scenario ω (MW).

$r^{\text{S,D}}_{sdt\omega}$ Down reserve capacity scheduled by storage unit s, characteristic day d, period t, and scenario ω (MW).

$r^{\text{S,DC}}_{sdt\omega}$ Down reserve capacity scheduled by storage unit s in charging mode, characteristic day d, period t, and scenario ω (MW).

$r^{\text{S,DD}}_{sdt\omega}$ Down reserve capacity scheduled by storage unit s in discharging mode, characteristic day d, period t, and scenario ω (MW).

$r^{\text{S,U}}_{sdt\omega}$ Up reserve capacity scheduled by storage unit s, characteristic day d, period t, and scenario ω (MW).

$r^{\text{S,UC}}_{sdt\omega}$ Up reserve capacity scheduled by storage unit s in charging mode, characteristic day d, period t, and scenario ω (MW).

$r^{\text{S,UD}}_{sdt\omega}$ Up reserve capacity scheduled by storage unit s in discharging mode, characteristic day d, period t, and scenario ω (MW).

v^{EV}_{kcm} Binary variable that is equal to 1 if the incentive offered to EV group k, in charging mode c, block m of the price-quota curve, being 0 otherwise.

$x^{\text{EV,C}}_{kc\omega}$ Per-unit number of controlled electric vehicles in group k, charging mode c and scenario ω (pu).

$\alpha^{\text{EV,C}}_{kbd\omega}$ Dual variable associated with constraint (10.12f) (€/pu).

$\alpha^{\text{I,G}}_{gd\omega}$ Dual variable associated with constraint (10.12c) (€/MW).

$\alpha^{\text{I,SE}}_{gd\omega}$ Dual variable associated with constraint (10.12d) (€/MWh).

$\alpha^{\text{I,SP}}_{gd\omega}$ Dual variable associated with constraint (10.12e) (€/MW).

β Auxiliary variable used to estimate the expected operation costs for given first-stage variables in the Benders' decomposition algorithm (€).

$\lambda^{\mathrm{EV}}_{kc}$ Incentive offered to electric vehicle group k and charging mode c (€).
$\lambda^{\mathrm{EV,A}}_{kcm}$ Auxiliary variable used to compute the incentive offered to electric vehicle group k, charging mode c, and block m (€).
$\Theta^{\mathrm{D}}_{bdt\omega}$ Voltage angle at bus b in the characteristic day d, period t and scenario ω (rad).

10.10 GAMS Codes

This section provides the GAMS [13] code used to solve the Illustrative Case Study described in Sect. 10.6.

```
$title GENERATION CAPACITY EXPANSION WITH ELECTRIC VEHICLES


*****************************************************************
*          DATA
*****************************************************************


SETS
w             'Scenarios'                                           /w1*w3/,
d             'Representative days'                                 /d1/,
t             'Time periods'                                        /t1*t3/,
k             'Electric vehicle groups'                             /k1*k2/,
m             'Blocks in price-quota curve'                         /m1*m3/,
c             'Control modes'                                       /c1*c2/,
g             'Candidate generating units'                          /g1*g3/,
s             'Candidate storages'                                  /s1*s3/,
b             'Buses'                                               /b1*b3/,
l             'Lines'                                               /l1*l3/,
GD(g)         'Dispatchable units'
GI(g)         'Intermittent units'
T0(t,k,d)     'Start of the charge of EVs'
TF(t,k,d)     'End of the charge of EVs'
TK(t,k,d)     'Periods in which vehicles can be charged'
GN(g,b)       'Location of generating units'
SN(s,b)       'Location of storages'
OL(b,l)       'Origin of line l'
FL(b,l)       'Destination of line l';

GD('g1')=yes; GD('g2')=yes;
GI('g3')=yes;

T0('t2','k1',d)=yes; T0('t1','k2',d)=yes;
TF('t3','k1',d)=yes; TF('t2','k2',d)=yes;
TK('t2','k1',d)=yes; TK('t3','k1',d)=yes;
TK('t1','k2',d)=yes; TK('t2','k2',d)=yes;

GN('g1','b1')=yes; GN('g2','b2')=yes; GN('g3','b3')=yes;
SN('s1','b1')=yes; SN('s2','b2')=yes; SN('s3','b3')=yes;

OL('b1','l1')=yes; OL('b2','l2')=yes; OL('b3','l3')=yes;
```

```
FL('b2','l1')=yes; FL('b3','l2')=yes; FL('b1','l3')=yes;


SCALARS

etaEV          'Charg/disch efficiency (EVs)' /0.98/,
PEVmax         'Charge rate of EVs' /0.015/,
EEVBmax        'Battery capacity of EVs' /0.04/,
gammaEVmin     'Minimum energy stored in EVs' /0.05/,
thetaDmax      'Maximum voltage angle deviation' /1.5707963/
gammaCUD       'Up reserve requirement from demand' /0.1/
gammaCUI       'Up reserve requirement from interm. units' /0.2/
gammaCDD       'Down reserve requirement from demand' /0.1/
gammaCDI       'Down reserve requirement from interm. units' /0.2/
CUD            'Unserved demand cost' /1000/;


PARAMETERS

WD(d)          'Period weight'
gammaNC(k)     'Non-clients of the supplier belonging to group k'
E0(k,w)        'Initial status of the batteries of EVs'
EE(k,w)        'Ending status of the batteries of EVs'
SOC_max(k)     'Capacity of the battery of each EV'
XHatEV(k,c,m,w)   'Per-unit number of EVs controlled'
lambdaHatEV(k,c,m) 'Incentive in price-quota curve'
gammaSmin(s)      'Per-unit energy that must remain in storages'
gammaS0(s,d)      'Per-unit energy in the first period (storage)'
gammaEV0(k,b,d)   'Per-unit energy in the first period'
gammaEVF(k,b,d)   'Per-unit energy in the last period'
etaS(S)           'Charge and discharge efficiency'
PEV(k,b,d,t,w)    'Demand EVs'

pi(w)             'Scenario probability'
/w1  0.25
w2   0.50
w3   0.25/,

PIG_max(g)        'Potential of candidate generating unit g'
/g1        400
g2         400
g3         600/,

EISE_max(s)       'Energy potential of candidate storing unit s'
/s1        100
s2         100
s3         100/,

PISP_max(s)       'Power potential of candidate storing unit s'
/s1        100
s2         100
s3         100/,

PG_up(g)          'Per-unit rump up limit of candidate unit g'
/g1        0.25
g2         0.25/,

PG_dw(g)          'Per-unit rump down limit of candidate unit g'
/g1        0.25
g2         0.25/,
```

```
PLmax                   'Lines capacity'
/l1         100
l2          100
l3          100/,

X(l)                    'Lines reactance'
/l1         0.01
l2          0.01
l3          0.01/,

gammaSEP(s)             'Relationship between energy and power'
/s1         3
s2          6
s3          9/;

TABLE AD(g,d,t) 'Availability of intermittent units'
            t1          t2          t3
g3.d1       0.4         0.9         0.6;

TABLE XEVHat_C1(k,m,w) 'Per-unit number controlled (mode c=1)'
            w1          w2          w3
k1.m1       0           0           0
k1.m2       0.3         0.3         0.3
k1.m3       0.4         0.4         0.4
k2.m1       0           0           0
k2.m2       0.3         0.3         0.3
k2.m3       0.4         0.4         0.4;

TABLE XEVHat_C2(k,m,w)  'Per-unit number controlled (mode c=2)'
            w1          w2          w3
k1.m1       0           0           0
k1.m2       0.2         0.2         0.2
k1.m3       0.3         0.3         0.3
k2.m1       0           0           0
k2.m2       0.2         0.2         0.2
k2.m3       0.3         0.3         0.3 ;

TABLE lambdaHatEV_C1(k,m) 'Incentive (mode c=1)'
            m1          m2          m3
k1          35          40          45
k2          35          40          45;

TABLE lambdaHatEV_C2(k,m) 'Incentive (mode c=2)'
            m1          m2          m3
k1          50          55          60
k2          50          55          60;

TABLE NEV(k,b,w) 'Number of EVs'
            w1          w2          w3
k1.b1       550         750         950
k1.b2       800        1000        1200
k1.b3       400         600         800
k2.b1       300         500         700
k2.b2       400         600         800
k2.b3       150         350         550;

TABLE PD(b,d,t,w) 'Demand'
            w1          w2          w3
b1.d1.t1    80          80          80
b1.d1.t2    100         100         100
b1.d1.t3    90          90          90
```

```
b2.d1.t1 220      220      220
b2.d1.t2 250      250      250
b2.d1.t3 230      230      230
b3.d1.t1 35       35       35
b3.d1.t2 50       50       50
b3.d1.t3 40       40       40;

TABLE PEV_pu(k,b,d,t,w) 'Demand of EVs'
                    w1      w2      w3
k1.b1.d1.t2        15.000  15.000  15.000
k1.b1.d1.t3         5.408   5.408   5.408
k1.b2.d1.t2        15.000  15.000  15.000
k1.b2.d1.t3         5.408   5.408   5.408
k1.b3.d1.t2        15.000  15.000  15.000
k1.b3.d1.t3         5.408   5.408   5.408
k2.b1.d1.t1        15.000  15.000  15.000
k2.b1.d1.t2         5.408   5.408   5.408
k2.b2.d1.t1        15.000  15.000  15.000
k2.b2.d1.t2         5.408   5.408   5.408
k2.b3.d1.t1        15.000  15.000  15.000
k2.b3.d1.t2         5.408   5.408   5.408;

TABLE CIG(g,w)  'Annualized investment cost (generating units)'
          w1      w2      w3
g1        2000    2000    2000
g2        6000    6000    6000
g3        3200    3200    3200;

TABLE CISE(s,w) 'Annualized energy investment cost (storages)'
          w1      w2      w3
s1        500     500     500
s2        500     500     500
s3        500     500     500;

TABLE CISP(s,w) 'Annualized power investment cost (storages)'
          w1      w2      w3
s1        3000    3000    3000
s2        3000    3000    3000
s3        3000    3000    3000;

TABLE CGD(g,w)  'Operation cost (generating units)'
          w1      w2      w3
g1        0.400   0.400   0.400
g2        0.250   0.250   0.250;

TABLE CSD(s,w)  'Discharge cost (storages)'
          w1      w2      w3
s1        0       0       0
s2        0       0       0
s3        0       0       0;

TABLE CSC(s,w)  'Charge cost (storages)'
          w1      w2      w3
s1        0       0       0
s2        0       0       0
s3        0       0       0;

TABLE CGRU(g,w) 'Up reserve capacity cost (gen. units)'
          w1      w2      w3
g1        0.25    0.25    0.25
g2        0.30    0.30    0.30;
```

```
TABLE CSRU(s,w) 'Up reserve capacity cost (storages)'
           w1       w2       w3
s1         0.20     0.20     0.20
s2         0.20     0.20     0.20
s3         0.20     0.20     0.20;


TABLE CGRD(g,w) 'Down reserve capacity cost (gen. units)'
           w1       w2       w3
g1         0.25     0.25     0.25
g2         0.30     0.30     0.30;

TABLE CSRD(s,w) 'Down reserve capacity cost (storages)'
           w1       w2       w3
s1         0.20     0.20     0.20
s2         0.20     0.20     0.20
s3         0.20     0.20     0.20;

WD(d)=365*8;
lambdaHatEV(k,'c1',m)=lambdaHatEV_C1(k,m);
lambdaHatEV(k,'c2',m)=lambdaHatEV_C2(k,m);
XHatEV(k,'c1',m,w)=XEVHat_C1(k,m,w);
XHatEV(k,'c2',m,w)=XEVHat_C2(k,m,w);
gammaSmin(s)=0.1;
gammaS0(s,d)=0.7;
gammaEV0(k,b,d)=0.5;
gammaEVF(k,b,d)=1;
etaS(S)=0.98;
PEV(k,b,d,t,w)=PEV_pu(k,b,d,t,w)*NEV(k,b,w)/1000;



*****************************************************************
*   DECLARATION OF VARIABLES AND MATHEMATICAL CHARACTERIZATION
*****************************************************************

VARIABLES

**        OBJECTIVE FUNCTION
z                  'Objective function'

**        FIRST-STAGE VARIABLES
pIG(g)             'Generation capacity installed'
eISE(s)            'Storage energy capacity installed'
pISP(s)            'Storage power capacity installed'
lambdaEV(k,c)      'Incentive offered to EVs'
lambdaEVA(k,c,m)   'Auxiliary variable to compute the incentive'
vEV(k,c,m)         'Selection of block of the price-quota curve'

**        SECOND-STAGE VARIABLES
xEVC(k,c,w)        'Per-unit number of EVs in mode c'
xEV(k,w)           'Per-unit number of controlled EVs'
pGD(g,d,t,w)       'Energy scheduled in the DA market'
rGU(g,d,t,w)       'Up reserve scheduled in the DA market'
rGD(g,d,t,w)       'Down reserve scheduled in the DA market'
pGDS(g,d,t,w)      'Intermittent power spillage'
pSC(s,d,t,w)       'Charge scheduled in the DA market by storages'
pSD(s,d,t,w)       'Discharge scheduled in the DA market by storages'
rSU(s,d,t,w)       'Up reserve scheduled in the DA market (storages)'
rSD(s,d,t,w)       'DW reserve scheduled in the DA market (storages)'
```

```
rSUC(s,d,t,w)      'Up reserve capacity provided by charging'
rSUD(s,d,t,w)      'Up reserve capacity scheduled by discharging'
rSDC(s,d,t,w)      'Down reserve capacity scheduled by charging'
rSDD(s,d,t,w)      'Down reserve capacity scheduled by discharging'
eSmin(s,d,t,w)     'Minimum energy stored (storages)'
eSmax(s,d,t,w)     'Maximum energy stored (storages)'
pEVC(k,b,d,t,w)   'Charge energy scheduled in the DA market by EVs'
pEVD(k,b,d,t,w)   'Discharge energy scheduled in the DA market by
     EVs'
rEVU(k,b,d,t,w)   'Up reserve scheduled in the DA market by EVs'
rEVD(k,b,d,t,w)   'Dw reserve scheduled in the DA market by EVs'
rEVU(k,b,d,t,w)   'Up reserve capacity by charging (EVs)'
rEVUD(k,b,d,t,w) 'Up reserve capacity by discharging (EVs)'
rEVDC(k,b,d,t,w) 'DW reserve capacity by charging (EVs)'
rEVDD(k,b,d,t,w) 'DW reserve capacity by discharging (EVs)'
eEVmin(k,b,d,t,w) 'Minimum energy stored (EVs)'
eEVmax(k,b,d,t,w) 'Maximum energy stored (EVs)'
pLD(l,d,t,w)       'Power flows'
thetaD(b,d,t,w)    'Voltage angles'

POSITIVE VARIABLES pIG(g),eISE(s),pISP(s),lambdaEV(k,c),lambdaEVA(k
     ,c,m),xEVC(k,c,w),xEV(k,w)
,pGD(g,d,t,w),rGU(g,d,t,w),rGD(g,d,t,w),pGDS(g,d,t,w),pSC(s,d,t,w),
     pSD(s,d,t,w),rSU(s,d,t,w),rSD(s,d,t,w),rSUC(s,d,t,w),rSUD(s,d,t
     ,w),rSDC(s,d,t,w),rSDD(s,d,t,w),eSmin(s,d,t,w),eSmax(s,d,t,w),
pEVC(k,b,d,t,w),pEVD(k,b,d,t,w),rEVU(k,b,d,t,w),rEVD(k,b,d,t,w),
     rEVUC(k,b,d,t,w),rEVUD(k,b,d,t,w),rEVDC(k,b,d,t,w),rEVDD(k,b,d,
     t,w),pUD(b,d,t,w),eEVmin(k,b,d,t,w),eEVmax(k,b,d,t,w);

BINARY VARIABLES
vEV(k,c,m)        'Binary variable to select financial incentives of
     EVs';


*****************************************************************
*   EQUATIONS
*****************************************************************

EQUATIONS

**        OBJECTIVE FUNCTION
ObjectiveFunction

**        INVESTMENT
InvestmentGen
InvestmentSto_01
InvestmentSto_02
InvestmentSto_03

**        FINANCIAL INCENTIVES
FinantialInc_01
FinantialInc_02
FinantialInc_03
FinantialInc_04
FinantialInc_05
FinantialInc_06
FinantialInc_07

**        GENERATING UNITS
GenUnit_01
GenUnit_02
```

```
GenUnit_03
GenUnit_04
GenUnit_05
GenUnit_06
GenUnit_07

**        STORAGES
StoUnit_01
StoUnit_02
StoUnit_03
StoUnit_04
StoUnit_05
StoUnit_06
StoUnit_07
StoUnit_08
StoUnit_09
StoUnit_10
StoUnit_11
StoUnit_12
StoUnit_13
StoUnit_14
StoUnit_15
StoUnit_16

**        ELECTRIC VEHICLES
EleVehicles_01
EleVehicles_02
EleVehicles_03
EleVehicles_04
EleVehicles_05
EleVehicles_06
EleVehicles_07
EleVehicles_08
EleVehicles_09
EleVehicles_10
EleVehicles_11
EleVehicles_12
EleVehicles_13
EleVehicles_14
EleVehicles_15
EleVehicles_16

**        DAY-AHEAD MARKET
DayAhead_01
DayAhead_02
DayAhead_03
DayAhead_04
DayAhead_05
DayAhead_06
DayAhead_07
DayAhead_08;

ObjectiveFunction..
     z =e= 0.001*sum(w,pi(w)*(
      sum(g,CIG(g,w)*pIG(g))+sum(s,CISE(s,w)*eISE(s)+CISP(s,w)*pISP(
         s)) +
     sum((k,b,c,m),NEV(k,b,w)*XHatEV(k,c,m,w)*lambdaEVA(k,c,m))+
     sum(d,WD(d)*sum(t,
         sum(g$GD(g),CGD(g,w)*pGD(g,d,t,w)+CGRU(g,w)*rGU(g,d,t,w)+
            CGRD(g,w)*rGD(g,d,t,w))+
```

```
            sum(s,CSD(s,w)*pSD(s,d,t,w)-CSC(s,w)*pSC(s,d,t,w)+CSRU(s,w
                )*rSU(s,d,t,w)+CSRD(s,w)*rSD(s,d,t,w))
             +sum(b,CUD*pUD(b,d,t,w))
              ))
      ));

InvestmentGen(g)..                      pIG(g)=l=PIG_max(g);
InvestmentSto_01(s)..                   eISE(s)=l=EISE_max(s);
InvestmentSto_02(s)..                   pISP(s)=l=PISP_max(s);
InvestmentSto_03(s)..                   eISE(s)=e=gammaSEP(s)*pISP(s);

FinantialInc_01(k,c,m)$(ord(m) gt 1)..              lambdaHatEV(k,c,m-1)
     *vEV(k,c,m) =l=lambdaEVA(k,c,m);
FinantialInc_02(k,c,m)..                lambdaHatEV(k,c,m)*vEV(k,c,m) =g=
     lambdaEVA(k,c,m);
FinantialInc_03(k,c)..                  sum(m,vEV(k,c,m)) =e= 1;
FinantialInc_04(k,c)..                  lambdaEV(k,c) =e= sum(m,lambdaEVA(
     k,c,m));
FinantialInc_05(k,c,w)..                xEVC(k,c,w) =e= sum(m, XHatEV(k,c,
     m,w)*vEV(k,c,m));
FinantialInc_06(k,w)..                  xEV(k,w) =e= sum(c,xEVC(k,c,w));
FinantialInc_07(k,w)..                  xEVC(k,'c2',w) =l= xEVC(k,'c1',w);

GenUnit_01(g,d,t,w)$GD(g)..             pGD(g,d,t,w)+rGU(g,d,t,w) =l= pIG(
     g);
GenUnit_02(g,d,t,w)$GD(g)..             0 =l= pGD(g,d,t,w)-rGD(g,d,t,w);
GenUnit_03(g,d,t,w)$GI(g)..             pGD(g,d,t,w)+pGDS(g,d,t,w) =e= AD(
     g,d,t)*pIG(g);
GenUnit_04(g,d,t,w)$(GD(g) and (ord(t) gt 1))..        (pGD(g,d,t,w)+
     rGU(g,d,t,w))-(pGD(g,d,t-1,w)-rGD(g,d,t-1,w)) =l=  PG_up(g)*pIG
     (g);
GenUnit_05(g,d,t,w)$(GD(g) and (ord(t) gt 1))..        (pGD(g,d,t-1,w
     )+rGU(g,d,t-1,w))-(pGD(g,d,t,w)-rGD(g,d,t,w)) =l=  PG_dw(g)*pIG
     (g);
GenUnit_06(g,d,t,w)$GI(g)..             rGU(g,d,t,w) =e= 0;
GenUnit_07(g,d,t,w)$GI(g)..             rGD(g,d,t,w) =e= 0;

StoUnit_01(s,d,t,w)..                   rSU(s,d,t,w) =e= rSUC(s,d,t,w)+
     rSUD(s,d,t,w);
StoUnit_02(s,d,t,w)..                   rSD(s,d,t,w) =e= rSDC(s,d,t,w)+
     rSDD(s,d,t,w);
StoUnit_03(s,d,t,w)..                   pSD(s,d,t,w)+rSUD(s,d,t,w) =l=
     pISP(s);
StoUnit_04(s,d,t,w)..                   pSD(s,d,t,w)-rSDD(s,d,t,w) =g= 0;
StoUnit_05(s,d,t,w)..                   pSC(s,d,t,w)+rSDC(s,d,t,w) =l=
     pISP(s);
StoUnit_06(s,d,t,w)..                   pSC(s,d,t,w)-rSUC(s,d,t,w) =g= 0;
StoUnit_07(s,d,t,w)$(ord(t) gt 1)..  eSmin(s,d,t,w) =e= eSmin(s,d,t
     -1,w)+etaS(s)*(pSC(s,d,t,w)-rSUC(s,d,t,w))-1/etaS(s)*(pSD(s,d,t
     ,w)+rSUD(s,d,t,w));
StoUnit_08(s,d,t,w)$(ord(t) gt 1)..  eSmax(s,d,t,w) =e= eSmax(s,d,t
     -1,w)+etaS(s)*(pSC(s,d,t,w)+rSDC(s,d,t,w))-1/etaS(s)*(pSD(s,d,t
     ,w)-rSDD(s,d,t,w));
StoUnit_09(s,d,t,w)$(ord(t) eq 1)..  eSmin(s,d,t,w) =e= gammaS0(s,d
     )*eISE(s)+etaS(s)*(pSC(s,d,t,w)-rSUC(s,d,t,w))-1/etaS(s)*(pSD(s
     ,d,t,w)+rSUD(s,d,t,w));
StoUnit_10(s,d,t,w)$(ord(t) eq 1)..  eSmax(s,d,t,w) =e= gammaS0(s,d
     )*eISE(s)+etaS(s)*(pSC(s,d,t,w)+rSDC(s,d,t,w))-1/etaS(s)*(pSD(s
     ,d,t,w)-rSDD(s,d,t,w));
StoUnit_11(s,d,t,w)..                   eSmin(s,d,t,w)=g= gammaSmin(s)*
     eISE(s);
```

```
StoUnit_12(s,d,t,w)..                   eSmax(s,d,t,w)=g= gammaSmin(s)*
    eISE(s);
StoUnit_13(s,d,t,w)..                   eSmin(s,d,t,w)=l= eISE(s);
StoUnit_14(s,d,t,w)..                   eSmax(s,d,t,w)=l= eISE(s);
StoUnit_15(s,d,t,w)$(ord(t) eq card(t))..              eSmin(s,d,t,w)=g=
    gammaS0(s,d)*eISE(s);
StoUnit_16(s,d,t,w)$(ord(t) eq card(t))..              eSmax(s,d,t,w)=g=
    gammaS0(s,d)*eISE(s);

EleVehicles_01(k,b,d,t,w)$TK(t,k,d)..                 rEVU(k,b,d,t,w) =e
    = rEVUC(k,b,d,t,w)+rEVUD(k,b,d,t,w);
EleVehicles_02(k,b,d,t,w)$TK(t,k,d)..                 rEVD(k,b,d,t,w) =e
    = rEVDC(k,b,d,t,w)+rEVDD(k,b,d,t,w);
EleVehicles_03(k,b,d,t,w)$TK(t,k,d)..                 pEVD(k,b,d,t,w)+
    rEVUD(k,b,d,t,w) =l= NEV(k,b,w)*xEV(k,w)*PEVmax;
EleVehicles_04(k,b,d,t,w)$TK(t,k,d)..                 pEVD(k,b,d,t,w)-
    rEVDD(k,b,d,t,w) =g= 0;
EleVehicles_05(k,b,d,t,w)$TK(t,k,d)..                 pEVC(k,b,d,t,w)+
    rEVDC(k,b,d,t,w) =l= NEV(k,b,w)*xEVC(k,'c1',w)*PEVmax;
EleVehicles_06(k,b,d,t,w)$TK(t,k,d)..                 pEVC(k,b,d,t,w)-
    rEVUC(k,b,d,t,w) =g= 0;
EleVehicles_07(k,b,d,t,w)$(TK(t,k,d) and not(T0(t,k,d)))..  eEVmin(
    k,b,d,t,w) =e= eEVmin(k,b,d,t-1,w)+etaEV*(pEVC(k,b,d,t,w)-rEVUC
    (k,b,d,t,w))-1/etaEV*(pEVD(k,b,d,t,w)+rEVUD(k,b,d,t,w));
EleVehicles_08(k,b,d,t,w)$(TK(t,k,d) and not(T0(t,k,d)))..  eEVmax(
    k,b,d,t,w) =e= eEVmax(k,b,d,t-1,w)+etaEV*(pEVC(k,b,d,t,w)+rEVDC
    (k,b,d,t,w))-1/etaEV*(pEVD(k,b,d,t,w)-rEVDD(k,b,d,t,w));
EleVehicles_09(k,b,d,t,w)$T0(t,k,d)..                 eEVmin(k,b,d,t,w)
    =e= gammaEV0(k,b,d)*NEV(k,b,w)*xEV(k,w)*EEVBmax+etaEV*(pEVC(k,b
    ,d,t,w)-rEVUC(k,b,d,t,w))-1/etaEV*(pEVD(k,b,d,t,w)+rEVUD(k,b,d,
    t,w));
EleVehicles_10(k,b,d,t,w)$T0(t,k,d)..                 eEVmax(k,b,d,t,w)
    =e= gammaEV0(k,b,d)*NEV(k,b,w)*xEV(k,w)*EEVBmax+etaEV*(pEVC(k,b
    ,d,t,w)+rEVDC(k,b,d,t,w))-1/etaEV*(pEVD(k,b,d,t,w)-rEVDD(k,b,d,
    t,w));
EleVehicles_11(k,b,d,t,w)$TK(t,k,d)..                 eEVmin(k,b,d,t,w)=
    g= gammaEVmin*NEV(k,b,w)*xEV(k,w)*EEVBmax;
EleVehicles_12(k,b,d,t,w)$TK(t,k,d)..                 eEVmax(k,b,d,t,w)=
    g= gammaEVmin*NEV(k,b,w)*xEV(k,w)*EEVBmax;
EleVehicles_13(k,b,d,t,w)$TK(t,k,d)..                 eEVmin(k,b,d,t,w)=
    l= NEV(k,b,w)*xEV(k,w)*EEVBmax;
EleVehicles_14(k,b,d,t,w)$TK(t,k,d)..                 eEVmax(k,b,d,t,w)=
    l= NEV(k,b,w)*xEV(k,w)*EEVBmax;
EleVehicles_15(k,b,d,t,w)$TF(t,k,d)..                 eEVmin(k,b,d,t,w)=
    g= gammaEVF(k,b,d)*NEV(k,b,w)*xEV(k,w)*EEVBmax;
EleVehicles_16(k,b,d,t,w)$TF(t,k,d)..                 eEVmax(k,b,d,t,w)=
    g= gammaEVF(k,b,d)*NEV(k,b,w)*xEV(k,w)*EEVBmax;

DayAhead_01(b,d,t,w)..    sum(g$GN(g,b),pGD(g,d,t,w))+sum(s$SN(s,b),
    pSD(s,d,t,w)-pSC(s,d,t,w))+sum(k$TK(t,k,d),pEVD(k,b,d,t,w)-pEVC
    (k,b,d,t,w))-sum(l$OL(b,l),pLD(l,d,t,w))+sum(l$FL(b,l),pLD(l,d,
    t,w))=e=
PD(b,d,t,w)+sum(k,(1-xEV(k,w))*PEV(k,b,d,t,w))-pUD(b,d,t,w);
DayAhead_02(l,d,t,w)..    pLD(l,d,t,w)=e=1/X(L)*(sum(b$OL(b,l),
    thetaD(b,d,t,w))-sum(b$FL(b,l), thetaD(b,d,t,w)));
DayAhead_03(l,d,t,w)..    pLD(l,d,t,w)=l=pLmax(l);
DayAhead_04(l,d,t,w)..    pLD(l,d,t,w)=g=-pLmax(l);
DayAhead_05(b,d,t,w)..    thetaD(b,d,t,w) =l= thetaDmax;
DayAhead_06(b,d,t,w)..    thetaD(b,d,t,w) =g= -thetaDmax;
DayAhead_07(d,t,w)..      sum(g$GD(g),rGU(g,d,t,w))+sum(s,rSU(s,d,t,
    w))+sum((b,k)$TK(t,k,d),rEVU(k,b,d,t,w)) =g= gammaCUD*sum(b,PD(
```

```
    b,d,t,w)+sum(k,  (1-xEV(k,w))*PEV(k,b,d,t,w)))+gammaCUI*sum(g$GI
    (g),pGD(g,d,t,w) );
DayAhead_08(d,t,w)..       sum(g$GD(g),rGD(g,d,t,w))+sum(s,rSD(s,d,t,
    w))+sum((b,k)$TK(t,k,d),rEVD(k,b,d,t,w)) =g= gammaCDD*sum(b,PD(
    b,d,t,w)+sum(k,  (1-xEV(k,w))*PEV(k,b,d,t,w)))+gammaCDI*sum(g$GI
    (g),pGD(g,d,t,w) );

*******************************************************************
*   MODEL AND SOLVING
*******************************************************************

MODEL GenExpansionEV /all/;

GenExpansionEV.optcr=0;

SOLVE GenExpansionEV USING mip MINIMIZING z;
```

References

1. Asensio, M., Contreras, J.: Stochastic unit commitment in isolated systems with renewable penetration under CVaR assessment. IEEE Trans. Smart Grid. **7**(3), 1356–1367 (2016)
2. Baringo, L., Conejo, A.J.: Wind power investment: a Benders decomposition approach. IEEE Trans. Power Syst. **27**(1), 433–441 (2012)
3. Bertsekas, D.P., Sandell, N.R.: Estimates of the duality gap for large-scale separable nonconvex optimization problems. In: Proceedings of the 21st IEEE Conference Decision and Control, vol. 21, pp. 782–785 (1982)
4. Cañas-Carretón, M., Carrión, M.: Generation capacity expansion considering reserve provision by wind power units. IEEE Tran. Power Syst. **35**(6), 4564–4573 (2020)
5. Carrión, M., Domínguez, R., Cañas-Carretón, M., Zárate-Miñano, R.: Scheduling isolated power systems considering electric vehicles and primary frequency response. Energy **168**, 1192–1207 (2019)
6. Carrión, M., Zárate-Miñano, R.: Operation of intermittent-dominated power systems with a significant penetration of plug-in electric vehicles. Energy **90**, 827–835 (2015)
7. Carrión, M., Domínguez, R., Zárate-Miñano, R.: Influence of the controllability of electric vehicles on generation and storage capacity expansion decisions. Energy **189**, 116156 (2019)
8. Chen, X., Lv, J., McElroy, M.B., Han, X., Nielsen, C.P., Wen, J.: Power system capacity expansion under higher penetration of renewables considering flexibility constraints and low carbon policies. IEEE Trans. Power Syst. **33**(6), 6240–6253 (2018)
9. Conejo, A.J., Carrión, M., Morales, J.M.: Decision Making Under Uncertainty in Electricity Markets. Springer, New York (2010)
10. Conejo, A.J., Castillo, E., Mínguez, R., García-Bertrand, R.: Decomposition Techniques in Mathematical Programming. Springer, Berlin (2006)
11. The ILOG CPLEX (2021). www.ilog.com/products/cplex/
12. Domínguez, R., Conejo, A.J., Carrión, M.: Toward fully renewable electric energy systems. IEEE Trans. Power Syst. **30**(1), 316–326 (2015)
13. GAMS (2021). www.gams.com
14. Gils, H.C., Simon, S.: Carbon neutral archipelago- 100% renewable energy supply for the Canary Islands. App. Energy. **188**, 342–355 (2017)

15. Gröwe-Kuska, N., Heitsch, H., Römisch, W.: Scenario reduction and scenario tree construction for power management problems. In: Proceedings of the IEEE Bologna Power Technology Conference, Bologna, Italy (2003)
16. Krukanont, P., Tezuka, T.: Implications of capacity expansion under uncertainty and value of information: the near-term energy planning of Japan. Energy **32**(10), 1809–1824 (2007)
17. Liu, Y., Sioshansi, R., Conejo, A.J.: Multistage stochastic investment planning with multiscale representation of uncertainties and decisions. IEEE Trans. Power Syst. **33**(1), 781–791 (2018)
18. Meneses de Quevedo, P., Muñoz-Delgado, G., Contreras, J.: Impact of electric vehicles on the expansion planning of distribution systems considering renewable energy, storage and charging stations. IEEE Trans. Smart Grid. **10**(1), 794–804 (2018)
19. Min, D., Ryu, J., Choi, D.G.: A long-term capacity expansion planning model for an electric power system integrating large-size renewable energy technologies. Comput. Oper. Res. **96**, 244–255 (2018)
20. Moghaddam, S.Z.: Generation and transmission expansion planning with high penetration of wind farms considering spatial distribution of wind speed. Int. J. Electr. Power Energy Syst. **106**, 232–241 (2019)
21. Morales, J.M., Pineda, S., Conejo, A.J., Carrion, M.: Scenario reduction for futures market trading in electricity markets. IEEE Trans. Power Syst. **24**(2), 878–888 (2009)
22. Nunes, J.B., Mahmoudi, N., Saha, T.K., Chattopadhyay, D.: Multi-stage co-planning framework for electricity and natural gas under high renewable energy penetration. IET Gener. Trans. Distrib. **12**(19), 4284–4291 (2018)
23. Pineda, S., Morales, J.M., Boomsma, T.: Impact of forecast errors on expansion planning of power systems with a renewables target. Euro J. Oper. Res. **248**, 1113–1122 (2016)
24. Qiu, T., Xu, B., Wang, Y., Dvorkin, Y., Kirschen, D.S.: Stochastic multistage coplanning of transmission expansion and energy storage. IEEE Trans. Power Syst. **32**(1), 643–651 (2017)
25. Ramírez, P.J., Papadaskalopoulos, D., Strbac, G.: Co-optimization of generation expansion planning and electric vehicles flexibility. IEEE Trans. Smart Grid. **7**(3), 1609–1619 (2016)
26. Sarid, A., Tzur, M.: The multi-scale generation and transmission expansion model. Energy **148**, 977–991 (2018)
27. Wogrin, S., Gayme, D.F.: Optimizing storage siting, sizing, and technology portfolios in transmission-constrained networks. IEEE Trans. Power Syst. **30**(6), 3304–3313 (2014)
28. Xu, B., Wang, Y., Dvorkin, Y., Fernández-Blanco, R., Silva-Monroy, C., Watson, J.P., Kirschen, D.S.: Scalable planning for energy storage in energy and reserve markets. IEEE Trans. Power Syst. **32**(6), 4515–4527 (2016)

Chapter 11
Transmission Expansion Planning Considering Renewable Energies and Electric Vehicles

This chapter studies the transmission expansion problem considering the presence of renewable energies and electric vehicles. The expected replacement of a relatively small number of high-capacity thermal units by a large number of renewable energy units with smaller capacity per unit may change substantially the utilization of the transmission network. Additionally, the incorporation of a large number of electric vehicle chargers may increase significantly the energy load at specific periods in some buses and may require the reinforcement of the transmission system. Then, a two-stage stochastic programming formulation is described in this chapter to decide the investments in transmission lines to face the increase of renewable power and the number of electric vehicles. A number of illustrative examples and a realistic case study are included in this chapter to show the performance of the described formulation. The realistic case study included at the end of the chapter is based on the isolated power system of Gran Canaria island in Spain.

11.1 Introduction

The objective of the transmission network is to conduct the electricity produced by large power plants to demand centers. Since electricity is typically transported throughout long distances, the transmission network operates at a high voltage to reduce power losses. For the sake of reliability and security, the transmission network has a mesh configuration that allows the supply of demand even if a fail in any element of the transmission system occurs. The main assets of the transmission system are the transmission lines, which connect power plants with substations. Substations connected to the transmission network can be considered interconnection buses where power is transformed decreasing voltage magnitudes to feed distribution networks. The distribution networks reach single consumers and operate at a lower

L. Baringo et al., *Electric Vehicles and Renewable Generation*,
Green Energy and Technology, https://doi.org/10.1007/978-3-031-09079-0_11

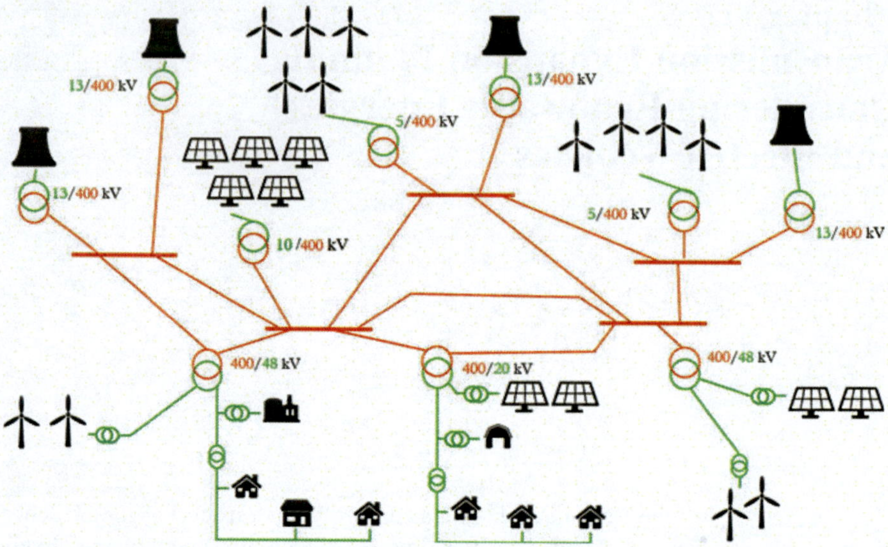

Fig. 11.1 Example of power system configuration

voltage than transmission networks. Voltages are stepped down in distribution networks sequentially in several transformation stations. In this manner, voltages reduce as the distance to the end users of electricity decreases. In contrast to transmission networks, distribution networks are typically operated using a radial configuration. Therefore, the structure of a distribution network is less redundant than that of the transmission network, which implies that failures in this infrastructure may cause outages for end users of electricity.

Figure 11.1 represents an example of a power system configuration. The transmission network is represented in red color. Observe that the operation voltage of this network is 400 kV. In real-world systems, the transmission network may operate at several voltage values in different parts of the network. The transmission system conducts power flows from power plants to the substations, where distribution lines are connected. Distribution lines are represented in green color. In this case, we consider two voltage levels for distribution lines, 20–48 kV. Observe that industrial, agricultural, or livestock facilities may be connected directly to a distribution line through a transformer. On the other hand, voltages are stepped-down at least one time before households are connected to the distribution line. Finally, observe that renewable units as wind and solar photovoltaic (PV) power plants can inject power directly into the distribution lines.

How to expand and reinforce the transmission network is key to ensure a reliable and safe operation of future power systems. Therefore, the impact of renewable energies and electric vehicles (EVs) in these networks must be analyzed thoroughly. Unlike traditional power plants that generate great amounts of energy in specific locations of the system, renewable energies are expected to be installed in a large number of locations in the network and their production is, in general, smaller than that of thermal and large hydro units. Additionally, the incorporation of a large number of EVs will increase the demand of the system and, thus, also the usage of the transmission system.

Considering the above, in this chapter we assume the role of a transmission network planner that desires to determine the optimal investments in transmission assets to accommodate existing networks to the incorporation of renewable energies and EVs. Uncertainties as the renewable energy capacity installed and the number of EVs are also considered in the developed mathematical problem.

Relevant references dealing with the transmission capacity expansion problem are [2–7, 10, 13, 14, 24]. A comprehensive description of generation and transmission investment models is provided in monograph [9].

Illustrative Example 11.1 Transmission expansion problem with electric vehicles

In this example we describe the problem faced by a transmission network planner that desires to determine the investments in new transmission lines for a future target year. The decisions that the transmission expansion planner has to make are the new transmission lines to install from a set of candidate transmission lines.

The most relevant input data that the transmission system planner must either know or estimate are:

- The existing transmission system topology, including the number of buses and transmission lines parameters.
- The technical description of the set of candidate transmission lines, including resistance, reactance, voltage, power capacity, and length.
- The capital costs of candidate transmission lines.
- The temporal characterization of the target period, for instance one year represented using ten characteristic days.
- An estimation of the generation capacity per bus that will be at operation at the target year.
- The estimation of the availability of wind and solar PV power for each hour of each characteristic day.
- The estimation of the hourly demand for each hour of each characteristic day.
- The estimation of the number of EVs and their charging patterns.

Observe that the input data described above are subject to different sources of uncertainties as the demand and the number of EVs in the target year, or the generation capacity installed in each bus.

Considering the above, the transmission system planner desires to determine the optimal transmission lines to install to satisfy the expected demand resulting in the minimum expected cost. This cost comprises the investment costs of new transmission lines and the operation cost of the system. □

This chapter is organized as follows. The decision framework is described in Sect. 11.2. The uncertainty characterization is described in Sect. 11.3. The formulation of the problem is provided in Sect. 11.4. In Sect. 11.5, an illustrative case study is solved to show the performance of the presented formulation. In Sect. 11.6, a realistic case study is solved and comprehensively discussed. Section 11.7 provides a summary of the chapter and includes some relevant conclusions. The main notation used in this chapter is defined in Sect. 11.8. Finally, Sect. 11.9 provides the GAMS [11] code used to solve the illustrative case study of Sect. 11.5.

11.2 Decision Framework

In this chapter, we assume the role of a transmission system planner that desires to determine the optimal transmission investment decisions for a future planning horizon. The considered investment decisions comprise the installation of new transmission lines and storage units. Observe that the installation of storage units by the transmission system planner may defer expensive investments in new transmission lines.

In order to simulate adequately the operation of the power system, the cost associated with the supply of the demand is taken into account in the transmission expansion problem. Then, a target year will be considered and represented by a set of characteristic days $d \in D$, which are divided into hourly periods denoted by index $t \in T$. Since hourly time periods are considered, power and energy terms are equivalent in the proposed model.

The demand of a large number of EVs will be also considered in the analyzed problem. In order to consider the most unfavourable case, it will be assumed that the charge of vehicles cannot be controlled by the power system operator. Considering this, the objective of the transmission expansion planer is to decide the optimal transmission investments that minimize the total cost.

Figure 11.2 represents graphically the decision-making process described above. The black point represents the time in which the transmission expansion decisions are decided. Note that these are *here-and-now* decisions that are made before knowing the actual realization of the uncertain parameters. The operation of the resulting power system is simulated in the second stage after knowing the actual realization of the uncertain parameters. For this reason, operation decisions are considered as *wait-and-see* decisions that can be different for each scenario.

11.3 Uncertainty Characterization

The transmission expansion problem is subject to different sources of uncertainty. The main sources of uncertainty for the transmission expansion planner are the number of EVs and the renewable generation capacity installed in the target year. Uncertain parameters are characterized as random variables and they are represented by the set of scenarios Ω. As stated in Chap. 3, this type of uncertain parameters

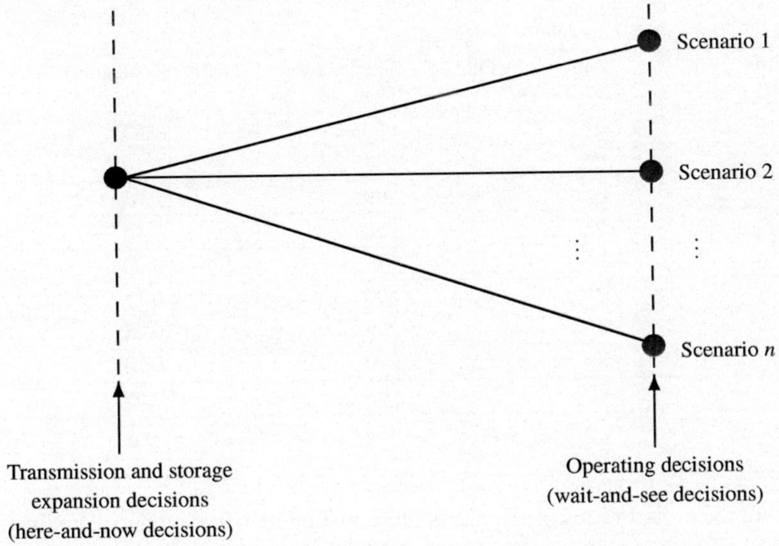

Fig. 11.2 Transmission expansion considering electric vehicles: scenario tree

Table 11.1 Illustrative Example 11.2: uncertainty

Parameters	Low		Medium		High	
	Value	Prob.	Value	Prob.	Value	Prob.
Capacity installed of solar PV (%)	10	0.20	30	0.60	50	0.20
Penetration of electric vehicles (%)	50	0.25	70	0.50	90	0.25

can be statistically characterized by expert judgment using information available in reliable technical reports.

Illustrative Example 11.2 Uncertainty

Let us consider two uncertain parameters: the future values of capacity installed of solar PV units and number of EVs. Table 11.1 lists three possible values (low, medium, and high) for each parameter. The capacity installed of solar PV is expressed in percentage with respect to the potential capacity of installation of this technology. In the same manner, the number of EVs is expressed in terms over the total number of vehicles.

A total number of $3^2 = 9$ scenarios can be generated considering the data included in Table 11.1. The resulting scenarios are included in Table 11.2. Assuming that both parameters are not correlated, the probability of a given scenario corresponds to the

Table 11.2 Illustrative Example 11.2: scenarios

Scenarios	Capacity installed of Solar PV (%)	Penetration of electric vehicles (%)	Probability
1	10	50	0.05
2	10	70	0.10
3	10	90	0.05
4	30	50	0.15
5	30	70	0.30
6	30	90	0.15
7	50	50	0.05
8	50	70	0.10
9	50	90	0.05

product of the probabilities of all parameters considered in that scenario. For instance, scenario 1 with a capacity installed of solar PV equal to 10% of the potential capacity (probability equal to 0.20) and a 50% penetration of EVs (probability equal to 0.25) has associated a probability equal to $0.20 \cdot 0.25 = 0.05$.

□

11.4 Formulation

This section presents the mathematical formulation of the transmission expansion planning problem described above. The description of the symbols used throughout this section is provided in Sect. 11.8.

11.4.1 Transmission Expansion Decisions

The investment decisions that the transmission system planner desires to determine are the following:

- Investments in transmission lines. It is considered that transmission line investments can be made from a set of candidate units $\ell \in L^{\mathrm{C}}$. Considering a planning horizon of one year, the investment costs are annualized as described in Chap. 2 using the Capital Recovery Factor (CRF), where $\mathrm{CRF} = \dfrac{i(1+i)^y}{(1+i)^y - 1}$, y is the expected lifetime of the line and i is the interest rate. The annualized costs are represented as $C_{\ell}^{\mathrm{I,L}}$.
- Investments in storage units. Likewise to the case of the installation of transmission lines, storage investments can be made over a set of candidate units $s \in S^{\mathrm{C}}$.

Investments in storages are made in terms of power and energy, $p_s^{\mathrm{I,SP}}$ and $e_s^{\mathrm{I,S}}$. That is, the term of power refers to the maximum power that can be charged or discharged from the storage, whereas the energy term represents the maximum amount of energy that can be stored in the facility. The maximum power and energy capacities that can be installed are positive values and they are bounded by the maximum capacities $P_{\mathrm{max},s}^{\mathrm{I,SP}}$ and $E_{\mathrm{max},s}^{\mathrm{I,SE}}$, respectively. The power and energy terms have associated different annualized costs, $C_s^{\mathrm{I,SP}}$ and $C_s^{\mathrm{I,SE}}$.

Then, the investment decisions on transmission lines and storage units are formulated as follows:

$$y_\ell^{\mathrm{L}} \in \{0, 1\}, \quad \forall \ell \in L^{\mathrm{C}}, \tag{11.1a}$$
$$0 \le e_s^{\mathrm{I,SE}} \le E_{\mathrm{max},s}^{\mathrm{I,SE}}, \ \forall s \in S^{\mathrm{C}}, \tag{11.1b}$$
$$0 \le p_s^{\mathrm{I,SP}} \le P_{\mathrm{max},s}^{\mathrm{I,SP}}, \ \forall s \in S^{\mathrm{C}}, \tag{11.1c}$$
$$e_s^{\mathrm{I,SE}} = \gamma_s^{\mathrm{S,EP}} p_s^{\mathrm{I,SP}}, \ \forall s \in S^{\mathrm{C}}. \tag{11.1d}$$

Constraints (11.1a) formulate investments in candidate transmission lines $\ell \in L^{\mathrm{C}}$ using binary variable $y_\ell^{\mathrm{L}} \in \{0, 1\}$, which is equal to 1 if transmission line ℓ is built, being equal to 0 otherwise. Constraints (11.1c)–(11.1d) model storage investment decisions. Constraints (11.1d) establish a prefixed relationship between energy and power capacities as a function of parameter $\gamma_s^{\mathrm{S,EP}}$.

11.4.2 Generating Units Constraints

Constraints (11.2a)–(11.2e) formulate the technical constraints used to characterize the behavior of generating units.

$$0 \le p_{gdt\omega}^{\mathrm{G,D}} \le P_{g\omega}^{\mathrm{I,G}}, \quad \forall g \in G^{\mathrm{D}}, \forall d \in D, \forall t \in T, \forall \omega \in \Omega, \tag{11.2a}$$
$$p_{gdt\omega}^{\mathrm{G,D}} + p_{gdt\omega}^{\mathrm{G,DS}} = A_{gdt}^{\mathrm{D}} P_{g\omega}^{\mathrm{I,G}}, \quad \forall g \in G^{\mathrm{I}}, \forall d \in D, \forall t \in T, \forall \omega \in \Omega, \tag{11.2b}$$
$$p_{gdt\omega}^{\mathrm{G,D}} - p_{gdt-1,\omega}^{\mathrm{G,D}} \le P_{\mathrm{up},g}^{\mathrm{G}} P_{g\omega}^{\mathrm{I,G}}, \quad \forall g \in G^{\mathrm{D}}, \forall d \in D, \forall t \in T, \forall \omega \in \Omega, \tag{11.2c}$$
$$p_{gdt-1,\omega}^{\mathrm{G,D}} - p_{gdt\omega}^{\mathrm{G,D}} \le P_{\mathrm{dw},g}^{\mathrm{G}} P_{g\omega}^{\mathrm{I,G}}, \quad \forall g \in G^{\mathrm{D}}, \forall d \in D, \forall t \in T, \forall \omega \in \Omega, \tag{11.2d}$$
$$\{p_{gdt\omega}^{\mathrm{G,D}}, p_{gdt\omega}^{\mathrm{G,DS}}\} \ge 0, \quad \forall g \in G, \forall d \in D, \forall t \in T, \forall \omega \in \Omega. \tag{11.2e}$$

Constraints (11.2a) and (11.2b) impose that the power generated by each dispatchable and intermittent generating unit has to be smaller than the installed capacity in each scenario, respectively. The power capacity of unit g is a random variable that depends on scenario index ω and is denoted as $P_{g\omega}^{\mathrm{I,G}}$. The availability of intermittent units $g \in G^{\mathrm{I}}$ in each period t and day d is modeled by parameter A_{gdt}^{D}. Up and down power ramps are characterized through constraints (11.2c) and (11.2d), respectively. Note that power ramps are expressed in per unit over the capacity installed of each unit in each scenario, $P^{\mathrm{I,G}}_{g\omega}$. Constraints (11.2e) define the non-negative nature of the variables characterizing the power produced and the power spillage.

11.4.3 Storage Units

The mathematical formulation of the operation of energy storage systems is provided below. In this formulation, it is considered simultaneously both existing and candidate storage facilities, which are denoted by sets S^{E} and S^{C}, respectively. The power charged and discharged by these storage units are characterized by $p^{\mathrm{S,C}}_{sdt\omega}$ and $p^{\mathrm{S,D}}_{sdt\omega}$, respectively.

Constraints (11.3a) and (11.3b) limit the amount of power that can be charged and discharged in each period in existing storage units. Constraints (11.3c) define the upper and lower limits of the energy stored in existing storage units according to the installed energy capacity, $E^{\mathrm{I,SE}}_s$. Constraints (11.3d) establish the initial status ($\gamma^{\mathrm{S,0}}_{sd}$) of the existing storage units for each day. Besides, constraints (11.3e) formulate the energy stored in existing storage units at the end of the day, i.e., at $t = 24$, which has to be at least greater than a desired value, $\gamma^{\mathrm{S,F}}_{sd}$. Constraints (11.3f) compute the level of energy in candidate and existing storage units in each period. Finally, constraints (11.3g)–(11.3k) are equivalent to (11.3a)–(11.3e) but for candidate storage units. Observe that the difference between existing and candidate storage units is that the energy and power capacities installed in existing storage units are known for each scenario, $P^{\mathrm{I,SP}}_s$ and $E^{\mathrm{I,SE}}_s$, whereas they are optimization variables in the case of candidate units, $p^{\mathrm{I,SP}}_s$ and $e^{\mathrm{I,SE}}_s$.

$$0 \le p^{\mathrm{S,D}}_{sdt\omega} \le P^{\mathrm{I,SP}}_s, \quad \forall s \in S^{\mathrm{E}}, \forall d \in D, \forall t \in T, \forall \omega \in \Omega, \tag{11.3a}$$

$$0 \le p^{\mathrm{S,C}}_{sdt\omega} \le P^{\mathrm{I,SP}}_s, \quad \forall s \in S^{\mathrm{E}}, \forall d \in D, \forall t \in T, \forall \omega \in \Omega, \tag{11.3b}$$

$$\gamma^{\mathrm{S}}_s E^{\mathrm{I,SE}}_s \le e^{\mathrm{S}}_{sdt\omega} \le E^{\mathrm{I,SE}}_s, \quad \forall s \in S^{\mathrm{E}}, \forall d \in D, \forall t \in T, \forall \omega \in \Omega, \tag{11.3c}$$

$$e^{\mathrm{S}}_{sdt\omega} = \gamma^{\mathrm{S,0}}_{sd} E^{\mathrm{I,SE}}_s, \quad \forall s \in S^{\mathrm{E}}, t = 0, \forall d \in D, \forall \omega \in \Omega, \tag{11.3d}$$

$$e^{\mathrm{S}}_{sdt\omega} \ge \gamma^{\mathrm{S,F}}_{sd} E^{\mathrm{I,SE}}_s, \quad \forall s \in S^{\mathrm{E}}, t = 24, \forall d \in D, \forall \omega \in \Omega, \tag{11.3e}$$

$$e^{\mathrm{S}}_{sdt\omega} = e^{\mathrm{S}}_{sdt-1,\omega} + \eta^{\mathrm{S}} p^{\mathrm{S,C}}_{sdt\omega} - \frac{1}{\eta^{\mathrm{S}}} p^{\mathrm{S,D}}_{sdt\omega},$$
$$\forall s \in S, \forall d \in D, \forall t \in T, \forall \omega \in \Omega, \tag{11.3f}$$

$$0 \le p^{\mathrm{S,D}}_{sdt\omega} \le p^{\mathrm{I,SP}}_s, \quad \forall s \in S^{\mathrm{C}}, \forall d \in D, \forall t \in T, \forall \omega \in \Omega, \tag{11.3g}$$

$$0 \le p^{\mathrm{S,C}}_{sdt\omega} \le p^{\mathrm{I,SP}}_s, \quad \forall s \in S^{\mathrm{C}}, \forall d \in D, \forall t \in T, \forall \omega \in \Omega, \tag{11.3h}$$

$$\gamma^{\mathrm{S}}_s e^{\mathrm{I,SE}}_s \le e^{\mathrm{S}}_{sdt\omega} \le e^{\mathrm{I,SE}}_s, \quad \forall s \in S^{\mathrm{C}}, \forall d \in D, \forall t \in T, \forall \omega \in \Omega, \tag{11.3i}$$

$$e^{\mathrm{S}}_{sdt\omega} = \gamma^{\mathrm{S,0}}_{sd} e^{\mathrm{I,SE}}_s, \quad \forall s \in S^{\mathrm{C}}, t = 0, \forall d \in D, \forall \omega \in \Omega, \tag{11.3j}$$

$$e^{\mathrm{S}}_{sdt\omega} \ge \gamma^{\mathrm{S,F}}_{sd} e^{\mathrm{I,SE}}_s, \quad \forall s \in S^{\mathrm{C}}, t = 24, \forall d \in D, \forall \omega \in \Omega. \tag{11.3k}$$

11.4.4 Transmission Lines

The power flow through the transmission lines is formulated using the DC approximation of the transmission network. This approximation is useful to approximate linearly the nonlinear relationships of active and reactive power flows with respect to the voltage magnitudes and angles, as described in Sect. 4.4.7 in Chap. 4.

DC power flows in existing lines are formulated through constraints (11.4a) and bounded by constraints (11.4b).The power flow in line ℓ depends on the reactance of the line, X_ℓ, and on the difference of voltage angles between the origin bus of the line, $O(\ell)$, and the destination bus, $F(\ell)$. The limits to voltage bus angles are set in constraints (11.4c).

$$p^{\mathrm{L,D}}_{\ell dt\omega} = \frac{1}{X_\ell}\left(\theta^{\mathrm{D}}_{O(\ell)dt\omega} - \theta^{\mathrm{D}}_{F(\ell)dt\omega}\right), \quad \forall \ell \in L^{\mathrm{E}}, \forall d \in D, \forall t \in T, \forall \omega \in \Omega, \qquad (11.4a)$$

$$-P^{\mathrm{L}}_{\max,\ell} \le p^{\mathrm{L,D}}_{\ell dt\omega} \le P^{\mathrm{L}}_{\max,\ell}, \quad \forall \ell \in L^{\mathrm{E}}, \forall d \in D, \forall t \in T, \forall \omega \in \Omega, \qquad (11.4b)$$

$$-\pi \le \theta^{\mathrm{D}}_{bdt\omega} \le \pi, \quad \forall b \in B, \forall d \in D, \forall t \in T, \forall \omega \in \Omega. \qquad (11.4c)$$

In the case of candidate lines, it has to be considered that power flows have to be limited by the capacity limits of the lines if they are built, but they have to be zero if the lines are not installed. Considering this, the power flows in candidate lines are formulated by constraints (11.5a) and (11.5b). Then, if binary variable $y^{\mathrm{L}}_\ell = 1$, transmission line ℓ is built, and the power flow $p^{\mathrm{L,D}}_{\ell dt\omega}$ is equal to $\frac{1}{X_\ell}\left(\theta^{\mathrm{D}}_{O(\ell)dt\omega} - \theta^{\mathrm{D}}_{F(\ell)dt\omega}\right)$ by constraints (11.5a), and the power flow is bounded by $P^{\mathrm{L}}_{\max,\ell}$ (constraints (11.5b)). On the other hand, if $y^{\mathrm{L}}_\ell = 0$, transmission line ℓ is not built, and the power flow $p^{\mathrm{L,D}}_{\ell dt\omega}$ is equal to 0 by constraints (11.5b). Note that the large value of parameter M causes that (11.5a) does not constraint the value of voltage angles.

$$\begin{aligned} &\frac{1}{X_\ell}\left(\theta^{\mathrm{D}}_{O(\ell)dt\omega} - \theta^{\mathrm{D}}_{F(\ell)dt\omega}\right) - \left(1 - y^{\mathrm{L}}_\ell\right) M \le p^{\mathrm{L,D}}_{\ell dt\omega} \\ &\le \frac{1}{X_\ell}\left(\theta^{\mathrm{D}}_{O(\ell)dt\omega} - \theta^{\mathrm{D}}_{F(\ell)dt\omega}\right) + \left(1 - y^{\mathrm{L}}_\ell\right) M, \\ &\forall \ell \in L^{\mathrm{C}}, \forall d \in D, \forall t \in T, \forall \omega \in \Omega, \end{aligned} \qquad (11.5a)$$

$$-P^{\mathrm{L}}_{\max,\ell} y^{\mathrm{L}}_\ell \le p^{\mathrm{L,D}}_{\ell dt\omega} \le P^{\mathrm{L}}_{\max,\ell} y^{\mathrm{L}}_\ell, \quad \forall \ell \in L^{\mathrm{C}}, \forall d \in D, \forall t \in T, \forall \omega \in \Omega. \qquad (11.5b)$$

Illustrative Example 11.3 Formulation of power flows in candidate transmission lines

Let us consider a candidate transmission line with power capacity equal to 100 MW and reactance equal to 0.002 pu (the power base is equal to 1 MW). If parameter M is equal to 1000, and the voltage angles at both ends of the line are 0 and −0.1 rad, the active power flow through this candidate line depending on whether it is installed or not is the following:

- Candidate line is built ($y^{\mathrm{L}}_\ell = 1$). In this case, constraints (11.5a) and (11.5b) are the following:

$$\frac{1}{0.002}(0+0.1)-(1-1)1000 \leq p^{\mathrm{L,D}} \leq \frac{1}{0.002}(0+0.1)+(1-1)1000,$$
$$-100 \leq p^{\mathrm{L,D}} \leq 100.$$

Indices ℓ, d, and t are removed in this example for the sake of clarity. Then, $p^{\mathrm{L,D}} = \frac{1}{0.002}(0+0.1)+(1-1)100 = 50$ MW by constraint (11.5a) and this value is within the lower and upper limits established in constraint (11.5b).

- Candidate line is not built ($y_\ell^{\mathrm{L}} = 0$). In this case, constraints (11.5a) and (11.5b) take the following values:

$$\frac{1}{0.002}(0+0.1)-(1-0)1000 \leq p^{\mathrm{L,D}} \leq \frac{1}{0.002}(0+0.1)+(1-0)1000,$$
$$0 \leq p^{\mathrm{L,D}} \leq 0.$$

In this case, $p^{\mathrm{L,D}} = 0$ MW by constraint (11.5b) and this value is within the lower and upper limits established in constraint (11.5a):

$$\frac{1}{0.002}(0+0.1)-(1-0)1000 = -950,$$
$$\frac{1}{0.002}(0+0.1)+(1-0)1000 = 1050.$$

□

11.4.5 Energy Balance Constraints

Constraints (11.6a) formulate the energy balance for each bus, period and scenario. The different terms in the energy balance constraint comprise the power output of generating units, the power discharged and charged from energy storage units, outgoing and incoming transmission power flows, nodal demand, the power charged from EVs, and the unserved demand. The total expected power charged by EVs belonging to group k in bus b, day d, period t, and scenario ω is characterized by $P^{\mathrm{EV}}_{kbdt\omega}$. Constraint (11.6b) states that the unserved demand has to be positive and lower than the inflexible demand.

$$\sum_{g\in G_b} p^{\mathrm{G,D}}_{gdt\omega} + \sum_{s\in S_b}\left(p^{\mathrm{S,D}}_{sdt\omega} - p^{\mathrm{S,C}}_{sdt\omega}\right) - \sum_{\ell\in L^{\mathrm{O}}_b} p^{\mathrm{L,D}}_{\ell dt\omega} + \sum_{\ell\in L^{\mathrm{F}}_b} p^{\mathrm{L,D}}_{\ell dt\omega}$$
$$= P^{\mathrm{D}}_{bdt\omega} + \sum_{k\in K_t} P^{\mathrm{EV}}_{kbdt\omega} - p^{\mathrm{UD}}_{bdt\omega},$$
$$\forall b\in B, \forall d\in D, \forall t\in T, \forall\omega\in\Omega, \tag{11.6a}$$
$$0 \le p^{\mathrm{UD}}_{bdt\omega} \le P^{\mathrm{D}}_{bdt\omega} + \sum_{k\in K_t} P^{\mathrm{EV}}_{kbdt\omega},$$
$$\forall b\in B, \forall d\in D, \forall t\in T, \forall\omega\in\Omega. \tag{11.6b}$$

11.4.6 Objective Function

The objective function is formulated in (11.7) and comprises the expected annual cost. This cost includes the annualized investment costs plus the operating cost during one year.

$$\sum_{\omega\in\Omega}\pi_\omega\Bigg[\sum_{\ell\in L^{\mathrm{C}}} C^{\mathrm{L}}_\ell y^{\mathrm{L}}_\ell + \sum_{s\in S^{\mathrm{C}}}\left(C^{\mathrm{I,SE}}_s e^{\mathrm{I,SE}}_s + C^{\mathrm{I,SP}}_s p^{\mathrm{I,SP}}_s\right)$$
$$+\sum_{d\in D}\sum_{t\in T} W_{dt}\left(\sum_{g\in G^{\mathrm{D}}} C^{\mathrm{G,D}}_{gdt\omega} p^{\mathrm{G,D}}_{gdt\omega} + \sum_{s\in S}\left(C^{\mathrm{S,D}}_{sdt\omega} p^{\mathrm{S,D}}_{sdt\omega} - C^{\mathrm{S,C}}_{sdt\omega} p^{\mathrm{S,C}}_{sdt\omega}\right)\right.$$
$$\left.+\sum_{b\in B} C^{\mathrm{UD}} p^{\mathrm{UD}}_{bdt\omega}\right)\Bigg], \tag{11.7}$$

where parameter π_ω indicates the probability of scenario ω, and W_{dt} is a parameter used to assign a weight to each period t in day d. Therefore, considering 24 hourly periods per day, it should be satisfied that $\sum_{\omega\in\Omega}\pi_\omega = 1$ and $\sum_{d\in D}\sum_{t=1}^{24} W_{dt} = 365\cdot 24 = 8760$.

11.4.7 Problem Formulation

Considering the previous definitions, the transmission expansion planning problem is finally formulated as the following mixed-integer linear programming problem:

$$\text{Minimize}_\Theta \quad (11.7) \tag{11.8a}$$

subject to:

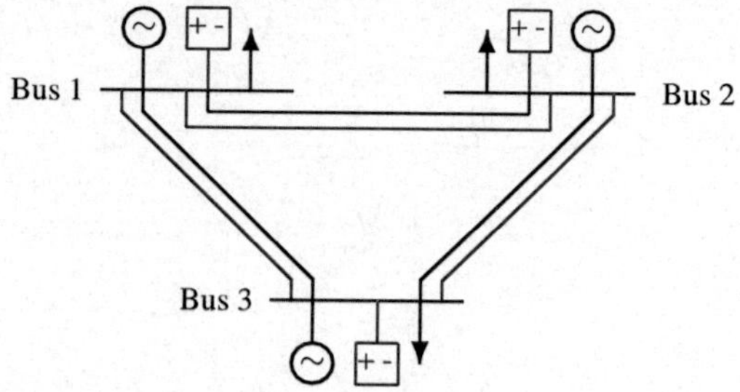

Fig. 11.3 Illustrative case study: three-node system

Table 11.3 Illustrative case study: demand (MWh)

Bus	Period 1	Period 2	Period 3
1	80	100	90
2	220	250	230
3	35	50	40

$$\text{Constraints} \quad (11.1){-}(11.6), \qquad (11.8b)$$

where set Θ comprises the decision variables.

11.5 Illustrative Case Study

We present an example to illustrate the formulation of the transmission expansion problem considering the presence of renewable energies and EVs explained in Sect. 11.4. We consider a planning horizon of one year characterized by a single day comprising three hourly periods. The weight of each hourly period, W_{dt}, is equal to $365 \cdot 8 = 2920$. The power system comprises three buses and three existing transmission lines. Three candidate lines are considered to reinforce the existing ones. Figure 11.3 represents the considered power system. It is also considered that a storage facility can be installed in each bus of the system. Candidate transmission lines and storage units are depicted in blue color. The number of EVs and the capacity installed for each generation technology in each bus are considered as uncertain parameters.

The hourly demand associated to each bus and period, with the exception of the electricity demanded by EVs, is provided in Table 11.3.

The parameters describing generating units are listed in Table 11.4, namely the location, the ramp capability, and the operating cost. Units 1 and 2 are considered to be dispatchable units, whereas unit 3 is an intermittent unit. The down power ramp, $P^{\mathrm{G}}_{\mathrm{dw},g}$ is assumed to be equal to the upper power ramp, $P^{\mathrm{G}}_{\mathrm{up},g}$. The power availability of unit 3 in each time period, A^{D}_{gdt}, is 0.4, 0.9, and 0.6, respectively.

Table 11.4 Illustrative case study: generating units data

Unit	Bus	$P^{\mathrm{G}}_{\mathrm{up},g}$ (pu)	$C^{\mathrm{G,D}}_{gdt\omega}$ (€/MWh)
1	1	0.25	0.25
2	2	0.25	0.40
3	3	–	0.00

Table 11.5 Illustrative case study: transmission lines data

Line	Type	Origin bus	Destination bus	X_ℓ (pu)	$P^{\mathrm{L}}_{\max,\ell}$ (MW)	C^{L}_ℓ (k€)
1	Existing	1	2	0.01	70	–
2	Existing	2	3	0.01	70	–
3	Existing	3	1	0.01	70	–
4	Candidate	1	2	0.01	70	200
5	Candidate	2	3	0.01	70	200
6	Candidate	3	1	0.01	70	200

Table 11.6 Illustrative case study: storage unit data

Storage	Bus	$E^{\mathrm{I,SE}}_{\max,s}$ (MWh)	$P^{\mathrm{I,SP}}_{\max,s}$ (MW)	γ^{SEP}_s (h)	$C^{\mathrm{I,SE}}_s$ (€/MWh)	$C^{\mathrm{I,SP}}_s$ (€/MW)
1	1	100	100	6	500	3000
2	2	100	100	6	500	3000
3	3	100	100	6	500	3000

The technical parameters of existing and candidate transmission lines are listed in Table 11.5. As stated above, three candidate lines can be built to reinforce the transmission network. The base power is 1 MVA.

The parameters describing the candidate storage facilities are included in Table 11.6. We assume that the cost of charging and discharging energy is null. The initial status of the storage units is 70% of the full capacity, and the minimum quantity of energy that must remain in the storage units in every period is 10% of the capacity. The efficiency of charging and discharging is equal to 0.98. The energy stored at the end of the last period has to be greater than or equal to the energy stored at the beginning of the planning horizon.

In order to characterize the uncertainty associated with the number of EVs and the generation capacity investments, six equiprobable scenarios are considered representing low, average, and high number of EVs and low and high investments in renewable technologies.

Two groups of EVs are considered for each bus. The number of vehicles in each bus and scenario is listed in Table 11.7. The expected charging demand of a single vehicle belonging to each group is included in Table 11.8. Vehicles belonging to

Table 11.7 Illustrative case study: number of electric vehicles

Group	Bus	Scenarios 1–2	Scenarios 3–4	Scenarios 5–6
1	1	550	750	950
	2	800	1000	1200
	3	400	600	800
2	1	300	500	700
	2	400	600	800
	3	150	350	550

Table 11.8 Illustrative case study: electric vehicle demand (kWh)

Group	Period 1	Period 2	Period 3
1	0	15	5
2	15	5	0

Table 11.9 Illustrative case study: generation investment plans

Unit	Scenarios 1, 3, and 5	Scenarios 2, 4, and 6
1	300	300
2	90	90
3	100	300

Table 11.10 Illustrative case study: total cost (k€)

Investment costs		Generation costs	Total
Transmission	Storage		
200	68.8	725.4	994.2

group 1 charge their batteries during periods 2 and 3, whereas vehicles in group 2 are charged in periods 1 and 2.

Table 11.9 shows the generation investment plans. Two different situations are considered. In the first one, the presence of non dispatchable units is low (100 MW, for scenarios 1, 3, and 5). For the rest of scenarios, the investments in intermittent units is much higher (300 MW).

Problem (11.8) is characterized by 940 constraints, 460 continuous variables, and 3 binary variables and is solved using CPLEX under GAMS [8].

The resulting expected cost is provided in Table 11.10. It is observed that investments in both transmission lines and storage units are made. However, the generation costs constitute the largest expenditure in this case.

The candidate lines and storage units that are installed are specified in Table 11.11. From this table we observe that candidate line 4 linking buses 1 and 2 is built. Additionally, a storage facility has been installed in bus 2. The reason for this result

Table 11.11 Illustrative case study: transmission and storage capacity investments

Candidate transmission lines			Storage units					
Line 4	Line 5	Line 6	Unit 1		Unit 2		Unit 3	
			(MWh)	(MW)	(MWh)	(MW)	(MWh)	(MW)
Yes	No	No	0	0	68.80	11.47	0.00	0.00

Table 11.12 Illustrative case study: power flows in existing lines (MW)

Scenario	Existing line 1			Existing line 2			Existing line 3		
	Periods			Periods			Periods		
	1	2	3	1	2	3	1	2	3
1	70.00	70.00	70.00	–36.38	–51.59	–43.92	–33.63	–18.41	–26.08
2	57.06	53.38	55.68	–69.90	–70.00	–70.00	12.85	16.62	14.32
3	70.00	70.00	70.00	–34.88	–49.55	–43.38	–35.13	–20.45	–26.62
4	57.34	53.52	54.32	–68.54	–70.00	–70.00	11.21	16.48	15.68
5	70.00	70.00	70.00	–33.38	–47.51	–42.84	–36.63	–22.49	–27.16
6	55.34	50.56	49.73	–66.04	–70.00	–70.00	10.71	19.44	20.27

Table 11.13 Illustrative case study: power flows in candidate lines (MW)

Scenario	Candidate line 4			Candidate line 5			Candidate line 6		
	Periods			Periods			Periods		
	1	2	3	1	2	3	1	2	3
1	70.00	70.00	70.00	0.00	0.00	0.00	0.00	0.00	0.00
2	57.06	53.38	55.68	0.00	0.00	0.00	0.00	0.00	0.00
3	70.00	70.00	70.00	0.00	0.00	0.00	0.00	0.00	0.00
4	57.34	53.52	54.32	0.00	0.00	0.00	0.00	0.00	0.00
5	70.00	70.00	70.00	0.00	0.00	0.00	0.00	0.00	0.00
6	55.34	50.56	49.73	0.00	0.00	0.00	0.00	0.00	0.00

is that, as observed in Tables 11.3 and 11.7, bus 2 has the highest demand and number of EVs.

Tables 11.12 and 11.13 provide the power flows in existing and candidate lines per period and scenario, respectively. In these tables we observe that the power flows in lines 1 and 4 reach the maximum capacity in some scenarios (1, 3, and 5), which correspond with low capacity of non dispatchable unit in bus 3. In these scenarios, the power flows go from bus 1 to buses 2 and 3, and from bus 3 to bus 2. On the contrary, in scenarios 2, 4, and 6 with a high capacity of the non-dispatchable unit located in bus 3, the power flow in the line connecting buses 3 to bus 1 is reversed, and the power flow of line 2 reaches its capacity in periods 2 and 3.

The operation of the installed energy storage is provided in Table 11.14. In this table we observe that the power capacity installed, 11.47 MW, coincides with the max-

Table 11.14 Illustrative Case Study: power charged/discharged in the storage installed in bus 2 (MW)

Scenario	Periods		
	1	2	3
1	0.23/0.00	0.00/0.22	0.00/0.00
2	11.47/0.00	0.00/11.47	0.47/0.00
3	0.02/0.00	0.00/2.05	2.11/0.00
4	11.47/0.00	0.00/11.47	0.47/0.00
5	0.00/0.00	0.00/4.24	4.42/0.00
6	11.47/0.00	0.00/11.47	0.47/0.00

Table 11.15 Illustrative case study: expected production (MW)

Period	Generating units			Storage units					
	Unit 1	Unit 2	Unit 3	Unit 1		Unit 2		Unit 3	
				(Ch.)	(Disch.)	(Ch.)	(Disch.)	(Ch.)	(Disch.)
1	225.97	56.80	80.00	0.00	0.00	5.77	0.00	0.00	0.00
2	237.20	79.21	119.20	0.00	0.00	0.00	6.82	0.00	0.00
3	221.96	56.80	95.00	0.00	0.00	1.32	0.00	0.00	0.00

imum power charged and discharged by the storage facility. Note that the operation of the storage is closely related to the presence of a large amount of non-dispatchable production. Therefore, the quantities of power charged and discharged in scenarios 2, 4, and 6 with a high amount of non-dispatchable production is much higher than in the rest of scenarios. The storage charges at period 1, characterized with a low demand, and discharges in period 2, where the demand is the highest.

Table 11.15 provides the expected energy provided by generating and storage units. From Table 11.15 it is observed that dispatchable generating unit 1, with the lowest operation cost, is the unit with highest power production. As mentioned above, storage unit 2 charges on low-demand periods, i.e., 1 and 3, and discharges in period 2.

11.6 Realistic Case Study

This section analyzes a case study based on the power system of Gran Canaria island in Spain. Gran Canaria is the largest island in the Canary Islands and it has available high potentials of wind and solar PV power.

The objective of this case study is to determine the optimal transmission and storage capacity investment decisions at year 2050. All mathematical problems in this section have been solved by CPLEX 12.6.1 under GAMS [8].

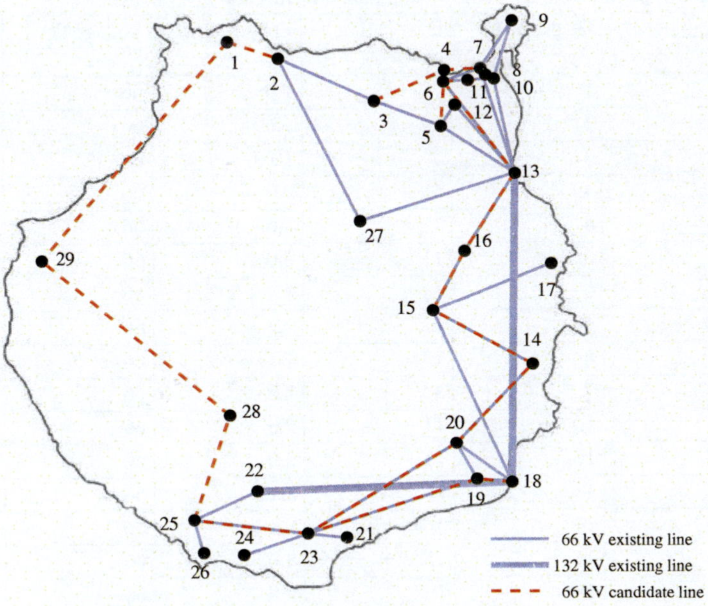

Fig. 11.4 Realistic case study: single-line diagram of Gran Canaria power system

11.6.1 Input Data

The characterization of the Canary Island power system is based on the representation included in [6]. This power system comprises 29 buses and 47 generating units. The technologies of generating units are Open-Cycle Gas Turbines (OCGT), wind, and solar PV. Figure 11.4 shows the power system topology. As observed in this figure, the transmission network is operated in two different voltage levels: 66 and 132 kV. Transmission lines have been represented using the DC model. Existing transmission lines are represented in blue color, whereas candidate lines are depicted in red color.

A set of 24 candidate lines are considered. The parameters that characterize the set of candidate lines are included in Tables 11.16 and 11.17. The power base is 100 MVA, and the voltage bases are 66 and 132 kV.

The target year is represented by a set of nine characteristic days. The weight of each day is included in Table 11.18. These days have been selected using the procedure proposed in [12] over the expected net demand in each single day. Therefore, a total number of $9 \cdot 24 = 216$ hourly time periods are considered. Figure 11.5 represents the hourly values of demand and wind and solar PV availabilities. The demand and wind and solar PV availabilities are obtained from actual values of Gran Canaria power system in 2018. It has been considered that the demand will grow by 25% from 2018 to the target year. The cost of unserved demand is 1000 €/kWh.

Table 11.16 Realistic case study: candidate transmission line parameters

Line	Origin bus	Destination bus	Length (km)	Voltage (kV)	Capacity (MW)
1	1	2	9.0	66	160
2	1	29	17.0	66	160
3	3	4	10.0	66	160
4	4	5	9.0	66	160
5	4	7	4.0	66	160
6	6	11	2.0	66	80
7	6	13	7.7	66	80
8	6	13	7.7	66	160
9	6	13	7.7	66	240
10	7	8	2.0	66	80
11	10	11	2.0	66	160
12	10	11	2.0	66	240
13	13	16	8.1	66	80
14	13	16	8.1	66	160
15	14	15	8.0	66	80
16	15	16	9.0	66	80
17	15	16	9.0	66	160
18	18	19	0.7	66	80
19	19	23	15.0	66	80
20	20	14	6.0	66	80
21	20	23	15.0	66	80
22	23	25	9.6	66	80
23	25	28	12.0	66	80
24	28	29	20.0	66	80

Table 11.17 Realistic case study: characteristics of candidate lines

Capacity (MW)	Voltage (kV)	Reactance (Ω/km)	Existing corridor	Cost (k€/km)
80	66	0.3616	No	373
			Yes	264
160	66	0.1808	No	621
			Yes	439
240	66	0.1205	No	812
			Yes	703

Table 11.18 Realistic case study: selection of characteristic days

Selected days	Weight
14 January 2018	15
16 January 2018	15
16 March 2018	199
6 May 2018	13
13 May 2018	13
14 May 2018	42
24 June 2018	18
13 July 2018	30
25 October 2018	20

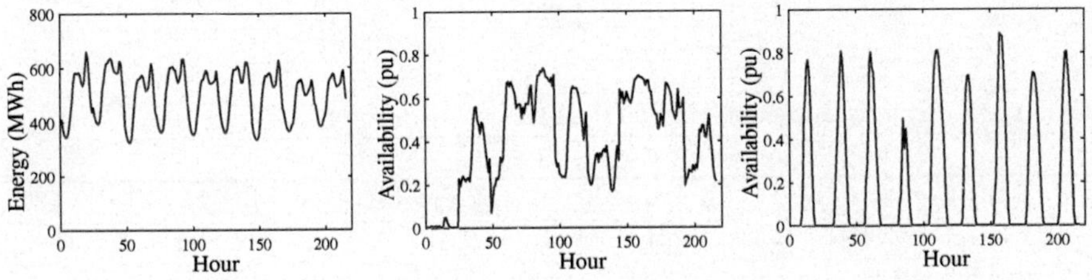

(a) Demand (b) Wind availability (c) Solar PV availability

Fig. 11.5 Realistic case study: energy demand and wind and solar PV availability

The energy demanded by EVs has been modeled for each bus and day considering that the energy consumption of vehicles is equal to 0.2 kWh/km. Four different groups of EVs, (1)–(4), are considered according to their charging patters: (1) between 00:00 and 7:00 h, (2) between 16:00 and 7:00 h of next day, (3) between 2:00 and 14:00 h, and (4) between 3:00 and 19:00 h. The percentage of vehicles in each group are 45, 35, 15, and 5%, respectively. The distance driven by each vehicle before charging is modeled using a Weibull distribution with mean equal to 140 km and scale and shape factors equal to 158.02 and 2.25. The starting charge hour for each group in each bus for each day is also modeled using a Weibull distribution with scale and shape factors equal to 2 and 1.5 for vehicles belonging to groups 2–4 and, and 1 and 1.5 for those in group 1. Considering that the penetration of EVs over the total number of vehicles may be 55, 75, or 95%, Fig. 11.6 represents the energy demanded by EVs in each hour of each considered day.

The allocation of the total demand per bus is performed using the share factors provided in Table 11.19.

Table 11.20 provides the considered potentials of generating units. The operation cost of OCGT units is 288.2 €/MWh, whereas it is equal to 0 for wind and solar PV units.

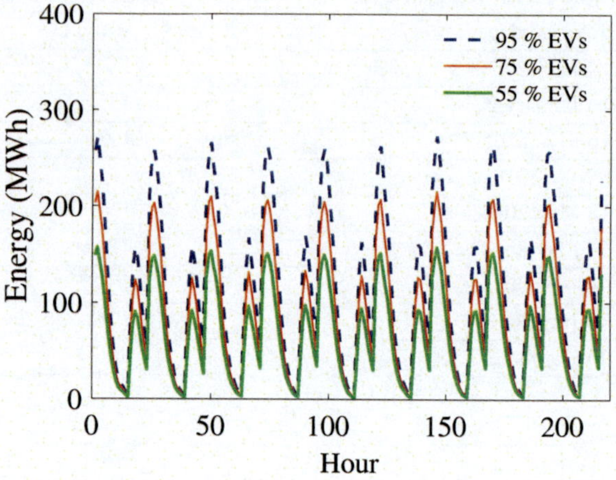

Fig. 11.6 Realistic case study: demand of EVs

Table 11.19 Realistic case study: demand share factor (%)

Bus																			
1	2	3	4	5	6	7	8	9	10	11	12	14	15	16	18	23	27	28	29
3.5	3.4	4.5	4.9	4.9	4.9	4.9	4.9	4.9	4.9	4.9	4.9	3.7	3.7	12.1	8.6	6.3	6.4	2.4	0.9

Table 11.20 Realistic case study: technical characteristics of generating units

Techn.	Unit	Node	Potential capacity (MW)	Techn.	Unit	Node	Potential capacity (MW)
OCGT	1–5	13	32.0	Solar PV	32	20	23.2
	6–10	13	32.0		33	22	235.6
	11–15	18	32.0		34	23	190.2
	16–20	18	32.0		35	24	302.9
Wind	21	9	22.8		36	9	22.3
	22	22	165.8		37	10	6.0
	23	23	2.4		38	11	38.4
	24	24	1.8		39	12	23.2
	25	26	111.1		40	13	23.2
	26	31	143.6		41	14	23.2
	27	37	0.6		42	15	23.2
Solar PV	28	16	23.2		43	26	12.0
	29	17	23.2		44	31	50.8
	30	18	23.2		45	35	8.7
	31	19	23.2		46	36	18.5
					47	37	5.4

Table 11.21 Realistic case study: scenarios

Scenario	Parameters		Probability
	Capacity installed of renewables over potentials (%)	Penetration of electric vehicles (%)	
1	25	55	1/9
2	25	75	1/9
3	25	95	1/9
4	50	55	1/9
5	50	75	1/9
6	50	95	1/9
7	75	55	1/9
8	75	75	1/9
9	75	95	1/9

Table 11.22 Realistic case study: technical characteristics of candidate storage units

Unit	Node	Potential capacity (MW)/(MWh)
5	13	300/1800
6	14	300/1800
7	18	300/1800
8	23	300/1800

Three different generation investment plans are accounted for modeling the uncertainty of the investments in generation capacity. It is considered that all OCGTs listed in Table 11.20 are operative in these expansion plans, and that 25, 50, and 75% of the potentials of wind and solar PV units are installed in each plan, respectively. Considering the three scenarios used to model the number of EVs, an the three generation investment plans, a total number of $3 \cdot 3 = 9$ scenarios is considered in this case study, whose main characteristics are summarized in Table 11.21.

Additionally to the installation of new transmission lines, we consider that storage facilities can be installed to increase the operation flexibility of the system. A total of four candidate Li-ion storage facilities is accounted for. A typical relationship between power and energy terms equal to 6 h is considered. The total investment cost of this type of storage units, including energy and power terms, is 148 €/MWh. The location of the candidate storage units and the maximum storage capacity, in power and energy units, is listed in Table 11.22. The cost of charge and discharge energy in the day-ahead market is null.

Table 11.23 Realistic case study: expected cost (million €)

Case	Transmission cost	Storage cost	Generation cost	Unserved demand	Total
Case *100%Sto*	4.56	63.7	677.2	0.00	745.4
Case *50%Sto*	4.56	54.4	689.5	0.00	748.5
Case *0%Sto*	4.31	0.0	828.3	22.78	855.4

The capital costs of transmission lines and storage units are annualized using a capital recovery factor equal to 10.95%, considering that the expected life of lines is 20 years and the interest rate is 9%.

11.6.2 Results

This section discuses the results obtained from solving three different cases:

- Case *100%Sto*: Problem (11.8) is solved considering the candidate storage units described in Table 11.22.
- Case *50%Sto*: This case is identical to Case *100%Sto* but considering only 50% of the storage potentials indicated in Table 11.22.
- Case *0%Sto*: This case is identical to Case *100%Sto* but eliminating the possibility of installing storage units.

The number of constraints, continuous and binary variables in Case *100%Sto* are $1.5{\cdot}10^6$, $0.7{\cdot}10^6$, and 24, respectively.

Table 11.23 includes the expected costs resulting from each case, considering transmission and storage investment costs, generation costs, and unserved demand costs. As expected, the smallest cost is obtained when the highest capacity of storage units can be installed. It is observed that the investments in transmission lines do not depend to a large extent on the capacity installed from storage units. By contrast, the generation cost increases significantly if the storage capacity is reduced. For instance, the generation cost increases by 22.3% with respect to Case *100%Sto* if storage units are not installed. It is also worth noting that there is unserved demand in Case *0%Sto*. This indicates that the system demand cannot be fully satisfied with the considered generation plans if storage units are not considered.

The investments in transmission lines for each analyzed case are summarized in Tables 11.24 and 11.25. As mentioned above, we notice that the investments in new transmission lines are relatively influenced by the available capacity of storage units. In this manner, we observe that the investments in transmission lines are identical in Cases *100%Sto* and *50%Sto*.

As an example, Tables 11.26 and 11.27 show respectively the average and maximum usage of every installed line for each considered scenario. In general, we

Table 11.24 Realistic case study: investments in transmission lines

Case	# Lines	Length (km)	Cost (k€)
Case *100%Sto*	13	104.60	4559.04
Case *50%Sto*	13	104.60	4559.04
Case *0%Sto*	12	85.50	4311.20

Table 11.25 Realistic case study: installed transmission lines

Case	Lines
Case *100%Sto*	2, 5, 6, 9, 10, 13, 14, 15, 16, 18, 20, 23, 24
Case *50%Sto*	2, 5, 6, 9, 10, 13, 14, 15, 16, 18, 20, 23, 24
Case *0%Sto*	1, 2, 5, 6, 9, 10, 14, 15, 17, 18, 20, 23

observe that the usage of lines increases as the number of EVs is higher. In this manner, the usage of transmission lines in scenarios 3, 6, and 9 is generally higher than in scenarios 1, 4, and 7, respectively. On the other hand, the influence of the capacity installed from renewable energies is uneven. For instance, let us consider the case of line 20, which links buses 14 and 20 (see Fig. 11.4). Tables 11.26 and 11.27 indicate that the usage of this line increases as the renewable capacity installed grows. The reason of this result is that in bus 14 it is installed a wind power unit with maximum capacity equal to 165.8 MW (see Table 11.20). Therefore, as the production of this wind power unit increases (scenario 7, for instance), also does the power flow in line 20. The same reasoning can be applied to the case of line 15 that links buses 14 and 15. On the contrary, power flows in lines linking buses with moderate values of renewable capacity, for instance, lines 5, 6, and 9, are not so sensible to the renewable capacity installed. Finally, observe that maximum usage of lines 13, 14, and 16 are close to their limits when the installed renewable capacity is high (scenarios 7, 8, and 9).

Table 11.28 provides the storage capacity installed in each case. It is worth noting that the limit to the storage capacity to install is active in Cases *50%Sto* and *0%Sto*.

Table 11.29 provides the expected energy provided by each generation technology. Note that the energy produced by OCGT units increases as the capacity of storage units decreases. The production of OCGT units increases by 22.3% from Case *100%Sto* to *0%Sto*. The opposite effect is observed for wind and solar PV units, whose productions decrease by 9.6 and 28.4%, respectively, if storage units are not installed. Observe that the power output profile of solar PV units is ideal for storage units, which charge in sunny hours at midday, and discharge during the night.

Figure 11.7 represents the production of each generation technology in each hour for scenarios 1, 5, and 9 in Case *100%Sto*. The description of each scenario is provided in Table 11.21. Observe that investment decisions on transmission lines and storage units are identical for each scenario. Figure 11.7a represents the case of scenario 1, which is characterized by low renewable capacity and low number of EVs. As

Table 11.26 Realistic case study: average usage of installed lines in Case *100%Sto* (%)

Scenario	Installed lines												
	2	5	6	9	10	13	14	15	16	18	20	23	24
1	10.0	17.8	35.7	47.3	35.8	13.8	13.8	24.9	30.6	4.3	25.7	25.8	26.2
2	10.6	18.6	37.4	49.4	37.5	14.0	14.0	25.9	31.7	4.4	27.1	27.3	27.6
3	11.2	19.4	39.0	51.6	39.2	14.2	14.2	26.9	32.6	4.5	28.6	28.8	29.1
4	7.2	16.4	32.9	43.5	33.0	23.2	23.2	32.3	41.3	7.0	32.0	21.3	19.3
5	7.8	17.2	34.5	45.7	34.7	23.1	23.1	33.3	42.1	7.1	30.5	22.6	20.7
6	8.3	18.0	36.1	47.8	36.3	23.7	23.7	34.2	43.1	6.8	32.8	23.9	22.1
7	7.6	15.3	30.0	41.8	31.2	31.9	31.9	43.0	50.2	10.0	40.4	24.2	19.6
8	8.1	16.1	31.6	44.0	33.0	32.9	32.9	44.2	50.8	9.9	40.7	26.2	20.8
9	8.5	16.9	33.3	46.0	34.2	32.4	32.4	45.3	51.0	9.9	39.6	26.5	22.1

Table 11.27 Realistic case study: maximum usage of installed lines in Case *100%Sto* (%)

Scenario	Installed lines												
	2	5	6	9	10	13	14	15	16	18	20	23	24
1	17.3	24.3	49.4	64.9	49.5	43.3	43.3	42.8	53.7	12.7	65.1	45.6	43.3
2	17.8	25.1	51.2	67.1	51.4	43.5	43.5	45.1	53.6	14.7	65.6	47.2	44.8
3	18.4	25.8	53.1	69.4	53.3	45.0	45.0	47.4	55.4	14.6	65.4	48.8	46.2
4	17.2	24.3	49.4	64.9	49.5	87.9	87.9	74.9	80.1	17.4	76.9	45.6	43.2
5	17.8	25.1	51.2	67.1	51.4	87.2	87.2	74.9	81.6	17.4	76.9	47.2	44.7
6	18.4	25.8	53.1	69.4	53.3	87.1	87.1	74.9	85.3	17.4	76.9	48.8	46.2
7	19.6	24.3	49.4	64.9	49.5	99.9	99.9	74.9	99.9	17.4	76.9	57.6	43.2
8	19.6	25.1	51.2	67.1	51.4	99.9	99.9	74.9	99.9	17.4	76.9	62.7	44.7
9	18.4	25.8	53.1	69.4	53.3	99.9	99.9	74.9	99.9	17.4	76.9	61.7	46.2

Table 11.28 Realistic case study: storage capacity installed

Case	Storage pow. (MW)	Storage ene. (MWh)
Case *100%Sto*	702.0	4212.0
Case *50%Sto*	600.0	3600.0
Case *0%Sto*	0.0	0.0

Table 11.29 Realistic case study: energy generation (GWh)

Case	OCGT	Wind	Solar PV	Storage
Case *100%Sto*	2350.1	1156.6	1801.4	603.8
Case *50%Sto*	2392.6	1135.8	1770.5	564.2
Case *0%Sto*	2874.4	1045.9	1289.7	0.0

a consequence of the low renewable capacity, most of the demand is provided by OCGT units. Additionally, storage units are rarely used in this scenario. Figure 11.7b represents scenario 5, in which average values of renewable capacity and number of EVs are considered. In this scenario it is observed that storage units are used more frequently. Unlike scenario 1, the power production of wind and solar PV units is higher than the demand in many periods. The exceeding renewable energy is used to charge storage units, which are discharged mainly at night hours. Finally, the generation in scenario 9 is depicted in Fig. 11.7c. The renewable capacity and the number of EVs are high in this scenario. Observe that the demand in this scenario is higher than in the other two analyzed scenarios. Additionally, the energy produced by OCGTs is significantly lower in this scenario.

11.7 Summary and Conclusions

This chapter has presented the formulation of a two-stage stochastic programming problem to model the transmission expansion planning problem of a power system considering the presence of renewable energies and a large number of EVs. The proposed model considers that the transmission system planner may install new lines and storage units to reinforce the transmission system. The installed capacity of generation technologies, the demand, and the number of EVs in future power systems are uncertain parameters for the transmission planner that has to decide at the present moment investment decisions. The mathematical formulation of this problem uses a binary variable to model the installation or not of candidate transmission lines. In order to make appropriate transmission investment decisions, the operation of the power system is modeled. The final mathematical problem is a large-scale mixed-integer linear programming problem that can be solved using commercial software.

The numerical results obtained from a realistic case study indicate that investment decisions in transmission lines are affected moderately by the presence of storage units. It has been also observed that the usage of transmission lines is highly dependent on the capacity installed of renewable energies and, at a lesser extent, on the number of EVs.

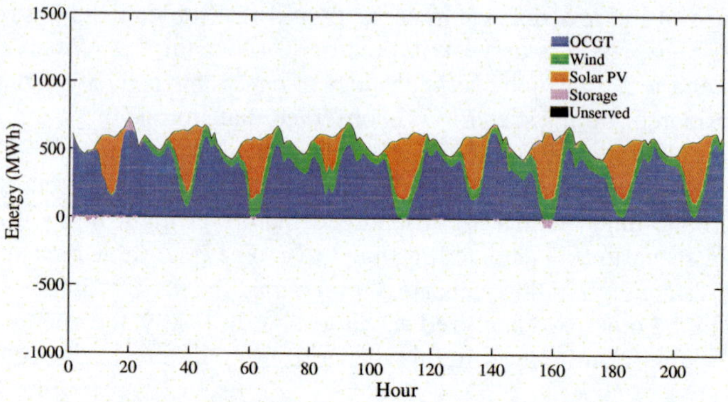

(a) Energy generation in scenario 1 (low renewable capacity, low number of EVs)

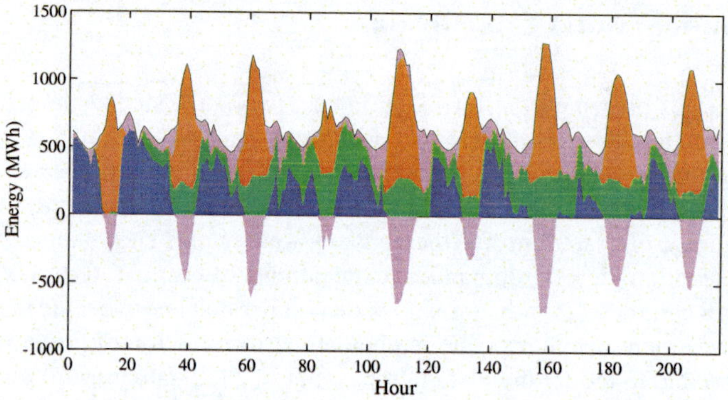

(b) Energy generation in scenario 5 (medium renewable capacity, medium number of EVs)

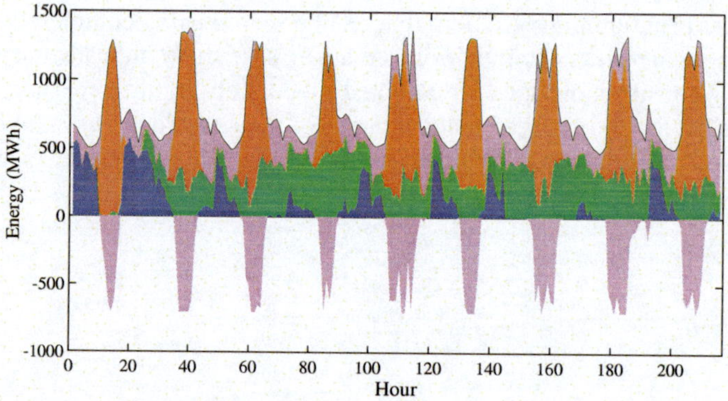

(c) Energy generation in scenario 9 (high renewable capacity, high number of EVs)

Fig. 11.7 Realistic case study: energy generation in selected scenarios

11.8 Notation

This section defines the main notation used in this chapter.

Acronyms

CRF Cost recovery factor.
DC Direct current.
EV Electric vehicle.

Sets and Indices

b Index of buses.
B Set of buses.
d Index of characteristic days.
D Set of characteristic days.
g Index of generating units.
G Set of generating units.
G_b Set of generating units located in bus b.
G^{D} Set of dispatchable generating units.
G^{I} Set of intermittent generating units.
k Index of electric vehicle groups.
K Set of electric vehicle groups.
K_t Set of electric vehicle groups that are connected to the grid in period t.
ℓ Index of transmission lines.
L Set of transmission lines.
L^C Set of candidate transmission lines.
L^E Set of existing transmission lines.
s Set of storage units.
S Set of storage units.
S^{C} Set of candidate storage units.
S^{E} Set of existing storage units.
S_b Set of storage units located at bus b.
t Index of time periods.
T Set of time periods.
T_k Set of time periods in which EV group k can be charged from or discharged to the grid.
ω Index of scenarios.
Ω Set of scenarios.

Parameters

A^{D}_{gdt} Availability of intermittent unit g in characteristic day d and period t (pu).
$C^{\mathrm{G,D}}_{gdt}$ Operation cost of generating unit g in characteristic day d and period t (€/MWh).
$C^{\mathrm{I,L}}_{\ell}$ Annualized capital cost of transmission line ℓ (€/MW).
$C^{\mathrm{I,SE}}_{s}$ Annualized capital cost of the energy component of storage unit s (€/MWh).
$C^{\mathrm{I,SP}}_{s}$ Annualized capital cost of the power component of storage unit s (€/MW).
$C^{\mathrm{S,C}}_{sdt}$ Consumption cost of storage unit s in characteristic day d and period t (€/MWh).
$C^{\mathrm{S,D}}_{sdt\omega}$ Offering cost of storage unit s in the day-ahead energy market in characteristic day d, period t, and scenario ω (€/MWh).
C^{UD} Cost of unserved demand (€/MWh).
$E^{\mathrm{I,SP}}_{s}$ Energy capacity of existing storage unit s (MWh).
$E^{\mathrm{I,SP}}_{\max,s}$ Maximum energy capacity that can be installed from candidate storage unit s (MWh).
M Large enough constant.
$N^{\mathrm{EV}}_{kn\omega}$ Number of EVs pertaining to group k and bus b in scenario ω.
$P^{\mathrm{D}}_{bdt\omega}$ Demand in bus b, characteristic day d, period t, and scenario ω (MWh).
$P^{\mathrm{EV}}_{kbdt\omega}$ Demand of EV group k at bus b and characteristic day d, period t, and scenario ω (MWh).
$P^{\mathrm{I,SP}}_{s}$ Power capacity of existing storage unit s (MW).
$P^{\mathrm{I,SP}}_{\max,s}$ Maximum power capacity that can be installed from candidate storage unit s (MW).
$P^{\mathrm{G}}_{\mathrm{up},g}$ Upper ramp factor of generating unit g (pu).
$P^{\mathrm{G}}_{\mathrm{dw},g}$ Down ramp factor of generating unit g (pu).
$P^{\mathrm{I,G}}_{g\omega}$ Capacity built of generating unit g in scenario ω (MW).
$P^{\mathrm{L}}_{\max,\ell}$ Capacity of transmission line ℓ (MW).
W_d Weight of characteristic day d (h).
X_ℓ Reactance of line ℓ (Ω).
$\gamma^{\mathrm{S,O}}_{s}$ Factor used to model the initial status of the storage unit s (pu).
$\gamma^{\mathrm{S,F}}_{s}$ Factor used to model the final status of the storage unit s (pu).
$\gamma^{\mathrm{S,EP}}_{s}$ Relationship between energy and power capacities in storage unit s (h).
$\gamma^{\mathrm{S,min}}_{s}$ Factor used to model the minimum energy that must contain storage unit s (pu).
η^{S} Efficiency of charging/discharging storage units (pu).
π_ω Probability of scenario ω (pu).

Variables

$p^{\mathrm{G,D}}_{gdt\omega}$ Power output of generating unit g in characteristic day d, period t, and scenario ω (MW).
$p^{\mathrm{G,DS}}_{gdt\omega}$ Power spillage of intermittent generating unit g in characteristic day d, period t, and scenario ω (MW).
$e^{\mathrm{I,SE}}_{s}$ Energy capacity built of storage unit s (MWh).
$p^{\mathrm{I,SP}}_{s}$ Peak power of storage unit s (MW).
$p^{\mathrm{L,D}}_{\ell dt\omega}$ Power flow through line ℓ in characteristic day d, period t, and scenario ω (MW).
$p^{\mathrm{S,C}}_{sdt\omega}$ Consumption power of storage unit s in characteristic day d, period t, and scenario ω (MW).
$p^{\mathrm{S,D}}_{sdt\omega}$ Discharged power of storage unit s in characteristic day d, period t, and scenario ω (MW).
y^{L}_{ℓ} Binary variable that is equal to 1 if candidate line ℓ is built, being 0 otherwise.
$\theta^{\mathrm{L,D}}_{bdt\omega}$ Voltage angle at bus b in characteristic day d, period t, and scenario ω (rad).

11.9 GAMS Codes

This section provides the GAMS [11] code used to solve the Illustrative Case Study included in Sect. 11.5.

```
$title TRANSMISSION CAPACITY EXPANSION WITH ELECTRIC VEHICLES


*****************************************************************
*          DATA
*****************************************************************


SETS
w        'Scenarios'                                  /w1*w6/,
d        'Representative days'                        /d1/,
t        'Time periods'                               /t1*t3/,
k        'Electric vehicle groups'                    /k1*k2/,
g        'Candidate generating units'                 /g1*g3/,
s        'Candidate storages'                         /s1*s3/,
b        'Buses'                                      /b1*b3/,
l        'Lines'                                      /l1*l6/,
EL(l)    'Existing lines'
CL(l)    'Candidate lines'
GD(g)    'Dispatchable units'
GI(g)    'Intermittent units'
TK(t,k,d)  'Periods in which vehicles can be charged'
GN(g,b)  'Location of generating units'
SN(s,b)  'Location of storages'
OL(b,l) 'Origin of line l'
FL(b,l) 'Destination of line l';
```

```
EL(l)$(ord(l) le 3)=yes;
CL(l)$(ord(l) gt 3)=yes;

GD('g1')=yes; GD('g2')=yes;
GI('g3')=yes;

TK('t2','k1',d)=yes; TK('t3','k1',d)=yes;
TK('t1','k2',d)=yes; TK('t2','k2',d)=yes;

GN('g1','b1')=yes; GN('g2','b2')=yes; GN('g3','b3')=yes;
SN('s1','b1')=yes; SN('s2','b2')=yes; SN('s3','b3')=yes;

OL('b1','l1')=yes; OL('b2','l2')=yes; OL('b3','l3')=yes;
OL('b1','l4')=yes; OL('b2','l5')=yes; OL('b3','l6')=yes;
FL('b2','l1')=yes; FL('b3','l2')=yes; FL('b1','l3')=yes;
FL('b2','l4')=yes; FL('b3','l5')=yes; FL('b1','l6')=yes;


SCALARS
etaEV          'Cha and disch efficiency (electric vehicles)' /0.98/,
EEVBmax        'Battery capacity of electric vehicles' /0.04/,
gammaEVmin     'Minimum energy stored in electric vehicles' /0.05/,
thetaDmax      'Maximum voltage angle deviation' /1.5707963/
CUD            'Unserved demand cost' /1000/
M              'Large enough constant' /1000/;



PARAMETERS
WD(d)               'Period weight'
gammaSmin(s)        'Per-unit energy that must remain in storages'
gammaS0(s,d)        'Per-unit energy in the first period (storage)'
etaS(S)             'Charge and discharge efficiency (storages)'
PEV(k,b,d,t,w)      'Demand electric vehicles'
pi(w)               'Scenario probability'

PIG_max(g)          'Potential of candidate generating unit g'
/g1       400
g2        400
g3        600/

EISE_max(s)         'Energy potential of candidate storing unit s'
/s1       100
s2        100
s3        100/

PISP_max(s)         'Power potential of candidate storing unit s'
/s1       100
s2        100
s3        100/

PG_up(g)            'Per-unit rump up limit of candidate unit g'
/g1      0.25
g2       0.25/

PG_dw(g)            'Per-unit rump down limit of candidate unit g'
/g1      0.25
g2       0.25/
```

```
PLmax                  'Lines capacity'
/l1        70
l2         70
l3         70
l4         70
l5         70
l6         70/

X(l)                   'Lines reactance'
/l1        0.01
l2         0.01
l3         0.01
l4         0.01
l5         0.01
l6         0.01/

CIL(l)                 'Transmission line cost'
/l4        200000
l5         200000
l6         200000/

gammaSEP(s)            'Relationship between energy and power'
/s1        6
s2         6
s3         6/;

TABLE AD(g,d,t) 'Availability of intermittent power units'
           t1        t2        t3
g3.d1      0.4       0.9       0.6;

TABLE NEV(k,b,w) 'Number of electric vehicles'
           w1        w2        w3        w4        w5        w6
k1.b1      550       550       750       450       950       950
k1.b2      800       800       1000      1000      1200      1200
k1.b3      400       400       600       600       800       800
k2.b1      300       300       500       500       700       750
k2.b2      400       400       600       600       800       850
k2.b3      150       150       350       350       550       550;

TABLE PIG(g,w) 'Capacity installed from each generation technology
     '
           w1        w2        w3        w4        w5        w6
g1         300       200       300       200       300       200
g2         90        90        90        90        90        90
g3         100       300       100       300       100       300;

TABLE PD(b,d,t,w) 'Demand'
           w1        w2        w3        w4        w5        w6
b1.d1.t1   80        80        80        80        80        80
b1.d1.t2   100       100       100       100       100       100
b1.d1.t3   90        90        90        90        90        90
b2.d1.t1   220       220       220       220       220       220
b2.d1.t2   250       250       250       250       250       250
b2.d1.t3   230       230       230       230       230       230
b3.d1.t1   35        35        35        35        35        35
b3.d1.t2   50        50        50        50        50        50
b3.d1.t3   40        40        40        40        40        40;
```

```
TABLE PEV_pu(k,b,d,t,w) 'Unitary demand of electric vehicles (kW)'
                  w1       w2       w3       w4       w5       w6
k1.b1.d1.t2     15.000   15.000   15.000   15.000   15.000   15.000
k1.b1.d1.t3      5.408    5.408    5.408    5.408    5.408    5.408
k1.b2.d1.t2     15.000   15.000   15.000   15.000   15.000   15.000
k1.b2.d1.t3      5.408    5.408    5.408    5.408    5.408    5.408
k1.b3.d1.t2     15.000   15.000   15.000   15.000   15.000   15.000
k1.b3.d1.t3      5.408    5.408    5.408    5.408    5.408    5.408
k2.b1.d1.t1     15.000   15.000   15.000   15.000   15.000   15.000
k2.b1.d1.t2      5.408    5.408    5.408    5.408    5.408    5.408
k2.b2.d1.t1     15.000   15.000   15.000   15.000   15.000   15.000
k2.b2.d1.t2      5.408    5.408    5.408    5.408    5.408    5.408
k2.b3.d1.t1     15.000   15.000   15.000   15.000   15.000   15.000
k2.b3.d1.t2      5.408    5.408    5.408    5.408    5.408    5.408;


TABLE CISE(s,w) 'Annualized energy investment cost (storages)'
         w1      w2      w3      w4      w5      w6
s1       500     500     500     500     500     500
s2       500     500     500     500     500     500
s3       500     500     500     500     500     500;

TABLE CISP(s,w) 'Annualized power investment cost (storages)'
         w1      w2      w3      w4      w5      w6
s1       3000    3000    3000    3000    3000    3000
s2       3000    3000    3000    3000    3000    3000
s3       3000    3000    3000    3000    3000    3000;

TABLE CGD(g,w)  'Operation cost (generating units)'
         w1      w2      w3      w4      w5      w6
g1       0.250   0.250   0.250   0.250   0.250   0.250
g2       0.400   0.400   0.400   0.400   0.400   0.400;

TABLE CSD(s,w)  'Discharge cost (storages)'
         w1      w2      w3      w4      w5      w6
s1       0       0       0       0       0       0
s2       0       0       0       0       0       0
s3       0       0       0       0       0       0;

TABLE CSC(s,w)  'Charge cost (storages)'
         w1      w2      w3      w4      w5      w6
s1       0       0       0       0       0       0
s2       0       0       0       0       0       0
s3       0       0       0       0       0       0;

pi(w)=1/card(w);
WD(d)=365*8;
gammaSmin(s)=0.1;
gammaS0(s,d)=0.7;
etaS(S)=0.98;
PEV(k,b,d,t,w)=PEV_pu(k,b,d,t,w)*NEV(k,b,w)/1000;

*******************************************************************
*   DECLARATION OF VARIABLES AND MATHEMATICAL CHARACTERIZATION
*******************************************************************

VARIABLES

**        OBJECTIVE FUNCTION
```

```
z                    'Objective function'

**        FIRST-STAGE VARIABLES
eISE(s)              'Storage energy capacity installed'
pISP(s)              'Storage power capacity installed'

**        SECOND-STAGE VARIABLES
pGDS(g,d,t,w)        'Intermittent power spillage'
pSC(s,d,t,w)         'Power charged from storages'
pSD(s,d,t,w)         'Power discharged from storages'
eS(s,d,t,w)          'Energy stored in storages'
pLD(l,d,t,w)         'Power flows'
thetaD(b,d,t,w)      'Voltage angles'
pUD(b,d,t,w)         'Unserved demand';

POSITIVE VARIABLES eISE(s),pISP(s),pGD(g,d,t,w),pGDS(g,d,t,w),pSC(
    s,d,t,w),pSD(s,d,t,w),eS(s,d,t,w),pUD(b,d,t,w);

BINARY VARIABLES  yL(l) Binary variable that is equal to 1 if line
     l is built;


*****************************************************************
*    EQUATIONS
*****************************************************************

EQUATIONS

**        OBJECTIVE FUNCTION
ObjectiveFunction

**        INVESTMENT
InvestmentSto_01
InvestmentSto_02
InvestmentSto_03

**        GENERATING UNITS
GenUnit_01
GenUnit_02
GenUnit_03
GenUnit_04

**        STORAGES
StoUnit_01
StoUnit_02
StoUnit_03
StoUnit_04
StoUnit_05
StoUnit_06
StoUnit_07

**        TRANSMISSION
Transmision_01
Transmision_02
Transmision_03
Transmision_04
Transmision_05
Transmision_06
```

```
Transmision_07
Transmision_08
Transmision_09

**        ENERGY BALANCE
EnergyBalance_01;

ObjectiveFunction..
     z =e= 0.001*sum(w,pi(w)*(
      sum(CL(l),CIL(l)*yL(l))+sum(s,CISE(s,w)*eISE(s)+CISP(s,w)*pISP(s)) +
      sum(d,WD(d)*sum(t,
           sum(g$GD(g),CGD(g,w)*pGD(g,d,t,w))+
           sum(s,CSD(s,w)*pSD(s,d,t,w)-CSC(s,w)*pSC(s,d,t,w))
            +sum(b,CUD*pUD(b,d,t,w))
             ))
     ));

**        INVESTMENT
InvestmentSto_01(s)..                    eISE(s)=l=EISE_max(s);
InvestmentSto_02(s)..                    pISP(s)=l=PISP_max(s);
InvestmentSto_03(s)..                    eISE(s)=e=gammaSEP(s)*pISP(s);

**        GENERATING UNITS
GenUnit_01(g,d,t,w)$GD(g)..              pGD(g,d,t,w)=l= PIG(g,w);
GenUnit_02(g,d,t,w)$GI(g)..              pGD(g,d,t,w)+pGDS(g,d,t,w) =e= AD(g,d,t)*PIG(g,w);
GenUnit_03(g,d,t,w)$(GD(g) and (ord(t) gt 1))..        pGD(g,d,t,w)-pGD(g,d,t-1,w) =l= PG_up(g)*PIG(g,w);
GenUnit_04(g,d,t,w)$(GD(g) and (ord(t) gt 1))..        pGD(g,d,t-1,w)-pGD(g,d,t,w) =l= PG_dw(g)*PIG(g,w);

**        STORAGES
StoUnit_01(s,d,t,w)..                                  pSD(s,d,t,w) =l= pISP(s);
StoUnit_02(s,d,t,w)..                                  pSC(s,d,t,w) =l= pISP(s);
StoUnit_03(s,d,t,w)$(ord(t) gt 1).. eS(s,d,t,w)  =e= eS(s,d,t-1,w)+etaS(s)*pSC(s,d,t,w)-1/etaS(s)*pSD(s,d,t,w);
StoUnit_04(s,d,t,w)$(ord(t) eq 1).. eS(s,d,t,w)  =e= gammaS0(s,d)*eISE(s)+etaS(s)*pSC(s,d,t,w)-1/etaS(s)*pSD(s,d,t,w);
StoUnit_05(s,d,t,w)..                                  eS(s,d,t,w)=g=gammaSmin(s)*eISE(s);
StoUnit_06(s,d,t,w)..                                  eS(s,d,t,w)=l=eISE(s);
StoUnit_07(s,d,t,w)$(ord(t) eq card(t))..              eS(s,d,t,w)=g=gammaS0(s,d)*eISE(s);

**        TRANSMISSION
Transmision_01(l,d,t,w)$EL(l)..    pLD(l,d,t,w)=e=1/X(L)*(sum(b$OL(b,l), thetaD(b,d,t,w))-sum(b$FL(b,l), thetaD(b,d,t,w)));
Transmision_02(l,d,t,w)$CL(l)..    pLD(l,d,t,w)=g=1/X(L)*(sum(b$OL(b,l), thetaD(b,d,t,w))-sum(b$FL(b,l), thetaD(b,d,t,w)))-(1-yL(l))*M;
Transmision_03(l,d,t,w)$CL(l)..    pLD(l,d,t,w)=l=1/X(L)*(sum(b$OL(b,l), thetaD(b,d,t,w))-sum(b$FL(b,l), thetaD(b,d,t,w)))+(1-yL(l))*M;
Transmision_04(l,d,t,w)$EL(l)..    pLD(l,d,t,w)=l=pLmax(l);
Transmision_05(l,d,t,w)$EL(l)..    pLD(l,d,t,w)=g=-pLmax(l);
```

```
Transmision_06(l,d,t,w)$CL(l)..   pLD(l,d,t,w)=l=pLmax(l)*yL(l);
Transmision_07(l,d,t,w)$CL(l)..   pLD(l,d,t,w)=g=-pLmax(l)*yL(l);
Transmision_08(b,d,t,w)..   thetaD(b,d,t,w) =l= thetaDmax;
Transmision_09(b,d,t,w)..   thetaD(b,d,t,w) =g= -thetaDmax;

**        ENERGY BALANCE
EnergyBalance_01(b,d,t,w)..   sum(g$GN(g,b),pGD(g,d,t,w))+sum(s$SN
    (s,b),pSD(s,d,t,w)-pSC(s,d,t,w))-sum(l$OL(b,l),pLD(l,d,t,w))+
    sum(l$FL(b,l),pLD(l,d,t,w))=e=PD(b,d,t,w)+sum(k,PEV(k,b,d,t,w)
    )-pUD(b,d,t,w);


*****************************************************************
*   MODEL AND SOLVING
*****************************************************************

MODEL TransExpansionEV /all/;

TransExpansionEV.optcr=0;

SOLVE TransExpansionEV USING mip MINIMIZING z;
```

References

1. Akbari T., Bina M. T.: A linearized formulation of AC multi-year transmission expansion planning: a mixed-integer linear programming approach. Electr. Power Syst. Res. **114**, 93–100 (2014)
2. Alvarez, J., Ponnambalam, K., Quintana, V.H.: Transmission expansion under risk using stochastic programming. In: Proceedings of the 9th International Conference on Probabilistic Methods Applied to Power Systems, PMAPS, Beijing, China, 16–20 October 2016, vol. 22, pp. 1369–1378 (2006)
3. Baringo, L., Conejo, A.J.: Transmission and wind power investment. IEEE Trans. Power Syst. **27**, 885–893 (2012)
4. Baringo, L., Baringo, A.: A stochastic adaptive robust optimization approach for the generation and transmission expansion planning. IEEE Trans. Power Syst. **33**, 792–802 (2018)
5. Bustos, C., Sauma, E., de la Torre, S., Aguado, J.A., Contreras, J., Pozo, D.: Energy storage and transmission expansion planning: substitutes or complements? IET Gener. Transm. Distrib. **12**(8), 1738–1746 (2018)
6. Cañas-Carretón, M., Carrión, M., Iov, F.: Towards renewable-dominated power systems considering long-term uncertainties: case study of las Palmas. Energies **14**(11), 3317 (2021)
7. Chen, X., Lv, J., McElroy, M.B., Han, X., Nielsen, C.P., Wen. J.: Power system capacity expansion under higher penetration of renewables considering flexibility constraints and low carbon policies. IEEE Trans. Power Syst. **33**, 6240–6253 (2018)
8. The ILOG CPLEX (2021). www.ilog.com/products/cplex/
9. Conejo, A.J., Baringo, L., Kazempour, S.J., Siddiqui, A.S.: Investment in Electricity Generation and Transmission, 1st edn. Springer, Berlin (2016). ISBN 978-3-319-29501-5
10. Domínguez, R., Conejo, A.J., Carrión, M.: Toward fully renewable electric energy systems. IEEE Trans. Power Syst. **30**, 316–326 (2015)
11. GAMS (2021). www.gams.com
12. Gröwe-Kuska, N., Heitsch, H., Römisch, W.: Scenario reduction and scenario tree construction for power management problems. In: Proceedings IEEE Bologna Power Technology Conference, Bologna, Italy (2003)

13. Liu, Y., Sioshansi, R., Conejo, A.J.: Multistage stochastic investment planning with multiscale representation of uncertainties and decisions. IEEE Trans. Power Syst. **33**, 781–791 (2018)
14. Majd, A., Farjah, E., Rastegar, M.: Composite generation and transmission expansion planning toward high renewable energy penetration in Iran power grid. IET Renew. Power Gener. **14**, 1520–1528 (2020)
15. Moeini-Aghtaie, M., Abbaspour, A., Fotuhi-Firuzabad, M.: Incorporating large-scale distant wind farms in probabilistic transmission expansion planning-part I: theory and algorithm. IEEE Trans. Power Syst. **27**(3), 1585–1593 (2012)
16. Moghaddam, S.Z.: Generation and transmission expansion planning with high penetration of wind farms considering spatial distribution of wind speed. Int. J. Electr. Power Energy Syst. **106**, 232–241 (2019)
17. Pandžic, H., Wang, Y., Qiu, T., Dvorkin, Y., Kirschen, D.S.: Near- optimal method for siting and sizing of distributed storage in a transmission network. IEEE Trans. Power Syst. **30**(5), 2288–2300 (2015)
18. Pandžic, K., Pandžic, H., Kuzle, I.: Coordination of regulated and merchant energy storage investments. IEEE Trans. Sustain. Energy. **9**(3), 1244–1254 (2018)
19. Pozo, D., Sauma, E., Contreras, J.: A three-level static MILP model for generation and transmission expansion planning. IEEE Trans. Power Syst. **28**, 202–210 (2013)
20. Qiu, T., Xu, B., Wang, Y., Dvorkin, Y., Kirschen, D.S.: Stochastic multistage coplanning of transmission expansion and energy storage. IEEE Trans. Power Syst. **32**(1), 643–651 (2017)
21. Rahmani, M., Vinasco, G., Rider, M.J., Romero, R., Pardalos, P.M.: Multistage transmission expansion planning considering fixed series compensation allocation. IEEE Trans. Power Syst. **28**(4), 3795–3805 (2013)
22. Roldán, C., Sánchez de la Nieta, A.A., García-Bertrand, R., Mínguez, R.: Robust dynamic transmission and renewable generation expansion planning: walking towards sustainable systems. Int. J. Electr. Power Energy Syst. **96**, 52–63 (2018)
23. Roh, J.H., Shahidehpour, M., Wu, L.: Market-based generation and transmission planning with uncertainties. IEEE Trans. Power Syst. **24**, 1587–1598 (2009)
24. Sahraei-Ardakani, M., Hedman, K.W.: A fast LP approach for enhanced utilization of variable impedance based facts devices. IEEE Trans. Power Syst. **31**(3), 2204–2213 (2015)
25. Silva, I.J., Rider, M.J., Romero, R., Murari, C.A.F.: Transmission network expansion planning considering uncertainty in demand. IEEE Trans. Power Syst. **21**(4), 1565–1573 (2006)
26. Ugranli, F., Karatepe, E.: Coordinated TCSC allocation and network reinforcements planning with wind power. IEEE Trans. Sustain. Energy. **8**(4), 1694–1705 (2017)
27. Zhang, H., Heydt, G.T., Vittal, V., Quintero, J.: An improved network model for transmission expansion planning considering reactive power and network losses. IEEE Trans. Power Syst. **28**(3), 3471–3479 (2013)
28. Zhang, X., Conejo, A.J.: Robust transmission expansion planning representing long- and short-term uncertainty. IEEE Trans. Power Syst. **33**(2), 1329–1338 (2018)
29. Ziaee, O., Choobineh, F.F.: Optimal location-allocation of tcsc devices on a transmission network. IEEE Trans. Power Syst. **32**(1), 94–102 (2017)

Chapter 12
Distribution Expansion Planning Considering Electric Vehicles

This chapter studies the distribution expansion problem considering the existence of a large number of chargers of electric vehicles connected to a low-voltage distribution network. The presence of electric vehicle chargers may change substantially the energy load at some locations and may require the reinforcement of the distribution system to preserve its adequate operation. We assume the role of a distribution system operator that desires to determine the distribution lines to reinforce to avoid voltage drops and congestion problems considering the additional power consumption of electric vehicle chargers. A number of illustrative examples are included to describe the performance of the formulation of the distribution expansion problem. Additionally, a case study based on a realistic distribution network is presented at the end of the chapter.

12.1 Introduction

The main objective of the distribution network is to deliver electricity to the end users of electricity. For doing that, the distribution network connects electricity consumers with the transmission grid through several steps. The connection between transmission and distribution networks is done in substations. Those substations connecting transmission and distribution networks are used to step the voltage down. For instance, voltages at transmission level (e.g., 200 or 400 kV) can be reduced up to 130, 60, or 45 kV in these substations. Other substations located in the distribution network are used to reduce again voltages up to 20 or 10 kV. Finally, transformer stations located close to the consumption points are used to step voltages down again to reach consumption voltage values, which may be, for instance, 380 V.

L. Baringo et al., *Electric Vehicles and Renewable Generation*,
Green Energy and Technology, https://doi.org/10.1007/978-3-031-09079-0_12

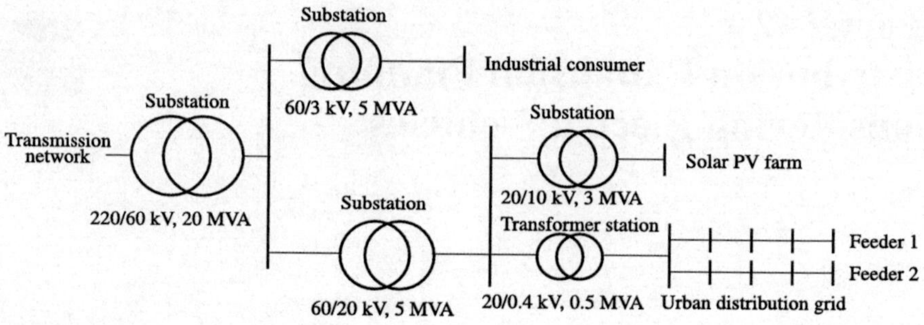

Fig. 12.1 Illustrative Example 12.1: distribution network

Illustrative Example 12.1 Example of distribution network configuration

Let us consider the distribution network represented in Fig. 12.1. This distribution network is connected to the transmission system through a substation that reduces the voltage from 220 to 60 kV. Two distribution lines are connected to this substation. One of these lines connects the former substation with another substation that further reduces voltage from 60 to 3 kV. A large industrial consumer is connected to this substation. Power intensive consumers may be connected to the distribution network at high voltage to reduce power losses in their facilities. The second line is connected to a substation that reduces the voltage from 60 to 20 kV. A solar photovoltaic (PV) farm is connected to this substation through another substation that reduces the voltage from 20 to 10 kV. Note that solar PV and wind power units can be connected to the transmission or to the distribution networks. Finally, a transformer station is connected to the 60/20 kV substation to deliver electricity to an urban grid. This transformer station steps voltage down from 20 to 0.4 kV. Households and commercial buildings are connected to the low voltage grid. Considering that these consumers have a relatively small consumption of energy and power, they usually have a single-phase connection to the distribution grid. Observe that urban consumers in this network are connected to the transformer station through two feeders. Traditionally, distribution grids are operated radially and a single path is used to feed the different consumptions. □

Distribution system operators (DSOs) are in charge of the electricity delivery to end users according to established regulations. It is frequent the presence of multiple DSOs in a power system, but the operation of the distribution network in a specific location (for instance a region or a large city) is responsibility of a single DSO. The tasks of DSOs consist in operating and maintaining the grid. One of the main objectives of DSOs is the long-term planning of the distribution grid. Then, decisions of DSO include the reinforcement of distribution lines, the construction of new corridors to feed new consumptions, installation or upgrading existing substations and transformation centers, the installation of capacitor banks, storage, or voltage control mechanisms.

The main challenges nowadays for DSOs are the massive incorporation of electric vehicle (EV) chargers and renewable distributed resources. In order to reduce charging periods, the power capacity of EV chargers is increasing and there exist commercial domestic EV chargers with rated power higher than 10 kW. This value may double or triple the peak power consumption of a typical home if power limiting devices are not used. This fact can cause voltage drops, phase unbalances and the congestion of distribution lines in specific periods. On the contrary, the installation of renewable distributed resources in the distribution network may cause undesirable voltage rises when the power output of renewable units is high. Over- and under-voltage situations are traditionally addressed through centralized control of tap changers of power transformers feeding the distribution network. However, local-voltage control mechanisms and storage units can be also used to address voltage problems.

Another problem that EVs and renewable distributed resources can cause is the existence of bidirectional flows in distribution lines. Traditionally, the distribution network has been characterized by unidirectional power flows, heading from the substation or transformation center to consumption points. However, the installation of renewable generation units in the distribution network, specially solar PV and wind power units, as well as the vehicle-to-grid capability of EVs may cause that the power flows in the distribution lines be bidirectional depending on the quantity of power injected into the grid by EVs and renewable units. This fact will require to adapt both the operation of the network and the design of the protection system of the different distribution network assets.

In this chapter we focus on the long-term planning problem of a DSO that desires to decide the reinforcement of the distribution network to allow the incorporation of a large number of EV chargers. The objective of the DSO is to determine the distribution lines to reinforce to avoid voltage drops and congestion problems considering the additional power consumption of EV chargers. This problem is formulated considering the AC model of distribution lines. As a result, the final problem is a non-linear mixed-integer linear programming problem that can be solved using commercial software.

Relevant references dealing with the expansion of the distribution network to accommodate renewable generation and EVs are [1–23].

The remainder of this chapter is organized as follows.

The decision framework is described in Sect. 12.2. The formulation of the problem is provided in Sect. 12.3. In Sect.12.4, an illustrative case study is analyzed to show the performance of the analyzed model. In Sect. 12.5, a realistic case study is solved and the results are analyzed. Section 12.6 provides a summary of the chapter and draws the most relevant conclusions. Section 12.7 defines the notation used in this chapter. Finally, Sect. 12.8 provides the GAMS [8] code use to solve the illustrative case study included in Sect.12.4.

12.2 Decision Framework

Let us consider the role of a DSO that desires to determine the optimal investment decisions to make to incorporate a high number of EV chargers in an urban distribution network. As an example, Fig. 12.2 represents an example of urban distribution network with radial configuration. The considered investment decisions comprise the installation of new distribution lines in existing or new corridors. Improvements of other distribution assets are not considered in this chapter. Therefore, new investments in transformers, capacitors, and storage units are not modeled. The objective of the DSO is to avoid unacceptable voltage drops and line congestion if a significant number of EV chargers are connected to the distribution network. We assume that EV chargers are installed in residential buildings. Considering this, it is assumed that the DSO has available a forecast of the location of EV chargers throughout the distribution network. Since the proposed approach is focused on high-power rate chargers, only three-phase chargers are considered.

In long-term planning problems, DSOs usually take a risk-averse position and try to reinforce the distribution network considering the worst case. In this problem, worst case happens when end users of electricity consume simultaneously their peak power consumption. This worst case can be relaxed by applying a simultaneity coefficient, assuming that it is unlikely that all end users of electricity consume its peak power demand simultaneously. Since the objective of the DSO is to adequate the distribution network against the worst scenario, the DSO does not need to solve a decision-making problem under uncertainty. Instead, the uncertain parameters as the peak demand of the consumers or the number of EV chargers are represented using their worst expected values.

Finally, an optimal power flow problem considering the AC formulation of the distribution lines has been formulated in order to analyze the steady state operation of the distribution network considering the presence of new charging points. As it is usual in long-term planning models, and considering that the starting point is an existing urban distribution network that operates properly under normal conditions, a single-phase equivalent analysis has been adopted. Additionally, observe that three-phase chargers do not cause power unbalances among phases.

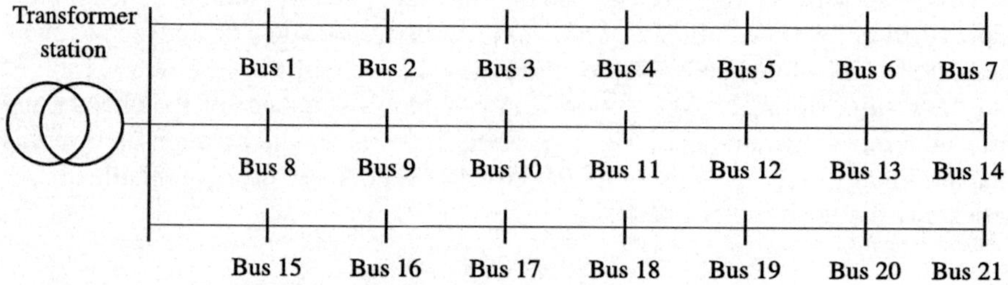

Fig. 12.2 Example of urban distribution network with radial configuration

12.3 Formulation of the Distribution Expansion Problem

This section presents the mathematical formulation of the distribution expansion problem described above.

12.3.1 Distribution Expansion Decisions

The investment decisions that the DSO desires to determine are investments in distribution lines. Distribution lines may consist in overhead lines or underground cables. Underground cables are more expensive but they are more frequent in urban zones. The investments in distribution lines are denoted by binary variable $y^{\mathrm{L}}_{\ell nm}$, which is equal to 1 if candidate line ℓ linking buses n and m is built, being 0 otherwise. The binary nature of variables $y^{\mathrm{L}}_{\ell nm}$ is stated in constraint (12.1a). Considering that line ℓ linking buses n and m also links these same buses in the opposite sense, it is necessary to impose constraint (12.1b).

$$y^{\mathrm{L}}_{\ell nm} \in \{0, 1\}, \quad \forall \{\ell, n, m\} \in L^{\mathrm{C}}, \tag{12.1a}$$

$$y^{\mathrm{L}}_{\ell nm} = y^{\mathrm{L}}_{\ell mn}, \quad \forall \{\ell, n, m\} \in L^{\mathrm{C}}, \tag{12.1b}$$

where L^{C} is the set of candidate distribution lines that can be built. Observe that if $\{\ell, n, m\} \in L^{\mathrm{C}}$, then it is also satisfied that $\{\ell, m, n\} \in L^{\mathrm{C}}$.

12.3.2 Voltage Magnitude Limits

Voltage magnitudes v_n are bounded by constraints (12.2a) and (12.2b). Parameters $V_n^{\max}$ and $V_n^{\min}$ indicate the established upper and lower voltage magnitude limits. As an example, these values can be equal to 1.1 and 0.9 times the nominal voltage values. In order to avoid infeasible solutions when voltages cannot be physically comprised between these two limits, the auxiliary variables $v_n^{\mathrm{aux},+}$ and $v_n^{\mathrm{aux},-}$ are considered. These variables model the excess and defect of nodal voltage values with respect to the fixed limits, respectively. It is important to note that the value of these auxiliary variables is penalized in the objective function (12.8) with the purpose of ensuring that they be different from zero only in those cases in which is not possible to obtain any feasible solution where voltage magnitudes be comprised between $V_n^{\max}$ and $V_n^{\min}$.

$$V_n^{\min} - v_n^{\mathrm{aux},-} \leq v_n \leq V_n^{\max} + v_n^{\mathrm{aux},+}, \quad \forall n \in N, \tag{12.2a}$$

$$v_n^{\mathrm{aux},+}, v_n^{\mathrm{aux},-} \geq 0, \quad \forall n \in N. \tag{12.2b}$$

Illustrative Example 12.2 Formulation of the voltage magnitude limits

Let us consider a single bus located in a distribution network whose voltage magnitude is denoted as v. The established upper and lower limits of single-phase voltage magnitudes are 240 and 220 V, respectively. Let us consider three different cases depending on the value of voltage magnitude v:

- Case 1: $v = 230$ V. In this case, voltage magnitude v is within the established limits, 220 and 240 V. Then, constraint (12.2a) is formulated as follows:

$$220 - v_n^{\text{aux},-} \leq 230 \leq 240 + v_n^{\text{aux},+}.$$

Since auxiliary variables $v_n^{\text{aux},+}$ and $v_n^{\text{aux},-}$ are penalized in the objective function, they will take the smallest possible values, which are equal to 0, satisfying the following expression:

$$220 \leq 230 \leq 240.$$

- Case 2: $v = 210$ V. In this case, voltage magnitude v is less than the lower limit, 220 V. Then, constraint (12.2a) is formulated as follows:

$$220 - v_n^{\text{aux},-} \leq 210 \leq 240 + v_n^{\text{aux},+}.$$

In order to satisfy the preceding constraints, auxiliary variable $v_n^{\text{aux},-}$ will take value equal to 10 V. Since v is less than 240, $v_n^{\text{aux},+}$ will be equal to 0. Then, the resulting expression is as follows:

$$220 - 10 \leq 210 \leq 240.$$

- Case 3: $v = 245$ V. In this case, voltage magnitude v is greater than the upper limit, 240 V. Considering this, constraint (12.2a) is formulated as follows:

$$220 - v_n^{\text{aux},-} \leq 245 \leq 240 + v_n^{\text{aux},+}.$$

In order to satisfy this constraint, auxiliary variable $v_n^{\text{aux},+}$ will be equal to 5 V. In this case, since v is greater than 220, $v_n^{\text{aux},-}$ will be equal to 0. Then, the resulting expression will be recast as:

$$220 \leq 245 \leq 240 + 5.$$

□

12.3.3 Transformer Station Capacity Constraints

Constraints (12.3) limit the power output of each distribution transformer s. Despite the fact that distribution networks are usually operated as radial systems, for the sake of generality, we have considered in this formulation that several distribution substations may feed the analyzed urban distribution system. Variables p_s^{S} and q_s^{S} represent the active and reactive power injected by transformer station s, respectively, while $S_s^{\mathrm{S,max}}$ is the rated apparent power capacity of the transformer station.

$$(p_s^{\mathrm{S}})^2 + (q_s^{\mathrm{S}})^2 \leq (S_s^{\mathrm{S,max}})^2, \quad \forall s \in S. \tag{12.3}$$

12.3.4 Active and Reactive Power Flow Constraints

The active and reactive power flows through existing distribution lines are expressed by constraints (12.4a) and (12.4b), respectively. The capacity of existing distribution lines is bounded by constraints (12.4c). As usual, distribution lines are represented by using the series impedance model.

$$p_{\ell nm}^{\mathrm{L}} = v_n^2 G_{\ell nm} - v_n v_m (G_{\ell nm} \cos(\theta_n - \theta_m) + B_{\ell nm} \sin(\theta_n - \theta_m)),$$
$$\forall \{\ell, n, m\} \in L^{\mathrm{E}}, \tag{12.4a}$$

$$q_{\ell nm}^{\mathrm{L}} = -v_n^2 (B_{\ell nm} - B_\ell^{\mathrm{S}}) - v_n v_m (G_{\ell nm} \sin(\theta_n - \theta_m) + B_{\ell nm} \cos(\theta_n - \theta_m)),$$
$$\forall \{\ell, n, m\} \in L^{\mathrm{E}}, \tag{12.4b}$$

$$(p_{\ell nm}^{\mathrm{L}})^2 + (q_{\ell nm}^{\mathrm{L}})^2 \leq \left(S_{\ell nm}^{\mathrm{L,max}}\right)^2, \quad \forall \{\ell, n, m\} \in L. \tag{12.4c}$$

Illustrative Example 12.3 Example of power flow constraints

Let us consider the π-model of a distribution line linking buses 1 and 2 that is represented in Fig. 12.3.

Let us assume that the series admittance is equal to $G + jB = 0.002 - j0.004\ \Omega$, the shunt admittance is $\mathrm{j}10^{-5}\ \Omega$, and single-phase voltage phasors of buses 1 and 2 are equal to $240\angle 0°$ and $220\angle -5°$ V, respectively.

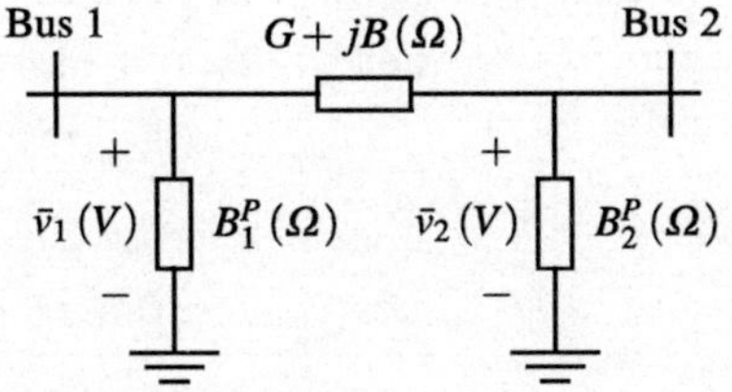

Fig. 12.3 Illustrative Example 12.3: π-model of the distribution line

Considering $-5° = -\frac{\pi}{36}$ rad, the active and reactive power flows from bus 1 to 2 are computed as follows:

$$\begin{aligned}
p_{12}^{\mathrm{L}} &= v_1^2 G - v_1 v_2 (G\cos(\theta_1-\theta_2) + B\sin(\theta_1-\theta_2))\\
&= 240^2 \cdot 0.002 - 240 \cdot 220(0.002\cos(0+\frac{\pi}{36}) - 0.004\sin(0+\frac{\pi}{36}))\\
&= 28.41\,\mathrm{W},\\
q_{12}^{\mathrm{L}} &= -v_1^2(B+B_1^{\mathrm{S}}) - v_1 v_2 (G\sin(\theta_1-\theta_2) - B\cos(\theta_1-\theta_2))\\
&= -240^2(-0.004+0.0001) - 240 \cdot 220(0.002\sin(0+\frac{\pi}{36}) + 0.004\cos(0+\frac{\pi}{36}))\\
&= 10.22\,\mathrm{var}.
\end{aligned}$$

In the same manner, the power flows from bus 2 to bus 1 are the following:

$$\begin{aligned}
p_{21}^{\mathrm{L}} &= v_2^2 G - v_2 v_1 (G\cos(\theta_2-\theta_1) + B\sin(\theta_2-\theta_1))\\
&= 220^2 \cdot 0.002 - 220 \times 240(0.002\cos(0-\frac{\pi}{36}) - 0.004\sin(0-\frac{\pi}{36}))\\
&= -26.81\,\mathrm{W},\\
q_{21}^{\mathrm{L}} &= -v_2^2(B+B_2^{\mathrm{S}}) - v_2 v_1 (G\sin(\theta_2-\theta_1) - B\cos(\theta_2-\theta_1))\\
&= -220^2(-0.004+0.0001)\\
&\quad -220 \cdot 240(0.002\sin(0-\frac{\pi}{36}) + 0.004\cos(0-\frac{\pi}{36}))\\
&= -8.08\,\mathrm{var}.
\end{aligned}$$

Negative values in p_{21}^{L} and q_{21}^{L} indicate that power flows actually go from bus 1 to bus 2. Line losses can be computed as follows:

$$\begin{aligned}
p_{12}^{\mathrm{L}} + p_{21}^{\mathrm{L}} &= 1.60\,\mathrm{W},\\
q_{12}^{\mathrm{L}} + q_{21}^{\mathrm{L}} &= 2.15\,\mathrm{var}.
\end{aligned}$$

□

The active and reactive power flows through candidate distribution lines depend on binary variable $y_{\ell nm}^{\mathrm{L}}$ and are expressed by constraints (12.5):

$$\begin{aligned}
&v_n^2 G_{\ell nm} - v_n v_m (G_{\ell nm}\cos(\theta_n-\theta_m) + B_{\ell nm}\sin(\theta_n-\theta_m))\\
&\quad -(1-y_{\ell nm}^{\mathrm{L}})M \le p_{\ell nm}^{\mathrm{L}} \le v_{ndt}^2 G_{\ell nm} - v_n v_m (G_{\ell nm}\cos(\theta_n-\theta_m)\\
&\quad + B_{nm}\sin(\theta_n-\theta_m)) + (1-y_{\ell nm}^{\mathrm{L}})M, \quad \forall\{\ell,n,m\}\in L^{\mathrm{C}}, \qquad (12.5a)\\
&-v_n^2(B_{\ell nm}+B_\ell^{\mathrm{S}}) - v_n v_m (G_{\ell nm}\sin(\theta_n-\theta_m) - B_{\ell nm}\cos(\theta_n-\theta_m))\\
&\quad -(1-y_{\ell nm}^{\mathrm{L}})M \le q_{\ell nm}^{\mathrm{L}} \le -v_n^2(B_{\ell nm}+B_\ell^{\mathrm{S}}) - v_n v_m (G_{\ell nm}\sin(\theta_n-\theta_m)\\
&\quad - B_{\ell nm}\cos(\theta_n-\theta_m)) + (1-y_{\ell nm}^{\mathrm{L}})M, \quad \forall\{\ell,n,m\}\in L^{\mathrm{C}}. \qquad (12.5b)
\end{aligned}$$

The capacity of candidate distribution lines is bounded by constraints (12.6).

$$(p^{\mathrm{L}}_{\ell nm})^2 + (q^{\mathrm{L}}_{\ell nm})^2 \leq \left(S^{\mathrm{L,max}}_{\ell nm}\right)^2 y^{\mathrm{L}}_{\ell nm}, \quad \forall\{\ell, n, m\} \in L^{\mathrm{C}}. \tag{12.6}$$

12.3.5 Active and Reactive Power Balance Constraints

Equations (12.7a) and (12.7b) respectively establish the active and reactive power balances in each bus for each time period. These equations enforce the active and reactive power balances in each bus of the distribution network considering the nodal consumption and distribution line flows. Parameter P^{EV}_k indicates the peak rate power of EV charger k. It is assumed that the power factor of EV chargers is equal to 1 and, therefore, they do not inject or consume reactive power.

$$\sum_{s\in S_n} p^{\mathrm{S}}_s - \sum_{\{\ell,m\}\in N_n} p^{\mathrm{L}}_{\ell nm} = P^{\mathrm{D}}_n + \sum_{k\in K_n} P^{\mathrm{EV}}_k, \quad \forall n \in N, \tag{12.7a}$$

$$\sum_{s\in S_n} q^{\mathrm{S}}_s - \sum_{\{\ell,m\}\in N_n} q^{\mathrm{L}}_{\ell nm} = Q^{\mathrm{D}}_n, \quad \forall n \in N. \tag{12.7b}$$

12.3.6 Objective Function

The objective function of the distribution expansion problem is formulated in (12.8) and comprises the investment cost of distribution lines and the penalization cost of auxiliary variables $v^{\mathrm{aux},+}_n$ and $v^{\mathrm{aux},-}_n$. Please, note that the penalization cost is a fictitious cost that is used to compute adequately the value of variables $v^{\mathrm{aux},+}_n$ and $v^{\mathrm{aux},-}_n$.

$$\sum_{\substack{\{\ell,n,m\}\in L^{\mathrm{C}} \\ n>m}} C^{\mathrm{L}}_{\ell nm} y^{\mathrm{L}}_{\ell nm} + \sum_{n\in N} C^{\mathrm{VP}} \left(v^{\mathrm{aux},+}_n + v^{\mathrm{aux},-}_n\right). \tag{12.8}$$

12.3.7 Problem Formulation

Considering the definitions of the previous sections, the formulation of the distribution expansion planning problem is finally formulated as the following mixed-integer nonlinear programming (MINLP) model:

$$\text{Minimize}_{\Theta} \quad (12.8) \tag{12.9a}$$

subject to

$$\text{Constraints (12.1a)–(12.7b)}, \tag{12.9b}$$

where set $\Theta = \left\{p^{\mathrm{L}}_{\ell nm}, p^{\mathrm{S}}_{s}, q^{\mathrm{L}}_{\ell nm}, q^{\mathrm{S}}_{s}, v_n, y^{\mathrm{L}}_{\ell nm}, v^{\mathrm{aux},+}_{n}, v^{\mathrm{aux},-}_{n}, \theta_n\right\}$ comprises the set of optimization variables of problem (12.9).

12.4 Illustrative Case Study

In this section we analyze an example to illustrate the formulation of the distribution expansion problem considering the presence of EVs presented in Sect. 12.3. Please note that the parameters used in this example are not representative of a realistic network. All mathematical problems analyzed in this section have been solved by DICOPT under GAMS [6].

Let us consider a radial distribution network where a single distribution transformer feeds the demand of four buses linked by a distribution feeder divided in four sections or lines. The objective of this example is to determine the investment in additional lines to accommodate the installation of an EV charging point in each considered bus. Each bus has associated the demand of a dwelling, 6 kW and 1 kvar per phase, plus the demand of an EV charging point, which is equal to 7 kW per phase. It is considered that the power factor of the EV charger is equal to one. The magnitudes of the impedance and resistance of each existing distribution lines are equal to 117.04 and 41.1 Ω, respectively. The parameters of candidate lines are equal to the existing ones. The nominal single phase voltage magnitude is 230 V. The minimum and maximum values of voltage magnitudes are 218.5 and 241.5 V, respectively. The penalization cost of voltages is 2 k€/V and the cost of building each new line is 10 k€.

Figure 12.4 represents graphically the single-line diagram of the considered network. Black lines represent existing lines, whereas blue lines characterize candidate lines. The consumption data of each bus is also included in this figure.

First, problem (12.9) is solved considering that investments in new distribution lines are not allowed. The obtained results are depicted in Fig. 12.5. This figure

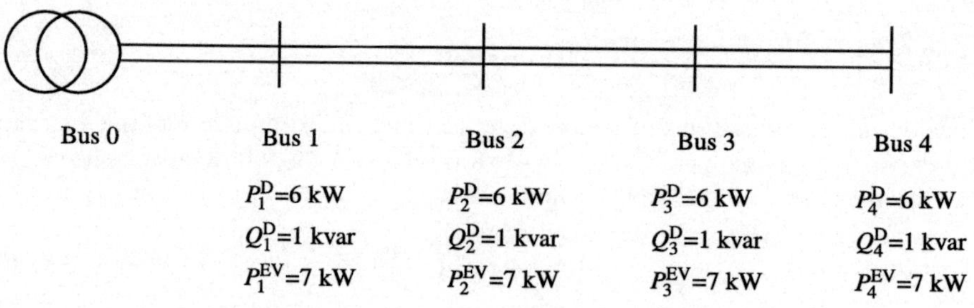

Fig. 12.4 Illustrative case study: single-line diagram of the distribution network

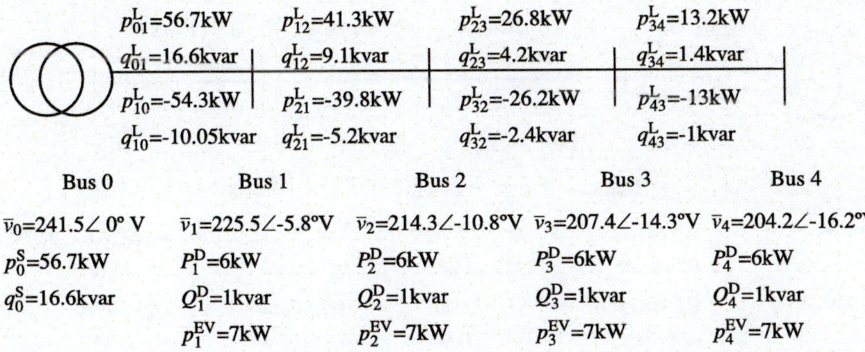

Fig. 12.5 Illustrative case study: power flows without considering network expansion

represents the voltage magnitudes at every bus, the active and reactive power injected by the power station at Bus 0, and the active and reactive power flows in the lines. For the sake of clarity, the consumption in each bus is also included. Observe that, even though the voltage at Bus 0 is the highest possible (241.5 V), voltage magnitudes at buses 2, 3, and 4 are under the minimum value, 218.5 V. Therefore, it can be concluded that the current configuration of the distribution network is not able to supply the demand of the consumers at the required quality level.

Next, problem (12.9) is solved considering the set of candidate lines depicted in Fig. 12.4. The obtained results are shown in Fig. 12.6. The installed lines are highlighted in blue color. For the sake of conciseness, power flows are represented in only one direction (from left to right in the figure). We observe that two sections of the distribution feeder are reinforced, from Bus 0 to Bus 1, and from Bus 1 to Bus 2. Since the parameters of new lines are identical to those in existing lines, the power flows in new and existing lines connecting the same buses are equal. In this case we notice that the voltage magnitudes in all buses are higher than the lower level, 218.5 V. Additionally, because of the reinforcement of two lines, line losses decrease significantly. The consequence of this is the reduction of the active and reactive power injected at Bus 0. In this manner the injected active power reduces by 4.1%, whereas the reactive power is reduced by 37.3%.

Finally, Table 12.1 provides the resulting costs in each solved case. If the distribution expansion is not allowed, the distribution investment cost is null, but the penalization cost of exceeding voltage limits is high. On the contrary it is observed that the penalization costs are zero in the case in which new distribution lines can be built.

A GAMS code to solve the above illustrative example is provided in Sect. 12.8.

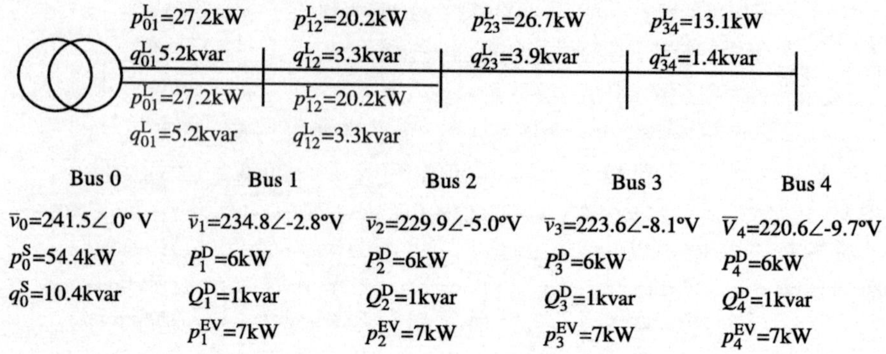

Fig. 12.6 Illustrative case study: power flows considering network expansion

Table 12.1 Illustrative case study: expected cost (thousand €)

Case	Distribution investment cost	Voltage penalization cost	Total cost
No expansion	0	349.5	349.5
Expansion	20	0	20

12.5 Realistic Case Study

This section analyzes a case study based on a realistic distribution system. We consider an urban district comprised by 26 buildings that is fed by a 10/0.4, 400 kVA power transformer.

The objective of this case study is to determine the optimal distribution investment decisions to accommodate the installation of a large number of EV charging points. All mathematical problems analyzed in this section have been solved by DICOPT under GAMS [6].

12.5.1 *Input Data*

Table 12.2 provides the information about the 26 lines comprising the distribution network. Lines are indicated by the buses they are connecting. As observed in Table 12.2, four branches with different number of lines leave the bus in which the power transformer is located (Bus 0). It is considered that the impedance per phase of each line is equal to $0.551 + j0.089\ \Omega/\text{m}$. The distribution network can be expanded by duplicating all lines, except those connecting buses 0 and 26, which can be triplicated. Observe that this is the longest branch and high voltage drops are foreseeable after connecting EV chargers. The power capacity of each line is 100 kVA and the cost of installing new lines in existing corridors is €100/m.

Table 12.2 Realistic case study: characteristics of distribution lines

Existing corridor	Length (m)	Number of candidate lines (#)	Existing corridor	Length (m)	Number of candidate lines (#)
0–1	65	1	0–13	70	1
0–2	18	1	13–14	10	1
2–3	34	1	14–15	10	1
3–4	19	1	15–16	10	1
4–5	16	1	16–17	10	1
5–6	15	1	17–18	10	1
0–7	50	1	18–19	10	1
7–8	8	1	0–20	95	2
8–9	8	1	20–21	10	2
9–10	8	1	21–22	10	2
10–11	10	1	22–23	10	2
11–12	7	1	23–24	10	2
			24–25	10	2
			25–26	10	2

The maximum power consumption and the number of EV chargers to install in each considered building are listed in Table 12.3. It is assumed a high penetration of EVs so that 44 EV charging points are expected to be connected to the considered distribution network. The power factor of residential demands is equal to 0.95 (lagging). Observe that the maximum power values indicated Table 12.3 do not consider the power associated with EV chargers. Note that all loads are assumed to have a three-phase grid connection and a nominal neutral-to-phase voltage magnitude equal to 230 V. The minimum and maximum voltage magnitudes are equal to 0.95 and 1.05 times the nominal value, i.e., 218.5 and 241.5 V, respectively. It is assumed that chargers are three-phase chargers with rate power equal to 11 kVA. The power factor of the charging points is 1.

Considering the above, the characterization of the distribution system is represented in Fig. 12.7. Candidate lines to be built are represented in blue color. Observe that all values are provided per phase.

12.5.2 Results

First, in order to analyze the behaviour of the existing network, the distribution expansion problem is solved without considering the installation of EV chargers. In other words, the power demanded by each EV charger, P_k^{EV}, is null. Therefore, only the maximum power associated with the households is considered in this analysis. Figure

Table 12.3 Realistic case study: maximum power and number of EV chargers per bus

Bus	Maximum power (kW)	Number of chargers (#)		Maximum power (kW)	Number of chargers (#)
1	6.90	1	13	6.90	3
2	6.90	1	14	6.90	2
3	6.90	2	15	6.90	1
4	6.90	1	16	5.75	1
5	6.90	1	17	6.90	2
6	6.90	2	18	9.2	2
7	5.75	1	19	5.75	1
8	6.90	3	20	4.60	1
9	5.75	1	21	6.90	1
10	6.90	1	22	6.90	2
11	6.90	2	23	6.9	2
12	8.1	2	24	5.8	3
			25	8.1	3
			26	6.9	1

12.8 represents the state of the distribution network in terms of voltage magnitudes per bus. The active and reactive power consumed per bus are also represented for the sake of clarity. Note that in the case of Bus 0, where the power transformer is located, the indicated power values refer to generated power.

If we analyze the voltage magnitudes of all buses, we can state that voltage limits are not exceeded in any bus. Observe that voltages in buses 19 and 26 (ending buses of the fourth and fifth branches, respectively) are equal to the minimum specified value, 218.5 V. However, the voltage at the power transformer bus is 231.9 V, which is less than the upper limit, 241.5 V. Therefore, the distribution network would be able to supply a certain additional power quantity, while maintaining voltages within specified values, by increasing the voltage at the power transformer bus.

Figure 12.9 represents the state of the distribution network when EV chargers are considered, but the reinforcement of the distribution network is not allowed. We represent in red color those voltages smaller than the lower limit, 218.5 V. In this case, we observe that 14 from the 27 buses of the distribution network have a voltage magnitude smaller than the lower limit. Therefore, the incorporation of EV chargers in the distribution network has the expected effect of reducing significantly the voltage magnitudes in those buses located far from the power transformer bus. Note also that the voltage magnitude at the power transformer bus is equal to the voltage upper limit, 241.5 V. Then, it is not possible to enlarge voltage magnitudes throughout the

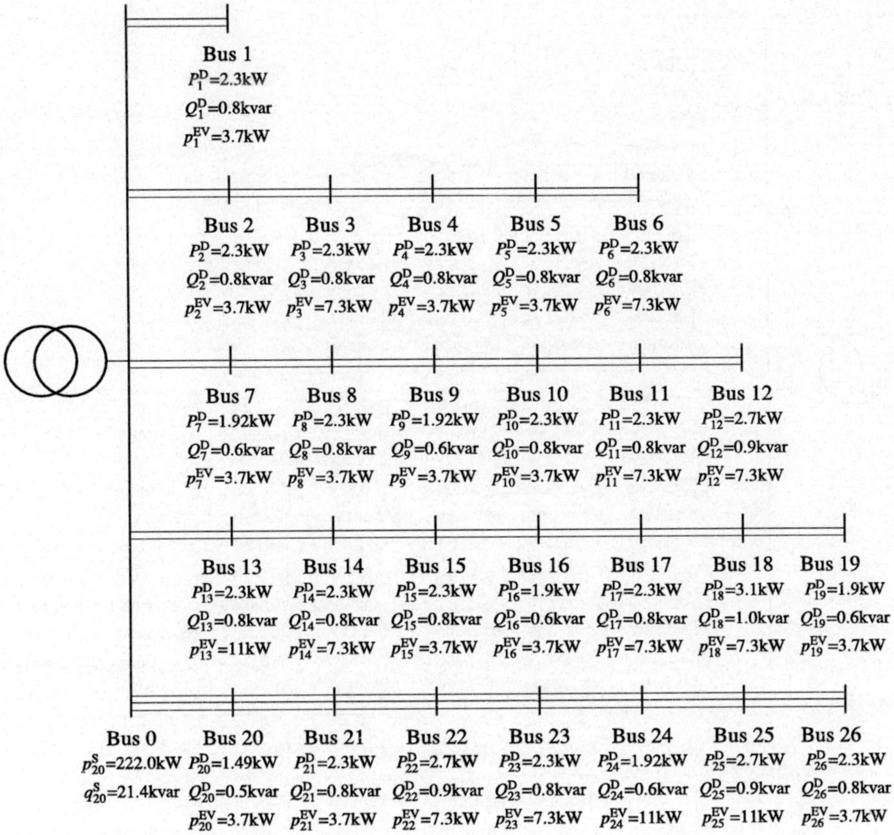

Fig. 12.7 Realistic case study: distribution network

distribution network by increasing the voltage at the power transformer bus. In this case, all lines are operating under its capacity, 100 kVA. As a conclusion, we can state that, if a large number of EV chargers is incorporated in the distribution network under study, a reinforcement of the network is needed to increase the voltage magnitudes at some specific buses.

Finally, Fig. 12.10 represents the resulting distribution network configuration after solving problem (12.9) formulated in Sect. 12.3. Installed lines are represented in blue color in the figure. First, observe that all voltages are within the specified limits. Therefore, the distribution expansion realized in this case is enough to ensure that steady-state voltages are acceptable. Note that the distribution expansion performed consists in the reinforcement of 17 existing corridors. All reinforced lines correspond to branches with 6 or more households. The total cost is equal to 20,100 €. The number of constraints, continuous variables and binary variables is equal to 888, 407, and 33, respectively. The total solution time is 30 s.

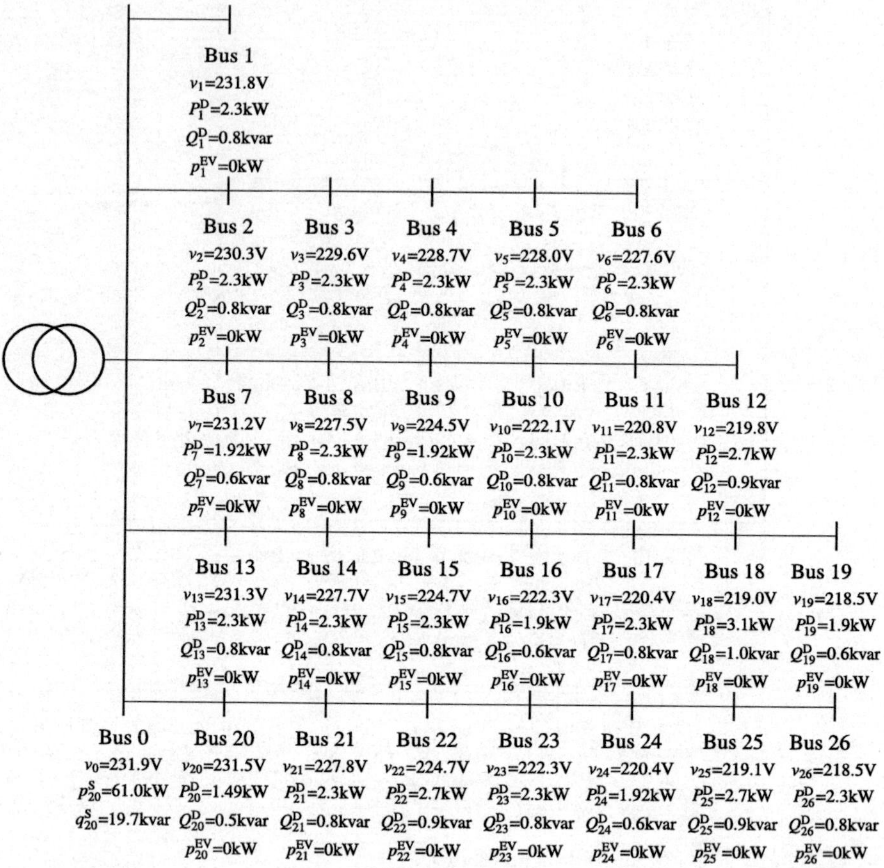

Fig. 12.8 Realistic case study: steady-state of the distribution network without considering EV chargers

12.6 Summary and Conclusions

This chapter has presented a formulation to model the distribution expansion problem considering the installation of new distribution lines in a low-voltage urban grid to accommodate the connection of a large number of EV chargers in the distribution network. The objective of the DSO is to avoid unacceptable voltage drops and line congestion. We assume that the DSO has a risk-averse position considering that end users of electricity consume simultaneously their peak power consumption.

The mathematical formulation of the installation of new distribution lines requires the usage of binary variables. Distribution lines are modelled considering the π-equivalent model of the lines and the power flows are computed using an AC model. As a consequence of this, the resulting mathematical problem is a mixed-integer non-linear programming problem. The proposed model can be expanded considering

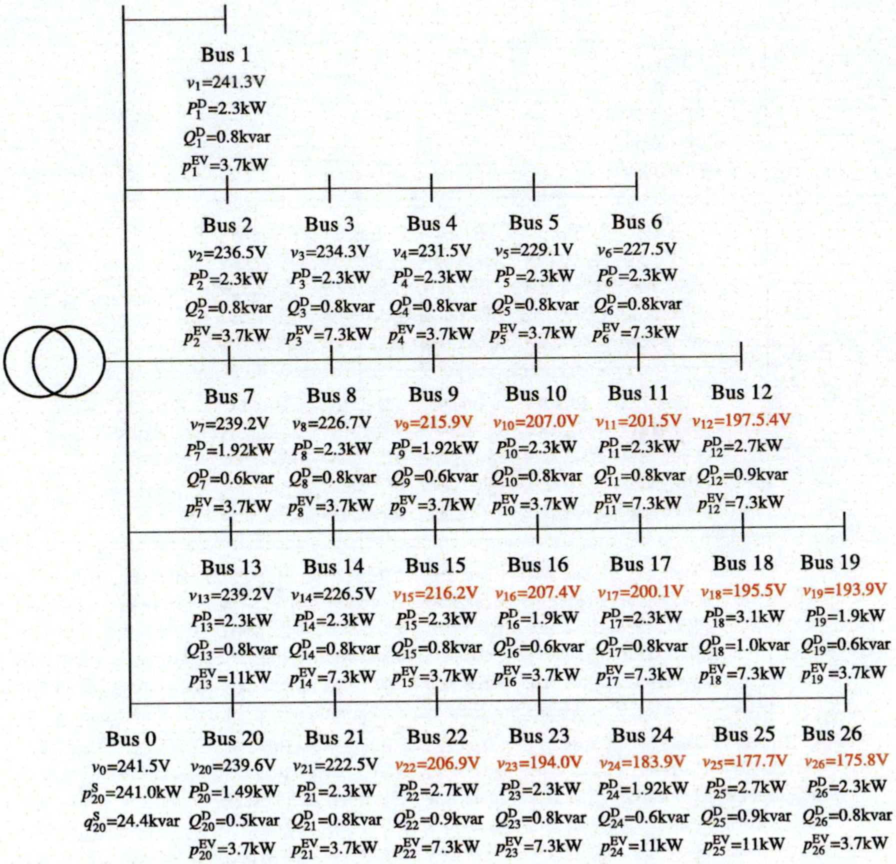

Fig. 12.9 Realistic case study: steady state of the distribution network considering EV chargers but no distribution expansion

single-phase chargers and demands. In this case the single-phase equivalent of the distribution network cannot be used, and the proposed formulation must consider the explicit modeling of each phase. This causes an increase of the computational size of the problem.

The numerical results obtained from the realistic case study solved in this chapter indicate that the presence of a large number of EVs decreases significantly the voltage magnitudes in the weakest feeders of the distribution network. It has been also verified that the proposed procedure is able to determine the distribution lines to reinforce to maintain voltages at appropriate levels.

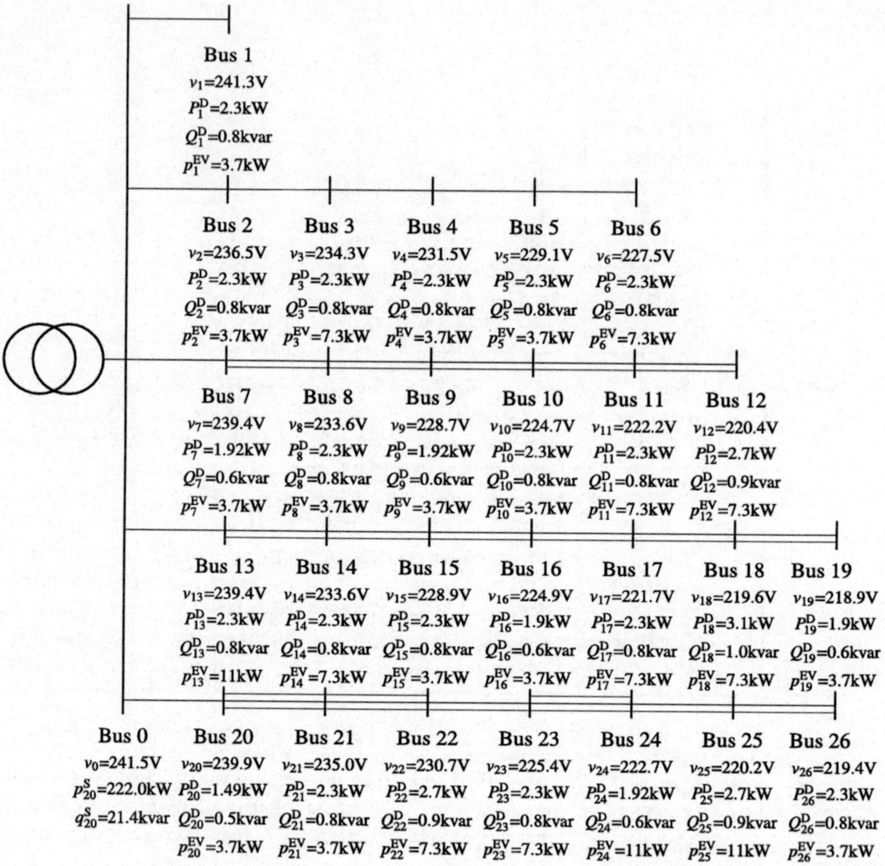

Fig. 12.10 Realistic case study: steady state of the distribution network considering EV chargers and distribution expansion decisions

12.7 Notation

This section defines the main notation used in this chapter.

Acronyms

AC Alternating current.
EV Electric vehicle.
PV Photovoltaic.

Sets and Indices

k	Index of electric vehicle chargers.
K	Set of electric vehicle chargers.
K_n	Set of electric vehicles chargers at usage at bus b and n.
ℓ	Index of distribution lines.
L	Set of distribution lines.
n	Index of buses.
m	Index of buses.
N	Set of buses.
N_n	Set of lines connected to bus n.
s	Index of distribution transformers.
S	Set of distribution transformers.
S_n	Set of distribution transformers at bus n.

Parameters

B^{S}_{ℓ}	Shunt susceptance of line ℓ (S).
$B_{\ell nm}$	Susceptance of line ℓ linking buses n and m (S).
$C_{\ell nm}$	Investment cost of line ℓ linking buses n and m (€).
C^{VP}	Penalization cost of voltage magnitude deviations (€/V).
$G_{\ell nm}$	Conductance of line ℓ linking buses n and m (S).
M	Large enough parameter.
P^{D}_n	Active power demand at bus n (kW).
P^{EV}_k	Active power demanded by charger k (kW).
Q^{D}_n	Reactive power demand in bus n (kvar).
$S^{\mathrm{L,max}}_{\ell nm}$	Capacity of line ℓ linking buses n and m (kVA).
$S^{\mathrm{S,max}}_s$	Capacity of distribution transformer s (kVA).
$V^{\max}_n$	Upper limit of the voltage magnitude at bus n (V).
$V^{\min}_n$	Lower limit of the voltage magnitude at bus n (V).

Variables

$p^{\mathrm{L}}_{\ell nm}$	Active power flow through line ℓ linking buses n and m (kW).
p^{S}_s	Active power supplied by distribution transformer s (kva).
$q^{\mathrm{L}}_{\ell nm}$	Reactive power flow through line ℓ linking buses n and m (kvar).
q^{S}_s	Reactive power supplied by distribution transformer s (kvar).
v_n	Voltage magnitude at bus n (V).

$y^{\mathrm{L}}_{\ell nm}$ Binary variable that is equal to 1 if candidate line ℓ linking buses n and m is built, being equal to 0 otherwise.

$v^{aux,+}_n$ Positive voltage magnitude deviation at bus n (V).

$v^{aux,-}_n$ Negative voltage magnitude deviation at bus n (V).

θ_n Voltage angle at bus n (rad).

12.8 GAMS Codes

This section provides the GAMS [8] code used to solve the Illustrative Case Study included in Sect. 12.4.

```
$title DISTRIBUTION NETWORK EXPANSION WITH ELECTRIC VEHICLES$


*****************************************************************
*         DATA
*****************************************************************


SETS
k              'Electric vehicles'          /k1*k4/,
n              'Buses'                      /n0*n4/
l              'Lines'                      /l1*l8/
s              'Substations'                /s1*s1/


KN(k,n)              'Set of electric vehicles connected to bus n'
SN(s,n)              'Set of substations in bus n'
EL(l)                'Set of existing lines'
CL(L)                'Set of candidate lines'
;

alias(n,m);

EL(l)$(ord(l) le 4)=yes;
CL(l)$(ord(l) gt 4)=yes;
SN('s1','n0')=yes;
KN('k1','n1')=yes;KN('k2','n2')=yes;
KN('k3','n3')=yes;KN('k4','n4')=yes;


PARAMETERS
S_Smax(s)            'Capacity of the transformer station'
G(l,n,m)             'Conductance'
B(l,n,m)             'Susceptance'
Bp(l,n)              'Shunt susceptance'
S_Lmax(l,n,m)        'Capacity lines'
P_EVC(k)             'Power charged'
P_D(n)               'Active power demand'
```

```
Q_D(n)            'Reactive power demand'
P_EV_max(k)       'Active power capacity of the charger of vehicle k'
V_max(n)          'Maximum voltage magnitude'
V_min(n)          'Minimum voltage magnitude'
MM                'Big enough constant'
CLin(l)           'Distribution line cost'
CVP               'Penalization cost'
S_Lmax_sqr(l,n,m)         'Square of the aparent power capacity of line l'
;


Table Lines(l,*)   'set of lines'
         O       F       G         B        Bp          Smax       Cost
* Existing lines
L1       0       1       0.005     -0.008   0.0001      1000
L2       1       2       0.005     -0.008   0.0001      1000
L3       2       3       0.005     -0.008   0.0001      1000
L4       3       4       0.005     -0.008   0.0001      1000
* Candidate lines
L5       0       1       0.005     -0.008   0.0001      1000       10
L6       1       2       0.005     -0.008   0.0001      1000       10
L7       2       3       0.005     -0.008   0.0001      1000       10
L8       3       4       0.005     -0.008   0.0001      1000       10;


CLin(l)=Lines(l,'Cost');
CVP=200;
MM=300;
P_EV_max(k)=7;
S_Smax(s)=5000;
V_max(n)=230*1.05;
V_min(n)=230*0.95;
Bp(l,n)=0;
P_EVC(k)=P_EV_max(k);
P_D(n)$(ord(n) gt 1)=6;
Q_D(n)$(ord(n) gt 1)=1;
SET ORI(l,n), FIN(l,n);
SET LN(n,m),   LNM(l,n,m);
ORI(l,n)$(ord(n) = Lines(l,'O') +1)=yes;
FIN(l,n)$(ord(n) = Lines(l,'F') +1)=yes;

loop(l,
         LN(n,m)$(ORI(l,n) and FIN(l,m))=yes;
         LN(m,n)$(ORI(l,n) and FIN(l,m))=yes;
         LNM(l,n,m)$(ORI(l,n) and FIN(l,m))=yes;
         LNM(l,m,n)$(ORI(l,n) and FIN(l,m))=yes;
         G(l,n,m)$(ORI(l,n) and FIN(l,m))= Lines(l,'G');
         B(l,n,m)$(ORI(l,n) and FIN(l,m))= Lines(l,'B');
         G(l,m,n)$(ORI(l,n) and FIN(l,m))= Lines(l,'G');
         B(l,m,n)$(ORI(l,n) and FIN(l,m))= Lines(l,'B');
         S_Lmax(l,n,m)= Lines(l,'Smax');
         S_Lmax(l,m,n)=Lines(l,'Smax');
);

G(l,n,m)$LNM(l,n,m)=0.003;
B(l,n,m)$LNM(l,n,m)=-0.008;
S_Lmax_sqr(l,n,m)= S_Lmax(l,n,m)**2;
```

```
***********************************************************************
*   DECLARATION OF VARIABLES AND MATHEMATICAL CHARACTERIZATION
***********************************************************************
VARIABLES
**        OBJECTIVE FUNCTION
obfun               'Objective function'
p_S(s)              'Active power substation s'
q_S(s)              'Reactive power substation s'
v(n)                'Voltage magnitude in bus n'
p_L(l,n,m)          'Active power flow'
q_L(l,n,m)          'Reactive power flow'
theta(n)            'Voltage angle'
;

POSITIVE VARIABLES
v_max_aux(n)       'Auxiliary voltage variable'
v_min_aux(n)       'Auxiliary voltage variable'
;


**        UPPER AND LOWER LIMITS OF THE VARIABLES
p_S.up(s)=100;p_S.lo(s)=0;
q_S.up(s)=100;q_S.lo(s)=0;
v.up(n)=300;v.lo(n)=200;
p_L.up(l,n,m)=100;p_L.lo(l,n,m)=-100;
q_L.up(l,n,m)=100;q_L.lo(l,n,m)=-100;
theta.l(n)=0;


BINARY VARIABLES
y(l)                'Binary variable that is equal to 1 if distribution line l is built'
;




***********************************************************************
*   EQUATIONS
***********************************************************************

EQUATIONS

**        OBJECTIVE FUNCTION
Objective_function

**        VOLTAGE MAGNITUDE LIMITS
Voltage_limit_01
Voltage_limit_02

**        TRANSFORMER STATION CAPACITY
Transformer_station_01

**        ACTIVE AND REACTIVE POWER FLOWS FOR EXISTING LINES
PowerFlow_existing_01
PowerFlow_existing_02
PowerFlow_existing_03

**        ACTIVE AND REACTIVE POWER FLOWS FOR CANDIDATE LINES
```

```
PowerFlow_candidate_01
PowerFlow_candidate_02
PowerFlow_candidate_03
PowerFlow_candidate_04
PowerFlow_candidate_05
PowerFlow_candidate_06
PowerFlow_candidate_07
PowerFlow_candidate_08
PowerFlow_candidate_09

**        ACTIVE AND REACTIVE POWER BALANCE CONSTRAINTS
Balance_01
Balance_02;



**        OBJECTIVE FUNCTION
Objective_function..      obfun =e=sum((l,n,m)$(LNM(l,n,m) and (ord(n) gt ord(m))),CLin(l)*y(l)) +CVP*sum(n,v_max_aux(n)+v_min_aux(n));

**        VOLTAGE MAGNITUDE LIMITS
Voltage_limit_01(n)..     v(n)  =l= V_max(n)+v_max_aux(n);
Voltage_limit_02(n)..     v(n)  =g= V_min(n)-v_min_aux(n);

**        TRANSFORMER STATION CAPACITY
Transformer_station_01(s)..         sqr(p_S(s))+sqr(q_S(s))=l= S_Smax(s)**2;

**        ACTIVE AND REACTIVE POWER FLOWS FOR EXISTING LINES
PowerFlow_existing_01(l,n,m)$(LNM(l,n,m) and EL(l))..     p_L(l,n,m)=e= ((v(n)**2)*G(l,n,m)- v(n)*v(m)*(G(l,n,m)*cos(theta(n)-theta(m))+B(l,n,m)*sin(theta(n)-theta(m))));
PowerFlow_existing_02(l,n,m)$(LNM(l,n,m) and EL(l))..     q_L(l,n,m)=e= -(v(n)**2)*(B(l,n,m)+Bp(l,n))- v(n)*v(m)*(G(l,n,m)*sin(theta(n)-theta(m))-B(l,n,m)*cos(theta(n)-theta(m)));
PowerFlow_existing_03(l,n,m)$(LNM(l,n,m) and EL(l))..     sqr(p_L(l,n,m))+sqr(q_L(l,n,m))=l= S_Lmax(l,n,m)**2;

**        ACTIVE AND REACTIVE POWER FLOWS FOR CANDIDATE LINES
PowerFlow_candidate_01(l,n,m)$(LNM(l,n,m) and CL(l))..    p_L(l,n,m)=g= ((v(n)**2)*G(l,n,m)- v(n)*v(m)*(G(l,n,m)*cos(theta(n)-theta(m))+B(l,n,m)*sin(theta(n)-theta(m)))) -S_Lmax_sqr(l,n,m)*(1-y(l));
PowerFlow_candidate_02(l,n,m)$(LNM(l,n,m) and CL(l))..    p_L(l,n,m)=l= ((v(n)**2)*G(l,n,m)- v(n)*v(m)*(G(l,n,m)*cos(theta(n)-theta(m))+B(l,n,m)*sin(theta(n)-theta(m))))+S_Lmax_sqr(l,n,m)*(1-y(l));
PowerFlow_candidate_03(l,n,m)$(LNM(l,n,m) and CL(l))..    q_L(l,n,m)=g= -(v(n)**2)*(B(l,n,m)+Bp(l,n))- v(n)*v(m)*(G(l,n,m)*sin(theta(n)-theta(m))-B(l,n,m)*cos(theta(n)-theta(m)))-S_Lmax_sqr(l,n,m)*(1-y(l));
PowerFlow_candidate_04(l,n,m)$(LNM(l,n,m) and CL(l))..    q_L(l,n,m)=l= -(v(n)**2)*(B(l,n,m)+Bp(l,n))- v(n)*v(m)*(G(l,n,m)*sin(theta(n)-theta(m))-B(l,n,m)*cos(theta(n)-theta(m)))+S_Lmax_sqr(l,n,m)*(1-y(l));
PowerFlow_candidate_05(l,n,m)$(LNM(l,n,m) and CL(l))..    sqr(p_L(l,n,m))+sqr(q_L(l,n,m))=l= y(l)*S_Lmax_sqr(l,n,m);
```

```
PowerFlow_candidate_06(l,n,m)$(LNM(l,n,m) and CL(l)).. p_L(l,n,m
    )=l= y(l)*S_Lmax_sqr(l,n,m);
PowerFlow_candidate_07(l,n,m)$(LNM(l,n,m) and CL(l)).. p_L(l,n,m
    )=g= -y(l)*S_Lmax_sqr(l,n,m);
PowerFlow_candidate_08(l,n,m)$(LNM(l,n,m) and CL(l)).. q_L(l,n,m
    )=l= y(l)*S_Lmax_sqr(l,n,m);
PowerFlow_candidate_09(l,n,m)$(LNM(l,n,m) and CL(l)).. q_L(l,n,m
    )=g= -y(l)*S_Lmax_sqr(l,n,m);

**        ACTIVE AND REACTIVE POWER BALANCE CONSTRAINTS
Balance_01(n)..  sum(SN(s,n),p_S(s))-sum((l,m)$(LNM(l,n,m)),p_L(l,
    n,m)) =E=   P_D(n)+  sum(KN(k,n),P_EVC(k));
Balance_02(n)..  sum(SN(s,n),q_S(s))-sum((l,m)$(LNM(l,n,m)),q_L(l,
    n,m)) =E=   Q_D(n);

*****************************************************************
*   MODEL AND SOLVING
*****************************************************************

MODEL DisExpansionEV /all/;

DisExpansionEV.optcr=0;

SOLVE DisExpansionEV USING minlp MINIMIZING obfun;
```

References

1. Abdalrahman, A., Zhuang, W.: QoS-aware capacity planning of networked PEV charging infrastructure. IEEE Open J. Veh. Technol. **1**, 116–129 (2020)
2. Arias, N.B., Hashemi, S.: Distribution system services provided by electric vehicles: recent status, challenges, and future prospects. IEEE Trans. Intell. Transp. Syst. **20**, 4277–4296 (2019)
3. Behravesh, V., Keypour, R., Akbari Foroud, A.: Control strategy for improving voltage quality in residential power distribution network consisting of roof-top photovoltaic-wind hybrid systems, battery storage and electric vehicles. Sol. Energy **182**, 80–95 (2019)
4. Carrión, M., Zárate-Miñano, R., Domínguez, R.: Integration of electric vehicles in low-voltage distribution networks considering voltage management. Energies. **13**, 4125 (2020)
5. Clement-Nyns, K., Driesen, J.: The impact of charging plug-in hybrid electric vehicles on a residential distribution grid. IEEE Trans. Power Syst. **25**, 371–380 (2010)
6. DICOPT (2022). https://www.gams.com/33/docs/S_DICOPT.html
7. Demirok, E., Casado-González, P., Frederiksen, K.H., Sera, D., Rodríguez, P., Teodorescu, R.: Local reactive power control methods for overvoltage prevention of distributed solar inverters in low-voltage grids. IEEE J. Photovoltaics **1**, 174–182 (2011)
8. GAMS (2021). www.gams.com
9. Humayd, A.S.B., Bhattacharya, K.: A novel framework for evaluating maximum PEV penetration into distribution systems. IEEE Trans. Smart Grid **9**, 2741–2751 (2018)
10. Kong, C., Jovanovic, R., Bayram, I.S., Devetsikiotis, M.: A hierarchical optimization model for a network of electric vehicle charging stations. Energies **10**(5), 675 (2017)
11. Leemput, N., Geth, F., Van Roy, J., Büscher, J., Driesen, J.: Reactive power support in residential LV distribution grids through electric vehicle charging. Sustain. Energy Grids Netw. **3**, 24–35 (2015)
12. Liu, Y., Xiang, Y., Tan, Y., Wang, B., Liu, J., Yang, Z.: Optimal allocation model for EV charging stations coordinating investor and user benefits. IEEE Access **6**, 36039–36049 (2018)

13. Knezović, K., Marinelli, M., Zecchino, A., Andersen, P.B., Traeholt, C.: Supporting involvement of electric vehicles in distribution grids: lowering the barriers for a proactive integration. Energy **134**, 458–468 (2017)
14. Munoz, E.R., Razeghi, G., Zhang, L., Jabbari, F.: Electric vehicle charging algorithms for coordination of the grid and distribution transformer levels. Energy **113**, 930–942 (2016)
15. Valsera-Naranjo, E., Sumper, A., Villafafila-Roble,s R., Martínez-Vicente, D.: Probabilistic method to assess the impact of charging of electric vehicles on distribution grids. Energies 1503–1531 (2012)
16. Sortomme, E., Hindi, M.M., MacPherson, S.D.J., Venkata, S.S.: Coordinated charging of plug-in hybrid electric vehicles to minimize distribution system losses. IEEE Trans. Smart Grid **2**, 198–205 (2011)
17. Sortomme, E., Negash, A.I., Venkata, S.S., Kirschen, D.S.: Voltage dependent load models of charging electric vehicles. In: Proceedings of the IEEE PES General Meeting, Vancouver, BC, Canada, 21–25(July), pp. 1–5 (2013)
18. Wang, J., Bharati, G.R., Paudyal, S., Ceylan, O., Bhattarai, B.P., Myers, K.S.: Coordinated electric vehicle charging with reactive power support to distribution grids. IEEE Trans. Ind. Informat. **15**, 54–63 (2018)
19. Wang, L., Sharkh, S., Chipperfield, A.: Optimal coordination of vehicle-to-grid batteries and renewable generators in a distribution system. Energy **113**, 1250–1264 (2016)
20. Xiang, Y., Yang, W., Liu, J., Li, F.: Multi-objective distribution network expansion incorporating electric vehicle charging stations. Energies **9**, 1–17 (2016)
21. Zárate-Miñano, R., Flores Burgos, A., Carrión, M.: Analysis of different modeling approaches for integration studies of plug-in electric vehicles. Int. J. Electr. Power Energy Syst. **114**, 105398 (2020)
22. Zeng, B., Feng, J., Zhang, J., Liu, Z.: An optimal integrated planning method for supporting growing penetration of electric vehicles in distribution systems. Energy **126**, 273–284 (2017)
23. Zhang, H., Moura, S.J., Hu, Z., Song, Y.: PEV fast-charging station siting and sizing on coupled transportation and power networks. IEEE Trans. Smart Grid **9**, 2595–2605 (2018)

Index

L. Baringo et al., *Electric Vehicles and Renewable Generation*, Green Energy and Technology, https://doi.org/10.1007/978-3-031-09079-0

U

V

W